INDUSTRIAL ENGINEERING
AND
PRODUCTION MANAGEMENT

[For Undergraduate, Postgraduate Courses and Diploma Programmes in Mechanical, Production and Industrial Engineering Students. A Useful Guide for IIE, Management Courses, Professional Engineers and Competitive Examinations for GATE and UPSC] and Engineering Services Examinations.

MARTAN[illegible] [illegible]LESANG
Professor and Head,
Department of Mechanical Engineering,
K.E. Society's Rajarambapu Institute of Technology,
Rajaramnagar (Sakharale)
(Distt. Sangli - Maharashtra)

S Chand And Company Limited
(AN ISO 9001 : 2008 COMPANY)
RAM NAGAR, NEW DELHI - 110 055

S Chand And Company Limited

(An ISO 9001:2008 Company)

Head Office: 7361, RAM NAGAR, NEW DELHI - 110 055

Phone: 23672080-81-82, 9899107446, 9911310888 Fax: 91-11-23677446

www.schandpublishing.com; e-mail: helpdesk@schandpublishing.com

Branches

Ahmedabad	:	Ph: 27541965, 27542369, ahmedabad@schandpublishing.com
Bengaluru	:	Ph: 22268048, 22354008, bangalore@schandpublishing.com
Bhopal	:	Ph: 4274723, 4209587, bhopal@schandpublishing.com
Chandigarh	:	Ph: 2625356, 2625546, chandigarh@schandpublishing.com
Chennai	:	Ph: 28410027, 28410058, chennai@schandpublishing.com
Coimbatore	:	Ph: 2323620, 4217136, coimbatore@schandpublishing.com (Marketing Office)
Cuttack	:	Ph: 2332580, 2332581, cuttack@schandpublishing.com
Dehradun	:	Ph: 2711101, 2710861, dehradun@schandpublishing.com
Guwahati	:	Ph: 2738811, 2735640, guwahati@schandpublishing.com
Hyderabad	:	Ph: 27550194, 27550195, hyderabad@schandpublishing.com
Jaipur	:	Ph: 2219175, 2219176, jaipur@schandpublishing.com
Jalandhar	:	Ph: 2401630, 5000630, jalandhar@schandpublishing.com
Kochi	:	Ph: 2378740, 2378207-08, cochin@schandpublishing.com
Kolkata	:	Ph: 22367459, 22373914, kolkata@schandpublishing.com
Lucknow	:	Ph: 4026791, 4065646, lucknow@schandpublishing.com
Mumbai	:	Ph: 22690881, 22610885, mumbai@schandpublishing.com
Nagpur	:	Ph: 6451311, 2720523, 2777666, nagpur@schandpublishing.com
Patna	:	Ph: 2300489, 2302100, patna@schandpublishing.com
Pune	:	Ph: 64017298, pune@schandpublishing.com
Raipur	:	Ph: 2443142, raipur@schandpublishing.com (Marketing Office)
Ranchi	:	Ph: 2361178, ranchi@schandpublishing.com
Siliguri	:	Ph: 2520750, siliguri@schandpublishing.com (Marketing Office)
Visakhapatnam	:	Ph: 2782609, visakhapatnam@schandpublishing.com (Marketing Office)

First Edition 1998

Subsequent Editions and Reprints 2002, 2004 (Twice), 2005, 2006 (Twice), 2007, 2008 (Twice), 2009, 2010, 2011, 2012, 2013 (Twice), 2014 (Twice), 2015, 2016

Reprint 2017

ISBN : 978-81-219-1773-5 **Code** : 1010A 197

PRINTED IN INDIA

By Nirja Publishers & Printers Pvt. Ltd., 54/3/2, Jindal Paddy Compound, Kashipur Road, Rudrapur-263153, Uttarakhand and published by S. Chand & Company Limited, 7361, Ram Nagar, New Delhi -110 055.

This book is dedicated to the
loving memory of my beloved brother
Late Krishna Telsang,
who met an untimely death.

Preface to the Second Edition

Valuable suggestions and constructive criticisms from number of colleagues at various institutions and the feedback from the students, as well as the rapid progress in manufacturing philosophies and technology have led to the second edition of the book Industrial engineering and production management."

It has become all the necessary on the part of organizations to enhance productivity and reduce cost, eliminate waste and non-value adding activities to gain competitive advantage. Various techniques of industrial engineering help to increase productivity and help to create working climate wherein each one will contribute positively towards achieving Organizational goals. The orientation of production management as a practicing discipline has changed significantly on global bases. All the important changes pertain to a fact that a customer is in very much focus and all the activities and operations should be centered around enhancing customer value. The advances in information technology have changed the way the business is being done and the business processes are re-engineered using the information technology leverage.

The book is reorganized in to two parts. Part I consists of various industrial engineering techniques and Part II discusses the various aspects of production management. This edition basically follows the same introductory nature, format, organization and balance as of the First edition.

Special Features of Second Edition

The second edition of the book retains all the good features of the First edition with respect to simplicity of the presentation and the format. It tries to bring in the flavor of changes that are taking place in the areas of manufacturing like new technologies, philosophies and concepts through suitable additions and modifications. These features include :

- Seven new chapters, which are important from the point of view of changes in curriculum of various universities, are added to enhance the usefulness of the book.
- Additional material to the existing chapters, new techniques and philosophies are added in the form of supplements.
- Additional numerical problems with solutions are provided to deepen the understanding of the concepts, analysis and decision-making.
- New and evolving concepts like Lean and Agile manufacturing, theory of constraints (TOC), Quality Function Deployment (QFD), Design for manufacturing and assembly (DFMA), Failure Mode Effect and Criticality Analysis (FMECA), Robust Design are added to bring substantial exposure to the readers.
- The style and language of presentation is intentionally kept simple in order to make students understand the concepts without much difficulty.

Audience

Like the First edition, this edition has been recommended for students in mechanical and industrial and production engineering programme at undergraduate and postgraduate level of various institutions and universities. It also forms a good reference for diploma programme in mechanical and production engineering, AMIE and AIIIE examinations and competitive examinations like GATE and UPSC.

Acknowledgements

This text together with the first edition represents a total of about six years effort. It couldn't have been written and produced without the help of many people. It gives me a great pleasure to acknowledge the support and assistance of the following people in bringing the second edition.

I take this opportunity to thank all those readers and faculty members who have given the valuable suggestions for the improvement of the book. I take this opportunity to express thanks to Principal Dr. B.R. Badwe for his inspiration, support and the appreciation for the work undertaken. I am very much indebted to my learned colleagues at the department of

Mechanical Engineering Rajarambapu Institute of Technology, Rajaramnagar. My special thanks to Prof. A.H. Mishrikoti for his contribution through constructive feed back during revision process. My sincere thanks are due to my Family members who bore the burden of the domestic chores during the period of writing. I express my sincere thanks to my wife Kshama and daughter Trupti for their moral support and inspiration. I seek your suggestions and criticisms to make this book more useful to the readers.

Rajaramnagar **Prof. Martand Telsang**
14.1.2003

CONTENTS

PART I : INDUSTRIAL ENGINEERING

PART III

ADVANCED TOPICS IN PRODUCTION MANAGEMENT

PART-I
INDUSTRIAL ENGINEERING

INDUSTRIAL ENGINEERING – TODAY

Ever since the beginning of the civilization, man has attempted to improve productivity of his limited resources in order to maximize creation of wealth, which is essential for survival and enjoyment of life. His first resources were manpower and animal power and then he created hand tools and later the machines and the machine tools to improve the productivity of manpower. The continued pursuit of higher productivity since then led to the dawn of industrial age and gave birth to the two disciplines. "Industrial Management and Industrial Engineering" during the mid twentieth century. Industrial Engineering seeks to maximize the performance of interactive man-machine-material systems. Systems integration that cuts across boundaries of functions within organizations and across boundaries of organizations that together make a whole enterprise.

It is seen that industrial engineering has evolved over the last century as a broad profession concerned with designing effective systems and developing the best processes with the purpose of integrating people, machine and material resources for improved overall effectiveness of organizations and delivering the products and services to the consumer. The focus is on manufacturing systems but still the other systems in the areas such as transportation, communication, finance etc. are given due importance. Industrial engineering discipline enables interface engineering facilities and their operations for converting resources into products and services, which are in turn delivered to the consumer. Thus, industrial engineering is referred appropriately as "Interface Engineering".

Due to increasing complexity of operations and need of customer responsiveness, there is a great need for teamwork between different functions such as marketing, product design, procurement, process design, manufacturing, logistics and distribution. Increasing competition is demanding a greater need to ensure the optimal utilization of all assets such as people, material, equipment and time in business systems. Thus, there is an established need to design and implement variety of planning and control systems in the various technology management interfaces such as

- Facilities Planning and Control
- Operations Planning and Management
- Plant Engineering and Management
- Quality Engineering and Management
- Manpower (HR) Management
- Logistics and Supply Chain Management
- Project Engineering and Management
- Environmental Engineering and Management

Thus, the scope of industrial engineering — Design, improve and install work systems (integrated systems of men, material and machines) have broadened and deepened to encompass all the activities in factories, fields, homes and offices, irrespective of purpose or size of the organizations. Today's Organizations work in constantly changing environment and has to accommodate new products, volume changes new technologies and management philosophies and there is a constant pressure to improve all work processes to meet the demands of changing environment. Industrial engineers play a pioneering role as change agents in the organization as system integrators with strong mathematical, statistical, technical and management background.

INDUSTRIAL ENGINEERING AND INFORMATION TECHNOLOGY

Information is the important element or key to achieve system integration. Industrial engineering as a discipline is now poised to break new frontiers in terms of enabling the new information technologies to serve man by guiding in the design, development and management of growing complex, interactive man-machine-material systems. All the capability of Information Technology (IT) help in improving the co-ordination and information access across organi-zational units enabling effective management of task interdependency.

Industrial engineers should understand the capabilities of IT and exploit the opportunities in design of integrated systems to achieve system optimization.

If we trace back the evolution of industrial engineering — 1950's to 1970's marked the quantitative management and this period expanded the knowledge base providing industrial engineering a firm base ir mathematics, which allowed for improved and better understanding of mathematical modelling.

80's IT as a New Business Tool

1980's saw the availability of high-speed, stored program digital computer and upward performance brought IT as a handy tool for industrial engineers. The benefits include a high speed calculating device computer and the ability to make comparison with previously stored data and the capability to simulate. With more sophisticated tools including IT, an industrial engineers of 80's had to specialize to a greater degree than even before. During this period, I.E. saw many specialty areas like

- Reliability Engineering
- Value Engineering
- Inventory Control and Production Control
- Human Engineering.

1990's IT as a Business Enabler

The alignment of IE with IT gave new dimensions to the way the business was conducted. The business process reengineering philosophy of radically overhauling and improving business processes (Michael Hammer) changed the way the business is done. IE have several assets that make them good candidate for BPR team membership. With new role of BPR team leaders, it was the first time that industrial engineering started getting more strategic and authoritative position in the organizations ranging from operations to banking, manufacturing to services, automotive to health care. ERP implementation and supply chain management and management consulting has made the industrial engineers suitable for posts in the areas of strategic management and supply chain managers.

Beyond 2000 — IT as a Business Driver

IT is all set to take a driving seat in business for the days to come. IT played a supporting role in the business so far, now the business would need to be designed as per the progress in the IT. IT and IE together can make a world of difference combining the efficiency of IE and accuracy if IT, we can meet the demands of the future. Thus, combined IE + IT can work wonders for the business and industry in the areas of risk management, industrial automation, web-enabled ERP and E-Engineering. Thus, Industrial Engineering discipline can make a unique contribution to society by improving the productivity of the economy and health, quality of work life (QWL) and human happiness by giving proper emphasis on abilities and limitations of human being in economic work situations and the technology.

This part starts with the introduction of Industrial Engineering functions and discuses the various tools and techniques of Industrial engineering in order to enhance the productivity and efficiency of the organizations.

Part I contains the following chapters.

Chapter 1 Introduction to Industrial Engineering
2 Productivity
3 Work Study
4 Method Study
5 Work Measurement
6 Value Engineering
7 Plant Location
8 Plant Layout
9 Material Handling
10 Job Evaluations and Merit Rating
11 Wages and Incentives
12 Ergonomics.

1

INTRODUCTION TO INDUSTRIAL ENGINEERING

• Definition • History and Development of Industrial Engineering • Contributions to Industrial Engineering • Activities of Industrial Engineering • Industrial Engineering Approach • Objectives of Industrial Engineering • Functions of Industrial Engineer • Place of Industrial Engineering in an Organisation • Industrial Engineering in Service Sector • Systems Approach.

1.1. DEFINITION

Present techno-economic scenario is marked by increasing competition in almost every sector of economy. The expectations of the customers are on the rise and manufacturers have to design, and produce goods in as many variety as possible (concept of economics of scale is no more talked off) to cater to the demands of the customers. Thus, there is a challenge before the industries to manufacture goods of right quality and quantity and at right time and at minimum cost for their survival and growth. This demands an increase in productive efficiency of the organisations. Industrial Engineering is going to play a pivotal role in increasing the productivity. Various industrial engineering techniques are used to analyse and improve the work methods, to eliminate waste and proper allocation and utilisation of resources.

Industrial engineering is a profession in which a knowledge of mathematical and natural sciences gained by study, experience and practice is applied with judgement to develop the ways to utilise economically the materials and other natural resources and forces of nature for the benefit of mankind.

American Institute of Industrial Engineers (AIIE) defines Industrial Engineering as follows:

Industrial Engineering is concerned with the design, improvement and installation of integrated system of men, materials and equipment. It draws upon specialised knowledge and skills in the mathematical, physical sciences together with the principles and methods of engineering analysis and design to specify, predict and evaluate the results to be obtained from such systems.

The prime objective of industrial engineering is to increase the productivity by eliminating waste and non-value adding (unproductive) operations and improving the effective utilisation of resources.

1.2. HISTORY AND DEVELOPMENT OF INDUSTRIAL ENGINEERING

History of industrial engineering dates back to industrial revolution and it has passed through various phases to reach the present advanced and developed stage. Though Frederick Taylor is named as father of scientific management and Industrial Engineering, there are many others who contributed to the industrial engineering field before Taylor and then they got associated with industrial engineering,

Adam Smith's concept of Division of labour through his book *The wealth of the nation* in 1776 is important as it influenced the factory system.

James Watt, Arkuwright, Boultin Mathew and Robinson obtained a place in the history of industrial engineering because of their progressive and scientific attitude towards the improvements in the performance of machines and industries.

Period between 1882-1912 was the critical period in the history of industrial engineering.

The important works during this period are:

1. Factory system and owner — engineer and manager concept.
2. Equal work, equal pay and incentive schemes.
3. Scheduling and Gantt charts.
4. Engineers started taking interest in cost control, and accounting. The most often quoted and acknowledged investigator that have lead to be discipline of industrial engineering in present form was F.W. Taylor, who took interest in human aspects of production and productivity.

The modern industrial engineering techniques had their origin during the period between 1940 to 1946. Predetermined time standards (PMTS), value analysis and system analysis are few prominent ones. They were expanded, refined and applied in subsequent years. Operation Research technique has brought a revolution and changed and expanded the scope of industrial engineering activities. The computers have added dimension to the industrial engineering activities.

Present State of Industrial Engineering

Industrial engineering has not remained restricted to manufacturing activities but has extended its services to service industries also. The development of techniques like

1. Value Engineering,
2. Operation Research,
3. CPM and PERT,
4. Human Engineering (Ergonomics),
5. Systems Analysis,
6. Advances in Information Technology and Computer Packages, and
7. Mathematical and Statistical Tools.

have expanded the scope of activities of industrial engineering. Thus industrial engineering has taken a firm position in the organisation and it is contributing maximum towards increasing productivity and efficiency in particular and Quality of Work Life (QWL) in general.

1.3. CONTRIBUTIONS TO INDUSTRIAL ENGINEERING

1. **Adam Smith (1776):** Adam Smith through his book titled *Wealth of Nations* laid foundation to scientific manufacturing. He introduced the concept of "division of labour". Through his concept of division of labour which included the skill development, time savings and the use of specialised machine was able to influence the factory system.

2. **James Watt (1864):** Steam engine advanced the use of mechanical power to increase productivity.

3. **Charles Babbage** an English mathematician worked on the same line as Adam Smith's division of labour and advocated specialisation as one more advantage of division of labour.

4. **Frederick Taylor (1859-1915):** Frederick Taylor is generally credited with being the father of industrial management and industrial engineering. Taylor was a mechanical engineer who initiated investigations of better work methods and went on to become the first individual to develop an integrated theory of management principles and methodologies. Taylor believed that a

scientific approach to management could improve labour efficiency. He proposed the following actions:

1. Collect data on each element of work and develop standardised procedures for workers.
2. Scientifically select, train and develop workers instead of letting them train themselves.
3. Strive for a spirit of cooperation between management and workers so that high production at good pay is fostered.
4. Divide the work between management and labour so that each group does the work for which it is best suited.

The above principles over the periods, developed into method study and work measurement, training, selection, placement and Industrial relations. So, the Taylor's contribution are:

1. Constitution of day's work,
2. Wage payment system,
3. Elimination of waste,
4. Training of workers,
5. Understanding between managers and workers.

5. **Henry L. Gantt (1913):** Gantt an engineering contemporary of Taylor, had a profound impact on the development of management thinking His contributions were:

1. Work in the area of motivation field and development of task and bonus plan, a highly successful incentive plan,
2. Measurement of management results by Gantt charts,
3. Recognition of social responsibility of business and industry,
4. Advocated training of workers by management.

6. **Frank and Lillian Gilbreth (1917):** The advancement of motion studies is a contribution by Gilbreth. Assisted by his wife, he developed method study as a tool for work analysis. Gilbreth emphasised the relationship between output and the effort of the worker. He developed micro motion study, a breakdown of work into fundamental elements called therbligs.

7. **Harrington Emerson (1913):** He developed his managerial concepts simultaneously with Taylor, Gantt and Gilbreth. Amongst his contributions is the Emerson's Efficiency Bonus Plan, an incentive plan which guarantees the base day rate and pays a graduated bonus.

Emerson's Twelve Principles of Efficiency

1. Clearly defined ideas,
2. Common sense,
3. Competent counsel,
4. Discipline,
5. Fair deal,
6. Reliable, immediate and adequate records,
7. Dispatching,
8. Standards and schedules,
9. Standardised conditions,
10. Standard operations,
11. Written standard practice instructions,
12. Efficiency reward.

8. **L.H.C. Tippet (1937):** He developed the concept of work sampling to determine the equipment and manpower utilisation and for setting performance standards for long cycle, heterogeneous jobs involving teamwork.

1.4. ACTIVITIES OF INDUSTRIAL ENGINEERING

The primary activities as spelled out by AIIE are:

1. Selection of processes and assembling methods.

2. Selection and design of tools and equipment.
3. Design of facilities including plant location, layout of buildings, machines and equipment, material handling system, raw materials and finished goods storage facilities.
4. Design and improvement of planning and control systems for production, inventory, quality and plant maintenance and distribution systems.
5. Developing a cost control system such as budgetary control, cost analysis and standard costing.
6. Development of time standards, costing, and performance standards.
7. Development and installation of job evaluation systems.
8. Installation of wage incentive schemes.
9. Design and installation of value engineering and analysis system.
10. Operation research including mathematical and statistical analysis.
11. Performance evaluation.
12. Organisation and methods (O and M).
13. Project feasibility studies.
14. Supplier selection and evaluation.

1.5. INDUSTRIAL ENGINEERING APPROACH

In carrying out the various activities (functions), the industrial engineering department uses the scientific approach, *i.e.,* the industrial engineer gathers and analyses facts, prepares the alternative solutions taking into consideration all the constraints both internal and external, and selects the best solution for implementation.

For example, an industrial engineering department selects the operation or job for improvement.

This is the stage referred to as problem identification or definition of the problem.

- All the details or facts about the job/operation are collected and recorded using various recording techniques like charts, diagrams or models and templates.
- All the recorded facts are subjected to critical examination by asking series of questions.
- Alternative ways of doing the operation and/or jobs is found out by using various techniques like brain storming, etc.
- Based upon the criteria fixed for evaluation, the best alternative is selected.

Only selection of best solution is not the end but industrial engineering department has the responsibility of preparing recommendation for implementation so that the organisation will get benefit out of this improved method.

To ensure that the actions it recommends are beneficial to the company, the industrial engineering department must operate with objectivity:

- In approaching the problem it has to listen to and evaluate objectively the viewpoints of the concerned departments affected.
- In making recommendations, its selected course of action should be supported by sound reasoning to prove that it offers the best solution.
- The industrial engineering department must be prepared to meet the prejudiced point of view and to treat them understandably but firmly.
- It should seek opinion of line management and it should not loose the sight of the fact that its first concern is to strengthen the overall operations of the company.

1.6. OBJECTIVES OF INDUSTRIAL ENGINEERING

The basic objectives of industrial engineering departments are:

1. To establish methods for improving the operations and controlling the production costs, and

2. To develop programmes for reducing those costs.

Industrial engineering department exists primarily to provide specialised services to production departments. The services offered depend on the type of organisation. Normally the services include such functions as—method study, establishing time standards, development of wage — incentive schemes, job evaluation and merit rating. In some cases, industrial engineering department is assigned to head projects.

1.7. FUNCTIONS OF AN INDUSTRIAL ENGINEER

1. Developing the simplest work methods and establishing one best way of doing the work. (Standard Method)
2. Establishing the performance standards as per the standard methods. (Standard Time)
3. To develop a sound wage and incentive schemes.
4. To aid in the development and designing of a sound inventory control, determination of economic lot size and work-in-process for each stage of production.
5. To assist and aid in preparing a detailed job description, and job specification for each job and to evaluate them.
6. Development of cost reduction and cost control programmes, and to establish standard costing system.
7. Sound selection of site and developing a systematic layout for the smooth flow of work without any interruptions.
8. Development of standard training programmes for various levels of organisation for effective implementation of various improvement programmes.

1.8. TECHNIQUES OF INDUSTRIAL ENGINEERING

The tools and techniques of industrial engineering aim at improving the productivity of the organisation by optimum utilisation of organisation's resources, *i.e.,* men, materials and machines. The various tools and techniques of industrial engineering are:

1. **Method study:** To establish a standard method of performing a job or an operation after thorough analysis of the jobs and to establish the layout of production facilities to have an uniform flow of material without back tracking.
2. **Time study (work measurement):** This is a technique used to establish a standard time for a job or for an operation.
3. **Motion economy:** This is used to analyse the motions employed by the operators to do the work. The principles of motion economy and motion analysis are very useful in mass production or for short cycle repetitive jobs.
4. **Financial and non-financial incentives:** These helps to evolve at a rational compensation for the efforts of the workers.
5. **Value analysis:** It ensures that no unnecessary costs are built into the product and it tries to provide the required functions at the minimum cost. Hence, helps to enhance the worth of the product.
6. **Production, planning and control:** This includes the planning for the resources (like men, materials and machines), proper scheduling and controlling production activities to ensure the right quantity, quality of product at predetermined time and pre established cost.
7. **Inventory control:** To find the economic lot size and the reorder levels for the items so that the item should be made available to the production at the right time and quantity to avoid stock out situation and with minimum capital lock-up.
8. **Job evaluation:** This is a technique which is used to determine the relative worth of jobs of the organisation to aid in matching jobs and personnel and to arrive at sound wage policy.

9. **Material handling analysis:** To scientifically analyse the movement of materials through various departments to eliminate unnecessary movement to enhance the efficiency of material handling.
10. **Ergonomics (Human Engineering):** It is concerned with study of relationship between man and his working conditions to minimise mental and physical stress. It is concerned with man-machine system.
11. **System analysis** is the study of various sub-systems and elements that make a system, their interdependencies in order to design, modify and improve them to achieve greater efficiency and effectiveness.
12. **Operation research techniques:** These techniques aid to arrive at the optimal solutions to the problems based on the set objective and constraints imposed on the problems. The techniques that are more often used are:
    (*i*) Linear programming problems,
    (*ii*) Simulation Models,
    (*iii*) Queuing models,
    (*iv*) Network analysis (CPM and PERT),
    (*v*) Assignment, sequencing and transportation models,
    (*vi*) Dynamic and integer programming,
    (*vii*) Games theory.
13. **The other techniques include:** Statistical process control techniques, group technology. organisation and methods (O & M).

1.9. PLACE OF INDUSTRIAL ENGINEERING IN AN ORGANISATION

With due consideration to nature of its activities, responsibilities and its working relationship within the company, industrial engineering department should report to the executive who has got the overall responsibility of planning, quality and sales, etc. A typical organisation structure is shown in Fig. 1.1.

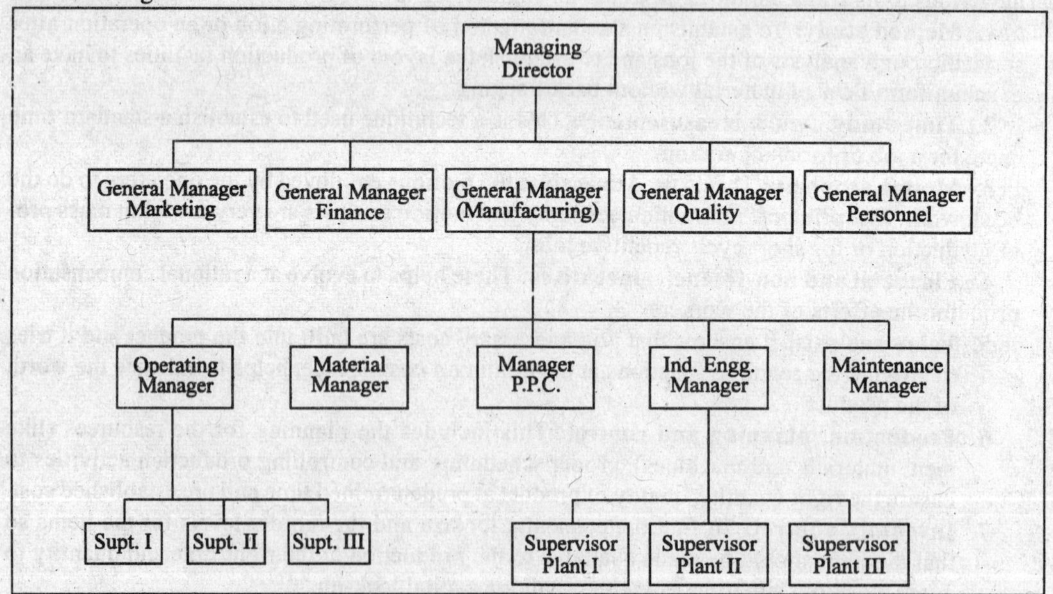

Fig. 1.1: Centralised industrial engineering organisation.

There is no hard and fast rule governing the place of industrial engineering in an organisation. The factors that influence its position in the organisation are:

1. The number of direct employees.
2. Scope of industrial engineering activities.
3. Complexity of manufacturing organisation.

Forms of Internal Organisation

1. **Small company (100 to 300 direct workers):** In a small company 2 to 3 industrial engineers may undertake all the responsibilities. The senior industrial engineer should be the man with a variety of capabilities as he is expected to handle large variety of assignments. Typically the emphasis will be on work methods, fixation of time and performance standards, plant layout, estimating, cost reduction, etc.

2. **Medium company (300 to 600 direct workers):** The organisation generally takes the form of individual sections for various activities of industrial engineering. Functions are well defined and titles like Time study engineer, Method study engineer, Process engineer, etc., are used to denote the activities performed by each specialist and there is a division of responsibility for industrial engineering service.

3. **Large company:** In a large company increased emphasis is placed on such functions as value engineering, operation research, training, wage programmes which is done in greater depth apart from regular specialised functions.

1.10. INDUSTRIAL ENGINEERING IN SERVICE SECTOR

Modern industrial engineering is a broad discipline encompassing the analysis, design and improvement of any and all productive elements of any enterprise.

As a part of the general improvement trend in the service industries, large number of industrial engineers are in demand and attracted to careers in exciting, challenging and rewarding new fields. Fields such as health care, education, etc., invite a creative talents, high degree of professional and technical competence, offer challenging opportunities. The various service industries are:

1. **Industrial engineering in health services:** A number of developments in health industry have resulted in a much wider acceptance and use of industrial engineers in hospitals, with a trend towards cooperative programmes in which group of hospitals share industrial engineering services. The advances in information technology and use of optimisation techniques are helpful to achieve the expected service levels in health care industries by optimum utilisation of resources.

2. **Industrial engineering in government organisations:** The range of activities encompassed by government is as extensive as the industrial engineering techniques themselves. Personnel activities, plant or office location problems, organisation and methods are so complex because of the integrated nature of government and their solution requires not only the application of normal industrial engineering techniques but an extremely broad-based new techniques.

3. **Industrial engineering in banking:** Banking is a large-scale business, dealing with the production and delivery of vital services throughout the economy. The computer, as a tool of industrial engineering is making a major impact on banking. The role of industrial engineering is:

(*a*) ***Training***: Training to make the employees technically competent to carry out the increasingly specialised trend in bank operations.

(*b*) ***Use of operation research***: One of the primary reasons for the slow growth of operations research in banking is that banks have rarely employed engineers in a technical capacity on their staff. Now with a changing trend, technically competent engineers are making a head way in applying OR techniques in banking.

(*c*) ***Cost System***: There is a need for engineers with conceptual awareness of interfacing amongst the various bank management functions. Engineers with ability to bring together the related functions of the bank in actual or simulated mode to present the management an objective appraisal of individual contribution of each unit. A well designed cost system provides the basis for such an evaluation.

(*d*) ***Information systems*:** Reports generated by information system require skilled and thoughtful design. The format and presentation of the data should be simple and should motivate the reader to action. Thus a competent industrial engineer is able to design the information system to integrate various banking activities.

4. **The other areas of application of industrial engineering:** Public utilities such as the companies which supply essential commodities such as water, gas, electricity telephone services.

1.11. SYSTEMS APPROACH

A system is defined as a collection of elements which are interdependent and independent to achieve objective *e.g.,* a production system.

The systems consists of many subsystems *e.g.,* in a production system we have many subsystems like planning system, inventory system, quality system, etc.

The Characteristics of the System

1. Every system has a specified objective to achieve.
2. Every system has got a boundary within which it operates.
3. System has got inputs which are processed into outputs.
4. There are restrictions or constraints imposed on the system by the factors which are internal to the system or external to the system.

The systems thinking or methodology is a wholistic approach which integrates the activities of subsystems and elements and optimises the system effectiveness. A simple system is shown in the Fig.1.2.

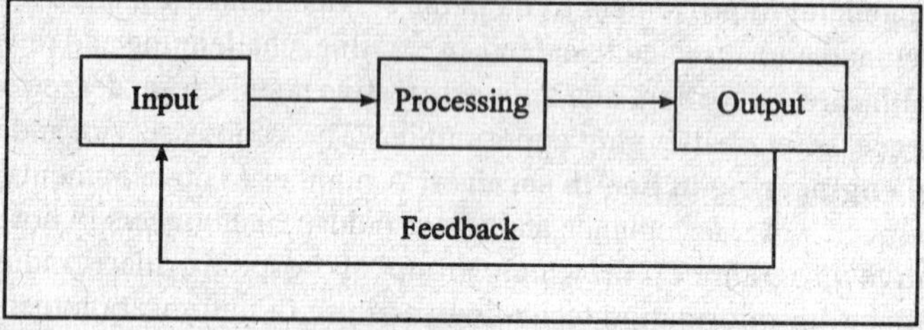

Surroundings (External to the system)

Fig. 1.2: Representation of a simple system.

The systems approach consists of systems analysis, systems engineering and systems management. It is a powerful tool for solving large, complex problems involving men and machines.

A key feature of the systems approach is its emphasis on analysing the interrelationships among various elements of the system. The systems approach takes the wholistic and integrated view of the problem and suggests solutions taking into consideration the influence of elements on the performance of the system. Systems approach consists of:

Systems analysis: Systems analysis includes investigation of system objectives, selection of criteria for evaluating alternative solutions, examination of the feasibility of proposed solutions, evaluation of feasible solution and selection of the optimal solution.

Systems engineering: It is a top level engineering process associated with the development or modification of the complex system. Systems engineering identifies and specifies the subsystem elements that will be assembled engineered, procured, developed, tested and evaluated in accordance with a development plan.

Systems management: Systems management includes the development of the procedures and the organisational structure for planning, directing and controlling systems engineering activities and operations throughout the life of the system.

Systems analysis consists of the following:

1. Problem definition.
2. System objectives.
3. System boundaries.
4. User requirements analysis.
5. System effectiveness measures.
6. Functional analysis.
7. Constraint evaluation.
8. Selection of feasible alternative.
9. Evaluation of the feasible solution against prefixed criteria.
10. Selection of the best alternative.

System Design

The system design is divided into two phases preliminary and detailed design.

Methodology for preliminary design consists of (As per Asimow):

- Screening of design alternatives.
- Development of mathematical model.
- Sensitivity analysis.
- Compatibility analysis.
- Stability analysis.
- Optimisation.
- Projection into future.
- Testing the design concept.
- Simplification of design.

Methodology for detailed design:

- Preparation for design.
- Establishing performance requirements for hardware and software.
- Design specifications preparation.
- Design of elements and subsystems.
- Design of parts.
- Preparation of assembly drawings.
- Experimental construction.
- System integration.
- Testing.
- Redesign.

For managing the systems effectively, the identification of checkpoints (milestones) is necessary which can be used to monitor schedule, budget, and technical performance. These milestones in addition to representing convenient points for monitoring project status, should be chosen to provide decision points for future course of action without affecting schedule and budget.

The major strength of the systems approach is that all the factors that influence the system throughout its life from the conception stage is considered.

Systems Approach in Industrial Engineering

Industrial engineering, relies heavily on systems approach in solving the problems. Instead of analysing the problem in isolation, at the elemental or subsystem level, an integrated view of the problem is considered for suggesting the solution keeping in mind the constraints imposed on the system from both inside and outside the system and the factors which influence the system. Thus,

this wholistic approach helps to achieve the system optimisation instead of analysing part by part which leads to sub-optimisation.

References for Further Reading

1. Maynard, H.B., *Industrial Engineering Handbook*, McGraw Hill Book Company, New York.
2. Vaughan Richard, C., *Introduction to Industrial Engineering*, Lowa State University Press, Ames.
3. Optner Standford, L., *Systems Analysis for Business Management*, Prentice Hall Inc. Englewood Cliffs, N.J. (1968).
4. Asimow, Morris, *Introduction to Design*, Prentice Hall Inc. Englewood Cliffs, N.J. (1962).
5. Chestnut Harold, *Systems Engineering Methods*, John Wiley and Sons. Inc., New York, (1967).

REVIEW QUESTIONS

1. Define industrial engineering? What is its importance?
2. Discuss the scope and objectives of industrial engineering.
3. Discuss the history and development of industrial engineering.
4. What are the functions of industrial engineering?
5. Describe the tools and techniques of industrial engineering?
6. Comment on the place of industrial engineering in an organisation.
7. How does industrial engineering help to increase the productivity of an organisation?
8. Write short notes on:
   (*a*) Industrial engineering approach.
   (*b*) Organisation structure for industrial engineering.
   (*c*) Qualities of an industrial engineer.
   (*d*) Productivity and I.E. techniques.
   (*e*) Industrial engineering in service sector.
   (*f*) Systems approach.
9. What are the phases involved in system methodology?
10. How systems approach is useful to problem-solving?

2

PRODUCTIVITY

• Introduction, concept and definitions • Production and productivity • Expectations from productivity, Benefits • Dynamics of productivity change • Measures of productivity • Productivity measurement models • Levels of productivity • Factors influencing productivity • Measures (Methods) to improve productivity.

2.1. INTRODUCTION

Productivity has now become an everyday watchword. It is crucial to the welfare of the industrial firm as well as for the economic progress of the country. High productivity refers to doing the work in a shortest possible time with least expenditure on inputs without sacrificing quality and with minimum wastage of resources.

Today the term productivity has aquired a wider meaning. Originally, it was used only to rate the workers according to their skills. The person who produced more either faster or harder were said to have higher productivity. Subsequently emphasis was laid to improve the hourly output by analysing and improving upon the techniques applied by different workers. A system of measurement was then evolved to compare the improvement made in relation to the rate of output and inorder to improve productivity further, machines were introduced. Manufacturers of machines started incorporating new features with the help of latest technological developments. Today we have machines that are completely controlled by computers. Computers have now become powerful tools towards improving productivity.

2.2. CONCEPT

Productivity is the quantitative relation between what we produce and what we use as a resource to produce them, *i.e.,* arithmetic ratio of amount produced (output) to the amount of resources (input). Productivity can be expressed as:

$$\text{Productivity} = \frac{\text{Output}}{\text{Input}}$$

Productivity refers to the efficiency of the production system. It is the concept that guides the management of production system. It is an indicator of how well the factors of production (Land, Capital, labour and energy) are utilised.

European Productivity Agency (EPA) has defined productivity as,

"Productivity is an attitude of mind. It is the mentality of progress, of the constant improvements of that which exists. It is the certainty of being able to do better today than yesterday and continuously. It is the constant adaptation of economic and social life to changing conditions. It is the continual effort to apply new techniques and methods. It is the faith in human progress."

A major problem with productivity is that it means many things to many people. Economists determine it from Gross National Product (GNP), Managers view it as cost cutting and speed up, engineers think of it in terms of more output per hour. But generally accepted meaning is that it is the relationship between goods and services produced and the resources employed in their production.

Table 2.1: Productivity as Viewed by Different People

ECONOMISTS	Ratio of output to input (partial productivity Measure and Total Productivity Measure)
ACCOUNTANTS	Financial Ratios, Budgetary Variances
BEHAVIOURAL SCIENTISTS	Labour Utilisation (Man days)
ENGINEERS	Capacity Utilisation, Production per Man hour, Manpower efficiency

2.3. DEFINITIONS OF PRODUCTIVITY

1. Productivity is a function of providing more and more of everything to more and more people with less and less consumption of resources.
2. The volume of output attained in a given period of time in relation to the sum of the direct and indirect efforts expended in its production.
3. Productivity is the measure of how well the resources are brought together in an organisation and utilised for accomplishing a set of objectives.
4. Productivity is concerned with establishing congruency between organisational goals with societal aspirations through input-output relationship.
5. Productivity is the multiplier effect of efficiency and effectiveness.

2.4. PRODUCTION AND PRODUCTIVITY

Production is defined as a process or procedure to transform a set of input into output having the desired utility and quality. Production is a value addition process. Production system is an organised process of conversion of raw materials into useful finished products represented as in Fig. 2.1.

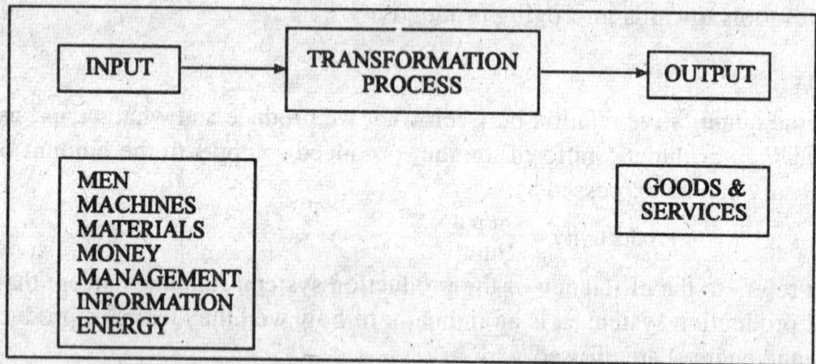

Fig. 2.1: Production system.

The concept of production and productivity are totally different production refers to absolute output whereas productivity is a relative term where in the output is always expressed in terms of inputs. Increase in production may or may not be an indicator of increase in productivity. If the production is increased for the same output, then there is an increase in productivity.

Productivity can be Increased

1. When production is increased without increase in inputs.

2. The same production with decrease in inputs.
3. The rate of increase in output is more compared to rate of increase in input.

***Illustration 1*:** A company produces 160 kg of plastic moulded parts of acceptable quality by consuming 200 kg of raw materials for a particular period. For the next period, the output is doubled (320 kg) by consuming 420 kg of raw material and for the third period, the output is increased to 400 kg by consuming 400 kg of raw material.

During the first year, production is 160 kg

$$\text{Productivity} = \frac{\text{Output}}{\text{Input}} = \frac{160}{200} = 0.8 \text{ or } 80\%.$$

For the second year, production is increased by 100%.

$$\text{Productivity} = \frac{\text{Output}}{\text{Input}} = \frac{320}{420} = 0.76 \text{ or } 76\% \downarrow$$

For third period, production is increased by 150%.

$$\text{Productivity} = \frac{\text{Output}}{\text{Input}} = \frac{400}{400} = 1.0, \textit{ i.e.}, 100\% \uparrow$$

***Comments*:** From the above illustration it is clear that, for second period, though production has doubled, productivity has decreased from 80% to 76%, for period third, production is increased by 150% and correspondingly productivity increased from 80% to 100%.

2.5. EXPECTATIONS FROM PRODUCTIVITY

Expectations differ amongst the various stakeholders, some of the expectations are quite contrast, *i.e.*, the workers expect more leisure time in contrast to managers expectation of hard work.

Table 2.2 shows the expectations of various groups interested in productivity.

Table 2.2: Expectations from Productivity

MANAGEMENT AND ENTREPRENEURS	High Return on Investment (ROI), Higher market share and corporate image.
MANAGERS	Maximum utilisation of resources, lower unit cost, higher quality.
WORKERS	Higher wages, safer work environment, increased quality of work life (QWL).
SUPPLIERS	Prompt payment, continuous order.
CUSTOMERS	Lower cost, quality, reliability, safety and timeliness of delivery.
GOVERNMENT	Economic development, employment generation, more exports.
SHARE HOLDERS	Higher dividends.

2.6. BENEFITS FROM PRODUCTIVITY

Always there is a misunderstanding about productivity in the minds of the workforce. To the workers, higher productivity means higher work load, higher efforts, more profits to owners and unemployment and threat to job security. These are not the correct observations.

Productivity integrates the objectives of owners and workers. Productivity contributes towards increase in production through efficient utilisation of resources and inputs rather than making workers to work hard. Productivity strives to minimise human hazards and human efforts with a view to utilise them to those areas where they can contribute maximum to the output.

2.7. DYNAMICS OF PRODUCTIVITY CHANGE

Productivity improvement results in lower cost per unit by effective utilisation of all the resources and reducing wastage. Lower cost per unit contributes to increased profit levels so that company can reinvest the surplus in new technology, equipment's and machines. This will result in further productivity increase and also there is a greater employment generation due to new investments. The productivity increase results in higher wages to employees as profit potential of the company increases thereby increasing purchasing power of workers. Thus productivity increase sets in a chain reaction as shown in Fig. 2.2.

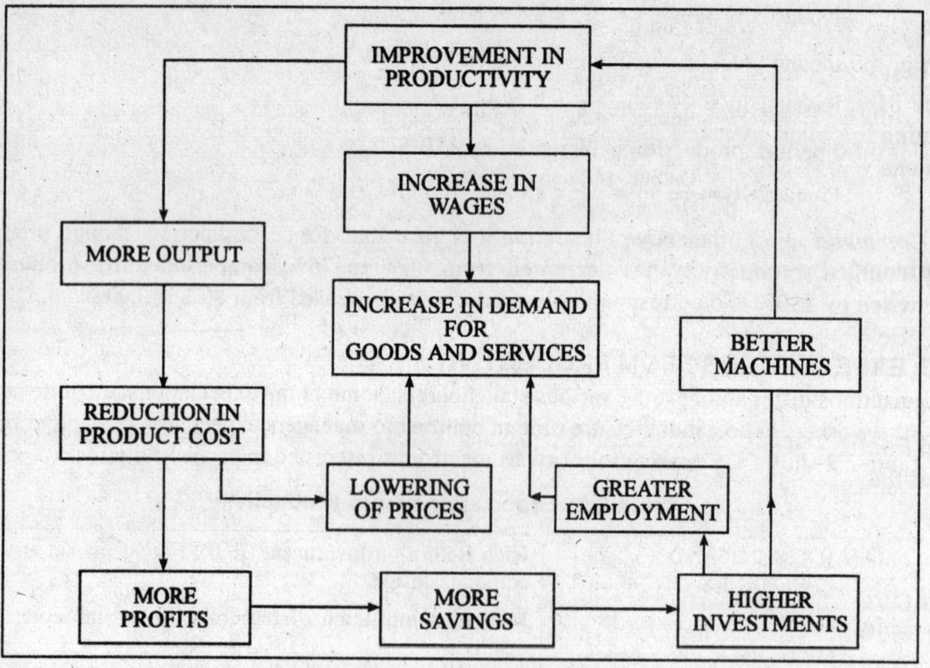

Fig. 2.2: Dynamics of productivity change.

2.8. PRODUCTIVITY MEASURES

Partial Productivity Measures (PPM)

Depending upon the individual input partial productivity measures are expressed as:

$$\text{Partial Productivity} = \frac{\text{Total Output}}{\text{Individual Input}}$$

1. $\text{Labour Productivity} = \frac{\text{Total Output}}{\text{Labour Input}}$

Labour input is measured in terms of man-hours.

2. $\text{Capital Productivity} = \frac{\text{Total Output}}{\text{Capital Input}}$
3. $\text{Material Productivity} = \frac{\text{Total Output}}{\text{Material Input}}$
4. $\text{Energy Productivity} = \frac{\text{Total Output}}{\text{Energy Input}}$

One of the major disadvantage of partial productivity measures is that there is an over emphasis on one input factor to the extent that other inputs are underestimated or even ignored. This cannot represent the overall productivity of the firm.

Total Productivity Measure (TPM)

It is based on all the inputs. This model can be applied to any manufacturing organisation or service company.

$$\text{Total Productivity} = \frac{\text{Total tangible output}}{\text{Total tangible input}}$$

Total tangible output = Value of finished goods produced + value of partial units produced + dividends from securities + interest + other income

Total tangible output = Value of (human + material + capital + energy + other inputs) used

The word tangible here refers to measurable.

The output of the firm as well as the inputs must be expressed in a common measurement unit. The best way is to express them in rupee value. To compare productivity, indices are be adjusted to the base year, and must be stated in terms of base year rupee value. This is referred to as deflating the input and output factors. Deflators are used to nullify the effect of changing price from one year to another.

$$\text{Deflator} = \frac{\text{Current year price}}{\text{Base year price}}$$

Features of Total Productivity Measures

1. Gives both firm level and detailed unit level index.
2. Helps to find out the performance and productivity of the operational unit.
3. Helps to plan, evaluate and control.
4. An important information to strategic planners regarding expansion or phasing out decisions.

Total Factor Productivity Measure (TFP)

It is the ratio of net output to the labour and capital (factor) input.

$$\text{Total Factor Productivity} = \frac{\text{Net output}}{\text{Labour} + \text{capital inputs}}$$

2.9. ADVANTAGES AND LIMITATIONS OF PRODUCTIVITY MEASURES

Advantages	*Limitations*
A. PARTIAL PRODUCTIVITY MEASURE	
1. Easy to understand and calculate.	1. Misleading if used alone.
2. A tool to pinpoint improvement.	2. No consideration of overall impact.
B. TOTAL PRODUCTIVITY MEASURE	
1. Easy and more accurate representation of the total picture of the company.	1. Difficulty in obtaining the data.
2. Easily related to total costs.	2. Requirement of special data collection system.
3. Considers all quantifiable outputs and inputs.	
C. TOTAL FACTOR PRODUCTIVITY MEASURE	
1. Data from company records is relatively easy to obtain.	1. No consideration for material and energy input.
2. Value added approach.	2. Difficult to relate value added approach to production efficiency.

***Illustration 2*:** The following information regarding the output produced and inputs consumed for a particular time period for a particular company given below:

Output = 10,000 Rs.

Human input = 3,000 Rs.
Material input = 200 Rs.
Capital input = 300 Rs.
Energy input = 100 Rs.
Other misc. input = 50 Rs.

The values are in terms of base year rupee value. Compute various productivity indices

***Solution*:**

Partial Productivity

1. Labour Productivity $= \dfrac{\text{Output}}{\text{Human Input}} = \dfrac{10{,}000}{3000} = 3.33$
2. Capital Productivity $= \dfrac{\text{Output}}{\text{Capital Input}} = \dfrac{10{,}000}{3000} = 3.33$
3. Material Productivity $= \dfrac{\text{Output}}{\text{Material Input}} = \dfrac{10{,}000}{2000} = 5.00$
4. Energy Productivity $= \dfrac{\text{Output}}{\text{Energy Input}} = \dfrac{10{,}000}{1000} = 10.00$
5. Other misc. expenses $= \dfrac{\text{Output}}{\text{Other misc. exp.}} = \dfrac{10{,}000}{500} = 20.00$
6. Total Productivity $= \dfrac{\text{Total Output}}{\text{Total Input}}$

$$= \frac{\text{Total Output}}{(\text{Human} + \text{Material} + \text{Capital} + \text{Energy} + \text{Other misc. expenses})}$$

$$= \frac{10{,}000}{3000 + 2000 + 3000 + 1000 + 500} = \frac{10{,}000}{9{,}500} = 1.053$$

7. Total Factor Productivity (TFP) $= \dfrac{\text{Net output}}{(\text{Labour} + \text{Capital})\ \text{input}}$

$$= \frac{\text{Total output} - \text{Material and services purchased}}{(\text{Labour} + \text{Capital})\ \text{input}}$$

Assume that the company purchases all its material and services including energy, m/c and equipment (leasing). Then

$$\text{Total Factor Productivity} = \frac{10{,}000 - (2000 + 3000 + 1000 + 500)}{3000 + 3000} = \frac{3500}{6000} = 0.583$$

***Illustration 3*:** Following information is given pertaining to a firms performance for the last four periods. Compute the partial productivity and total productivity indexes for the company for each of the four periods. Present the results in a tabular form

Assume period-1 as a base year.

Particulars	*Period-1*	*Period-2*	*Period-3*	*Period-4*
A. OUTPUT				
1. Finished goods produced	2500	2200	2800	3200
2. Work-in-process	1200	1600	1000	4000
% of completion	60	50	30	15
Price per unit (Rs.)	1000	1200	1500	1700
3. Dividend from securities	12000	15000	28000	29000
Deflator for item (3)	1	1.11	1.12	1.5
B. INPUTS				
1. Skilled labour (hrs)	[illegible]0,000	12,000	12,000	10,000

(*Contd.*)

Average Wage Rate	60	70	75	75
2. Unskilled labour (hrs)	5000	8000	5000	3000
Wage Rate (Rs.)	30	40	40	50
3. Materials				
Raw materials (tonnes)	20	18	23	25
Price per tonne	1200	1600	2000	2000
4. Total plant hrs worked	1800	2400	2500	2500
Plant hour rate	650	650	1500	2400
5. Energy				
(*i*) Oil used (lts)	5000	3000	2000	1500
Price/litre	4	6	8	12
(*ii*) Coal (tonnes)	200	150	50	–
Price/tonne	1200	1800	2800	–
(*iii*) Electricity (kWH)	15000	18000	22000	30000
Rate/kWH	2.5	3.2	4.2	6.7
6. Other expenses				
(*i*) Consulting fees	20000	–	40000	200000
(*ii*) Information expenses	10000	15000	28000	50000
Deflator for item (6)	1	1.2	1.3	1.3

***Solution*: Sample calculations for period-2.** Assume period-1 as base period.

Deflator for period-2 $= \dfrac{\text{Current year price}}{\text{Base year price}}$

Total output for period-2

$$= \frac{2200 \times 1200}{1.2} + \frac{1600 \times 0.5 \times 1200}{1.2} + \frac{15000}{1.11} = 3013513.50$$

Inputs for period-2

1. Labour input $= \dfrac{12000 \times 70}{1.16} + \dfrac{8000 \times 40}{1.33} = 964739.43$

2. Material input $= \dfrac{18 \times 1600}{1.33} = 21654.13$

3. Capital input (Plant hr cost) $= \dfrac{2400 \times 650}{1} = 156000$

4. Energy input $= \dfrac{3000 \times 6}{1.5} + \dfrac{150 \times 1800}{1.5} + \dfrac{18000 \times 3.2}{1.28} = 237000$

5. Other expenses $= \dfrac{1500}{1.2} = 12500$

Total Input for period-2 (1 + 2 + 3 + 4 + 5) = 2870893.6

Calculation of productivity measures for period-2

1. Total Productivity Measure $= \dfrac{\text{Total Output}}{\text{Total Input}} = \dfrac{3013513.5}{964739.43} = 1.04$

1. Labour Productivity $= \dfrac{\text{Output}}{\text{Labour Input}} = \dfrac{3013513.5}{964739.43} = 3.12$

2. Capital Productivity $= \dfrac{\text{Output}}{\text{Capital Input}} = \dfrac{3013513.5}{156000} = 1.93$

3. Material Productivity $= \dfrac{\text{Output}}{\text{Material Input}} = \dfrac{3013513.53}{21654.13} = 139.64$

4. Energy Productivity $= \dfrac{\text{Output}}{\text{Energy Input}} = \dfrac{3013513.5}{2370000} = 12.71$

5. Other misc. expenses $= \dfrac{\text{Output}}{\text{OtherMisc. exp.}} = \dfrac{3013513.5}{2500} = 241.08$

Productivity measures and productivity indexes are represented in the Tables 2.3 and 2.4.

Table 2.3: Productivity Ratios

Particulars	*Period-1*	*Period-2*	*Period-3*	*Period-4*
T.P.R.	1.36	1.04	1.10	1.37
L.P.R.	4.3	3.12	3.59	5.53
M.P.R.	134.6	139.2	112.8	126.8
C.P.R.	2.76	1.93	1.91	2.34
E.P.R.	10.86	12.71	25.22	47.15
OTHER EXP. PROD. RATIO	107.73	241.8	59.74	19.86

T.P.R. = Total productivity ratio
L.P.R. = Labour productivity ratio
M.P.R. = Material productivity ratio
C.P.R. = Capital productivity ratio
E.P.R. = Energy productivity ratio

Table 2.4: Productivity Index

Particulars	*Period-1*	*Period-2*	*Period-3*	*Period-4*
T.P.I.	1	0.76	0.81	1.01
L.P.I.	1	0.73	0.83	1.02
M.P.I.	1	1.04	0.84	0.94
C.P.I.	1	0.70	0.69	0.85
E.P.I.	1	1.17	2.32	4.34
OTHER EXP. PROD. INDEX	1	2.23	0.56	0.19

T.P.I. = Total productivity index
L.P.I. = Labour productivity index
M.P.I. = Material productivity index
C.P.I. = Capital productivity index
E.P.I. = Energy productivity index

2.10. PRODUCTIVITY MEASUREMENT MODELS

1. **Craig and Harris model:** This model points out the inadequacy of partial productivity measure. It is also called as "Service flow model" because physical inputs are converted into rupees that are payments for services provided by inputs. Productivity is viewed as efficiency of conversion process.

Total productivity is expressed as,

$$P = \frac{O}{L + C + R + Q}$$

where P = Total productivity
L = Labour input factor
C = Capital input factor
R = Raw materials and purchased parts
Q = Other misc, goods and services.

2. **Taylor-Davis model:** Contrary to Craig and Hariss total productivity model, they defined a total factor productivity (TFP) Model.

$$\text{Total Factor Productivity (TFP)} = \frac{S + C + MP - E}{(W + B) + [(KW + KF)\ Fb*df]}$$

where S = Net sales adjusted (*i.e.*, deflated to base year)

C = Inventory change (Raw materials, finished goods and WIP)

MP = Manufacturing plant (Unsaleable products like jigs and fixture, SPM)

E = Exclusions (Materials and services purchased from outside + depreciation of buildings + plant + equipment + rentals)

W = Wages and salary

B = Benefits

KW = Working capital

KF = Fixed capital

Fb = investors contribution (expressed as %)

df = price deflator.

In this model, raw material was not considered as input on the basis that raw material is the result of some other labour and effort.

3. **APC model:** American productivity centre (APC) has developed a comprehensive measure which distinguishes among profitability, price recovery and productivity. It can be utilised to measure productivity changes in labour, materials, energy and capital. It also measures the corresponding effect each one has on profitability.

APC model is based on the premise that profitability is a function of productivity and price recovery. Productivity relates to quantities of output and quantities of inputs, while price recovery relates to price of output and costs of inputs. Price recovery can be thought of as the degree to which input cost increases are passed on to the customers in the form of higher output price.

Relationship between productivity, profitability and price recovery are represented as,

$$\text{Profitability} = \frac{\text{Revenue}}{\text{Cost}}$$

$$= \frac{\text{Output Quantities} \times \text{Sales Price}}{\text{Input Quantities} \times \text{Unit Cost}}$$

$$= \frac{\text{Output Quantities}}{\text{Input Quantities}} \times \frac{\text{Sales Price}}{\text{Unit Cost}}$$

$$\text{Profitability} = \text{Productivity} \times \text{Price recovery}$$

The model compares data from one period (base period) with the data from the current period.

2.11. FACTORS INFLUENCING PRODUCTIVITY

Factors influencing productivity can be classified broadly into two categories: (*a*) Controllable or internal factors and (*b*) Non-controllable or external factors.

S.No	*Controllable (Internal Factors)*	*Uncontrollable (External Factors)*
1.	Product	Structural Adjustments (economic and social)
2.	Plant and Equipment.	Natural Resources
3.	Technology	Government Policy
4.	Materials	Infrastructure
5.	Human Factors	
6.	Work Methods	
7.	Management Style	
8.	Financial Factors	
9.	Sociological Factors	

A. Controllable Factors (Internal Factors)

Product factor: In terms of productivity means the extent to which the product meets output requirements. Product is judged by its usefulness. The cost benefit factor of a product can be enhanced by increasing the benefit at the same cost or by reducing cost for the same benefit.

Plant and equipment: These play a prominent role in enhancing the productivity. The increased availability of the plant through proper maintenance and reduction of idle time increases the productivity. Productivity can be increased by paying proper attention to utilisation, age, modernisation, cost, investments, etc.

Technology: Innovative and latest technology improves productivity to a greater extent. Automation and information technology helps to achieve improvements in material handling, storage, communication system and quality control. The various aspects of technological factors to be considered are

(*i*) Size and capacity of the plant.
(*ii*) Timely supply and quality of inputs.
(*iii*) Production planning and control.
(*iv*) Repairs and maintenance.
(*v*) Waste reduction.
(*vi*) Efficient material handling systems.

Material and energy: Efforts to reduce materials and energy consumption brings about considerable improvement in productivity.

The factors that are to be considered are:

1. Selection of quality material and right material,
2. Control of wastage and scrap,
3. Effective stock control,
4. Development of sources of supply,
5. Optimum energy utilisation and energy savings.

Human factors: Productivity is basically dependent upon human competence and skill. Ability to work effectively is governed by various factors such as education, training, experience aptitude, etc., of the employees. Motivation of employees will influence productivity.

Work methods: Improving the ways in which the work is done (methods) improves productivity. Work study and industrial engineering techniques and training are the areas which improve the work methods which in term enhances the productivity.

Management style: This influence the organisational design, communication in organisation, policy and procedures. A flexible and dynamic management style is a better approach to achieve higher productivity.

B. External Factors

Structural adjustment includes both economic and social changes. Economic changes that influence significantly are:

1. Shift in employment from agriculture to manufacturing industry,
2. Import of technology,
3. Industrial competitiveness.

Social changes such as women's participation in the labour force, education, cultural values, attitudes are some of the factors that play a significant role in the improvement of productivity.

Natural resources: Manpower, land and raw materials are vital to the productivity improvement.

Government and infrastructure: Government policies and programmes are significant to productivity practices of government agencies, transport and communication power, fiscal policies (interest rates, taxes) influence productivity to the greater extent.

2.12. PRODUCTIVITY IMPROVEMENT TECHNIQUES

The basic productivity improvement techniques are represented in the Fig. 2.3.

Technology Based

1. **Computer Aided Design (CAD), Computer Aided Manufacturing (CAM), and Computer Integrated Manufacturing System(CIMS):** CAD refers to design of products, processes or systems with the help of computers. The impact of CAD on human productivity is significant for the advantages of CAD are:
   - Speed of evaluation of alternative designs
   - Minimisation of risk of functioning
   - Error reduction.

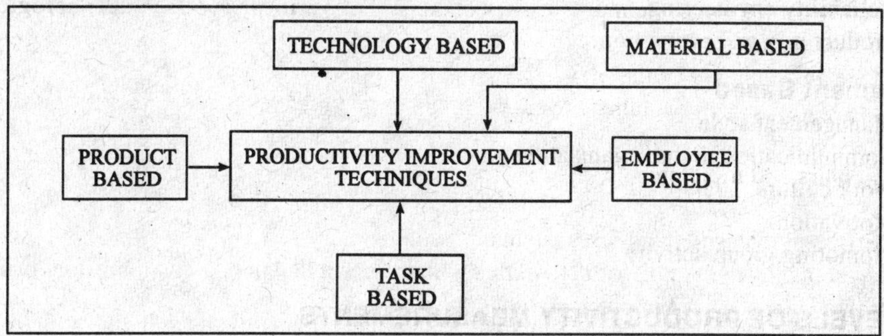

Fig. 2.3: Productivity improvement techniques.

   CAM is very much useful to design and control the manufacturing system. It helps to achieve the effectiveness in production system by Line Balancing.
   - Production planning and control
   - Capacity Requirements Planning (CRP) Manufacturing Resource Planning (MRP II) and Materials Requirement Planning (MRP)
   - Automated inspection.

   Computer integrated manufacturing (CIM) is characterised by automatic line balancing, machine loading/scheduling and sequencing, automatic inventory control and inspection.
2. Robotics.
3. Laser technology.
4. Modern maintenance techniques.
5. Energy technology.
6. Flexible manufacturing system (FMS).

Employee Based

1. Financial and non-financial incentives at individual and group level.
2. Employee promotion.
3. Job design, job enlargement, job enrichment and job rotation.
4. Worker participation in decision-making.
5. Quality circles (QC), Small Group Activities (SGA).
6. Personal Development.

Material Based

1. Material planning and control.
2. Purchasing, logistics.
3. Material storage and retrieval.
4. Source selection and procurement of quality material.
5. Waste elimination.

6. Material recycling and reuse.

Process Based

1. Methods engineering and work simplification.
2. Job design, job evaluation, job safety.
3. Human factors engineering.

Product Based

1. Value analysis and value engineering.
2. Product diversification.
3. Standardisation and simplification.
4. Reliability engineering.
5. Product mix and promotion.

Management Based

1. Management style.
2. Communication in the organisation.
3. Work culture.
4. Motivation.
5. Promoting group activity.

2.13. LEVELS OF PRODUCTIVITY MEASUREMENTS

1. International Level	Development of indexes to compare the growth and competitive position of competing countries.
2. National Level	Developing economic indicators to enable the country to plan its resources on a rational basis.
3. Industry (Sector) Level	Developing measures to compare each others performance to plan its manpower requirements, to compare performance of companies which comprise the industry in a sector.
4. Company Level	Measures to enable them compare themselves in terms of performance experience. To measure trends of productivity improvements to plan effectively company resources.
5. Individual Resource Level	To develop measures to compare performance of each resource amongst one another. To plan the future requirement of these resources.

References for Further Reading

1. David, J. Sumanth, *Productivity Engineering and Management,* Tata McGraw Hill, New Delhi (1990).
2. Adam, E.E. Jr. and R.J. Ebert., *Production and Operations Management,* Prentice Hall Englewood Cliff N.J. (1978).
3. Riggs J.L., *Production System, Planning Analysis and Control,* John Wiley and Sons, New Delhi.
4. *Introduction to Work-Study* (Third Revised Edition) International Labour Office (ILO), Geneva.

REVIEW QUESTIONS

1. What is productivity and what is its relationship with production?
2. "Mere increase in production may or may not contribute to increase in productivity". Comment.
3. What is the relationship between work-study and productivity?

4. Explain partial productivity measures and total productivity measure and what are the advantages and limitations of both.
5. Explain the dynamics of productivity change and how it brings about chain reaction throughout the society.
6. What is deflator? Why it is essential to deflate outputs and inputs?
7. Explain the various models of productivity and compare each model.
8. What are the benefits and expectations of various stakeholders from productivity.
9. Enlist the factors influencing productivity. Explain how each factor will affect productivity.
10. Explain the various tools and techniques to improve productivity.
11. Explain basic work content and excess work content. What are the reasons for excess work content?
12. Mention the tools to reduce the excess work content.
13. Write short notes on:
    (*i*) Level oı productivity measurements,
    (*ii*) Productivity measurement models,
    (*iii*) Measuring productivity of a firm,
    (*iv*) Techniques to improve productivity.

3

WORK-STUDY

• Introduction • Importance of work-study • Advantages of work-study • Work-study procedure • Work simplification and work-study • Human consideration in work-study • Work-study and management • Work-study and supervisor • Work-study and the workers • Work-study man • Influence of method and time study on production activities. • Concept of work content • Reasons for excess work content • Techniques to reduce work content • Work-study as a tool to improve productivity.

3.1. INTRODUCTION

Work-study forms the basis for work system design. The purpose of work design is to identify the most effective means of achieving necessary functions. Historically, this work-study aims at improving the existing and proposed ways of doing work and establishing standard times for work performance.

"Work-study is a generic term for those techniques, method study and work measurement which are used in the examination of human work in all its contexts. And which lead systematically to the investigation of all the factors which affect the efficiency and economy of the situation being reviewed, in order to effect improvement."

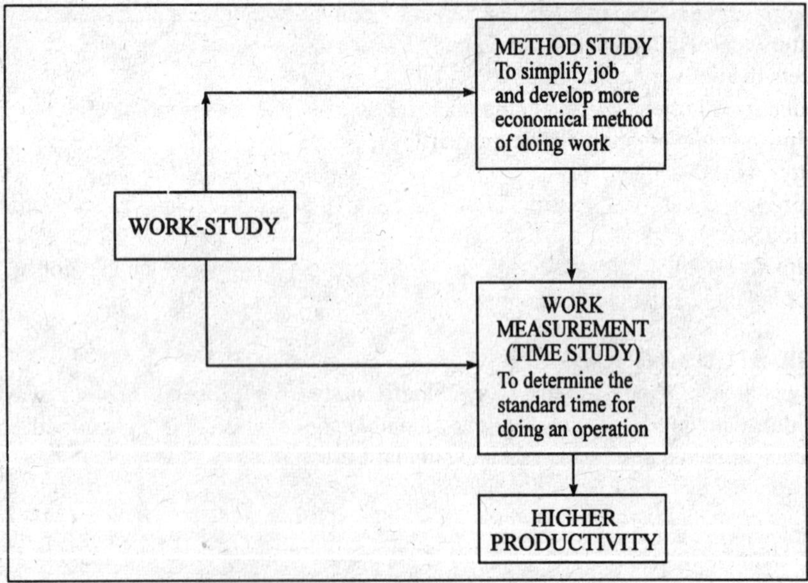

Fig. 3.1: Components of work-study.

Work-study is not new. Since the beginning of human civilisation, there has been a tendency to bring about improvements in the activities the people performed. But this has been organised into a technique and got recognition in the later stages.

Work-study is encompassed by two techniques, *i.e.*, method study and time measurement as shown in Fig. 3.1. **"Method study is the systematic recording and critical examination of existing and proposed ways of doing work, as a means of developing and applying easier and more effective methods and reducing costs."**

"Work measurement is the application of techniques designed to establish the time for a qualified worker to carry out a specified job at a defined level of performance."

There is a close link between method study and work measurement. Method study is concerned with the reduction of the work content and establishing the one best way of doing the job where as work measurement is concerned with investigation and reduction of any ineffective time associated with the job and establishing time standards for an operation carried out as per the standard method.

3.2. IMPORTANCE OF WORK-STUDY

1. Work-study is a means of enhancing the production efficiency (productivity) of the firm by elimination of waste and unnecessary operations.
2. It is a technique to identify non-value adding operations by investigation of all the factors affecting the job.
3. It is the only accurate and systematic procedure oriented technique to establish time standards.
4. It is going to contribute to the profit as the savings will start immediately and continue throughout the life of the product.
5. It has got universal application.

3.3. ADVANTAGES OF WORK-STUDY

1. It helps to achieve the smooth production flow with minimum interruptions.
2. It helps to reduce the cost of the product by eliminating waste and unnecessary operations.
3. Better worker-management relations.
4. Meets the delivery commitment.
5. Reduction in rejections and scrap and higher utilisation of resources of the organisation.
6. Helps to achieve better working conditions.
7. Better workplace layout.
8. Improves upon the existing process or methods and helps in standardisation and simplification.
9. Helps to establish the standard time for an operation or job which has got application in manpower planning, production planning.

3.4. WORK-STUDY PROCEDURE

Work-study is a procedure oriented and systematic study to establish the one best way (standard) method of doing an operation by investigation and analysis of all the details regarding the job or operation carried out as per the established standard method.

Steps Involved in Work-Study

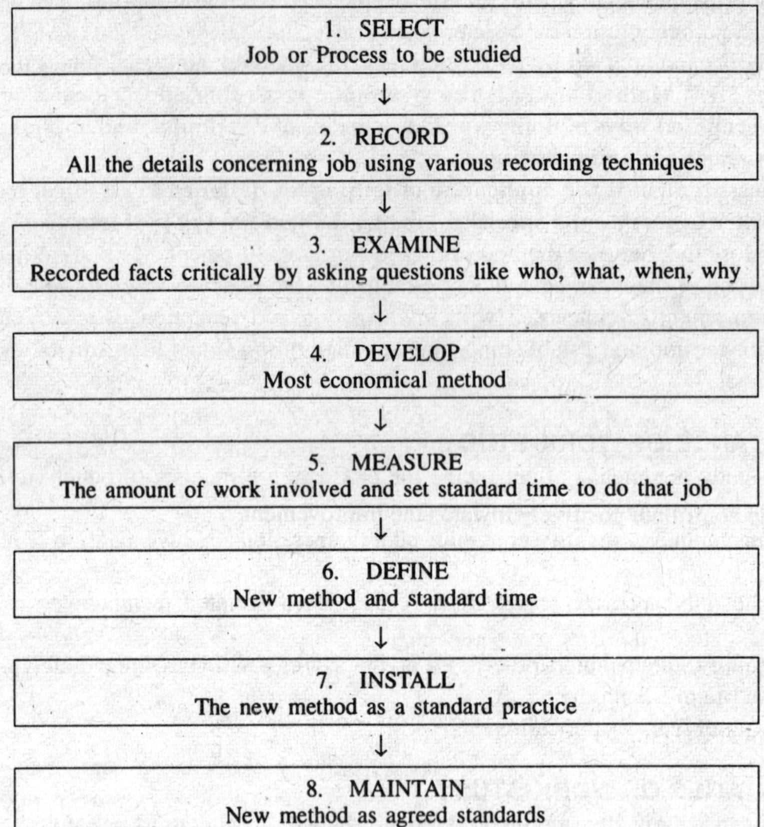

3.5. WORK SIMPLIFICATION AND WORK-STUDY

Any production system is characterised by the coordination of machines and equipments, materials and men. Rapid change in technology and introduction of new technologies are making the processes and methods more complex. Human factor has become all the more important though automation and computer controls are catching up.

The process management is key to the success of the product and company. Method study aims to identify the key processes and process parameters. A detailed investigation is carried out to get all the necessary details in order to analyse the existing process and break the process into parts (operations) which helps to plan and control. A detailed analysis with respect to process inputs (men, material, money) and also the process parameters is carried out to improve the process and to get the desired level of output both in terms of quality and quantity.

The work simplification starts with the analysis of the product and a detailed evaluation with regards to whether it can be changed in such a way as to make it easier to produce by reducing the waste, eliminating non-value adding operations, design modification, etc.

Thus work-study is a powerful tool to make work simplification.

3.6. HUMAN CONSIDERATIONS IN WORK-STUDY

Work-study will become the powerful management tool to improve productivity only if a good relationship is established between the managers, supervisors, employees (workers). Due considerations should be given to everyone concerned as an individual, and should see that no one will perceive a threat to his security and self-respect.

Thus it is essential to consider the relationship between work-study and workers, supervisors and the management as work-study is a participative tool for investigation of the work being performed.

3.7. WORK-STUDY AND THE MANAGEMENT

Managements have to take a positive outlook and attitude towards all the concerned, the supervisor and workers. Traditionally the management should change the preoccupied belief that workers always does not want to work and too much stress on only increasing the labour productivity (making workers to work hard) without giving due considerations to other aspects like working environment, technology and motivation to workers. The management should realise the fact that the workers are the key contributors to the productivity and success of the organisation. The workers should be compensated fairly for contribution and also provide an opportunity to get involved in the affairs concerning them and they should be given an opportunity to participate in the decision-making concerning their work and work related problems. Thus the management should create a climate of mutual trust and confidence in which every individual should feel to contribute positively towards the improvement.

Thus work-study being worker centred, due considerations should be given to their needs, motivation and problems to get maximum benefit out of this technique. Thus management can gain a lot from work-study if it is able to convince the workers and unions regarding objectives of this study and there should a free and open communication between the management and employees to get maximum benefit out of work-study.

3.8. WORK-STUDY AND SUPERVISOR

The work-study man is going to face difficult problems because of supervisor or foreman's attitude. Foreman is a manager on the shop floor to workers and the success of work-study in all its phases depends on him as it is he who is going to cooperate with the work-study man.

Before the work-study begins, the whole purpose of work-study and the procedures involved in the work-study must be carefully explained to the foreman so that he understands exactly what is being done. This facilitates the work of work-study man as he is going to convey and convince the workers regarding the purpose and benefits of work-study. Thus the work-study man has to establish trust and friendship and sell his idea to get the acceptability of the foreman

3.9. WORK-STUDY AND THE WORKERS

Work-study brings about the improvements through changes in the methods, procedures and also some habits. This change, the workers always perceive it as a threat to their job security and their familiarity. Any change is always resisted by human beings as there is lot of uncertainty (probability) is associated. Thus the management and the work-study man should be able to gain the acceptability and confidence of the workers by making them understand the need for the change and how this change is going to benefit both the workers and their organisation. Now, the workers attitude towards work is changing fast workers no more tolerate boredom and monotony on the work. Because of longer years of education and exposure, they want to be master of their own and wants to take decision concerning their work by themselves. So a greater responsibility now rests on the management to constructively channelise their efforts into constructive outlets by providing them an opportunity and climate where in workers will feel affiliated, work to their full potential. In this context, work-study from the workers point of view is gaining special attention.

3.10. WORK-STUDY MAN

There are some qualities and qualifications expected from work-study man. He must be educated to the level that he can grasp the problems. Preferably university degree in engineering is preferred if he is expected to take other responsibility in the area of production management. Basic requirements are:

1. Exposure and experience to the various production systems.
2. A good knowledge of methods and systems of work-study.
3. Objective approach to shopfloor problems.
4. A strong believer of improvement of work methods through work-study.
5. Mentally suited to the work.

Personal Qualities

1. Sincerity and honesty.
2. Enthusiasm.
3. Interest in and sympathy with people.
4. Good appearance and self-confidence.
5. Tact in dealing with people.

3.11. INFLUENCE OF METHOD AND TIME STUDY ON PRODUCTION ACTIVITIES IS SHOWN IN FIG. 3.2

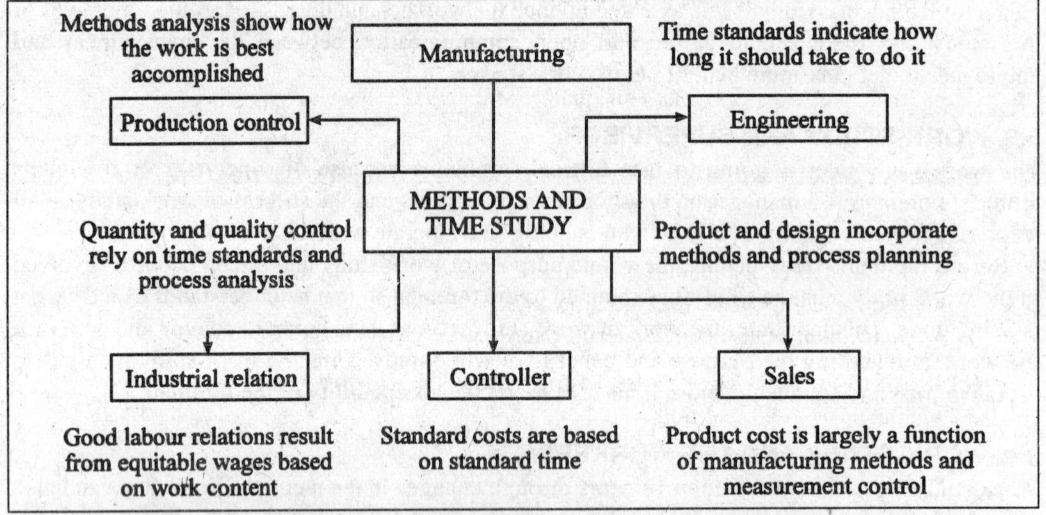

Fig. 3.2: Influence of method and time study on production activities.

The basic objective of production management is to manufacture the right quantity and quality of goods at the predetermined time and pre-established cost. Work-study is tool to achieve this objective. During the product design and process design, the methods of manufacture are fixed and process planning is done using the standard times and standard method. Methods analysis guide with respect to how the work is to be best accomplished and time standards indicate how long will it take to complete the job.

Process analysis and standard times, helps to have a control on quality and quantity manufactured. Based upon the standard times, standard cost are determined and this helps the analysis of variance between actual and standard costs. Product cost which is a function of method and standard time and cost control is very much essential to be in competition. Standard time form the basis for compensation. This helps to link wages and the work content. Thus work-study applied in right spirit helps to accomplish the production objectives.

3.12. CONCEPT OF WORK CONTENT

The amount of work contained in a given job is referred to as work content. For a given job work content is measured in terms of man-hours or machine-hours.

Work content has two constituents:

(*a*) **Basic work content:** Which is the minimum time theoretically required to do an operation or job. This cannot be reduced. Basic work content will result in the following conditions:

- The design and the specification are perfect.
- Process of manufacture is exactly followed.
- No loss of working time due to any of the reasons.

Thus, the basic work content represents an ideal condition which is not possible to achieve.

(*b*) **Excess work content:** The actual time required to complete an operation or job is more than the basic time in practical situations. This additional portion of the work content is called excess work content.

3.13. REASONS FOR EXCESS WORK CONTENT

In a manufacturing company, the excess work content gets added because of the following:

(*a*) Work content added due to defects in design or specification of a product

Typical causes under this classification are

- Bad design of the product.
- Lack of standardisation of components.
- Incorrect specifications and quality standards.
- Faulty design of components.

(*b*) Work content added due to inefficient methods of manufacture

- Improper selection of a manufacturing process/machine.
- Wrong selection of tools.
- Lack of process standardisation.
- Improper layout of the shop/factory.
- Inefficient methods of material handling.

(*c*) Ineffective time added due to shortcomings of the management

- Bad working conditions.
- Frequent production interruptions due to breakdowns.
- Poor production planning and control.
- Lack of safety measures.
- Lack of quality mindedness.
- Improper communication (lack of instructions).
- Frequent changes in set-ups (smaller lot size).
- Lack of performance standards.
- Shortage of materials/tools.

(*d*) In effective time added due to reasons attributed to work man

- Unauthorised absence from work.
- Substandard performance.
- Carelessness in working.
- Unnecessary wastage of time (Idleness).

Figure. 3.3 shows how manufacturing time is made up of.

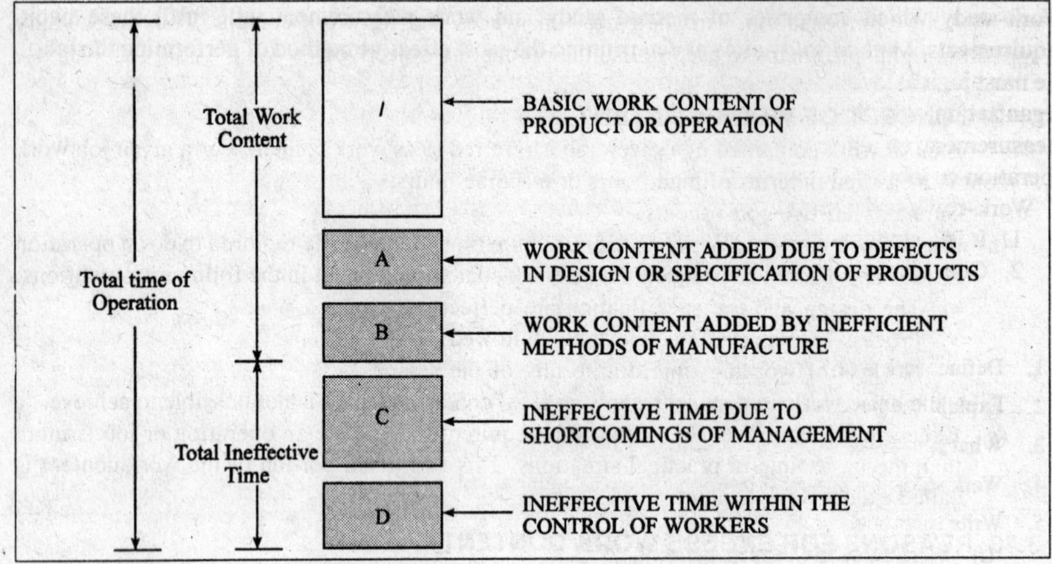

Fig 3.3: How manufacturing time is made up of.

3.14. TECHNIQUES TO REDUCE WORK CONTENT

1. Management techniques to reduce work content due to product

(*a*) Product development.
(*b*) Standardisation (variety reduction)
(*c*) Value analysis.
(*d*) Market research/consumer research.

2. Management techniques to reduce work content due to process or methods

(*a*) Process planning.
(*b*) Methods study.

3. Management techniques to reduce ineffective time due to management

(*a*) Product standardisation and simplification.
(*b*) Product specialisation.
(*c*) Standardisation of component.
(*d*) Production planning and control.
(*e*) Materials control.
(*f*) Plant maintenance.
(*g*) Safety measures and improved working condition.

4. Management techniques to reduce ineffective time within control of the workers

(*a*) Sound personnel policies.
(*b*) Operators training.
(*c*) Safety training.
(*d*) Financial incentives.

3.15. WORK-STUDY AS A TOOL TO IMPROVE PRODUCTIVITY

The important functions of production management are setting up the most effective method of performing the operation (standard method) and control or effective utilisation of resources.

Work-study which comprises of method study and work measurement will fulfil these two requirements. Method study aims at determining the most effective method of performing the job, the most logical layout for manufacturing facilities, uninterrupted flow of materials throughout the organisation, will help to complete the job in the least possible time and at optimum cost. Work measurement on the other hand determine the time required by an operator to complete the operation or job for the standard method at the defined level of performance.

Work-study is the most effective tool to enhance productvity because of the fact that:

1. It is a straight-forward way of increasing the productive efficiency of the organisation.
2. Considers all the factors influencing productivity.

Review Questions

1. Define work study. What are the components of work study ?
2. Explain the work study procedure.
3. What constitutes excess work content ? What are the techniques of reduce the work contents ?
4. Work study is a powerful tool for improving productivity.
5. Write short notes on :
   (*i*) Work study and the management.
   (*ii*) Work study and the workers.
   (*iii*) Work study and the supervisors.

4

METHOD STUDY

• Introduction • Objectives of method study • Scope of method study • Steps involved • Selection of job for method study • Recording the facts • Method study symbols • Recording techniques • Charts • Operation process chart • Flow process chart • Two handed process chart • Multiple activity chart • Diagrams—Flow diagram and String diagram • Micro motion study, Simo chart • Memo motion study, Cycle graph and Chronocycle graph • Critical examination • Development and selection of new method • Principles of motion economy • Installation of the proposed method • Maintaining the proposed method.

4.1. INTRODUCTION

Method study enables the industrial engineer to subject each operation to systematic analysis. The main purpose of method study is to eliminate the unnecessary operations and to achieve the best method of performing the operation.

Method study is also called methods engineering or work design. Method engineering is used to describe collection of analysis techniques which focus on improving the effectiveness of man and machines.

According to British Standards Institution (BS 3138):

"Method study is the systematic recording and critical examination of existing and proposed ways of doing work as a means of developing and applying easier and more effective methods and reducing cost."

Fundamentally method study involves the breakdown of an operation or procedure into its component elements and their systematic analysis. In carrying out the method study, the right attitude of mind is important. The method study man should have:

1. The desire and determination to produce results.
2. Ability to achieve results.
3. An understanding of the human factors involved.

Method study scope lies in improving work methods through process and operation analysis. Such as;

(*i*) Manufacturing operations and their sequence.
(*ii*) Workmen.
(*iii*) Materials, tools and gauges.
(*iv*) Layout of physical facilities and work station design.
(*v*) Movement of men and material handling.
(*vi*) Work environment.

4.2. OBJECTIVES OF METHOD STUDY

Method study is essentially concerned with finding better ways of doing things. It adds value and increases the efficiency by eliminating unnecessary operations, avoidable delays and other forms of waste.

The improvement in efficiency is achieved through:

1. Improved layout and design of workplace.
2. Improved and efficient work procedures.
3. Effective utilisation of men, machines and materials.
4. Improved design or specification of the final product.

The objectives of method study techniques are:

(*i*) To present and analyse true facts concerning the situation.
(*ii*) To examine those facts critically.
(*iii*) To develop the best answer possible under given circumstances based on critical examination of facts.

4.3. SCOPE OF METHOD STUDY

The scope of method study is not restricted to only manufacturing industries. Method study techniques can be applied effectively in service sector as well. It can be applied in offices, hospitals, banks and other service organisations.

The areas to which method study can be applied successfully in manufacturing are:

1. To improve work methods and procedures.
2. To determine the best sequence of doing work.
3. To smoothen material flow with minimum of back tracking and to improve layout.
4. To improve the working conditions and hence to improve labour efficiency.
5. To reduce monotony in the work.
6. To improve plant utilisation and material utilisation.
7. Elimination of waste and unproductive operations.
8. To reduce the manufacturing costs through reducing cycle time of operations.

4.4. STEPS INVOLVED IN METHOD STUDY

The detailed procedure for conducting the method study is shown in Fig. 4.1.

Steps in Method Study

SELECT **The job to be analysed.**
RECORD **All relevant facts about present method.**
EXAMINE **The recorded facts critically.**
DEVELOP **The most efficient, practical and economic method.**
DEFINE **The new method.**
INSTALL **The method as a standard practice.**
MAINTAIN **That standard practice.**

4.5. SELECTION OF THE JOB FOR METHOD STUDY

Cost is the main criteria for selection of a job, process, department for methods analysis. To carry out the method study, a job is selected such that the proposed method achieve one or more of the following results:

(*a*) Improvement in quality with lesser scrap.
(*b*) Increased production through better utilisation of resources.
(*c*) Elimination of unnecessary operations and movements.
(*d*) Improved layout leading to smooth flow of material and a balanced production line.
(*e*) Improved working conditions.

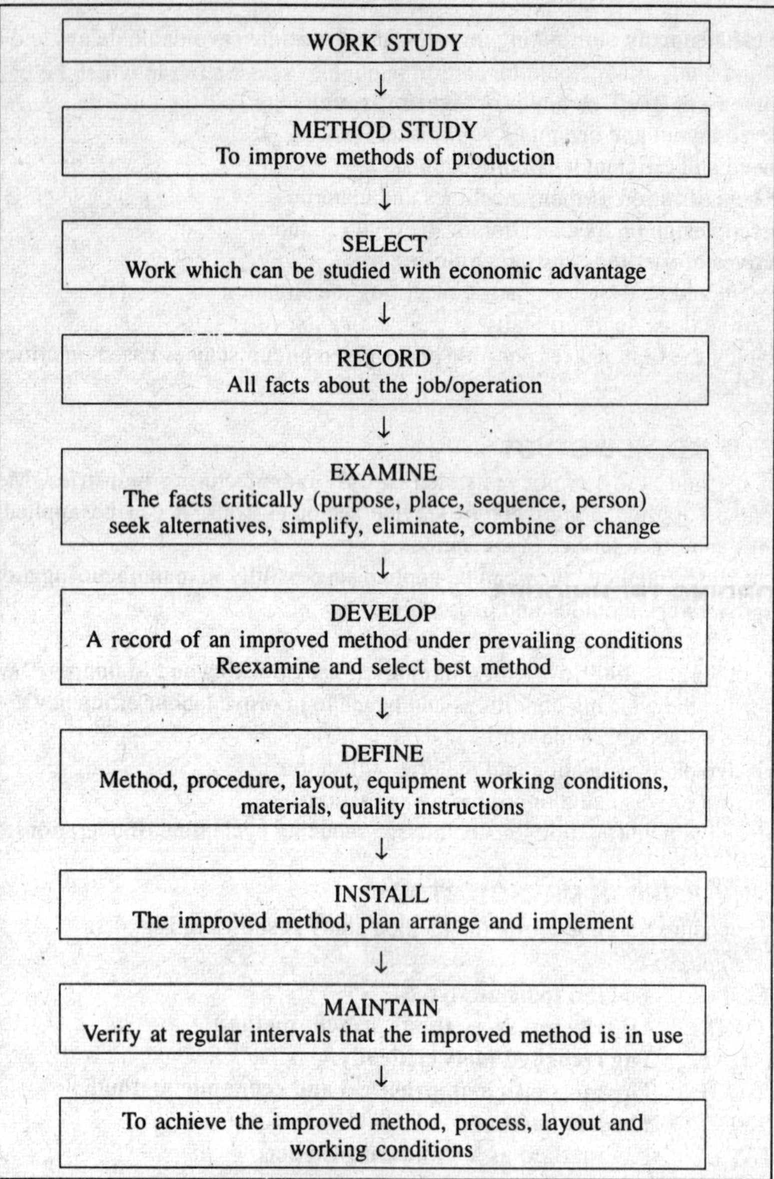

Fig. 4.1: Method study procedure.

The job should be selected for the method study based upon the following considerations: 1. Economic aspect, 2. Technical aspect, and 3. Human aspect.

1. Economic Aspects

The method study involves cost and time: If sufficient returns are not attained, the whole exercise will go waste. Thus the money spent should be justified by the savings from it. The following guidelines can be used for selecting a job:

(*a*) Bottleneck operations which are holding up other production operations.

(*b*) Operations involving excessive labour.

(*c*) Operations producing lot of scrap or defectives.

(*d*) Operations having poor utilisation of resources.

(*e*) Backtracking of materials and excessive movement of materials.

2. Technical Aspects

The method study man should be careful enough to select a job in which he has the technical knowledge and expertise. A person selecting a job in his area of expertise is going to do full justice.

Other factors which favour selection in technical aspect are:

1. Job having in consistent quality.
2. Operations generating lot of scraps.
3. Frequent complaints from workers regarding the job.

3. Human Considerations

Method study means a change as it is going to affect the way in which the job is done presently and is not fully accepted by workman and the union. Human consideration play a vital role in method study. These are some of the situations where human aspect should be given due importance:

1. Workers complaining about unnecessary and tiring work.
2. More frequency of accidents.
3. Inconsistent earnings.

4.6. RECORDING TECHNIQUES

The next step in basic procedure, after selecting the work to be studied is to record all facts relating to the existing method. In order that the activities selected for investigation may be visualised in their entirety and in order to improve them through subsequent critical examination, it is essential to have some means of placing on record all the necessary facts about the existing method. Records are very much useful to make before and after comparison to assess the effectiveness of the proposed improved method.

The recording techniques are designed to simplify and standardise the recording work.

Graphical method of recording was originated by Gilberth. In order to make the presentation of the facts clearly, without any ambiguity and to enable to grasp them quickly and clearly, it is useful to use symbols instead of written description.

Method Study Symbols:

Symbol	Activity
○	OPERATION
□	INSPECTION
⇨	TRANSPORTATION
D	DELAY
▽	STORAGE

1. Operation ○

An operation occurs when an object is intentionally changed in one or more of its characteristics (physical or chemical). This indicates the main steps in a process, method or procedure.

An operation always takes the object one stage ahead towards completion

Examples of operation are:

- Turning, drilling, milling, etc.
- A chemical reaction.
- Welding, brazing and riveting.
- Lifting, loading, unloading.
- Getting instructions from supervisor.
- Taking dictation.

2. Inspection □

An inspection occurs when an object is examined and compared with standard for quality and quantity. The inspection examples are:

- Visual observations for finish.
- Count of quantity of incoming material.
- Checking the dimensions.

3. Transportation ⇨

A transport indicates the movement of workers, materials or equipment from one place to another.

Ex.–Movement of materials from one work station to another.
Workers travelling to bring tools.

4. Delay D

Delay (Temporary Storage)

A delay occurs when the immediate performance of the next planned thing does not take place.

Ex.–Work waiting between consecutive operations.
Workers waiting at tool cribs.
Operators waiting for instructions from supervisor.

5. Storage ▽

A storage occurs when the object is kept in an authorised custody and is protected against unauthorised removal. For example, materials kept in stores to be distributed to various work centres.

4.7. RECORDING TECHNIQUES

According to the nature of the job being studied and the purpose for which the record is required the techniques fall into following categories:·

1. Charts.
2. Diagrams.
3. Templates and models.

CHART:	
1. OPERATION PROCESS CHART (outline process chart)	Gives bird's-eye view of process and records principal operations and inspecting.
2. FLOW PROCESS CHART	
• Man type	Sequence of activities performed by worker.
• Material type	Sequence of activities performed on materials.
• Equipment type	Sequence of activities performed by equipment.
3. MULTIPLE ACTIVITY CHART	Charts activities of men and/or machines on a common time scale.
4. TWO HANDED PROCESS CHART	Activities performed by worker's two hands.
5. TRAVEL CHART	Movement of materials and/or men between departments.
6. SIMO CHART	Activities of worker's hands, legs and other body movements on common time scale.
DIAGRAM:	
7. FLOW AND STRING DIAGRAMS	Path of movement of men and materials.
8. MODELS AND TEMPLATES	Work place layout.
9. CYCLE GRAPH AND CHRONOCYCLE GRAPH	High speed, short cycle operation recording.

4.7.1. CHARTS

This is the most popular method of recording the facts. The activities comprising the jobs are recorded using method study symbols. A great care is to be taken in preparing the charts so that the information it shows is easily understood and recognised. The following information should be given in the chart:

(*a*) Adequate description of the activities.
(*b*) Whether the charting is for present or proposed method.
(*c*) Specific reference to when the activities will begin and end.
(*d*) Time and distance scales used wherever necessary.
(*e*) The date of charting and the name of the person who does charting.

1. Operation Process Chart

It is also called outline process chart. An operation process chart gives the bird's-eye view of the whole process by recording only the major activities and inspections involved in the process. Operation process chart uses only two symbols, *i.e.*, operation and inspection. Operation, process chart is helpful to:

- Visualise the complete sequence of operations and inspections in the process.
- Know where the operation selected for detailed study fits into the entire process.
- In operation process chart, the graphic representation of the points at which materials are introduced into the process and what operations and inspections are carried on them are shown.

Construction of the Chart

A start is made by drawing an arrow to show the entry of the main materials, writing above the descriptions of the components, and below the line the description of the condition. As each operation, inspection takes place, the symbol is entered and numbered in sequence, with a brief description on the right hand side and the time required for the operation on the left side. During assembly process, the major process is charted towards the right hand side of the chart and the subsidiary process on its left hand side. These are joined to each other and to the main trunk at the place of entry of the material or subassembly. The chart does not show where the work takes place, or who performs it. An illustration of operation process chart is shown in Fig. 4.2.

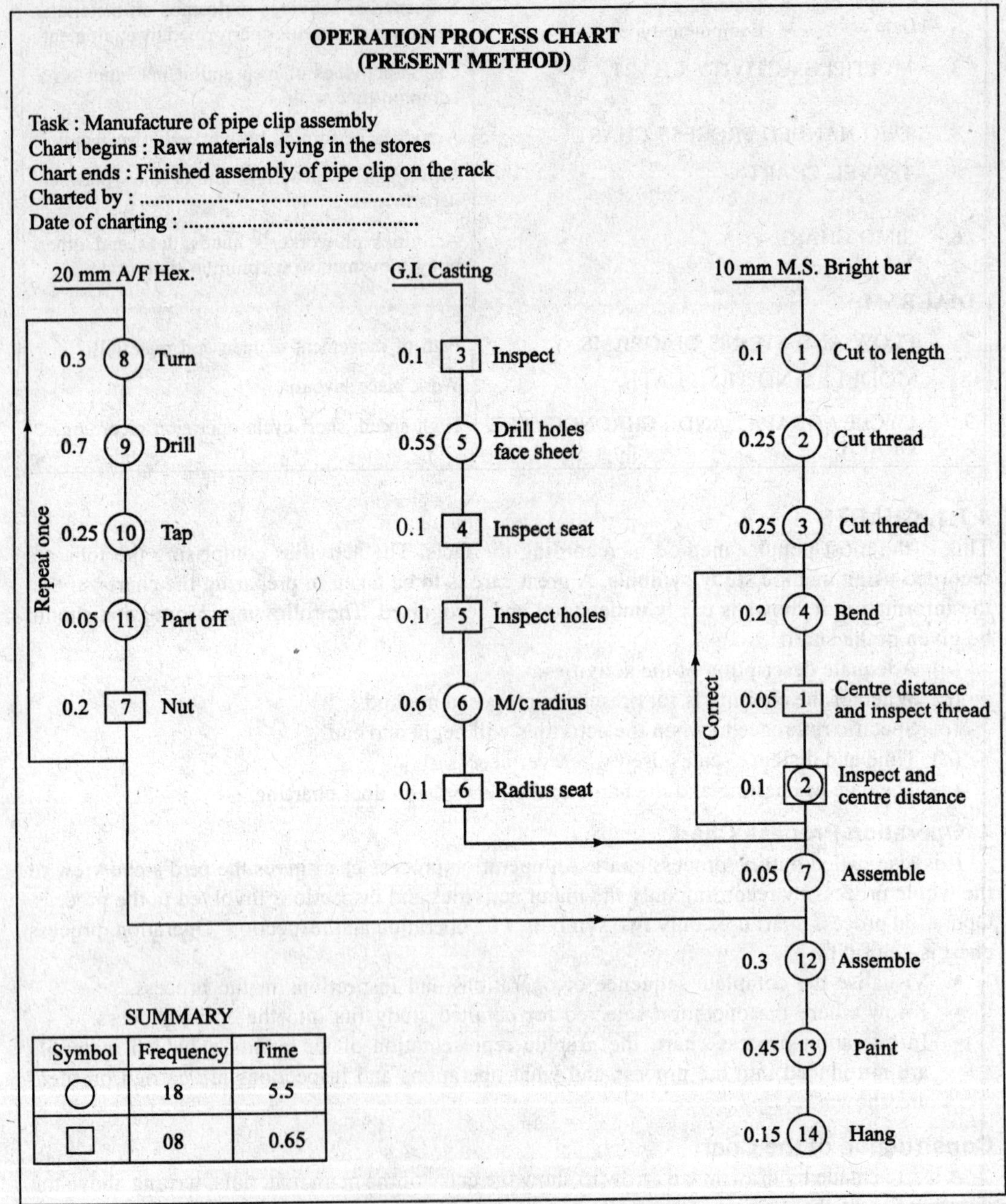

Symbol	Frequency	Time
○	18	5.5
□	08	0.65

Fig. 4.2: Operation process chart.

FLOW PROCESS CHART (Material type)
(PRESENT METHOD)

Task : Machining of the component
Chart begins : Component lying in the stores
Chart ends : The machined component lying in the stores
Charted by : ....
Date of charting : ....

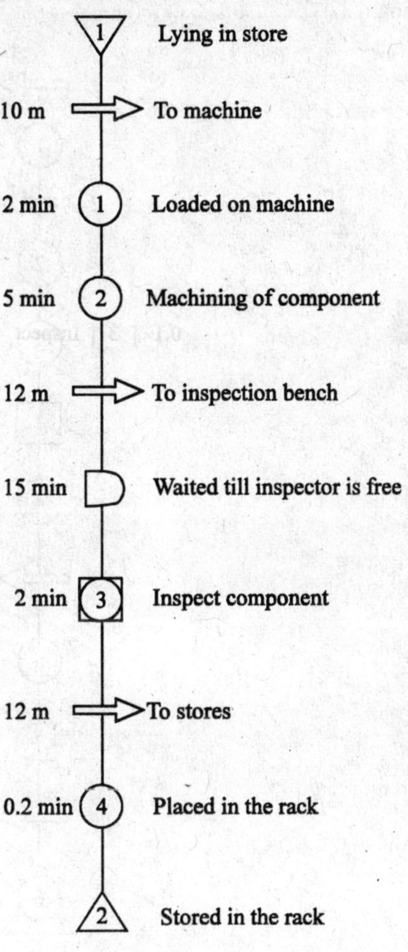

SUMMARY

Symbol	Frequency	Time	Distance
○	4	9.2 min	–
⇨	3	–	34 m
□	1	2 min	–
D	1	15 min	–
▽	2	–	–

Fig. 4.3: Flow process chart (material type).

FLOW PROCESS CHART (Man type)
(PRESENT METHOD)

Task : Writing a letter
Chart begins : Typist in his chiar at his office
Chart ends : Typist puts letter in "out tray"
Charted by : ....
Date of charting : ....

1 ⇨ To officers cabin
(1) Take dictation
2 ⇨ To his own seat
(2) Prepare for typing
(3) Types letter
[1] Checks for mistakes
(4) Place in file for signature
3 ⇨ To officer's cabin
(5) Places file for signature
D1 During checking and signature
4 ⇨ Back to own seat
(6) Type envelope
(7) Put letter in envelope
(8) Keep letter in "out" tray

SUMMARY

Symbol	○	⇨	□	▽	D
Frequency	08	04	01	–	01

Fig. 4.4: Flow process chart (man type).

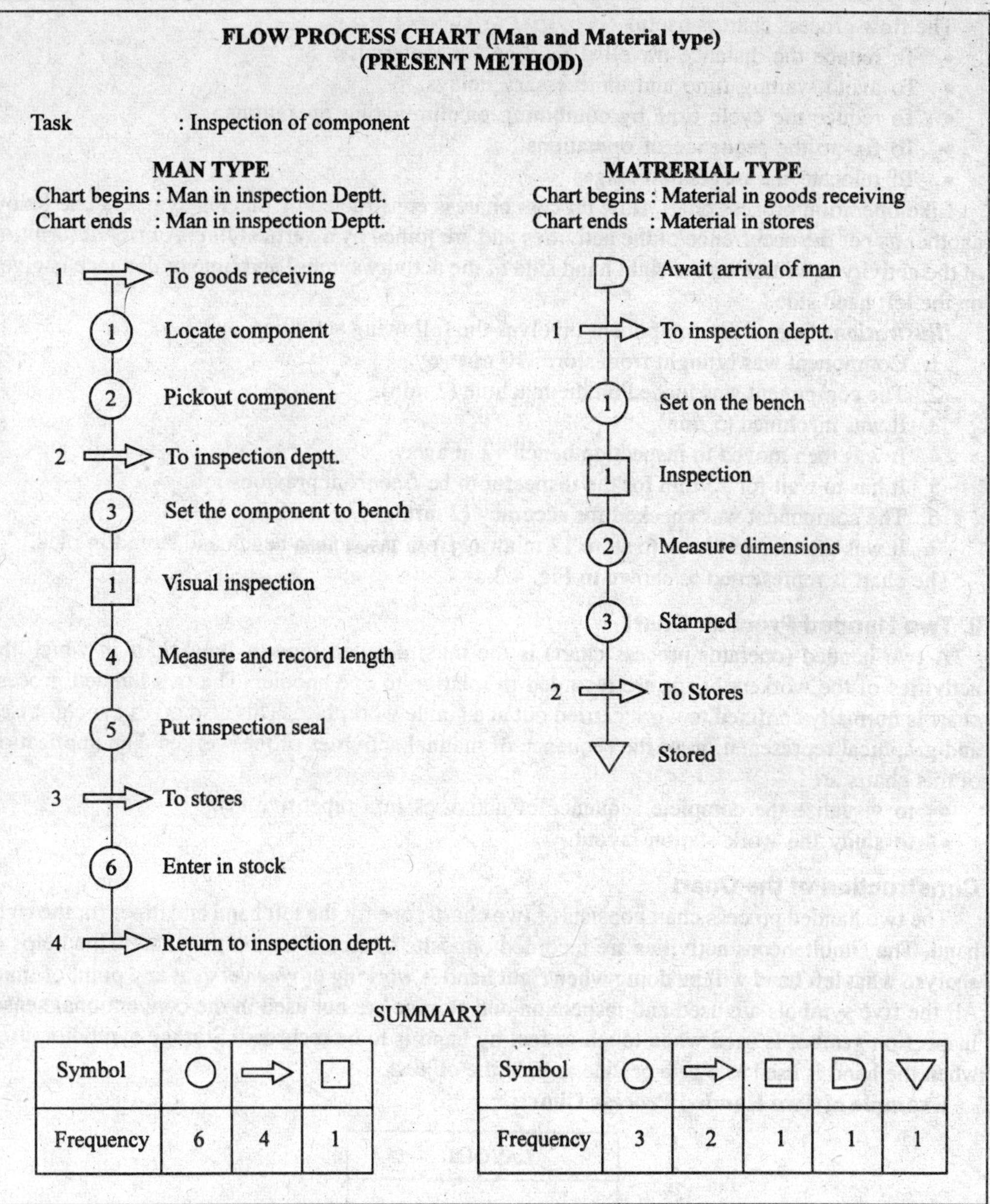

Symbol	○	⇨	□
Frequency	6	4	1

Symbol	○	⇨	□	D	▽
Frequency	3	2	1	1	1

Fig. 4.5: Flow process chart (man and material type).

2. Flow Process Chart

Flow process chart gives the sequence of flow of work of a product, or any part of it through the work centre or the department recording the events using appropriate symbols. It is the amplification of the operation process chart in which operations, inspection, storage, delay and transportation are represented. Flow process charts are of three types:

- Material type—Which shows the events that occur to the materials.
- Man type—Activities performed by the man.
- Equipment type—How equipment is used.

The flow process chart is useful

- To reduce the distance travelled by men (or materials).
- To avoid waiting time and unnecessary delays.
- To reduce the cycle time by combining or eliminating operations.
- To fix up the sequence of operations.
- To relocate the inspection stages.

Like operation process chart, flow process chart is constructed by placing symbols one below another as per the occurrence of the activities and are joined by a vertical line. A brief description of the activity is written on the right hand side of the activity symbol and time or distance is given on the left hand side.

***Illustration*:** One of the component involves the following activities.

1. Component was brought from stores 10 m away.
2. The component was loaded on the machine (2 min).
3. It was machined (5 min).
4. It was then moved to inspection bench 12 m away.
5. It has to wait for 15 min for the inspector to be free from previous job.
6. The component was checked for accuracy (2 min).
7. It was then moved back to store 12 m away from inspection bench and stored in rack.

The chart is represented as shown in Fig. 4.3.

3. Two Handed Process Chart

A two handed (operator process chart) is the most detailed type of flow chart in which the activities of the workers hands are recorded in relation to one another. The two handed process chart is normally confined to work carried out at a single workplace. This also gives synchronised and graphical representation of the sequence of manual activities of the worker. The application of this charts are:

- to visualise the complete sequence of activities in a repetitive task.
- to study the work station layout.

Construction of the Chart

The two handed process chart consists of two charts, one for the left hand and other for the right hand. The simultaneous activities are recorded opposite to each other on the chart. This helps to analyse what left hand will be doing when right hand is working or *vice versa* at any point of time. All the five symbols are used and inspection and storage are not used in the conventional sense. Inspection symbol is used when touch or feel by hand is to be recorded. Storage symbol is used when the hand is used as a grip or vice to hold the object.

Example of Two Handed Process Chart:

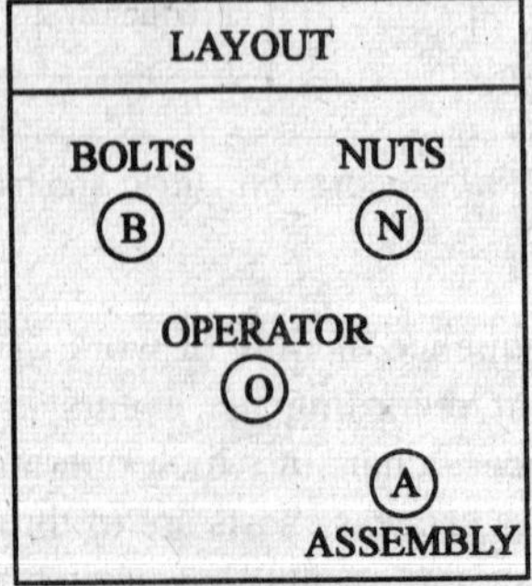

Fig. 4.6: Work bench layout.

4. Multiple Activity Chart

It is a chart where activities of more than subject (worker or equipment) are each recorded on a common time scale to show their inter-relationship. Multiple activity chart is made to

- Study idle time of the man and machines
- Determine number of machines handled by one operator
- Determine number of operators required in teamwork to perform the given job.

Construction of the Chart

A multiple activity chart consists of a series of bars (columns) placed against a common time

TWO HANDED PROCESS CHART
(PRESENT METHOD)

Task : Assembly of nut and bolt
Chart begins : Both hands free before assembly
Chart ends : Both hands free after assembly
Charted by : ....
Date of charting : ....

LEFT HAND		RIGHT HAND	
Description	*Symbol*	*Symbol*	*Description*
Reach for bolt	1 ⇨	1 ⇨	Reach for nut
Grasp bolt head	(1)	(1)	Grasp nut
Carry to central position	2 ⇨	2 ⇨	Carry to central position
Hold bolt	▽	(2)	Place nut on bolt
		(3)	Screw nut
Hold bolt	▽	(4)	Grasp assembly
		3 ⇨	Carry to box
Transfer assembly to right hand	(2)		
	D	(4)	Release assembly
		4 ⇨	Return hand to central position
	D		

SUMMARY

Symbol	○	⇨	▽	D
Frequency (R.H.)	5	4	–	–
Frequency (L.H.)	2	2	2	2

Fig. 4.7: Two handed process chart.

scale. Each subject is allocated one bar and the activities related to the subjects are represented in this bar. The columns are placed against a common time scale which starts at zero and ends at cycle time of the job. The task to be recorded is broken into smaller elements and time for each element is measured with the helps of a stop watch. The activities are then recorded in the chart in their respective columns.

Two symbols are used in the chart—One representing working and other idle. Working is represented by hatched column and Idle is represented by blank as shown in Fig. 4.8.

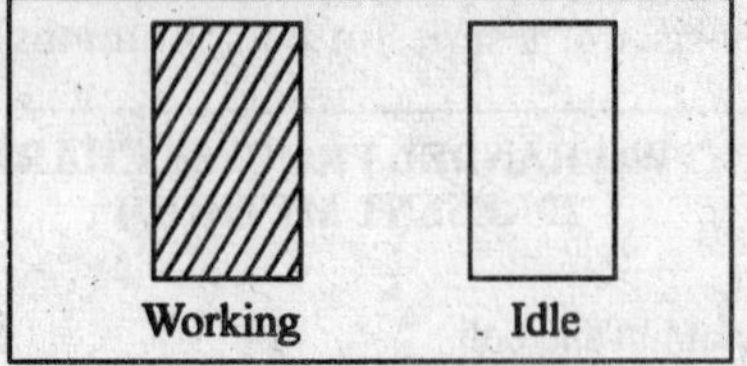

Fig. 4.8: Symbols used in multiple activity chart.

The multiple activity chart is extremely useful in organising teams of operatives on mass production work. This is also used in maintenance. It is used to determine the number of machines which an operator can handle. It is useful in:

- Reducing idle time of machines and operators.
- Combine or eliminate some of the operations.
- It helps to explore ways to increase utilisation of men and machines.

MULTIPLE ACTIVITY CHART
(Present Method)

Task : Machining of a component
Chart begins : The part to be machined lying near machine
Chart ends : Machined part lying in the container
Charted by : ....
Charting date : ....

	Operator			*Machine*		
0	Description	T	S		T	S
0.20	LOAD JOB	0.2		IDLE		
0.28	SWITCH 'ON'	0.08		IDLE		
0.36	SWITCH 'ON'	0.08		IDLE		
1.86	IDLE			MACHINING OF PART "Autocycle"	1.5	
1.91	PICKUP PART	0.05		IDLE		
1.96	KEEP IN TRAY	0.05		IDLE		

Subject	*Cycle time (min)*	*Time worked per cycle*	*Percentage utilisation*
OPERATOR	1.96	0.46	23.4
MACHINE	1.96	1.5	76.6

Fig. 4.9: Multiple activity chart.

An Illustration of Multiple Activity Chart

The operator engaged on the machine performs the following operations:

1. Pick up the job, place it between the jaws of a hydraulic vice (0.2 min).
2. Make the switch 'ON' to tightly hold the part (0.08 min).
3. Make the switch 'ON' start automatic cycle of the operation (0.08 min).
4. Machining of the part on auto cycle (1.5 min).
5. Wait till the vice opens automatically (0.08).
6. Pickup the machined job from the vice (0.05).
7. Keep it in the tray (0.05).

Construct the multiple activity chart for the machining operation.

Multiple activity chart is shown in Fig. 4.9.

4.7.2. DIAGRAMS

The flow process chart shows the sequence and nature of movement but it does not clearly show the path of movements. In the paths of movements, there are often undesirable features such as congestion, back tracking and unnecessary long movements. To record these unnecessary features, representation of the working area in the form of flow diagrams, string diagrams can be made:

(*i*) To study the different layout plans and thereby select the most optimal layout.
(*ii*) To study traffic and frequency over different routes of the plant.
(*iii*) Identification of back tracking and obstacles during movements

Diagrams are of two types: 1. Flow Diagram and 2. String Diagram.

1. Flow Diagram

Flow diagram is a drawing, substantially to scale, of the working area, showing the location of

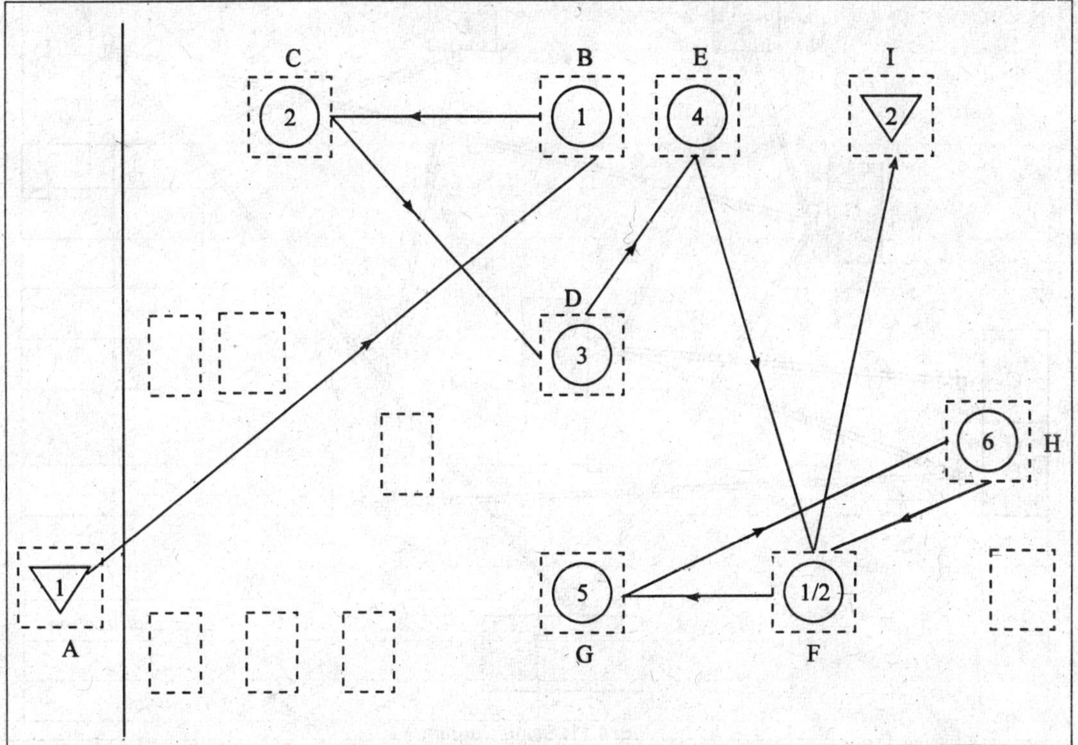

Fig. 4.10: A simple flow diagram.

the various activities identified by their numbered symbols and are associated with particular flow process chart either man type or material type.

The routes followed in transport are shown by joining the symbols in sequence by a line which represents as nearly as possible the paths or movement of the subject concerned.

The procedure to make the flow diagram:

1. The Layout of the workplace is drown to scale.
2. Relative positions of the machine tools, work benches, storage, inspection benches are marked on the scale.
3. Path followed by the subject under study is traced by drawing lines.
4. Each movement is serially numbered and indicated by arrow for direction.
5. Different colours are used to denote different types of movements.

A Simple Flow Diagram is shown in Fig. 4.10.

2. String Diagram

The string diagram is a scale layout drawing on which length of a string is used to record the extent as well as the pattern of movement of a worker working within a limited area during a certain period of time. It is especially valuable where the journeys are so irregular in distance and frequency to see exactly what is happening.

The primary function of a string diagram is to produce a record of an existing set of conditions so that the job of seeing what is actually taking place is made as simple as possible.

One of the most valuable features of the string diagram is the way it enables the actual distance travelled during the period of study to be calculated by relating the length of the thread used to the scale of the drawing. Thus it helps to make a very effective comparison between different layouts or methods of doing job in terms of the travelling involved.

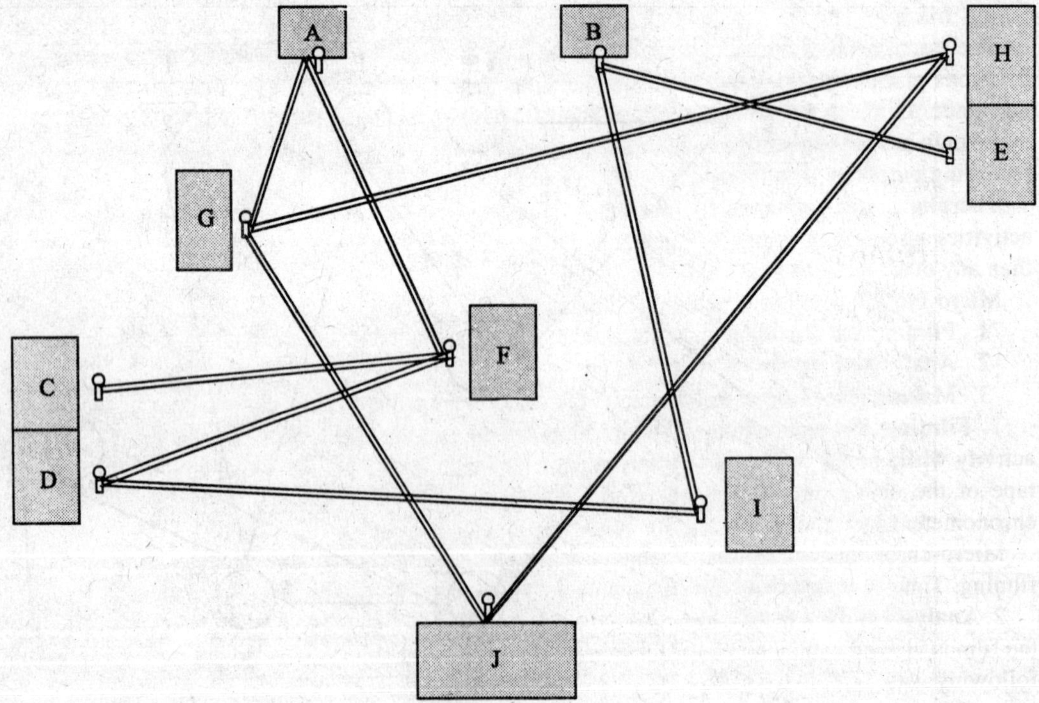

Fig. 4.11: String diagram.

The main advantage of string diagram compared to flow diagram is that repetitive movements between work stations which are difficult to be traced on the flow diagram can be conveniently shown on string diagram.

Procedure to draw string diagram:

1. A layout of the workplace or factory is drawn to scale on a soft board.
2. Pins are fixed into boards to mark the locations of work stations, pins are also driven at the turning points of routes.
3. A measured length of thread is taken to trace the movement (path).
4. The distance covered by the object is obtained by measuring the remaining part of the thread and subtracting it from the original length.

Figure 4.11. shows a string diagram.

4.8. MICRO-MOTION STUDY

Micro-motion study provides a technique for recording and timing an activity. Micro-motion study is a set of techniques intended to divide the human activities in a groups of movements or Micro-motions (called as therbligs) and the study of such movements helps to find for an operator one best pattern of movement that consumes less time and requires less effort to accomplish the task. Therbligs were suggested by Frank B. Gilbreth, the founder of motion study. Micro-motions study was originally employed for job analysis but new uses have been found for this tool. The applications of micro-motion study include the following:

1. Is an aid in studying the activities of two or more persons on a group work.
2. As an aid in studying the relationship of the activities of the operator and the machine as a means of timing operations.
3. As an aid in obtaining motion time data for time standards.
4. Acts as a permanent record of the method and time of activities of the operator and the machine.

The micro-motion group of techniques is based on the idea of dividing human activity into divisions of movements or groups of movements (therbligs) according to purpose for which they are made. Gilbreth differentiated 17 fundamental hand or hand and eye motions to which an eighteenth has subsequently been added. Each therblig has a specific colour, symbol and letter for recording purposes. The therbligs are shown in Table 4.1.

Therbligs refer primarily to motions of human body at the workplace and to the mental activities associated with it. They permit a much more precise and detailed description of the work than any other recording techniques.

Micro-motion study involves the following steps:

1. Filming the operation to be studied.
2. Analysis of the data from the films.
3. Making recording of the data.

1. **Filming the operation:** Micro-motion study consists of taking motion pictures of the activity while being performed by an operator. The equipment required to make a film or video tape of the operation consists of 16 mm movie camera, 16 mm film, wink counter (micro-chronometer) and other usual photographic aids.

Micro-chronometer (or wink counter) is a timing device placed in the field of view while filming. Time is recorded in winks. (1 wink =1/2000 of a minute)

2. **Analysis of data from films:** Once the operation has been filmed and film is processed, then the film is viewed with help of projector for analysis of micro-motions. The film is analysed in the following way:

Film is run at normal speed so as to get familiar with the pattern of movement involved.

Table 4.1: Therbligs

SYMBOL	*CODE*	*NAME*	*DESCRIPTION*	*COLOUR*
	SH	SEARCH	Locate an article	BLACK
	F	FIND	Mental reaction at end of search	GRAY
	ST	SELECT	Selection from a number	LIGHT GRAY
	G	GRASP	Taking hold	RED
	H	HOLD	Prolonged grasp	GOLD OCHRE
	TL	TRASPORT LOADED	Moving an article	GREEN
	P	POSITION	Placing in a definite location	BLUE
#	A	ASSEMBLE	Putting parts together	VIOLET
	U	USE	Causing a device to perform its function	PURPLE
	DA	DISASSEMBLE	Separating parts	LIGHT VIOLET
	I	INSPECT	Examine or test	BURNT OCHRE
	PP	PREPOSITION	Placing an article ready for use	PALE BLUE
	RL	RELEASE LOAD	Release an article	CARMINE RED
	TE	TRASPORT EMPTY	Movement of a body member	OLIVE GREEN
	R	REST	Pause to overcome fatigue	ORANGE
	UD	UNVOIDABLE DELAY	Idle - outside person's control	YELLOW

- A typical work cycle is selected from amongst the filmed cycles.
- Film is run at a very low speed and is usually stopped or reversed frequently to identify the motions (therbligs).

Therbligs after identification are entered in analysis sheet.

3. Recording of data is done using SIMO chart.

4.8.1. SIMO CHART

Simultaneous motion cycle chart (SIMO chart) is a recording technique for micro-motion study. A simo chart is a chart, based on the film analysis, used to record simultaneously on a common time scale the therbligs or a group of therbligs performed by different parts of the body of one or more operators.

It is the micro-motion form of the man type flow process chart. To prepare simo chart, an elaborate procedure and use of expensive equipment are required and this study is justified when the saving resulting from study will be very high.

The format for SIMO chart is shown in Fig. 4.12.

SIMO CHART

Operation : .... Film No. : ....
Part drawing No. : .... Chart No. : ....
Method : .... Present/Proposed Date : ....
Operation No. : .... Charted by: ....

Wink counter Reading	*Left hand description*	*Therbligs*	*Time*	*Time in 200/m*	*Time*	*Therbligs*	*Right hand description*

Fig. 4.12: Format for simo chart.

4.9. MEMO MOTION STUDY

Memo motion photography is a form of time-lapse photography which records activity by the use of cine camera adapted to take picture at longer intervals than normal (time interval normally lies between 1/2 sec to 4 sec).

There are many jobs that have activities which does not need to be examined in fine detail, and are still too fast or intricate to be recorded accurately without the help of a film. There are cases where Micro-motion study is not justified like smaller production quantities, job cycle may exceed 4 minute duration.

The filming of these various classes of works can be performed efficiently and economically by a method of time lapse cine photography known as memo motion.

This is carried out by attaching an electric time lapse unit to the cine camera so that a picture is taken at an interval of time set at any convenient unit between 1/2 sec to 4 sec in frequency. A camera is placed with a view over the whole working area to take pictures at the rate of one or two per second instead of 24 frames a second. The result is that the activities of 10 or 20 minutes may be compressed into one minute and a very rapid survey of the large movements giving rise to wasted efforts can be detected and steps are taken to eliminate them. It is economical compared to Micro-motion study.

4.10. CYCLE GRAPH AND CHRONOCYCLE GRAPH

These are the photographic techniques for the study of path of movements of an operators hands, fingers, etc. These are used especially for those movements which are too fast to be traced by human eye.

A cycle graph is a record of path of movement usually traced by a continuous source of light on a photograph. A small electric bulb is attached to hand, finger or other part of the body of the operator performing the operation. A photograph is taken by still camera and the light source shows the path of the motion and the path of the photograph is called "cycle graph".

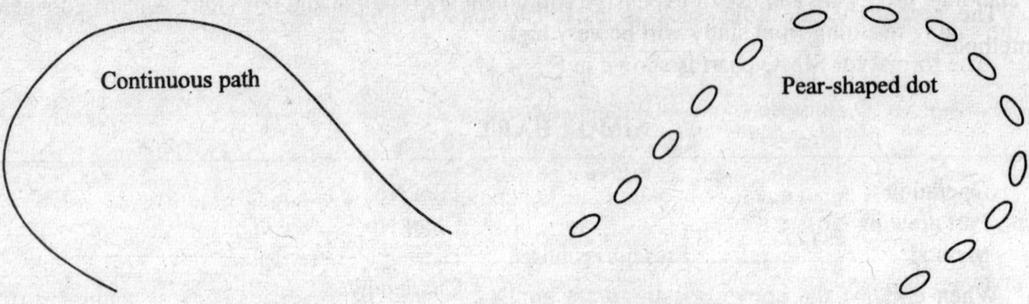

Fig. 4.13: Cycle graph and chronocycle graph.

Chronocycle Graph

Cycle graph has a limitation. It will not give the direction or the speed of movements. This limitation is overcome by chronocycle graph.

The chronocycle graph is a special form of cycle graph in which the light source is suitably interrupted so that the path appears as a series of pear-shaped dots, the pointed end indicating the direction of movement and the spacing indicating the speed of movement. The time taken for the movement can be determined by knowing the rate at which the light source is being interrupted and by counting the number of dots. Cycle graph and chronocycle graph are shown in Fig. 4.13.

4.11. CRITICAL EXAMINATION

The objective of critical examination of recorded facts of an existing or proposed method is to determine the true reasons underlying each event and to make a systematic list of all the possible improvements for later development in a new and improved method.

The principles to be followed during critical examination are:

1. Facts should be examined as they are, not as they appear to be or they should be.
2. Pre-conceived ideas, which often colour the interpretation of facts should be avoided.
3. Hasty judgements should be avoided.
4. All aspects of the problem must be approached with a challenging and sceptical attitude.
5. Every detail must be examined logically and no answer should be accepted until it has been proved correct.

Critical examination is conducted through a systematic and methodical questioning process. The examination is achieved by means of two sets of detailed questions:

- Primary questions.
- Secondary questions to indicate the alternatives and consequently the means of improvements.

Primary Questions

The following are the primary questions under their respective headings:

1. **Purpose:** Analyses whether the job/activity is essential (what is achieved?).
2. **Means:** Analyses whether the activity is being done using right material, right measuring devices and guages. (How is it done? why that way?).
3. **Sequence:** Analyses the sequence of activities in relation to other activities is challenged by asking "when it is done" and why then?
4. **Place:** Analyses whether the activity is done at right place by asking "where is it done" and why there?
5. **Person:** The final question is regarding person performing the activity by asking "who does it" and why that person.

Secondary Questions

The secondary questions seek to establish suitable alternatives to existing or proposed methods.

Under	*It is asked*
Purpose -	What else could be done?
Means -	How else could it be?
Place -	Where else could be?
Sequence -	When else it could be?
Person -	Who else could do it?

When each of the above questions are applied to any event, it may suggest a number of possibilities. When these have been established, it is necessary to ask.

What should be done?
How should it be done?
Where should it be done?
When should it be done?
Who should do it?

The answers to the last five questions indicate the line along which a new method for the overall process should be developed.

Table 4.2: Questions for Critical Examination

Primary Questions			*Secondary Questions*	
			Possible alternatives	*Selected alternatives*
PURPOSE	What is achieved?	Is it necessary? Why?	What else could be done?	What should be done?
PLACE	Where is it done?	Why there? Advantages: Disadvantages:	Where else could it be? A: D:	Where should it be done?
SEQUENCE	When it is done? After: Before:	Why then? A: D:	When else could it be? A: D:	When should it be?
PERSON	Who does it?	Why that person? A: D:	Who else could do it? A: D:	Who should do it?
MEANS	How is it?	Why that way? A: D:	How else could it be? A: D:	How should it?

4.12. DEVELOPMENT AND SELECTION OF NEW METHOD

Critical examination gives rise to number of creative ideas. Since all the ideas are not practicable, some of the ideas are required to be discarded and others are to be refined and developed.

Development involves the analysis of the three phases–evaluation, investigation and selection. For development of the new method, the following approaches can be considered:

1. Eliminate all unnecessary operations.
2. Combine operations and elements.
3. Change the sequence of operations.
4. Simplify the necessary operations.

Steps in development and selection:

(A) **Evaluation:** Evaluation phase tests the true worth of each alternative and thereby decide whether an idea should be pursued or discarded. It is therefore, an exercise to short list the creative ideas.

(B) **Investigation**: Investigation explores as to how the ideas cleared at the evaluation stage suitably can be converted into practical suggestions. Investigation usually involves preparation of drawings, making prototypes, conducting trial runs. The aim is to test each idea for its economic and technical feasibility so that each suggestion is definite and supported by evidence of practicability. Investigation involves the testing technical and economic feasibility.

(C) **Selection**: Each alternative needs to be evaluated against a set of specific factors. The most commonly selected factors are investment required, production rate, manufacturing cost per piece, return on investment. Using point system, weights are then assigned to each of the factors, performance of each factor is then predicted for each alternative. This step is followed by evaluation process of each alternative against each specific factor.

To select a preferred alternative, the points scored by each alternative against each specific factors are added. The alternative scoring the maximum is selected.

4.13. PRINCIPLES OF MOTION ECONOMY

There are a number of principles concerning the economy of movements which have been developed as a result of experience and which forms the basis for the development of improved methods at the workplace. These are first used by Frank Gilbreth, the founder of motion study and further rearranged and amplified by Barnes, Maynard and others.

The principles are grouped into three headings:

(*a*) Use of the human body.
(*b*) Arrangement of workplace.
(*c*) Design of tools and equipment.

A. Uses of Human Body

When possible:

1. The two hands should begin and complete their movements at the same time.
2. The two hands should not be idle at the same time except during periods of rest.
3. Motions of the arms should be made simultaneously.
4. Hand and body motions should be made at the lowest classification at which it is possible to do the work satisfactorily.
5. Momentum should be employed to help the worker, but should be reduced to a minimum whenever it has to be overcome by muscular effort.
6. Continuous curved movements are to be preferred to straight line motions involving sudden and changes in directions.

7. "Ballastic" (*i.e.,* free swinging) movements are faster, easier and more accurate than restricted or controlled movements.
8. Rhythm is essential to the smooth and automatic performance of a repetitive operation. The work should be arranged to permit easy and natural rhythm wherever possible.
9. Work should be arranged so that eye movements are confined to a comfortable area, without the need for frequent changes of focus.

B. Arrangement of the Workplace

10. Definite and fixed stations should be provided for all tools and materials to permit habit formation.
11. Tools and materials should be pre-positioned to reduce searching.
12. Gravity fed, bins and containers should be used to deliver the materials as close to the point of use as possible.
13. Tools, materials and controls should be located within a maximum working area and as near to the worker as possible.
14. Materials and tools should be arranged to permit the best sequence of motions.
15. "Drop deliveries" or ejectors should be used wherever possible, so that the operative does not have to use his hands to dispose of finished parts.
16. Provision should be made for adequate lightning, and a chair of type and height to permit good posture should be provided. The height of the workplace and seat should be arranged to allow alternate standing and seating.

C. Design of Tools and Equipments

17. The colour of the workplace should contrast with that of work and thus reduce eye fatigue.
18. The hands should be relieved of all work of "holding" the work piece where this can be done by a jig or fixture or foot operated device.
19. Two or more tools should be combined where possible.
20. Where each finger performs some specific movement, as in type writing, the load should be distributed in accordance with the inherent capacities of the fingers.
21. Handles such as those used on screw drivers and cranks should be designed to permit maximum surface of the hand to come in contact with the handle.
22. Levers, cross bars and wheel bars should be in such position that operator can manipulate them with least body change and with greatest mechanical advantage.

4.14. INSTALLATION OF THE PROPOSED METHOD

Installation refers to the implementation of the proposed method and it serves the following objectives:

- Preparation of change proposal to management.
- Steps to prepare its implementation on acceptance of proposal.
- To get formal approval from management.
- To implement the accepted proposal.

Installation is composed of two steps:

- Recommendation phase.
- Implementation phase.

In the recommendation phase: The formal written report should be prepared for the changed method, present the recommendations to the management. Also provide information on implementation plan and get the approval of the management.

Implementation phase: The entire study effort will go waste if the proposal is not implemented. Though the responsibility of implementation is that of top management yet the assistance of method study man is required to :

1. to tackle problems at implantation phase.
2. minimise delay in implementation process.
3. the proposal is implemented in its entirety.

4.15. MAINTAIN THE PROPOSED METHOD

Method changes does not end up with implementation of the proposal. Follow-up after the implementation is equally important. The maintenance of the proposed method involves:

1. Monitoring and control.
2. Audit of the savings.
3. Review of the approach.
4. Evaluation of effectiveness of proposed method.

References for Further Reading

1. Barnes, R.L., *Motion and Time Study, Design and Measurement of Work*, 7th Edn., John Wiley and Sons, New York (1980).
2. Mundel, M.E., *Motion and Time Study*, 5th Edn., Prentice Hall, Englewood Cliffs, N.J. (1978).
3. Niebel, B.W., *Motion and Time Study*, 6th Edn., Richardo Irwin Inc. Homewood Ill, (1976).
4. Curie, R.M.; *Work-Study*, 4th Edn., ELBS and PITMAN (1977).
5. Jhamb, L.C., *Work-Study and Ergonomics*.
6. ILO Work-Study.

REVIEW QUESTIONS

1. Define method study. What are its objectives? Differentiate between work measurement and method study.
2. Justify why method study should proceed work measurement.
3. Explain the procedure for method study.
4. How is the job selected for method study?
5. Differentiate between method study and value engineering.
6. Explain the significance, construction and applications of the following recording techniques:
   (*a*) Operation (outline) process chart,
   (*b*) Flow process chart,
   (*c*) Multiple activity chart,
   (*d*) Two handed process chart,
   (*e*) String diagrams and flow diagrams,
   (*f*) Templates and models,
   (*g*) Travel chart.
7. Construct a two handed process chart for an operator washing glass bottles with brush whose normal cycle is as follows:
   Pick up two glass bottles with left hand
   Put them on the fixture
   Press the fixture with right hand towards the brush
   Extend left hand and pick up two more bottles
   Wait few seconds
   When washing is complete with draw fixture with right hand
   Life two washed bottles on side table and simultaneously load
   Fixture with two bottles ready for washing.

8. What are therbligs? When it is used? What are the advantages of micro-motion study?
9. What are therbligs? Give any five therbligs with symbols.
10. Critical examination forms the basis for methods improvement explain.
11. What are primary and secondary questions? How they are useful?
12. What are principles of motion economy?
13. Write notes on
    (*i*) Development of preferred methods,
    (*ii*) Implementation and maintenance of new methods.
14. What is memo motion study? Discuss this application.
15. Write a note on cycle graph and chronocycle graph.

5

WORK MEASUREMENT

• Definition • Objectives of time study • Techniques of time measurement • Steps in making a time study • Selecting job for time study • Obtaining and recording information • Breaking the jobs into elements • Reasons for breaking jobs into elements • Types of elements • Guidelines for breaking jobs into elements • Number of cycles to be timed • Measuring duration of each element • Extending observed time into normal time • Time study equipment • Selection of worker for time study • Performance rating methods • Allowances • Computation of standard time • Work sampling • Synthetic data • PMTS (Predetermined Motion and Time Study) • MTM (Method Time Measurement) • MOST(Maynard's Operation Sequence Technique).

5.1. DEFINITION

Work measurement is also called by the name "Time study". Work measurement is absolutely essential for both the planning and control of operations. Without measurement data, we cannot determine the capacity of facilities or it is not possible to quote delivery dates or costs. We are not in a position to determine the rate of production and also labour utilisation and efficiency. It may not be possible to introduce incentive schemes and standard costs for budget control.

Time study has been defined by British standard Institution as **"The application of techniques designed to establish the time for a qualified worker to carry out a specified job at a defined level of performance."**

5.2. OBJECTIVES OF WORK MEASUREMENT

The use of work measurement as a basis for incentives is only a small part of its total application.

The objectives of work measurement are to provide a sound basis for:

1. Comparing alternative methods.
2. Assessing the correct initial manning (manpower requirement planning).
3. Planning and control.
4. Realistic costing.
5. Financial incentive schemes.
6. Delivery date of goods.
7. Cost reduction and cost control.
8. Identifying substandard workers.
9. Training new employees.

5.3. TECHNIQUES OF WORK MEASUREMENT

For the purpose of work measurement, work can be regarded as:

1. **Repetitive work:** The type of work in which the main operation or group of operations

repeat continuously during the time spent at the job. These apply to work cycles of extremely short duration.

2. **Non-repetitive work:** It includes some type of maintenance and construction work, where the work cycle itself is hardly ever repeated identically.

Various techniques of work measurement are:

1. Time study (stop watch technique),
2. Synthesis,
3. Work sampling,
4. Analytical estimating,
5. Predetermined motion and time study.

Time study and work sampling involve direct observation and the remaining are data based and analytical in nature.

Time study: A work measurement technique for recording the times and rates of working for the elements of a specified job carried out under specified conditions and for analysing the data so as to determine the time necessary for carrying out the job at the defined level of performance.

Synthetic data: A work measurement technique for building up the time for a job or parts of the job at a defined level of performance by totalling element times obtained previously from time studies on other jobs containing the elements concerned or from synthetic data.

Work sampling: A technique in which a large number of observations are made over a period of time of one or group of machines, processes or workers. Each observation records what is happening at that instant and the percentage of observations recorded for a particular activity, or delay, is a measure of the percentage of time during which that activities delay occurs.

Predetermined motion time study (PMTS): A work measurement technique whereby times established for basic human motions (classified according to the nature of the motion and conditions under which it is made) are used to build up the time for a job at the defined level of performance. The most commonly used PMTS is known as Methods Time Measurement (MTM).

Analytical Estimating

A work measurement technique, being a development of estimating, whereby the time required to carry out elements of a job at a defined level of performance is estimated partly from knowledge and practical experience of the elements concerned and partly from synthetic data.

The work measurement techniques and their applications are shown in Table 5.1.

Table 5.1: Work Measurement Techniques and their Application.

Techniques	*Applications*	*Unit of Measurement*
1. Time study	Short cycle repetitive jobs. Widely used for direct work	Centi minute (0.01 min)
2. Working sampling	Long cycle jobs/heterogeneous operations.	Minutes
3. Synthetic Data	Short cycle repetitive jobs	Centi minutes
4. MTM	Manual operations confined to one work centre	TMU (1 TMU = 0.006 min)
5. Analytical estimating	Short cycle non-repetitive job	Minutes

Steps in Making Time Study

Stop watch time is the basic technique for determining accurate time standards. They are economical for repetitive type of work. Steps in taking the time study are:

1. Select the work to be studied.
2. Obtain and record all the information available about the job, the operator and the working conditions likely to affect the time study work.

3. Breakdown the operation into elements. An element is a distinct part of a specified activity composed of one or more fundamental motions selected for convenience of observation and timing.
4. Measure the time by means of a stop watch, taken by the operator to perform each element of the operation. Either continuous method or snap back method of timing could be used.
5. At the same time, assess the operators effective speed of work relative to the observer's concept of "Normal" speed. This is called performance rating.
6. Adjust the observed time by rating factor to obtain normal time for each element

$$\text{Normal time} = \frac{\text{Observed time} \times \text{Rating}}{100}$$

7. Add the suitable allowances to compensate for fatigue, personal needs, contingencies, etc., to give standard time for each element.
8. Compute allowed time for the entire job by adding elemental standard times considering frequency of occurrence of each element.
9. Make a detailed job description describing the method for which the standard time is established.
10. Test and review standards where necessary. The basic steps in time study are represented by a block diagram in Fig. 5.1.

STEPS IN TIME STUDY

SELECT
The job to be timed

↓

OBTAIN & RECORD
Details Regarding method, Operator, Job and Working Conditions

↓

DEFINE
The elements, Break the job into elements convenient for timing

↓

MEASURE
Time duration for each element and assess the rating

↓

EXTEND
Observed time into normal time (Basic time)

↓

DETERMINE
Relaxation and personal allowances

↓

COMPUTE
Standard time for the operation for defined job or operation

Fig. 5.1: Steps in time study.

1. Selecting Job for Time Study

The reasons for which time study may be done:

(*a*) The job in question is new one or not previously carried out.

(*b*) Change in the method of existing time standard.

(*c*) Complaint received from workers or unions regarding the time standard.

(*d*) A particular operation becomes bottle-neck operation which holds up number of subsequent activities.

(*e*) Change in the management policy regarding how time standards are used, *i.e.,* General purpose or wage incentive plans.

The general guidelines for selecting the job for time study:

(*a*) Bottle-neck operations.

(*b*) Repetitive jobs.

(*c*) Jobs using a greater deal of manual labour.

(*d*) Jobs with longer cycle time.

(*e*) Sections/department frequently working overtime.

2. Obtaining and Recording Information

During this step, all the relevant and necessary information regarding the method, operator and details of working conditions are recorded:

- The accuracy of time standards depends upon the correctness of the method employed by the operators. So wrong methods should not be timed. The method is to be standard and the time required to carry out the job as per the standard method is to be timed.
- The selection of an operator refers to choosing an operator amongst many operators doing the same job. He should be a representative worker with a normal pace neither too fast nor too slow. So the details of the operator is essential to be recorded before starting actual time study.
- Information to enable the identification details such as. Part number and name, machine No. speed and feed, materials, operator details, etc.
- Working conditions under which an operator carries out the job like temperature, dust, smoke, vibrations, noise, etc.
- Working position such as standing, sitting, bending, etc., and weights handled, protective clothing, etc.

3. Breaking the Jobs into Elements

Once the recording of the basic information regarding the job and, operator are done, the next step is breaking job into elements.

Element is a distinct part of a specified job selected for convenience of observation, measurement and analysis.

Work cycle is a complete sequence of elements necessary to perform a specified activity or job to yield one unit of production. It may also include the elements which do not occur with every cycle.

Reasons for Breaking the Jobs into Elements

1. To ensure that productive time is separated from unproductive activities (separating effective time and ineffective time).
2. To permit the rate of performance to be assessed more accurately than would be possible if the assessment were made over a complete cycle.
3. To enable different types of elements to be identified and distinguished so that each element is given an appropriate treatment.
4. To ensure elements involving a high degree of fatigue to be isolated and to make the allocation of fatigue allowances more accurately.

5. To enable the detailed work specification to be produced.
6. To enable machine elements to be distinguished from 'human' elements.
7. To enable time standards to be checked or modified at later date, omissions and errors to be rectified.
8. For accuracy of rating.
9. To enable time values for frequently recurring elements, such as the loading/unloading of jobs into fixture, machine adjustment to be extracted and used in the compilation of standard data.

5.4. TYPES OF ELEMENTS

1. **A repetitive element** is an element which occurs in every work cycle of the job. *Examples*, Picking up part for assembly, element of locating a work piece in a holding device.
2. **An occasional element** is one that does nor occur in every work cycle of the job or which may occur at regular intervals. *Examples,* tool changing after sometime, adjusting tension or machine setting, instruction from supervisor. Occasional element is useful work to be included in standard time.
3. **A constant element** is an element for which the basic time remains constant whenever it is performed. *Examples*, Switch on machine, measure diameter, insert cutting tools.
4. **A variable element** is an element for which the basic time varies in relation to some basic characteristics of the product, equipment or process. *Examples*, Dimensions, weight, quality, etc.
5. **A manual element** is an element performed by a worker.
6. **A machine element** is an element automatically performed by a power driven machine. *Examples*, Press working parts, annealing tubes.
7. **Governing element** is an element occupying a longer time than that of any other element which is being performed concurrently example Gauge dimensions while turning diameter (turning diameter will be a governing element).
8. **A foreign element** is one that is observed during study but do not form part of the given activity of the cycle. *Example*, Dropping work on the floor, operator talking to his colleague.

Guidelines for Breaking Jobs into Elements

1. Elements should be easily identified.
2. Each element should have a definite beginning and end.
3. Manual elements should be separated from variable elements.
4. Occasional elements should be timed separately.
5. Elements should be as short as can be conveniently timed by a trained observer.
6. Elements should be chosen so that they represent naturally unified and recognisably district segments of the operation.

Example of Standardised Element Breakdown

DRILLING WITH TWIST DRILL IN BENCH TYPE MACHINE

Element	*Break Point (end)*	*Remarks*
Pick up piece, place in jig	Moving jig towards spindle start	
Place under spindle, Advance drill	Machine feed engages Tool starts cutting	Machine feed Manual feed
Drill	Tool finishes cutting	–
Lift up spindle	Hand leaves lever handle	–

Element	*Break Point (end)*	*Remarks*
Take out of jig place aside	Noise when piece reaches bottom of container	Small pieces
Take out of jig	The moment when the piece is separated from jig	Pieces which cannot be thrown away
Place aside	(*a*) Piece released from hand grip (*b*) Crane hook released after reaching floor container	Pieces which cannot be thrown away

How Many Cycles to be Timed ?

1. The number of cycles through which any particular job should be observed varies directly as the amount of variations in the times of the elements of the job.
2. The number of cycles to be observed will depend on the degree of accuracy desired. This in turn will depend on the length of run of the job and the number of people engaged on it.
3. The study should be continued through a sufficient number of cycles to ensure that occasional elements such as handling boxes of finished parts, periodical cleaning of machines, etc., can be observed several times.
4. Where more than one operator is engaged on the same job it is preferable to take a short study on each of several operators rather than timing too long on a single operator.

The number of observations at 95% confidence level and accuracy of ± 5 per cent is given by the statistical formula

$$n = \left[40 \frac{\sqrt{n' \, \Sigma x^2 - (\Sigma X^2)}}{\Sigma x} \right]^2$$

where n' = number of preliminary readings, Σx = sum of preliminary set of observations
n = sample size (number of observations)

(Measure) Duration of Each Element

When elements have been selected, the next step is starting the timing of operations. There are two principal methods of timing with the stop watch: (*a*) Cumulative timing, and (*b*) Fly back timing.

In *cumulative,* the watch runs continuously throughout the study. It is started at the beginning of the first element of the first cycle to be timed and is not stopped until the whole study is completed. At the end of each element the watch reading is recorded and individual element times are obtained by successive subtractions after the study is completed.

In *fly back timing,* the hands of the stop watch are returned to zero at the end of each element and allowed to start immediately, the time for each element is obtained directly.

While recording the time of the elements, operators speed of working is assessed and recorded on the observation sheet. Rating is the time study engineers assessment of the operator's pace of working in relation to the concept of standard or normal. Rating is used to convert observed time into normal time.

Extend Observed Time into Normal Time

The representative time established from the observation data is the time which an operator has taken while working at a certain pace. The observed time is converted into basic or normal time by multiplying it by rating factor.

$$\text{Normal time} = \frac{\text{Observed time} \times \text{performance rating (\%)}}{100}$$

Determine relaxation and other allowances

Normal times of elements added together give normal time for the operation. But this will not be equal to standard time as the operators cannot work continuously. Some additional time is added to normal time to arrive at the standard time. The additional time is needed to:

(*i*) To provide the operator to attend to his personal needs (relaxation allowances).
(*ii*) Interference allowances.
(*iii*) Contingency allowance.
(*iv*) Policy allowance.

Calculate Standard Time for the Job

The various allowances are added to the normal time as applicable to get the standard time. Thus basic constituents of standard time are:

1. Elemental (observed time).
2. Performance rating to compensate for difference in pace of working.
3. Relaxation allowance.
4. Interference and contingency allowance.
5. Policy allowance.

5.5. TIME STUDY EQUIPMENTS

Basic time study equipment required to make the time study are: (1) Time study board, (2) Stop watch, and (3) Time study forms.

Time study board: Time study board is simply a flat board, usually of plywood or of any suitable plastic sheet and it should have fittings to hold a stop watch and time study forms. The use of board provides support and resting face while writing observations on the shop-floor and makes the hands free to write and operate stop watch.

Stop watch: Stop is the measuring instrument to observe the elemental timings and usually a decimal watch is used.

A decimal minute stop watch has two hands. The small hand represent minutes on dial and completes one revolution in 30 minutes. The large hand represents centi minutes (1/100th minute) and completes one revolution in one minute and each division on large dial represents 0.01 minute. Two commonly used types of stop watches are:

Cumulative stop watch: The watch is started by pressing the winding knob located on the head of the watch and is stopped by pressing the winding knob. Pressing winding knob third time snaps the hands back to zero. Once started it will run until required number of cycles have been timed.

Fly back stop watch: This is most commonly used watch. In this type of watch the movement is started and stopped by a slide (A) at the side of the winding knob (B). Pressure on the top of the winding knob causes both the hands to fly back to zero without stopping the mechanism from which point they move forward immediately. This type of watch is used for either fly back or cumulative timing method. The stop watch is shown in Fig. 5.2.

Electronic timers are most widely used timing devices for time study. The electronic timer which performs the same function as the stop watch is sometimes referred to as electronic stop watch.

Electronic data collectors and computers and motion picture camera (with constant speed motor drive) are also used for the purpose.

Time study forms: Time study forms are usually printed forms of standard size. The use of standard forms is desirable as the constant informations, such as part number and part name, operation description, observers name and other description are pre-printed on the top of the form which eliminates the possibility of any details being missed. As the size of the forms are standardised they can be easily filed for future referencing. Time study forms are shown in Fig. 5.3 and Fig. 5.4.

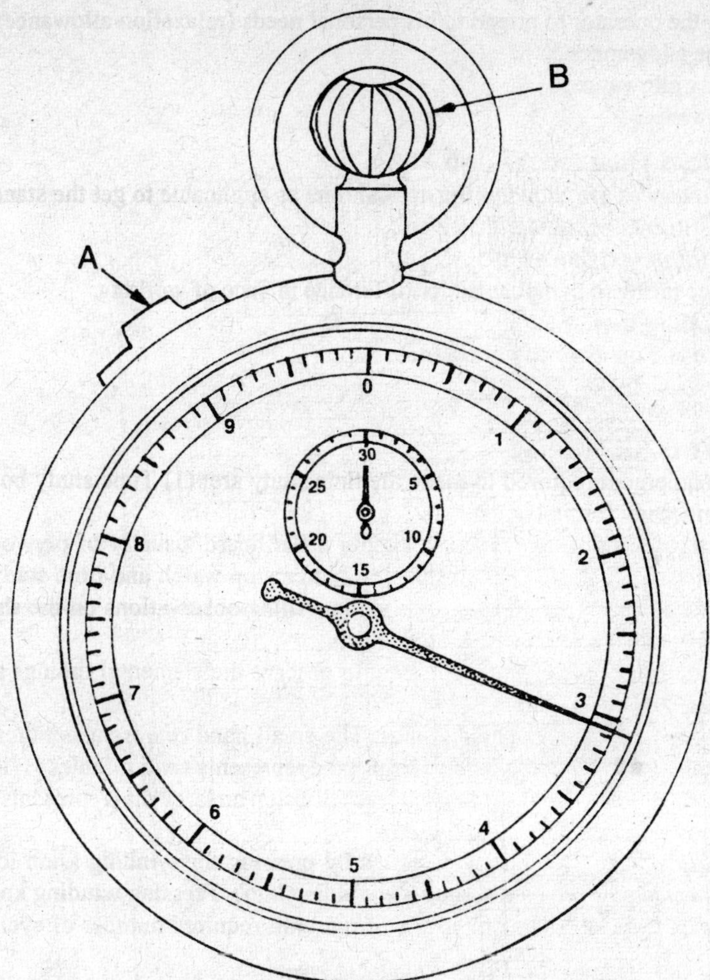

A = Slide for stopping and starting the movement
B = Winding knob. Pressure on this knob returns both the hands to zero.

Fig. 5.2: Decimal-minute stop watch.

Selection of Worker for Time Study

If a choice of workers is available, it is good policy to consult the foreman and the workers representatives to suggest the one most suitable to be studied first. The worker selected should be skilled and should have a good temperament. His pace of performance should be close to the average so that observed times are near to the normal times.

When a large number of workers are working on the job, it is a good policy to take studies on more than one qualified worker.

The distinction is made between a qualified worker and representative worker. A representative worker is one whose skill and performance is the average of the group under consideration. A representative worker may not be necessarily be a qualified worker.

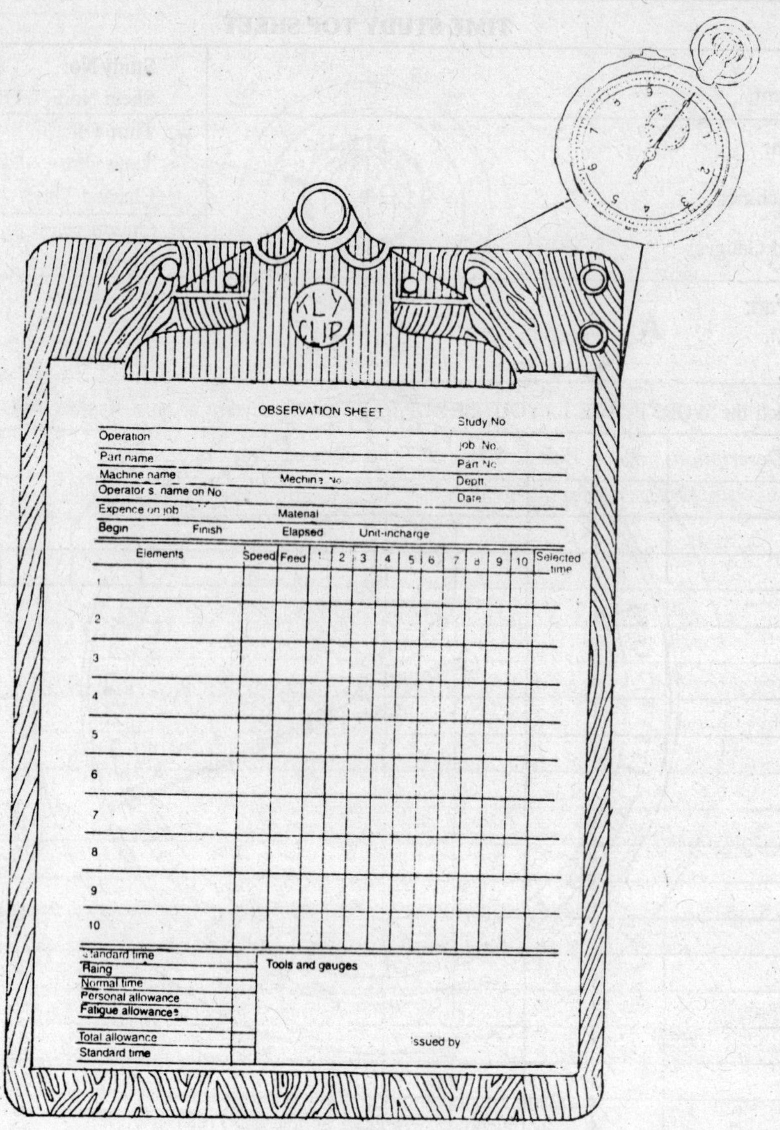

Fig. 5.3: Time study board.

A Qualified Worker as Defined by ILO

A qualified worker is one who is accepted as having the necessary physical attributes, who possesses the required intelligence and education, and who has acquired necessary skill and knowledge to carry out the work in hand to satisfactory standards of safety quantity and quality.

The time study man should not make any attempt to time the operative without his knowledge. from a concealed position or with the watch in the pocket.

5.6. PERFORMANCE RATING

Performance rating is the process of adjusting the actual pace of working of an operator by comparing it with the mental picture of pace of an operator working at normal speed.

TIME STUDY TOP SHEET

Department:		Study No: Sheet No.: Of
Operation:	M.S. No.:	Time Off: Time On:
Plant/Machine:	No.:	Elapsed Time:
Tools and Gauges:		Operative: Clock No.:
Product/Part: Dwg. No.: Meterial Quality:	No.	Studies By: Date
		Checked:

N.B. Sketch the WORKPLACE LAYOUT/SET-UP/PART on the reverse or on a separate sheet and attach

Element Description	*R.*	*W.R.*	*S.T.*	*B.T.*	*Element Description*	*R.*	*W.R.*	*S.T.*	*B.T.*

N.B. R. = Rating. W.R. = Watch Reading. S.T. = Subtracted Time. B.T. = Basic Time.

Fig. 5.4 (*a*) General purpose time study top sheet

Study No.:	TIME STUDY CONTINUATION SHEET					Seet No. of			
Element Description	*R.*	*W.R.*	*S.T.*	*B.T.*	*Element Description*	*R.*	*W.R.*	*S.T.*	*B.T.*
N.B. Reverse side similar, but without upper line of heading.									

Fig. 5.4 (*b*) **Continuation sheet for general-purpose time study (front)**

$$\text{Performance rating} = \frac{\text{Observed Time}}{\text{Normal Time}} \times 100$$

In other words, rating is a levelling factor to convert observed timings into normal timings.

Factors Affecting Performance Rating

There exists a variation from element to element and even among the elements in the same operation. This is due to the inconsistency in the speed of the working of the operator.

Each worker by nature has different temperament and attitudes towards the work. Some workers by their nature are fast (above the speed of the average worker) and some are by nature slow. Both these workers will not represent a normal worker.

The variation in actual times for a particular element may be due to the factors both internal and external. The external factors which are not in control of work study man are:

- Variation in the quality or other characteristics of the material used even though it is in prescribed tolerance limit.
- Changes in the operating efficiency of tools and equipment within their useful life.
- Unavoidable changes in methods or conditions of operations.
- Change in working conditions like heat, light, dust, etc.

Factors which are within control (Internal Factors) are:

- Acceptable variation in the quality of the product.
- Variation due to operators ability.
- Variation due to his attitude of mind.

The various methods of performance rating are: (1) Speed rating, (2) Westing house system of rating, (3) Synthetic rating, and (4) Objective rating.

Speed rating: In this technique the speed of the movements of the operator is the only factor considered for performance rating. The speed rating is found by the observer by comparing pace of operators working with his own concept of normal pace. An average worker is rated at 100%, better than average worker is rated at a figure higher than 100 and below average worker will be rated below 100. If a worker is rated at 125% it means that the speed is 25% higher than the observers concept of normal and rating of 80% means the worker is working 20% below the observers concept of a normal worker.

In speed rating, the process of rating is confined to the comparison of speed of movements with a concept of normal speed. On the basis of this assumption, the rating process is made simpler and with training in developing the concept of normal pace, the observer can become quite proficient in his judgement.

Westing house method of rating: Westing house system utilises a set of criteria to measure the performance of the operators. The factors are:

- Skill,
- Effort,
- Consistency,
- Conditions.

1. ***Skill*:** Measures the workers proficiency in adhering to a given method, coordination of proper hand and eye movements, rhythm of the movements. The skill has been classified into six degrees, each degree indicating a specified class of skill within which an operator performs the task.
2. ***Effort*:** Measures the speed with which the skill is applied. The effort is also divided into six degrees.
3. ***Consistency*:** Measures factors which affect the consistency of the operator to perform the work cycle repeatedly within the same time. Elements which affect the consistency are—variations in materials, hard spots, presence of foreign elements. Consistency is subdivided into six classes.

4. ***Conditions***: Measure the extent to which the conditions like temperature, vibrations, light and noise affect the operator's performance.

The Westing house system of classification of skill, effort, consistency and conditions are shown in Table 5.2.

As per this system, the time study observer assign rating for a criteria of particular task. Numerical values are than obtained from Table 5.2 and establishes the performance rating by adding the four values and adding the levelling factor to normalise the observed time.

It is applied to the cycle time in case of a manual time rather than to the individual elemental times.

Table 5.2: Performance Rating Table. (Westing house method)

Skill			*Effort*		
+ 0.15	A1	Superskill	+ 0.13	A1	Excessive
+ 0.13	A2		+ 0.12	A2	
+ 0.11	B1	Excellent	+ 0.10	B1	Excellent
+ 0.08	B2		+ 0.08	B2	
+ 0.06	C1	Good	+ 0.05	C1	Good
+ 0.03	C2		+ 0.02	C2	
0.00	D	Average	0.00	D	Average
– 0.05	E1	Fair	– 0.04	E1	Fair
– 0.10	E2		– 0.08	E2	
– 0.16	F1	Poor	– 0.12	F1	Poor
– 0.22	F2		– 0.17	F2	
Conditions			*Consistency*		
+ 0.06	A	Ideal	+ 0.04	A	Perfect
+ 0.04	B	Excellent	+ 0.03	B	Excellent
+ 0.02	C	Good	+ 0.01	C	Good
0.00	D	Average	0.00	D	Average
– 0.03	E	Fair	– 0.02	E	Fair
– 0.07	F	Poor	– 0.04	F	Poor

Illustration of Westing House Method: An observed time for an operation is 0.05 minutes and the ratings are as follows:

Skill (Excellent) B2

Effort (Good) C2

Condition (Good) C

Consistency (Good) C

The values for the ratings are assigned from Table 5.2.

Criteria	*Rating*	*Numerical Value*
Skill	B2	+ 0.08
Effort	C2	+ 0.02
Condition	C	+ 0.02
Consistency	C	+ 0.01
	Total	+ 0.13

(*a*) Performance rating factor = 1 + 0.13 = 1.13 = 113% = 113%

(*b*) Normal time of operation = Observed time × performance rating

$= 0.05 \times 1.13$

$= 0.0565$ minutes

Synthetic rating: The performance rating under this method is established by comparing observed time of some of the manual elements with those of known time values of the elements from predetermined motion and time studies (PMTS).

The procedure is to make the time study in a usual manner and then compare the actual time for the elements with predetermined time values for the same elements.

A ratio is computed between predetermined time value for the element and actual time value for the element.

This ratio is the performance index or rating factor for the operator for the particular element.

Performance rating factor, (R) is given by

$$R = \frac{P}{A}$$

P = Predetermined time for elements (minutes)

A = Average actual time value (selected time) for the same element 'P' (minutes)

Objective rating: In this method, the operator's speed is rated against a single standard pace which is independent of job difficulty. The observer merely rates speed of movement or activity, paying no attention to job itself. After the pace rating is made, an allowance or a secondary adjustment is added to the pace rating to take care of job difficulty.

Job difficulty is divided into six classes, and percentage is provided for each of these factors.

The job difficulties as per the founder of this system—M.E. Mundel have been categorised into six classes as follows:

1. Amount of body used.
2. Foot pedals.
3. Bi-manualness.
4. Eye-hand coordination.
5. Handling requirements.
6. Weight.

5.7. ALLOWANCES

The normal time for an operation does not contain any allowances for the worker. It is impossible to work throughout the day even though the most practicable, effective method has been developed. Even under the best working method situation, the job will still demand the expenditure of human effort and some allowance must therefore be made for recovery from fatigue and for relaxation. Allowances must also be made to enable the worker to attend to his personal needs. The allowances are categorised as: (1) Relaxation allowance, (2) Interference allowance, and (3) Contingency allowance.

Relaxation Allowance

Relaxation allowances are calculated so as to allow the worker to recover from fatigue.

Relaxation allowance is a addition to the basic time intended to provide the worker with the opportunity to recover from the physiological and psychological effects of carrying out specified work under specified conditions and to allow attention to personal needs. The amount of allowance will depend on nature of the job.

Relaxation allowances are of two types—fixed allowances and variable allowances.

Fixed allowances constitute:

(*a*) Personal needs allowance. It is intended to compensate the operator for the time necessary to leave, the workplace to attend to personal needs like drinking water, smoking, washing hands. Women require longer personal allowance than men. A fair personal allowance is 5% for men and 7% for women.

(b) Allowances for basic fatigue. This allowance is given to compensate for energy expended during working. A common figure considered as allowance is 4% of the basic time.

Variable Allowance

Variable allowance is allowed to an operator who is working under poor environmental conditions that cannot be improved, added stress and strain in performing the job.

The variable fatigue allowance is added to the fixed allowance to an operator who is engaged on medium and heavy work and working under abnormal conditions. The amount of variable fatigue allowance varies from organisation to organisation.

Interference Allowance

It is an allowance of time included into the work content of the job to compensate the operator for the unavoidable loss of production due to simultaneous stoppage of two or more machines being operated by him. This allowance is applicable for machine or process controlled jobs.

Interference allowance varies in proportion to number of machines assigned to the operator. The interference of the machine increase the work content.

Contingency Allowance

A contingency allowance is a small allowance of time which may be included in a standard time to meet legitimate and expected items of work or delays, the precise measurement of which is uneconomical because of their in frequent or irregular occurrence.

This allowance provides for small unavoidable delays as well as for occasional minor, extra work.

Some of the examples calling for contingency allowance are:

- Tool breakage involving removal of tool from the holder and all other activities to insert new tool into the tool holder.
- Power failures of small duration.
- Obtaining the necessary tools and gauges from central tool store. Contingency allowance should not exceed 5%.

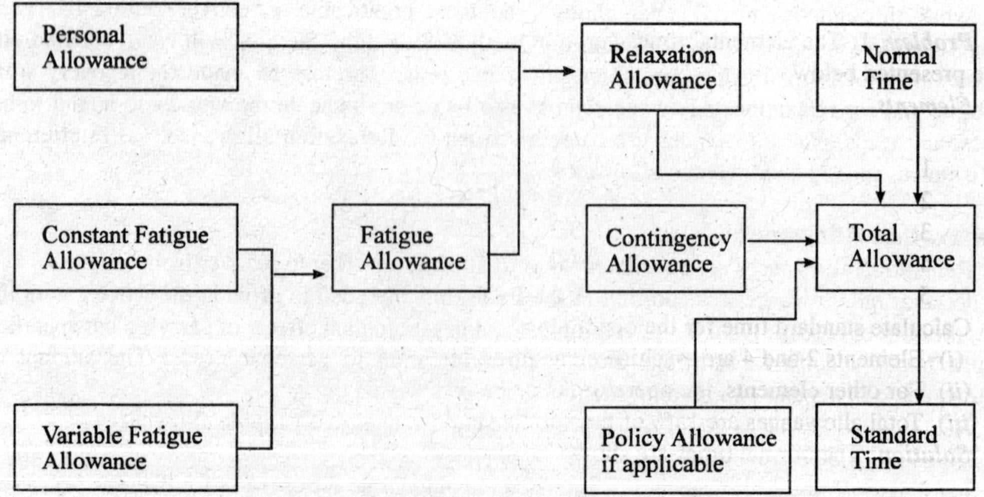

Fig. 5.5: Various allowances to build standard time.

Policy Allowance

Policy allowances are not the genuine part of the time study and should be used with utmost care and only in clearly defined circumstances.

The usual reason for making the policy allowance is to line up standard times with requirements of wage agreement between employers and trade unions.

The policy allowance as defined by ILO:

"A policy allowance is an increment, other than bonus increment, applied to a standard time (or to some constituent part of it, *e.g.,* work content) to provide a satisfactory level of earnings for a specified level of performance under exceptional circumstances. Policy allowance are sometimes made as imperfect functioning of a division or part of a plant."

Various allowances used to build the standard time is shown in Fig. 5.5.

5.8. COMPUTATION OF STANDARD TIME

Standard time is the time allowed to an operator to carry out the specified task under specified conditions and defined level of performance.

The basic constituents of standard time are as shown in Fig. 5.6.

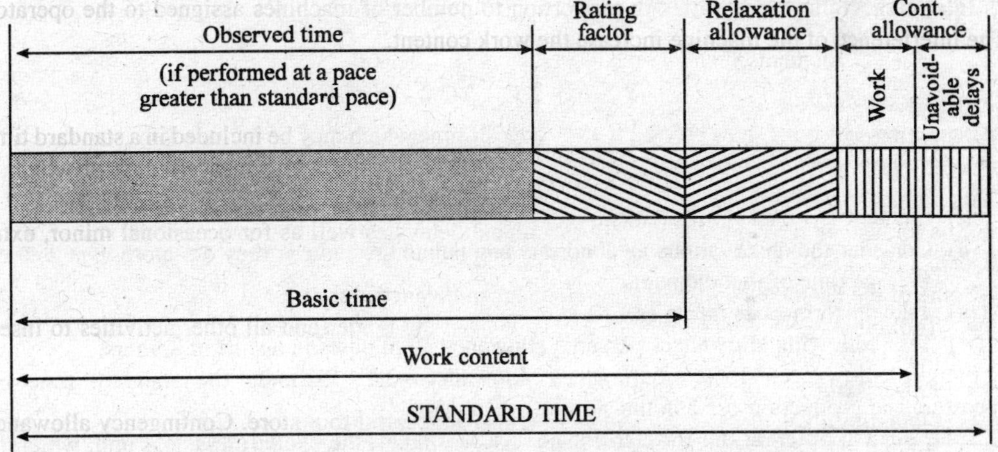

Fig. 5.6: How the standard time for a simple manual job is made up.

***Problem 1*:** The elemental times (in minutes) for 4 cycles of an operation using a stop watch are presented below:

Elements	*Cycle time in minutes*			
	1	2	3	4
1	1.5	1.5	1.3	1.4
2	2.6	2.7	2.4	2.6
3	3.3	3.2	3.4	3.4
4	1.2	1.2	1.1	1.2
5	0.51	0.51	0.52	0.49

Calculate standard time for the operation if

(*i*) Elements 2 and 4 are machine elements

(*ii*) For other elements, the operator is rated at 110%

(*iii*) Total allowances are 15% of the normal time.

***Solution*:** The normal times are shown in Table 5.3.

Table 5.3: Computation of Normal Time

Element No.	*Cycle Time (min.)*				*Avg. cycle time (3)*	*Rating (4)*	*Normal time = Avg. time × rating*
	1	2	3	4			
1	1.5	1.5	1.3	1.4	1.425	110%	1.425 × 1.1 = 1.568
2	2.6	2.7	2.4	2.6	2.575	m/c elem.	= 2.575
3	3.3	3.2	3.4	3.4	3.325	110%	3.325 × 1.1 = 3.658
4	1.2	1.2	1.1	1.2	1.175	m/c elem.	= 1.175
5	0.51	0.51	0.52	0.4	0.505	110%	0.505 × 1.1 = 0.555

Normal time for the cycle = 1.568 + 2.575 +3.658 + 1.175 + 0.555 = 9.531

Standard time = (9.531 + (0.15 × 9.531))

= **10.484 minutes**

Problem 2: The following data refers to the study conducted for an operation. Table shows actual time for elements in minutes

Cycle— Elements	1	2	3	4	5
1	2.5	2.1	2.2	5.4	2.5
2	6.2	6.00	6.1	5.9	5.9
3	2.3	2.0	2.1	2.1	2.2
4	2.4	2.1	2.8	3.0	2.3

(*i*) Element 2 is a machine element.

(*ii*) Consider the observations as abnormal and delete the same if they are more than 2 % of average time of that element.

(*iii*) Take performances rating as 120.

(*iv*) Take following allowances personal allowance 30 minutes in a shift of 8 hours.

Fatigue allowance—15%, contingency allowance—2%. Estimate the standard time of operation and production per 8 hours shift.

Solution: On observation, for element no. 1, cycle no. 4, the cycle time is 5.4 min, which is more than 25% of the average time for that element.

∴ **The cycle time 5.4 is neglected.**

Element. No.	*Cycle*					*Avg. Time*
	1	2	*3*	*4*	*5*	
1	2.5	2.1	2.2	*5.4	2.5	2.325
2	6.2	6.00	6.1	5.9	5.9	6.02
3	2.3	2.0	2.1	2.1	2.2	2.14
4	2.4	2.1	2.8	3.0	2.3	2.52
				Total observed cycle time		13.005

Normal time for the cycle = Observed time/cycle × Rating

= 13.005 × 1.2

= 15.606 min

Total Allowance = Fatigue allowance + contingency allowance

(∴ Personal allowance per shift is given)

Standard time = 15.606 (1 + 0.15 + 0.2)

= 18.259 minutes

(ii) Production Rate Per Shift

Total time per shift of 8 hrs = 8×60 = 480 min.
Less personal allowance 30 min.
Effective production time 450 min.

$$\text{Production in 8 hrs shift} = \frac{\text{Time availablefor production}}{\text{Std. Time}}$$

$$= \frac{450}{18.259}$$

$$= 24.64$$

$$= \textbf{25 jobs.}$$

Problem 3: The following table shows a time study data. The times shown are continuous watch readings in minutes. Initial setting of stop watch is at 0.00.

S.No.	*Element*	*Cycle Time*			*Performance rating*
		1	*2*	*3*	
1.	Get two cases	0.5	4.2	8.6	1.05
2.	Put parts into cases	1.5	5.7	9.9	1.15
3.	Clamp two parts in position	3.8	8.1	12.6	0.95

Take relaxation allowance as 15% and find the standard time.

Solution: This is a cumulative timing method the cycle times are tabulated as shown in table.

S.No.	*Element*	*Cycle Time*			*Avg. time*	*Rating*	*Normal time*
		1	*2*	*3*			
1.	Get two cases	0.5	0.4	0.5	0.466	1.05	0.49
2.	Put parts into cases	1.0	1.5	1.3	1.266	1.15	1.456
3.	Clamp two parts in position	2.3	2.4	2.7	2.466	0.95	2.343

Standard time = Normal time (1 + 0.15)
= (0.49 + 1.456 + 2.343) (1.15)
= **4.932 min.**

Problem 4: The observed times and the performance ratings for the five elements are given compute the standard time assuming rest and personal allowance as 15% and contingency allowance as 2% of the basic time.

Element	1	2	3	4	5
Observed time (min)	0.2	0.08	0.50	0.12	0.10
Performance rating	85	80	90	85	80

Solution: The normal time is computed from the observed time as shown in the table below:

Element	*Observed time (minute)*	*Rating*	*Normal time (min)*
1	0.20	85	$0.2 \times 85/100 = 0.170$
2	0.08	80	$0.8 \times 0.8 = 0.064$
3	0.50	90	$0.50 \times 0.9 = 0.450$
4	0.12	85	$0.12 \times 0.85 = 0.102$
5	1.10	80	$1.10 \times 0.8 = 0.080$
		Total	0.866

Normal time per piece = 0.866 min.

Rest and personal allowance = $0.15 \times 0.866 = 0.01299$

Contingency allowance = $0.02 \times 0.866 = 0.0173$ min.

Standard time per piece = Normal time/piece + Rest and personal allowance + contingency allowance

$= 0.866 + 0.1299 + 0.0173$

= 1.0732 minutes.

Problem 5**:** A worker operating on a machine performs the following elements. The description of element, their observed time and ratings are given. Compute the standard time for the component.

Element	*Description*	*Observed time*	*Rating*	*Relaxation Allowance*
A	Position the job	0.25	80%	10%
B	Switch 'ON' and lower drill	0.09	100	11%
C	Drill hole	2.8	90	12%
D	Raise drill and switch 'OFF'	0.05	80	10%
E	Remove job from jig	0.15	110	11%

Solution**:** The relaxation allowance is given separately for each element so the elementwise standard time is to be computed and then added to together to get the standard time for the job.

The computation of standard time is shown the table:

Element	*Observed time*	*Rating*	*Normal time*	*Relaxation allowance*	*Standard time (min.)*
A	0.25	80	0.25×0.8	.10	0.22
B	0.09	100	0.09×1	0.11	0.099
C	2.80	90	0.80×0.9	0.9	2.746
D	0.05	80	0.05×0.8	0.8	0.043
E	0.15	110	0.15×1.1	1.1	0.183

∴ Standard time for the job = $0.22 + 0.099 + 2.746 + 0.043 + 0.183$

= 3.251 minutes

Problem 6**:** The elemental timings are given below along with the respective ratings. Assuming rest and personal allowance as 12% and contingency allowance of 2%, calculate the standard time for the operation.

Element	*Observed time*	*Rating*	*Remark*
A	0.2	90	–
B	0.05	80	–
C	0.03	100	–
D	0.78	100	–
E	0.06	100	–
F	0.05	100	–
G	0.02	85	Once in 5 pieces
H	0.06	80	–
I	0.10	90	–
J	0.04	90	Once in 20 pieces

***Solution*:** The normal time for each component is computed as shown in Table 5.4.

Table 5.4: Computation of Normal Time

Element (1)	*Observed time (2)*	*Rating (3)*	*Frequency (4)*	*Normal time (5)*
A	0.2	90	1	0.080
B	0.05	80	1	0.040
C	0.03	100	1	0.030
D	0.78	100	1	0.7800
E	0.06	100	1	0.0600
F	0.05	100	1	0.0500
G	0.02	85	1/5	0.0340
H	0.06	80	1	0.0480
I	0.10	90	1	0.0900
J	0.04	90	1/20	0.0180

Normal time for the operation = 1.330
Standard time = Normal time + allowance
= 1.330 (1 + 0.14)
= **1.5162 min.**

***Problem 7*:** Turning gear blanks on centre lathe involves the following elements. The stop watch data is given. Assuming the rest and the personal allowance as 13% and contingency allowance of 2%, calculate the standard time.

Element	*Description*	*Observation*					*Rating*
		1	2	3	4	5	
1	Pick and place	0.2	1.46	5.22	6.49	14.25	90
2	'ON' M/c and tool approach	0.3	1.55	5.30	13.10	14.35	110
3	Turn diameter	1.05	2.31	6.05	13.84	15.10	110
4	Withdraw tool and stop Machine	1.13	2.38	6.14	13.92	15.17	110
5	Release part and place it aside	1.28	2.54*	6.29	14.06	15.32	95%

Note: Element No. 1 Observation No. 4, Foreign element 2.54 to 5.02
Element No. 5, Observation No. 2, Foreign element 6.29 to 12.98

***Solution*:** Calculate the individual element times considering the stop watch readings. The times for the foreign elements are excluded from the corresponding elemental times. The individual timing are shown in table:

Element No.	*Elemental time value in min.*					*Average time (minutes)*
	1	2	3	4	5	
1	0.20	0.18	0.20	0.2	0.19	0.1940
2	0.10	0.09	0.08	0.12	0.10	0.0980
3	0.75	0.76	0.75	0.74	0.75	0.7500
4	0.08	0.07	0.09	0.08	0.07	0.0780
5	0.15	0.16	0.15	0.14	0.016	0.1500

Normal times are computed by multiplying observed time by rating factor.

Element	*Observed time (min)*	*Rating*	*Normal time*
1	0.1940	90	0.1746
2	0.0980	110	0.1078
3	0.7500	100	0.7500
4	0.0780	110	0.0858
5	0.1500	95	0.1425
		Total	1, 2607

Basic (Normal) time per job = 1.2607 min.

Rest and allowance and contingency allowance = 15%

∴ Standard time = 1.2607 (1 + 0.15) = **1.4498 min.**

***Problem 8*:** A time study was conducted on a job consisting of three elements. Stop watch readings in hundredth of a minute are given. Using cumulative timing method along with rating factors.

Calculate the standard time if allowance is 12%.

Element	*Stop watch readings*					*Rating*
	1	*2*	*3*	*4*	*5*	
A	10	73	139	203	266	80
B	25	88	155	218	280	100
C	64	128	193	257	320	110

***Solution*:** This stop watch readings are cumulative. The individual timings for elements are computed by subtracting proceeding reading from successive figure as shown in table below:

Example 25 – 10 = 15 for element *B*, 1st observation

Element	*Individual elemental timings*					*Avg. time (min)*	*Rating*	*Normal time*
	1	*2*	*3*	*4*	*5*			
A	10	09	11	10	09	0.098	80	0.0784
B	15	15	16	15	14	0.150	100	0.1500
C	39	40	38	39	40	0.392	110	0.4312
							Total	0.6596

Standard time = 0.6596 (1 + 0.12) = **0.7387 minutes.**

5.9. COMPARISON OF VARIOUS TECHNIQUES

Sr. No.	*Creteria*	*Work Sampling*	*Predetermined time Standards*	*Stop Watch Timing*
1	Speed, time required to measure and establish standards	Average to fast	Slow to average	Average
2	Training and skill required, supervision	Low to moderate	High	Moderate to high
3	Cost employee time, equipment, etc.	Average	Fairly high	Average
4	Assistance in methods improvement	Low to moderate	High	Good
5	Accuracy, subjective, objective, degree or distortion	Fair to good	Very high	Good to high
6	Acceptability: employee, supervisor	Fair	Good	Fair to good

Sr. No.	Creteria	Work Sampling	Predetermined time Standards	Stop Watch Timing
7	Interruption of work operations	Moderate	Low	Fairly high
8	Applicability: for physical, clerical, professional work	Very good	Average	Average
9	Savings: How quickly how much	Average to high	High	Average to high
10	Usability: In Scheduling production, evaluating performance	Average to high	High	High
11	Reporting requirements difficulty of furnishing data	Average	Average	Average

5.10. WORK SAMPLING

Work sampling was originally developed by L.H.C. Tippett in Britain in 1934 for the British Cotton Industry Research Board. Work sampling is a fact finding tool.

Work sampling is defined as:

"A technique in which a statistically competent number of instantaneous observations are taken, over a period of time, of a group of machines, processes or workers. Each observations recorded for a particular activity or delay is a measure of the percentage of time observed by the occurrence."

Work sampling has three main applications

1. **Activity and delay sampling:** To measure the activities and delays of workers or machine. *e.g.,* the percentage of time in a day, a person is working and the percentage that a person is not working.
2. **Performance sampling:** To measure working time and non-working time of a person on a manual work, and to establish a performance index or performance level for a person during his working time.
3. **Work measurement:** Under certain circumstances, to measure a manual task, that is, to establish a time standard for an operation.

Procedure for Conducting a Work Sampling Study

The following steps are involved in making sampling study:

1. **Decide on the objective of the study:** It is very important to first set the objectives of study as the duration of the study, number of observations, the design study sheet and elemental breakdown depends upon the objective.
2. Obtain the approval of the supervisor of the department in which work study is to be conducted. Make sure that the operators to be studied and the other people in the department understand the purpose of the study. Obtain their cooperation.
3. **Decide upon work and delay elements:** Work and delay elements represent the headings under which the observations are to be recorded. The nature of the work and delay elements differ from company to company depending upon the objective of the study and the work.
4. **Decide upon the duration of the study:** The duration of study depends upon the objective, number of observers, the accuracy desired and the frequency of occurrence of the activity.
5. **Determine the desired accuracy of results:** This may be stated as the standard error of a percentage or desired accuracy. The confidence level is also to be stated.
6. Make a preliminary estimate of the percentage occurrence of the activity or delay to be measured.

7. **Design the study:**
   (*a*) Determine number of observations to be made.
   (*b*) Determine number of observers needed.
   (*c*) Determine the number of days or shifts needed for the study.
   (*d*) Make the detailed plans for taking observations.
   (*e*) Design the observation form.
8. Make the observations according to the plan, analyse and summarise the data.
9. Check the accuracy or precision of the data at the end of the study
10. Prepare the report and state conclusions.

Design of Work Sampling Study

1. Determination of Required Number of Observations

The number of observations depends upon:

- Activity percentage (P)
- Limits of accuracy (A)
- Confidence level (C)

Number of observations at a confidence level of 95% is given by

$$N = \frac{4\,(1-P)}{A^2,\,P}$$

Setting Performance Standards with Work Sampling

Procedure to develop performance standards are detailed below:

1. **Taking the study:** A work sampling study is carried out for the operation whose standard time is to be determined. Observations are made at random intervals of time and are noted. Whether subject under study is working or idle are noted. Reasons for delays and interruption are recorded. The observations of production activity (working) are divided into machine working and hand working. Operators pace of performance is noted down when manual working is observed.
2. **Rating index:** Individual performance ratings are averaged out to obtain an overall rating index.
3. **Production quantity:** Number of pieces produced during the period of study are determined from production reports.
4. **Overall time per unit (T_o):** It is calculated by dividing production time (duration of study) by number of pieces produced.
5. **Effective time per unit (T_e):** Overall time per piece includes even the time spent on unproductive activities. Overall time is multiplied by percentage of productive activities to get the effective time per piece.

Let T_o = Overall time per piece
N = Total number of observations.
N_p = Observations of Production activity.
$= N_m + N_h$
N_m = Observations of machine controlled work.
N_h = Observations of hand controlled work.

∴ Effective time per piece (T_e) = Overall time / unit × production activity %

$$= T_o \frac{\times N_p}{N}$$

The effective time per piece (T_e) can be compared to observed time of stop watch study. To get the normal time, effective time is broken down into manual and machine controlled time.

Machine controlled effective time (T_m) $= T_e \times \dfrac{N_m}{N_p}$

$$= T_e \times \frac{N_m}{N_m + N_h}$$

and Manual (hand) controlled time $(T_h) = T_e \times \dfrac{N_h}{N_m + N_h}$

or $(T_e - T_m)$

Normal time per piece

Let R be the performance rating index

T_h = Hand controlled portion of effective time

T_m = Machine controlled effective time

$$\text{Normal time} = \frac{T_h \times R}{100} + T_m$$

Rating is applied to only manual (hand controlled) elements

Standard time per piece

Standard time is calculated by adding relevant allowances to the normal time.

Control Charts in Work Sampling

Control charts are used in work sampling to continuously keep track of particular activity. A chart is employed where the proportions of activities obtained from work sampling are represented as a function of time. The points in the chart gives the idea of the trend and the presence of out of control condition if exist.

An investigation is made if the point falls outside the control limits. Control chart is useful aid to a work sampling man.

Advantages of Work Sampling Compared to Time Study

1. Many operations or activities which are impractical or costly to measure by time study can be measured by work sampling.
2. A simultaneous work sampling study of several operators or machines may be made by a single observer.
3. It usually requires lesser man-hours and costs less to make a work sampling study instead of making a continuous time study.
4. Observations may be taken over a period of days or weeks thus reducing the chances of day-to-day variations affecting results.
5. Any interruption during study will not affect the results.
6. Work sampling measurements may be made with a pre-assigned degree of reliability.
7. Work sampling studies are preferred to continuous time studies by the operators being studied.
8. A stop watch is not needed for work sampling studies
9. Work sampling studies cause less fatigue and are less tedious.

Disadvantages of Work Sampling

1. Work sampling is uneconomical for short cycle jobs.
2. It is also uneconomical for studying a single workman or even small group of workmen or machines.
3. Time study permits a finer breakdown of activities and delays than is possible with work sampling study.
4. Workman may change their normal pattern of working on seeing the observer, making the sampling study of very little value.
5. Insufficient observations are likely to produce inaccurate results.
6. It does not normally account for speed of the operator.

***Problem 1*:** A job order shop has 12 general purpose machines. A work sampling study has been designed to know the ineffective time of the entire shop. The study conducted revealed that the ineffective time is to the extent of 30%.

Compute the number of observations that are required to have the accuracy of 5% with confidence level of 95%.

***Solution*:** Number of observations $(N) = \dfrac{4\,(1-P)}{A^2\,P}$

where P = Activity percentage
= 0.30 (in the problem it is the ineffective time)
A = Accuracy required
= 0.02

Substituting the values in the equation,

$$N = \frac{4 \times (1-0.3)}{(0.02)^2 \times 0.3}$$

= **23333** observations.

***Problem 2*:** A work sampling study was conducted to establish the standard time for an operation. The observations of the study conducted is given below:

Total number of observations = 160
Manual (hand controlled work) = 14
Machine controlled work = 106
Machine idle time = 40
Average performance rating = 80%
No. of parts produced = 36
Allowance for personal needs and fatigue = 10%
Study conducted for 3 days
Available working hours/day = 8 hrs

Calculate the standard time per piece.

***Solution*:** Total number of observations (N) = 160
No. of observations of production activity (N_p) = 120
Observations of machine controlled work (N_m) = 106
Observations of hand controlled work (manual) N_h = 14
Average performance rating R = 80%
Duration of study = 3 days
= 1440 minutes
(Each day 8 hour)
No. of parts produced/day = 36

(*i*) **Overall time per unit (T_o)**

$$T_o = \frac{\text{Duration of study}}{\text{No. of pieces produced / days}}$$

$$= \frac{1440}{36} = 40 \text{ minutes}$$

(*ii*) **Effective time per piece (T_e)**

T_e = Overall time/piece × % of time spent on productive activity.

$$= T_o \times \frac{N_p}{N}$$

$$= T_o \times \frac{N_p + N_m}{N}$$

$$= 40 \times \frac{(106 + 14)}{160}$$

$$= 30 \text{ min}$$

(iii) Breakdown of effective time per piece into machine controlled portion and hand controlled portion

T_m = machine controlled portion of the effective time per piece

$$T_m = T_e \times \frac{N_m}{N_p}$$

$$= T_e \times \frac{N_m}{N_m + N_h}$$

$$= 30 \times \frac{106}{106 + 14} = 26.5 \text{ min.}$$

Hand controlled portion of effective time per piece (T_h)

$$T_h = T_e \times \frac{N_h}{N_m + N_h}$$

$$= 30 \times \frac{14}{120}$$

$$= 3.5 \text{ minutes}$$

Normal time per piece

Normal time per piece = Machine controlled portion of effective time per piece + Normal time of hand controlled portion of effective time/piece

$$= T_m + (T_h \times R)$$

$$= 26.5 + 3.5 \times 0.8$$

$$= 29.3 \text{ min}$$

Standard time per piece = Normal time + allowances

$$= 29.3\,(1 + 0.1)$$

$$= \mathbf{32.23} \text{ min.}$$

Problem 3: A work study was conducted in a machine shop. The data has been recorded.

Total number of observations = 2000
No activity = 500
The ratio between manual to machine portion of the activities = 3 : 1
Average performance rating = 85%
Total number of pieces produced during study = 120
Duration of the study = 60 hrs.

Calculate the standard time/ piece assuming 15% relaxation allowance.

Solution: **(i) Overall time per unit (T_o)**

$$T_o = \frac{\text{Duration of study}}{\text{No. of pieces produced during study}}$$

$$= \frac{60 \times 60}{120} = 30 \text{ minutes}$$

(ii) Effective time per piece (T_e)

T_e = Overall time per piece × production of productive observations to total observations.

$$= T_o \times \frac{N_p}{N}$$

$$= 30 \times \frac{2000}{2500}$$

$$= 24 \text{ min.}$$

As the rating is to be applied to only human controlled activities break the effective time into machine controlled and human controlled.

∴ Machine controlled time per piece = $(T_m) = 24 \times 1/4 = 6$ min.

= Hand controlled time/piece $(T_h) = 24 \times 3/4 = 18$ min.

= Normal time per piece = $T_m + T_h \times R$

= $6 + 18 \times 0.85$

= **21.3** min.

Standard time per piece (T_s) T_s = Normal time + allowances

= $21.3 + (1 + 0.15)$

= **24.5** min.

Problem 4: The following data refers to a sampling study of production of one component.

1. Duration of data collection 5 days @ 8 hours per day
2. Number of operators = 10
3. Allowances given for the process = 15%
4. Production quantity in 5 days = 6000 components
5. Sampling data collected

Days	1	2	3	4	5
No. of observations	230	240	200	180	225
Occurrence of activity	200	190	170	150	210

Calculate standard time of production of the component if average performance rating of the operator is 120% and the entire operation is manual.

Solution : No. of observations $(N) = 1075$

No. of observations $(N_p) = 920$
(working)

$$\text{Overall time per piece} = T_o = \frac{\text{Total time worked}}{\text{No. of units produced}}$$

$$= \frac{5 \times 8 \times 10 \times 60}{6000} \text{ min.}$$

= 40 min.

$$\text{Effective time per piece } (T_e) = T_o \times \frac{N_p}{N}$$

$$= 40 \times \frac{920}{1075}$$

= 34.23

Normal time = Observed time × Rating

= 34.23×1.2

= 41.07 min.

Standard time = Normal time (1 + allowances)

= 41.07 (1 + 0.15)

= **47.24** min.

5.11. SYNTHETIC DATA

A work measurement technique for building up time for the job at a defined level of performance by totalling elemental times obtained previously from time studies on other jobs containing the elements concerned, form synthetic data.

The steps involved in synthetic data:

1. Collect all the details about the job (dimensions, tools, methods, conditions).

2. Analyse jobs into constituent elements (activity grouping to enable synthetic elements to be applied if relevant).
3. Select appropriate basic times from synthetic data covering contingent factors.
4. Select and apply synthetic data covering contingent factors.
5. Verify details of elemental analysis for job method and condition.
6. Total the basic times, rating and allowances to compute standard time for the job.

Advantages

Although synthesis was originally developed to establish the work content for short batch production and jobbing work, it can be used in place of time study to determine times for many other types of work including repetitive work provided necessary data is available.

The main advantage of synthesis is the reduced cost of application. By means of synthesis it is possible to establish times, which are equally satisfactory for planning and production control purpose.

5.12. PREDETERMINED MOTION TIME ANALYSIS (PMTS)

A standard time for a job or an operation may be established by time study, by work sampling or by the usé of predetermined times.

A predetermined time system consists of a set of time data and a systematic procedure which analyses and subdivides any manual operation of human task into motions, body motions, or other elements of human performance, and assigns to each the appropriate time value. This system of time data was originally developed from extensive studies of all aspects of human performance through measurement, evaluation and validation procedures.

Predetermined times are the tabulated values of normal times required to perform individual movements such as moving an arm from one position to another, etc. The total times needed to perform the operation is the sum of the times needed for basic motions. By arranging the basic motions and aggregating associated times, an existing task can be analysed or a proposed operation can be timed without actually performing it.

Factors to be Considered While Using PMTS

- Application of PMTS requires that an operation which is to be measured is divided into basic motions as per the system selected. Each system has its own specific rules and procedures which must be followed exactly.
- Most PMTS do not include allowances, so these are added as in stop watch study.
- At the time of application of PMTS for the first time in a company the adjustment should be made if necessary, in order to match company's performance level which is one time activity PMTS can be classified as to accuracy level, time required for application and the extent of method description.

Types of PMTS

1. Methods time analysis (MTA): A. B. Segur of Oak Park Illinois was one of the first to establish the relationship between the time element and the motion itself. Segur stated that the method must be well defined before an attempt is made to time-analyse the motions involved. He developed a table of improvement principles involving many of his basic motions such as hold, grasp, preposition, position, avoidable delay and balance delay. The improvement principle involved here is in the elimination of the left hand as a holding device. In MTA, motion values are given up to fifth decimal.

2. Work factor system (WF): This is first system of PMTS to have a general use with the work factor system it is possible to determine the work factor time for manual tasks by the use of predetermined data. A detailed analysis of each of the task is made based upon the identification of major variables of work and the use of work factor as a unit of measure. Then the standard time from the table of motion values is applied to each motion.

Four major variables of work factor system are:

1. Body member.
2. Distance.
3. Manual control.
4. Weight or resistance.

This system is applicable to highly repetitive system.

Methods Time Measurement (MTM)

Methods Time Measurement procedure is defined as:

"A procedure which analysis any manual operation or method into the basic motions required to perform it and assigns to each motion a predetermined time standard which is determined by the nature of the motion and the conditions under which it was made."

The primary objective of MTM is to improve methods of operation and it establishes methods accurately before production starts by determining correct times and operations.

Basic Motion Time study (BMT)

Basic motion time study was developed and is thought by J.P. Woods and Gordon Limited, Toronto, Canada. Like other predetermined motion time system, all manual activity has been divided into basic motions.

A basic motion, according to Woods and Gordon, is defined as "Any motion which starts from rest, moves through space, and ends at rest."

(Type 1) Reach

(Type 2) Move

(Type 3) Turn

The body motion and symbols are very similar to the body motions employed by MTM. The only difference lies in the step, where the distance measured is the distance the foot travels.

Advantages of PMTS

1. Short cycle jobs can be timed accurately.
2. Rating, the most difficult part of time study is not necessary.
3. The results obtained are consistent.
4. A reasonable estimate of work content can be obtained before the task is actually carried out.

METHOD TIME MEASUREMENT (MTM)

The objective of MTM is the establishment of tangible, understandable and acceptable data for the scientific measurement of human effort.

Method Time Measurement is defined as:

"A procedure which analyses any manual operation or method into the basic motions required to perform it and assigns to each motion a predetermined time standard which is determined by the nature of the motion and the conditions under which it is made."

USES OF MTM

1. Developing effective methods and plans in advance of beginning production.
2. Improving existing methods.
3. Establishing time standards.
4. Cost estimating.
5. Training supervisors to become method conscious.
6. Research in the areas like operating methods, performance rating.

MTM procedure recognises:

- Eight manual movements.
- Nine pedal and trunk movements.
- Two ocular movements.

Thus there are nineteen fundamental motions to be considered in the establishment of any motion pattern. The time for each of these motions are determined not only by the physical conditions involved in the motions performance but also by the nature of the conditions under which it is made. Thus, the time for a given motion is affected by a combination of physical and mental conditions.

Unit of MTM is TMU. One TMU = 0.0006 minutes.

CONVENTIONS FOR RECORDING MTM DATA

To simplify recording individual MTM methods, a system of MTM conventions has been developed. By using this system, every detail of the motion can be easily recorded. For example:

Reach: Reach is the basic element when the predominant purpose is to move the hand or finger to a destination. The time for making a reach varies with (1) condition (nature of destination), (2) length of the destination, (3) type of reach.

Classes of reach: There are five classes of reach. The time to perform a reach is affected by the nature of the object towards the reach is made.

Case A Reach: to object in fixed location or to object in other hand or on which other hand rests.

Case B Reach: to object whose general location known. Location may vary slightly from cycle to cycle.

Case C Reach: to object jumbled with other objects in group.

Case D Reach: to very small object or where accurate grasp required.

Case E Reach: to indefinite location to get hand into position for body balance or next move or out of the way.

The length of a motion is the true path, not just the straight line distance between the two terminal points.

For example, R 8 C represent Reach 8 inches, case C.
R 12 A represent Reach 12 inches, case A.

The values of TMUS for these symbols are obtained from MTM tables.

Similarly, the details for other symbols:

Move, Turn, Apply pressure, Grasp, Position, Release, Disengage.

Eye motions, Body leg and foot motions are obtained from the MTM tables (published).

MTM VERSIONS

MTM-1 is the most accurate. Provides the most detailed method detailed description but requires the longest time for analysis.

MTM-2 was developed by constructing motion combinations from basic motion of MTM-1. The analysis can be done more quickly than MTM-1.

MTM-3 is the simplest of the MTM systems and is intended for use with long cycle short run operations.

Speed of analysis is seven times faster than MTM-1.

MTM-3 should not be used for analysing manual motions with a frequency higher than 10 or sequence of eye motions.

The MTM should be used with caution. A sufficient training is essential to take up the MTM measurement.

INTRODUCTION TO "MAYNARD OPERATION SEQUENCE TECHNIQUE" (MOST)

H.B. Maynard and Company has introduced MOST system and this new system was brought into practice in the United States in 1975. It has gained a wide recognition as a major contribution to the body of Industrial Engineering. This techniques has a wide application and can be successfully applied in all industries ranging from ship building to electronics, automobile, textile. Application have been made in offices, assembly shops, materials handling, maintenance and other such operations.

Levels of MOST and their Applications:

Maxi MOST: At the highest level, maxi-MOST is used to analyse operations that are likely to be performed lesser than 150 times per week. An operation in this category is less than 2 minutes to more than several hours in length.

Basic MOST: At the intermediate level, operations that are likely to be performed more than 150 times but lesser than 150 times per week should be analysed with basic MOST.

Mini MOST: At the lowest level, mini-MOST provides the most detailed and precise methods analysis. In general, this level of detail and precise is required to analyse any operation likely to be repeated more than 1500 times per week.

DEFINITIONS OF SOME TERMS RELATING TO MOST SYSTEM

1. **Operation:** It is a job task consisting of one or more work element usually done essentially in one location or the performance of any planned work.
2. **Sub-operation:** A sub-operation is desecrate, logical and measurable part of an operation. The content of such a sub-operation may vary depending on type of operation requirements and application area.
3. **Time standard:** It is the total allowed time including manual time, process time and allowance that it should take to perform a task or do a job.
4. **Activity:** It is defined as the series of logical events that take place when an object is moved, observed or treated by hand, a tool or transportation device.
5. **Method step:** A method step is a descriptive formulation of an activity one or more steps organised in sequence according to the applied method will constitute an operation or sub-operation.
6. **Sequence model:** A sequence model is a multi-character representation of a single activity.
7. **Sub-activity:** It is defined as discrete sub-division of an activity or sequence model.
8. **Parameters:** It is a character representation of a sub-activity.
9. **Most analysis:** Most analysis is a computer study of an operation consisting of one or several methods steps and corresponding sequence models as well as appropriate parameters time values and total normal time for the operation or sub-operation.

Consequently only three Basic, Most activity sequence are needed for describing manual work plus a fourth for measuring the movements of object with manual cranes.

- The GENERAL MOVE SEQUENCE (for spatial movement of an object freely through air).
- The CONTROLLED MOVE SEQUENCE (for the movement of an object when it remains in control with a surface or it is attached to another object during the movement).
- The TOOL USE SEQUENCE (for the use of common hand tools).

Application of MOST

This technique finds its application for method improvement. It helps to established the standards and also for determining the production delays and labour performance index.

References for Further Reading

1. Barnes, R.L., *Motion and Time Study Design and Measurement of Work.*, 7th Edn., John Wiley and Sons, New York (1980).
2. Mundel, M.E., *Motion and Time Study*, 5th Edn., Prentice Hall, Englewood Cliffs, N.J. (1978).
3. Niebel, B.W., *Motion and Time Study*, 6th Edn., Richardo Irwin Inc Homewood Ill. (1976).
4. Curie, R.M., *Work Study*, 4th Edn., ELBS and PITMAN (1977).
5. Jhamb, L.C., *Work Study and Ergonomics.*
6. ILO Work Study.

REVIEW QUESTIONS

1. Define time study and explain its objectives.
2. Explain the various steps involved in time study.
3. What is performance rating? Why it is required to rate the worker? What are different rating methods?
4. Explain the terms qualified worker, and normal worker.
5. Explain the Westinghouse method of performance rating.
6. Why the job as a whole cannot be timed?
7. Explain the various types of elements with examples for each.
8. Explain the principle techniques of work measurement and their applications.
9. What are the advantages of stop watch technique?
10. Explain various timing methods in stop watch study.
11. Explain the effect of following on standard time:
    (*a*) Skill of the operator,
    (*b*) Variation in work,
    (*c*) Pace of performance,
    (*d*) Working conditions.
12. Why it is necessary to give allowances? What are different types of allowances?
13. How standard time is computed?
14. What is the relationship between observed time, normal time and the standard time?
15. Write a note on time study equipment.
16. What are the precautions to be taken during stop watch study?
17. What are the different standard forms used for making stop watch study?
18. What is work sampling? What are its merits and limitations?
19. Where work sampling can be useful in the area of production?
20. Explain briefly the steps in work sampling study.
21. How do you determine number of observation to be taken?
22. Write short notes on:
    (*i*) Synthetic rating and analytical estimating,
    (*ii*) Synthetic data and analytical estimating,
    (*iii*) Predetermined motion time study (PMTS),
    (*iv*) Methods time measurement (MTM),
    (*v*) Maynard's operation sequence technique (MOST),
    (*vi*) Allowances,
    (*vii*) Work sampling,
    (*viii*) Work content.

PROBLEMS

1. The elemental times for the elements is shown below

Element	*Observed time*	*Rating*	*Frequently*
A	0.15	80	1
B	0.06	100	1
C	0.12	100	1
D	0.05	80	1
E	8.00	85	1
F	1.05	90	1
G	0.05	110	1
H	10.00	80	1/150
I	6.00	100	1/500

The relaxation allowance is 12% calculate the work content of each element and work content of the job. Also calculate the standard time for the job contingency allowance of 2% is to be allowed.

2. The following data refers to the time study of a drilling process. (Time is given in seconds).

			Cycle timed				
S. No.	*Element*	*Rating %*	*1*	*2*	*3*	*4*	*5*
1	*A*	90%	50	45	46	48	47
2	*B*	105%	30	35	34	29	132*
3	*C*	110%	80**	20	30	25	26
4	*D*	115%	105	110	115	100	103
5	*E*	110%	20	23	24	18	19
5	*F*	90%	40	45	39	44	46

* Switch problem ** Drill replaced

Take suitable allowances with proper reasoning and establish the standard time for this operation.

3. Following data is obtained from a manufacturing unit:
   (1) Number of working days in a week = 6
   (2) Daily shifts –2 of 8 hours duration.
   (3) Number of operators working/shift =10
   (4) Standard hours per unit of production = 4
   (5) Number of units produced in a week = 150
   (6) Absentee man hours = 96
   (7) Ideal time due to power cut off = 108 hrs

Calculate

(*i*) Absenteeism percentage.
(*ii*) Labour utilisation index.
(*iii*) Productive efficiency of labour.
(*iv*) Overall productivity of labour in terms of units produced per week per employee.

4. In a machine shop work sampling study was conducted for 160 hrs in order to estimate the standard time. Total number of observation recorded were 3500. There were 600 no working activities. Ratio between manual to machine element was 2 : 1. Average rating factor was 1 : 2 and total number of jobs produced during the study were 8000. Rest and personal allowances taken together will be 17% of normal time determine the standard time per job.

6

VALUE ENGINEERING

• Origin of value engineering • Meaning of value • Definition of value engineering • Value analysis and value engineering • Uses of V.E. • When to apply value analysis • Reasons for unnecessary costs • Difference between value analysis and other cost reduction techniques • Steps in value analysis • Phases and constituents elements of each phase • FAST technique • Ten commandments of V.E.

6.1. ORIGIN OF VALUE ENGINEERING

The technique of value engineering gained considerable momentum during the Second World War. During the year 1947, Lawrence Miles, who was working for the American General Electric Company (GEC) became increasingly aware of the limitations of traditional cost reduction techniques. As a design engineer involved in cost reduction activities, was able to distinguisl between what is good value and bad value and realised that very few people had a real grasp o good and bad value. He established that this is because of altogether a different approach. He evolved a set of tests that could be applied to any product either projected or being currently produced. Then he persuaded the management to try this technique which was already subjected to cost reduction exercises many times. The result was encouraging and soon the G.E. Company started using this to their products.

The technique soon spread to other American companies and later to Europe. It is now generally accepted by management as a powerful tool to ensure that unnecessary costs are not built into their products. The principle behind value analysis is that it is a functionally oriented method for improving product value by relating the various elements of product worth to their corresponding elements of cost. Thus value analysis allows the required functions to be performed at the minimum cost. To apply principles of value analysis effectively it is recommended that a systematic and rigorous approach is to be followed to ensure that no steps are omitted. Value engineering is a team approach. A team of people drawn from various specialist areas is found to be the best way to handle V.E. projects. The participating team members are drawn from design, purchase, production, accounts and marketing coordinated by an engineer belonging to any one of the above departments.

6.2. MEANING OF VALUE

Value means different things to different people. Value is the word used by individuals without being clearly understood. Even different departments of the same company have different opinions of the value of the product that their own company manufactures. The designer relates value with reliability, purchase men with price paid for item, production person equates the value with what it costs to manufacture and sales person with what customer is willing to pay.

The Value is of Four Types

1. **Cost value:** It is sum of all costs incurred in manufacturing the product. The cost value is

the sum of raw material cost, labour cost, tooling cost and other overhead costs to produce the product.

2. **Exchange value:** It is the measure of all the properties, and features of the product which make the product possible of trading (exchanging) with other products.
3. **Use value:** It is that value of the product which constitutes the amount of its cost included to make the product work.
4. **Esteem value:** is that amount of cost included into the product to enhance its customer appeal (desire to own).

Example: For a motor car bumper, its use value is very much less than its esteem value.

Gudgeon pin has got 100% use value but esteem value is nil. Value of the product varies according to the place and time of consideration. A balance between use and esteem value is to be obtained.

Value is the relationship between cost and quality and can be expressed by a simple equation,

$$\text{Value} = \frac{\text{Quality (need perfromance } (P))}{\text{Cost } (C)}$$

Thus, the value can be improved by several means.

1. Decreasing costs, while ensuring same level performance same level of performance.

$$V\uparrow = \frac{P\rightarrow}{C\downarrow}$$

2. Enhancing performance at same cost. $V\uparrow = \frac{P\uparrow}{C_{\rightarrow}}$

3. Decreasing cost and increasing performance. $V\uparrow = \frac{P\uparrow}{C\downarrow}$

4. Both performance and costs increasing. But performance increases faster than cost. $V\uparrow = \frac{P\uparrow}{C\uparrow}$

6.3. DEFINITION OF V.E.

According to the Society of American Value Engineers (SAVE) **"Value analysis is the systematic application of recognised techniques which identify the function of a product or service, establish a monetary value for the function and provide the necessary function reliably at that lowest overall cost."**

Principle behind value analysis is that it is a functionally oriented method for improving product value by relating various elements of product worth to their corresponding elements of cost.

V.E. is an organised creative approach to ensure that essential functions of a product or service are provided at minimum overall cost without sacrificing quality and reliability.

The objective of value analysis is not to degrade or cheaper the products but to improve its value by reducing cost. The elimination of unnecessary cost cause no adverse effect on quality and reliability of the product. Thus V.E. is a cost preventive technique which eliminates unnecessary cost built up into the product.

6.4. VALUE ANALYSIS AND VALUE ENGINEERING

Though the two terms are used synonymously and philosophy underlying is same, *i.e.*, unnecessary cost elimination, there is a difference between the two. The difference lies in time and the phase of product life-cycle at which the technique is applied.

Value Analysis is applied to the existing product with a view to improve its value. It is analysis after the fact and it is a remedial procedure.

Value Engineering is applied to the product at the design stage and thus ensures prevention rather than elimination.

6.5. USES OF VALUE ENGINEERING

1. It is a cost prevention as well as cost elimination technique thus reducing cost of the product.
2. Helps employees for better understanding of their jobs and orients them towards creative thinking.
3. Balance the cost and performance.
4. Prevents over design of components.
5. Motivates employees to come out with creative ideas.
6. Increases the profits and deflates costs.
7. Helps to satisfy the customer with company's products.

6.6. WHEN TO APPLY VALUE ANALYSIS

One can expect very good results from application of value analysis at the proper time and correct phase of product. Life cycle. V.E. should be applied in case of the following indications:

- Company's products are loosing in the market and there is a decline in sales.
- Company's products are priced higher than the competitors.
- New design of products being undertaken.
- Symptoms of disproportionate increase in cost of production.
- Decreasing profitability and return on investment (ROI).
- Company failing to meet its delivery commitment.

6.7. REASONS FOR UNNECESSARY COSTS

No designer or operations manager builds costs knowingly into the product/process though some how the unnecessary costs do creep in without one's knowledge.

In any product, if one carefully analyses, approximately 20-30% of the product cost is unnecessary. The reasons are:

1. Lack of relevant information—leads to wrong decisions which increase costs.
2. Honest and wrong beliefs—These results from accepting opinions, beliefs and theories without justification and verification.
3. Temporary circumstances.
4. Habits and attitudes.
5. Lack of ideas.

6.8. DIFFERENCE BETWEEN V.E. AND OTHER COST REDUCTION TECHNIQUES

There is a basic difference in approach between conventional cost reduction techniques and value analysis *e.g.*, a cost reduction team examining a component may come to the conclusion that the process is right, purchase price seems to be O.K. and all that could be done is to reduce the size (diameter or thickness) slightly which would achieve 5% reduction in cost.

- Generally traditional techniques are post-production oriented and concentration is on the part or element in contrast to value analysis which is function oriented.
- Other techniques try to reduce the cost starting from existing method where as V.A. never considers any process, method or element granted and it analyses by function rather than part or element.
- Cost reduction can reduce the costs up to 10% whereas 30-40% cost reduction is possible with V.A. without sacrificing its quality and performance.

- Value analysis questions the very existence of part or process by analysis of its design specifications, materials and process of manufacturing.
- The cost reduction is cost oriented and concentrates on finding the cheaper cost whereas value analysis is function oriented and try to produce the product at lowest cost without sacrificing its quality and performance.

6.9. STEPS IN VALUE ANALYSIS

Value analysis is a systematic approach and it follows a planned and disciplined procedure to analyse the job. The steps in V.A. are shown in Fig. 6.1.

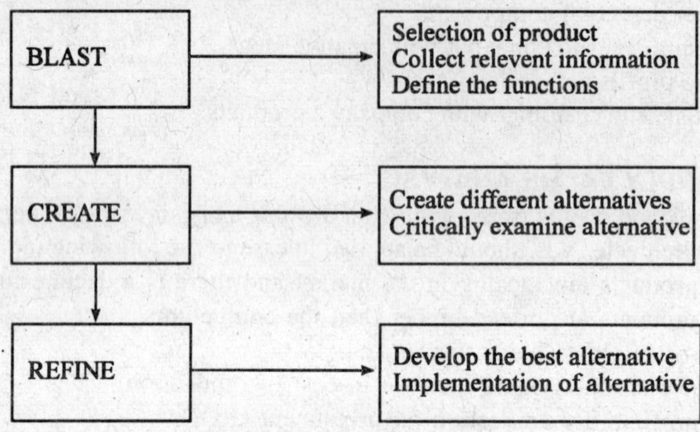

Fig. 6.1: Steps in value analysis.

6.9.1. SELECTION OF PRODUCT

High volume product will give maximum returns (benefits) from value analysis. The factors influencing the selection of the product are:

(*a*) Market research estimated product life.
(*b*) Expected variation in labour and material cost throughout the expected life of the product.
(*c*) Changes in trading conditions.
(*d*) Time and cost to modify tooling, carry out field tests.
(*e*) Cost effect of increased inventory, interchangeability and spares.
(*f*) Overall cost to make the change and likely savings.

It is essential for the value engineering team to justify the selection of V.E. project. Criteria used for justification are:

1. There should not be any major changes in the product or item under study.
2. The functions of components are definable and understandable.
3. The proposed changes after the recommendations are likely to be implemented.
4. A large saving potential.
5. Components are large in number (minimum 8-10) and appear to be large, more complex and heavier than necessary.

6.9.2. INFORMATION GATHERING

All the relevant information is to be collected before the team meets for the session. People will be very much tempted to make decision on mere opinion, if all the facts are not available. This goes against the very basis of value analysis.

The following information will be needed:

- All drawings and specifications.
- All cost details.

- Machining and assembly details.
- Features preferred by customers.
- Supply details.
- Failure details and service records.
- Quantities involved.
- All manufacturing details.
- Any other relevant information pertaining to the product.

6.9.3. DEFINITION AND ANALYSIS OF FUNCTIONS

This step differentiates V.E. from all other cost reduction techniques. Functional analysis is the heart of V.E.

Function is that which a product does in order to work or sell.

Miles set the rules for defining functions. The expression of all the functions must be accomplished in just two words, a VERB and NOUN.

Product	*Function*	
	VERB	*NOUN*
Newspaper	Provide	Information
Shaft	Transmit	Torque
Pen	Communicate	Ideas
Bolt	Secure	Part

The two words definition of functions helps to state with clarity the function performed by a product or part. This helps to break the problems into simplest elements. Faulty communication and misunderstanding is reduced by defining the function in two words.

All the functions can be divided into two levels as per the importance BASIC and SECONDARY.

Basic (or primary) function is the basic purpose for which the product exists, *i.e.,* the Basic function of pen is to communicate idea. Or A bottle holding fruit juice is "contain liquid"

Secondary functions can be more than one and they support the primary function.

They are basically a result of a specific design approach.

Examples.

Product	*Basic function*	*Secondary function*
Bottle of cold drink	Contain liquid	Facilitate drinking, Preserve liquid
Pen	Communicate Idea (Make mark)	Improve look provide, grip, control flow

The functional approach is applied to the whole product, subassemblies and then individual components. Any deviation from this sequence leads to the analysis of only part of the function. The analysis of subassembly may lead to the elimination or combination elsewhere.

In the function phase,

First step is to identify the functions and their levels as basic and secondary.

Second step is to allocate proper costs to the selected functions.

6.10. FUNCTION ANALYSIS SYSTEMS TECHNIQUE (FAST)

This technique is developed by Charles Bytheway and it is used to establish the relationship between functions in the analysis of an entire system, process or complicated assembly and gives a better understanding of interrelationships of function and their cost.

This technique basically finds answers to three questions about each function performed by a product or service.

FAST Diagram is shown in Fig. 6.2.

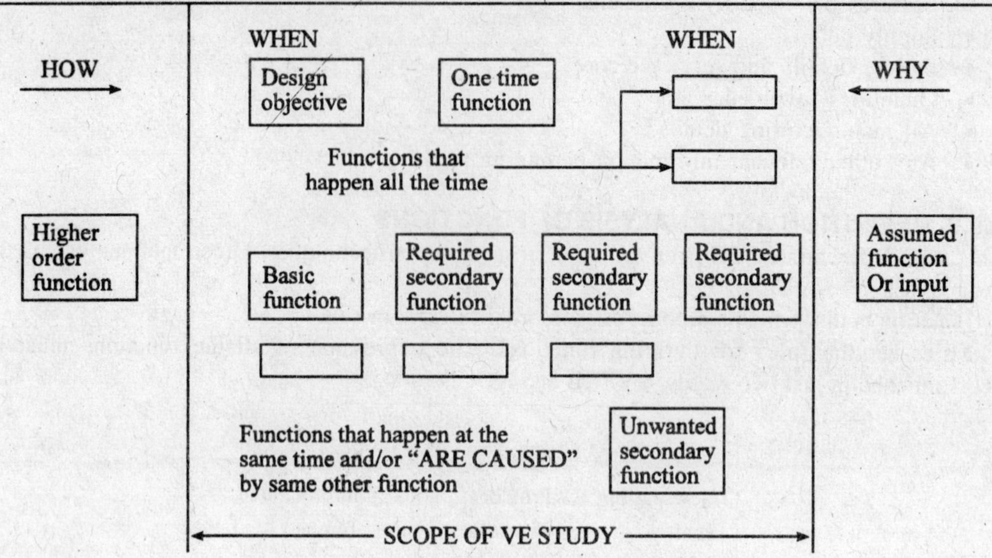

Fig. 6.2: Function analysis systems technique (FAST).

Construction of a FAST Diagram for a ball point pen

The fast diagram is represented in Fig. 6.3.

The first step is to identify all the functions of the parts.

The parts of the ball pen and functions of parts are shown in Table 6.1.

Table 6.1: Parts and Functions of Parts of Ball Pen.

Part	*Function*	*Remarks*
Tip	Improves look, locate part	Chromium plating
Plastic neck	Hold part, provide grip, facilitate maintenance	Step in tip
Spring	Provide tension	
Refill	Contain ink, control flow, restrict movement	
Ink in refill	Make marks	
Plastic barrel	Enclose part, identify company, air flow	Hole in barrel, name in barrel

Basic function—Provide impression or MAKE MARK

Secondary functions—Improve look
Provide grip
Control flow

To start with, MAKE MARK (basic function).

The question "why do we need this function is asked".

The answer is "to communicate ideas". This function is shown on the left side of FAST diagram.

How do you make marks?

The answer is by performing the functions "contain ink and control flow" which is represented on the right side of basic function.

Next question is how do we control flow and contain Ink?

Then the answer is to position refill, *i.e.,* by making use of spring whose function is to provide tension. Which is represented in the figure 6.3.

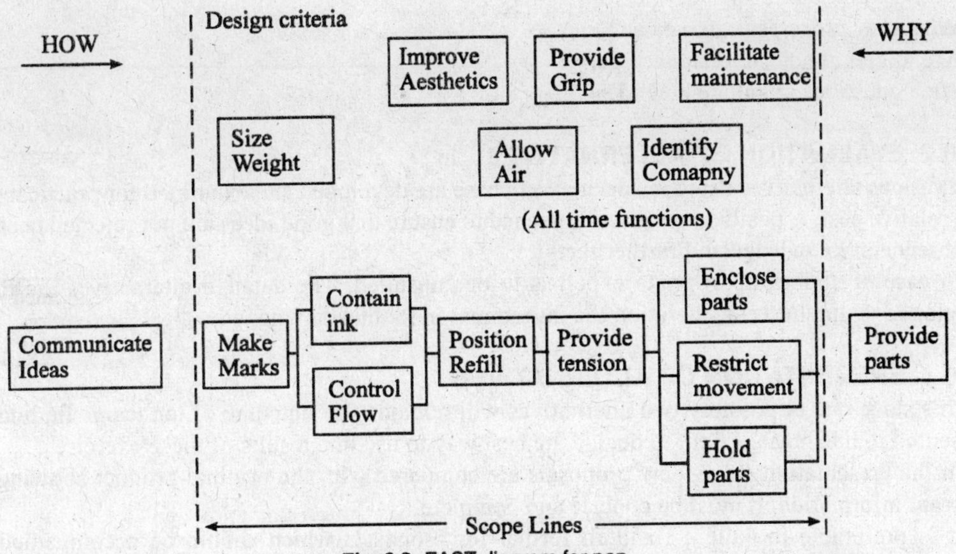

Fig. 6.3: FAST diagram for pen.

Then the other functions are to be considered.

The question "when do we improve look"? All the time. The all time functions are placed on the right hand top corner of the diagram. The other functions are represented as shown in order to determine the scope of the problem, the FAST diagram includes scope lines. (Vertical broken lines).

The left scope line is placed between the basic function under consideration (make marks) and highest order basic function (Communicate Ideas). The Function on the left of the scope line is the next higher order function which will be satisfied by several means.

The scope line on the right hand is determined by drawing a vertical broken line to the left of the function that is suitable input to the system "In this case, provide part is the suitable input." Design parameters are placed inside the scope lines.

In the next phase, (creative phase) the generation of alternative ways of performing desired function is taken up. By observation of adjacent functions on the FAST diagram many new ideas may be found to combine, modify the functions. This leads to better performance of function.

6.10.1. SPECULATION OF ALTERNATIVES

The objective here is to generate alternative ways of achieving functions defined in the previous step. Ideas should be suggested by team members regardless of practicability and cost. Brain storming technique is used to generate ideas.

It is important that alternatives are only listed and are not discussed or evaluated. People will not usually give ideas as there is a fear of being ridiculed.

The following questions often lead to creative thinking

Can we do without it?

Why this particular shape?

(If it is flat) can it be round? or *vice versa*

Can two parts be combined?

Can one component be made into two?

Why does it need to be so thick?

Can standard part be used?

Can we buy it at cheaper rate?

What new processes can we use?
Are tolerances closer than necessary?
Such questions stimulate new ideas.

6.10.2. EVALUATION OF ALTERNATIVES

The various alternatives listed in speculation phase are developed and examined for practicability and relative cost. A positive approach is needed to ensure that good ideas are not rejected because of absence of knowledge and further facts.

In case of doubts, appropriate expert is to be consulted. The detailed alternatives are to be developed taking into consideration the investment in tooling, saving, etc.

6.10.3. PRESENTATION OF ALTERNATIVES

At this stage one or possibly two alternatives will remain. The outcome of the teams findings is presented to line management to decide the best way to use line results of these exercises.

In the presentation stage, new proposals are compared with the original product showing all relevant information. It must be concise and complete.

It is preferable to have a standard format for proposals which should be accompanied by drawing or sketches as appropriate.

The format should contain the following information:

Title and scheme number.
Description of change.
Part number.
Old and new costs with systematic break ups.
Savings, Annual Quantity and Annual Savings.
Tooling and other implementation costs.
Then implementation strategy is to be worked out.
Hence, a firm backing from top management is essential. If V.A. is to succeed.

6.10.4. PHASES AND CONSTITUENT ELEMENTS OF EACH PHASE

1. **ORIENTATION PHASE**
- SELECT - Project to be handled by V.E. Team
- ESTABLISH - Priorities
- PLAN - Specific product
- APPOINT - The team
- BRIEF - Members regarding the project
- REVIEW - Methodology to be followed for investigation
- PREPARE - Terms of reference for selected project
- FIX - Responsibility for Data collection

↓

2. **INFORMATION PHASE**
- LOCATE - Data from different sources
- IDENTIFY - FACTS
- ARRANGE - Facts into required form

3. FUNCTIONAL ANALYSIS PHASE
- LIST - The components or parts
- PREPARE - Verb-Noun description of functions
- ESTABLISH - Cost of BASIC FUNCTIONS
- ESTIMATE - Worth of each Basic function
- DETERMINE - Value improvement potential

↓

4. SPECULATION OR CREATIVE PHASE
- CONDUCT - Creative problem solving
- GENERATE - Ideas, Combine and/or rearrange them to provide different ways to accomplish basic functions

↓

5. EVALUTION PHASE
- SHORT LIST - Creative ideas
- DEVELOP - Ideas functionally
- EVALUATE - Ideas according to the prefixed criteria
- RANK - Alternatives according to the cost reduction potential
- SELECT - One or more alternatives for development

↓

6. DEVELOPMENT PHASE
- ANALYSE - The alternatives for technical requirement
- CONFIRM - The economic feasibility
- ARRANGE - The necessary test runs
- SELECT - One or two alternatives for recommendations

↓

7. RECOMMENDATION PHASE
- PREPARE - The reports of V.A. proposals
- MAKE - Presentation to the decision-making authority
- SUBMIT - A definite plan for action
- SECURE - Approval for implementation

↓

8. IMPLEMENTATION PHASE
- OBTAIN - Approval to the proposal
- FIX - Responsibility for implementation
- AUTHORISE - Change to convert recommendations into actions
- EXPEDITE - Implementation
- RESOLVE - Problem identified during implementation
- TRAIN - Workmen
- RETIME - Jobs

9. FOLLOW-UP PHASE	
• AUDIT	- Savings and compare results with original expectations
• SUBMIT	- Cost savings achievement reports to the management
• EVALUATE	- The effectiveness
• RECOMMEND	- Changes/corrective actions and awards for contributors to value engineering proposals

6.11. TEN COMMANDMENTS (PRINCIPLES) OF VALUE ANALYSIS

1. Do not use a component or part that does not contribute to the value of the product.
2. Do not use a component or part whose cost is not proportional to its usefulness.
3. Do not provide any features to the component or finished product that are not absolutely necessary.
4. Accept the change if part of the required quality can be made out of cheaper and easily available material.
5. If the part of required quality is made by a process or method costing less, then do use the alternative process or method.
6. Use standard parts wherever possible.
7. Use proper toolings and manufacturing methods taking into consideration the quantities.
8. The cost of the component used should be proportional to its use or function.
9. Use the material, part best suited for the purpose.
10. Purchase the part instead of manufacturing in house, if suitable supplier can provide the part of good quality at the reasonable price.

References for Further Reading

1. Miles, Lawrence, *"Techniques of Value Analysis and Engineering"*, McGraw Hill Book Company, New York (1961).
2. Harald, Tufty, *Compendium of Value Engineering.*
3. ASTME, *Value Engineering in Manufacturing.*, Prentice Hall Inc. Englewood, Cliff, N.J. (1967).

REVIEW QUESTIONS

1. Define value. What are different types of values? Gives example for each type of value.
2. Define value engineering. Discuss the various fields of application of value engineering.
3. What are the steps involved in value analysis?
4. What is functional analysis? Explain with respect to V. E.
5. What is the difference between value engineering and value analysis?
6. What is use value and esteem value? Explain with an example for each.
7. How value engineering helps to improve productivity?
8. Explain the creative phase of value analysis.
9. What is the difference between V.E. and method study?
10. Explain FAST with respect to V.E.
11. "Value engineering prevents unnecessary cost build up into the product." Explain.
12. "V.E. is a cost reduction technique but with a difference." Comment.

7

PLANT LOCATION

• Introduction • Need for selection of a location • Importance of location • Systems view of location • Plant location, problem • Advantages and limitations of urban, semi-urban and rural locations • Comparison between rural and urban location • Factors influencing the plant location • Quantitative method for evaluating plant location.

7.1. INTRODUCTION

The facilities location problem is an important strategic level decision-making for an organisation. One of the key features of a conversion process (manufacturing system) is the efficiency with which the product (services) are transferred to the customers. This fact will include the determination of where to place the plant or facility.

Holmes defines plant location problem as one of determining **"That location which, in consideration of all factors affecting products delivered-to-customers cost of product(s) to be manufactured, will afford the enterprise the greatest advantages obtained by virtue of location."**

The selection of location is a key-decision as large investment is made in building plant and machinery. It is not advisable or not possible to change the location very often. So an improper location of plant may lead to waste of all the investments made in building and machinery, equipment.

Before a location for a plant is selected, long range forecasts should be made anticipating future needs of the company.

The plant location should be based on the company's expansion plan and policy, diversification plan for the products, changing market conditions, the changing sources of raw materials and many other factors that influence the choice of the location decision. The purpose of the location study is to find an optimum location one that will result in the greatest advantage to the organisation.

7.2. NEED FOR SELECTING A SUITABLE LOCATION

The need for selecting a suitable location arises because of two situations.

1. When starting a new factory.
2. In case of existing factory.

The existing firms will seek new locations in order to expand the capacity or to place the existing facilities. The increase in demand for the company's products can give rise to following decisions:

1. Whether to expand the existing capacity and facilities.
2. Whether to look for new locations for additional facilities.

3. Whether to close down existing facilities to take advantage of some new locations.

The probable reasons for replacement of existing facilities to new locations is shown in Table 7.1.

Table. 7.1: Reasons for Replacement of Existing Facilities to New Locations

Changes in location of demand
Changes in availability of materials
Changes in availability of transport
Changes in the cost and/or supply of labour
Changes in regulations and law
Changes in availability of raw materials
Changes in policy of industries to relocate on which the firm is dependent

7.3. PLANT LOCATION PROBLEM

The locations of the facilities is carried out in three stages:

First stage—Selection of a general territory (or region).

Second stage—Selection of a Community.

Third stage—Selection of specific site.

1. **Selection of a region:** This refers to the selection of a particular geographical zone or state taking into consideration such factors as nearness to market and sources of raw materials, basic infrastructure facilities available, climatic conditions and taxation and laws.
2. **Selection of a community:** This refers to the selection of the specific locality within the selected region. The factors that influence the selection of community are, availability of labour, community attitude, social structure and service facilities.

   Generally the following alternatives are available:

   (1) Urban area.

   (2) Rural area.

   (3) Semi-urban area near the urban area.
3. **Selection of a particular site:** This refers to the selection of specific site within the community. The factors that influence the site selection are the cost of the land, availability and suitability of the land. The type of manufacturing process may dictate the site selection. The conditions that govern the particular types of community are as follows:

   A. Condition that demand city (Urban) location

   (*i*) Highly skilled labour requirement

   (*ii*) Manufacturing dependent on urban utilities

   (*iii*) Excellent communication and transportation facility

   (*iv*) Concentrated suppliers.

   B. Conditions that demand sub-urban location

   (*i*) Semi-skilled or female workforce required

   (*ii*) Large space availability for future expansion

   (*iii*) Community close to large population centre.

   C. Conditions demanding rural location

   (*i*) Large site required for future expansion

   (*ii*) Requirement of unskilled labour

   (*iii*) Manufacturing process is dangerous and objectionable

   (*iv*) Low wage structure

   (*v*) Lower property tax rates

   (*vi*) Lower cost of land.

7.4. ADVANTAGES OF URBAN, SUBURBAN, RURAL LOCATIONS

A. URBAN AREA

Advantages

1. Excellent communication network.
2. Good transportation facilities for material and people.
3. Availability of skilled and trained manpower.
4. Factory in the vicinity of the market hence high local demand.
5. Excellent sourcing (subcontracting) facilities.
6. Good educational, recreational and medical facilities.
7. Availability of service of consultants, training institutes and trainers.

Disadvantages

1. High cost of land compared to rural area.
2. Sufficient land is not available for expansion.
3. Labour cost is high due to high cost of living.
4. Industrial unrest due to trade union activities.
5. Management labour relations are much influenced by union activities.
6. Municipal and other authority restrictions on buildings etc.
7. High labour turnover.

B. RURAL AREA

Advantages

1. Cheaper and ample availability of site.
2. Cheaper labour rates.
3. Less turnover of labours because of limited mobility.
4. No municipal restrictions.
5. Good industrial relations.
6. Scope for expansion and diversification.
7. No slums and environmental pollution.

Disadvantages

1. Poor transportation network.
2. No good communication facilities.
3. Sourcing of components and materials should be from outside.
4. Far away from market.
5. High absenteeism during harvest season.
6. No educational, medical and recreational facilities.

C. SUBURBAN AREA

Advantages

1. Land available at cheaper rate compared to urban location.
2. Infrastructure facilities are developed by promotional agencies.
3. Because of nearness to city availability of skilled manpower.
4. Educational, medical facilities are available because of nearness to city.

Limitations

1. Due to concentration the suburban area will become crowdy and will become urban in turn within short period.

2. High mobility of workers and hence higher labour turnover.
3. Government incentive and subsides to promote industries.

7.5. IMPORTANCE OF LOCATION

Location decisions are important and require management's careful attention for several reasons. Three important reasons are:

1. **Competition:** A company's location affects its ability to compete. In manufacturing firms, location affects direct cost by, to and from transportation from the location and cost of labour and others. Thus a good location helps to deliver the product at a cheaper price and hence helps to combat competition.
2. **Cost:** Failures to make a good location become too expensive for the company and have long lasting effects. Plant location is a major investment decision. If selected wrongly, it is going to affect the performance of the company and location once selected will be difficult to change in near future.
3. **Indirect benefits:** These are not visible directly and will not be reflected in company's accounts.

7.6. SYSTEMS VIEW OF LOCATION

A board systems view is necessaryfor considering location of facilities. The problems of selection will encompass many interrelated factors.

The manufacturing company is a part of the larger system—Logistic chain.

Manufacturing companies depend on suppliers for their inputs and need to supply their outputs to consumers.

Thus several company's are linked together by logistic chain as shown in Fig. 7.2.

The location of one component in logistic chain depends on the suppliers consumers and other facilities involved in manufacturing and distribution system.

The systems view evaluates all the components and their interrelationships to arrive at an optimal location for all the components of the chain.

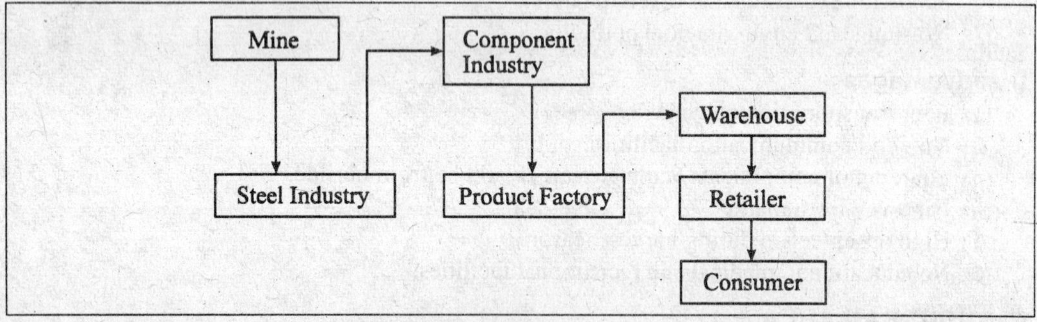

Fig 7.1: Logistic chain for a product.

7.7. LOCATION FACTORS

Location factors involve many factors that influence the revenues, costs or both and they affect profits.

In general, location factors are grouped into:

1. **Market related factors:** Location of demand and competition.
2. **Tangible cost factors:** Labour, materials, transportation utilities, site cost, taxes, etc.
3. **Intangible factors:** Legal aspects, environmental factors, climate, schools, hospital and recreational facilities, community attitude, etc.

Each one of these factors are discussed in detail.

7.8. COMPARISON BETWEEN URBAN AND RURAL LOCATIONS

	Factor	*Urban Location*	*Rural Location*
1.	Market	Local demand is high so less distribution and transportation cost	Market far away which increase the distribution and transportation cost
2.	Labour	Adequate availability of skilled labour	Difficult to get the skilled labour
3.	Cost and availability of land	Cost of land will be high and difficult to get land for future expansion	Adequate land will be available at the lower cost
4.	Transport	Good transport facility is available	Non-availability of good transport system
5.	Communication facilities	Availability of good communication network	Communication facilities are not available
6.	Municipal and other civil amenities	Available	Not available
7.	Pollution control	Strictly measures are to be taken to control pollution	Lower efforts are required for control of pollution measures
8.	Labour turnover	Higher labour turnover due to better opportunities in the vicinity of the area	Low labour turnover
9.	Union activities	High union activities resulting in strikes	Least disturbances due to strikes and lockouts
10.	Community services (School, hospital)	Community services like schools services and recreation facilities are available	Not available. It is to be created by the firm on its own
11.	Cost of living	High	Low
12.	Storage facilities	Adequate storage facilities including cold storages	Warehouse and storage facilities are not available

7.9. FACTORS INFLUENCING PLANT LOCATION

1. **Proximity to markets:** Every company is expected to serve its customers by providing goods and services at the time needed and at reasonable price organisations may choose to locate facilities close to the market or away from the market depending upon the product. When the buyers for the product are concentrated, it is advisable to locate the facilities close to the market.

Locating nearer to the market is preferred if

(*i*) The products are delicate and susceptible to spoilage.

(*ii*) After sales services are promptly required very often.

(*iii*) Transportation cost is high and increase the cost significantly.

(*iv*) Shelf life of the product is low.

Nearness to the market ensures a consistent supply of goods to customers and reduces the cost of transportation.

2. **Supply of raw material:** It is essential for the organisation to get raw material in right qualities and time in order to have an uninterrupted production. This factor becomes all the important if the materials are perishable and cost of transportation is very high.

General guidelines suggested by Yaseen regarding effects of raw materials on plant location is:

(*a*) When a single raw material is used without loss of weight, locate the plant at the raw material source, at the market or at any point in between.

(*b*) When weight loosing raw material is demanded, locate the plant at the raw material source.

(*c*) When raw material is universally available, locate close to the market area.

So the general guideline is " Nearness to source of raw material is of special importance when the material is bulky in relation to its value and when the volume and weights are significantly reduced during the processing." The place of production is likely to be at the place of consumption where the final product is more expensive to carry because it is more bulky, fragile and perishable.

If the raw materials are processed from variety of locations, the plant may be situated so as to minimise total transportation costs.

Nearness to raw material is important in case of industries such as sugar, cement, jute and cotton textiles.

3. **Transport facilities:** Speedy transport facilities ensure timely supply of raw materials to the company and finished goods to the customers. The transport facility is a prerequisite for the location of the plant. There are five basic modes of physical transportation, air, road, rail, water and pipeline. Goods that are mainly intended for exports demand a location near to the port or large airport. The choice of transport method and hence the location will depend on relative costs, convenience, and suitability. Thus transportation cost to value added is one of the criteria for plant location.

4. **Infrastructure availability:** The basic infrastructure facilities like power, water and waste disposal, etc., become the prominent factors in deciding the location.

Certain types of industries are power hungry *e.g.,* aluminium and steel and they should be located close to the power station or location where uninterrupted power supply is assured throughout the year. The non-availability of power may become a survival problem for such industries. Process industries like paper, chemical, cement, etc., require continuous supply of water in large amount and good quality, and mineral content of water becomes an important factor. Waste disposal facilities for process industries is an important factor which influences the plant location.

5. **Labour and wages:** The problem of securing adequate number of labour and with skills specific is a factor to be considered both at territorial as well as at community level during plant location. Importing labour is usually costly and involve administrative problem. The history of labour relations in a prospective community is to be studied. Prospective community is to be studied. Productivity of labour is also an important factor to be considered. Prevailing wage pattern, cost of living and industrial relation and bargaining power of the unions forms the important considerations.

6. **Law and taxation:** The policies of the state governments and local bodies concerning labour laws, building codes, safety, etc., are the factors that demand attention.

In order to have a balanced regional growth of industries, both central and state governments in our country offer the package of incentives to entrepreneurs in particular locations.

The incentive package may be in the form of exemption from a sales tax and excise duties for a specific period, soft loan from financial institutions, subsidy in electricity charges and investment subsidy. Some of these incentives may tempt to locate the plant to avail these facilities offered.

7. **Suitability of land and climate:** The geology of the area needs to be considered together with climatic conditions (humidity, temperature). Climate greatly influence human efficiency and behaviour. Some industries require specific climatic conditions *e.g.,* textile mill will require humidity.

Now, with the development in air-conditioning facilities the climatic condition can be controlled but at a high cost.

8. **Supporting industries and services:** Now-a-days the manufacturing organisation will not make all the components and parts by itself and it subcontracts the work to vendors. So, the source of supply of component parts will be the one of the factors that influences the location.

The various services like communications, banking services professional consultancy services and other civil amenities services will play a vital role in selection of a location.

9. **Community and labour attitudes:** Community attitude towards their work and towards the

prospective industries can make or mar the industry. Community attitudes towards supporting trade union activities is an important criteria.

Facility location in specific location is not desirable even though all factors are favouring because of labour attitude towards management which brings very often the strikes and lock-outs.

10. **Social infrastructure:** Availability of community facilities like—(1) Housing facilities, (2) Recreational facilities, (3) Educational facilities, and (4) Medical facilities—are to be considered while selecting a location.

The Factors influencing the location are listed as shown in Table 7.2.

Table 7.2: Factors Influencing Plant Location

Location Factor	*Territory Selection*	*Community Selection*	*Site Selection*
1. Proximity to market	✓		
2. Supply of raw material	✓		
3. Transport facilities	✓	✓	
4. Infrastructure availability	✓	✓	
5. Labour and wages	✓	✓	
6. Law and taxation	✓	✓	
7. Suitability of land and climate	✓	✓	
8. Supporting industries and services		✓	✓
9. Community and labour attitudes		✓	✓

7.10. QUANTITATIVE METHOD FOR EVALUATION OF PLANT LOCATION

1. **Dimensional analysis:** Dimensional analysis involves calculation of the relative merits or cost ratios for each of the factors, giving each of the cost factor an appropriate weightage by means of an index to which the cost ratio is raised and multiplying these weighted ratios in order to arrive at a figure on the relative merits of alternative sites.

Let $C_{m1}, C_{m2}, C_{m3}, \ldots\ldots C_{mz}$ are the costs associated with site M for different cost factors.

$C_{n1}, C_{n2}, C_{n3}, \ldots\ldots C_{nz}$ are the costs associated with site N on Z different factors.

$W_1, W_2, W_3, \ldots\ldots W_z$ are the weightages for various factors

Merit of location $M = (C_{m1})W_1 \times (C_{m2})W_2 \times \ldots\ldots (C_{mz})W_z$

Merit of location $N = (C_{n1})W_1 \times (C_{n2})W_2 \times \ldots\ldots (C_{nz})\ W_z$

The relative merits of sites M and N are given by

$$\frac{\text{Merit of } M}{\text{Merit of } N} = \left[\frac{C_{m1}}{C_{n1}}\right]^{W_1} \times \left[\frac{C_{m2}}{C_{n2}}\right]^{W_2} \times \ldots\ldots \left[\frac{C_{m2}}{C_{nz}}\right]^{W_2}$$

If the value of $\dfrac{\text{Merit of } M}{\text{Merit of } N} > 1$, select site N

Otherwise, *i.e.*, < 1, select M

The advantages of this method are—it compares both subjective and objective (tangible and intangible) factors and gives a quantitative figure to the decision maker.

2. **Method based on the economics of various sites:** The ideal location for the plant should be such that the cost of procuring the materials and processing them into products and the cost of distribution of finished goods to the customer should be minimum.

This method is illustrated through an example.

An ABC company intends to select one of the three locations. Both tangible and intangible factors collected by the expert is given below:

PARTICULARS		SITE		
		A	B	C
(*a*)	Total investments in land, building, plant and m/c (Rs. '000)	250	325	270
(*b*)	Revenues ('0000)	410	515	360
(*c*)	Expenses on raw materials ('0000)	89	100	98
(*d*)	Distribution cost ('0000)	40	60	30
(*e*)	Expenses on utilities ('0000)	50	40	25
(*f*)	Wages and salaries ('0000)	25	30	28
(*g*)	Community facilities	Indifferent	Good	Bad
(*h*)	Community attitude	Indifferent	Good	Indifferent

Here the criteria used to evaluate the suitability of the sites is rate of return on investment. Rate of return is computed as

$$\text{R.O.I.} = \frac{\text{Total Revenues} - \text{Total expenses}}{\text{Total investment}}$$

The table shows the calculation of R.O.I. for the three alternatives.

PARTICULARS		SITE		
		A	B	C
1.	Total investments	250	325	270
2.	Total sales	410	515	360
3.	Cost of raw material	89	100	98
4.	Cost of distribution	40	60	30
5.	Expenses on utilises	50	40	25
6.	Salaries and wages	25	30	28
7.	Total expenses	204	230	181
8.	Rate of return (%)	82.4%	87.6%	66.29%

As the R.O.I. for site 'B' is higher compared to others, it is the best choice.

References for Further Reading

1. Raymond, Mayer, *Production and Operations Management,* McGraw Hill, New York (1982).
2. Ray-wild, *Production and Operations Management*—Principle and Practice, Halt Rinehart and Winston, London (1980).
3. Magee, *"Industrial Logistics"*, McGraw Hill, New York (1968).
4. Love, R.F. *et. al.*, *"Facilities Location Models and Methods"*, Elsevier (1986).
5. James, Riggs., *Production System-Planning, Analysis and Control,* 4th Edition, John Wiley and Sons, New York.

REVIEW QUESTIONS

1. Explain why plant location decisions are important to the organisation?
2. What are the factors that influence the selection of location for a plant?
3. Explain the advantages and disadvantages of urban, semi-urban and rural locations.
4. Explain the quantitative methods available for plant location.
5. Explain the modern trend in plant location.
6. Define plant location problem.

7. How the Government policy affects the selection of location?
8. Explain the factors that influence the location for the following products. Justify your answer:
   (*a*) Textile (cotton) industries,
   (*b*) Steel,
   (*c*) Cement,
   (*d*) Aluminium,
   (*e*) Food processing Industries,
   (*f*) Sugar Industries,
   (*g*) Paint Industries.

8

PLANT LAYOUT

• Definition • Plant layout problem • Objectives of plant layout • Principles of plant layout • Advantages of plant layout • Factors influencing plant layout • Types of manufacturing systems—Make to stock—Make to order—Job type, Batch and Mass production • Types of layouts • Material flow pattern • Symptoms of bad layout • Plant layout procedure • When to use, process, product and fixed position layout • Tools and techniques of plant layout • Computer packages for layout analysis • Systematic layout planning.

8.1. DEFINITION

Plant layout refers to the physical arrangement of production facilities. It is the configuration of departments, work centres and equipment in the conversion process.

According to Moore **"Plant layout is a plan of an optimum arrangement of facilities including personnel, operating equipment, storage space, material handling equipment and all other supporting services along with the design of best structure to contain all these facilities."**

The overall objective of plant layout is to design a physical arrangement that meets the required output quality and quantity most economically.

8.2. PLANT LAYOUT PROBLEM

Each one in the organisation is connected with the plant layout some way or the other. So the need for the plant layout change arise because of the following reasons:

1. Changes in the product design or introduction of the new product.
2. Changes in the volume of demand for the company's product.
3. Increasing frequency of accidents because of existing layout.
4. Plant and machinery becomes outdated and is to be replaced by new one.
5. Poor working environment affecting worker efficiency and productivity.
6. Change in the location or markets.
7. Minimising the cost through effective facilities location.

8.3. OBJECTIVES OF PLANT LAYOUT

The primary goal of the plant layout is to maximise the profit by arrangement of all the plant facilities to the best advantage of total manufacturing of the product.

Thus the objective of plant planning is the best relationship between output, space and manufacturing cost.

The objectives of plant layout are:

1. Stremline the flow of materials through the plant.
2. Facilitate the manufacturing process.

3. Maintain high turnover of in process inventory .
4. Minimise materials handling.
5. Effective utilisation of men. Equipment and space.
6. Make effective utilisation of cubic space.
7. Flexibility of manufacturing operations and arrangements.
8. Provide for employee convenience, safety and comfort.

8.4. PRINCIPLES OF PLANT LAYOUT

1. **Principle of integration:** A good layout is one that integrates men, materials, machines and supporting services and others in order to get the optimum utilisation of resources and maximum effectiveness.
2. **Principle of minimum distance:** This principle is concerned with the minimum travel (or movement) of man and materials. The facilities should be arranged such that, the total distance travelled by the men and materials should be minimum and as far as possible straight line movement should be preferred.
3. **Principle of cubic space utilisation:** The good layout is one that utilise both horizontal and vertical space. It is not only enough if only the floor space is utilised optimally but the third dimension, *i.e.*, the height is also to be utilised effectively.
4. **Principle of flow:** A good layout is one that makes the materials to move in forward direction towards the completion stage, *i.e.*, there should not be any backtracking.
5. **Principle of maximum flexibility:** The good layout is one that can be altered without much cost and time, *i.e.*, future requirements should be taken into account while designing the present layout.
6. **Principle of safety and security and satisfaction:** A good layout is one that gives due consideration to workers safety and satisfaction and safeguards the plant and machinery against fire, theft, etc.
7. **Principle of minimum handling:** A good layout is one that reduces the material handling to the minimum.

8.5. ADVANTAGES OF PLANT LAYOUT

1. **Advantages to the worker**: A good layout will reduce the effort of the workers and minimises the manual material handling. It reduces the number of accidents and provide better working conditions.
2. **Advantages to the management**: Effective plant layout reduce the labour costs and enhances the productivity thus ultimately reducing the cost per unit. This helps the management to gain competitiveness in manufacturing.
3. **Advantages to manufacturing**: Minimises the movement between work centres and also results in reduced manufacturing cycle.
4. **Advantages to production control**: A good layout facilitates production through uniform and uninterrupted flow of materials and helps to carry out production activities within the predetermined time period and with effectiveness.

8.6. FACTORS INFLUENCING PLANT LAYOUT

1. Type of production—Engineering industry, process industry.
2. Production system—job shop, batch production, mass production.
3. Scale of production.
4. Availability of the total area.
5. Arrangement of material handling system.
6. Type of building—single storey or multi-storey.

7. Future expansion plan.
8. Type of production facilities—dedicated or general purpose.

8.7. TYPES OF MANUFACTURING SYSTEMS

The production system (facility, equipment and operating methods) that a company uses depends upon the type of the product that is offered to the customer and the strategy that it employees to serve its customers.

1. **Make to stock production:** In this type, the products are manufactured and placed in stock before the customers order is received. The product is despatched to the customer "Off the shelf" from finished goods inventory after receipt of customer order.

Examples—Manufactures of standard items like bearings, nut and bolts, etc., produce and keep the finished goods inventory.

2. **Make to order production:** Some companies make the product to order and manufactures the product after receipt of customer order. Here the lead time to deliver the item to the customer will be more as the production activity starts only after the receipt of firm order.

Types of Production

According to volume and standardisation of the production of the products the manufacturing systems are classified as: (*a*) Job type production, (*b*) Batch production, and (*c*) Continuous production.

A. Job Type Production

It is characterised by manufacturing of one or few quantities of products designed and produced as per the specifications of the customers within the prefixed time and cost, *i.e.,* this type of production is distinguished by high variety and low volume.

Characteristics of Job Type Production

1. High variety and low volume.
2. General purpose machines and equipment to perform wider range of operations.
3. The flow of materials and components between different work station is highly discontinuous due to imbalance in work content.
4. Manufacturing cycle time is more.
5. Highly skilled workforce is required.
6. Highly competent and qualified supervisors are required.
7. Very large work-in-process inventory.
8. Flexible material handling system with a capability to move objects of various sizes and shapes along widely varying paths.
9. Difficulty in planning, scheduling and coordinating the productions of numerous components of wide variety.

B. Batch Production

Batch production is characterised by manufacture of limited number of products produced at regular intervals and stocked at warehouses awaiting sales

Examples– Pharmaceutical industry, chemical industry, assembly shops such as machine tools, subcontractors who take component for processing from large manufacturer.

Characteristics of Batch Production

1. Short production runs.
2. The plant and machineries set up is used for limited number of parts and then it is used to make different product.

3. More number of set-ups.
4. The workers are expected to posess skill in one particular manufacturing operation.
5. The amount of supervision required is less compared to job type.
6. Plant and machineries are flexible.
7. Manufacturing cycle time is comparatively lower than job production.
8. Large work-in-process inventory.
9. Flexible material handling system.

C. Repetitive (Mass) Production

D. This is characterised by high volume and low variety. This manufactures several standard products produced and stocked in the warehouses as finished goods awaiting to be despatched. Examples of mass production are plastic goods, manufacture and assembly shops of automobiles, etc.

Characteristics

1. Flow of material is continuous.
2. Special purpose machines are used.
3. Material handing system is mechanised most of the time by conveyers, etc.
4. Relatively lower skilled persons can manage work.
5. Shorter cycle time.
6. Work-in-process is comparatively low because of line balancing.
7. Higher inventory of raw materials.
8. Less flexibility of equipment and machines.

The types of Manufacturing Systems are shown in Fig. 8.1.

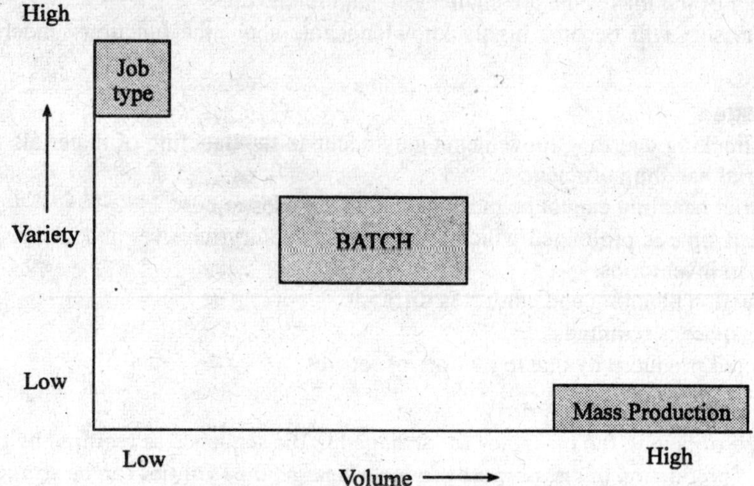

Fig 8.1: Types of manufacturing systems.

8.8. TYPES OF LAYOUT

1. Functional Layout (Process Layout)

This layout is recommended for batch production. All machines performing similar type of operations are grouped at one location in the process layout *e.g.*, all lathes, milling machines, etc., are grouped in the shop will be clustered in like groups.

Thus in process layout the arrangement of facilities are grouped together according to their functions.

A typical process layout is shown in Fig. 8.2. The flow paths of material through the facilities from one functional area to another vary from product to product.

Usually the paths are long and there will be possibility of backtracking.

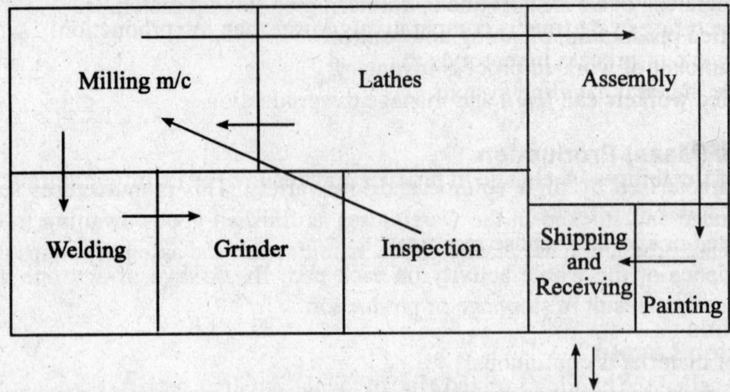

Fig 8.2: Process layout.

Advantages

1. Flexibility of equipment and personnel.
2. Lower investment on account of comparatively less number of machines and lower cost of general purpose machines.
3. Higher utilisation of production facilities.
4. Greater flexibility with regards to work distribution to machineries and workers.
5. Variety of job makes the job challenging and interesting.
6. Supervisors will become highly knowledgeable about the functions under their department.

Disadvantages

1. Backtracking and long movements may occur in the handling of materials thus reducing material handling efficiency.
2. Material handling cannot be mechanised which adds to cost.
3. Process time is prolonged which reduce the inventory turnover and increases the investment in inventories.
4. Production planning and control is difficult.
5. More space is required.
6. Lowered productivity due to number of set-ups.

2. Product Layout (Line Layout)

In this type of layout, the machines are arranged in the sequence as required by the product. If the volume of production of one or more products is large, the facilities can be arranged to achieve efficient flow of materials and lower cost per unit. Special purpose machines are used which perform the required function quickly and reliably. The equipment is closely placed along the sequence in which the item is processed.

The Product Type Layout is shown in Fig. 8.3.

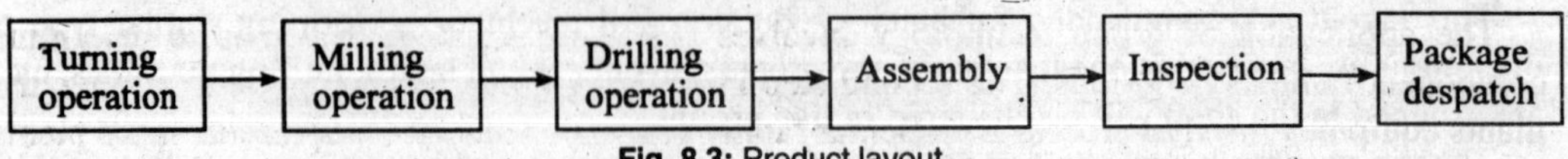

Fig. 8.3: Product layout.

Advantages

1. Reduced material handling cost due to mechanised handling systems and straight flow.
2. Perfect line balancing which eliminates bottlenecks and idle capacity.
3. Manufacturing cycle is short due to uninterrupted flow of materials.
4. Simplified production, planning and control.
5. Small amount of work-in-process inventory.
6. Unskilled workers can learn and manage the production.

Disadvantages

1. Lack of flexibility—A change in product may require the facility modification.
2. Large capital investment.
3. Dedicated or special purpose machines.
4. Dependence of the whole activity on each part. Breakdown of any one machine in the sequence may result in stoppage of production.

3. Combination Layout

This is also called the hybrid or mixed type of layout usually a process layout is combined with the product layout. For example, refrigerator manufacturing uses a combination layout. The process or functional layout is used to produce various operations like stamping, welding, heat treatment are carried out in different work centres as per the requirement. The final assembly of the product is done in a product type layout.

Thus for manufacturing various component parts process layout is used and for assembly product layout is used.

4. Fixed Position Layout

This is also called the project type of layout.

In this type of layout, the material, or major components remain in a fixed location and tools, machinery, men and other materials are brought to this location.

This type of layout is suitable when one or few pieces of identical heavy products are to be manufactured and when the assembly consists of large number of heavy parts, the cost of transportation of these parts is very high. The major advantages of this type of layout are:

1. Helps in job enlargement and upgrades the skills of the operators.
2. The workers identify themselves with a product in which they take interest and pride in doing the job.
3. Greater flexibility with this type of layout.
4. Layout capital investment is lower.

5. Group Layout

There is a trend now to bring an element of flexibility into manufacturing system as regards to variation in batch sizes and sequence of operations. A grouping of equipment for performing a sequence of operations on family of similar components or products has become all the important.

Group technology (GT) is the analysis and comparisons of items to group them into families with similar characteristics. GT can be used to develop a hybrid between pure process layout and pure flow line (product) layout. This technique is very useful for companies that produce variety of parts in small batches to enable them to take advantage and economics of flow line layout.

The application of group technology involves two basic steps, first step is to determine component families or groups. The second step in applying group technology is to arrange the plants equipment used to process a particular family of components. This represents small plants within the plants. The group technology reduces production planning time for jobs. It reduces the set up time.

Comparison between Product and Process Layout is shown in Table 8.1.

Table 8.1: Comparison between Process and Product Layout

	Characteristic	*Product Layout*	*Process Layout*
1.	Nature	A sequence of facilities as per processing requirement of products	All similar facilities are grouped together
2.	Application	High volume, few products	High volume and high variety
3.	Product	Standardised, stable rate of output	Diversified or variety of products using common operations
4.	Work flow	Straight line, same sequence of operations for all product	Variable flow for each product type
5.	Material Handling	Flow Predictable and systematic can be automated easily	Cannot be automated as flow depends upon the product type
6.	Inventory	High turnover of raw materials and WIP inventory	Low turnover of both raw material and WIP
7.	Breakdowns	Breakdown in any one machine stops production line	Can tolerate breakdowns
8.	Production centre	Simple	Complex
9.	Flexibility	Low	High
10.	Space utilisation	Efficient	Low
11.	Product cost	High fixed cost, Low variable cost	Low fixed cost, High variable cost

8.9. MATERIAL FLOW PATTERNS

The pattern of material flow is an important consideration in the plant layout decision because good layout aims at minimising flow of materials. The flow pattern of materials helps in eliminating bottle-necks, rushing and backtracking and ensures good supervision and control.

The various flow patterns of materials is shown in Fig. 8.4. The material flow systems can be classified on the basis of the availability of floor space as.

1. **Horizontal flow system:** Usually devised for a single story building when the flat floor area is available.
2. **Vertical flow system:** This system is used in case of multi-storey buildings and limited area is available.

Characteristics of different Flow Patterns

(*a*) **Straight line**

1. Shortest route and must have roads on both sides.
2. Plant area has long length and narrow width.
3. Unsuitable for longer production lines.

(*b*) **U-type**

1. Less difficulty in returning empty containers.
2. Suitable for longer production lines.
4. Requires square like floor area.
5. One side road link will be required.

(*c*) **Serpentine (Inverted S-Shape)**

1. Requires roads on both sides.
2. Suitable for longer production lines.
3. Difficulty in returning empty containers.
4. Requires square like floor area.

(*d*) **Comb or dendrite arrangement**

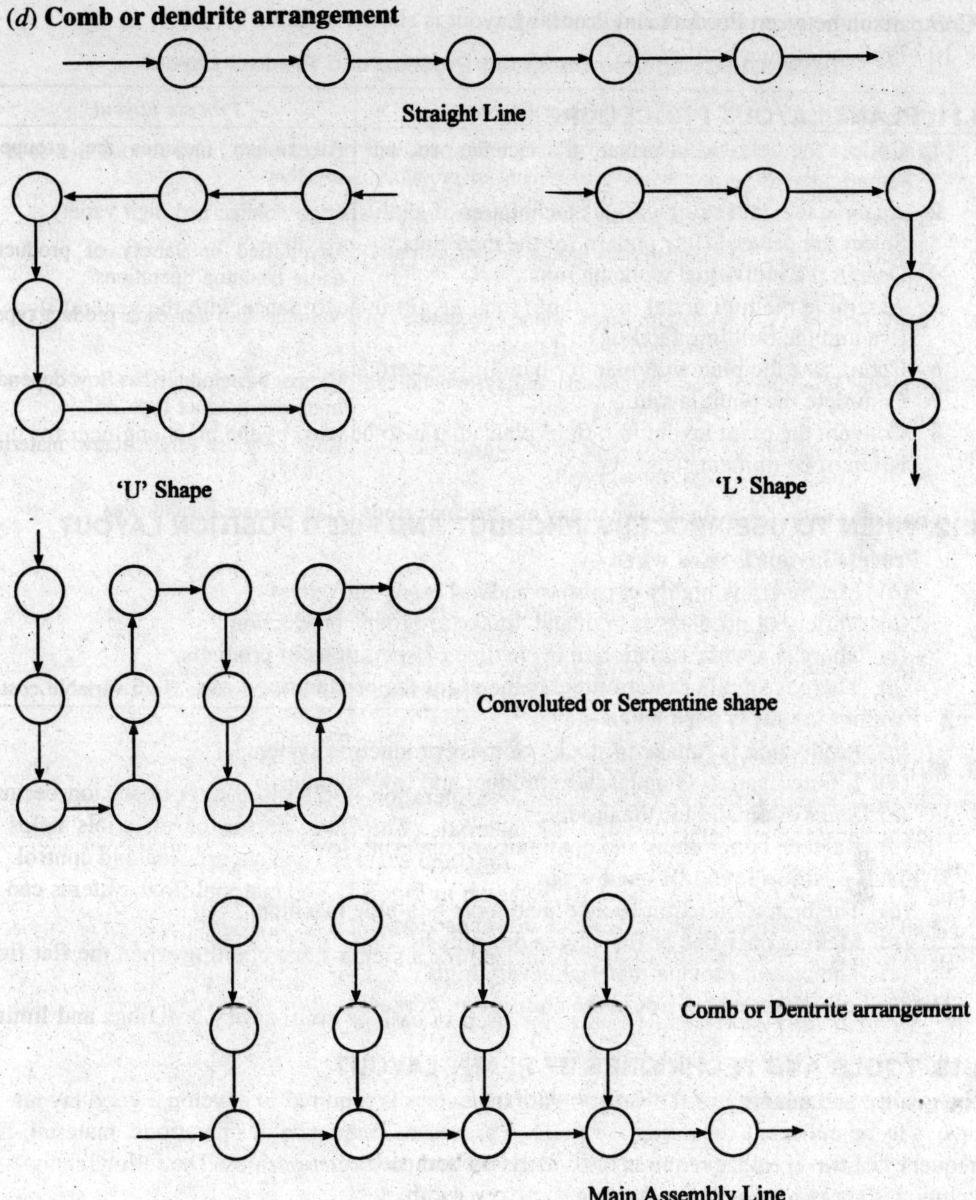

Fig. 8.4: Flow pattern.

8.10. SYMPTOMS OF BAD LAYOUT

1. Long material flow lines and backtracking (rehandling).
2. Poor utilisation of space.
3. Congestion for movement of materials and men.
4. Large amount of work-in-process.
5. Long production cycles.
6. Excessive handling of materials.
7. More frequent accidents.
8. Difficult to supervise and control.

9. Spoilage of products during handling.
10. Production line bottlenecks.

8.11. PLANT LAYOUT PROCEDURE

1. Collect the detailed information about the product, process, etc., and record the data systematically.
2. Analyse the data using various techniques of analysis.
3. Select the general flow pattern for the materials.
4. Design the individual work stations.
5. Assemble the individual layout into total layout in accordance with the general flow pattern and the building facilities.
6. Coordinate the plan with plan for handling materials.
7. Complete the plant layout.
8. Convent the plant layout into floor plans that is to be used by the plant engineer for installation of equipment.

8.12. WHEN TO USE PROCESS, PRODUCT AND FIXED POSITION LAYOUT

1. **Process layout is used when –**
   (*a*) Machinery is highly expensive and not easily moved.
   (*b*) Variety of products is high and "make to order" production.
   (*c*) There is a wide variation in cycle time of operations of products.
   (*d*) There is small or intermittent demand for the products.
2. **Product layout is used when –**
   (*a*) Production is "make to stock" or mass production system.
   (*b*) Limited variety, standardised product and low volume.
   (*c*) Steady demand for the product.
   (*d*) Balanced operations and continuity of material flow.
3. **Fixed position layout is used when –**
   (*a*) The operation requires only hand tools or single machine.
   (*b*) Making only one or few pieces of product.
   (*c*) The cost of moving material is very high.

The skill of workmanship lies in the abilities of workers.

8.13. TOOLS AND TECHNIQUES OF PLANT LAYOUT

The quality and quantity of the data on various factors is required to develop a good layout. The data is to be collected regarding the various processes, sequence of operations, material, flow, frequency of travel, space requirements, activities and their relationships. The following tools and techniques are used to analyse the data.

1. Process charts—(operation process charts, flow process charts)
2. Travel Chart
3. Diagrams—(flow diagrams and string diagrams)
4. REL—(Relationship chart)
5. Templates
6. Scaled models

The process charts and diagrams are discussed in chapter on method study.

TRAVEL CHART

The flow of material between functional areas of the plant is recorded on a From—To chart. It records the distance and frequency of movements between various pairs of departments. This chart is used to determine the degree of closeness between the departments. It forms the basis for layout design that seeks to minimise the total material handling costs.

The advantages of travel charts are:

1. It is a useful tool for movement analysis.
2. It is helpful to locate the activities and backtracking.
3. Useful for comparing alternative flows.
4. Can be easily computerised as quantitative data is involved.

Procedure to draw the Travel Chart

Departments or work centres are listed both rowwise and columnwise, in the same sequence. Each intersecting square (cell) is used to record data from one department to another. The entries in the chart are scattered on both sides of the diagonal.

A typical travel chart is shown in Fig. 8.5.

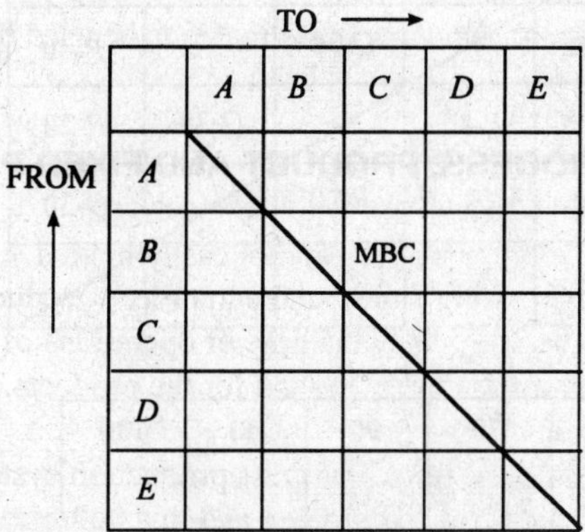

Fig 8.5: Travel chart.

MBC represents number of moves from *B* to *C*.

The entries below the diagonal represent backtracking.

Problem 1: A company manufactures three products *P*, *Q*, and *R* using the same manufacturing facilities arranged in six departments *A*, *B*, *C*, *D*, *E* and *F*. The material handling is done by a forklift. The containers can carry 300, 400 and 600 pieces of the products *P*, *Q* and *R* respectively. The annual demand for each product is 12000 units. Sequence of operations of product movement are given below.

Product	*Movement*
P	$A \longrightarrow E \longrightarrow B \longrightarrow D \longrightarrow C \longrightarrow F$
Q	$A \longrightarrow B \longrightarrow C \longrightarrow D \longrightarrow E \longrightarrow F$
R	$C \longrightarrow B \longrightarrow A \longrightarrow E \longrightarrow D \longrightarrow F$

Construct the travel chart.

Solution: Annual demand for product *P* = 12000 per annum

Capacity of the containers to carry *P* = 300

$$\text{Number of trips (frequency) of forklift} = \frac{12000}{300} = 40 \text{ trips}$$

Material movement for product *P* is *A* to *E*, *E* to *B*, *B* to *D*, *D* to *C*, *C* to *F*

Annual demand for product *Q* = 12000 per annum

Similarly, capacity of the containers to carry *Q* = 400

$$\text{Number of trips (frequency) of forklift} = \frac{12000}{400} = 30 \text{ trips}$$

Material movement for product Q is A to B, B to C, C to D, D to E, E to F

Annual demand for product $R = 12000$ per annum

Capacity of the containers to carry $R = 600$

Number of trips (frequency) of forklift $= \frac{12000}{600} = 20$ trips

Material movement for product R is C to B, B to A, A to E, E to D, D to F

These are represented on the From — To chart as shown in Fig. 8.6.

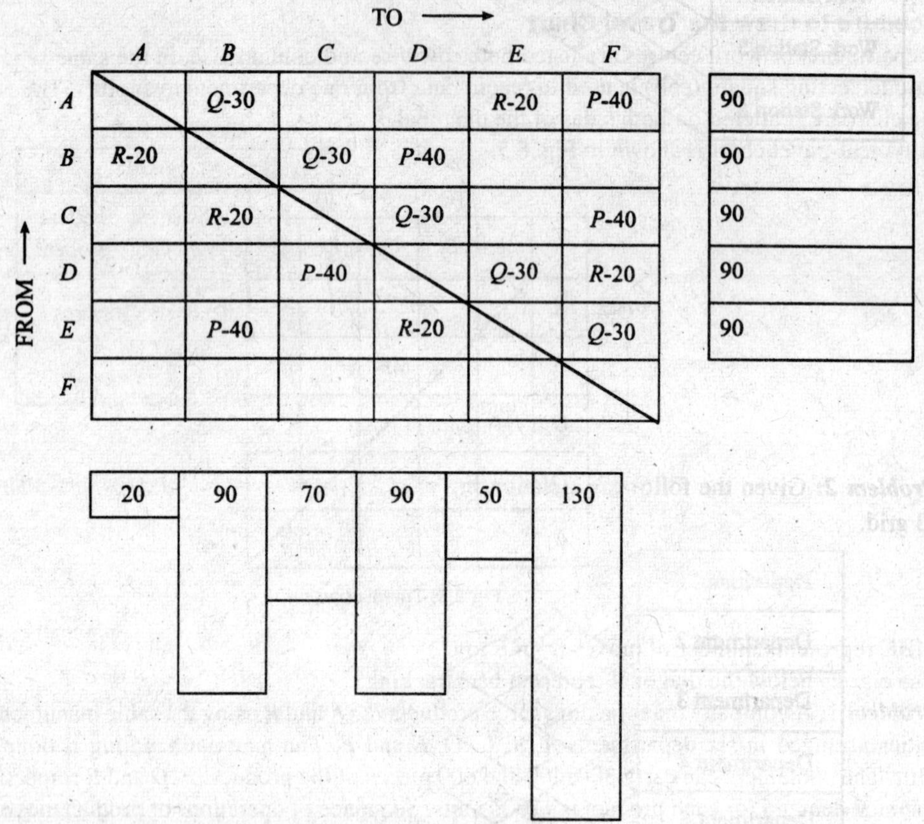

Fig. 8.6: Travel chart.

REL (RELATIONSHIP CHART)

The relative importance of having one department near another is displayed in Relationship chart. Robert Muther developed the REL chart. Each diamond-shaped cell in the chart shows the relationship if any between two functional plant areas. Two entries come into the cells. The top entry is a letter that indicates the degree of closeness of the relationship and the number below represents the important reason for relationship.

The relationship diagrams recognise the need for exploring "Relationship" rather than calculating exact flows and costs. The requirements of closeness are expressed on a scale.

These pair-wise interdepartmental "closeness requirement ratings" are then used to develop a suitable layout which satisfies as many pairwise relationship as possible and to the extent possible. The relationship chart is shown in the Fig. 8.7.

Reasons behind closeness value

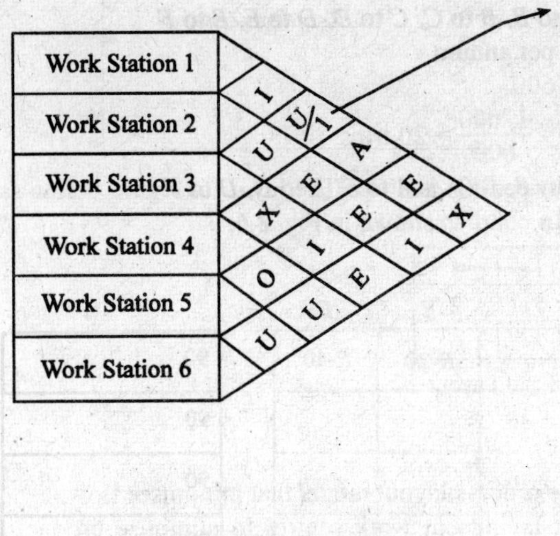

Code	Reason
1	....
2	....
3	....
4	....
5	....

Closeness Rating

Value	Closeness
A	Absolutely essential
E	Especially important
I	Important
O	Ordinary closeness
U	Unimportant
X	Not desirable

Fig. 8.7: Relationship chart.

***Problem 2*:** Given the following Relationship chart, arrange the work centres into a suitable 2×3 grid.

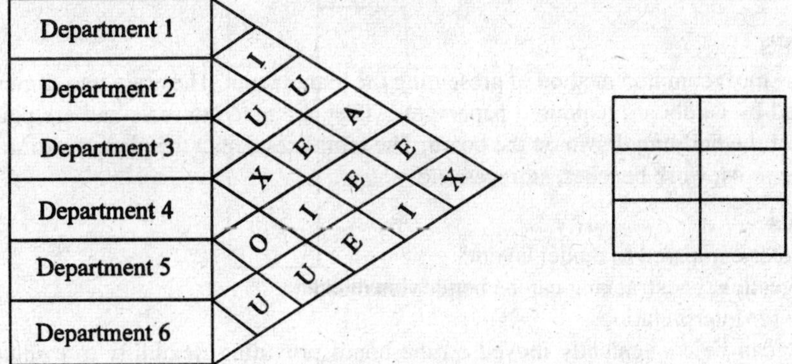

This problem can be solved using intuitive approach. In the intuitive approach, prime importance is given to *A* relationship (closeness is absolutely essential).

Department (1) and department (4) requires such relation.

∴ Departments (1) and (4) should be as close as possible. Then, another extreme relationship *i.e., X* (closeness undesirable). This is found between Departments (1) and (6).

So, Departments (1) and (6) should be as far as possible from each other.

This is shown in figure below

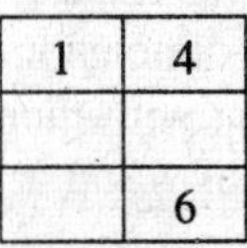

Layout step I

This requirement is to be checked between (4) and (6). Whose relationship is 'U' (Unimportant) so the figure is okay.

The remaining set of Departments are located to satisfy the requirements of nabours. I

Department	*Closeness Type*	*Requirement with*
1	*E*	5
4	*E*	2
6	*E*	3

and there is an undesirable relationship between (3) and (4). The layout is figure below satisry these requirements. Which is the final layout.

1	4
5	2
3	6

Final Layout

LOAD DISTANCE MODEL

This is most commonly used quantitative process layout model that minimises flow.

Load distance model—an algorithm for laying out work centres to minimise product flow, based on the number of loads moved and the distance between each pair of work centres.

The model can be mathematically expressed as—

$$C = \left(\sum_{i=1}^{n} \sum_{j=1}^{n} L_{ij}\, d_{ij}\right) \times K$$

where C is the cost to be minimised

'n' is the number of work centres

L = no. of loads moved between work centres i and j

D_{ij} = distance between work centres i and j

K = cost to move one load by one unit distance is constant

TEMPLATES

This is the most common method of presenting the plant layout. These are two-dimensional cut outs prepared by cardboard, coloured papers, etc. They are made to scale and are placed on the scaled plan of the building drawn on the board. The templates depict the plan of various activities like machine tools, work benches, storages, etc.

Advantages

1. Cheaper compared to model layout.
2. Congestion, backtracking can be better visualised.
3. Easy for interpretation.
4. They can be conveniently moved on the board providing flexibility to evaluate various feasible arrangements.

Limitations

1. Interpretation is difficult for non-technical persons.
2. It is difficult to visualise the impact of overhead facilities on the plant and process.

SCALE MODELS

Three-dimensional models are the scaled models of the facilities. The models of the plant and equipment with their appropriate scale are used in preparing the model layout. They are usually made up of wood or plastic. These are costly.

Advantages

1. Easy to visualise and explain.

2. It facilitates the study of overhead structures and devices like lighting, ventilation, safety features, etc.

8.14. COMPUTER PAKAGES FOR LAYOUT ANALYSIS

The major advance in layout planning is computerised analysis. Computer programmes have been developed to assist the layout analyst in identifying layouts that appear to meet some specified criterion. The user can specify a matrix of the anticipated number of trips between departments.

CRAFT

Gordon Armour and Elwood Buffa presented a computerised layout programme called CRAFT (Computerised Relative allocation of facilities). Input data includes material flow per unit time, cost per unit per distance moved and space requirements in the form of an initial layout.

It is an improvement algorithm
Uses heuristic method
Minimises costs
Limitations: Maximum number of departments = 40
Maximum layout size = 30 × 30

CRAFT considers exchange between locations repeatedly until no further significant costs reductions are possible. The programme output is in the form of facilities in a basic rectangular form that is close to the lowest cost layout.

CRAFT programme works on the assumption that movement between departments occur along straight line between the centroids of the departments. It assumes that the costs vary linearly with distance. This model is applicable to both manufacturing and non-manufacturing organisations.

The flowchart for CRAFT is shown in Fig. 8.8.

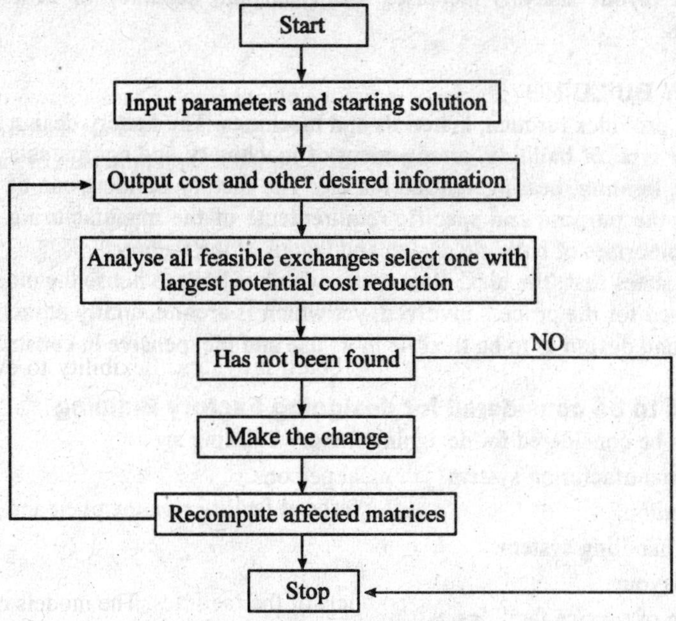

Fig. 8.8: Flowchart for CRAFT.

CORELAP

Computerised Relationship Layout Planning (CORELAP) uses the *A-E*, *I*, *O*, *U* closeness ratings, space requirements and maximum building length to width ratio to develop a layout.

- **The inputs are—**
  No. of departments and area requirements
  Relationship chart
  Weightages for REL chart
  Location of pre-assigned department on layout
  Maximum building length to width ratio

Mechanism

Compute Total Closeness Rating (TCR)
Highest TCR is selected and located
Next locate "*A*" of the relationship
Next locate "*E*" of the relationship, etc.

To break the tie, choose higher TCR.

ALDEP

Automated layout design programme requires input data for building specifications and preference matrix of location relationships. It has the capability of laying out up to 63 departments. It uses a matrix of letter codes to specify nearness priorities. These rankings are converted to a qualitative scale to facilitate the evaluation of trade off. The programme can deal with multi-storey facilities.

There are number of other techniques like—
PLANET (Plant Layout Analysis and Evaluation Technique)
COFAD (Computerised facilities Design)
PREP (Computerised Relationship Layout Planning)
FLAG (Facility Layout Algorithm using Graphics)

Computerised layout analysis increases the speed and accuracy of evaluation of various alternatives.

8.15. FACTORY BUILDING

Factory building provides for men, materials and machines. The factory design refers to the plan for the particular type of building, arrangement of machinery and equipments and provision of service facilities, lighting, heating ventillation etc. The factory building can of different designs depending upon the purpose and specific requirements of the manufacturing process. Thus, a factory design comprises of building design and layout of the factory.

James Lundy states that "the ideal plant is one which is built to house the most efficient layout that can be devised for the process involved, yet which is architecturally attractive and of such a standard, shape and design as to be flexible in its use and inexpensive in construction.

8.15.1. Factors to be considered for designing Factory Building

The factors to be considered for designing factory building are :

(*i*) Type of manufacturing system.
(*ii*) Plant location.
(*iii*) Material handling system.
(*iv*) Factory layout.
(*v*) Provision of service facilities.
(*vi*) Future expansion plans.
(*vii*) Safety and security.
(*viii*) Convenience and facilities to employees.

(*xi*) Flexibility.
(*x*) Aesthetic aspect.

8.15.2. Types of Factory Buildings

There are basically two types of factory buildings :

(*i*) Single storey buildings
(*ii*) Multi-storey buildings

Single storey buildings are suitable

(*a*) when land is available in abundance and at the lower cost.
(*b*) when the product to be manufactured is bulky and heavy.
(*c*) where the frequent changes in layout are desired.
(*d*) where the material handling is complex and difficult.

Multi-storey buildings are preferred when

(*a*) the cost of land is very high.
(*b*) product to be manufactured is light in weight.
(*c*) the availability of land is limited.

The advantages of single storey buildings include - greater flexibility, low cost of construction, use of natural light and ventillation, effective utilisation of floor space and facilities expansion.

The advantages of multi-storey buildings include - use of gravity flow for material handling, more cubic space utilisation and more storage space.

References For Further Reading

1. Apple, J.M., *Plant Layout and Material Handling,* 3rd Edn., John Wiley and Sons, New York (1977).
2. Moore, J.M., *Plant Layout and Design,* Macmillan, New York (1962).
3. Muthur, R., *Practical Plant Layout,* McGraw Hill (1955).
4. Chary, S.N., *Production and Operations Mgt.,* TMH, New Delhi (1996).
5. Maynard., *Industrial Engineering Handbook,* McGraw Hill, New York.

REVIEW QUESTIONS

1. Define plant layout. What are the objectives of good plant layout?
2. What are the various types of layout? Explain the application of each.
3. Compare product layout and process layout.
4. What is the significance of group layout?
5. What are the various flow patterns? Explain with a diagram for each.
6. What are the principles of plant layout?
7. What are the factors that influence the plant layout?
8. How the plant layouts are related to type of production?
9. Explain the various tools and techniques of plant layout.
10. Explain the procedure for plant layout.
11. What are the advantages of computer packages in plant layout?
12. Explain the steps involved in systematic layout planning (SLP) with the help of a block diagram.

13. Write short note on:
    (*i*) Type of production systems,
    (*ii*) Process and product type layout,
    (*iii*) Travel chart,
    (*iv*) Relationship chart,
    (*v*) Dimensional Analysis,
    (*vi*) Single storey *vs*. multi-storey buildings,
    (*vii*) Symptoms of bad layout,
    (*viii*) Flow pattern.

9

MATERIAL HANDLING

• Introduction. • Objectives of M.H. • Elements of material handling • Material handling activities and functions • Relationship between plant layout and M.H. • Principles of M.H. • Symptoms of bad material handling • Selection of M.H. equipment's • Types of material handling equipment's • Unit load concept in M.H. • Systematic handling analysis • Economics of material handling.

9.1. INTRODUCTION

In order to convert the raw materials into finished products, it is essential that one of the three basic elements of production, *i.e.,* material, men or machines should move. In majority of the industrial processes it is the material that moves from raw material stage to the stage of when it becomes the finished goods. Because the material is more widely moved rather than the men or machines. Hence the name "Material Handling"

Haynes defines **"Material handling embraces the basic operations in connection with the movement of bulk, packaged and individual products in a semi-solid or solid state by means of gravity manually or power-actuated equipment and within the limits of individual producing, fabricating, processing or service establishment."** Material handling does not add any value to the product but adds to the cost of the product and hence it will cost the customer more. So the handling should be kept at minimum.

Material handling amounts to 15 to 25% of total cost of product according to American material handling society. Material handling is the art and science involving the movement, handling and storage of materials during different stages of manufacturing.

Out of the total time spent for manufacturing a product, 20% of the time is utilised for actual processing on them while the remaining 80% of the time is spent in moving from one place to another, waiting for the processing or storage (temporary).

9.2. OBJECTIVES OF MATERIAL HANDLING

1. Minimise cost of material handling.
2. Minimise delays and interruptions by making available the materials at the point of use at right quantity and at right time.
3. Increase the productive capacity of the production facilities by effective utilisation of capacity and enhancing productivity.
4. Safety in material handling through improvement in working condition.
5. Maximum utilisation of material handling equipment.
6. Prevention of damages to materials.
7. Lower investment in inprocess inventory.

9.3. ELEMENTS OF MATERIAL HANDLING

1. **Motion:** Move in most economic, safe and efficient manner.
2. **Time:** Provide materials on time.
3. **Quantity:** Ensure supply of correct quantity continuously at each manufacturing organisation.
4. **Space:** Ensure optimum use of cubic space.

9.4. MATERIAL HANDLING ACTIVITIES AND FUNCTIONS

In manufacturing organisation, the handling activity encompass —

1. Transportation and handling at suppliers end.
2. Material handling at manufacturing plant.
3. Transportation and handling from warehouse to customer (physical distribution).

But we are more concerned here with handling within the plant.

The material handling activity in the plant starts with the unloading of the material after receipt from suppliers, and extends throughout the processing from raw material stage till it is manufactured and stored in the warehouse to be dispatched to the customer.

The various activities and functions are shown in Fig. 9.1.

(In Manufacturing Organisation)

• Packaging at customers end • Loading and transportation at suppliers end • External plant handling	• Unloading • Receiving and temporary storage location • Storing • Issuing • Workplace handling • In process handling and storage • Inter-department • Intra plant • Packaging • Warehousing of finished goods • Loading and shipping	• Transportation to consumers • Inter-plant handling

Fig. 9.1: Material handling activities and functions.

Relationship between Plant Layout and Material Handling

There is a close relationship between plant layout and material handling. A good layout ensures minimum material handling and eliminate rehandling.

- Material movement does not add any value to the product so, the material handling should be kept at minimum though not avoid it. This is possible only through the systematic plant layout. Thus a good layout minimises handling.
- The productive time of workers will go without production if they are required to travel long distance to get the material tools, etc. Thus a good layout ensures minimum travel for workman thus enhancing the production time and eliminating the hunting time and travelling time.

- Space is an important criteria. Plant layout integrates all the movements of men, material through a well designed layout with material handling system.
- Good plant layout helps in building efficient material handling system. It helps to keep material handling shorter, faster and economical. A good layout reduces the material backtracking, unnecessary workmen movement ensuring an effectiveness in manufacturing.

Thus a good layout always ensure minimum material handling.

9.5. PRINCIPLES OF MATERIAL HANDLING

1. **Planning principle:** All handling activities should be planned.
2. **Systems principle:** Plan a system integrating as many handling activities as possible and coordinating the full scope of operations (receiving, storage, production, inspection, packing, warehousing, supply and transportation).
3. **Space utilisation principle:** Make optimum use of cubic space.
4. **Unit load principle:** Increase quantity, size, weight of load handled.
5. **Gravity principle:** Utilise gravity to move a material wherever practicable.
6. **Material flow principle:** Plan an operation sequence and equipment arrangement to optimise material flow.
7. **Simplification principle:** Reduce combine or eliminate unnecessary movement and/or equipment.
8. **Safety principle:** Provide for safe handling methods and equipment.
9. **Mechanisation principle:** Use mechanical or automated material handling equipment.
10. **Standardisation principle:** Standardise method, types, size of material handling equipment.
11. **Flexibility principle:** Use methods and equipment that can perform a variety of task and applications.
12. **Equipment selection principle:** Consider all aspect of material, move and method to be utilised.
13. **Dead weight principle:** Reduce the ratio of dead weight to pay load in mobile equipment.
14. **Motion principle:** Equipment designed to transport material should be kept in motion.
15. **Idle time principle:** Reduce idle time/unproductive time of both MH equipment and manpower.
16. **Maintenance principle:** Plan for preventive maintenance or scheduled repair of all handling equipment.
17. **Obsolescence principle:** Replace obsolete handling methods/equipment when more efficient method/equipment will improve operation.
18. **Capacity principle:** Use handling equipment to help achieve its full capacity.
19. **Control principle:** Use material handling equipment to improve production control, inventory control and other handling.
20. **Performance principle:** Determine efficiency of handling performance in terms of cost per unit handled which is the primary criteria.

9.6. SYMPTOMS OF BAD MATERIAL HANDLING

1. Frequent interruption in production due to delay in handling and supplying materials to the point of use.
2. Skilled labour performing duties like storing, movement and handling of materials.
3. Damages to materials in handling.
4. Accumulation of work-in-process and materials in different locations.
5. Reworking and rejections due to handling defects.

6. Crowded floor space with scrap and materials.
7. Congestion at receipt, production and inspection areas.
8. Long waiting for material handling equipment to pick up and deliver materials.

9.7. SELECTION OF MATERIAL HANDLING EQUIPMENTS

Selection of MH equipment is an important decision as it affects both cost and efficiency of handling system.

The following factors are to be taken into account while selecting material handling equipment:

1. **Nature of operations**
   (*i*) Whether handling is temporary or permanent.
   (*ii*) Whether the flow is continuous or intermittent.
   (*iii*) Material flow pattern—vertical or horizontal.
   (*iv*) Type of layout—process layout, product layout or combination layout.
2. **Material to be handled**
   (*i*) Size and shape of the material.
   (*ii*) Quantity and weight of the material
   (*iii*) Material characteristics.
   (*iv*) Susceptibility to damage during handling.
3. **Distance over which the material is to be moved**
   (*i*) Fixed distance.
   (*ii*) Long distance.
   (*iii*) Work station.
4. **Installation and operating costs**
   (*i*) Initial investment.
   (*ii*) Operating and maintenance costs.
5. **Plant facilities**
   (*i*) Types of buildings.
   (*ii*) Floor load capacity.
6. **Safety considerations**
7. **Engineering factors**
   (*i*) Door and ceiling dimensions.
   (*ii*) Floor conditions and structural strength.
   (*iii*) Traffic safety.
8. **Equipment reliability**
   (*i*) Use of standard components.
   (*ii*) Service facilities.
   (*iii*) Supplier reputation.

9.8. TYPES OF MATERIAL HANDLING EQUIPMENTS

The material handling equipment are classified based on:

(*i*) Types of services required: (1) Lifting, (2) Moving, (3) Stacking, and (4) Positioning
(*ii*) Types of equipment
(*iii*) Relative mobility of equipment: (*a*) Travel between fixed points, and (*b*) Travel over wide areas
(*iv*) Movement of equipment.
   1. On the floor.
   2. Above the floor.
   3. Overhead.
   4. Underground.

Categories of equipment

(*i*) Conveyers.
(*ii*) Cranes and hoists.
(*iii*) Industrial trucks.

1. Conveyers

Conveyers primarily perform the movement of uniform loads between fixed points. They occupy space continuously except when they are of portable type. They reduce handling.

Types of conveyers are:

1. Belt conveyers.
2. Roller conveyers.
3. Screw conveyers.
4. Pipeline conveyers.
5. Monorail.
6. Trolley conveyers.

Conveyers are useful when –

(*a*) Loads are uniform.
(*b*) Materials move continuously.
(*c*) Routes do not vary.
(*d*) Movement rate is relatively fixed.
(*e*) Movement is from one point to another point.

2. Cranes and Hoists

Cranes are overhead devices capable of moving materials vertically and laterally in area of limited length and width and height. Cranes are employed for lifting and lowering heavy objects and moving them from one point to another.

Cranes find their application in heavy engineering industries and in intermittent type of production.

Types of cranes are:

1. Overhead travelling cranes.
2. Jib crane.
3. Gantry crane.

Hoists are used for loading and unloading of heavy objects and they are also used for raising and lowering heavy and long objects.

Type of hoists are:

1. Chain hoists.
2. Pneumatic hoists.
3. Electric hoists.

The hoists and cranes are most commonly used when –

(*a*) Movement is within fixed area.
(*b*) Moves are intermittent.
(*c*) Loads vary in size and weight.
(*d*) Loads handled are not uniform.

3. Industrial Trucks

Hand or powered vehicles are used for movement of mixed or uniform loads intermittently over varying paths which have suitable running surfaces and clearances and where the primary function is transporting. These are various types of material handling trucks.

Types of industrial trucks are:

1. Forklift truck.

2. Platform truck.
3. Tractor trailer.

Industrial trucks are generally used when:

1. Materials are moved intermittently.
2. Movement is through changing routes and distances.

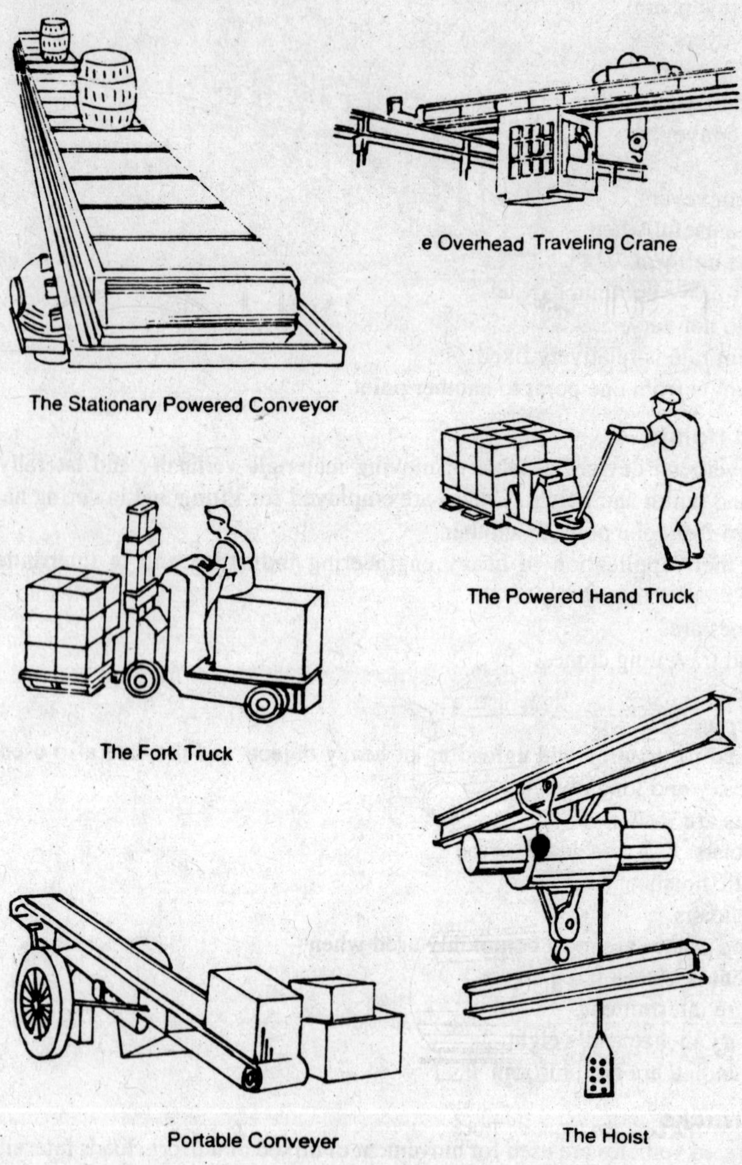

Fig. 0.2: Motorial handling oquipmonto.

3. Loads are uniform mixed is size and weight.
4. Materials can be put into unit loads.

Material handling equipment are shown in Figs. 9.2 and 9.3.

9.9. UNIT LOAD CONCEPT

Material handling efficiency is proportional to the size of the load handled, *i.e.*, number of units handled per unit time. It is economical and faster to handle small parts by grouping them into one

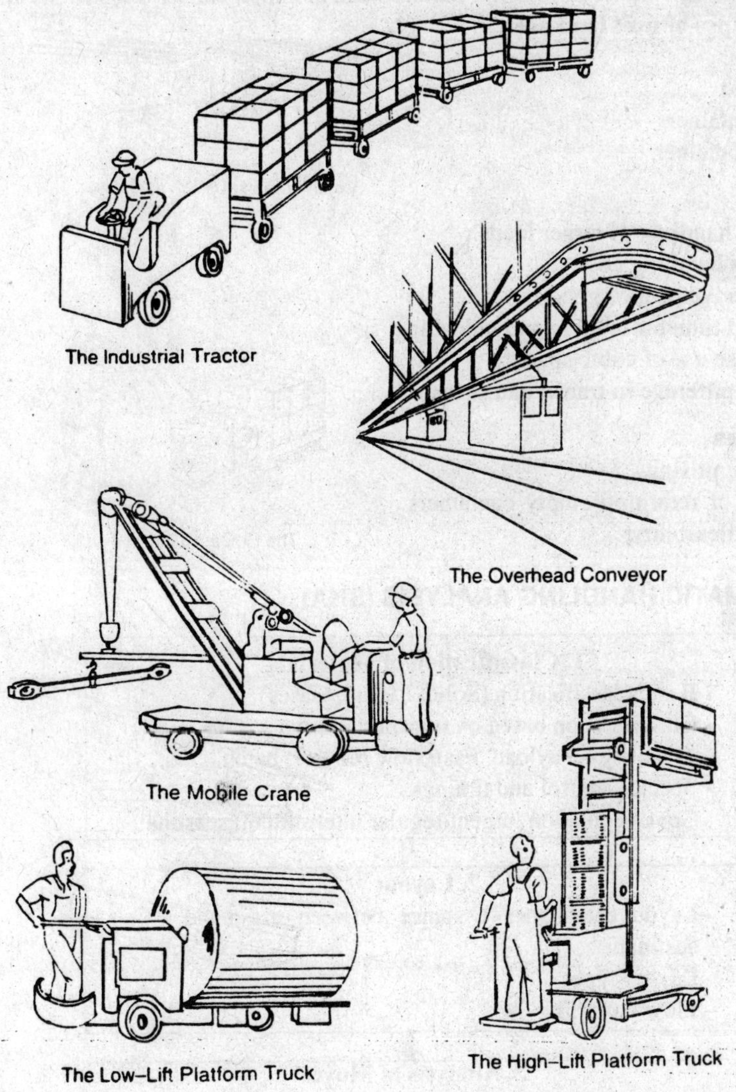

Fig. 9.3: Material handling equipments.

unit (called unit load) than moving them individually. James R. Bright defines unit load as—**"A number of items or bulk material, so arranged or restrained that the mass can be picked up and moved as a single object too large for manual handling and which upon being released will return its initial arrangement for subsequent movement."**

Unit load should–

- Perform a minimum number of handling and eliminate manual handling.
- Assemble materials into unit load for economy of handling and storage.
- Make the unit load as large as possible considering the limitations of building, handling equipment.

Production areas, volume of material required and common carrier dimensions and capacity.

Common types of unit load are:

- Part bins.
- Pallet box.
- Bulk container.
- Cargo container.

Advantages

1. Permits handling of larger loads.
2. Reduces handling cost.
3. Faster movement of goods.
4. Reduced time for loading and unloading.
5. Maximise use of cubic space.
6. Reduce pilferage in transit and storage.

Disadvantages

- Cost of unitising.
- Problem of returning empty containers.
- Lack of flexibility.

9.10. SYSTEMATIC HANDLING ANALYSIS (SHA)

1. Classification of Materials
- Basic classification (Solid, Liquid, Gases)
- Classification based on transportability.
  Quantity – Payload, Fast/Slow movers, batch.
- Special control and timings.
  Govt. regulation, urgent/regular intermittent, seasonal.

↓

2. Layout
- Layout establishes 'Distance' between origin and destination.
- Layout types.
- Flow patterns.

↓

3. Analysis of Moves
- Flow analysis – Intensity and Condition of Flow.
- Process charting – Route chart.

↓

↓

4. Visualisation of Move
- Flow diagram, Distance intensity. chart – Distance for each move.

↓

5. Understanding MH Method
- Movement system
- MH equipment
- Unit load, palletisation.
- Packaging method, Calculation of requirements.

↓

6. Preliminary handling plan
- Systematic engineering .
- Determination of method.
- Conventions.
- Visualisation.
- Development of more than one plan.

↓

7. Modifications and limitations
- Procedural problem.
- People problems.
- Other modifications.

↓

8. Evaluation of alternatives
- Investment, operating costs.
- Intangible factors.

↓

SELECTED M.H. SYSTEM.

9.11. ECONOMICS OF MATERIAL HANDLING

The American Society for Mechanical Engineers (ASME) has developed the formulae for estimating the economies with the application of certain equipment to a material handling problem.

Let

A = Percentage allowance on investment
B = Percentage allowance for insurance and taxes
C = Percentage allowance for maintenance
D = Percentage allowance for depreciation
E = Annual cost of power, supplies and others
S = Yearly saving in direct labour cost in Rs.
U = Yearly savings or earnings through increased production (Rs.)
T = Yearly Savings in fixed charges and operating charges (Rs.)
I = Initial cost of equipment
X = Percentage of year during which the equipment is used

1. Maximum justifiable investment —

$$Z = \frac{(S + T + U - E)\,X}{A + B + C + D}$$

2. Yearly cost of maintaining the equipment

$$Y = I\,(A + B + C + D)$$

3. Yearly profit from the operation of the equipment above the sample intent

$$V = [(S + T + U - E)X] - Y$$

4. Estimated rate of Profit

$$P = \frac{V}{I} + A$$

5. No. of year required for amortisation of investments out of earning

$$H = \frac{100}{P + D}$$

References for Further Reading

1. Apple, James M., *Material Handling Systems Design,* John Wiley and Sons, New York, (1972).
2. Muther, T. and K. Haganas, *Systematic Handling Analysis,* CBI Publishing Co., Boston, (1973).
3. Adam E., Jr. and Ebert, *Production and Operations Management,* 5th Edition, Prentice Hall of India, New Delhi, (1993).
4. Bolz, H.A. and G.E. Hagerrann, *Material Handling Hand Book,* Ronald Press, New York.
5. James, Riggs, *Production System Planning Analysis and Control,* John Wiley and Sons, New York, (1987).

REVIEW QUESTIONS

1. Define material handling. State objectives of material of handling.
2. "Material handling is considered necessary evil." Comment.
3. What are the principles of M.H.?
4. What are different types of material handling equipment?
5. What are the symptoms of bad layout?
6. Explain the elements of material handling.
7. What is the relationship between plant layout and M.H. system?
8. What is unit load concept in material handling?
9. What is economics of material handling?
10. What are the factors that influence the selection of material handling equipment?
11. Explain the steps involved in Systematic Handling Analysis (SHA).
12. Material handling does not add any value to the product instead adds to its cost comment.
13. State for what applications the following M.H. equipment are used:
    (*i*) Forklift truck,
    (*ii*) Jib crane,
    (*iii*) Belt conveyer,
    (*iv*) Roller conveyer.

10

JOB EVALUATION AND MERIT RATING

• Introduction • Definitions • Objectives of job evaluation • Job evaluation procedure • Job analysis — application, stages • Job evaluation systems — Ranking method • Job classification method • Factor comparison method • Point rating system • Merit rating — Introduction • Benefits of merit rating • Requirements of sound merit-rating system • Methods of merit rating.

10.1. INTRODUCTION

The basic reason today in industrial disputes is regarding the wages. It is the tendency of the employees to compare their wages and salaries in relation to those of others in the same organisation or working in the similar jobs in other organisations. Always the dissatisfaction is created amongst the employees if there is a difference in wages that other employees are getting for the same type of work performed.

Industries and business whether medium or large are facing the problem of basic wage payments.

Because of the difference in wage structure for the same type of jobs, the employee turn over is on the rise.

Thus employers are also facing the challenge to retain and maintain the workforce under changing techno-economic scenario and wage parity amongst the organisations because of the entry of multinational organisations. One of the prime objectives of sound wage and salary administration is to eliminate inequalities and see that comparable jobs should be paid the same wage.

Even both employees and employers accept the fact that jobs having the same amount of work content, same level of difficulty should be paid same amount. But this is possible only, if wage structure is based on the classification of jobs as per the difficulty.

Thus, job evaluation is a technique to systematically determine the worth of each job and help in establishing basic wage rates of jobs.

10.2. DEFINITIONS

Job evaluation is a process to determine in a systematic manner and analytically the worth of each job in the organisation based upon the set of carefully selected factors such as skill, effort and responsibility demanded by the job and translating these worth of jobs into monetary terms (*i.e.,* pay and wages).

It is an attempt to determine and compare the demands which the normal performance of the particular jobs makes on normal workers without taking account of the individual abilities or performance of workers concerned. It is a job rating method and not the job ranking method.

Job evaluation aims to provide a means of establishing a wage structure acceptable to both workers and management.

10.3. OBJECTIVES OF JOB EVALUATION

1. To establish a sound wage and salary system by determining the worth of each job in the factory in relation to various factors like skill required, effort and responsibility involved.
2. To eliminate the wage inequalities.
3. To establish a general wage level for a given factory.
4. To clearly define the line of authority and responsibility.
5. To formulate an appropriate and uniform wage structure.
6. To provide a sound base for recruitment, selection, promotion and transfer of employees.
7. To identify the training needs of the employees so as to prepare them for future positions.
8. A sound base for individual performance measurement.
9. To promote a good employee-employer relations.

10.4. PROCEDURE FOR JOB EVALUATION

The steps involved in job evaluation are shown in Fig. 10.1.

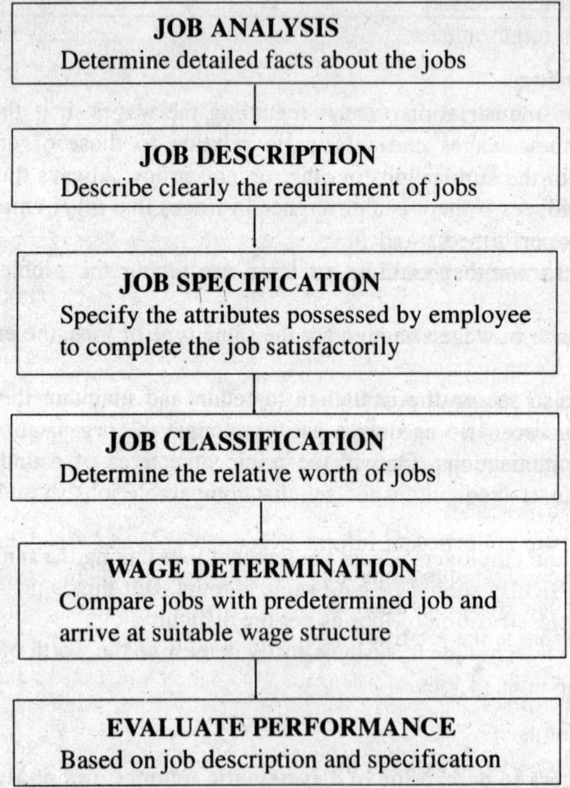

Fig. 10.1: Steps in job evaluation.

10.5. JOB ANALYSIS

Job analysis is the process of determining the facts relating to the jobs. It involves a systematic examination of the job to find out:

1. Nature of tasks performed by the workers.
2. Purpose or objectives of the tasks.
3. Working conditions under which the tasks are carried out.

4. Responsibility, skill required to perform the tasks.
5. Relationship between various jobs done in the department/organisation.

Job analysis programmes are usually tailor made as the nature of the information to be collected will depend on the organisation and purpose for which it is undertaken.

The information gathered through job analysis is useful for:

- Job evaluation.
- Personnel and general management decisions, recruitment, selection, promotion, transfer of staff in the organisation.
- Performance review and appraisal.
- Manpower planning.
- Design of training programme.

STAGES IN JOB ANALYSIS

Stage I. Job identification.

Stage II. Job information collection.

Stage III. Qualification requirements.

Stage I. Job Identification

Study and gather general information on the organisation with a view to locate each job in its overall context.

The information can be sought through:

- Organisational chart.
- Diagram of production process and functional relationship between jobs.
- Other sources of information should be consulted to construct or update the organisational charts/process diagrams.

The list of jobs to be analysed must be identified. During this some problems may arise like:

(*i*) Job titles may not be an indicative of job content.

(*ii*) Need for judicious sampling of post to be analysed.

(*iii*) Same job title may cover two or more basically different posts.

(*iv*) Similar work may be done in posts with different job titles.

Stage II. Information Collection

At this stage, a systematic collection of information on all jobs is carried out using a standard job analysis questionnaire. The questions are carefully designed to uncover the essential characteristics of the job. The questions such as:

Who does the work? What is the job title?

What are the essential tasks?

How are the tasks performed?

What are the equipment used?

What is the relationship between tasks of the job and tasks of other jobs?

What are job holders responsibilities towards his colleagues and towards the machines and equipment?

Under what working conditions the task is performed? (hours of work, noise, temperature, lighting, etc.).

Methods of Information Collection

1. Questionnaire method—A carefully designed questionnaire is to be filled by the worker and his/her supervisor.
2. Interview with the worker and his supervisor.
3. Direct observations at the workplace.

Stage III. Qualification Requirements for Satisfactory Performance of the Job

1. Knowledge.
2. Level of education.
3. Skills including experience.
4. Physical ability.
5. Mental ability.
6. Aptitude (initiative, tact, etc.).

The qualification requirements must consider only those which are essential to do the job.

A typical job analysis form is shown in Fig. 10.2.

10.6. JOB DESCRIPTION

Job description follows the job analysis. It gives all essential facts about the job like duties, responsibilities, working conditions and other required facts.

Job description is composed of three parts:

(*i*) Job identification containing the details like job title, department, section, job code, names of supervisor and other details to identify the job.

(*ii*) Job summary gives the overall picture of the duties performed.

(*iii*) Work performed gives the details of both regular as well as occasional tasks performed machines and tools used, working conditions and hazards. A typical job description is shown in Fig. 10.3 and example sheet in Fig. 10.5.

10.7. JOB SPECIFICATION

Job specifications are prepared from the data collected during job analysis. It is the statement of qualities and capabilities that an employee must possess to perform the job satisfactorily. Job specification describe the extent to which each of the job factor such as education experience, physical effort, responsibility for others work, materials machines and equipment, etc., present in the job and the degree of difficulty present. The job descriptions and job specifications both form the basic for job evaluation and so it is essential to make it sure that the facts are presented correctly. The typical job specification form is shown in the Fig. 10.4.

10.8. JOB EVALUATION SYSTEMS

The job evaluation system should be simple and readily understood by all the concerned. There are four accepted systems of job evaluation:

1. Ranking method.
2. Job classification system.
3. Factor comparison method.
4. Point rating system.

1. Ranking System

In this system, jobs are ranked in terms of their importance (with respect to level of duties and responsibilities) from the lowest to the highest. For ranking job is considered as a whole and it is not broken down into various elements or tasks. The rank is decided by the committee. This is the easiest and most simple method.

The major steps involved in ranking system are:

1. Selection of the jobs.
2. Job analysis.

JOB ANALYSIS DATA SHEET	
Job Title ------------------------- Proposed Title ------------------- Department --------------------- Person Interviewed ------------- Date of Analysis ---------------	Job code ------------- Reports to -----------
1. Job summary	
2. Work performed (*a*) Regular duties (*b*) Occasional duties (*c*) Periodic duties	
3. Equipment (*a*) Machines (*b*) Tools	
4. Education	
5. Experience (type and amount)	
6. Responsibilities for (*a*) Product and materials (*b*) Equipment and machinery (*c*) Work of others	
7. Physical demands— (*a*) Physical effort (*b*) Mental effort (*c*) Monitory (*d*) Hazards	

Fig. 10.2: Job analysis sheet.

JOB DESCRIPTION	
Job Title ----------------- Department --------------	Job Code No. ------------ Section --------------------
1. JOB SUMMARY	
2. RESPONSIBLE FOR	
3. EQUIPMENT AND MACHINE USED	
4. WORKING CONDITIONS AND HAZARDS	

Fig 10.3: Typical job description format.

JOB SPECIFICATION		
Job Title (present) -------------------- Job Title (proposed) -----------------	Job Code No. ----------------- Deptt./section -------------------	
FACTOR	DESCRIPTION	DEGREE/POINTS
1. Education and knowledge		
2. Experience		
3. Responsibility for machinery and equipment		
4. Responsibility for work of others		
5. Responsibility for safety of others		
6. Physical effort		
7. Mental effort		
8. Responsibility for products and materials		
9. Working conditions and hazards		

Fig. 10.4: Typical job specification sheet.

JOB DESCRIPTION

Job Title (Existing) : Turner
Deptt./section : Tool Room
Job Code : 04 – B
Employee Interviewed
Date : 7–5–1997

Job Summary

Operates the centre lathe machine

A. **Regular duties**

1. Preparation and setting of the machine (tools and work holding devices).
2. Selection of cutting parameters, *i.e.,* speed, feed depth of cut using charts.
3. From the drawings decides the sequence of operations.
4. Carries out all the operations on the job (turning, facing, taper turning, knurling, etc., as required).
5. Checks dimension of finished parts using micrometers, verniers, dial gauges slip gauges, etc.).
6. Grinds and prepares cutting and form tools.

B. **Casual Tasks**

(*a*) Occasionally does maintenance/small repairs of the m/c.
(*b*) Occasionally does assembly work.
(*c*) Works on other machines if instructed.

C. **Equipment and Machine used**

Precision Centre Lathe.

D. **Working Conditions**

Works within the tool room. Chances of injuries to limbs while loading and unloading of jigs/fixtures, work holding devices, chips may enter the eyes or burrs.

Fig. 10.5: Example of job description.

3. Choosing the committee for ranking jobs.
4. Ranking of the jobs.
5. Grouping the job into groups by the committee.

The advantage of this method lies in its simplicity, less time consuming and less paperwork (forms). The main limitations being the lack of standard criteria for ranking jobs and hence less accurate. Applicable where there are only limited jobs to be analysed.

2. Job Classification System

The job classification system is the process of allocating jobs to grades which are predefined. The rater is provided with the yardstick. The grade differences are defined in terms of differences in the levels of duties, responsibilities and requirements of special skills.

The job evaluation by this method involves the following steps:

1. Decide and describe number of grades.
2. Write grade level descriptions
3. Listing of jobs to be evaluated
4. Assigning the jobs to grades based upon grade and job level description.

Major Advantages

1. Method is easily understood by all and easy to use.
2. The results are fairly accurate.

The limitation being the system gets complicated as the number of jobs are increased and grade descriptions are relatively difficult to write.

3. Factors Comparison Method

It is a qualitative method of job evaluation. This method involves the detailed analysis of the jobs which are then ranked in respect of predetermined jobs. Five factors are considered for evaluation of jobs. They are:

1. Mental effort.
2. Skill requirement.
3. Physical requirement.
4. Responsibility.
5. Working conditions.

Steps involved in factor comparison method.

1. The key jobs are to be identified which can be accurately described. Key jobs are so chosen that they represent each major level of duties, responsibilities and skill, etc., covering the entire range of jobs.
2. Key jobs are analysed for each of the five factors defined.
3. The salary paid for each key job is apportioned amongst the factors.
4. The wage paid for each key job is apportioned amongst the factors in proportion to their importance in the job. Thus a money rating scale for each of the factors is obtained.
5. Each of the jobs are evaluated for each of the factors on its money rating scale of key jobs Monitory value of the job is obtained by adding up the individual money values for each one of the factors depending upon the importance of the job.

Advantages

1. Job-to-job comparison ensures that jobs are compared on a common comparable scale.
2. Directly the money value of the job is obtained.
3. It takes into account all the factors that constitute a job.
4. This method is applicable even for large organisations.

Limitations

(*a*) Selection of a wrongly paid job is likely to introduce error in the considerable way.

(*b*) Lot of subjectivity is existing in constructing the rating scales for money value.
(*c*) It is difficult to explain the rating scale to employees.
(*d*) It is more time consuming.

4. Point Rating Method

This the most popular and quantitative method designed by Merill. R. Lott. This method evaluates jobs based on the carefully selected factors such as education, experience, physical effort, responsibility for machines and materials which are common to majority of the jobs. The major steps involved in point rating system are:

1. **Decide the type of jobs to be evaluated:** During this step, the jobs which are to be evaluated are listed and only jobs having different work contents are to be recorded.
2. **Select the factors:** The following factors are generally used for evaluating jobs (with special reference to factory employees):
   - Education.
   - Experience.
   - Judgement and initiative.
   - Physical effort.
   - Mental effort.
   - Responsibility for equipment, materials.
   - Responsibility for process.
   - Responsibility for the work of others.
   - Working conditions.
   - Hazards.
   - Adaptability.
   - Creative ability.
   - Aptitude for learning.
   - Initiative.

The factors selected must be rateable. Only important factors are to be considered. There should not be any overlap of meaning of factors selected. The factors should be as few as possible selected factors should be acceptable to both workers and management.

3. **Definition of factors:** There should not be confusion in meaning of the factor. Each factor should be defined such that everyone has the same interpretation. Simple language should be used and complicated words are to be avoided, *e.g.*, Education. This factor is used in the context of academic or technical training necessary to acquire competence to understand the work being done and responsibilities associated with it. It does not include the practical trade knowledge acquired on the job.
4. **Define the degrees for each factor:** Each factor is to be divided into degrees. The degrees provide a rating scale for each factor.

For example, degrees for the education factor are defined as:

Degree	*Description*
1	Requires the ability to understand simple verbal instructions.
2	Requires the ability to read and write. Requires trade knowledge equivalent to use hand tools or/and operation of simple machines.
3	Requires the knowledge to be able to operate the machines or repetitive jobs with few set-ups. Knowledge of the part and tools. Be able to do arithmetic calculations.
4	Specialised knowledge regarding the engineering drawing mathematics and English.
5	The factors (5, 6 .... etc.) are defined with respect to further educational requirements.

5. **Assigning weightages to the factors:** The weightages are assigned to each factor on a percentage basis depending upon the proportionate contribution of each factor to the job. Weightages are arrived by consensus at the joint committee of management and union.
6. **Assigning points to the degrees:** Once the weightages are decided, the next step is to assign points to the degrees. The most common method is to allow the percentage weightage of the factor to represent the point value of the first in each factor and let the remaining values of the degrees to follow an arithmetic progression, using the weightage of the factor itself as the difference between two degrees.

For example, if the percentage weightage for education factor is 20%, and the factor is split into degrees, the point for the degrees is

Degree	*Points*
1	20
2	40
3	60
4	80
5	100

7. **Preparing the job evaluation manual:** In this manual, the job factors and their degree are defined. It thus serves as a mechanism for determining the relative values of the existing and proposed jobs.
8. **Preparing job descriptions and job specifications.**
9. **Rating of the job (determining the points).**

10. **Placing jobs into grades:** Once the jobs have been rated in terms of points, they are placed into grades.

Advantages

(*i*) More accurate and reliable compared to other methods.
(*ii*) More objectivity due to quantification of factors.
(*iii*) It is logical and practicable.
(*iv*) Good acceptability.

Limitations

(*i*) Requires a large experience and expertise to define degrees and points.
(*ii*) Points allocated to degree are arbitrary.

10.9. MERIT RATING

Job evaluation evaluates the job and the merit rating assess the worth of a person performing the job. Merit rating is also called the performance appraisal. It evaluates, controls and reviews the performance. Both job evaluation and performance appraisal are aimed at systematically determining the wage rates paid to the employees.

Benefits of Merit Rating

(*i*) Useful in rewarding the person and the reward can be linked to the performance.
(*ii*) Helps to identify the person's potential to perform the assigned jobs and to decide the future positions he can take up.
(*iii*) To identify the training needs of the employees.
(*iv*) Helps in counselling employees regarding their strengths and weaknesses.
(*v*) It motivates employees to perform better.
(*vi*) Acts as a constructive performance appraisal system.

Requirements of a sound performance appraisal system:

(*i*) The merit rating system should be transparent in the sense that it should be known to everyone.

(*ii*) The criteria should be fixed and known to the rater as well as to the ratee.
(*iii*) There should not be any bias or ambiguity.
(*iv*) The rating should be done at the prefixed intervals.
(*v*) It should be related to the job related behaviour only.
(*vi*) It should act as a basic for sound reward system.

Methods of Merit Rating

1. **Ranking method:** This is the conventional and easy method. Normally the employees are ranked in the order from best to worst. This is applicable to industries where number of people are few. The limitation of this method is that it cannot indicate specific strengths and weaknesses. Ranking becomes difficult as the number of employees increase.
2. **Paired comparison method:** In this method, the rater compares each employee in a group with all the remaining employees. The performance is the only parameter for comparison. This also becomes difficult to compare if the group is large.
3. **Forced choice method:** In this method, for each trait or behaviour number of statements are given and the rater is required to select only one statement which describes the particular behaviour of the employee being evaluated. This method is called forced choice because the rater is forced to check only one statement and is not allowed to describe behaviour in his own words. This is most popular method used for rating of lower cadre staff.
4. **Check list method:** These are the lists made up of series of questions or statements which are concerned about the important aspects of employees performance on the job. The process of rating simply consists of checking those questions concerned to ratee and answering the question in "YES" or "NO". It is easy to compare the employees by this method.
5. **Scale plan:** This is widely accepted method in industries. The scale is constructed to define the various degrees of the traits.
   There are two types of scale plans:
   (*i*) **Continuous scale:** Here the scale is constructed to represent the highest to lowest degree of required trait: (*a*) Numerical scale, (*b*) Description scales.
   (*ii*) **Discontinuous scales:** This is the scale which gives elaborate description of facts needed for rating.

References for Further Reading

1. Bim, "*Job Evaluation A Practical Guide*", British Institute of Management, London, (1961).
2. Maynard, H.B., "*Industrial Engineering Handbook*", McGraw Hill, 3rd Edition, New York, (1971).
3. Husband, T.M., "*Work Analysis and Pay Structure*", McGraw Hill, New York, (1976).

REVIEW QUESTIONS

1. What is job evaluation? What are its objectives?
2. How does job analysis differ from job description?
3. Describe various methods of job evaluation giving their advantages and limitations.
4. Explain the steps involved in point rating method of job evaluation.
5. What is merit rating and how it helps the industries?
6. Write short notes on:
   (*i*) Importance of job evaluation and merit rating,
   (*ii*) Job analysis, job description and job specification,
   (*iii*) Merit rating methods,
   (*iv*) Job evaluation systems.

11

WAGES AND INCENTIVES

• Introduction. • Definitions • Minimum wage • Need for rational wage policy • Factors influencing wage system • Characteristics of a good wage system • Types of wage payments—Wage payment by time basis •Piece rate system • Incentive schemes—Financial and non-financial incentives • Individual and group incentive schemes • Characteristics of a good incentive system • Incentive plans.

11.1. INTRODUCTION

The study of work measurement leads to wage payments. Theoretically the wages that worker gets is proportional to the amount of work he does. Wages are supposed to increase effective motivation to work hard and better. Wages constitute the principle source of income for the workers for the service rendered. A rational wage policy is essential to compensate the workers for their efforts rendered.

The compensation to the employees involves the following issues:

1. Determination of wage structure/levels for different positions in the organisation.
2. Determining wage for each individual employee occupying the position.
3. Determining the method of wage payment.

11.2. DEFINITIONS

1. **Wages:** These are the payments made by the employer to the efforts put in by the workers towards production. A wage determines the standard of living and it should represent a fair return for the effort of the worker and also wages should be able to satisfy the primary and secondary needs of the workers. They should be enough to provide him a reasonable standard of living.
2. **Nominal wages:** It is the amount of money paid to the worker in cash for the efforts of the worker towards production and no other benefits are given to the worker. This is called money wage. The rates of wages vary from one place to another depending upon the demand and supply of labour and the necessities of life.
3. **Real wages:** It represents the amount of necessaries, comforts, luxuries and cash payment a worker gets in return for his efforts. Some organisations provide their employees certain essential commodities, housing with free electric and water charges, uniforms and other such facilities in addition to the money in cash. If all these amounts are considered for wages, it becomes the real wage.
4. **Living wage:** When the wage rates are such that they are going to fulfil some of the requirements of a family like foods, cloths, education and insurance against misfortune along with other basic necessities, they are referred to as living wages.
5. **Fair wage:** It is a wage which is to be considered as a fair amount of return for the efforts of the employees and should be able to cover the other necessities of life, apart from basic

necessities like food, clothes and shelter for his family. The rate for the fair wage lies between real wage and minimum wage.

6. **Minimum wage:** Minimum wage may be defined as the wage, which not only provides for basic subsistence but something more than this. It should be able to keep the employees motivated and it should provide for some measure of education medical facilities and other essential requirements. It should also consider the cost of living.

11.3. MINIMUM WAGE

Wage cannot be paid beyond the paying capacity of the industries or factories.

The minimum wage is fixed taking into consideration the factors such as cost of living, maintaining the efficiency of the workforce, keeping them motivated and paying capacity of the industries.

Main objectives of the minimum wages:

(*i*) To protect the sections of working population whose wages are very low.
(*ii*) To prevent exploitation of the workers.
(*iii*) To improve general standard of life.
(*iv*) Satisfactory compensation towards efforts expended by the worker.

11.4. NEED FOR A RATIONAL WAGE POLICY

A sound wage policy should be aimed at social justice and the workers should get their due share for their efforts.

The rational wage policy should consider the following aspects:

1. Fixing minimum wages.
2. Fixing ceilings on wages.
3. Wage structure.
4. Price stability and price index.

The rationalised wage policy should aim at reducing the relative propensity and improve the living conditions of working class.

11.5. FACTORS INFLUENCING WAGE SYSTEM

It is very much complex to arrive at a wage which may be considered satisfactory for both workers and management. The various factors that determine the wage level are:

1. Labour market, *i.e.*, demand and supply of labour.
2. Legal and statutory restrictions. (Minimum Wage Act, Payment of Bonus Act 1965, Employees Provident Fund, Family Pension Fund Act 1952, Factory Act 1948, Employees State Insurance Act 1948, Payment of Gratuity Act 1992).
3. Organisation's ability and willingness to pay.
4. Bargaining capacity of the employer and the employees.
5. Prevailing wage structure in the specific sector or industry.
6. Workers skill, knowledge and experience.
7. Wage levels in the specific sector or industry.
8. Cost of living.

11.6. CHARACTERISTICS OF A GOOD WAGE SYSTEM

1. A good wage system should be acceptable to both employees and management.
2. It should guarantee a minimum wage to the employee.
3. To should be able to keep the worker motivated.
4. It should provide a scope for employees to get reward for their additional or extra effort (incentives).

5. It should be consistent and should not be altered frequently.
6. Should believe in equal work and equal pay.
7. The system should be simple and understood by all concerned.
8. Proper encouragement should be given to utilise the full potential of employees.
9. Should make the work challenging and interesting.

11.7. TYPES OF WAGE PAYMENTS

There are two basic methods of wage payments:

1. Wage payment on time basis.
2. Wage payment on output basis.

1. **Wage payment on time basis:** Under this method, wages are paid to the employee based on the time for which he works. The wage rates are predetermined in terms of Rs. per month, Rs./day or Rs./hour of employment. In this system the workers are paid on time they work irrespective of output produced.

This system is applicable where output is not quantifiable and it is not the criteria of payment, where work is of not repetitive type.

Advantages

1. System is easy to understand and simple to operate.
2. Reduces the problems of industrial relations.
3. The quality of the work is maintained as employees are not in a hurry to increase quantity.
4. The workman can show his efficiency and workmanship without loss to himself.
5. There is a scope for improvement in work methods.

Disadvantages

1. Does not provide any incentive to ambitious and more efficient employees.
2. The output will be lowered in the absence of strict supervision.
3. The basis for wage is time and not the output or efficiency. So it happens that less efficient workers are paid equal to efficient workers.
4. Employer will gain or loose by increase or decrease in output.

2. **Wage payment on the basis of output (piece rate system):** In this system, wages are paid to the employees in relation to the output produced. This method is very convenient where each individual worker is capable of performing his work without any dependence on the other individual and the output produced will be quantifiable.

This method can be applied where output is standardised. The work is of repetitive nature.

Advantages

1. It provides incentives to efficient workers.
2. Cost of supervision is low compared to time based system.
3. Higher speed increases the production rate and hence reduces the cost per unit.
4. Increases the utilisation of production facilities.
5. Workers innovates new ways and methods of doing the work in order to reduce the time per unit.
6. Motivates workers to produce more.

Disadvantages

1. Workers in order to increase their wages through faster working, may neglect the quality.
2. Because of speed, worker may be prone to accident as it is possible that he may neglect precautions and safety measures.
3. The security for the workers is low and this may seriously punish the aged and inefficient workers.

11.8. INCENTIVE SCHEMES

Incentive schemes are intended to increase workers motivation by allowing them to earn proportionately higher returns from greater efforts.

American Society of Mechanical Engineers (ASME) defines a wage incentive plan as **"a method of payment which directly relates earning to production. A system which enables workmen to increase their earning by maintaining or exceeding an established standard of performance."**

These are the tools of the management to stimulate the production by encouraging workers to produce more than average in accordance with their productivity.

Incentive plans are of two basic types:

1. **Financial incentives:** These are the rewards paid to the employees efforts in cash.
2. **Non-financial incentives:** These are non-monetary incentives (other than cash). These may include gift items, discount coupons, special holidays, etc.

Some of the non-financial incentives are:

1. Management may create a climate of competition amongst the employees to contribute constructively towards the organisation.
2. This incentive promotes creativity and idea generation.
3. The incentive include provision for good housing, with all modern amenities, recreation facilities, medical facilities, etc.
4. Promotion to employees and facilities for personal growth.
5. Foreign business or educational trips.

11.9. INDIVIDUAL AND GROUP INCENTIVE SCHEMES

Under individual incentive scheme, individual employee is paid incentive on the basis of the individual performance or output. This incentive is regardless of the output or performance of the department or organisation. The employers are liable to pay incentive to those employees who are producing more than the standard output.

Under group incentive scheme, each employee is paid incentive on the basis of collective performance of his group to which he belongs. This group incentive scheme is preferred by the management as in turn they are getting an output from the group. Within the group, each employee is going to get equal share of the incentive. Highly competent or productive employees are not in favour of this scheme.

11.10. CHARACTERISTICS OF A GOOD INCENTIVE SYSTEM

1. The plan should be simple to understand and easy to operate. The employee should be able to calculate his earnings.
2. The incentive scheme should be consistent. Once installed, the incentive scheme should not be alterated to often.
3. The incentive scheme should be such that it should motivate the employee to produce more.
4. There should be direct relation between the effort and the reward.
5. The incentive system should not create disharmony amongst the employees.

11.11. INCENTIVE PLANS

(*a*) **Differential piece rate system**

(*i*) Taylor's Differential piece rate system.

(*ii*) Merrick's Differential piece rate system.

(*b*) **Time and piece rate system**

Gantt task and bonus scheme.

(*c*) **Premium bonus schemes**
  (*i*) Halsey plan.
  (*ii*) Rowan plan.
  (*iii*) Bedaux plan.
(*d*) **Efficiency based plans**
  (*i*) Emerson's efficiency plan.
(*e*) **Group incentive schemes**

1. Taylor's Differential Piece Rate System

This plan was developed by F.W. Taylor, father of scientific management.

Under this system, a standard output for the day is fixed based upon the accurate time and method study. The employee's who achieve this standard or produce more than the standard are given higher rates than those who fail to reach the standard. It considers both time and output, the differential rates are paid based upon the quantity produced.

Characteristics of this system

1. The day's wage or earning is not guaranteed.
2. Two piece rates are applicable—lower piece rate for output below standard output and higher piece rate for output exceeding the standard output.
3. Lower piece rate is considered for below standard performance workers and higher piece rate for standard and above standard performance.

Advantages

1. Worker is paid as per his efficiency and he is paid only for production time and not for the idle time.
2. Clerical work is kept at minimum and worker knows how much he should produce to get the required wage.
3. Reduce the supervision to the minimum.
4. Encourage the worker to produce more to earn more.
5. It ensures better utilisation of the resource.

Disadvantages

1. Workers in order to earn more will not care much for quality of the work. Hence more wastages and rejections.
2. Quantity orientation makes the worker to work fast, in due course he may neglect safety measures.
3. The decrease in demand may have an adverse effect on the employees.

Illustration.

Computation of earnings by this method

Output of worker *A* = 700 units
Output of worker *B* = 900 units
Standard production = 100 No's/hour
Labour rate = Rs. 10 per hour
differential rate 80% of the standard piece rate for below standard performance 120% of the standard piece rate for standard and above standard performance.

compute the earnings of *A* and *B*.

Solution

Standard output per hour = 100
Labour rate/hour = Rs.10

$$\text{Standard piece rate} = \frac{10}{100} = 0.10 \text{ Rs.}$$

Piece rate for below performance = $0.10 \times 0.8 = 0.08$

Higher piece Rate (above std. performance) = $0.1 \times 1.2 = 0.12$

Earning of operator $A = 700 \times 0.08 = 56$ Rs.

Earning of operator $B = 900 \times 0.12 = 108$ Rs.

Note: Assuming the working hours of 8, the standard output per day is 800 pieces. Hence operator A is below standard performance, (Lower piece rate applicable) and operator B is above standard performance, (so higher wage rate is applicable).

2. Merrick's Differential Piece Rate System

This system is similar to Taylor's differential piece rate system. It provides three piece rates as compared to two piece rates given by Taylor.

The prevailing piece rates under this plan.

Up to 83% of standard output	Basic rate per piece
Over 83% to 100%	$1.10 \times$ Basic rate per piece
Over 100%	$1.2 \times$ Basic rate per piece

Characteristics of the System

1. Minimum wage is not guaranteed.
2. Standard rate is established after a through analysis of each job.
3. It encourages the workers to produce more and if they come up to 83% of standard performance, they are going to get the piece rate above standard piece rate.

3. Gantt Task and Bonus Scheme

This scheme was proposed by Henry Gantt.

In this system a minimum wage is guaranteed for those workers who cannot reach bonus level. The standard output is fixed based upon the careful study. If the worker achieves the standard output or produces more than the standard output, he is provided with an extra allowance between 25% to 50% of the hourly rates.

Characteristics of the System

1. Minimum day's wage is guaranteed irrespective of the performance of the worker.
2. The actual output of the worker is compared with the standard output to calculate efficiency.
3. Time wages without bonus are paid below standard performance.
4. Time wages plus an increase in wage rate is considered at standard performance.
5. A high piece rate is considered in place of time wages above standard performance.

Computation of Earnings

Illustration.

The following information is given for an operator's output on a milling machine for six days.

Day	Monday	Tuesday	Wed.	Thu.	Fri.	Sat.
Output	50	55	48	40	35	45

The incentive scheme is as follows:

Output below standard	Time wages
Output at standard	Time wages +20% bonus.
Output above standard	Higher piece rate on entire production.

The hourly rate of the operator is Rs. 10. The std. time is 12 minutes per piece of the job and higher rate is Rs. 2.5 per piece. Calculate day wise earnings of the operator.

Solution

Assuming 8 hours of working per day and standard time of 12 minutes.

$$\text{Standard output per day} = \frac{80 \times 60}{12} = 40 \text{ Nos}$$

The earnings are computed as shown in table:

Day	*Actual Output*	*Actual output as a % of std. Output*	*Basis of Payment*	*Day's Earnings (Rs)*
MON	50	125%	High Piece Rate	50 × 2.5 = 125.0
TUE	55	137%	High Piece Rate	55 × 2.5 = 137.5
WED	48	120%	Higher Piece Rate	120.00
THU	40	100%		80 + 20% of 80 = 96.00
FRI	35	87.5%		8 × 10 = 80.00
SAT	45	112.5%	Higher Piece Rate	45 × 1.5 =112.5

4. Halsey Premium Plan

It is a gain sharing plan. In this plan, standard times are established based upon the past experience. The plan guarantees to worker a certain base rate and in addition a certain percentage of time he has saved on the job.

Earnings in this scheme are computed as follows:

$$\text{Bonus earned} = \frac{1}{2}(T_s - T_t) \times R$$

where

R = hourly labour rate (Rs./hr)

T_s = std. time

T_t = Time actually taken

$T_s - T_t$ = Time saved.

$$\text{Earnings} = \frac{1}{2}(T_s + T_t) \times R$$

Advantages

1. It guarantees the fixed wage to slow workers and at the same time additional payment to efficient workers.
2. Production cost saving.
3. Simple in operation and worker can compute his days earning without any difficulty.

Disadvantages

1. The plan does not give the bonus on the time saved but it gives bonus only on a percentage of time saved.
2. Any mistake in computing standard time affects the workers earnings.
3. Incentive is not much attractive to workers.

Illustration: A company is operating 50-50 Halsay premium plan. Compute the total wages of the worker, per hour workings in the factory based on the following information:

Time Rate = Rs. 6/hour

Time allowed = 100 hrs

Time taken = 75 hrs

Time saved = 25 hrs

$$\text{Earning} = \frac{1}{2}(T_s + T_t) \times R$$

$$= \frac{1}{2}(100 + 75) \times 6$$

= 525 Rs.

Rs. 525 earning for 75 hrs.

$$\text{Earnings per hour} = \frac{525}{75} = 7 \text{ Rs/hr.}$$

5. Rowan Premium Plan

This plan was devised by James Rowan of Glasgow in the year 1901.

The plan provides each workman a guaranteed minimum wage plus bonus for certain portion of the time saved, the proportion of time payable for incentive being the ratio of actual time taken to the standard time.

Earnings under this scheme are,

$$\text{Bonus} = \frac{T_t}{T_s}(T_s - T_t)\, R$$

$$\text{Earnings} = R \times T_t + \frac{T_s - T_t}{T_s} \times R \times T_t$$

$$= T_t + R\left(1 + \frac{T_s - T_t}{T_s}\right)$$

Advantages

1. Encouragement to slow workers and trainees.
2. Employer will get the partial benefit of increased output.
3. A good protection against loose standards.

Disadvantages

1. It is more complex compared to other incentives schemes.
2. The sharing concept is not favoured by workers and so unpopular amongst workers.

6. Bedaux Point Premium Plan

This system is applicable when the manufacturing involves veriety of short cycle jobs of varying work contents.

Characteristics of this system

1. Based upon the work measurement study, the standard time is established for each job in terms of points. One point equals one minute, *i.e.,* a standard hour consists of 60 points.
2. Guaranteed wage rate is paid up to standard performance.
3. Guaranteed wage rate plus incentive bonus is paid for above std. performance.
4. The incentive bonus is paid at the fixed proportion for the time saved. As per this plan, 75% of the points saved multiplied by one sixtieth (1/60th) of the workers hourly rate is paid as incentive.

Advantages

1. Plan is simple in design and easy to operate.
2. It really gives incentive to higher production.
3. This brings in a competitive spirit amongst the workers to produce more.

7. Emerson's Efficiency Plan

Under this plan,

(*i*) Days wages irrespective of the output performance is guaranteed.

(*ii*) A standard output based on past performance is taken as 100% efficiency.

(*iii*) Actual output of the worker is compared with standard output to calculate the efficiency.

$$\text{Efficiency \%} = \frac{\text{Actual output for the period}}{\text{Standard outputfor the period}} \times 100$$

(*iv*) An incentive bonus is paid to the workman of his output exceeds. 2/3rd (66.67%) of the standard performance (100%)

Earnings in this method are computed as under.

1. Efficiency Below 66.67%	No bonus, only time wages.
2. Between 66.67% to 100%	Time wages + Bonus at different percentages increasing rapidly to 20% at 100% efficiency.
3. Over 100%	Time wages + 20% bonus + 1% bonus for each 1% increase in efficiency beyond 100%.

8. Scanlon Plan

This is group incentive plan applicable to labour intensive firms. Characteristics of this system are:

(*i*) This is applicable to employees throughout the organisation.

(*ii*) All employees recieve a guaranteed time wages.

(*iii*) Ratio of labour cost to total sales value is set as an index of total labour effectiveness.

(*iv*) Actual labour cost to the total sales is calculated for the assessment period.

(*v*) An incentive bonus, equivalent to % reduction in labour to sales ratio is paid to each employee.

Incentive Bonus Computation.

$$\text{Labour cost during base month} = LB$$

$$\text{Total sales during base month} = SB$$

$$\text{Standard (base) labour to sales ratio} = \frac{LB}{SB}$$

$$\text{Labour cost for assessment month} = LA$$

$$\text{Total sales for the assessment month} = SA$$

$$\text{Actual labour to sales ratio for assessment year} = \frac{LA}{SA}$$

$$\text{Percentage reduction in the labour index} = \frac{\left(\frac{LB}{SB}\right) - \left(\frac{LA}{SA}\right) \times 100}{\frac{LB}{SB}}$$

9. Priestman Production Bonus

This is also a group incentive plan. The important features of this plan are:

1. All employees are paid guaranteed time wages.
2. A standard output to be achieved (by factory or deptt.) is agreed upon by employer and union representative. Output is expressed in units. In case of single product and points in case of multiple products.
3. To determine standard output, number of workers in employment is taken into account.
4. Actual output is compared with standard output. If actual output exceeds the standard output, each employee is paid production bonus in proportion to increase in output per worker.

Computation of Bonus

$$\text{Percentage productivity bonus payable to each employee} = \frac{\frac{PA}{WA} - \frac{PB}{WB} \times 100}{\frac{PB}{WB}}$$

$\frac{PA}{WA}$ = Actual production per workman during assessment period.

$\frac{PB}{WB}$ = Std. production per workman during base period.

PB = Std. production in base period.
PA = Actual production in assessment period.
WB = No. of workman employed during base month.
WA = No. of workman employed during assessment month.

References for Further Reading

1. Patten, T.H., *Pay, Employee Compensation and Incentive Plans*, Macmillan Publishing Co. Inc., New York, (1977).
2. Shwinger, P., *Wage Incentive Systems*, John Wiley and Sons Inc., New York, (1975).
3. Raymond, R., *Mayer, Production and Operations Management*, McGraw Hill International Book Company, Japan, (1982).
4. Friedman, B., *Effective Staff Incentives*, Kogan Page, (1990).
5. Backkan, J., *Wage determination—An analysis of wage Criteria*, Princeton N.J., Van Nostrand, (1960).

REVIEW QUESTIONS

1. Explain the factors that influence payment of wages to the employees.
2. What are the two basic methods of payment of wages? Explain the merits and demerits of the two methods.
3. What is incentive? What are the different types of incentives?
4. Explain the characteristics of a good wage system.
5. Explain the various incentive plans.
6. How does the Taylor's differential piece rate system differ from the Merrick's differential piece rate system?
7. Compare the various premium bonus schemes.
8. Write short notes on:
   (*a*) Minimum wage policy,
   (*b*) Financial and non-financial incentives,
   (*c*) Individual and group incentives,
   (*e*) Halsey premium plan,
   (*f*) Non-financial incentives.
9. What are the salient features of the following incentive plans? How earnings are computed in these plans:
   1. Emerson's efficiency plan,
   2. Scanlon plan,
   3. Priestman production plan,
   4. Badaux plan.

12

ERGONOMICS

• Introduction and definition • Objectives of human engineering • Ergonomics as a multidisciplinary approach • Ergonomics, productivity and working environment • Areas of human engineering • Man-machine system—types • Man in a closed loop system • Aspects of man-machine system • Design of information displays • Design of controls • Noise, vibration, ventilation and heat • Anthropometry • Manual material handling (MMH) • Physiological aspects of muscular work • Work place design.

12.1. INTRODUCTION AND DEFINITION

The word **Ergonomics** has its origin in two Greek words **Ergon** meaning work and **Nomos** meaning laws. So it is the study of the man in relation to his work. The word ergonomics is used commonly in Europe. In USA and other countries it is called by the name **"human engineering or it is also called human factors engineering"**. ILO defines human engineering as—**"The application of human biological sciences along with engineering sciences to achieve optimum mutual adjustment of men and his work, the benefits being measured in terms of human efficiency and well-being."** The human factors or human engineering is concerned with man-machine system. Thus another definition which highlights the man-machine system is: **"The design of human tasks, man machine system, and effective accomplishment of the job, including displays for presenting information to human sensors, controls for human operations and complex man-machine systems."**

Human engineering focuses on human beings and their interaction with products, equipment facilities and environments used in the work. Human engineering seeks to change the things people use and the environment in which they use the things to match in a better way the capabilities, limitations and needs of people.

12.2. OBJECTIVES OF HUMAN ENGINEERING

Human engineering (ergonomics) has two broader objectives:

1. To enhance the efficiency and effectiveness with which the activities (work) is carried out so as to increase the convenience of use, reduced errors and increase in productivity.
2. To enhance certain desirable human values including safety, reduced stress and fatigue and improved quality of life.

Thus in general the scope and objective of ergonomics is **"designing for human use and optimising working and living conditions".**

Thus human factors (ergonomics) discovers and applies information about human behaviour, abilities and limitations and other characteristics to the design of tools, machines, systems, tasks, jobs and environment for productive, safe, comfortable and effective human use.

12.3. ERGONOMICS IS MULTIDISCIPLINARY

The various disciplines that are going to have influence on human factors are:

1. **Engineering:** Design of work system suitable to worker
2. **Physiology:** Study of man and his working environment
3. **Anatomy:** Study of body dimensions and relations for work design
4. **Psychology:** Study of adaptive behaviour and skills of people
5. **Industrial hygiene:** Occupational hazards and workers health.

12.4. ERGONOMICS, PRODUCTIVITY AND WORKING ENVIRONMENT

Productivity is a powerful tool to improve the standards of living of people and to enhance the quality of work life (QWL). Ergonomics is concerned with man and his working conditions. Ergonomics aims at providing comfort and improved working conditions so as to channelise the energy, skills of the workers into constructive productive work. This accounts for increased productivity, safety and reduces the fatigue. This helps to increase the plant utilisation. Thus the benefits of ergonomics are shown in Fig. 12.1.

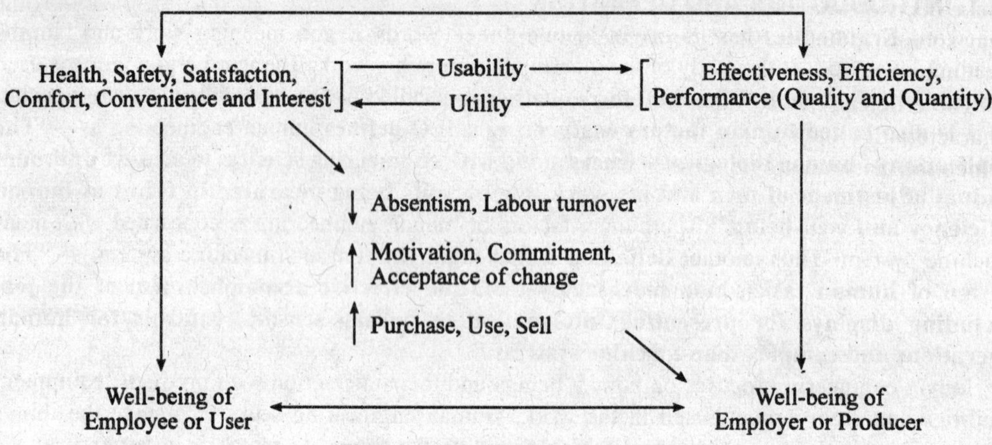

Fig. 12.1: Benefits of ergonomics.

12.5. STUDY OF HUMAN ENGINEERING AREAS

1. Anthropometry and bio mechanics.
2. Control of physical work environment
3. Design of man-machine system.
4. Design of controls and displays.
5. Accidents fatigue and safety.
6. Work place design.

12.6. MAN-MACHINE SYSTEM

The fundamental concept in human engineering is the system. A system is composed of humans, machines and other things that work together (or interact) to accomplish some goal which these same components could not produce independently.

The man-machine system is a combination of one or more human beings and one or more physical components interacting to bring about, from given inputs, some desired output.

Man-machine system consists of any type of physical object, device, equipment, facility and activities performed by man.

One way to characterise man-machine system is the degree of manual versus machine control.

The three broad categories of systems are:

Manual system: A manual system consists of hand tools and other aids which are coupled by a human operator who controls the operation. Operators use their own physical energy as the power source.

Mechanical system: These system's are semiautomatic, consist of well integrated physical part such as powered machine tools. They are generally designed to carry out their functions with little variation. Power is typically provided by a machine and operators function is mainly of control by the use of controlling devices.

Automated system: When the system is fully automated, it performs all operational functions without human intervention. Automated system require human to install, programme, reprogramme and maintain the system.

Man-machine system is a closed loop system. A simple man-machine system is shown in Fig. 12.2.

The man will receive certain information from the machine either from dials, displays, etc., designed for that purpose or by observation of machine itself. He will process this information and make decisions on what action to take and manipulate controls or attend machine in some other way so as to affect its behaviour in the required manner.

Environmental factors will have an influence on the working of the system.

The efficiency with which the man functions depends on environmental factors, on his own characteristics such as age, motivation, training and experience as well as the efficiency with which the machine provides the information feedback and accepts control measures.

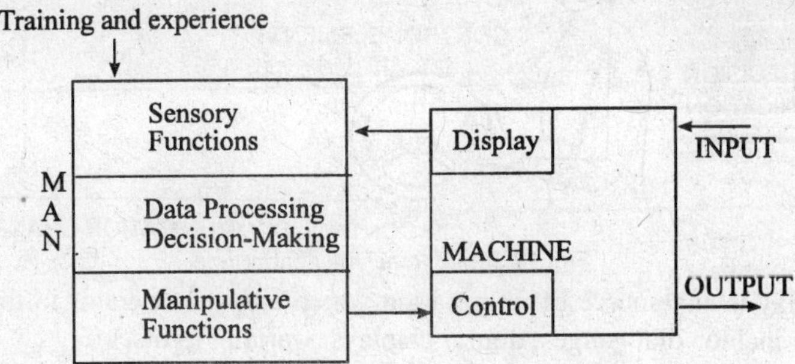

Fig. 12.2: Man-machine system.

The Functions Performed by Man and Machines

It is very essential to know as to which functions are performed by man and which functions does machines better perform.

Man is unique or better.

1. At discriminating relevant from irrelevant signals.
2. At innovation and creative in problem-solving.
3. In reasoning.
4. Ability to select his own inputs.
5. In improving, adopting flexible procedures, exercising judgement based on minimal information.
6. Sensitive to wide variety of stimuli.

7. In selective recall of old information.

Machines are unique and better:

1. Routine processing and storage of large amounts of facts and details.
2. For repetitive and monotonous operation.
3. For monitoring men and machines.
4. In operating under conditions that are stressful.
5. Rapid response to signals.
6. For rapid and complex situations.
7. For concurrent operations.
8. In sensing stimuli beyond the range of human sensitivity.

12.7. THREE ASPECTS OF A MAN-MAHCINE SYSTEM

1. Design of information displays.
2. Design of controls.
3. Environmental factors.

Man in a Control Loop System

Except for the automated systems, the man-machine system must include human as an operator and a system with machine and operator forms the close loop system.

A typical man-machine control loop is shown in Fig. 12.3. In display communication channel, information is sent to the operator from a display element through display communication channel. A display is any source of information which aids the operator in the control process. Typical displays include dial gauges, digital displays, warning lights, etc.

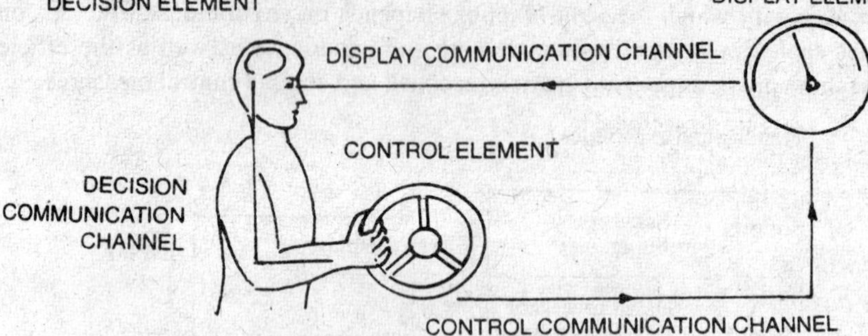

Fig. 12.3: Man-machine control loop.

Decision Communication Channel

The information from a display is passed on to the control mechanism of the brain via the nervous system where it is processed to arrive at a decision in relation to the required performance. This decision is communicated to mechanical leverage system of the human bone and muscular system which makes up the decision communication channel.

Control Communication Channel

A control is any device which regulates the action of a machine. Typical control includes hand wheels, levers, pedals, control knobs, push buttons, etc. The contact between the operator and the machine takes place at two channel only, *i.e.,* display communication channel and control communication channel. A poor design of display and control may cause an error in the system. Hence the displays and controls must be designed ergonomically.

12.8. DISPLAY DESIGN

Displays are necessary extensions to man's senses and provide both prime and supplemental information needed by operators in making decisions and in effecting control responses.

Information presented by displays can be considered dynamic or static. Dynamic information continuously changes or is subjected to change through time *e.g.*, traffic signals, charts or graphs.

The more detailed classification:

1. **Quantitative information:** Display presentation that reflect the approximate value of same variable such as temperature, speed, etc.
2. **Quantitative information:** Display presentations that reflect the approximate value, trend, rate of change, direction of change.
3. **Status information:** Display presentation that indicate the condition or status of the system such as ON-OFF condition, etc.
4. **Warning and signal indicators:** Display the emergency, unsafe condition.
5. **Representational information:** Pictorial or graphic representation.
6. **Identification information.**
7. **Time phased information.**

Types of Displays

1. **Visual Dispalys:** Depending upon the use visual displays classified as:

(*a*) **Quantitative display:** (To read a precise numeric value *e.g.*, display for pressure measurements, display for speed measurement, angle measurement). Conventional quantitative displays are mechanical devices of the following types:

- Fixed scale with moving points
  e.g., pressure gauge, Automobile speedometer.
- Moving scale with fixed pointer
  e.g., weighing machine to measure human weight.
- Digital display
  e.g., digital counters of tape recorder.
- The modern technology has made it possible to present electronically generated features.
  e.g., digital displays
  Types of displays are shown in the Fig. 12.4 (*a*) and Fig.12.4 (*b*).

(*b*) **Qualitative display:** The display is used to read an approximate value or to indicate rate of change, change in direction, etc.
e.g., the increase or decrease in pressure.

(*c*) **Check display:** The display gives information about the parameters whether they are normal.

2. **Auditory displays:** As compared to visual displays, auditory displays can make monitoring performance superior. So these devices are suitable as warning devices. Following are some of the situations in which auditory displays are more suitable:

- When the message is simple and short.
- When the message calls for immediate attention.
- When the receiver moves from one place another.
- When continuously changing in information of some type is presented.

Considerations in Display Design

1. What is the information to be transmitted? What is its purpose of function?
2. What type of display is to be used?
3. Nature of the visual environment in which information is to be transmitted.
4. Detailed design characteristics of the type of the display chosen.

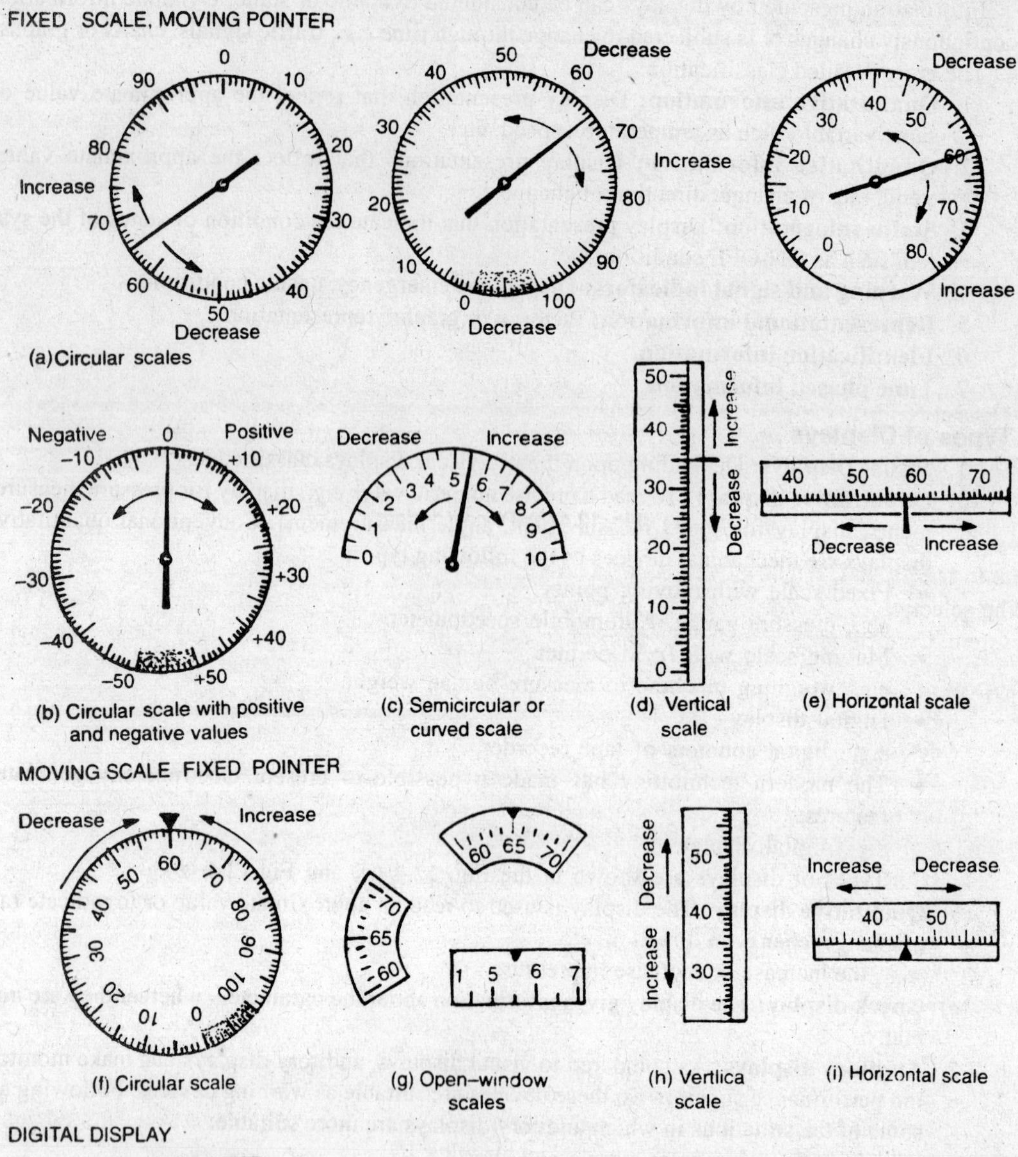

Fig. 12.4 (*a*): Types of displays.

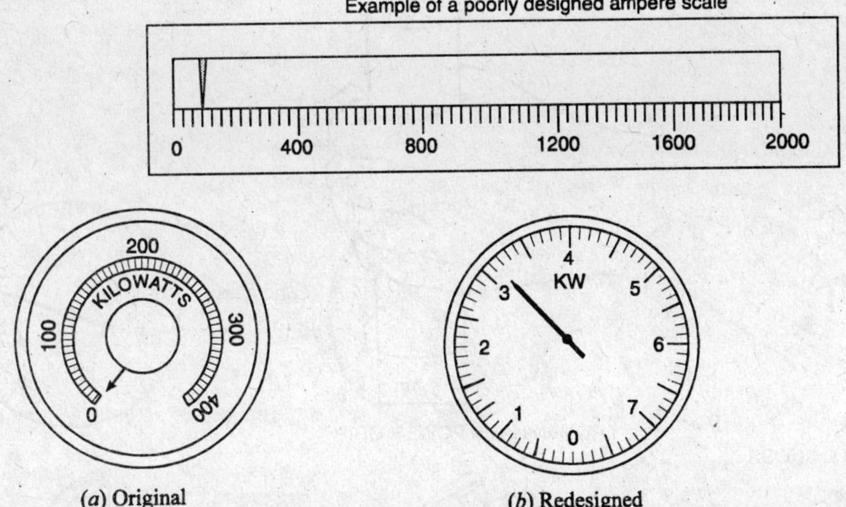

Meter at the right would be easier to read because it is bolder and less cluttered than one at the left. It has fewer graduations and the double arc line has been eliminated. The scale length is increased by placing the markers closer to the perimeter, although this requires that the numerals be placed inside the scale, the clear design and the fact that the numerals are up right probably would partially offset this disadvantage.

Fig. 12.4 (*b*): Dispaly design.

12.9. DESIGN OF CONTROLS

The selection of control should be considered with regard to the functional requirements of the system. Controls are the means by which information is transmitted to the machine from the man.

Types of Control

1. **Hand controls:** The anthropometric data for the human hand can be used as an aid to design dimensions of hand controls.
2. **Hand levers:** Levers give a quick control action and can accommodate large forces. They are not suitable for fine adjustments. Levers can provide efficient ON/OFF or step by step control.
3. **Hand wheels:** Hand wheel provide a controlling torque via both hands and they are used for heavy loads. They can provide good accuracy of adjustments.
4. **Cranks:** Cranks are intended to provide torque via one hand. Smaller cranks are use for fast control.
5. **Knobs:** Rotating knobs are recommended for light loading control with either fingers or with whole hand. Knobs are typically use in applications such as instrument control panels.
6. **Push buttons and toggle switches:** These are essentially used as light load ON/OFF controls and are normally designed for operation by one finger.
7. **Joysticks:** This is a type of hand control now extensively use in computer applications including CNC machines.
8. **Foot pedals:** These are used for fast action control with medium or heavy loading capacity. They lack in accuracy and range which may be obtained with hand controls. These are used in sitting position.

The Fig. 12.5 shows the optimum size for human power grip.

Specific Control Recommendations

The various functional requirements and design features are given below:

1. Push buttons should have positive, snap (click) action for operator feedback and concave or rough surface top to aid in fingering.

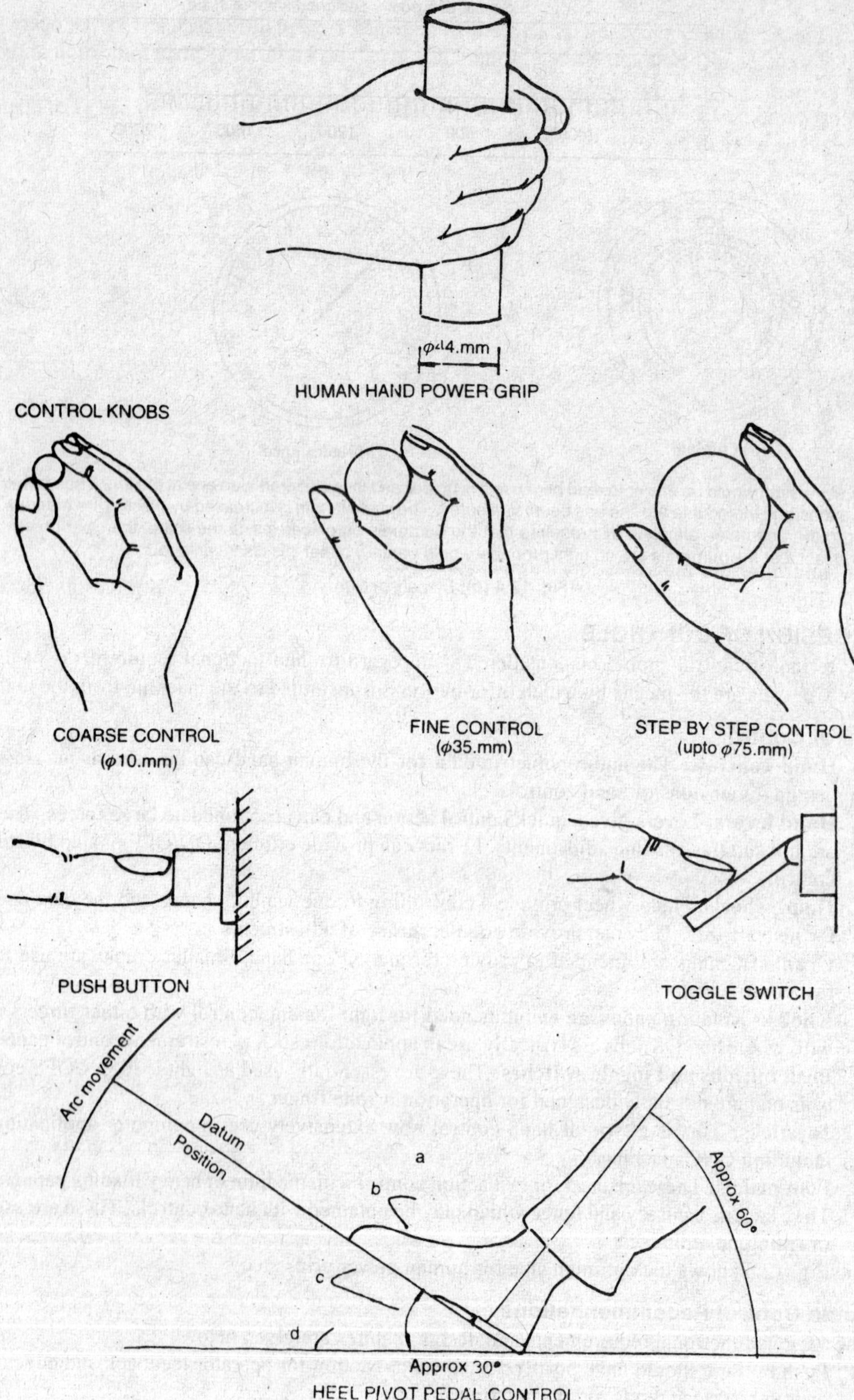

Fig. 12.5: Optimum size for human grip.

2. Toggle switches provide a quick mode of response, require little space and can be operated simultaneously with others in a group. They give both visual and tactual indication of their state.
3. Knobs for continuous control can be shape and colour coded effectively and with appropriate gearing. Provide a considerably flexible choice of adjustments.
4. Levers are commonly used as gear shifts, throttles, or joysticks in positioning or tracking tasks.
5. Cranks are used for high rates of adjustments over long distances position feedback is poor.
6. Steering wheels (hand wheels) are useful where large rotary forces must be applied but they require more space.
7. Pedals have characteristics, similar to hand wheels good for large force applications. They do not permit precision but are useful in distributing the work.

12.10. ENVIRONMENTAL FACTORS

The role that ergonomics plays in the environmental man-machine interface is essentially three fold.

- First, identifying the effects that the environment has on man's physiological and psychological process.
- Second, ensuring that work patterns, equipment and machine interfaces are designed to minimise the individual variation in performance.
- Third, ensuring that all the necessary protective systems are designed to take an account of physiological and psychological variations in man.

The environmental factors that affect the performance are:

Illumination: When human activities are carried out indoors or at night, it is necessary to provide some sort of artificial illumination the type of lighting or illumination depends upon the type of work being performed, the size of the objects, accuracy, speed and duration of the work, etc.

The lighting system should provide

- Sufficient brightness.
- Uniform illumination.
- A contrast between brightness of the job and of background.
- No direct or reflected glare.

Factors Affecting Visual Performance

1. Normal performance depends upon the eye/brain system, system at its optimum efficiency and be able to provide normal vision.
2. Colour blindness.
3. Visual adaptation to light levels is important physiological mechanism.
4. External factors affecting include luminance, contrast, quality and amount of illumination, time of observation. Movement size of objects and glare.

Principles Involved in Improving Visual Performance

1. Since time is required for an eye to adjust, design should avoid the need to identify visual information at widely different distances or in slightly different illumination levels.
2. Minimise the extent of eye movement required to resolve visual information by arrangement of displays rather than by reducing size of the details.

Noise: Noise has been defined as unwanted sound and it has been shown to have both short and long-term effects on human performance. These effects may be internal and physiological in nature, resulting in the auditory system being unable to perceive sound. The amount of having loss is related to the level of the noise to which the operator is exposed and also it depends upon the exposure time for high frequency intensities. Sudden load sound which are entirely unexpected can use an increase in blood pressure, sweating, heart rate respiration and mascular contraction and repeated exposure may affect person's digestion.

Recommendations

1. Where the problem is the talker, speech training may be necessary to correct those aspects that are at fault.
2. Where noise is present and has been identified as a problem it should be reduced at source.
3. Impact equipment and excessively noisy machines and equipment and operations should be isolated by constructing per enclosures that amount of noise transmits beyond the enclosure is reduced.
4. The use of baffles, sound absorbers and acoustical treatment of walls, ceilings and floors can also reduce sound reflection.
5. In server noise situations, workers should be provided with personal protective devices such as earplugs, earmuffs and helmets.

Vibration: Usually vibrations of the air are detected as sound but air vibrations below 20 Hz are not heard but can be felt. Vibrations can affect the performance on target tracking. Ideally vibrations should be minimised at source. Normally protection from residual vibration is achieved by reducing the forces transmitted, by converting vibration energy into thermal energy by using mechanical or hydraulic dampers and by altering body position and body support.

Thermal conditions: (Temperature, Humidity and Air Flow): Poor heat and humid conditions produce thermal stresses in the workers which affect their efficiency, concentration and dexterity of their members of the body. Working temperature of 60-65°F is considered as normal but it varies according to the nature of work. Humidity and heat are related to each other both affect comfort and tolerance of the body to the heat. If humidity is high, evaporation of the sweet is reduced which results in dissipated which results in dryness of mouth, throat and nose. Humidity as a general rule should not be allowed exceed 70 per cent. The effect of heat can be minimised by:

- Shielding, isolating heat sources to reduce direct transmission by radiation of heat between body and the heat source.
- Installation and provision for adequate local ventilation to get rid of smokes, fumes, etc.
- Permit rest pauses in cool, extreme hot conditions.

Ventilation: Ventilation is the process of displacement of state air of the building by fresh air to reduce the presence of bad odour, CO_2 concentration, humidity and temperature.

A good ventilation system provides fresh air.

Most common methods of ventilation are:

- Windows and ventilators provide natural ventilation.
- Exhaust fans extract stale air and creates low pressure area to be filled by fresh air.

12.11. ANTHROPOMETRY

Anthropometry deals with the measurements of the dimensions and certain other physical characteristics of the body such as volumes, centre of gravity, inertial properties and masses of body segments. There are two primary types of body measurements: Static and dynamic (functional). Static dimensions are measurements taken when the body is in a fixed (static) position. These consist of:

- Skeletal dimensions (between dimensions of joints).
- Contour dimensions (skin surface dimensions).

Body measurements vary as a function of age, sex and for different countries. There are differences in anthropometrics of male and female.

The specific body features measured are shown in Fig. 12.6.

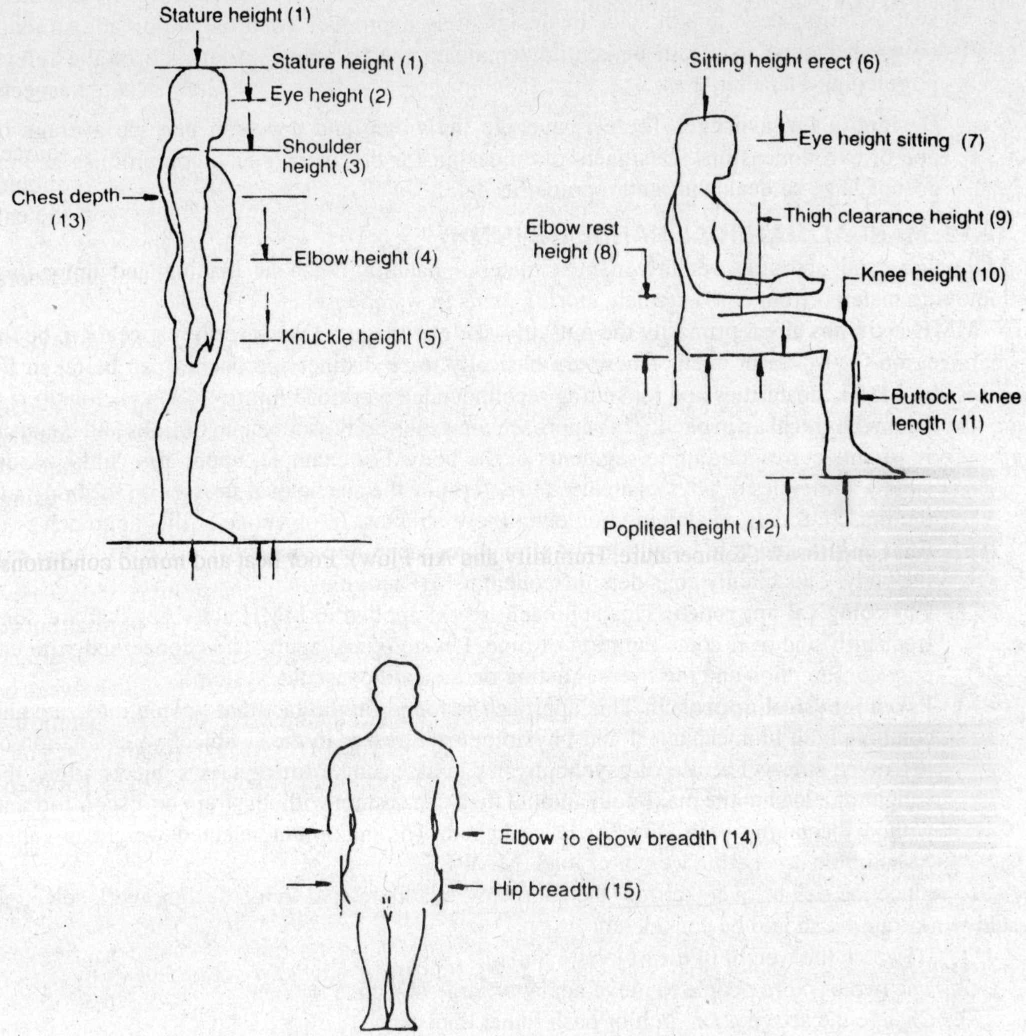

Fig. 12.6: Anthropometric data (specific body features).

Dynamic (functional) dimensions are taken under conditions in which the body is engaged in some physical activity.

Use of Anthropometric Data

It is very much essential in the design when items are designed for specific groups such as adult males, children, etc., the data used should be specific for such groups in the country or culture in question.

Principles in the Application of Anthropometric Data

- **Design for extreme individuals:** Designing for maximum population value is the recommended strategy if a given maximum (high) value of some design feature should accommodate of doors. Designing for minimum population value is an appropriate strategy if a given minimum (low) value of some design feature has to accommodate all.

  Example. Distance of control from the operator and force required to operate the control.

Designing for adjustable range: In the design features of equipment or facilities the provision for adjustment should be there for the individual who use them *e.g.*, automobile seats, chairs, desk height, etc. In design it is a produce to provide for adjustment to cover the range from 5th percentile female to the 95th percentile male of the relevant population characteristics.

Designing for average: There is average individual and a person may be average on one or two dimensions. Designers often design for the average as a compromise as they do not have to deal with anthropometric data.

12.12. MANUAL MATERIAL HANDLING (MMH)

Many jobs and activities require manual material handling such as loading and unloading, removing material from conveyer belt, storing items in warehouse, etc.

MMH activities affect primarily the musculo-skeletal system. Over exertion appears to be the main reason for MMH problem. These are basically three distinct approaches can be taken for assessing MMH capabilities and for setting recommended workload limits.

1. **Biomechanical approach**: This approach views the body as a system of links and connecting joints corresponding to segments of the body. For example, upper arm (link), elbow (joint). Principles of physics are used to determine the mechanical stresses on the body and the muscle forces needed to counteract these stresses. In objective if this approach is to limit task demands to be within the strength capacity and compressive force tolerance of the body. This mainly considers infrequent MMH activities.
2. **Physiological approach**: This approach is best applied to MMH activities that are done frequently and over some duration of time. Physiological approach is concerned with energy consumption and the stresses acting on the cardiovascular system.
3. **Psychophysical approach**: This approach is found on the fact that people integrate and combine both biomechanical and physiological stresses in their subjective evaluation of perceived stress. The use of psychophysics in assessing a lifting task subjects adjust the weight of a load to the maximum amount they can sustain without strain or discomfort and without becoming unusually tired, weakened. The maximum selected weight is called "Maximum acceptable weight of load (MAWL)".

To reduce the risk of overexertion, the following factors related to job design, work selection and work training should be considered:

1. Decrease the weight of the objects handled.
2. Use two or more people to move heavy or large objects.
3. Change the activity, *i.e.*, pull or push rather than carry.
4. Minimise carrying distance.
5. Stack materials below shoulder height.
6. Keep heavy objects at knuckle height.
7. Job rotation to less strenuous jobs.
8. Design containers with handles that can be held close to the body.
9. Select the person best suited by screening tests.
10. Assign the job within his capacity.
11. Train the workers properly in techniques of lifting.
12. Train them to adopt safe practices.

12.13. PHYSIOLOGICAL ASPECTS OF MASCULAR WORK

Physiological aspects of ergonomics can be best studied by work physiology. During muscular work physiological function change from the resting level. Heart rate, blood pressure, cardiac output, respiration, pulmonary ventilation, oxygen intake, CO_2 production, chemical composition of blood and urine, body temperature, rate of perspiration, etc., increase and they come back to resting level when the work stops. The period during which the work continue is called "work cycle" and period during which there is no work (work stops) is called recovery period. By measuring one or more physiological variable during activity it is possible to determine in what degree to working level differs from the resting level. This gives an estimate of physiological stress experienced in performing the given task. When the activity stops it is possible to determine the duration of the recovery period at the end of which the individual has returned to his pre-activity physiological equilibrium.

In order to evaluate the total physiological expenditure, one must consider physiological reactions both during work and during recovery period. A complete work cycle includes physiological cost of work plus the physiological cost of recovery.

Methods of Measurement of Work Output

Basically there are two methods:

1. Direct calorimetry in which the heat produced by the subject is measured directly in a human calorimeter.
2. Indirect calorimetry in which the subjects respiratory exchange (O_2 consumption and CO_2 production) is determined and used as a basis for calculating energy output.

Data required: 1. Oxygen consumption during the work

2. Calorie equivalent of oxygen.

For work, Bicycle ergometer, or tread mill can be used.

12.14. WORKPLACE DESIGN

The ideal design of any workplace should begin with the operator in mind. The design should ensure that the operator will have adequate and comfortable posture that he can see what he must and he can operate his controls in an effective manner. If the workplace is not properly adapted to his dimensions and to his typically human characteristics, he will not be able to perform his work with maximum efficiency.

Most of the production jobs in industry oblige the worker to remain sitting or standing for long periods of time while performing a given series of tasks. The workplace design is influenced by ergonomic considerations. Inadequate design of the workplaces will inhibit the ability of the worker to perform his tasks and may result the injuries, strain or fatigue, a reduction in quality or output, etc.

Determination of workplace requirements will involve an examination of the work elements which constitute the work cycle and an examination of the body measurements, reach and movement capacities of the worker.

1. **Dimensions of working surface:** The normal and maximum working area for an operator is shown in the Fig. 12.7. The circumscribed line defines the maximum reach area for the operator. Any control or object that is to be grasped must be located within this area. It is the greatest distance from which from which small objects can be procured. Large and heavy objects will have to be located even closer to the body.
2. **Dimensions of the working envelope:** Operator not only performs the jobs in horizontal planes, work is often done above the horizontal working plane. This is particularly important in man-machine system when controls are to be manipulated. To determine the location of where controls are to be placed, it is necessary to visualise a complex three-dimensional envelope of the space in front of the operator.
3. **Workplace height:** The correct working height depends on the nature of the task being done. Many manual tasks are performed when the work is at elbow height. If the job requires the perception of fine visual detail, it will be necessary to raise the work above elbow height and bring it closer to the eye.

Sit-Stand and Standing Workplace

In general, a sit-stand workplace is more desirable than either sit or a stand workplace. If a sit-stand workplace is to be suitable for use of the operators, it must be provided with an adjustable height chair and an adjustable foot rest. However, the workplace for a standing job will be greatly improved if it is made adjustable in height. When this is done, the distance from the floor to the top of the working surface, should be variable from 36 to 42 inches for females and 40 to 46 inches for males.

4. **Selection of chairs:** Many production workers spend entire day sitting at a workplace. The

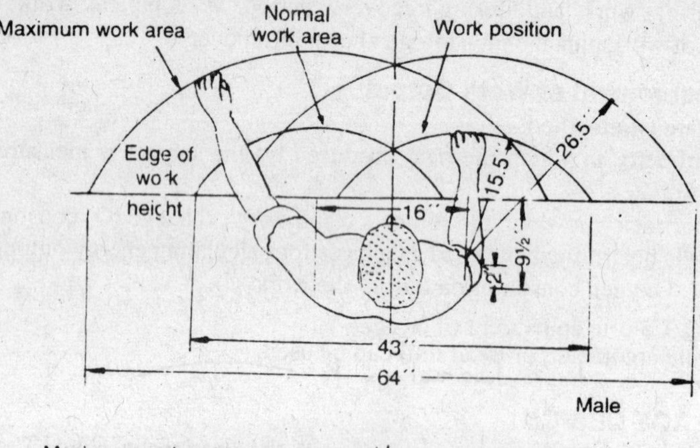

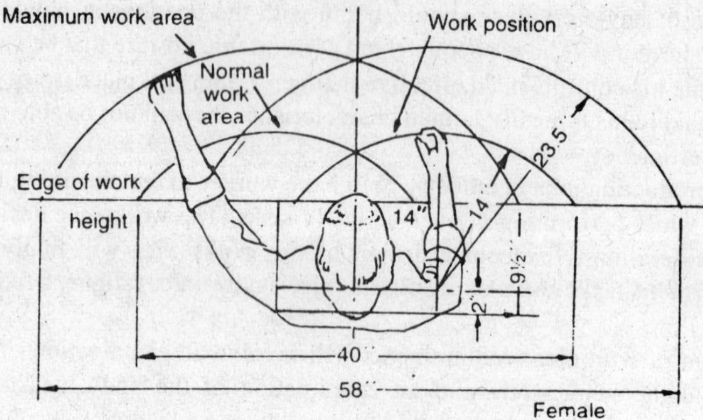

Fig. 12.7: Maximum and normal working area.

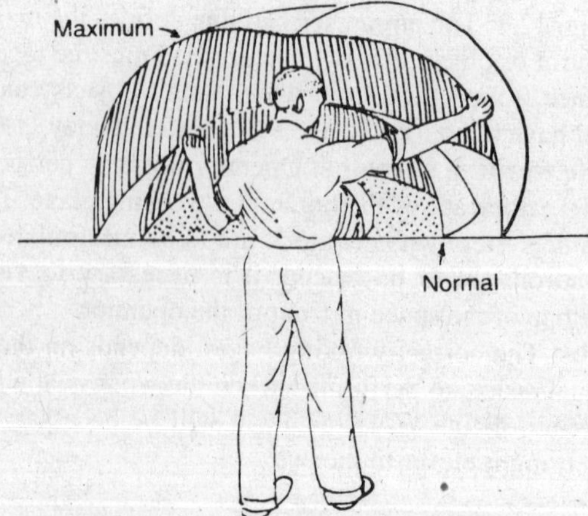

Fig. 12.8: Normal and maximum working space in three dimensions.

chair they are sitting in and along with the foot rest is one of the most important elements of the workplace design.

The Various aspects of dimensions of chairs like the width of the chair, depth of the seat, the back rest dimension should be considered in designing the chair.

The general guidelines for designing the work place are:

(*a*) Materials and tools should be located as far as possible within the normal working area of the operator.

(*b*) The materials and tools should be located in the order which they are used in assembly.

(*c*) Gravity should be employed, wherever possible, to make the raw materials reach the operator and deliver it to next work station.

(*d*) The height of the chair and other dimensions should be designed in such a way as to give maximum comfort to the operator.

(*e*) Foot pedals should be used wherever possible.

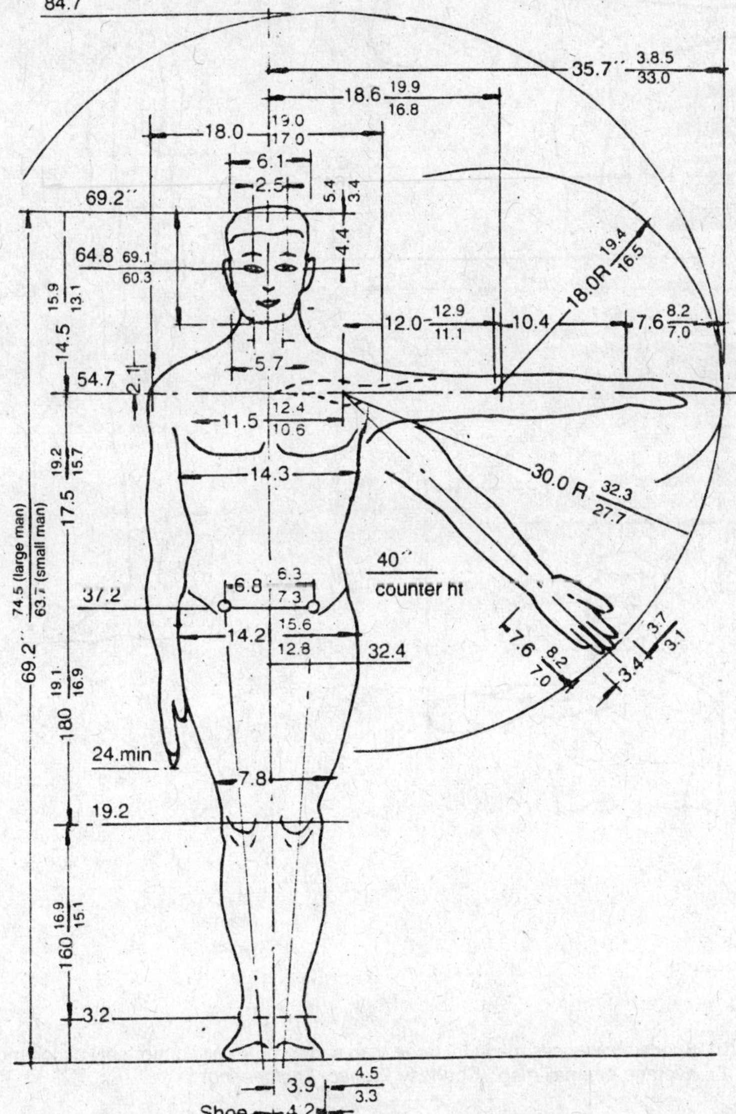

Fig. 12.9: Human dimensions of the average adult male, including the 2-1/2 to 97-1/2 percentile range of the subjects measured. (Courtesy *Product Engineering*)

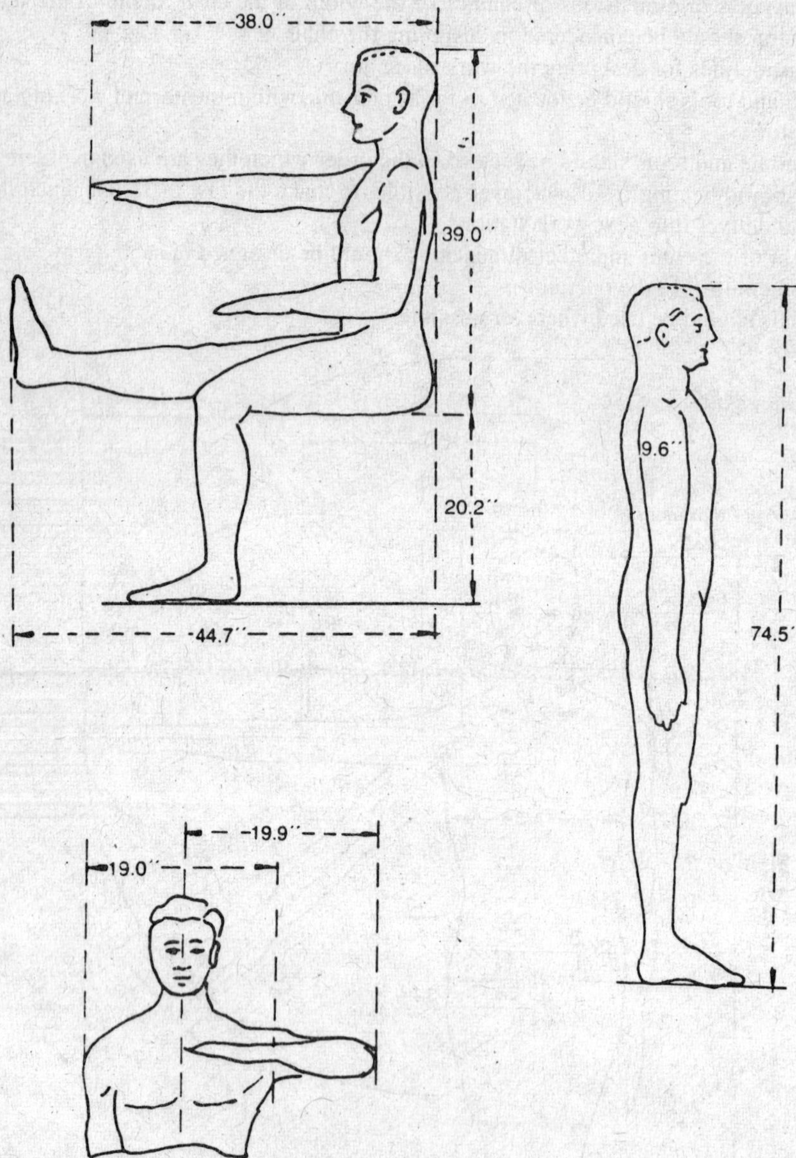

Fig. 12.10: Space allowances made for large man shown here will comfortably accommodate the average or small man. (Courtesy *Product Engineering*)

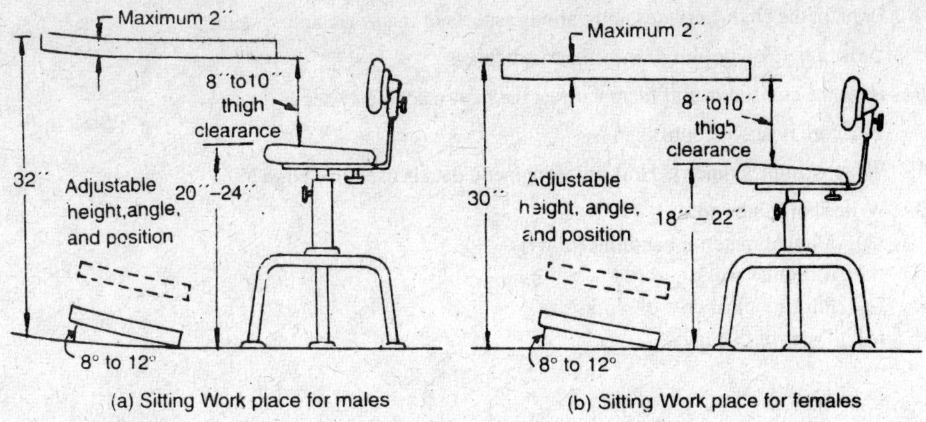

(a) Sitting Work place for males (b) Sitting Work place for females

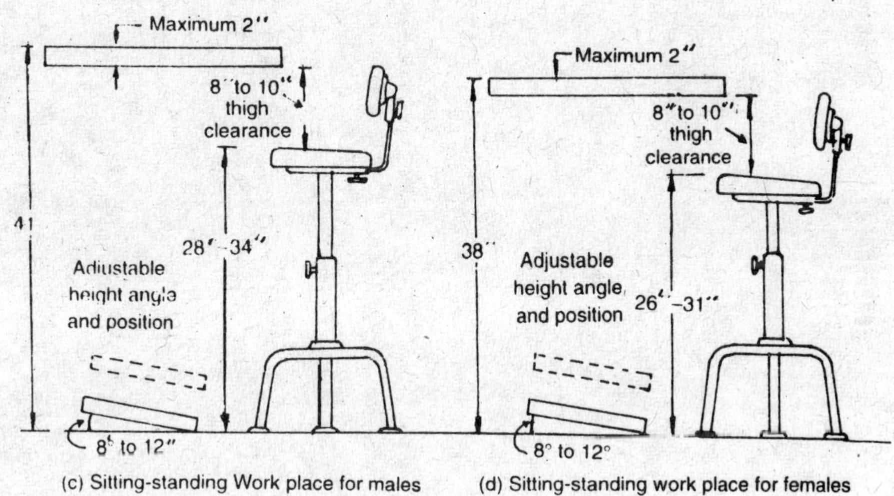

(c) Sitting-standing Work place for males (d) Sitting-standing work place for females

Fig. 12.11: Sitting and sitting-standing workplace.

References for Further Reading

1. Mark, S. Sanders and McCormick, *Human Factors in Engineering and Design*, 7th Edition, McGraw Hill Company, New York, (1992).
2. Maynard, *Industrial Engineering Handbook,* McGraw Hill Company, New York.
3. Barnes, R.M., *Motion and Time Study,* Wiley, New York.
4. Grandjeane, *Fitting task to the Man*, 4th Edition, Tayler and Francis London, (1988).
5. Tiechauer, E.R., *The Biomechanical Basis of Ergonomics,* Wiley, New York, (1978).

REVIEW QUESTIONS

1. Define ergonomics and what are its objectives.

2. Mention the areas of application of ergonomics.
3. What is system? Explain man-machine system.
4. Explain the characteristics and various aspects of man-machine system.
5. Write a note on design of information displays.
6. How the environmental factors affect the man-machine system.
7. What are types of controls in use?
8. Write is anthropometry. How anthropometic data is used in design?
9. Write short notes on:
   (*a*) Manual material handling (MMH),
   (*b*) Muscular work,
   (*c*) Physiological cost of work,
   (*d*) Types of displays.

PART II
PRODUCTION MANAGEMENT

13

INTRODUCTION TO PRODUCTION / OPERATIONS MANAGEMENT

Production/Operations Function • Production system • Objectives of Production Management • History and Development of Production Management •Functions and scope of production department • Production Management Frame work • Types of Production • Classification of Production System • Production Interface with other Functional Areas of Business • Production Interrelation with Sub Functional Areas of Production • Organisation Structure For Production Function

13.1. PRODUCTION/ OPERATIONS FUNCTION

Strategic growth and competitiveness of organizations are depending upon the effective utilization of the critical productive resources of the organization. Production/ Operations Function is concerned with design and control systems responsible for the productive use of raw materials, human resources, equipment and facilities in the development of a product or service. The words production and operations are used synonymously.

Production is a creation of utility. The production function creates utility by providing form, time and place utilities for the produced goods. Manufacturing provides form utility while physical distribution (a service function) provides the time and place utilities. Operation is often defined as a transformation process. In operations management, we try to ensure that the transformation process is transformed efficiently and that the output is of greater value than some of the inputs. Thus, the role of operations is to create value. The transformation process itself can be viewed as a series of activities along the value chain extending from supplier to customer. Any activities that do not add value are wastes (superfluous) and should be eliminated. The input transformation (process) - output process is a characteristic of a wide variety of operation systems. In an automobile factory, steel sheet is formed into different shapes, painted and finished and then assembled with thousands of component parts to produce a working automobile. In an Aluminum factory, various grades of bauxite are mixed, heated and cast into ingots of different sizes. In a hospital, patients are helped to become healthier individuals through special care, meals, medication, laboratory tests and surgical procedures.

Obviously, operations can take many different forms. The transformation process can be

- Physical: as in manufacturing operations.
- Locational: as in transportation or warehouse operations.
- Exchange: as in retail operations.
- Physiological: as in health care.
- Psychological: as in entertainment.
- Informational: as in communication.

Operations management may be defined as a process, which combines and transforms various resources used in the production operations subsystem of the organisation in to value added Products/Services in a controlled manner as per the policies of the organisations. Thus, production/operations functions are a part of an organisation, which is concerned, with the transformation of a range of inputs into required (product and services) outputs having a requisite quality level.

A set of various activities, which are involved in manufacturing certain products, is named as "Production Management". If the same concept is extended to service management, then the set of various management activities are called "Operations Management". In general, the concept of manufacturing products/offering services is called Production/Operations Management.

Activities in Production/operations management include, organising work, selecting process. Arranging layouts, locating facilities, designing jobs, measuring performance, controlling quality, scheduling work, managing inventory and planning production. The managers need good technical, conceptual and behavioral skills. Their activities are closely interrelated with other functional areas of the firm.

13.2. PRODUCTION SYSTEM

The production system (function) of an organisation is that part which produces the organisations products. Production is the basic activity of all organisations and all the other activities revolve around production activity. The output of production is the creation of goods or services, which satisfy the needs of the customer.

In some organisations the product is a physical (tangible) good. E.g. Refrigerators motorcars, Television, tooth paste etc., while in others it is a service (insurance, health care etc.)

The production system has the following characteristics.

1. Production is an Organised activity, so every production system has an objective.
2. The system transforms the various inputs (men, material, machines, information, energy) in to useful Outputs (Goods or Services.)
3. Production system does not operate in isolation from the other organizational systems such as finance, marketing etc.
4. There exists a feedback about the activities, which is essential to control and improve system performance.

Production system is shown in the figure (13.1)

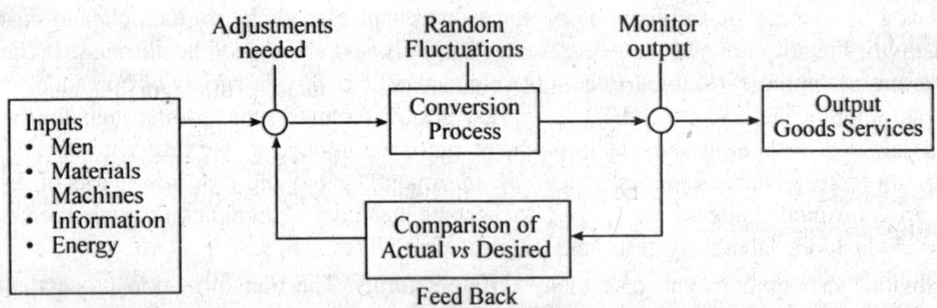

Fig. 13.1: Production system

Production and Production Management

Production is defined as the step-by-step conversion of one form of material in to another form through chemical or mechanical process to create or enhance the utility of the prouuct to the user. Thus economists define production as an activity by which form utility is created or enhanced.

For example, the iron ore exists in the nature. It will be converted in to steel by a chemical process, which is put to various uses like, making bars, pipes, angles, channels etc.

Thus production is a value addition process. At each stage of stage of processing, there will be a value addition.

Elwood Buffa defines production as " a process by which goods and services are created.

Production management is a process of planning, organising directing and controlling the activities of the production function.

Elwood Buffa defines production management as follows :

Production management deals with decision-making related to production processes so that the resulting goods or services are produced according to specifications, in the amounts and by the schedule demanded and out minimum cost.

13.3. OBJECTIVES OF PRODUCTION MANAGEMENT

Production is an Organised activity and each Organised activity has its objective. Which helps to evaluate its performance against the set objectives.

The objective of the production management is stated as:

To produce goods services of right quality and quantity at the predetermined time and pr established cost.

Thus the objective of production management are reflected in

1. Right Quality
2. Right Quantity
3. Predetermined time
4. Pre established cost (Manufacturing cost)

1. Right Quality

The quality of the product is established based upon the customers needs. Customer's needs are translated in to product specifications by the design or engineering department. The manufacturing department then translates these specifications in to measurable objectives.

Thus the cost quality trade off decides the final quality of the product. Thus a proper balance must be obtained such that the product quality offered to the customer should be within the pre-established manufacturing cost.

2. Right Quantity

The manufacturing organisation should produce the products at the right number.

If the products are produced in quantity excess of demand the capital will block up in the form of inventory and if it is produced in quantity short of demand, these will be shortages products. Thus a decision is to be taken regarding how much to produce. (Right quantity)

3. Manufacturing costs

Manufacturing costs are established before the product is actually manufactured. The manufacturing department has to manufacture the products at the pre-established cost in any case, any variation between the actual costs and the standard (pre established) should be kept at minimum.

4. Manufacturing schedule.

Timeliness of delivery (schedule) is one of the important parameter to judge the effectiveness of production department. There are many reasons like non-availability of materials at right time, absenteeism, machine break down etc. Which affect the timely completion of the products. So the manufacturing department should organize its activities in such a way that the products will be manufactured.

To achieve the above objective, the manufacturing production department has to make the optimum utilizations of various inputs like men, material and machine. So to have as better utilization of resources, the production department has to achieve the other objectives, which are lower in the hierarchy.

These objectives are called intermediate objectives are going to optimize the utilization of resources.

INTERMEDIATE OBJECTIVES:

The intermediate objectives can be stated in terms of

1. Machinery and equipments

The objective concerned to these areas is that the machine and equipment should be such tat they should be able to produce the products as per the specifications and accuracy required. The total cost of procurement and running cost should be minimum. Once the machines are procured and put to productive use, and then the next objective is to utilize these resources to the maximum extent.

2. Materials

The materials should be made available when required as per the specifications (shape, size, quality etc) and at the most economical price. The production department should aim at maximum utilization of the material with minimum wastage and scrap.

3. Manpower

Manpower is an important resource or input to production and the success of production depends to a greater degree upon the type of manpower an organisation have.

Thus these should be a perfect matching between the workers & jobs and the manufacturing department climate should be such that the potential skills and energies for the workers should be channalised in to constructive outputs. The objectives are set with respect to productivity per worker. Labour turnover rate, safety and industrial relations etc.

4. Supporting Services

This helps indirectly to achieve the other objectives and adequate provision of the services helps to utilise other inputs effectively. The objectives should be set for each of the services like water steam power, material handling etc.

Thus intermediate objectives are supporting to the primary objectives. The achievement of these objectives helps the company to satisfy the customer needs and increase the market share resulting in increased profitability.

13.4. HISTORY AND DEVELOPMENT OF PRODUCTION MANAGEMENT

Production systems have existed since the earliest days of civilization as evidenced by the pyramids of Egypt, the Great Wall of China etc. The wide spread production of consumer goods and thus production management did not begin until the Industrial Revolution in the 1700s. Prior to that time, the skilled craftsmen and their apprentices fashioned goods for individual customers. The typical production facility has a handful of apprentice workers under supervision of master craftsman cum owner. Its technology was embedded in minds and hands rather than equipment. Product design and production process were united in the person of the owner, materials and quality control, personnel scheduling were done from experience using simple rules. Markets were small and distribution was uncomplicated.

Crafts production, the process of handcrafting product/services for individual customers. Every piece was unique, hand fitted and made entirely by one person.

The availability of coal, iron ore and steam power set into motion a series of industrial inventions that revolutionized the work was performed. Mechanically powered machines replaced the labour as primary factor of production and brought workers to a central location to perform tasks under the direction of a supervisor in a place called 'Factory'. During the same time, Adam Smith's Wealth of Nations (1776) proposed the division of labour in which the production process was broken down into a series of small tasks, each performed by different workers. The specialisation of worker an limited, respective tasks made him an expert on the tasks and further encouraged the development of specialised machinery.

The introduction of interchangeable parts by Eli Whitney (1790s) allowed the manufacturing of firearms; watches, sewing machines and other goods to shift from customised one at a time production to volume production of standardised parts. Thus, the system of measurements and inspection, a standard method of production and supervisors to check the quality of the worker's production had started. Advances in technology continued through the 1800s. Cost accounting and other control systems were developed but the management theory and practice virtually non-existent.

In the beginning of 1900s, Frederick W. Taylor, approached management of work as a science, based on observation, measurement and analysis, he identified the best method of doing each job i.e. the methods are standardised for all workers and economic incentives are established to encourage workers to follow the standards. Taylor's philosophy became famous as "Scientific Management". This ideas was extended by efficiency experts Frank and Lillian Gilbreth and Henry Grant, among other.

Mass Production

American manufactures became adept at mass production over the next five decades and almost dominated manufacturing worldwide. The mass production refers to high volume production of a standard product for a mass market.

The human relations movement of 1930s, led by Mayo Elton and Hawthorne studies, introduced the idea that worker motivation as well as technical aspects of work, affected productivity. Theories of motivation were developed by Herzberg, Maslow, McGregor and others, Quantitative models and techniques by the operations research groups of world war II continued to develop and were applied successfully to manufacturing and services. Computers and automation led another break through in technological advancements applied to operations.

Japanese Production System

During 1970s, Japanese companies became significant players in the world economy, especially in products such as automobiles and consumer electronics. During the 1980s, they started dominating many industries. The robustness of the Japanese economy and their business success has caused operations managers throughout the world to study how Japanese companies were able to develop their production system dominance. Japanese production philosophy is embedded in three major principles or goals.

- Quality comes first
- Improve product and process quality continuously
- Eliminate all forms of waste.

Production/Operations Management - Today

Mass production can produce large volumes of goods quickly, but it cannot adapt very well to changes in demand. Today's consumer market is characterized by Product Proliferation, Shortened Product life cycles, Shortened product development times, changes in technology, Customised products and Segment markets. Mass production doesn't fit the environment of mass production. Using a concept known as just in time, Japanese manufacturers changed the rules of production from mass production to lean production characterized by flexibility and quality. Total quality management has since spread across the globe and is an underlying force for successful operations today. Technology with changing political and economic conditions prompted an era of industrial globalization in which companies competed worldwide for both market access and production resources.

Service Economy and Environmental Awareness

Since 1960, Service industries have grown rapidly compared to manufacturing service operations face almost all the problems confronted in manufacturing. So, operations management has become all the important in service sector.

A new challenge facing operation managers is to make production system environmentally compatible yet efficient. The opportunities for doing this include - reducing the production of harmful by-products, recycling waste materials and energy, recycling of water, green packaging and environmental friendly products.

Operations Management in E-Business Environment

The emergence of the Internet has energized the trend of globalization. Trade that occurs over the internet (or any computer network) is called Electronics Commerce or e-Commerce. E - commerce can take the form of trade between business, between consumers, between business and consumers.

- Business-to-Business (B2B) trade typically involves companies and their suppliers.
- Business to Consumer (B2C) trade can take the form of online retailing.
- Consumer to Business (C2B) transactions reverses the normal flow of trade by having customers post what they want and having business acceptor reject their offers.

Operations management skills in order fulfillment, warehousing, logistics and distribution enable B2C e-commerce. B2B e-commerce plays an important role in streamlining the operations. Function in the areas such as, product design, procurement and supply chain management.

13.5. FUNCTIONS AND SCOPE OF PRODUCTION DEPARTMENT

The activities of production department of an organisation are grouped into two broad categories.

1. The activities that convert the available capital in to physical resources required for production .
2. The activities that convert the physical resources in to saleable goods and services.

In carrying out the above activities, the production department must fulfill the following activities.

A. Production of goods at the right time and in sufficient quantity to meet the demand

B. Production of goods at minimum possible cost.

C. Production of goods of acceptable quality.

Thus, the functions of production personnel are:

- Forecasting the demand for the products and using the forecast to determine the requirements of various factors of production.
- Arranging for the procurement of required factors of production.
- Arranging for the services such as maintenance, store keeping material handling, inspection and quality control etc that would be required to attain the targeted level of production.
- Utilizing effectively the factors of production and service facilities available to produce the product.

Scope of production Management

The objectives of production management are aimed at satisfying the needs of the customers through offering organisations products / services.

The scope of production management can be considered from the point of view of both strategic decisions influencing the production system. And the operation level.

The strategic level decisions are mainly concerned with the design of product and production system. These decisions involve decisions, which have long terms implications.

The strategic level decisions are;

1. New product identification and Design

The success of an organisation depends upon the product mix that it offers to the customer. These exists a demands for the products if the product has good market acceptability. The products should be designed in such way as to meet the expectations of customers. The tools like value analysis should be applied at the design stage to avoid unnecessary cost building up in to the product.

2. Process design and planning.

This involves the appropriate technology for conversion of raw materials in to products. The choice of technology depends upon several factors such as demand, investment capability labour availably degree of automation required.

This is followed by selection of the process of conversion and determining the workstations and the flow of work. At this stage macro level process planning is done

3. Facilities location and layout planning

The facilities location is a strategic decision and facilities once located will not be altered in near feature so due considerations should be given to all the facte4s that affect the location.

4. Design of material handling system

As per the principle of Material handling, the handling should be kept at minimum though it is not possible to avoid handling. The selection of particular flow pattern and material handling equipment is dependent on the distance between the workstations, intensity of flow or traffic and size, shape and nature of materials to be handled.

5. Capacity planning

This decision is concerned with the procurement of fixed assets like plant and machineries. The decision regarding the size of the plant, output etc are decided at this stage

The capacity planning activity is again a function of volume of demand.

The operational level decisions are short-term decisions. These are mainly concerned with planning and control of production activities. The operational level decisions are.

1. Production planning

It is concerned with determining the future course of action regarding production to achieve the organisation objectives.

2. Production control

It is a management technique, which aims to see that the activities are carried out as per the plan. Production control activity is output and to take corrective action if these exists a deviation between actual and standard.

3. The other activities include

Inventory control, maintenance and replacement cost reduction and cost control and work system design.

13.6. PRODUCTION MANAGEMENT FRAME WORK

The division of production management functions in to 5 p's (product, plant, programme, processes and people) will provide useful conceptual framework for the various activities performed by production or operations manager.

The five P's

*** The product**

Product is the link between production and marketing. It is not enough that a customer requires product but the organisation must be capable of producing the product.

As per the product policy of the organisation an agreement is reached between the various functions on the following a aspects of the product,

1. Performance
2. Quality and reliability
3. Aesthetics and ergonomics
4. Quantity and selling price
5. Delivery schedule

To arrive at the above, the external and the internal factors which affect the various aspects such as market needs, existing culture and legal constraints and the environmental demands should be given due consideration. Thus the major policy decision regarding variety of product mix is going to affect the producing system

*** The plant**

The plant accounts for major investment (fixed assets)

The plant should match the needs of the product; market, the worker and the organisation.

The plant is concerned with;

– Design and layout of building and offices

– Reliability, perfect, maintenance of equipments

– Safety of operations

– The financial constraint

Plant layout deals with physical arrangement of plants are machineries within the selected site. The layout should be such that it should allow for smooth movement of men and material with minimum back tracking. The type of the layout is dependent on production type, volume of demands etc.

*** The Process**

These are always number of alternations methods of creating a product. But it is required to select the one best method, which attains the objectives.

In deciding about the process it is necessary to examine the following factors.

1. Available capacity
2. -Manpower skills available
3. -Type of production
4. Layout of plant
5. Safety
6. Maintenance required
7. Manufacturing costs

The programme

The Programme here refers to the timetable of production. Thus the programmes prepares schedules for

1. Purchasing
2. Transforming
3. Maintenance
4. Cash
5. Storage and transport

*** The people**

Production depends upon people. The people vary in their attitudes, skill and expectations from the work. Thus to make best are of available human resource, It is required to have a good match between people and jobs which may lead to job satisfaction. The producing manager should be involved in issues like

- Wages/salary administration
- Conditions of work/safety
- Motivation
- Training of employees
- Training of employees

Thus production management encompasses these 5 P's. The areas of 5 Ps are overlapping.

13.7. TYPES OF PRODUCTION

The production system (facility, equipments and operating methods) that a company use depends upon the type of the product that is offered to the customer and the strategy that it employs to serve it customers. Basically, production systems can be categorized as :

13.7.1. Make to Stock Production

In this system, manufacture stocks the finished goods (products) in inventory for immediate shipment. This system ensures immediate delivery of good quality, reasonably priced, off the shelf standard products. For example, automobile bearings, ready to wear garments, nuts and bolts, motors, televisions etc. Normally, the customer does not accept delay in delivery and the management is required to maintain adequate stock of finished products. Thus system implies the manufacture of products based on a well-known and predictable demand pattern. Operations management focuses entirely on replenishment of inventory, actual customer orders cannot be identified in the production process. The production volume of each sales units tends to be high and customers delivery time is usually determined by the availability of finished goods inventory. The finished goods inventory acts as a buffer against uncertain demand and stock out situations. The main advantage of this system being the short delivery time and the limitations being high costs of inventory and inability to express customer for the design of the product.

Situations for Make to Stock Production are :

1. Fairly constant and predictable demand.
2. Products are few and they are standardised.
3. Shorter delivery time expected by the customers.
4. Products having higher shelf life.

Information needed to make a production plan is as follows;

5. Forecasted demand for the planning period.
6. Starting inventory level.
7. Desired ending inventory level.
8. Any previous orders to be fulfilled (back orders).

Total Production	= Forecasted demand + Back orders
	+ Ending inventory - Opening Inventory

Make to stock items are generally mass consumed and pass through multiple channels before reaching the end user. Most of the data about customers is not known and hence, feed back from distribution channel will act as an important source of information. Demand is also calculated from these channel members and aggregated for production purpose.

This system is characterized by less complex production, process and product standardization and fairly constant production rate. Distribution system is critical and integration of production and distribution is essential to keep a stable flow of products at the point of consumption and should be responsive to any change. As there is no one contact between producer and the customer, distribution system acts as eyes and ears of organisation to support demand forecasting and demand analysis.

13.7.2. Make to Order

Some companies make or manufacture products after the receipt of the firm order from the customer. Here the production activities will be initiated only after the confirmation of the orders and the products are not supplied from the stock and hence the lead time (the time between ordering the product and delivery) is long.

Make to order production system describes a manufacturing facility in which the final product is usually made from parts/ components already designed but may include some custom designed components also. If has many of the base components available along with the engineering designs but the product is not completely specified. The order processing cycle begins when the customer specifies his requirements of the product. The manufacturer also some times assists the customer to prepare product specifications.

Made to order is a demand responsive strategy, and only the product and component designs and some standard raw material and components are held in stock. Examples are, custom tailored clothing, special purpose machinery and product made to customer specifications. Very expensive products are usually made to order.

Situations for make to order productions are;

1. Products are manufactured to customer specifications.
2. Customer can wait till the order is being processed (longer delivery schedule).
3. Product is non standard and expensive to store.
4. When there are several product options available to store.

In make to order production system, there is a direct interaction with customers during all the stages of but it is extensive during engineering phase. Manufacturer quotes delivery schedule and price and there is a discussion among the customer and producer regarding alternatives to reduce cost, reduce time to deliver.

In make to order situations, production schedule changes with changes in customer orders from one period to another. In this system, producers build large capacities in anticipation and capacity utilization is lower as compared to make to stock situation. Capacity requirements planning and shop floor control are critical and distribution is less complicated.

13.7.3. Assemble-To-Order Production System

When number of alternative combinations or options is available to customers as in automobiles, consumer electronics and computers and customer is not ready to wait until product is made, manufacturers produce and stock standard component parts. When the customer places the order, the customer does the assembly from the parts/ components selected. Since the components are manufactured and stocked, the only the time to assemble is needed before delivering product to the customer. The modular parts approach strategy is normally used here. The assemble to order system aims to combine product customization/ variety of make to order system with low cost and shorter lead-time.

13.8. CLASSIFICATION OF PRODUCTION SYSTEM

The production activities are classified according to the volume of production (quantity) and product standardization as follows.

- Job Shop Production
- Batch Intermittent Production
- Continuous Production
- Cellular Production

Table 13.1 : Comparison between Make to Stock, Make to order and Make to Assemble Systems

Particulars	*Make to Stock*	*Make to Order*	*Assemble to Order*
1. Product range	Low	High	Medium/high
2. Production volume	High	Low	Medium
3. Lead time	Low	High	Medium
4. Customer producer interface	Limited / distant	High at sales & design level	High at sales level
5. Handling of fluctuations in demand	safety stock of product units	Planning of excess capacity & raw material stock	Planning of standard modules & parts
6. Basis of planning	End item forecasts	Back logs & marketing intelligence reports	Backlogs and trend analysis
7. Inventory level	High F. G. Inventory & associated inventory carrying costs	Low inventory level & associated carrying cost	Major modules / parts held in inventory
8. Product category and cost	Standardised products with lower price/unit	Special products (high variety) & high cost/unit	Modular parts/ sub assembly medium/ high price/units

13.8.1. Job Shop Production

Job Shop Production/ industries are characterized by manufacturing of one or few quantity of products designed and produced as per the specification of customers within prefixed time and cost. The distinguishing feature of job shop is low volume and high variety of products

Examples of products manufactured by job shop industry include space vehicles, aircraft, machine tools, special purpose machines, tooling and jigs and fixtures, custom clothing, prototypes, large turbo generators, material handling machines, construction equipments and the like.

A job shop typically comprises of general-purpose machines arranged in to different departments. Each job demands an unique technological requirements, demands processing on machines in a certain sequence. Because of high variety, scheduling becomes complicated. Planning for job order involves deciding the order or priority for jobs for the jobs waiting to be processed in order to achieve the desired objectives. Job order type production applies to situations where products are to be manufactured against specific requirements of the customer.

Characteristics of Job Order Production

1. High variety of products and low volume.
2. Use of general purpose machines and facilities.
3. Highly skilled operators who can take up each job as a challenge because of its uniqueness.
4. Frequently changing set ups.
5. Process (or functional) type layouts for arrangement of facilities.
6. Large inventor of materials, tools and parts.
7. Movement of material is long and interrupted.
8. Relative imbalance of work loads of different departments and labour.
9. Functional departmentation exists.
10. Required numerous job instructions.
11. Detailed planning will evolve around sequencing requirements for each product, capacities for each work centre and order priorities, because of this, scheduling is relatively complicated in comparison to repetitive line manufacture.

12. Many products are run throughout the plant and material handling has to be modified and adjusted to suit different types of products.

Advantages of Job Order Production

- Because of general-purpose facilities, variety of products can be processed.
- Operators will become more skilled and competent that each job gives them learning opportunity.
- Utilisation of full potential of operators.
- Opportunity to use innovate ideas and creative methods.

Limitations

1. Higher set up and tooling up cost due to frequent set up changes.
2. High inventory level of raw material and in process and hence higher inventory costs.
3. Production planning is complicated.
4. Highly competent and skilled manpower is demanded.
5. Product cost comes to be high.
6. High cost of material handling and larger space requirement.

13.8.2. Project Industries

The key feature of the project type industries is that the materials, tools, equipment and personnel are brought to the location where the product is to be fabricated. Project is defined as "an endeavor" with a specific objective to be met within the prescribed time and cost limitations and that has been assigned for definition or execution. In engineering industries context, a project is a one at a time major task such as construction, shipbuilding, erection and commissioning of plants and design of new product.

The project industries are characterized by the site type of an organisation (project organisation). It is based on a single product, generally of large size, which takes months or some times years for completion. Each product is unique in the sense that it can be considered as a prototype as the structure is modified in case to meet the requirements of the customer.

The characteristics of project industries are;

1. Project is one-off job, which may not repeat in exactly the same manner.

1. Project has a definite start and finish i.e. it is executed in a time bound schedule.
2. It needs resources and skills of diversified nature.
3. It has a definite and definable goals or end results that can be defined in terms of costs, schedule and performance requirements.
4. Project passes through distinct activities, which constitute Project Life Cycle

Planning and scheduling of activities and procurement function are critical to the success of the project. Inventory is mainly concentrated in product itself and material always flow in the same direction. Reducing inventory simply means shortening the delivery time and is equivalent of using equipment to the maximum of its capacity. The objectives of manufacturing, namely minimizing delivery lead time (lead time refers to between start and end of the project), minimizing inventory investment and maximizing equipment utilization which usually are considered conflicting will coincide in this case.

The Programme Evaluation and Review Technique (PERT) and Critical Path Method (CPM) are typical of the scheduling techniques developed for project execution. The key variable in the project "time" is controlled more effectively using these PERT and CPM techniques.

13.8.3 Intermittent (Batch) Production

Intermittent production is defined by APICS (American Production and Inventory Control Society) as a form of manufacturing in which the job pass through the functional departments in lost or batches and each lot may have a different routing.

Batch production is characterized by the manufacture of limited number of products produced at regular intervals and stocked awaiting sales. Material tends to bė more complex. The facilities can be used for producing another batch of product between two successive productions runs. Batch production is represented as shown in fig (13.2).

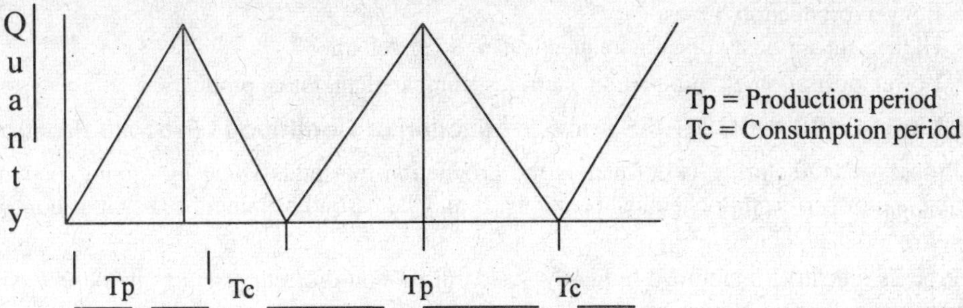

Fig 13.2. Batch production

Batch production is justified when the production rate exceeds the demand rate. The quantities in batch production are decided based on the balancing of two costs i.e. set up cost and inventory carrying cost.

Batch production aims at satisfying the continuous customer demand for an item. However, the plant is capable of production rate that exceeds demand rate. The shop produces the products to build an inventory and it changes over to other products. When the stock of the first item becomes depleted, the production is repeated to build up the inventory again. The intermittent production caters to make to stock, assemble to order and also make to order manufacturing environments.

Manufacturing equipment machines used in batch production is general-purpose machines, which are designed for higher production rate. For examples, Turret lathes capable of holding several cutting tools are used rather than engine lathes. The machine tools used in batch production are usually combined with specially designed jigs and fixtures. Which increase the production rate. Batch production plants include machine shops, foundries, plastic moulding units and press shops. This also includes some kinds of chemical and pharmaceutical units. Batch quantities may very from few units to many thousands of units.

Characteristics of Batch Production

i. Shorter production runs.

ii. Plant and machinery set up is used for the production of items in a batch and set up required to be changed for processing next batch of items.

iii. More number of set ups and hence higher set up cost.

iv. Amount of supervision required is less compared to job order.

v. Plant and machinery are flexible.

vi. Manufacturing lead time and also cost are lower as compared to job order production.

vii. Higher level of work in process inventory.

Advantages of Batch Production

- Bettcr utilisation of plant and machinery.
- Promotes functional specialisation.
- Cost per unit is lower as compared to job order production.
- Lower investment in plant and machinery.
- Flexibility to accommodate and process number of products.

- Job satisfaction exists for operators.

Limitations of Batch Production

- Material handling is complex because of irregular and longer flows.
- Production planning and control is complex.
- Work in process inventory is higher compared to continuous production.
- Longer production times.
- Higher set up costs due to frequent changes in set up.
- Lower utilisation of production facilities compared to mass production.

13.8.3. PROCESS INDUSTRIES (Flow Production or Continuous Process Production)

Process manufacturing is defined as the production that adds value by mixing, separating, forming and/or performing chemical reactions. It may be carried out in either batch or continuous mode.

Process industries manufacture highly standardised non-discrete products in extremely large volumes using a continuous process. Products that flow into continuous streams fall in this category. Process industries develop extensive long range resource requirement and they are more capital intensive. The capital budgeting decision is crucial to financial performance in process industries. Plant location, plant capacity, long range plan for materials, manpower, energy and waste disposal are the important aspects in process industries. Continuous process scheduling will be capacity controlled and no other shop floor control is required.

Maintenance planning is proved to be critical parameter, which affects the performance of process industries to the larger extent. Flow manufactures tend to sell high volumes and product differentiation is not very obvious. Thus, the marketing strategy focuses on product availability and price. Usually the output (product) from process industries have few design changes, low unit values and relatively high transportation costs. Using high volume, dedicated equipment arranged in production lines, minimizes cost per unit.

In flow production, the production process generally follows a specific and fixed sequence of operations (routing). High automation is possible with sophisticated controls for processes. Material handling can also be fully automated. Today, many process industries run unmanned.

Characteristics of Process Industry (Flow Production)

1. Dedicated plant and equipments with zero flexibility.
2. Material handling is fully automated.
3. Wet or dry product flow measurable by weight or volume.
4. Shorter lead-times.
5. Component materials cannot be readily identified with final product and the product cannot be disassembled.
6. The end product is not naturally divisible and hence an additional step such as bottling or canning is needed to get the products into saleable units.
7. Larger investment and usually capital-intensive units.
8. Process follows a predetermined sequence of operations.
9. Maintenance is an important aspect.
10. Unit cost is lower due to high volume.
11. Planning and scheduling is a routine action.
12. Persons with (semiskilled) limited skills, can be used on the production line.

Special features of process industries include;

- Variations in raw materials quality due to different sources of supply.
- Processing technology variations.
- Variations in the yields of output.

- Demand for by-products and joint products.
- Shelf life of raw materials.
- Product differentiation is limited.
- Plant availability depends on preventive maintenance system.

In flow type or process industries products are typically bulk commodities, which may be packaged to order. Steel, ores, cloth pharmacy products, gases, chemicals, petroleum products, cement, rubber, paint, paper processing etc. are some of examples of process production.

13.8.4. Mass Production

Manufacture of discrete parts or assemblies using a continuous process are call Mass production or Repetitive production. When the volumes are very large, a fixed assignment of resources, which otherwise would be risky, is justified. The machineries are arranged in a line or product layout. Specialised departments will disappear and their place is taken by assembly lines. Material handling systems can automated, thus bringing down the cost of material handling per unit. Automobile assembly line is a typically example of mass production. Product and process standardization exists and typically, all outputs follow the same path. Standardization provides for known and fixed through put time, giving managers easier control of the system and more reliable delivery dates.

Characteristics of Mass Production

1. Standardization of product and process sequence and hence line layout (product layout) is recommended.
2. Dedicated, special purpose machines (SPM) having higher production capabilities and output rates.
3. Large volumes of products.
4. Shorter cycle time of production.
5. Lower in process inventory.
6. Perfectly balanced production lines.
7. Flow of materials, components and parts is continuous and without any back tracking.
8. Production planning and control is easy.
9. Extent of supervision required is less.
10. Material handling can be fully automated.
11. Because of high volume, cost per unit is low.

Advantages of Mass Production

12. Higher rate of production with reduced cycle time.
13. Higher capacity utilization due to line balancing.
14. Less skilled operator can man the process.
15. Low in process inventory.
16. Production cost per unit will come down due to economics of scale.

Limitations of Mass Production

17. breakdown of one machine will stop an entire production line.
18. Line layout needs major adjustments/changes with the changes in the product design.
19. High investment in production facilities.
20. Supervision is general rather than specific.
21. Work for operators is monotonous without much challenge.
22. The cycle time is determined by the slowest operation.

13.8.5. Mass Production with Process Layout

This type of production is called "Quantity Production" . This system involves the mass production of single parts on fairly standard machine tools such as automatic screw machines, injection moulding machines, and punch presses. These standard machines are adapted to the production of the particular part by means of special tools designed such as die sets, moulds and form cutting tools. The production equipment is fully dedicated to meet the demand of the product. Examples of items in this type production include components for assembled products that have high demand rate e.g. automobile components/ subassembly, plastic moulded parts, nuts, screws, light bulbs etc.

13.8.6. Mass Production with Product Layout

It is referred to as "Continuous Production". Here the production facilities are arranged as per the sequence of production operations from first operation to the finished product. Here the items are made to flow through the sequence of operations through material handling devices such as conveyors, transfer devices etc. Examples of continuous production include automated transfer machines for production of complex discrete parts and manual assembly lines for assembly of complex products. The production line is dedicated to the product.

13.8.7. Cellular Production

The cellular production system is based on group technology, which seeks to achieve superior performance (efficiency) by exploiting similarities inherent in the parts. In cellular production, groups of parts that have a similar processing requirements are grouped into part family. After the parts are divided into families, a cells created that includes all the equipments, facilities and human skills required produce a part family.

In cellular manufacturing, a team is completely responsible for organising work within each cell. Based on the due date, the team members schedule the work.

Cellular manufacturing combines the advantages of job shop to obtain high variety possible with job and the reduced costs and response times available with mass production.

Cellular processes are most commonly used as substitutes for job shop processes that need-increased productivity. Increasingly, however they are being used in place of flow processes to obtain greater flexibility. They are also becoming a popular way to organise service operations.

Advantages of Cellular Production

1. Reduced material handling and transportation because workstations are spatially close and often are operated as repetitive flow processes.
2. Set up times are reduced because the jobs processed at the same cell often have similar characteristics that require less change over from job to job.
3. Reduction in throughput time because of reduction in waiting times between production stages and also waiting for transportation.
4. Lower in process inventories because of more efficient scheduling.
5. Less space is needed because the machines in cells are located close together and less storage of in process inventory.
6. Simplified shop floor control and reduction in defects due to increased accountability.
7. Lower total investment due to higher productivity and efficiency.

Limitations of Cellular Production

1. breakdown of a single machine, halts production of entire cell.
2. In situations like lowering of demand for family, breaking up the cell and redistribution of equipments may be necessary.
3. Implementation of this system requires a considerable amount of work and expertise to characterise and classify the products and then design appropriate work cells.

Table 13.2 : Comparison between Various Types of Production

Characteristics	*Job order production*	*Batch Production*	*Mass Production*
1. Volume of production (quantity)	One or few jobs	Limited number of small lots	Large quantity
2. Product variety	Larger variety	Medium or few variety	One or few standard products
3. Layout	Process or functional layout	Process or functional layout	Product or line layout
4. Set up time	High	High and frequent	Low
5. Manufacturing cycle time	Large	Medium	Low
6. Material flow	Discontinuous, non-uniform, travel long distances.	Discontinuous	Uniform and uninterrupted flow
7. Equipment & machinery	General purpose	General purpose with high production rate	Special purpose and dedicated
8. Flexibility	High	High	Very low
9. Production planning & Control	Complex	Complex	Simple & routine
10. Work in Process inventory	High	Medium	Low
11. Cost per unit	High	Medium	Low
12. Skill of labour	Highly skilled	Skilled	Semiskilled or unskilled
13. Investment	Low	Medium	Highly capital intensive
14. Material handling	Manual	Manual or semi automatic	Automated
15. Plant utilisation & productivity	Low	Medium	High

13.9. PRODUCTION INTERFACE WITH OTHER FUNCTIONAL AREAS OF BUSINESS

Production systems are one of the subsystems of larger organisational system, which has many other subsystems like marketing, finance, personal etc. Each organisational sub-systems are independent at the same time they are interdependent. They are independent in the sense that each functional subsystem has its own objectives and goals and at the same time they are functionally related with the other subsystems of the organisation. The goals and objectives of the subsystems should be supportive and they should help to achieve the goals of the organisation. The conflicts or mismatch between the goals of subsystems will work contrary towards the achievement of organisational goals and results in sub-optimisation of the goals. The typical organisational system, which contains the various subsystems, is shown in fig (13.3).

The organisational system shown has an objective and various elements that constitute the organisational systems are finance, production, marketing and personal subsystem, are;

- Production : To manufacture the products of right quality and right quality at predetermined cost at pre-established time.

Production department aims to offer the customers which products and or services that satisfy the needs of the customer at the affordable cost, and at the same time enhancing the production efficiency (i.e. productivity).

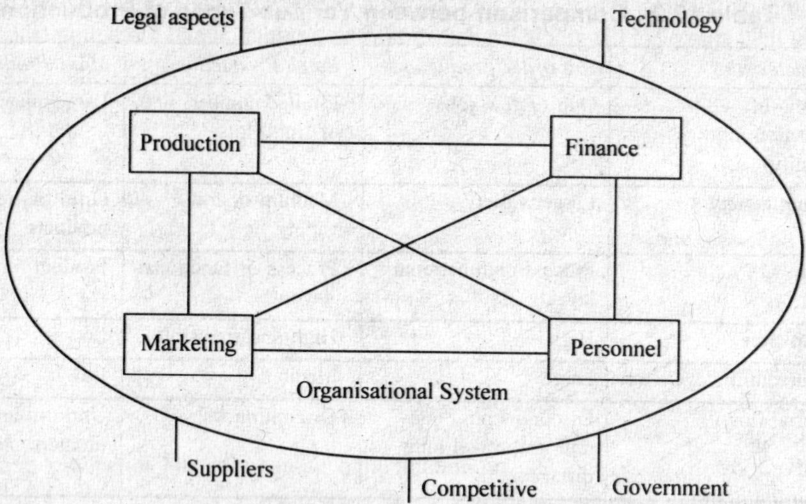

Fig 13.3 : Subsystems of organisational system

- Marketing : To create the demand for the company's products and/or services and satisfy the needs of the customer through company's products through various activities like market research, marketing planning, sales administration and advertising.
- Finance: To plan and allocate the finance to various activities of the organisation and to meet the long term and short-term financial requirements of the enterprise. The activities include financial planning, budgets, general and cost accounting etc.
- Personnel : The objective of the personnel function is to match the jobs and skills of the personnel and create a harmonious climate where in each and every individual in the organisation contributes positively towards the achievement of organizational objectives. The functions include - Recruitment, placement, compensation, promotion and training.

13.9.1. Interface between Production and Marketing

The production function is that which covers the analysis, supply and transformation of the facilities. The product identification and forecasting demand and the distribution are closely

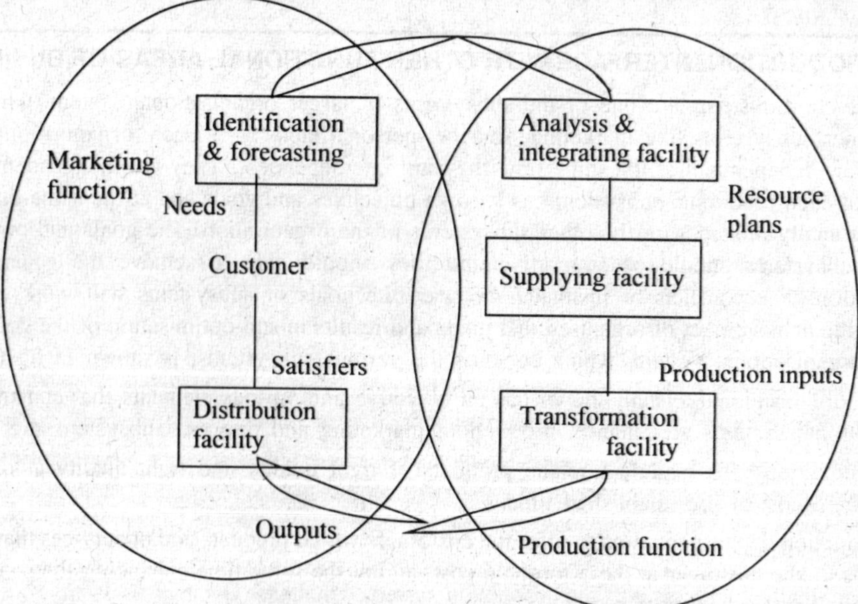

Fig 13.4 : Production and marketing tasks

linked to the marketing. The relationship between these two major functions is crucial to the success i.e. the survival of the whole organisation. The production and marketing tasks are shown in fig 13.4.

There is a strong interdependency between the two departments. The production department seeks the following information from marketing department.

1. Needs of the customer with respect to the company's products and services.
2. The demand for the products and the likely market trend for the future period.
3. The special features required by customer regarding the products, (feedback regarding the customer acceptability, performance, change in tastes and preferences etc.)
4. The delivery requirements of the products from production as per the customer requirements.

The marketing department requires the following information from production department.

5. Production status of the products.
6. Performance characteristics of products.
7. Delivery schedules as per the production plan based upon the forecasted demand.
8. Product features and specifications.

Most of the time production and marketing objectives are conflicting. Marketing department in order to be efficient and to achieve it goals, require variety of products and they want all the products to be on the production line (and or in ready stock) so that they will be able to achieve maximum sales through satisfying needs of the customer.

Contrary to this, the manufacturing efficiency is maximum when there is a minimum variety and same product is manufactured continuously (minimum set ups). This will reduce the set up cost and cost per unit will be minimum in case of continuous production.

Thus, a compromise is to be struck between two conflicting objectives so as to optimize the overall objectives of the organisation. Thus based upon the feedback received from the marketing, production depart should plan, organize and execute production activities which helps to achieve the objectives of the organisation.

13.9.2. Production and Finance

The finance is bloodline of business and it is the finance, which sets the wheels of production rolling. The production department has to invest in physical facilities, requires raw materials and component parts, have to pay wages and salaries and pay for utilities. Thus, the finance department has to make provisions for both long term and short-term requirements of funds to make the smooth running of production.

The production department has to furnish the detailed production budgets to finance department so that the funds will be released as per the plan.

13.9.3. Production and Personnel

The success of the production programmed depends upon the quality, attitude and skills of the people. The personnel department has its role throughout the organisation. The responsibility of matching the job and the person lies with personnel department.

The personnel department has to keep records of the development of workers, identify their training needs, manpower utilization etc.

It is only through skilled , committed and loyal workers the production objectives can be fulfilled. Thus in order to achieve production goals in particular and organizational goals in general, personnel department has to channelise the skills and efforts of the workforce into constructive outlets to achieve the set objectives.

13.10. PRODUCTION INTERRELATION WITH SUB FUNCTIONAL AREAS OF PRODUCTION

The production efficiency of the production system is stated in terms of its ability to produce the products with required quality and quantity at pre determined cost and pre-established time. This goals of production can be achieved through

- Automated production systems
- Flexible/human cantered production (Mass customisations)
- High value added production
- Defect free production
- Manufacturing for customer satisfaction.

Production efficiency or productivity depends on the various sub functions of production such as process design and process planning, product design, production planning and quality and maintenance functions.

13.10.1. Production and Materials Management

The production programme (planning) based upon the forecasted demand spells out the requirement of materials (quantity and time of requirements). The success of production depends upon the smooth flow of materials between the various workstations. The materials function to a greater extent contributes to the success of the production function by making available the materials and tools at the quantity required and at the required time.

The materials functions that influence production functions are;

- Selection of sources of supply (Vendor selection) this directly affects the quality of materials.
- Vendor rating.
- The purchase procedure.
- Inventory control.

Thus, the failure on the part of materials department in supplying materials in right quantity and right time, will lead to production interruptions or delays which may result in non availability of finished goods at the time required by the customers resulting in loss of sales. The excess storage of materials will make the inventory costs to rise, this the whole manufacturing economics will be disturbed. By using an inventory control systems a balance is struck and the materials will be made available in right quality and quantity at right time to increase production efficiency.

13.10.2. Production and Maintenance

There is a direct relationship between production and maintenance function. The efficiency of the maintenance function enhances the production efficiency.

A good or effective maintenance is characterised by;

- Less breakdowns and hence less down time.
- Increase in equipment/machine availability.
- No stoppage of production.
- Higher utilisation of machines/equipments.

Effective maintenance reduce the break downs and hence less production interruption and stoppages. Higher utilisation of facilities indicates higher production rates. Reduction of breakdowns keeps the delivery schedule promises. The modern maintenance practices like condition monitoring and total productive maintenance (TPM) will more effective in terms of reducing different types of losses such as;

- **Down Time**
  1. Equipment failure from break downs.
  2. Set up and adjustments.
- **Speed Losses**
  1. Idling and minor stoppages.
  2. Reduced speed.

• **Defect**

1. Process defects.
2. Reduced yields.

Thus, the effective maintenance system ensures the higher utilisation and availability of production machines and hence supports the production objective.

13.10.3 Production and Production Planning and Control (PPC)

Production planning and control can be defined as "the direction and co-ordination of firms resources towards attaining the prefixed goals". PPC helps to achieve uninterrupted flow of materials through production line by making available the materials at right time in required quantity. The production planning deals with activities such as process selection, process planning, loading, scheduling and sequencing. Production plan prepares the timetable of production. Manufacturing or production carries out the production as per the plan given by planning department of PPC. Every thing will not be perfect and predictable. There are many contingencies/obstacles like non availability of materials, unauthorized absenteeism machine breakdown rush orders etc. will try to deviate the output from the plan.

The control function will be so designed as to sense these deviations between planned and actual production. Thus, the production functions effectiveness depends on the accuracy of production planning and control function.

13.10.4 Production and Research & Development

Research and Development department develops new products/services and improves the design of existing products in order to enhance their functional utility, customer appeal. Design department's output will be in the form of drawings. Bill of materials specifications etc. i.e. they are still on the paper.

Actual manufacturing translates the design into physical products, which meets the customer requirements for which it is designed. A success of the product in the market i.e. customer acceptability depends upon the accuracy with which the product specifications are transformed into a physical product.

The designer cannot sit in isolation and design the product. The interaction between the various functions is a must for design to be floor proof. The manufacturing aspects of product will be considered at the design stage itself. The concepts like design for production and Design For Manufacturing and Assembly (AFMA) consider the manufacturing aspects at the design stage. Thus, the manufacturing will be able to produce goods/services if the ease with which the product is manufacture is considered at the design stage. Thus, the co-ordination between the production, materials management, marketing and engineering department is a must and they should work as a team.

13.11. ORGANISATION STRUCTURE FOR PRODUCTION FUNCTION

Organising function of the production management brings together human and physical resources in an orderly manner and arranges them in a co-ordinated pattern to accomplish planned objectives. Organisation structure gives the details of positional hierarchy duties and responsibilities and authority. A typical organisational structure is represented in fig (13.4).

The managing director (or CEO) is the overall head of the organisation that manages all the affairs of the organisation. Each functional heads i.e. vice presidents are going to look after the overall functions of respective functions and depending upon the size and type of the industry, product management style, the sub functional areas are decided and various positions are created in the structure.

Typical, the sub functional areas in operations or manufacturing include :

• Materials management

- Engineering department (R & D)
- Quality management
- Production.

The production sub functional areas include;

- Maintenance
- Production, planning and control (PPC)
- Industrial engineering

The number of tiers or layers in the management structure, number of positions, number in each positions will be depend upon the many factors. For effectiveness, the organisation structure should be flexible and adaptive.

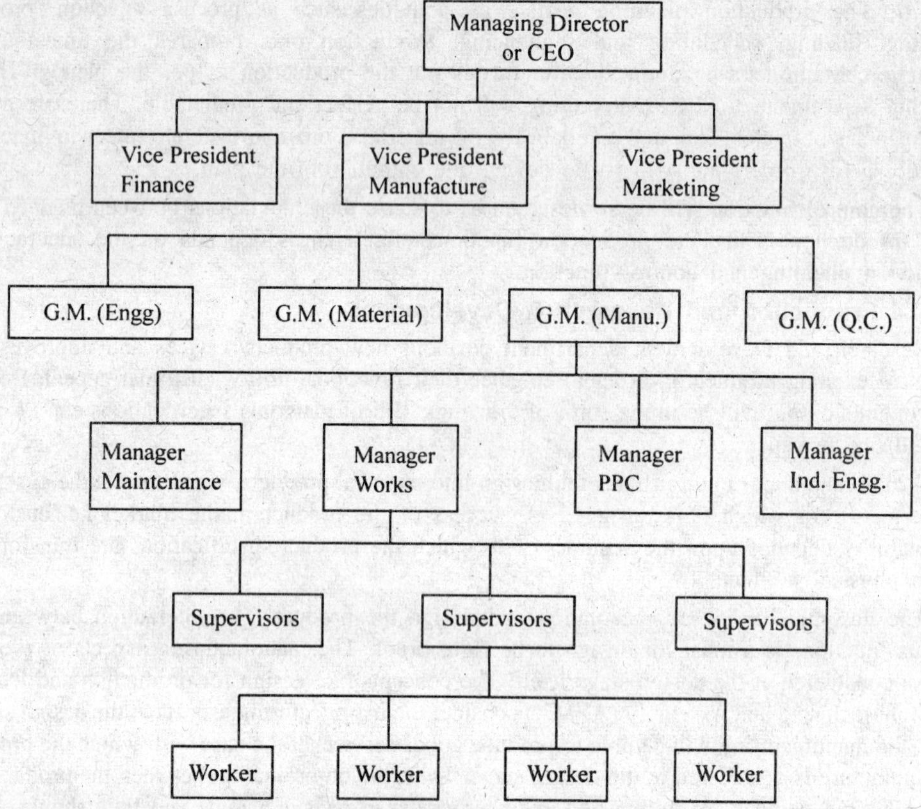

Fig 13.5 : A typical organizational structure for production functional area

References For Further Reading

1. Pr Alan Muhlemann And Others "Production And Operation Management" 6th Edn ELBS Pitman Publishing (1994)
2. Ray Wild "Production And Operations Management Holt, Rinehart And Winston London(1980)
3. Adam E And Ebert R "Production An Operations Management 5 Th Edition Prentice Hall Of India New Delhi (1993)
4. James Riggs "Production Systems Planning Analysis & Control 4th Edition Wiley New York.

Review Questions

1. "Production is a value addition process". Comment with an example?
2. "Production management is process of planning, organising and controlling the activities of production function". Explain.
3. Discuss the scope and objectives of production management.

4. Explain the frame work for production management.
5. Explain the factors that affect the performance of production function.
6. Explain the salient features, characteristics and when to use the following types of production system.
   (*i*) Make to order
   (*ii*) Make to stock
   (*iii*) Make to assemble.
7. Distinguish between "Make to stock" and "Make to order" production systems.
8. Explain the various functional areas of business.
9. Explain the various interrelationship between the following functional areas
   (*i*) Production and Marketing
   (*ii*) Production and Finance
   (*iii*) Production and Personnel.
10. Explain the sub functional areas of production function. Explain the objectives and functions of each.
11. Distinguish between job order production and mass production.
12. Give the classification of production system (Types of production) based on quantity (or volume) of production.
13. Give the characteristic features of the following types of production.
   (*i*) Job order and project type.
   (*ii*) Batch production
   (*iii*) Mass production and flow production
   (*iv*) Cellular manufacturing system.
14. Give the typical organizational structure for production management.

14

NEW PRODUCT DESIGN

• Introduction and need for new product • Product life-cycle (PLC) • Product policy of the organisation • Selection of a profitable product • Product design process • Product analysis • Market analysis • Product characteristics—Functional, operational, durability and dependability, aesthetic and economic aspects of product design • Standardisation • Production aspects of product design.

14.1. INTRODUCTION

Every organisation has to design, develop and introduce new products as a survival and growth strategy. Organisations objective of achieving growth of business is only through introduction of new products. Organisation's are required to design the new products for the following reasons:

- To be in business for a long time believing the fact that business is a long lasting institution.
- To satisfy unfulfilled needs of the customers.
- The company's existing product line becomes saturated and the sales is on the decline.
- To enter into new prospective businesses through diversification (related or unrelated).
- Too much competition in the existing product line.
- The profit margin is on the decline.

14.2. PRODUCT LIFE-CYCLE

The product once introduced into the market will undergo definite phases. The various phases of life-cycle of a product are represented in Fig. 14.1.

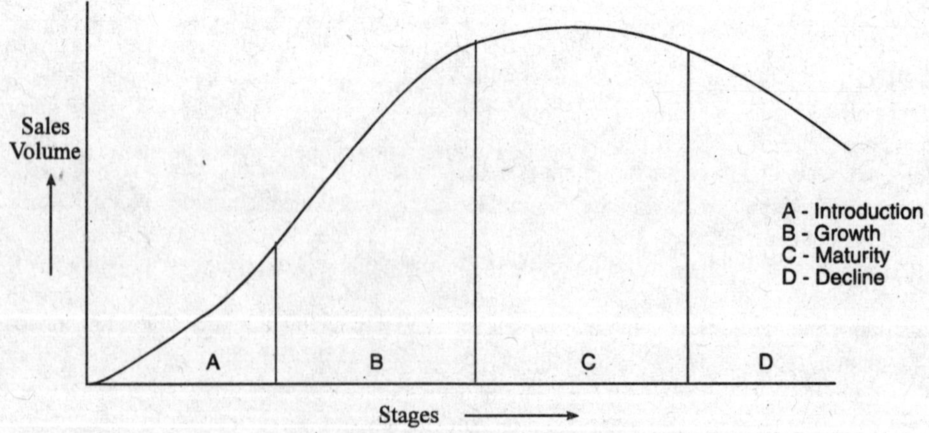

Fig. 14.1: Product life-cycle.

Characteristics of Phases in Product Life-cycle (PLC)

The demand for a product generally tends to follow a predictable pattern called product life-cycle (PLC). Products go through a series of stages beginning with start-up or introduction of product followed by rapid growth, maturity or saturation and finally the decline of demand. The time spans of stages of these products vary considerably across industries. These time spans vary from few weeks or months (for novelty and fashion goods) to years.

Introduction stage: This stage marks the introduction of the product into the market. It may be an entirely new product in the market or old product to the new market. The demand is low as customers do not know much about the product. So the organisations have to invest heavily in advertisement to make the product familiar to the customer. The volume of sales will be low and if proper care is not taken, the chances of product failures are high.

Growth : Once the product passes through the introduction stage, the sales starts increasing because of the acceptability of the product by the customer. The sales growth rate is high because of limited or no competition.

Maturity (Saturation): The sales growth reaches a point above which it will not grow. This is due to the market share taken by the competitor's products. Thus, the sales will be maintained for some period.

Decline: The competitors will enter the market with better product features, advanced technology and reduced prices. This is a threat to the very existenoe of product and sales start declining. If proper care like addition of special features, design changes are not incorporated there comes a time when the products are to be taken back from the market.

Characteristics of Phases of PLC are shown in Table 14.1.

Table 14.1: Characteristics of Phases of PLC

Particulars	*Introduction*	*Growth*	*Maturity*	*Decline*
1. Product Variety	High Variety	Increasing Standardisation	Dominant Design feature of product	High standard commodity
2. Volume	Low Volume	Increasing Volume Consolidation	High volume	Decreasing volume
3. Industry Structure	Small Competition	Beginning of Competition	Few large companies	Survivors
4. Form of Competition	Product Characteristics	Product Quality and availability	Price and dependability	Price

14.3 PRODUCT POLICY OF AN ORGANISATION

Product policy is the top management (Strategic) decision. Every organisation has its own product strategies or policies which form the basis of competing in the market. They become the unique selling proposition (USP) of the company. As per the requirements of the company, it may choose product policies. The same company can opt for different policies for the different products. The various product policies are:

1. **Lowest price:** The company will be the price leader and the company is going to offer the product at the cheapest price than its competitors. Price becomes the criteria used to compete in the market. Though the profit per unit is less, the company is going to make the substantial profit by the large volume.
2. **Highest quality:** Some organisations offer highest quality products irrespective of the cost. They are catering to the needs of special class of customers who value quality as the only criteria to purchase the product.
3. **Compromise between cost and quality:** Some organisations in order to capture the larger sections of the customers, offer products with the optimum blend of quality and cost. The

products are reasonably of good quality in proportion to its price. These organisations try to give good value to the customers for his money.

4. **Safety:** Some organisations give maximum importance to safety. Safety is the criteria on which they compete in the market. For example all home appliances, electrical gadgets, etc.

   Thus, organisations have to choose the policies suitable for them. This policy is going to influence the design to the large extent.

14.4. SELECTION OF A PROFITABLE PRODUCT

It is the product that makes or mar the fortune of the company. So utmost care should be taken in the identification of the product.

Before selecting a product, organisations have to carryout SWOT analysis in order to know their strength areas, weaknesses or limitations, opportunities before the organisation and the perceived threat. The organisations have to explore the opportunities (products) which fall under their strength areas so that they are able to cash on the opportunities. Product selection is a team effort.

IDENTIFICATION OF PROFITABLE PRODUCT

1. **By chance:** It is impossible to ignore the effects of chance. A meeting in a train with a stranger, sight of a new device, an attendance at a social gathering can create the idea which may lead to a successful product.
2. **Desire to utilise idle resources:** Many organisations have idle resources like excess cash, unused plant and equipment, unutilised management talent, surplus distribution channels and the management may conceive an idea of putting these resources to productive use which gives rise to new product.
3. **Demand supply gap:** If the gap between demand and supply is big, then the products are selected to bridge this gap.
4. Need to support existing range.
5. Forward and/or backward integration.
6. By spreading the risk.
7. To supplement a declining income.
8. To keep pace with changing fashion and customer preferences and tastes.
9. To exploit special skills.
10. To attract prestige.
11. To exploit special assets.

No project should be undertaken until it is viewed objectively from all aspects and a detailed feasibility study is carried out. The personal preference and immediate excitement should not be given any room.

14.5. PRODUCT DESIGN (DEVELOPMENT) PROCESS

The stages in the product design are shown in Fig. 14.2.

Stage I: CONCEPTION

The draft specifications for the product are laid down incorporating the user requirements at this stage. This stage provides the basis for all subsequent design activities. The specifications of the proposed product must be prepared by marketing department in as much details as possible. The following minimum information on design specifications should be furnished:

- The performance requirements
- The appearance or styling requirements.

- The estimated quantity which will be sold.
- The maximum price within which the product should be offered.
- The probable date of introduction of the product into the market.

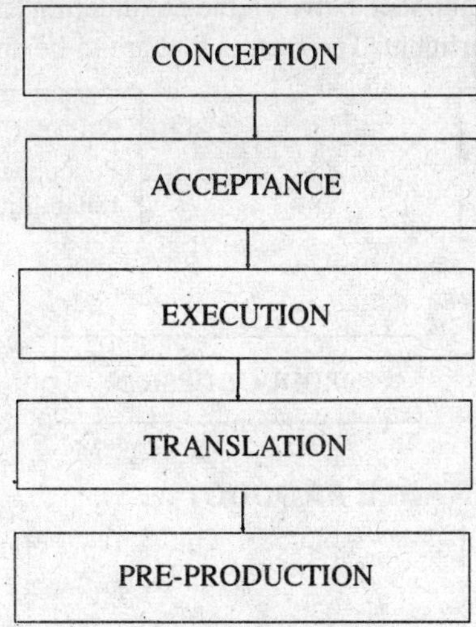

Fig. 14.2: Stages in product design.

Stage II: ACCEPTANCE

This is a stage where the design activity of the product begins after the feasibility analysis and model making and calculations of the product is accepted.

Stage III: EXECUTION

On general design considerations, a model is prepared as per the acceptance of specifications in stage II. The models should conform to specifications. The model is going to reveal the detailed feasibility aspects of proposed designs and special considerations. The cost of the product is built at this stage which the production engineers always try to stick to it. Now the advanced techniques like rapid prototyping technique (RPT) and experimental stress analysis techniques are available for prototype, modelling and testing.

Stage IV: TRANSLATION

At this stage the production engineering department is involved in design work. The manufacturing feasibility is tested at this stage. The final manufacturing drawings are prepared.

Stage V: PRE-PRODUCTION

In large scale production, it is recommended to carryout a pilot run under production conditions. This will consist of completely assembling quantity of production from parts or components made by normal production method and using the same degree of skill in the operatives which will be found in the final manufacture. The pre-production run will ensure the quality, reliability of product as per the specifications before the production will be started on commercial basis. Thus the pre-production stage will check:

1. Drawings.
2. Final tools.
3. Production techniques and estimates.
4. Specifications.

14.6. PRODUCT ANALYSIS

Once the profitable product is selected and an organisation decides the product policy for the proposed product, the next step is the detailed analysis of the product with respect to various factors that influence the product design.

The coordination is to be established between the engineering (design) department, production and industrial engineering department. The factors that are to be analysed are shown in Fig. 14.3.

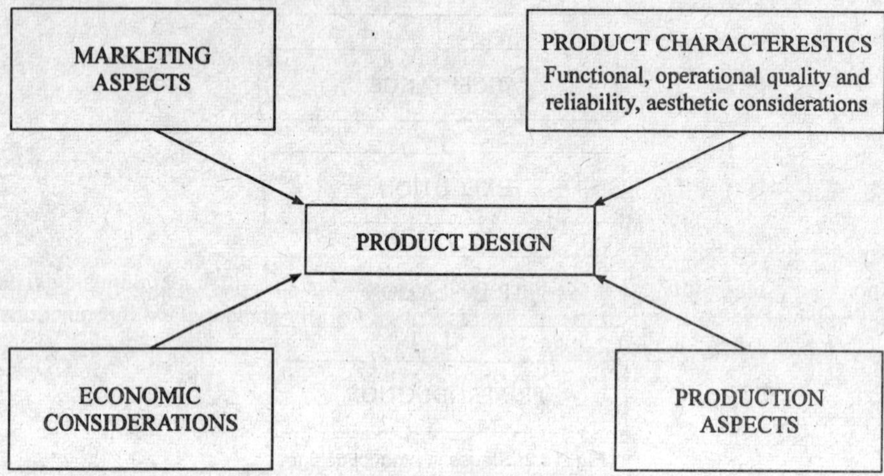

Fig. 14.3: Factors that influence product design.

14.6.1. MARKETING ASPECTS

Once the product is selected, then it is very important to know the marketability of the product. All further steps are dependent upon the demand for the proposed product and customer acceptability to the product. If there is no potential market, then it is a wasteful exercise to design and manufacture the product. The marketing analysis for the proposed product seeks to give answers to the following questions:

1. Whether the functions that are offered by the product are desirable and acceptable to the customers?
2. What will be the status of the product in the market
   (*i*) Product is already existing?
   (*ii*) Entirely new product about which the customer is totally unaware?
3. What will be the expected demand for the product both short-term and long-term?
4. The factors that influence the demand for this proposed product?
5. What is the level of competition and basis of competition for product?
6. What are the distinguishing features offered by competitors?
7. The price prevailing in the market.
8. Distribution system, etc.

The answers to the above question will furnish the designer with lot of information about the product.

It is easy to estimate the demand for the existing product. But if the product is entirely new, it is offered to the market first time, a detailed market survey is to be carried out to estimate the demand for the product.

The demand for the product depends on many factors. Some of which are related to local conditions and sometimes difficult to define. So, an organisation has to be in close touch with the target market it wants to capture so that it is able to feel the trend of the market.

A product that is offered to the customer should be of infinite variety if it is aimed at 100 per cent customer satisfaction. This is practicably impossible. Thus, in practice, product design is an outcome of some sort of compromise between infinite variety and designer's concept of ideal design. In selling this compromise to customers, the management resorts to advertising. The advertisement campaign is dependent upon the compromised design.

Advertising serves many purposes such as:

- Providing general information about the general existence of product.
- Technical information like its functional utility and characteristics.
- Drawing attentions of the customers to special features targeted at the customer.
- To create demand among passive population of customer.
- Educating the customer or telling him what he should want.

Thus, marketing aspects which analyses the factors that influence the demand for the product is an important step in product analysis.

14.6.2. FUNCTIONAL ASPECTS

Once the marketing feasibility exists for the proposed product, *i.e.*, a sufficient demand exists for the product, the functional scope of the product is to be carefully analysed and the functions are to be defined properly. The functional objectives are to be fixed with respect to the product like:

- What are the functions the product is expected to perform?
- Whether we should give a single function or multiple functions are to be incorporated.
- Cost considerations due to offering multiple functions.

The functional analysis helps in analysing the importance and worth of each function to be built into the product and, hence, affects the design of the product.

Example: A tape recorder has a well defined objective, *i.e.*, to play the cassettes and should give a good quality sound as output. But it this will not state all aspects. The functional analysis is to be done with respect to:

- Whether only tape recorder or should include the additional utilities like radio and/or clock.
- Whether should use single cassette or twin cassettes.
- Whether it should be provided with auto reverse facility.
- What should be the built in safety measures.
- Whether portable or stationary

and many of the other factors need to be analysed. Now, the trend is towards offering a functional versatility of the product which aims at increasing the range of applications of product to the customer.

For example, a kitchen mixer allows for large number of operations to be performed by additional attachments. Even functional versatility is offered to engineering equipment like machine tools, which are going to perform number of operations and with a minimum of investment and savings in space. The provision of multifunction will create a psychological satisfaction of owning more than one product. For example, a two-in-one tape recorder serves the twin objectives of tape recorder and radio.

14.6.3. OPERATIONAL ASPECTS

Once the functions expected to be serviced by the product are established then operational aspects of the products are to be determined. The product is not only expected to perform its functions satisfactorily but it should be easy to handle and operate at the customers end.

The product is used at different operational conditions and the customers vary with respect to skill and knowledge and the designer's problem becomes complicated with addition of more functions.

The designer has to analyse the preparation time (set-up time), operation time and put away time with respect to the customer. Thus, the operational aspect becomes all the important as it is the customer who is going to operate at his place and care should be taken to see that the product should have ease of operation.

14.6.3. DURABILITY AND DEPENDABILITY

These two factors define the quality and reliability of the product. Durability refers to the length of the active life of the product under given working conditions. Dependability refers to the reliability with which the product serves its intended function. Thus, quality of the product is directly proportional to the quality of inputs (materials, men, etc.), the process of manufacture. Thus, it is a function of cost.

Depending upon the management product policy, a compromise is to be struck between quality and cost. To be in competition the organisations have to give better quality products at reasonable price. Due consideration should be given to various aspects of quality measures, safety and maintenance aspects.

14.6.4. AESTHETIC ASPECT

Aesthetic aspect refers to the "external look good' aspect of the product and it is concerned with moulding the final shape around the basic skeleton. Aesthetic aspects help the sell function of the product by attracting the customers and creating the first impression about the product.

For consumer goods aesthetics is the dominant factor in creating the demand for the product. Styling becomes all the important factor in product design in situations such as changes in fashion and taste, evolution of form and introduction of new ideas to quickly outdate the old ones. If the market is ready to accept creative product and eager to discard old ones in favour of new ones, the styling becomes the main aspect of saleability of product. Designers use variety of tools to build aesthetic characteristics into the products such as:

1. **Use of Special materials:** For housing the components (body) or additional decorations. For example, use of chromium strips, plastics, wood, glass, fabric materials.
2. **Use of Colour:** Natural colour of the material concerned or by use of paints, plating, spraying.
3. **Texture as a supplement to colour:** Shape denoted by outer contours and similarity to familiar objects.
4. **Use of lines.**
5. **Scaling the product.**
6. **Packaging.**

Aesthetics has been recognised as an integral part of design and the scope of aesthetics is not limited to only consumer goods but now it has been extended to engineering goods also like machinery and equipment.

14.6.5. ECONOMIC ANALYSIS

An economic analysis is the key to the management decision in product design policy. Once the sufficient information on the marketability of the product is obtained and various aspects of the product like functional, operational, quality and aesthetic are studied about the proposed product. Then the next steps is the economic analysis which seeks to answer the following questions:

- What will be the amount of investment needed to manufacture the new product?
- What are the estimated production costs per piece?
- What will be the expected profit margin?

- Whether the prices (cost + profit) proposed to be offered by the company are competitive?
- What is the expected volume of sales?

Now, the concept of the product pricing is undergoing a change. In a competitive market, the price is fixed by the market and the manufacturer is expected to offer the products to the customers at that price if at all he wants remain in the competition. Now, the trend is working backwards towards the cost at which the product is to be manufactured.

For example, the market price of the product is '*X*'. The manufacturer is expected to offer the product at this price say, if he expects the profit margin of '*Y*'. Then total cost within which he should manufacture the product is $X - Y$ say $-Z$. Now, the manufacturer should see that the manufacturing cost will not exceed '*Z*', so that he can compete in the market. Techniques like Break-even Analysis, Cost-volume Profit Analysis are used to establish relationship between cost volume and profit.

14.6.6. STANDARDISATION

Standardisation is a tool for variety reduction—**"Standardisation is a process of defining and applying the conditions necessary to ensure that given range of requirements can normally be met with a minimum of variety and in a reproducible and economic manner on the basis of the best current techniques."**

The other two words associated with standardisation are:

Simplification—"a process of reducing types of products within a definite range".

Specialisation—"a process where in particular firms concentrate on the manufacture of the limited number of product types".

The three *S* are usually linked together and develop a logical sequence.

Objectives of Standardisation

- Interchangebility of parts, components, etc.
- Keeping the variety minimum.
- Helps to achieve a better control due to reduced variety.

Advantages

- Reduction of waste and obsolescence.
- Reduction in inventory.
- Reduced effort in book keeping and accounting.
- Standardisation reduces the price because of economy of scale.
- Ease in procurement because of availability.
- Reduction in maintenance and repair costs.

Standardisation covers wide variety of activities. The standardisation can be described by the following main categories:

- Physical dimensions and tolerance of components with a definite range.
- Rating of machines and equipment (energy, speed, voltage, temperature, etc.).
- Methods of testing characteristics.
- Performance standards.
- Specification of properties of material (Physical and chemical).
- Methods of safety considerations.
- Methods of installation to comply with precautionary measures and convenience of use.

Simplification is a point of disagreement between the production and marketing departments.

Production department is in favour of minimum of variety as it reduces the number of set ups and, hence, utilisation of production facilities is increased and can be able to produce the component at the reduced cost compared to more variety.

Marketing advocates that the higher variety helps them to sell more as it helps them to meet the needs of the larger customer groups.

Thus, a balance is struck between the two in order to arrive at the optimum variety of types of products to be manufactured.

The advantages of simplification (minimum variety) are:

1. Reduce inventories of materials and component parts.
2. Reduced investments in plant and machinery.
3. Reduced space requirements of storage.
4. Ease of planning and control.
5. Reduction in selling price.
6. Simplification of inspection and control.

The disadvantages are:

- Not able to meet the needs of wide range of customer preferences.
- Possibility of loosing orders to competitors.
- Creates a constant source of conflict between marketing and production.

14.6.7. PRODUCTION ASPECTS OF PRODUCT DESIGN

The design will be converted into reality in the production shop where it will be transformed into a physical product to be offered to the customer. So successful transformation of design into a saleable product is a challenge to the organisation. This needs a close coordination of production and design department at all stages of manufacturing the product. Thus, a designer cannot design the product in isolation and an increased interaction is need by production and other department to produce design that works and that can be manufactured without any problems.

Thus the concept of "DESIGN FOR PRODUCTION" is the current trend.

The following aspects of production should be evaluated:

(A) Selection of Suitable Process

1. Production quantity (volume of production).
2. Information regarding utilisation of equipment, capacity of equipment, accuracy, etc.
3. Selection of tooling such as jigs and fixtures.
4. Sequence of operations and methods of assembly.
5. Possibility of applications of new techniques, processes.

(B) Utilisation of Materials and Components with a View of

(*i*) Selecting the materials conforming to specifications.
(*ii*) Selection of method to reduce waste and scrap.
(*iii*) Usage of standard components and parts.
(*iv*) Interchangebility of components and assemblies within the product.

(C) Selecting the proper tolerances and work method to achieve the specified quality standard through available processes and equipment. The specification of quality is going to influence the selection of a process.

To achieve a successful transformation of the design, a designer engineer should work in close coordination with production and methods engineer to specify the best available process of transformation keeping in mind the limitations of the production.

References for Further Reading

1. Samuel A. Eilon, *Elements of Production Planning and Control*, Universal Publications, Mumbai (1994).

2. Joseph Monks, *Operations Management*, McGraw Hill Book Company, New York.
3. Evertt E. Adam Jr. and Ebert, *Production and Operation Management Concepts* , Model and Behaviour, 5th Edn., Prentice Hall of India (1993).

REVIEW QUESTIONS

1. What is the need for organisations to design and develop new products?
2. Explain with a graph various stages of product life-cycle.
3. What are the sources that help to identify a new product?
4. Explain the steps in new product design.
5. Why it is nèccessary to analyse the product? What are the various analysis that are carried out?
6. Explain the functional and operational aspects of product design.
7. What do you mean by Design for production? Explain in detail the production aspects of Product design.
8. Write a note on Standardisation.
9. Explain the quality and reliability aspects of product design.
10. Write short notes on:
    (*i*) Product policy of an organisation.
    (*ii*) Competition analysis.
    (*iii*) Identification of new product.
    (*iv*) Aesthetic aspects of product design.
    (*v*) Preferred numbers.
    (*vi*) Economic analysis.

15

DEMAND FORECASTING

• Introduction • Forecasting and prediction • Need for forecasting • Long-term and short-term forecasts • Classification of forecasting methods • Judgemental techniques • Time series analysis • Least square method of forecasting • Moving average method • Exponential smoothing method • Casual forecasting method • Regression and correlation • Forecast error • Steps in forecasting process • Requirements of a good forecasting method • Cost and accuracy of forecasts.

15.1. INTRODUCTION

Forecasting plays a crucial role in the development of plans for the future. It is essential for the organisations to know for what level of activities one is planning before investments in inputs, *i.e.*, men, machines and materials be made.

Before making an investment decision, many questions will arise like:

- What should be the size or amount of capital required?
- How large should be the size of the work force?
- What should be the size of the order and safety stock?
- What should be the capacity of the plant?

The answers to the above questions depends upon the forecast for the future level of operations. Modern production activities are becoming more complex technologically and the basic inputs are becoming expensive and there are lot of restrictions on them. The planning of production activity is, therefore, essential so that the resources are put to their best use.

Planning is a fundamental activity of management. Forecasting forms the basis of planning and forecasting enables the organisation to respond more quickly and accurately to market changes.

Forecasting as defined by American Marketing Association is: **"An estimate of sales in physical units (or monetary value) for a specified future period under proposed marketing plan or programme and under the assumed set of economic and other forces outside the organisation for which the forecast is made."** Forecasting is thus looked upon as a projection based on the past data. Forecasting is not a guess work. It is the inference based on large volume of data on past performance.

Forecasting is an important component of strategic and operational planning. It establishes a link between planning and controlling. It is essential for planning scheduling and controlling the system to facilitate effective and efficient output of goods and services.

15.2. FORECASTING AND PREDICTION

Prediction is an estimate of future event through subjective considerations other than just the past data. For prediction, a good subjective estimation is based on managers skill, experience and judgement.

There is an influence of one's own perception and bias in prediction. So it is less accurate and

has low reliability. Forecasting is an estimate of future event achieved by systematically combining and casting forward in a predetermined way data about the past.

Forecasting is based on the historical data and it requires statistical and management science techniques. When we refer to forecasting, usually it is some combination of forecasting and prediction.

15.3. NEED FOR DEMAND FORECASTING

1. Majority of the activities of the industries depend upon the future sales.
2. Projected demand for the future assists in decision-making with respect to investment in plant and machinery, market planning and programmes.
3. To schedule the production activity to ensure optimum utilisation of plant's capacity.
4. To prepare material planning to take up replenishment action to make the materials available at right quantity and right time.
5. To provide an information about the relationship between demand for different products in order to obtain a balanced production in terms of quantity required of different products as a function of time.
6. Forecasting is going to provide a future trend which is very much essential for product design and development.

Thus, in this changing and uncertain techno-economic and marketing scenario, forecasting helps to predict the future with accuracy.

15.4. LONG-TERM AND SHORT-TERM FORECASTS

Depending upon the period for which the forecast is made, it is classified as long-term forecasting and short-term forecasting. Forecasts which cover the periods less than one year is termed as short-term forecasting, and which cover the period over one year (5 years or 10 years) future are termed as long-term forecasts.

Short-term forecasts are made for the purpose of materials control, loading and scheduling and budgeting. Long-term forecasts are made for the purposes of product diversification, sales and advertising budgets, capacity planning and investment planning.

15.5. CLASSIFICATION OF FORECASTING METHODS

A large number of forecasting techniques with various degrees of complexity are available to the forecaster these days. The availability of computer programmes have facilitated the task of the forecaster. The methods can be classified as subjective depending upon the individual judgements and objective method after analysing the quantitative information. In general, forecasting methods are classified as:

1. Judgemental techniques.
2. Time Series methods.
3. Causal methods (Econometric Forecasting).

The judgemental technique is a method which relies on the art of human judgement. This is in practice since long time. The other two techniques are relatively new and are heavily use statistics for analysing the past data.

In econometric forecasting, the analyst tries to establish cause and effect relationships between sales and some other parameter that are related to sales, *i.e.*, the demand for cement depends upon the projected growth of construction industry.

The objective in this method is to establish a cause and effect relationship between changes in the sales level of the product and set of relevant explanatory variable. It utilises regression and correlation analysis.

Time series analysis, identifies the historical pattern of demand for the product and project or

extrapolates this demand into the future. The important method of making inference about future on the basis of what has happened in the past is called time series analysis.

15.6. JUDGEMENTAL TECHNIQUES

These are commonly used techniques in business and industries. This is a subjective method where in there is a heavy reliance on the past experience of the person and skill. The various judgemental techniques are:

1. **Opinion survey method:** Opinion method is relatively simple and practicable method for forecasting demand for the new products. In this method, opinions are collected from the prospective buyers regarding, why they buy a particular product, what they expect from the product. The sampling technique is used to survey the customers. From the representative sample it is possible to forecast how the targeted population responds to the product.

2. **Executive opinion method:** In this method, the experts opinion is sought on the future demand for the product. It is biased and subjective as it is not backed by any scientific method or statistical data. The accuracy of the predicted demand depends upon the skill, expertise and experience of the person making the forecast. This method is used for demand forecasting of established products.

3. **Customer and distributor surveys:** The individuals who have bought a product can be asked the reasons for making the purchase. The questionnaire can be given along with the guarantee cards. Estimates of expected sales (distributor surveys) can be requested from retail outlets and company's sales force. Many companies heavily rely on judgements made by their sales personnel.

4. **Marketing trials:** This method is specially applicable to new products. If the product is entirely new to the customer or market, then it is very difficult to know the acceptability of the product by the market. In this case it is advisable to expose the product to the limited market trial. Such a trial is like a controlled experiment in which the market area and the method of presentation are carefully selected and controlled. Its cost is very high. This method is normally recommended for consumables like cosmetics, tooth pastes, etc.

5. **Market research:** This method can be used for new products or existing products. Usually the work is assigned to external marketing agencies. This is recommended if extensive data is needed. The purpose of the research is to gather information regarding the nature of consumption. The details about various factors which influence the demand like location, buyer occupation, prices quantity, quality, consumer income, etc., are collected and the factors are related to get the forecast.

6. **Delphi technique:** To make the judgemental forecasts more realistic by minimising bias, this method is used. In this method, a panel of experts are asked sequential questions in which the response to one questionnaire is used to produce next questionnaire. The information available to some experts are made available to the other experts. This technique is an iterative process.

The Delphi method is a qualitative forecasting method in which opinions are collected from experts to arrive at a reliable consensus. A series of questionnaire is sent to a panel composed of experts from selected technical fields. Each questionnaire demands a written opinion about specific subjects and reasons for justifying the opinions. These reasons are summarised in each iteration and returned for inspection by the whole panel. Through this series of exchanged view, the consensus is reached. This method can be extended to several decades into future.

15.7. TIME SERIES ANALYSIS

Time series refers to the past data arranged in a chronological order as a dependent variable and time as an independent variable.

For example, the sales of TV sets for the last four years in a particular geographical region are:

Year	1993–94	1994–95	1995–96	1996–97
No. of TV sets ('000)	20	30	40	58

The time series methods does not study the factors that influence the demand and in this method all the factors that shape the demand are grouped into one factor-time and demand is expressed as a series of data with respect to time.

The time series analysis consists of determining the trend underlying the demand and extrapolate the future trend. Statistical methods are used to determine the trend. Time series consists of four components:

1. **Trend (T):** The long period tendency of the data to increase or decrease is referred to as a secular trend. It is the long run historical component of the time series which indicates overall growth or decline of the business overtime.
2. **Cyclical fluctuations (C)** are the oscillating patterns of increase or decrease around the trend. Cyclical fluctuations account for some of the variance between the trend line and the data points. The magnitude, timing and pattern of cyclic fluctuations vary so widely and the awareness of the effect of cyclic variation should influence the forecaster.
3. **Seasonal variations (S):** This seasonal variation occurs with some degree of regularity in a span of one year and are annually repetitive. These variation are caused by climatic conditions, social customs, festivals, *e.g.*, the sales of paint is highest during Diwali Festival, sale of umbrella and rain coat during monsoon, and warm clothes during winter.
4. **Irregular variations (R):** These occur without any fixed pattern. These are chance variations that are not explained by trend, cyclic or seasonal variations. These may account for natural calamities like flood, draught, earthquake, etc.

The most commonly used expression for a time series forecast is:

$$Y = TCSR$$

where Y = Forecasted value

T = Secular trend

C = Cyclic variations

S = Seasonal variations

R = Irregular fluctuations

15.8. LEAST SQUARE METHOD OF FORECASTING (Regression Analysis)

This is the mathematical method of obtaining the "the line of best fit between the dependent variable (usually demand) and an independent variable. This method is called least square method as the sum of the square of the deviations of the various points from the line of best fit is minimum or least. It gives the equation of the line for which the sum of the squares of vertical distances **between the actual values and the line values are at minimum**.

In a simple regression analysis, the relationship between the dependent variable y and some independent variable x can be represented by a straight line

$$y = a + bx$$

where b is the slope of the line

a is the y-intercept.

The values of the constants a and b are determined by the two simultaneous equations.

$$\Sigma y = Na + b\Sigma x \quad \text{... (1)}$$

$$\Sigma xy = a\Sigma x + b\Sigma x^2 \quad \text{... (2)}$$

These two equations are called normal equations.

To compute the values of a and b

(*i*) Calculate the deviation (x) for each period and also the sum of deviations.

(*ii*) Find the value of Σx^2

(*iii*) Find the value of Σxy

(*iv*) Calculate the values of a and b

(*v*) Make the sum of deviations $\Sigma x = 0$

Substituting the value of $\Sigma x = 0$ in equations (1) and (2)

We get, $\Sigma y = Na$

$\Sigma xy = b\Sigma x^2$

which gives the values of a and b as

$$a = \Sigma y/N$$

$$b = \Sigma xy/\Sigma x^2$$

Note: (*i*) If the Time Series consists of odd number of years to make $\Sigma x = 0$, the middle value of the time series is taken as the Origin.

(*ii*) If the time series consists of even number of years, the midway period between two middle periods is taken as origin to make $\Sigma x = 0$.

***Problem 1*:** The following data gives the sales of the company for various years. Fit the straight line. Forecast the sales for the year 1998 and 1999.

Year	*Sales (000)*
1989	13
1990	20
1991	20
1992	28
1993	30
1994	32
1995	33
1996	38
1997	43

***Solution*:**

Year	*Sale (y)*	*Deviation (x)*	x^2	*xy*
1	13	– 4	16	– 52
2	20	– 3	9	– 60
3	20	– 2	4	– 40
4	28	– 1	1	– 28
5	30	0	0	0
6	32	1	1	32
7	33	2	4	66
8	38	3	9	114
9	43	4	14	172
$N = 9$	$\Sigma y = 257$	$\Sigma x = 0$	$\Sigma x^2 = 60$	$\Sigma xy = 204$

Now, substituting the values in the equations to get

$$a = \Sigma y/n = 257/9 = 28.56$$

$$b = \Sigma xy/x^2 = 204/60 = 3.4$$

The equation of the straight line of best fit is

$$y = 28.56 + 3.4x$$

(*i*) Sales for the year 1998

$$y_{98} = 28.56 + 3.4 \times 5$$

$$= 45.56 = 45560$$

(*ii*) Sales for the year 1999

$$y_{99} = 28.56 + 3.4 \times 6 = 49000$$

Problem 2: The past data regarding the sales of SPMS for the last five years is given. Using the least square method, fit a straight line, Estimate the sales for the year 1996 and 1997.

Year	1991	1992	1993	1994	1995
Sales ('00)	35	56	79	80	40

Solution:

Year	*Sales (y)*	*Deviation (x)*	x^2	*xy*
1991	35	– 2	4	– 70
1992	56	– 1	1	– 56
1993	79	0	0	0
1994	80	1	1	80
1995	40	2	4	88
N = 5	$\Sigma y = 290$	$\Sigma x = 0$	$\Sigma x^2 = 10$	$\Sigma xy = 34$

In this case, as the number of periods is odd, to make the $\Sigma x = 0$, the deviations are calculated from the middle period, *i.e.*, 1993.

Now, to fit a straight line $y = a + bx$ the values of a and b are to be computed using the formula

$a = \Sigma y/n = 290/5 = 58$

$b = \Sigma xy/\Sigma x^2 = 34/10 = 3.4$

$y = 58 + 3.4\,x$ ... (1)

Deviation for 1996 is $x = 3$, and 1997 = 4

Substituting the values of $x = 3$ and $x = 4$ we get the forecast for 1996,1997

Year 1996 = $58 + 3.4 \times 3 = 68.2$

Therefore Forecast for 1996 = 6820 units

Year 1997 = $58 + 3.4 \times 4 = 71.6$

Therefore Forecast for 1997 = 7160 units

Problem 3: The sales for the domestic water pumps manufactured by Ajit Manufacturing Company is given forecast the demand for the pumps for the next three years using least square method.

Year	1986	1987	1988	1989	1990	1991	1992	1993	1994	1995
Sales ('000)	30	33	37	39	42	46	48	50	55	58

Solution:

Year	*Sales (y)*	*Deviation (x)*	x^2	*xy*
1986	30	– 4.5	20.25	– 135
1987	33	– 3.5	12.25	– 45.5
1988	37	– 2.5	6.25	– 92.5
1989	39	– 1.5	2.25	– 58.5
1990	42	– .5	0.25	– 21
1991	46	0.5	0.25	23
1992	48	1.5	2.25	72
1993	50	2.5	6.25	12.5
1994	55	3.5	12.25	192.5
1995	58	4.5	20.25	261
N = 10	438	$\Sigma x = 0$	$\Sigma x^2 = 82.5$	$\Sigma xy = 251$

As the number of periods is even, sum of the deviations is made equal to Zero ($Ex = 0$) by making midway period between two middle periods is made as origin. The 1990 and 1991 are the two middle periods and midway period between 1990 and 1991 is selected as Origin.

i.e., 1990 – 0.5

--------- 0

1991 + 0.5

and the deviations of the other periods are calculated as shown in the table.

$$a = \Sigma y/n = 438/10 = 43.8$$

$$b = \Sigma xy/\Sigma x_2 = 251/82.5 = 3.04$$

The regression line is

$$y = 43.8 + 3.04x$$

Forecast for the pumps for next three years

$$y_{96} = 43.8 + 3.04 \times 5.5$$
$$= 60.52 = 60520$$
$$y_{97} = 43.8 + 3.04 \times 6.5$$
$$= 63.56 = 63560$$
$$y_{98} = 43.8 + 3.04 \times 7.5$$
$$= 66.6 = 66600$$

Least square method when the sum of the deviations is not zero ($Ex = 0$)

For the straight line $y = a + bx$

The constants a and b can be calculated from the expression—

$$a = \frac{(\Sigma y \cdot \Sigma x^2) - (\Sigma x \cdot \Sigma xy)}{(N\Sigma x^2) - (\Sigma x)^2}$$

$$b = \frac{N\Sigma xy - (\Sigma x \cdot \Sigma y)}{(N\Sigma x^2 - (\Sigma x)^2}$$

or Alternatively

The values of a and b can be computed by solving the following two equations:

$$\Sigma y = Na + b\Sigma x \quad \text{... (1)}$$

$$\Sigma xy = a\Sigma x + b\Sigma x^2 \quad \text{... (2)}$$

***Problem 4*:** A company manufacturing washing machines establishes a fact that there is a relationship between sale of washing machines and population of the city. The market research carried out reveals the following information:

Population (million)	5	7	15	22	27	36
No. of washing machines demanded ('000)	28	40	65	80	96	130

Fit a linear regression equation and estimate the demand for washing machines for a city with a population of 45 million.

***Solution*:**

Population (x)	*No. of Washing m/c's demanded (y)*	x^2	*xy*
5	28	25	140
7	40	49	280
15	65	225	975
22	80	484	1760
27	96	729	2592
36	130	1296	4680
$\Sigma x = 112$	$\Sigma y = 439$	$\Sigma x^2 = 2808$	$\Sigma xy = 10427$

To find the regression equation, the values of a and b are computed as follows:

$$a = \frac{(\Sigma y \cdot \Sigma x^2) - (\Sigma x \cdot \Sigma xy)}{(N\Sigma x^2 - (\Sigma x)^2}$$

$$= \frac{(439 \times 3808) - (112 \times 10427)}{6 \times 2808 - (112)^2}$$

$$= \frac{1232712 - 1167824}{16848 - 12544} = 15.07$$

$$b = \frac{N\Sigma xy - (\Sigma x \cdot Ey)}{N\Sigma x^2 - (\Sigma x)^2}$$

$$= \frac{6 \times 10427 - 112 \times 439}{6 \times 2808 - (112)^2}$$

$$= \frac{62562 - 49168}{16864 - 12544}$$

$$= \frac{13394}{4304}$$

$$= 3.11$$

(*i*) The regression equation is

$$y = 15.07 + 3.11x$$

(*ii*) The demand for the washing machine when population of city is 45 million, *i.e.*, $x = 45$

$$y = 15.07 + 3.11 \times 45$$
$$= 155.02$$

Therefore, the number of washing machines demanded when $x = 45$ is 155020 Nos. (**Ans**).

15.9. MOVING AVERAGE (MA) FORECASTING

The past data of sales of a company can have fluctuations (high or low) because of the seasonal variations and random variations. Simple averaging of demand for previous periods is going to hide the trend and it is meaningless since trend is an important factor. Moving Average (MA) consists of series of arithmetic means calculated from overlapping groups of successive elements of time series.

Moving average is a simple statistical method to extrapolate and establish trend of past sales. This method uses a past data and calculates a rolling average for a constant period. At each period, fresh average is computed at the end of each period by adding the demand of the most recent period and deleting the data of the old-period since the data in this method changes from period to period, it is called moving average method.

A simple moving average is calculated as follows:

$$\text{MA} = \frac{\text{Sum of demands for periods}}{\text{chosen number of periods}}$$

For example, for the time series of values D_1, D_2, D_3 . . . . . D_n , etc., for different periods, the moving average for n periods, is given by,

First value moving average

$$= 1/n(D_1 + D_2 + D_3 + \ldots\ldots D_n)$$

Second value moving average

$$= 1/n(D_2 + D_3 + D_4 + \ldots\ldots D_{n+1})$$

Third value of moving average

$$= 1/n(D_3 + D_4 + \ldots\ldots + D_{n+2})$$

Period n of moving average should be carefully selected. The wrongly selected period will distort the data and gives wrong picture of the trend. Larger the period of moving average, the greater is the smoothing effect.

A three months MA has a weightage of 1/3rd and 5 months MA has a weightage of 1/5th. Larger the value of n (period of moving average), the smaller is the effect of random variation and a higher smoothing effect. The value of n depends upon the speed at which the pattern of demand changes. If the demand pattern is stable a high value of n is selected. If the pattern is not stable, a small value of n should be selected.

Centring of the Moving Average

A three period moving average is located against the second period, a five months moving average is located against third month.

In general, n months moving average is located against $\frac{n+1}{2}$.

Weighted Moving Average (WMA)

Sometimes the forecaster wants to use a moving average but does not want all the n periods equally weighted as in simple MA method. But some organisations base their forecasts on a weighted moving average.

In simple MA, equal weightage is given to 1st month, 2nd month and 3rd month in a three month moving average. But the organisation wants to attach more weightage to the third month and least to the first month.

For example, depending upon the importance it assigns weightages, *e.g.*, 0.2 to 1st period, 0.3 to second and 0.5 to the third. The sum of these weights should be equal to one.

Problem 5: The past data on the load on the weaving machines is shown below:

Month	*Load (HRS)*
May-96	–
June-96	585
July-96	610
Aug.-96	675
Sep.-96	750
Oct.-96	860
Nov.-96	970

(*a*) Compute the load on the weaving machine centre using 5th moving average for the month of December 1996.

(*b*) Compute a weighted three months moving average for December, 1996 where the weights are 0.5 for the latest month, 0.3 and 0.2 for the other months respectively.

Solution: (*a*) Five months moving average forecast for Dec. 1996

$$= \frac{D\text{ Nov.} + D\text{ Oct.} + D\text{ Sept.} + D\text{ Aug.} + D\text{ July}}{5}$$

$$= \frac{970 + 860 + 750 + 675 + 610}{5}$$

= 773 hrs.

(*b*) A three month weighted moving average forecast for Dec.96

= (W Nov. D Nov.) + (W Oct. + D Oct.) + (W Sept. D Sept.)

= $970 \times 0.5 + 860 \times 0.3 + 750 \times 0.2$

= 947.8 machine hours. (Ans)

Problem 6: The data given below represents sales figures of ABC company for the 12 months of the year 1996.

1. Compute 3 months moving average (ignoring decimal values)
2. Forecast the demand for the month of Jan.1997
3. If the actual demand for the month of Jan.1997 is 905 units, what should be the forecast for the month of Feb. 97

Month	Jan.	Feb.	Mar.	Apr.	May	June	July	Aug.	Sept.	Oct.	Nov.	Dec.
Sales Rs. (000)	400	490	570	500	640	680	710	800	820	910	860	950

Solution: (*i*) The 3 month moving average is shown in table below:

Month	*Sales*	*Moving Total*	*Moving Average*
Jan.	400	–	–
Feb.	490	1460	487
Mar.	570	1560	520
Apr.	500	1710	570
May	640	1820	607
June	680	2030	677
July	710	2190	730
Aug.	800	2330	777
Sept.	820	2530	844
Oct.	910	2590	864
Nov.	860	2720	907
Dec.	950	–	–

(*ii*) Forecast for the month of January is 907.
(As the secular trend for Jan.97 is not available, the forecast equals last moving average)

(*iii*) Forecast for Feb.97

$$= \frac{\text{Last moving total} + \text{demand for Jan. 97} - \text{demand for Oct. 96}}{3}$$

$$= \frac{2720 + 905 - 910}{3} = \frac{2715}{3}$$

= 905 units.

15.10. EXPONENTIAL SMOOTHING METHOD

One of the disadvantages of the moving average forecasting is the laborious operation of maintaining the data for all the previous years. Exponential smoothing method requires only the current demand and the forecasted demand for the current month. Simple moving average method gives the equal weightage to the all the periods. Exponential smoothing is distinguished by the fact that it assigns weight to all the previous data and the pattern of weights assigned are of exponential form.

Demand for the most recent data is given more weightage and the weights assigned to older periods decrease exponentially. Thus exponential forecasting ensure that the forecast made by this method keeps pace with changing business trend.

Exponential Smoothing Model

Simple exponential smoothing model estimates the average forecast for the next period by using the actual and the forecasted demand for the previous period.

Forecast for the period t (F_t)

$$= \text{Forecasted demand for the last period} + \alpha \begin{bmatrix} \text{Actual demand} & - & \text{Forecasted demand} \\ \text{for the last period} & & \text{for the last period} \end{bmatrix}$$

$$\therefore \quad F_t = F_{t-1} + \alpha \; [D_{t-1} - F_{t-1}]$$

where α is called smoothing constant. ... (1)

The basic advantage of this method over the moving average method is that, one needs only two figures - one for the old forecast and another for the actual sales observation.

The formula given in equation (1) can be expanded as follows. Equation (1) can be written as

$$F_t = \alpha \cdot D_{t-1} + (1 + \alpha) \cdot F_{t-1} \quad ...(2)$$

In the similar way, the forecast for F_{t-1} can be expanded as

$$F_{t-1} = \alpha \cdot D_{t-2} + (1 + \alpha) \cdot F_{t-2} \quad ...(3)$$

The value of F_{t-1} is substituted from equation (3) into equation (2).

$$\therefore \quad F_t = \alpha \, (D_{t-1}) + (1 - \alpha) \cdot [\alpha \cdot D_{t-2} + (1 - \alpha) \cdot F_{t-2}] \quad ...(4)$$

Rearranging the terms in equation (4), we get,

$$F_t = \alpha \cdot D_{t-1} + \alpha \, (1 - \alpha) \cdot D_{t-2} + (1 - \alpha)^2 \cdot F_{t-2} \quad ...(5)$$

In a similarly way, F_{t-2} can be written as

$$F_{t-2} = \alpha \cdot . \, D_{t-3} + (1 - \alpha) \cdot F_{t-3}$$

On substituting the value of F_{t-2} from equation (5) and rearranging equation (5).

$$F_t = \alpha \cdot D_{t-1} + \alpha \, (1 - \alpha) \cdot D_{t-2} + \alpha \, (1 - \alpha)^2 \, D_{t-3} + (1 - \alpha)^3 \cdot F_{t-4} \quad ...(6)$$

A general equation can be written as

$$F_t = \alpha \cdot D_{t-1} + \alpha \, (1 - \alpha) \, D_{t-2} + \alpha \, (1 - \alpha)^2 \, D_{t-3} \, ...$$

$$- \, ... + \alpha \, (1 - \alpha)^{t-1} \, D_{t-t} + (1 - \alpha)^t \cdot F_o$$

where D_{t-t} the starting forecast

$= F_o$, if an initial forecast cannot be made.

The weightage for each of the demands in the past is discounted by a factor of $(1 - \alpha)$. The last term is negligible for a very large value of period. The value of the α selected is small (0.05 to 0.1), if the demand pattern is smooth or stable and large value of α is used for the fluctuating demand.

Problem 7 :

The demand for the disposable plastic tubing for a general hospital is 300 units and 350 units for September and October respectively. Using 200 units as demand for September, compute the forecast for the month of November. Assume the value of α as 0.7.

Solution:

(*i*) Forecast (Ft) can be written as,

$$F_t = \alpha \cdot D_{t-1} + (1 - \alpha) \cdot F_{t-1}$$

In this case, D_{t-1} = Demand for September (300 units)

F_{t-1} = Forecast for September (given) 200 units.

$\therefore$ Forecast for October can be computed as,

$$(F_t) = 0.7 \times 300 + (1 - 0.7) \times 200$$

$$= 210 + 60 = 270 \text{ units.}$$

(*ii*) Forecast for the month of November

$$F_{t+1} = \alpha \text{ (Demand for Oct.)} + (1 - \alpha). \text{ Forecast for October}$$

$$= 0.7 \times 350 + (1\text{-}0.7) \times 270$$

$$= 326 \text{ units.}$$

Problem 8 :

The demand for the particular product is given for the last 8 periods. Compute the exponentially smoothed forecast for the periods taking $\alpha = 0.1$ and 0.3. Which of these forecast is better

Period	1	2	3	4	5	6	7	8
Demand	10	18	29	15	30	12	16	8

Solution.

The forecast for the initial period is not given and it is assumed equal to the demand for the first period. The forecasts with $\alpha = 0.1$ and 0.3 are computed and the details are given in the table below.

Period	*Demand* D_t	*Forecast* F_t at $\alpha = 0.1$	*Error* $D_t - F_t$	*Forecast* F_t at $\alpha = 0.3$	*Error* $D_t - F_t$
1	10	15	– 5.0	15	– 5.0
2	18	14.5	3.5	13.5	4.5
3	29	14.85	14.15	14.85	14.15
4	15	16.26	– 1.26	19.09	– 4.09
5	30	16.14	13.86	17.86	12.14
6	12	17.52	– 5.52	21.50	– 9.50
7	16	16.97	– 0.97	18.65	– 2.65
8	8	16.87	– 8.87	17.85	– 9.85

Problem 9 :

Estimate the sales forecast for the year 2000, using exponential smoothing forecastor. Take $\alpha = 0.5$ and the forecast for the year 1995 as 160×10^5 units. Compare the forecast with least square method.

Year	1995	1996	1997	1998	1999
Sales Rs. ($\times 10^5$)	180	168	159	170	188

Solution.

The demand for the year 2000 can be computed using least square method as follows.

Year	*Sales* ($\times 10^5$)	*Deviation (x)*	x^2	*xy*
1995	180	– 2	4	– 360
1996	168	– 1	1	– 168
1997	159	0	0	0
1998	170	+ 1	1	170
1999	188	+ 2	4	376
$N = 5$	$\Sigma y = 505$	$\Sigma x = 0$	$\Sigma x^2 = 10$	$\Sigma x^2 = 18$

Substituting these values in the equations,

$$a = \frac{\Sigma y}{N} = \frac{505}{5} = 101$$

$$b = \frac{\Sigma xy}{\Sigma x^2} = \frac{18}{10} = 1.8$$

The regression equation is

$$y = 100 + 1.8x$$

Now, the sales for year 2000 *i.e.*, for $x = 3$, we have,

$$y = 101 + 1.8 \times 3 = 106.4$$

The demand for year 2000 $= 106.4 \times 10^5$ Rs.

The forecasted demand using the smoothing constant $\alpha = 0.5$ is computed as shown in the table using the formula.

$$F_t = \alpha \cdot D_{t-1} + (1 - \alpha) \cdot F_{t-1}$$

Period	*Demand*	*Forecasted demand for α = 0.5*	F_t
1995	180	160 (Given)	
1996	168	$0.5 \times 180 = (1 - 0.5) \times 160$	170.0
1997	159	$0.5 \times 168 = (0.5) \times 170$	169.0
1998	170	$0.5 \times 159 = (0.5) \times 169$	164.0
1999	188	$0.5 \times 170 = (0.5) \times 164$	167.0
2000		$0.5 \times 188 = (0.5) \times 167$	177.5

The forecast demand for the year 2000 with $\alpha = 0.5 = 177.5$.

15.11. CAUSAL FORECASTING METHOD

Casual methods try to identify the factors which causes the variation of demand and try to establish a relationship between the demand and these factors.

In the method, the analyst tries to identify those factors that best explain the level of sales of the product. This process is called econometric forecasting.

The objective of this method is to establish a cause and effect relationship between the changes in the sales level of the product and set of relevant explanatory variable.

Some of the casual methods are,

(*a*) Regression and Correlation analysis.

(*b*) input-output analysis.

(*c*) End use analysis.

Regression and Correlation

Regression analysis is a forecasting technique that establishes the relationship between variables. Historical data establishes a functional relationship between the two variables.

Correlation analysis is determining the degree of closeness of the relationship between two variables.

In general, the sales of any product is influenced by external or internal factors. Sales is thus a dependent variable and is a function of one or more independent variable. A regression models (simple or multiple) establish the relationship between the independent and dependent variable.

Types of Correlation

(*i*) **Positive and negative correlation:** The direction of variation of variables determine whether the relationship is positive (direct) or negative (inverse or indirect). If the increase in value of one is accompanied by increase in value of the other it is called positive correlation, *i.e.*, both variables vary in the same direction. Correlation is said to be negative if increase (or decrease) in one variable results in decrease (or increase) in another variable.

(*ii*) **Simple and multiple correlation:** Correlation is said to be simple when only two variables are involved *e.g.,* price and demand for a product. If more than two variables are studied at the same time, the correlation is said to be multiple. *e.g.*, Price, demand and supply of a product.

(*iii*) **Linear and non-linear correlation:**

Co-efficient of correlation

The degree of relationship is called correlation. It is a single figure which expresses the degree and direction of correlation is called co-efficient of correlation.

Karl pearsons co-efficient of correlation

The formula for calculation the co-efficient of correlation (γ) is given by

$$\gamma = \frac{\Sigma dx.dy}{N\sigma \bar{x}\, \sigma y}$$

where $dx = x - \bar{x}$ = deviation of items of variable x from its arithmetic mean

$dy = y - \bar{y}$ = deviation of items of variable y from its arithmetic mean

σx = standard deviation of variable x.

σy = standard deviation of variable y

The positive value of $\Sigma dx\ dy$ indicates +ve value of co-efficient of correlation.

The value of γ is between + 1 and –1.

The formula can be written in a simple way as

$$\gamma = \frac{\Sigma dx \cdot dy}{\sqrt{\Sigma dx\, \Sigma dy^2}}$$

where $dx = x - \bar{x}$ and $dy = y - \bar{y}$.

The value of γ lies between +1 and –1. The value of one represents perfect correlation and value of zero indicates that there is no correlation between the two variables.

Regression Analysis

The co-efficient of regression gives the value by which one variable changes when there is a change in other variable.

The co-efficient of regression of y on x gives the value by which y varies for a unit change in x. It is expressed as $\frac{\gamma \cdot \sigma y}{\sigma x}$

similarly, co-efficient of regression of x on y is expressed as $\frac{\gamma \cdot \sigma x}{\sigma y}$

If y is independent variable and x is dependent variable, Then regression co-efficient of x on y is given by

$$\frac{\gamma \cdot \sigma x}{\sigma y} = \frac{\Sigma dx \cdot dy}{\Sigma (dy)^2}$$

If the deviations are taken from assumed means then the expression for regression of x on y is given as

$$\frac{\gamma \cdot \sigma y}{\sigma x} = \frac{\Sigma dxdy - \Sigma dx \cdot \Sigma dy/n}{\Sigma (dy)^2 - (\Sigma dy/n)^2}$$

If x is independent variable and y is dependent variable

Regression co-efficient of y on x is

$$\frac{r \cdot \sigma y}{\sigma x} = \frac{\Sigma dx \cdot dy}{\Sigma (dx)^2}$$

or $$\frac{\gamma \cdot \sigma y}{\sigma x} = \frac{\Sigma dx \cdot dy - \dfrac{\Sigma dx \cdot \Sigma dy}{n}}{\Sigma (dx)^2 - \left(\dfrac{\Sigma dx^2}{n}\right)}$$

Problem 10: The following data relates the cost of production and sales prices

	1986	1987	1988	1989	1990	1991	1992	1993	1994
Costs	203	216	223	239	248	253	279	301	311
Prices	225	242	250	271	275	277	255	318	329

Establish the co-efficient of correlation between costs and prices.

Solution:

Costs X	*Prices Y*	$X = x - \bar{x}$	$Y = y - \bar{y}$	X^2	XY	Y^2
203	225	49.6	– 50.8	2460	2520	2581
216	242	– 36.6	– 33.8	1340	1237	1142
223	250	– 29.6	– 25.8	876	764	666
239	271	– 13.6	– 04.8	185	65	23
248	275	– 4.6	– 0.08	21	4	1
253	277	– 0.4	0.2	0	0	1
279	295	26.4	19.2	697	507	369
301	318	48.4	42.2	2343	2042	1781
311	329	58.4	53.2	3411	3107	2830

$\Sigma x = 2273$ $\quad \Sigma y = 2482$ $\quad \Sigma x^2 = 11333$ $\quad \Sigma xy = 10246$ $\quad \Sigma y^2 = 9393$

$\bar{x} = 252.6$ $\quad \bar{y} = 275.8$

Co-efficient of correlation between variable x and y is given by

$$\gamma = \frac{\Sigma xy}{\sqrt{\Sigma x^2, \Sigma y^2}} = \frac{10246}{11333\ ,\ 9393} = 0.99$$

i.e., close correlation between costs and prices.

15.12. FORECAST ERROR

The demand for the product is forecasted using many forecasting methods. It is essential to have a good measure of effectiveness of the methods.

Forecast error is the numerical difference between the forecasted demand and the actual demand. The error should be minimum as far as possible.

Mean Absolute Deviation (MAD)

It is a measure of forecast error and it measures the average forecast error without direction.

It is calculated as the sum of the absolute value of the forecast error for all periods divided by the total number of periods.

$$\text{MAD} = \frac{\text{Sum of absolute value of forecast error for all periods}}{\text{number of periods}}$$

$$\sum_{j=1}^{n} = \frac{(\text{forecast error})}{N}$$

$$\sum_{j=1}^{n'} = \frac{(\text{Forecasted demand} - \text{Actual demand})}{n}$$

where n is the number of periods.

In MAD, errors are measured without considering signs.

Thus it expresses the magnitude but not the direction of error. This absolute value is referred as Mean Absolute Deviation.

BIAS

This measures the forecast error with regard to direction and shows any tendency to over forecast or under forecast.

This is also calculated as sum of forecast error for all periods divided by the total number of periods.

$$\text{BIAS} = \frac{\text{Sum of forecast error for all periods}}{\text{number of periods.}}$$

$$\sum_{j=1}^{n} = \frac{\text{Forecast Error}}{n}$$

$$\sum_{j=1}^{n} = \frac{(\text{Forecast demand} - \text{Actual demand})}{n}$$

Bias indicates the directional tendency of forecast errors. If the forecast repeatedly overestimates actual demand Bias will have a positive value and underestimation will be indicated by a negative bias.

Problem 11: A dealer for electrical appliances forecasts the demand for the Geyser at the rate of 500 per month for the next three months. The actual demands turned out to be 400, 560 and 700. Calculate the forecast error and bias, comment on the same.

Solution:
$$\text{MAD} = \sum_{j=1}^{n} \frac{(\text{Forecasted demand} - \text{Actual demand})}{3}$$

$$= (500-400) + (500-560) + (500-700)$$

$$= \frac{100+60+200}{3} = 120 \text{ units}$$

$$\text{Bias} = (500\text{-}400) + (500-560) + (500-700)$$

$$= \frac{100-60-200}{3} = -053 \text{ units}$$

In this case MAD is 120 units and bias is –53 units

Since MAD increases absolute error, we can conclude that it has very high degree of average error, *i.e.*, 24 per cent of the forecasted value of Geysers.

The Bias – The Forecasting method has a tendency to under estimate the forecast since bias is – ve, *i.e.*, it is about 9.6 per cent under forecast.

15.13. COSTS AND ACCURACY OF FORECASTS

There is a cost accuracy trade off while selecting a forecasting method. The more sophisticated methods involve high costs of implementation and maintenance, but they provide accurate forecasts which brings operational economy.

Thus, it is expected to balance the cost and accuracy and try to optimise the costs. The cost/accuracy trade off is given in Fig. 15.1.

Steps in Forecasting Process

1. Determine the purpose of the forecast.
   (The use to which the forecasted data is to be put)
2. Select the time horizon (period) for which the forecast is to be made.
   (Short term (up to one year) or Long-term)

3. Determine the forecasting methods to be used depending upon the accuracy required, availability of past data, cost.
4. Rather all the relevant information to be used in the forecast.
5. Make the forecast.

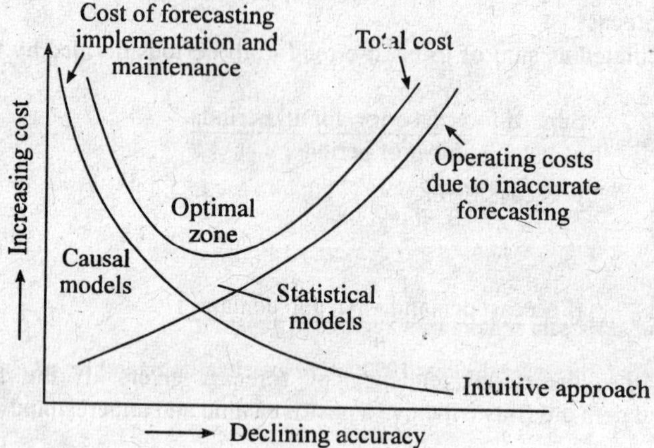

Fig. 15.1: Cost/accuracy of forecasts.

Requirements of a Good Forecasting Method

1. The method should be easy to understand and should be simple to use.
2. The method selected should give minimum forecast errors at the optimal cost.
3. The cost to make the forecast should be small.
4. The method selected should be stable in the sense that the changes should be minimum.

References For Further Reading

1. Buffa, Edwood, *Production and Operations Mgt.*
2. Granger C.W.J., *Forecasting in Business and Economics.*
3. Marks, *Production and Operations Management.*
4. Montgomery D.C. and L.A. Johnson, *Forecasting and Time Series Analysis*, McGraw Hill, New York. (1976).

REVIEW QUESTIONS

1. Explain the need for sales forecasting.
2. What is sales forecasting?
3. Explain the basis for classifying forecasts into short-term and long-term forecasting.
4. What are different methods of sales forecasting discuss the merits and limitations of sales forecasting method?
5. What is time series analysis? What are the components of time series? How the forecast is made from the time series.
6. What are seasonal variations? Why they are important for sales forecast?
7. How forecasting is useful in production scheduling?
8. Explain the Least Square method of forecasting.
9. When causal forecasting method is used?
10. When judgemental techniques are used for forecasting?
11. What are the disadvantages in using causal method?
12. Contrast forecasting and predicting. Give an example for each.
13. Discuss forecast errors.

14. Explain the cost accuracy trade off in forecasting.
15. Compare Moving Average and exponential smoothing method.
16. What are limitations of moving average method?

PROBLEMS

1. Following data refers to the past sales of one product:

Year	1982	1983	1984	1985	1986	1987	1988	1989	1990
Sales in Rs. (× 10000)	39	54	62	73	85	100	95	105	120

use least square method and estimate sales forecasting of year 1992.

2. The data given below refers to the past sales of production unit for last eleven years. Using least square method estimate sales forecast for next two years.

Year	1981	1982	1983	1984	1985	1986	1987	1988	1989	1990	1991
Sales Rs. (10000)	35	50	48	47	53	58	68	79	92	85	96

Plot the graph of sales data and draw the regression line on it.

3. Estimate the sales forecast for the year 1992, using exponential smoothing take $\alpha = 0.3$

Year	1984	1985	1986	1987	1988	1989	1990	1991
Sales Rs. (× 10)	180	168	159	170	188	205	190	210

Take the sales forecast of year 1984 as Rs. 16,00,000. Plot the actual sales and forecasted values on a suitable graph.

4. A company manufacturing tractors finds that there exists a relationship between sales of tractors and index of the agricultural income. The following data has been collected by the company for the last five years.

Years	1988	1989	1990	1991	1992
Demand ('000)	100	112	130	150	280
Demand index	125	140	180	190	220

(*a*) Fit a regression line.
(*b*) Estimate the sales of tractors for the year 1993. For the given index.

Ans. (*a*) $y = 14.62 + 1.137\,x$
(*b*) 29892 nos.

5. The following information gives the sales of the company for ten months:

Month	*Sales*
Jan.	700
Feb	1280
Mar.	840
Apr.	920
May	1020
June	900
July	1276
Aug.	1440
Sept.	1610
Oct.	1500

If the smoothing factor of 0.5 is used, forecast the demand for November. **(Ans.1475)**

6. The past sales of the three months is given below for the year 1996.

October	November	December
300	350	400

The forecasted demand for October was 315 units.
Forecast the sales for Jan. 1997 assuming $\alpha = 0.4$.

16

PRODUCTION PLANNING AND CONTROL

• Introduction • Need for production planning and control • Objectives of PPC • Functions of PPC • Comparison of production planning and production control • Information requirement of PPC • Production procedure • Organisation for PPC • Centralised and decentralised PPC.

16.1. INTRODUCTION

Production activity constitutes the transformation of materials into a desirable output (Products). Production consists of a series of sequential operations to produce a desirable product acceptable to customer and meets the customer demand, with respect to the quality and intended function. Production is an organised activity which has got specific objectives. "The efficiency of production system is stated in terms of its ability to produce the products with required quantity and specified quality at predetermined cost and pre-established time." Production planning and control is a tool available to the management to achieve the stated objectives. Thus, a production system is encompassed by the four factors, *i.e.*, quantity, quality, cost and time. Production Planning starts with the analysis of the given data, *i.e.*, demand for products, delivery schedule, etc., and on the basis of the information available, a scheme of utilisation of firms resources like machines, materials and men are worked out to obtain the target in the most economical way.

Once the plan is prepared, then operations (execution of plan) are performed in line with the details given in the plan. Production control comes into action if there is any deviation between the actual and planned. The corrective action is taken so as to achieve the targets set as per plan by using control techniques.

Thus production planning and control can be defined as the "**direction and coordination of firms resources towards attaining the prefixed goals.**" Production planning and control helps to achieve uninterrupted flow of materials through production line by making available the materials at right time and required quantity.

16.2. NEED FOR PPC

The present techno-economic scenario of India emphasize on competitiveness in manufacturing. To be in competition, Indian industries have to streamline the production activities and attain the maximum utilisation of firms resources to enhance the productivity. Production planning and control serves as a useful tool to coordinate the activities of the Production system by proper planning and control system.

Production system can be compared to the nervous system with PPC as a brain. Production Planning and Control is needed to achieve:

1. Effective utilisation of firms resources.

2. To achieve the production objectives with respect to quality, quantity, cost and timeliness of delivery.
3. To obtain the uninterrupted production flow in order to meet customers varied demand with respect to quality and committed delivery schedule.
4. To help the company to supply a good quality products to the customer on the continuous basis at competitive rates.

16.3. PRODUCTION PLANNING AND PRODUCTION CONTROL

Production planning is a pre-production activity. It is the pre-determination of manufacturing requirements such as manpower, materials, machines and manufacturing process.

Ray wild defines "Production planning is the determination, acquisition and arrangement of all facilities necessary for future production of products. It represents the design of production system. Apart from planning the resources, it is going to organise the production.

Based on the estimated demand for company's products, it is going to establish the production programme to meet the targets set using the various resources.

Production Control

Inspite of planning to the minute details, yet always (most of the time) it is not possible to achieve production 100 per cent as per the plan. There may be innumerable factors which affect the production system and because of which there is a deviation from the actual plan. Some of the factors that affect are—

1. Non-availability of materials (due to shortage, etc.).
2. Plant, equipment and machine breakdown.
3. Changes in demand and rush orders.
4. Absenteeism of workers.
5. Lack of coordination and communication between various functional areas of business.

Thus, if there is a deviation between actual production and planned production, the control function comes into action. Production control through control mechanism tries to take corrective action to match the planned and actual production. Thus production control reviews the progress of the work, and takes corrective steps in order to ensure that programmed production takes place. The essential steps in control activity are:

- Initiating the production.
- Progressing.
- Corrective action based upon the feedback and reporting back to the Production Planning.

16.4. OBJECTIVES OF PPC

1. Systematic Planning of production activities to achieve the highest efficiency in production of goods/services.
2. To organise the production facilities like machines, men, etc., to aehieve stated production objectives w.r.t. quantity and quality time and cost.
3. Optimum Scheduling of resources.
4. Coordinate with other departments relating to production to achieve regular balanced and uninterrupted production flow.
5. To conform to delivery commitments.
6. Materials planning and control.
7. To be able to make adjustments due to changes in demand and rush orders.

16.5. FUNCTIONS OF PPC

The functions of Production Planning and controlling are depicted in the Fig. 16.1. Pre-planning is a macro level planning and deals with analysis of data and is an outline of the planning policy based upon the forecasted demand, market analysis and product design and development. This stage is concerned with process design (new processes and developments, equipment policy and replacement and work flow (Plant layout). The preplanning function of PPC is concerned with decision-making with respect to methods, machines and work flow with respect to availability, scope and capacity.

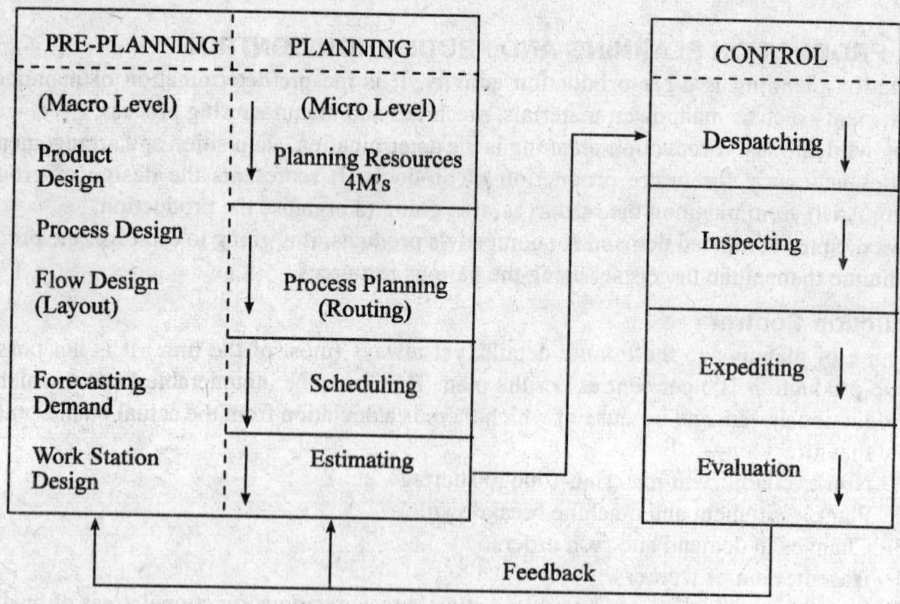

Fig. 16.1: Functions of Production Planning and Control.

The planning function starts once the task to be accomplished is specified, with the analysis of 4 M's, *i.e.*, Machines, Methods, Materials and Manpower. This is followed by process planning (routing). Both short-term (near future) and long-term planning are considered. Standardisation, simplification of products and processes are given due consideration.

Control phase is effected by despatching, inspection and expediting Materials control, analysis of work-in-process. Finally, evaluation makes the PPC cycle complete and corrective actions are taken through a feedback from analysis. A good communication, and feedback system is essential to enhance and ensure effectiveness of PPC.

The Main Functions of PPC

1. **Materials function:** Raw materials, finished parts and bought out components should be made available in required quantities and at required time to ensure the correct start and end for each operation resulting in uninterrupted production. The function includes the specification of materials (quality and quantity) delivery dates, variety reduction (standardisation) procurement and make or buy decisions.

2. **Machines and equipment:** This function is related with the detailed analysis of available production facilities, equipment down time, maintenance policy procedure and schedules. Concerned with economy of jigs and fixtures, equipment availability. Thus, the duties include the analysis of facilities and making their availability with minimum down time because of breakdowns.

3. **Methods:** This function is concerned with the analysis of alternatives and selection of the best method with due consideration to constraints imposed. Developing specifications for processes is an important aspect of PPC and determination of sequence of operations.

4. **Process planning (Routing):** It is concerned with selection of path or route which the raw material should follow to get transformed into finished product. The duties include,

(*a*) Fixation of path of travel giving due consideration to layout.

(*b*) Breaking down of operations to define each operation in detail.

(*c*) Deciding the set up time and process time for each operation.

5. **Estimating:** Once the overall method and sequence of operations is fixed and process sheet for each operation is available, then the operations times are estimated. This function is carried out using extensive analysis of operations along with methods and routing and a standard time for operation are established using work measurement techniques.

6. **Loading and scheduling:** Scheduling is concerned with preparation of machine loads and fixation of starting and completion dates for each of the operations. Machines have to be loaded according to their capability of performing the given task and according to their capacity. Thus the duties include:

(*a*) Loading the machines as per their capability and capacity.

(*b*) Determining the start and completion times for each operation.

(*c*) To coordinate with sales department regarding delivery schedules.

7. **Dispatching:** This is the execution phase of planning. It is the process of setting production activities in motion through release of orders and instructions. It authorises the start of production activities by releasing materials, components, tools, fixtures and instruction sheets to the operator. The activities involved are,

(*a*) To assign definite work to definite machines, work centres and men.

(*b*) To issue required materials from stores.

(*c*) To issue jigs, fixtures and make them available at correct point of use.

(*d*) Release necessary work orders, time tickets, etc., to authorise timely start of operations.

(*e*) To record start and finish time of each job on each machine or by each man.

8. **Expediting:** This is the control tool that keeps a close observation on the progress of the work. It is a logical step after dispatching which is called "follow-up" or "progress". It coordinates extensively to execute the production plan. Progressing function can be divided into three parts, *i.e.*, follow up of materials, follow up of work-in-process and follow up of assembly. The duties include:

(*a*) Identification of bottlenecks and delays and interruptions because of which the production schedule may be disrupted.

(*b*) To devise action plans (remedies) for correcting the errors.

(*c*) To see that production rate is in line with schedule.

9. **Inspection:** It is a major control tool. Though the aspects of quality control are the separate function, this is of very much important to PPC both for the execution of the current plans and its scope for future planning. This forms the basis for knowing the limitations with respects to methods, processes, etc., which is very much useful for evaluation phase.

10. **Evaluation:** This stage though neglected is a crucial to the improvement of productive efficiency. A thorough analysis of all the factors influencing the production planning and control helps to identify the weak spots and the corrective action with respect to preplanning and planning will be effected by a feedback. The success of this step depends on the communication, data and information gathering and analysis.

16.6. COMPARISON BETWEEN PRODUCTION PLANNING AND PRODUCTION CONTROL

	Production Planning	*Production Control*
1.	Production Planning is a pre-production activity.	Production control will be in action when production activity begins.
2.	Planning involves the collection, maintenance and analysis of data with respect to time standards, materials and their specification, machines and their process capabilities.	Control is concerned with communication of their information and producing reports like output reports, productivity, rejection rate, etc.
3.	Planning is useful to anticipate the problems and devising remedial measure in case the problem arises.	Control involves in taking corrective steps in case of error to match actual performance against the planned performance.
4.	Planning is a centralised activity and includes functions like materials control, tool control, process planning and control.	Control is a widespread activity. Includes functions such as dispatching programming and inspection, etc.
5.	• Planning sees that all the necessary resources are available to make the production at right quality and time.	Control keeps track of the activities and sees whether everything is going as per schedule or not.

16.7. INFORMATION REQUIREMENT OF PPC

The effectiveness of production planning and control depends to a greater extent upon the accuracy of the information it gets from other departments. The following information is vital to the success of PPC function in an organisation. The information required and its sources and the department responsible for providing the information is given below in Table 16.1.

Table 16.1: Information Required for PPC

	Information Required	*Sources of Information*	*Department*
1.	Production Programme • Quantity to be produced • Delivery date • Variety and different models and special features	The sales order or the order accepted by the marketing deptt.	Marketing department.
2	Quality Standards–Specifications and tolerances.	Engineering or design department who translate the customers needs into specifications.	Engineering purchase and stores.
3.	Production materials • Types of materials • Quality and quantity • Procurement lead time • Stock Position	Drawing and Bill of materials (BOM) Material Stock cards.	PPC.
4.	Toolings	Standard and special toolings.	PPC Deptt.
5.	Operational Details Sequence of operations Process capability of machines and equipment's Jigs and fixtures needed Cutting parameters or Process parameters	Process sheets Load charts Process capability studies	Industrial Engineering.
6.	Standard times for operation and set up time.	Work measurement data.	PPC.
7.	Starting and finishing dates.	Machine load and schedule charts.	Production.
8.	Progress of work (Status of work).	Production reports.	Production.

16.8. PRODUCTION PROCEDURE

Always there exists a difference between two plants or manufacturing units even though they are manufacturing the same product. The difference may be because of many reasons viz., the technology, manufacturing processes, types of layouts, organisation structure and management policies. This difference is the root cause for different practices of production, planning and control systems. The PPC function is not equally important in all the different industries. The PPC function changes from industry to industry. The PPC functions depend upon the type of manufacturing system like jobbing, batch, mass and flow production systems. It is the job shop unit which requires all the function of PPC.

Production procedure (cycle) starts with the customer and ends up with satisfying the needs of the customer by delivering products. Production Procedure consists of:

1. **Sales forecast:** The marketing or sales department after a thorough analysis and market research comes out with details like acceptability of the product by customers, consumers reactions to new modifications and designs. Based upon the analysis of the data, sales department prepares a sales forecast with breakdown of products and models as a function of time periods. Detailed forecast is submitted to the management.
2. **Preparation of production budget:** The production budget is prepared by the finance department in consultation with production department. The management reviews the forecast and the budget to take decision regarding annual quantities to be produced.
3. **Engineering department to prepare details:** The engineering department is instructed to prepare drawing, B.O.M (Bill of Materials) specifications or to check and modify the existing ones.
4. **Production Planning activity:** Production planning activity begins as soon as the technical information is received from the engineering department. The production planning activity results in a schedule or time table of production. The inventory levels are checked in order to initiate procurement activity of materials. Make or buy decision is made. The production planning section supplies the compete data on methods, process sheets, machine loading and production schedule to the dispatching section.
5. **Dispatching.** Detailed production orders are dispatched to the shop specifying what, how and when and where the operations are to be performed.
6. **Progressing.** Control action is exercised throughout the manufacturing period and progress is continuously compared with planned schedule so that suitable corrective steps are taken in case of difference between planned and actual production.
7. **Inspeection.** Inspections are carried out and quality control ensures that the desired specifications are in conformance with the actual.
8. **Evaluation.** Evaluation is carried out after and before production so that corrective actions are devised to improve methods, down times and finally the management gets reports from both production and financial departments.
9. The finished product is transferred to stock.
10. Finally, the product is delivered to the customer—Thus, the production procedure requires the coordinated effort of all the functional departments of the organisation. The production procedure is shown in Fig. 16.2.

16.9. ORGANISATION FOR PPC

The status of production planning and control department in the organisation depends upon many factors such as degree of centralisation desirable, composition of internal structure of PPC, manufacturing system and management policy.

1. In case of highly repetitive type of production with higher degree of automation, the planning work is carried out by line staff. There may not be a formal planning department.

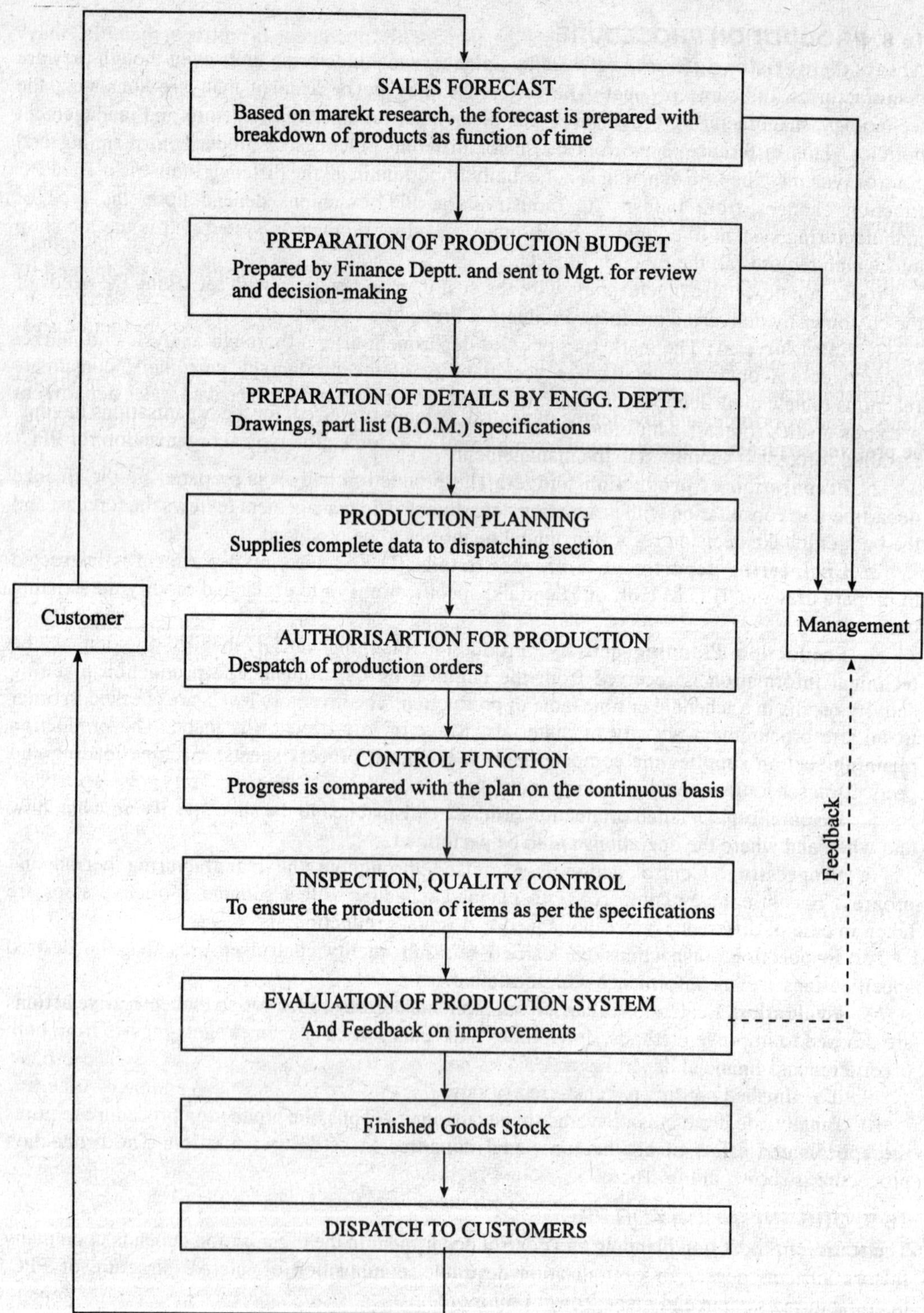

Fig. 16.2: Production procedure.

2. For the product type of layout where the sequence of operations is prefixed, then PPC may be the part of manufacturing department itself.
3. An independent PPC department is to be set-up if each product requires varying capacity, product variety is more, requires different sequence of operations with a functional layout. A separate department with a PPC head reporting directly to the works manager should exist.

Centralised and Decentralised PPC

In centralised planning the functions of production planning are controlled by a staff specialist. Centralised planning is more effective in case of multi-product, multi-plant organisations and it takes away the burden of planning from line function to allow them to concentrate on manufacturing. The decentralised planning involves the line staff in planning the production and this is going to take away the majority of their time in performing functions.

Higher degree of centralisation is recommended for multi-product (large variety), multiple plants, large workforce, and low degree of centralisation is preferred, for the organisations having the prefixed sequence of operation and having small workforce. The typical organisation for PPC in an organisation is shown in Fig. 16.3.

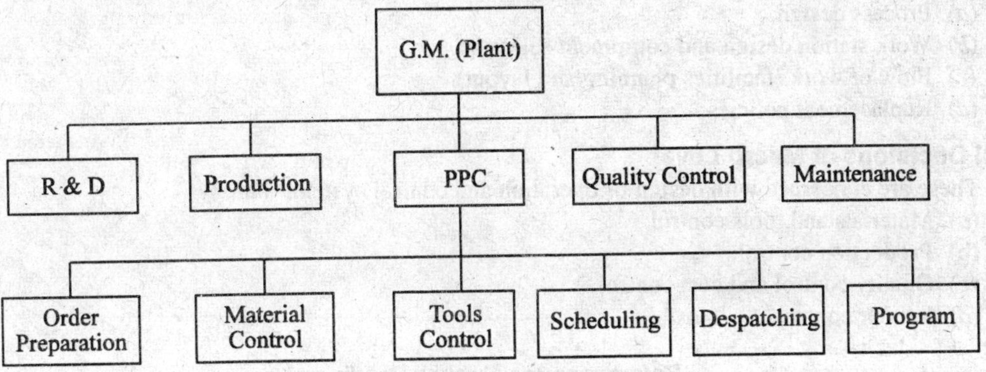

Fig. 16.3: Typical organisation for PPC in a firm.

16.10. MANUFACTURING METHODS AND PPC

1. Job Production–Functions of PPC

(*a*) Materials are purchased on receipt of the order.

(*b*) Standard tools are stocked and special tools are either made in house or purchased from outside.

(*c*) Process planning activity normally does not exist. Based upon the drawings and specification supervisor decides the work methods, fixes up the machines and estimates time for completion of the operation.

(*d*) Schedule is prepared to mark the beginning and finish of each activity. The day-to-day scheduling is at the descretion of the supervisor.

(*e*) Progressing is through the meeting with the supervisors.

2. Batch Production–Functions of PPC

Functions are more complex.

(*i*) Materials control and tools control are more important and systematic stock replenishment system is essential.

(*ii*) Detailed route sheets (Process sheets) are prepared.

(*iii*) Loading and scheduling are to be worked out with greater details.

(*iv*) Progressing function is crucial as the detailed data is to be collected on the progress of the work.

3. Continuous Production—PPC Functions

(*a*) Materials function is critical.

(*b*) No tools control because of nature of the plant.

(*c*) No process planning activity.

(*d*) Scheduling is restricted to final quantity required.

(*e*) Progressing requires only recording of final production quantity.

16.11. PROBLEMS OF PRODUCTION PLANNING AND CONTROL

The production planning and control has two problem areas–One is at macro level and long-term (design of system) and second is the control (short-term and at micro level). The classification of problem area are:

(A) Decisions at Macro Level

These are the decisions regarding the design of production system.

(*a*) Process design.

(*b*) Work station design and equipment selection.

(*c*) Flow of work (facilities planning and layout).

(*d*) Replacement policies.

(B) Decisions at Macro Level

These are concerned with design of operation and control systems namely:

(*a*) Materials and tools control.

(b) Production control.

(*c*) Quality control and cost control.

(*d*) Labour control.

References for Further Reading

1. Samuel Eilon, *Elements of Production Planning and Control*, Universal Publishing Corporation, Mumbai.
2. Ray Wild, *Production and Operations, Management*, Holt, Rinehart and Winston, London (1980).
3. James Riggs, *Production Systems, Planning, Analysis and Control*, John Willey & Sons .
4. Joseph Monks, *Operations Management—Theory and Problems*, 3rd edn., McGraw Hill Book Company, New York, (1987).

REVIEW QUESTIONS

1. What is the need for production planning and control in the organisation?
2. Explain the functions of PPC.
3. Planning the production is the pre-operation activity.
4. Explain w.r.t. to the manufacturing organisations.
5. What are objectives of PPC and how they are achieved?
6. Compare production planning and production control.
7. What information is essential for efficient working of PPC?
8. Explain the relationship of PPC with other departments.

9. Describe production cycle (Production procedure).
10. Explain centralised and decentralised PPC.
11. Production planning and control is the nerve centre of the organisation. Explain.
12. State the production problem area.
13. PPC contributes to effective utilisation of firms resources. Comment.
14. State functions of PPC for various types of manufacturing methods.

17

CAPACITY PLANNING

Introduction. • Measurement of capacity • Measures of capacity • Capacity planning • Estimating future capacity needs • Factors influencing effective capacity • Factors favouring over capacity and under capacity • Aggregate planning • Aggregate planning guidelines • Linear programming approach to aggregate planning • Master production schedule • Aggregate planning methods, advantages and limitations.

17.1. INTRODUCTION

Design of the production system involves planning for the inputs, conversion process and outputs of production operation. The effective management of capacity is the most important responsibility of production management. The objective of capacity management (*i.e.*, planning and control of capacity) is to match the level of operations to the level of demand.

Capacity planning is to be carried out keeping in mind future growth and expansion plans, market trends, sales forecasting, etc. It is a simple task to plan the capacity in case of stable demand. But in practice the demand will be seldom stable. The fluctuation of demand creates problems regarding the procurement of resources to meet the customer demand. Capacity decisions are strategic in nature. Capacity is the rate of productive capability of a facility. Capacity is usually expressed as volume of output per period of time.

Production managers are more concerned about the capacity for the following reasons:

- Sufficient capacity is required to meet the customers demand in time.
- Capacity affects the cost efficiency of operations.
- Capacity affects the scheduling system.
- Capacity creation requires an investment.

Capacity planning is the first step when an organisation decides to produce more or new products.

17.2. MEASUREMENT OF CAPACITY

It is easy and simple to measure the capacity of the unit manufacturing homogeneous tangible products which can be counted. The capacity of such units can be expressed in number of units of output per period. For example, the capacity of an automobile unit is expressed as number of vehicles produced per month. The capacity of steel plant is expressed as millions of tonnes per annum. Capacity of textile mill is expressed as metres of cloth per day.

It is difficult to express capacities when the company manufactures multiple products and some of the products requiring common facilities and others specialised facilities. In this situation measuring capacity is more complicated. In such situations, the capacity is not expressed as output per period of time but usually expressed as man-hours, machine hours or sometimes in terms of applicable resources.

Examples

- A job shop can measure its capacity in machine hours and/or man-hours.
- For hospitals it is expressed as bed days per month.
- The transport system is expressed in seat kms per month.

17.3. MEASURES OF CAPACITY

1. **Design capacity:** Designed capacity of a facility is the planned or engineered rate of output of goods or services under normal or full scale operating conditions.

For example, the designed capacity of the cement plant is 100 TPD. (Tonnes per day)

Capacity (designed) of the sugar factory is 150 tonnes of sugar cane crushing per day.

2. **System capacity:** System capacity is the maximum output of the specific product or product mix the system of workers and machines is capable of producing as an integrated whole. System capacity is less than design capacity or at the most equal it because of the limitation of product mix, quality specification, breakdowns. The actual is even less because of many factors affecting the output such as actual demand, downtime due to machine/equipment failure, unauthorised absenteeism.

The capacities and output relationship is shown in Fig. 17.1.

The system capacity is less than design capacity because of long range uncontrollable factors.

The actual output is still reduced because of short-term effects such as breakdown of

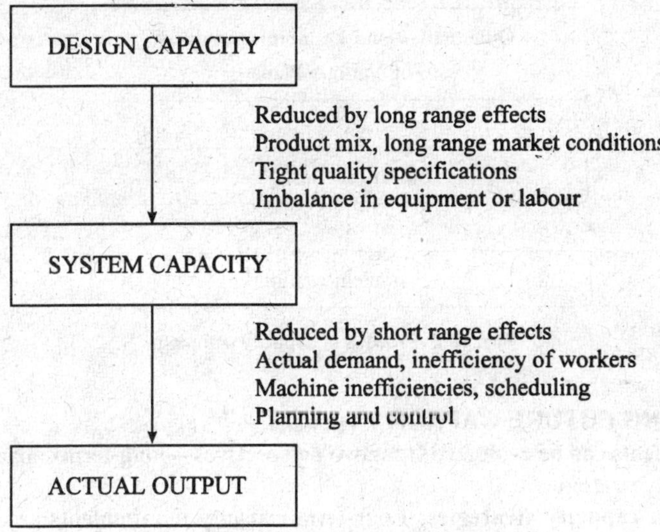

Fig. 17.1: Capacity and output relationship.

equipment, inefficiency of labour. The system efficiency is expressed as ratio of actual measured output to the system capacity.

$$\text{System efficiency (SE)} = \frac{\text{Actual Output}}{\text{System capacity}}$$

3. **Licensed capacity:** Capacity licensed by the various regulatory agencies or government authorities. This is the limitation on the output exercised by the government.

4. **Installed capacity:** The capacity provided at the time of installation of the plant is called installed capacity.

5. **Rated capacity:** Capacity based on the highest production rate established by actual trials is referred to as rated capacity.

17.4. CAPACITY PLANNING

Capacity planning is concerned with defining the long-term and the short-term capacity needs of an organisation and determining how those needs will be satisfied.

Capacity planning decisions are taken based upon the consumer demand (market) and this is merged with the human, material and financial resources of the organisation. The process of capacity planning is shown in Fig. 17.2.

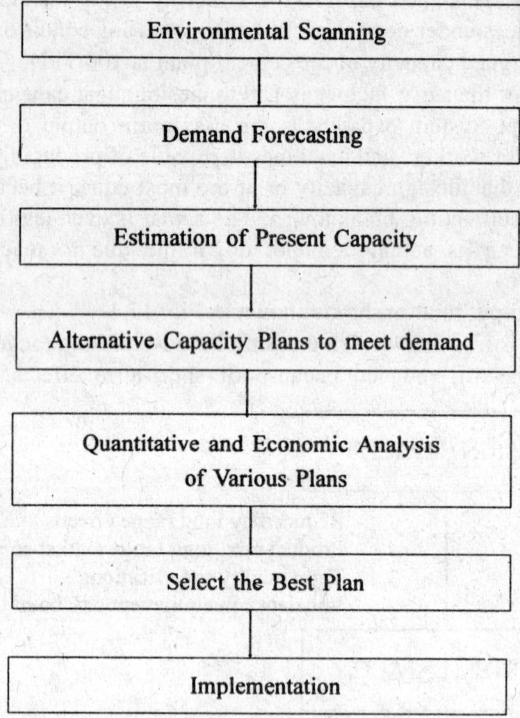

Fig. 17.2: Process of capacity planning.

17.5. ESTIMATING FUTURE CAPACITY NEEDS

Capacity requirements can be evaluated from two perspectives—long-term capacity strategies and short-term capacity strategies.

1. **Long-term capacity strategies:** Long-term capacity requirements are more difficult to determine because the future demand and technology are uncertain. Forecasting for Five or Ten years into the future is more risky and difficult. Even sometimes company's today's products may not be existing in the future. Long range capacity requirements are dependent on marketing plans, product development and life-cycle of the product. Long-term capacity planning is concerned with accommodating major changes that affect overall level of the output in long-term (more than one year). Marketing environmental assessment and implementing the long-term capacity plans in a systematic manner are the major responsibilities of management.

- **Multiple products:** Company's produce more than one product using the same facilities in order to increase the profit. The manufacturing of multiple products will reduce the risk of failure. Having more than on product helps the capacity planners to do a better job. Because products are in different stages of their life-cycles, it is easy to schedule them to get maximum capacity utilisation.

- **Phasing in capacity:** In high technology industries, and in industries where technology developments are very fast, the rate of obsolescence is high. The products should be brought into the market quickly. The time to construct the facilities will be long and there is no much time as the products should be introduced into the market quickly. Here the solution is phase in capacity on modular basis. Some commitment is made for building funds and men towards facilities over a period of 3-5 years. This is an effective way of capitalising on technological breakthrough.
- **Phasing out capacity:** The outdated manufacturing facilities cause excessive plant closures and down time. The impact of closures are not limited to only fixed costs of plant and machinery. Thus, the phasing out here is done with humanistic way without affecting the community. The phasing out options makes alternative arrangements for men like shifting them to other jobs and/or to other locations, compensating the employees, etc.

2. **Short-term capacity strategies:** Managers often use forecasts of product demand to estimate the short-term workload the facility must handle. Managers looking ahead up to 12 months, anticipate output requirements for different products, and services. Managers then compare requirements with existing capacity and then take decisions as to when the capacity adjustments are needed.

For short term periods of up to one year, fundamental capacity is fixed. Major facilities will not be changed. Many short term adjustments for increasing or decreasing capacity are possible. The adjustments to be required depend upon the conversion process like whether it is capital intensive or labour intensive or whether product can be stored as inventory.

Capital intensive processes depend on physical facilities, plant and equipment. Short-term capacity can be modified by operating these facilities more or less intensively than normal. In labour intensive processes short-term capacity can be changed by laying off or hiring people or by giving overtime to workers. The strategies for changing capacity also depends upon how long the product can be stored as inventory.

The short-term capacity strategies are represented in Table. 17.1.

Table 17.1: Short-term capacity strategies

Short term capacity strategies—

- **Inventories:** Stock finished goods during slack periods to meet the demand during peak periods.
- **Backlogs.** During peak periods, the willing customers are requested to wait, and their orders are fulfilled after the peak demand period.
- **Employment level (hiring and firing):** Hire additional employees during peak demand period and layoff employees as demand decreases.
- **Employee training:** Develop multi-skilled employees through training so that they can be rotated among different jobs. The multi-skilling helps as an alternative to hiring employees.
- **Workforce utilisation:** Employees are made to work overtime during peak hours and work fewer hours during slack hours (flexible work hours).
- **Subcontracting:** During peak periods, hire the capacity of other firms temporarily to make the component parts or products.
- **Process design:** Change job content by redesigning the job.
- **Maintenance:** Temporarily discontinue routine maintenance so that the this time can be utilised for production.

17.6. FACTORS INFLUENCING EFFECTIVE CAPACITY

The effective capacity is influenced by—(1) Forecasts of demand, (2) Plant and labour efficiency, (3) Subcontracting, (4) Multiple shift operation, (5) Management policies.

1. **Forecasts of demand:** Demand forecast is going to influence the capacity plan in a significant way. As such, it is very difficult to forecast the demand with accuracy as it changes significantly with the product life-cycle stage, number of products. Products with long life-cycle usually exhibit steady demand growth compared to one with shorter life-cycle. Thus the accuracy of forecast influences the capacity planning.

2. **Plant and labour efficiency:** It is difficult to attain 100 per cent efficiency of plant and equipment. The efficiency is less than 100 per cent because of the enforced idle time due to machine breakdown, delays due to scheduling and other reasons. The plant efficiency varies from equipment to equipment and from organisation to organisation. Labour efficiency contributes to the overall capacity utilisation. The standard time set by industrial engineer is for a representative or normal worker. But the actual workers differ in their speed and efficiency. The actual efficiency of the labour should be considered for calculating efficiency. Thus plant and labour efficiency are very much essential to arrive at realistic capacity planning.

3. **Subcontracting:** Subcontracting refers to off loading, some of the jobs to outside vendors thus hiring the capacity to meet the requirements of the organisation. A careful analysis as to whether to make or to buy should be done. An economic comparison between cost to make the component or buy the component is to made to take the decision.

4. **Multiple shift operation:** Multiple shifts are going to enhance the firms capacity utilisation. But specially in the third shift the rejection rate is higher. Specially for process industries where investment is very high it is recommended to have a multiple shifts.

5. **Management policy:** The management policy with regards to subcontracting, multiplicity of shifts (decision regarding how may shifts to operate), which work stations or departments to be run for third shift, machine replacement policy, etc., are going to affect the capacity planning.

17.7. FACTORS FAVOURING OVER CAPACITY AND UNDER CAPACITY

It is very difficult to forecast demand as always there is an uncertainty associated with the demand. The forecasted demand will be either higher or lower than the actual demand. So always there is a risk involved in creating capacity based on projected demand. This gives rise to either over capacity or under capacity.

The over capacity is preferred when:

(*a*) Fixed cost of the capacity is not very high.
(*b*) Subcontracting is not possible because of secrecy of design and/or quality requirement.
(*c*) The time required to add capacity is long.
(*d*) The company cannot afford to miss the delivery, and cannot afford the loose the customer.
(*e*) There is a economic capacity size below which it is not economical to operate the plant.

The under capacity is preferred when :

(*a*) The time to build capacity is short.
(*b*) Shortage of products does not affect the company (*i.e.*, lost sales can be compensated).
(*c*) The technology changes fast, *i.e.*, the rate of obsolescence of plant and equipment is high.
(*d*) The cost of creating the capacity is prohibitively high.

17.8. AGGREGATE PLANNING

Aggregate planning is an intermediate term planning decision. It is the process of planning the quantity and timing of output over the intermediate time horizon (3 months to one year). Within this range, the physical facilities are assumed to be fixed for the planning period. Therefore, fluctuations in demand must be met by varying labour and inventory schedule. Aggregate planning seeks the best combination to minimise costs.

Aggregate Planning Strategies

The variables of the production system are labour, materials and capital. More labour effort is required to generate higher volume of output. Hence, the employment and use of overtime (OT) are the two relevant variables. Materials help to regulate output. The alternatives available to the company are inventories, back ordering or subcontracting of items.

These controllable variables constitute pure strategies by which fluctuations in demand and uncertainties in production activities can be accommodated.

1. **Vary the size of the workforce:** Output is controlled by hiring or laying off workers in proportion to changes in demand.
2. **Vary the hours worked:** Maintain the stable workforce, but permit idle time when there is a slack and permit overtime (OT) when demand is peak.
3. **Vary inventory levels:** Demand fluctuations can be met by large amount of inventory.
4. **Subcontract:** Upward shift in demand from low level. Constant production rates can be met by using subcontractors to provide extra capacity.

Strategies to Meet Non-uniform Demand

1. **Strategy:** Absorb demand fluctuations by varying inventory level, back ordering or Shifting demand.

Method	*Costs*
• Produce in earlier period and hold until product is demanded.	• Cost of holding inventory.
• Offer to deliver the products later when capacity is available.	• Delay in receipt of revenue lost sales and customer dissatisfaction.
• Special marketing efforts to shift the demand to slack period.	• Cost of advertising, discounts or promotional programmes.

2. **Strategy:** Change only the production rate in accordance with non-uniform demand.

Method	*Costs*
• Work additional hours without changing the workforce size.	• Overtime premium wages.
• Increase workforce size for high production so that overtime is avoided.	• Excess wages during slack period.
• Subcontract work to other firms.	• Reduce company overheads and increase subcontractors profit.
• Revise make or buy decisions to purchase items when capacity is fully loaded.	• Waste of company skills, tooling and equipment unutilised in slack periods.

3. **Strategy:** Change the size of the workforce to vary production level in accordance with demand.

Method	*Cost*
• Hire additional personnel as demand increases.	• Employment costs for advertising and recruitment, cost of additional shift, if shift is added
• Lay off personnel as demand decreases.	• Cost of compensation to workers for lay off.

17.9. AGGREGATE PLANNING GUIDELINES

1. Determine corporate policy regarding controllable variables.
2. Use a good forecast as a basis for planning.
3. Plan in proper units of capacity.
4. Maintain the stable workforce.
5. Maintain needed control over inventories.
6. Maintain flexibility to change.
7. Respond to demand in a controlled manner.
8. Evaluate planning on a regular basis.

***Problem*:** ABC company produces toilet soaps at their works in Mumbai. Aggregate planning measures used by ABC is tonnes of soap which includes making and packaging of the soap. The planning is done for a time horizon of one year and for 4 quarters.

Quarter	I	II	III	IV
Demand (tonnes)	40	60	50	45

The company has a regular workforce which can produce 35 tonnes of output per quarter. If the workers are allowed to work overtime with a restriction that the extra time cannot be more than 20 per cent of the regular time in any time. The output rate is 25 per cent higher than regular time during overtime but the overtime expenses are 40 per sent more than that of regular time. The company subcontracts the soap making and packaging operation to a SSI unit but only at the cost of 50 per cent premium than the cost of regular production. The regular time production costs are Rs. 10,000 per tonne.

No shortages are allowed as per company policy. Inventory carrying costs are Rs. 5,000 pe tonne per annum.

Design the cost efficient aggregate plan assuming zero starting inventory. Compute total production cost.

***Solution*:** Production planning options are:

- Regular time production.
- Overtime production.
- Subcontracting.
- Carrying inventory.

The regular time production with a combination of remaining options (one or more) can be used to make the production plan.

There is a limitation on the regular time production. It cannot be more than 35 tonnes/quarter but the demand exceeds this in each of the four quarters. So along with (RT) regular time, other options are to be used and initial inventory is zero.

In order to prioritise these options, compute the cost per unit of production.

The costs of different options are,

1. Regular time production = 10,000 Rs. per tonne
2. Overtime production = $10,000 + 0.40 \times 10,000 = 14000$ Rs./tonne
3. Subcontracting = $10,000 + 0.50 \times 10,000 = 15,000$ Rs./tonne
4. Produce in OT and carry the inventory for 1 quarter = $14,000 + 5000/4 = 15,250$ per tonne

The options are preferred based upon the cost.

The maximum allowable products are produced in regular time. Then the preference is given to OT production but on this also there is a restriction.

Subcontracting can be done to any extent there is no limit and hence 4th option is not at all preferred.

In any quarter, the maximum amount that can be produced by overtime = 0.20 × 35 × 1.25 = 8,750 tonne.

In any quarter, the maximum amount that can be produced in a regular plus overtime is: 35 + 8.750 = 43,750 tonne.

In any quarter, if the demand exceeds 43,750 tonnes, then it is met by subcontracting.

Since there are no shortages are allowed, the entire demand in any quarter will have to be met in that quarter.

Quarter	*Output*			
	Regular Production (tonnes)	*Overtime (tonnes)*	*Subcontracting*	*Total (tonnes)*
I	35	5.00	–	40
II	35	8.75	16.25	60
III	35	8.75	6.25	50
IV	35	8.75	1.25	45

Production costs are computed for these plans are as follows –

Quarter I (35 × 10,000) + 5 × 14,000 = 4,20,000

Quarter II (35 × 10,000) + 8.75 × 14,000 + 16.25 × 15,000) = 7,16,250

Quarter III (35 × 10,000) + 8.75 × 14,000 + 6.25 × 15,000 = 5,66,250

Quarter IV (35 × 10,000) + 8.75 × 14,000 + 1.25 × 15000 = 4,91,250

Total Production Costs Rs. 21,93,750 **(Ans)**

***Example*:** The forecasted demand for a product for 6 months cycle is shown. Each unit requires 10 man hours and labour cost is Rs. 6 per hour regular time and Rs. 9 per hour overtime. The total cost per unit is estimated to be Rs. 200 and can be subcontracted at the cost of Rs. 208/unit. Currently there are 20 workers employed and hiring and training costs for additional workers are Rs. 300 per person where as lay off costs are Rs. 400 per person. Company policy is to retain a safety stock equal to 20 per cent of the monthly forecast and each months safety stock becomes the beginning inventory for the next month. There are currently 50 units in stock carried at a cost of Rs. 2 per month. Stock out cost is Rs. 20 per unit per month.

	Jan.	Feb.	Mar.	Apl.	May	June
Forecasted demand	300	500	400	100	200	300
Work days	22	19	21	21	22	20
Worker hr at 8/day	176	152	168	168	176	160

Three aggregate plans are proposed.

Plan 1: Vary the workforce size to accommodate demand.

Plan 2: Maintain a constant workforce of 20 and use overtime and idle time to meet demand.

Plan 3: Maintain constant workforce of 20, and build inventory or incur stockout cost. The firm must begin January with the 50 unit inventory on hand. Compared the costs of three plans.

***Solution*:** Determine what will be the production requirements as adjusted to include a safety stock of 20 per cent of the next months forecast. Starting with January inventory of 50 units, each subsequent months inventory reflects the difference between the forecasted demand and the production requirements of the previous month.

Month	*Forecasted demand*	*Cumulative demand*	*Safety Stock (20% of forecast)*	*Beginning Inventory*	*Production required forecast + s.s – beginning inventory*
January	300	300	60	50	300 + 60 – 50 = 310
February	500	800	100	60	500 + 100 – 60 = 540
March	400	1200	80	100	400 + 80 – 100 = 380
April	100	1300	20	80	100 + 20 – 80 = 40
May	200	1500	40	20	200 + 40 – 20 = 220
June	300	1800	60	40	300 + 60 – 40 = 320

Plan 1 (Vary the workforce size)

Sr. No.	*Particulars*	*Jan.*	*Feb.*	*March*	*April*	*May*	*June*	*Total*
1.	Production required	310	540	380	40	220	320	
2.	Production hrs. required (sr. no. 1 × 10)	3100	5400	3800	400	2200	3200	
3.	Hours available per worker at 8/day	176	152	168	168	176	160	
4.	No. of workers required (sr no 2 - sr no 3)	18	36	23	3	13	20	
5.	No. of worker hired	–	18	–	–	10	7	
6.	Hiring cost (sr no. 5 × 300 Rs.)	–	5400	–	–	300	2100	10500
7.	No. of workers laid off	2	–	13	20	–	–	
8.	Layoff cost (sr no. 7 × 400)	800	–	5200	8000	–	–	14000

Plan 2 (Use overtime and idle time)

Sr. No.	*Particulars*	*Jan.*	*Feb.*	*March*	*April*	*May*	*June*	*Total*
1.	Production required	310	540	380	40	220	320	
2.	Production hrs required (sr. no 1 × 10)	3100	5400	3800	400	2200	3200	
3.	Hours available per worker at 8/day	176	152	168	168	176	160	
4.	No. of hrs available (sr. no. 3 × 20)	3520	3040	3360	3360	3520	3200	
5.	No. of overtime (OT) hrs required (sr no. 2 – sr no. 4)	–	2360	440	–	–	0	–
6.	Over time premium (sr no. 5 × 3)	–	7080	1320	–	–	–	8400
7.	No. of idle time hrs.	420	–	–	2960	1320	–	–
8.	(sr no. 4 – sr no 2) idle time cost (sr no. 7 × 6)	3520	–	–	17.760	7920		28200

Plan 3 (Use inventory and stock out based on constant 20 worker force)

Sr. No.	*Particulars*	*Jan.*	*Feb.*	*March*	*April*	*May*	*June*	*Total*
1.	Production required	310	540	380	40	220	320	
2.	Production hrs required (sr. no. 1 × 10)	3100	5400	3800	400	2200	3200	
3.	Total hours available at 20 workers	3520	3040	3360	3360	3520	3200	
4.	Units product (sr no. 3 – 10)	352	304	336	336	352	320	
5.	Cumulative Production	352	656	992	1328	1680	2000	
6.	Units shorts (sr no. 2 × sr no. 5)	–	194	238	–	–	–	
7.	Shortage cost (sr no. 6 × 20 Rs.)	–	3880	4760	–	–	–	8640
8.	Excess units (sr no. 5 – s. no. 2)	42	–	–	58	190	190	–
9.	Inventory cost (sr No. 8 × 2)	84	–	–	116	380	380	960

The costs for the three plans

Plan 1: 10,500 (hiring) + 14,000 (Lay off) = 24,500

Plan 2: 8,400 (OT) + 28,200 (Idle time) = 36,600

Plan 3: 8,640 (stock out) + 96 (Inventory) = 9,600

Because plan 3 is economical, it is preferred.

17.10. LINEAR PROGRAMMING APPROACH TO AGGREGATE PLANNING

If aggregate planning problem is viewed as one that of the allocating capacity (supply) to meet forecast (demand) requirement. It can be solved in a linear programming format. Transportation method can be used to solve the problem. In this case the supply consists of the inventory on hand and units that can be produced through regular time (RT), overtime (OT) and subcontract (SC). Demand consists of individual month (or period) requirements plus any desired ending inventory. Costs associated with producing units in the given period or producing them and carrying them in inventory until a later period. Then solve the problem.

The transportation linear programming approach can be used to include back order costs. Production in a later month can be allocated to supply a back ordered demand from an earlier month at whatever stock out cost premium the firm chooses to assign.

***Problem*:** The supply, demand, cost and inventory data for a company which has a constant workforce is given. The company wants to meet all the demand (with no back orders). Allocate the production capacity to satisfy demand at minimum cost.

Supply Capacity (units)

Period	*RT*	*OT*	*Subcontract*
1	60	18	1000
2	50	15	1000
3	60	18	1000
4	65	20	1000

Demand Forecast

Period	*Demand*
1	100
2	50
3	70
4	80

Initial inventory = 20

Final inventory = 25

Regular time cost/unit = Rs.100

O.T. Cost/unit = 125

Subcontracting cost/unit = 130

Carrying cost/unit/period = Rs. 2

***Solution*:**

1. **Initial inventory:** Initial inventory of 20 units is available at period 1 at no additional cost. Carrying cost is Rs. 2 per unit per period. If it is retained for period 2, period 3 and so on. If it is unused, it costs Rs. 8/unit.
2. **Regular time (RT) production:** Cost per unit is Rs. 100 if the units are used in the same month when they are produced. A carrying cost of Rs. 2/unit/month is added on for each month for the units retained. Unused regular time costs the firm 50 per cent *i.e.*, Rs. 50/unit.
3. **Over time (OT):** Cost per unit is Rs. 125/unit if the units are used in the same month of their production. Otherwise a carrying cost of Rs. 2/unit/month is applicable. Unused OT has zero cost.
4. **Subcontracting (SC):** Cost per unit is Rs. 130. There is no cost for unused capacity, hence the initial allocation are made so as to use regular time as fully as possible. Overtime and subcontracting amounts can also be allocated on a minimum cost basis.
5. **Final inventory** requirement is (25 units) and must be available at the end of period 4 and has been added to period and demand of 40 units to obtain a total of 105.

The Tables 17.1. and 17.2 show initial matrix and matrix after allocation.

Table 17.1: Linear Programming Format

Supply units from		*Demand units for*				*Capacity*	
		Period 1	*Period 2*	*Period 3*	*Period 4*	*Unused*	*Total available*
Initial Inventory		0	2	4	6	8	20
Period 1	RT	100	120	104		50	60
	OT	125	127	129	131	0	18
	SC	130				0	1000
Period 2	RT		100	102	104	5	50
	OT		125	127	129	0	15
	SC		130			0	1000
Period 3	RT			100	102	50	60
	OT			125	127	0	18
	SC			130		0	1000
Period 4	RT				100	50	65
	OT				125	0	20
	SC				130	0	1000
Demand		100	50	70	105	4001	4323

Table 17.2: Allocation Matrix (Least Cost)

Supply units from		*Demand units for*				*Capacity*	
		Period 1	*Period 2*	*Period 3*	*Period 4*	*Unused*	*Total availble*
Initial Inventory		(20) 0	2	4	6	8	20
Period 1	RT	(60) 100	120	104		50	60
	OT	(18) 125	127	129	131	0	18
	SC	(2) 130				(998) 0	1000
Period 2	RT		(50) 100	102	104	50	50
	OT		125	127	(12) 129	(3) 0	15
	SC		130	130	130	(1000) 130	1000
Period 3	RT			(60) 100	102	50	60
	OT			(10) 125	(8) 127	0	18
	SC			130	130	(1000) 0	1000
Period 4	RT				(65) 100	50	65
	OT				(20) 125	0	20
	SC				0 130	(1000) 0	1000
Demand		100	50	70	105	4001	4326

◯ Indicate allocations.

17.11. MASTER PRODUCTION SCHEDULE (MPS)

Master scheduling follows aggregate planning. It expresses the overall plans in terms of specific end items or models that can be assigned priorities. It is useful to plan for the material and capacity requirements.

Functions of MPS

Master production schedule (MPS) gives a formal details of the production plan and converts this plan into specific material and capacity requirements. The requirements with respect to labour, material and equipment is then assessed.

Main Functions of MPS are:

1. **To translate aggregate plans into specific end items:** Aggregate plan determines level of operations that tentatively balances the market demands with the material, labour and equipment capabilities of the company. Master schedules translates this plan into specific number of end items to be produced in specific time period.
2. **Evaluate alternative schedules:** Master schedule is prepared by trial and error. Many computer simulation models are available to evaluate the alternate schedules.
3. **Generate material requirement:** It forms the basic input for material requirement planning (MRP).
4. **Generate capacity requirements:** Capacity requirements are directly derived from MPS. Master scheduling is thus a prerequisite for capacity planning.
5. **Facilitate information processing:** By controlling the load on the plant, master schedule determines when the delivery should be made. It coordinates with other management information systems such marketing, finance and personnel.
6. **Effective utilisation of capacity:** By specifying end item requirements, master schedule establishes the load and utilisation requirements for machines and equipment.

Time Intervals and Planning Horizon

Time interval used in master scheduling depends upon the type, volume, and component lead times of the products being produced. Normally weekly time intervals are used. The time horizon covered by the master schedule also depends upon product characteristics and lead times. Some master schedules cover a period as short as few weeks and for some products it is more than a year.

Master schedules often have firm and flexible portions.

Changes in master schedule affect lead times, work schedules, machine set-ups.

"Firm" portion will generally encompass the minimum lead-times necessary for components and cannot be changed.

Aggregate plan and Master schedule are shown in Fig. 17.3.

Aggregate Plan

Month	J	F	M	A	M	J	J	A	S
No.of Motors	30	45	50	30	60	30	30	40	40

Master Schedule

Month	**J**	**F**	**M**	**A**	**M**	**J**	**J**	**A**	**S**
Ac Motors (5 HP)	5	5	10	5	15	6	10	–	10
20 HP	10	7	10	5	10	4	5	–	20
D.C. Motor 20 HP	5	10	15	10	15	10	–	5	10
FHP Motors (1/2 HP)	10	23	15	10	20	10	–	35	–

Fig. 17.3. Aggregate plan and Master schedule.

An Example of Master Production Schedule

The quarterly time blocks are divided into months and then weeks. The quantity of each product family to be produced in a quarter, is divided by the weeks in that quarter. The total quantity of a product family to be made in the week is divided among individual product models in that family based on the mix of those products in the current demand pattern and adjusted or current inventory level of each product. MPS is reviewed each week to check that the proper mix of products is to be produced, particularly for the week that is about to enter the time fence.

17.12. AGGREGATE PLANNING METHODS, ADVANTAGES AND LIMITATIONS

	Method	*Advantages*	*Limitations*
1.	Trial and Error	Simple to understand and apply	Approximates with an aggregate product. Does not guarantee optimum solution.
2.	Rough cut capacity planning	Can deal with variety of products and can use weekly time blocks	Relies on judgement to determine the measurable master schedule, i.e., does not guarantee optimal solution.
3.	Linear Programming	Optimal solution to the stated problem is found	Approximates with an aggregate product actual problem may not fit the linear model.
4.	Linear decision Rule (LDR)	Optimal solution can be found	Actual problem does not fit in the quadratic model.
5.	Computer search techniques	Does not contain mathematical form of problems	Search may end with selecting local minimum instead of global minimum.

Production Plan

Television Production	Year 1 Q1	Q2	Q3	Q4
Portables	800	900	1000	900
Consoles	4000	5000	50000	4000

MASTER PRODUCTION SCHEDULE
WEEKS

Portables	1	2	3	4	5	6	7	8	9	10	11	12	13	14	15	16
Model 101	10	10	10	12	12	12	12	10	10	10	6	6	6	6	6	6
Model 106	25			25			25			30			30			30
Model 110	100			100		100										
Consoles																
Model 103																

References for Further Reading

1. Shore Barry, *Operations Management*, McGraw Hill, New York (1973).
2. Holt C., F. Modigliani, et al., *Production Planning Inventories and Workforce*, Prentice Hall Englewood Cliffs N.J. (1960).
3. Joseph Monks, *Operations Management Theory & Problems*, 3rd Edition, McGraw Hill Book Company, New York (1987).
4. James Dilworth, *Production and Operations Management*, 5th Edition, McGraw Hill Company, New York (1993).

REVIEW QUESTIONS

1. Define capacity and capacity planning.
2. Explain with examples how capacity is measured.
3. Define
   (*i*) Design capacity
   (*ii*) System capacity

(*iii*) Installed capacity
(*iv*) Licensed capacity
(*v*) Rated capacity

4. Explain the long-term and short-term capacity strategies.
5. Discuss the various factors influencing capacity planning.
6. Explain the steps involved in capacity planning.
7. Discuss the factors that favour over and under capacity.
8. How do you determine machine requirements?
9. Define aggregate planning and master productions schedule.
10. Distinguish between aggregate planning and master scheduling.
11. What are the variables associated with aggregate planning?
12. What are the various strategies associated with aggregate planning?
13. Explain pure strategies and mixed strategies.
14. Master production schedule is to drive the entire production system. Identify the MPS Functions.
15. Explain MPS with an example.
16. Discuss the advantages and disadvantages of various aggregate planning methods.
17. Discuss advantages and disadvantages of relying on overtime work to meet significant portion of demand.

PROBLEMS

1. The lead time to procure the raw material from a supplier is 4 weeks. The present stock is 54 kg of the material. There is also a scheduled receipt of 4.5 kg of it in 4 weeks. The production requirements over the period of next 9 weeks are

Week	1	2	3	4	5	6	7	8	9
Amount required (kg)	24	–	29	11	–	5	19	27	18

Of order quantity is 45 kg, find the planned order releases.
(**Ans.** Planned order release in 4th week for 45 kg)

2. Set up and solve the following aggregate planning problem through transportation linear programming matrix method.

	Regular	*Overtime*	*Subcontracting*
Production capacity/period	8000 units	2000 units	2000 units
Production Cost/unit	Rs. 7	Rs. 9	Rs. 10

Inventory initial = 1000 units, carrying cost = Rs 1/unit-period

Period	1	2	3	4
Demand (units)	6,000	18,000	3,000	10,000

Back orders are not allowed and unused regular time has a cost of Rs. 4 per unit:
(*a*) Show your solution
(*b*) Tabulate the total cost of your plan.

18

MATERIAL REQUIREMENT PLANNING (MRP)

• Introduction • MRP Objectives • Functions served by MRP • MRP terminology • MRP system • Master production schedule (MPS) • Inventory status file • Bill of material (BOM) • MRP logic • MRP outputs • Management information from MRP • Lot Sizing considerations • Manufacturing resource planning—MRP II • Capacity requirements planning (CRP).

18.1. INTRODUCTION

Material requirement planning (MRP) refers to the basic calculations used to determine component requirements from end item requirements. It also refers to a broader information system that uses the dependence relationship to plan and control manufacturing operations.

MRP is a technique of working backward from the scheduled quantities and needs dates for end items specified in a master production schedule to determine the requirements for components needed to meet the master production schedule. The technique determines what components are needed, how many are needed, when they are needed and when they should be ordered so that they are likely to be available as needed. The MRP logic serves as the key component in an information system for planning and controlling production operations and purchasing. The information provided by MRP is highly useful in scheduling because it indicates the relative priorities of shop orders and purchase orders.

"Materials Requirement Planning (MRP) is a technique for determining the quantity and timing for the acquisition of dependent demand items needed to satisfy master production schedule requirements."

MRP is one of the powerful tools that, when applied properly, helps the managers in achieving effective manufacturing control.

18.2. MRP OBJECTIVES

1. **Inventory reduction:** MRP determines how many components are required, when they are required in order to meet the master schedule. It helps to procure the materials/components as and when needed and thus avoid excessive build up of inventory.

2. **Reduction in the manufacturing and delivery lead times:** MRP identifies materials and component quantities, timings when they are needed, availabilities and procurements and actions required to meet delivery deadlines. MRP helps to avoid delays in production and priorties production activities by putting due dates on customer job orders.

3. **Realistic delivery commitments:** By using MRP, production can give marketing timely information about likely delivery times to prospective customers.

4. **Increased efficiency:** MRP provides a close coordination among various work centres and

hence helps to achieve uninterrupted flow of materials through the production line. This increases the efficiency of production system.

18.3. FUNCTIONS SERVED BY MRP

1. **Order planning and control:** When to release orders and for what quantities of materials/components.

2. **Priority planning and control:** How the expected date of availability is compared to the need date for each component.

3. Provision of a basis for planning capacity requirements and developing a broad business plans.

The following questions are addressed in MRP processing:

1.	What do we want to produce, and when.	–	Provided by Master Production Schedule
2.	What component are required to make it and how many.	–	Bill of Materials (BOM)
3.	How many are already scheduled to be available in each future period.	–	Inventory status file
4.	How many more we need to obtain for each future period.	–	Difference in required and available
5.	When to order these amounts so that they will be available when needed.	–	Planned order release

18.4. MRP TERMINOLOGY

1. **Dependent demand:** The demand for an item depends on another item. The demand dependency is the degree to which the demand for one item is associated with demand for another item.

2. **MRP:** A technique for determining the quantity and timing of dependent demand items.

3. **Lot size:** The quantity of items required for an order.

4. **Time phasing:** Scheduling to produce or receive an appropriate amount (Lot) of material so that it will be available in the time periods when required.

5. **Time bucket:** The time period used for planning purposes in MRP.

6. **Gross requirements:** The overall quantity of an item needed at the end of the period to meet the planned output levels.

7. **Net requirements:** The net quantity of an item that must be acquired to meet the scheduled output for the period. It is calculated as, Gross requirements minus scheduled receipts for the period minus amounts available from the previous period.

8. **Requirements explosion:** The breaking down of (exploding) parent items into component parts that can be individually planned and scheduled.

9. **Scheduled receipts:** The quantity of an item that will be received from suppliers as a result of orders that have been placed.

10. **Planned order receipts:** The quantity of an item that is planned to be ordered so that it will be received at the beginning of the period to meet net requirements for the period. The order has not yet been placed.

11. **Planned order release:** The quantity of an item that is planned to be ordered or it is a plan (quantity and date) to initiate the purchase or manufacture of materials so that they will be received on schedule after the lead time offset.

12. **Lead time offset:** The supply time or number of time buckets between releasing an order and receiving the materials.

18.5. MRP SYSTEM

Fig. 18.1 shows the MRP system components.

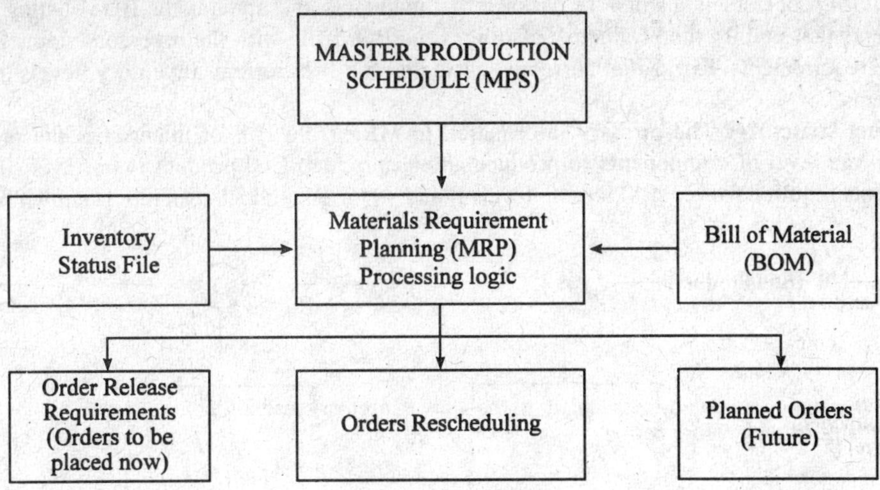

Fig. 18.1: MRP system.

The inputs to the MRP systems are: (1) A master production schedule, (2) An inventory Status file, and (3) Bill of material (BOM).

Using these three information sources, the MRP processing logic (Computer Programme) provides three kinds of information (Output) for each product component: Order release requirements order rescheduling and planned orders.

Master Production Schedule (MPS)

MPS is a series of time phased quantities for each item that a company produces, indicating how many are to be produced and when. MPS is initially developed from firm customer orders or from forecasts of demand before MRP system begins to operate. The MRP system accepts whatever the master schedule demands and translates MPS end items into specific component requirements. Most systems then make a simulated trial run to determine whether the proposed master schedule can be satisfied.

Inventory Status File

Every inventory item being planned must have an inventory status file which gives complete and up to date information on the on hand quantities, gross requirements, scheduled receipts and planned order releases for the item. It also includes planning information such as lot sizes, lead times, safety stock levels and scrap allowances.

Basically, the job of the inventory status file is to keep data, about the projected use and receipts of each item and to determine the amount of inventory that will be available in each time bucket. If the projected available inventory is not adequate to meet the requirement in a period, the MRP programme will recommend that this item be ordered.

Bill of Materials (BOM)

To schedule the production of an end product, the MRP system must plan for all the materials, parts and subassemblies that go into the end product. The Bill of Material file in the computer provide this information. BOM file identifies each component by unique part number and helps processing by a process which 'explodes' end item requirements into component requirements.

Thus BOM identifies how each end product is manufactured, specifying all subcomponents items, their sequence of build up, their quantity in each finished unit and the work centres performing the build up sequence. This information is obtained from product design documents, work flow analysis and other standard manufacturing information.

The BOM processor is a software package that maintains and updates the BOM listing of all components that go into the product. It also links the BOM file with the inventory status file so that the requirements explosion correctly accounts for the current inventory levels of all components.

Product structure: The primary information to MRP from Bill of material is the product structure, the level of components to produce an end product. End product is on level '0'; the components required for level '0' are on level 'l' and so on. A product structure is shown in Fig. 18.2.

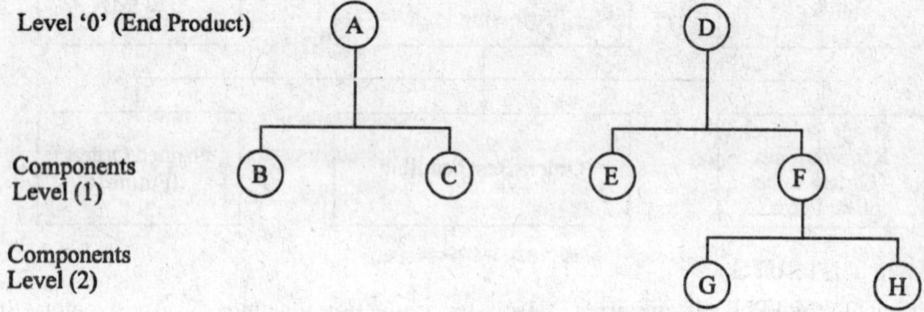

Fig. 18.2: Product structure for two assembled products.

One unit of end product *A* requires one unit each of components *B* and *C*. One unit of enα product *D* requires one unit of component *E* and one unit of component *F*. The component *F* ir urn requires one unit of component *G* and one units of component *H*.

ſo facilitate MRP processing, each component at every level of the BOM must have a unique part number for its identification. The separate identifications enable computer to find any parent item and to determine all the components needed to make it. Determining all the lower level components needed to make a parent is called exploding the requirement by the BOM.

Example of Product Structure

The product structure tree for a three drawer file cabinet without fasteners is represented in Fig. 18.3.

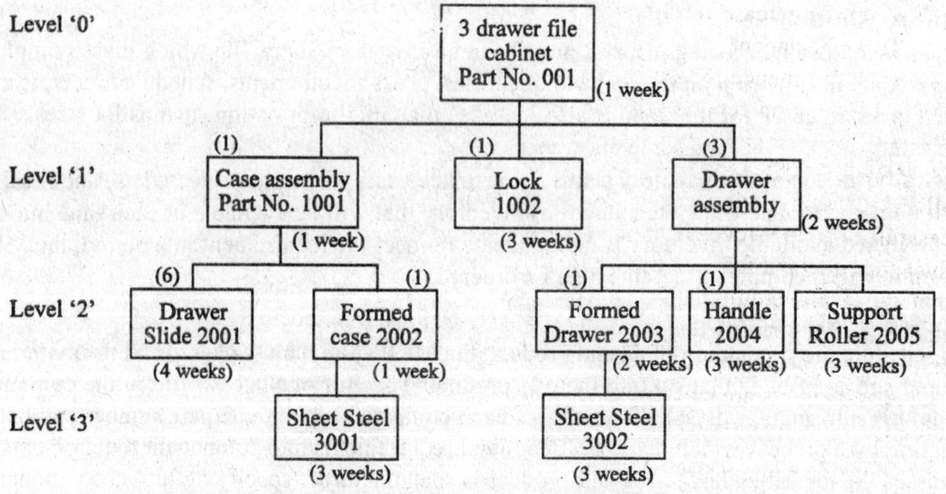

Fig. 18.3: Product Structure file for 3 drawer file cabinet.

Number	Description
001	File cabinet
1001 (1)	Case assembly
1002 (1)	Lock
1003 (3)	Drawer assembly
1001	Case assembly
2001 (6)	Drawer slides
2002 (1)	Formed case
1003	Drawer assembly
2003 (1)	Formed drawer
2004 (1)	Handle
2005 (2)	Support rollers
2002	Formed case
3001 (1)	Sheet steel
2003	Formed drawer
3002 (1)	Sheet Steel

18.6. MRP OUTPUTS

The most visible outputs are the actual and planned order releases that go to purchasing and in house production shops.

A variety of reports can be generated from the information made available by an MRP Program.

TO MPS Planners'

- Simulation of proposed MPS
- Researching information for open orders (due to cancellation, delays, shortages)

To Purchasing and Production

- Changes to keep priorities valid.
- Order releases (Purchase and shop orders)
- Planned order releases

To capacity Requirements Planning

- Order release information for load profiles, delays, shortages

To Management

- Performance measurement of (vendors, cost, forecast accuracy)
- Exception reports (on due dates BOM file, etc.)

18.7. MRP LOGIC

MRP processing logic accepts the master schedule and determine the components schedules successfully for low-level items of the product structures. It calculates for each item in each product structure and for each time period (typically one week) in the planning horizon how many of that items are required (Grass requirements) how many units from inventory are already available, the net quantity that must be planned (planned order receipts) and when ordered for new shipments must be planned (planned order releases) so that all materials arrive when needed.

Net requirements are calculated by adjusting for existing inventory items already, on order as recorded in inventory status file.

$$\text{Net requirements} = \begin{matrix}\text{Projected gross}\\ \text{requirements}\end{matrix} - \left[\begin{matrix}\text{Inventory}\\ \text{on hand}\end{matrix} + \begin{matrix}\text{Scheduled}\\ \text{receipts}\end{matrix}\right]$$

Order releases are planned for components in a time phased manner (using lead time data from the inventory file) so that materials will arrive precisely when needed. This is referred to as planned order receipt. When the orders are actually issued to vendors or shops the planned receipt becomes the scheduled receipt.

18.8. MANAGEMENT INFORMATION FROM MRP

MRP output includes a report as shown in Fig. 18.4 for one component. The detailed identification (code), lead time is given in the report.

The report shows that:

- 400 units are needed in week 4 and another 500 needed in week 8. (Gross requirements)
- No outstanding orders were previously placed, and no scheduled receipt of this items.
- There are 50 units of the item already available as inventory and this will meet the 4 weeks requirements.
- Thus, net requirements are 350 units for week 4 (400-50) and 500 for week 8.

To meet these net requirements, we should plan to receive 350 units in fourth week and 500 units in eighth week. Since the lead time for an item is three weeks, the first order must be placed in first week and second in fifth week.

This report clearly identifies the procurement actions required to keep production on schedule. Any change in end item demand with time, modifications in the MPS will dictate corresponding adjustment of lower level requirements.

Order quantity :		*WEEK*							
Lead time = 3 weeks		*1*	*2*	*3*	*4*	*5*	*6*	*7*	*8*
Gross requirements					400				500
Scheduled receipts									
Available for next period	50	50	50	50					
Net requirements					350				500
Planned Order receipts					350				500
Planned Order releases		350				500			

Fig. 18.4: An MRP report for one item.

18.9. LOT SIZING CONSIDERATIONS

The released orders should specify a discrete lot size for purchasing or manufacturing. The various lot sizing techniques available are given below:

1. **Fixed Order Quantity** —Each time an order is placed, the quantity remains same.
2. **Economic Order Quantity (EOQ).**
3. **Lot for Lot (LFL)**—Crder thc exact net requirements each period.
4. **Least Unit Cost (LUC)**
5. **Least Total Cost (LTC)** —Order the quantity that minimises the total set-up and carrying costs during the planning horizon.
6. **Part Period Algorithm (PPA)** —Use the ratio of ordering and carrying costs to derive a part period number and use thc number as a creation for cumulating requirements.
7. **Fixed Period Requirements** —Order a supply for a given number of periods each time (*e.g.*, a two months supply).

***Problem 1*:** Complete the material requirements plan for an item *X* shown below. The item has an independent demand and a safety stock of 40 is maintained.

Order Quantity = 70 Lead time = 4 weeks Safety stock = 40	*WEEK*											
	1	2	3	4	5	6	7	8	9	10	11	12
Projected requirement	20	20	25	20	20	25	20	20	30	25	25	25
Receipts		70			70						70	
On hand at the end of period (65)												
Planned order release												

Solution.

Order Quantity = 70 Lead time = 4 weeks Safety stock = 40	*WEEK*											
	1	2	3	4	5	6	7	8	9	10	11	12
Projected requirement	20	20	25	20	20	25	20	20	30	25	25	25
Receipts		70			70			70			70	
On hand at the end of period (65)	45	95	70	50	100 ~~30~~	75	55	105 ~~35~~	75	50	95 ~~25~~	70
Planned order release	70			70			70					

***Problem 2*:** A small-scale unit manufactures a product and it is expected to supply 80 units in week 1, 120 in week 4, 120 in week 6, and 100 in week 8. Each product is made of 2 housings, a shaft assembly and one wheel. For these components order quantities, lead times and inventories on hand at the beginning of period 1 are given below.

Part	*Order Qty.*	*Lead time*	*Inventory on hand*
Housings	600	2 weeks	200
Shaft assembly	400	3 weeks	440
Wheel	800	1 week	100

Apart from the above requirement, another 180 shaft assembly required for another customer 600 units of housing are already scheduled to be received at the beginning of week 2. Complete the material requirement plan for housing, shaft and wheel and show what quantities of orders must be released and when they must be released in order to satisfy the MPS.

***Solution*:** End item master schedule is shown in table 18.1.

Table 18.1: End item Master Schedule

Week Nos.	1	2	3	4	5	6	7	8
Requirement	80			120		120		100

Table 18.2: Components and Subcomponents Materials Plan

Component Material Plan—Housing

Order Quantity = 600 Lead time = 2 weeks	WEEK							
	1	2	3	4	5	6	7	8
Projected requirement	160			240		240		200
Receipts		600						600
On hand at the end of period (200)	40	640	640	400	400	160	160	560 −40
Planned order release						(600)		

(Place order in week 6 as in week, 8, there is negative stock)

Component Material Plan—Shaft Assembly

Order Quantity = 400 Lead time = 3 weeks	WEEK							
	1	2	3	4	5	6	7	8
Projected requirement	80			120	180*	120		100
Receipts						(400)		
On hand at the end of period (440)	360	360	360	240	60	340 −60	340	240
Planned order release			(400)					

Sub-component Material Plan—Wheel

Order Quantity = 800 Lead time = 3 weeks	WEEK							
	1	2	3	4	5	6	7	8
Projected requirement			400					
Receipts			(800)			400		
On hand at the end of period (100)	100	100	500 −300	500	500	500	500	500
Planned order release		(800)						

* Requirement from another product.

The MRP master schedule and component part schedule is shown in the tables.

Each unit of product requires two housing, one shaft assembly and one wheel. Each unit of product requires two housing, the projected material requirements for housing are double the number of end products. The projected requirements of 160 housings in period 1 are adequately satisfied by the 200 units on hand at the beginning of period 1, leaving 40 on hand at the end of period 1. On hand materials can be calculated with the following equation:

$$\begin{matrix}\text{On hand at} \\ \text{end of period}\end{matrix} = \begin{matrix}\text{On hand at end} \\ \text{of previous period}\end{matrix} + \text{Receipts} - \begin{matrix}\text{Projected} \\ \text{requirements}\end{matrix}$$

With the receipt of 600 housings in period 2, the on hand inventory will be adequate until week

8, which at first instant will be 40 units short. To overcome this a planned order release for 600 quantity has been scheduled for week 6 because housing has 2 weeks lead-time. The planned receipt of 600 in week 8 will thus result in an end of period inventory of 560 units.

Each unit of product requires one shaft assembly and 180 shaft assembly are needed for another product in week 5 are incorporated into the requirements. The on hand stock is adequate until week 6, when quantities will drop to – 60 unless a planned order is released in week 3.

The wheel is a sub-component of shaft assembly. Planned order of 400 units from the above shaft assembly plan shows up as a projected requirement for 400 wheels in week 3 on the sub-component plan. Since on hand inventory is inadequate to supply this need, a planned order release is scheduled for week 2. It should ensure that an order of 400 wheels will be available by the beginning of week 3.

18.10. MANUFACTURING RESOURCE PLANNING (MRP-II)

Manufacturing Resource Planning (MRP-II) is an integrated information system that synchronise all aspects of the business. MRP-II system coordinates sales, purchasing, manufacturing, finance and engineering by adopting a focal production plan and by using one unified data base to plan and update the activities in all the systems.

A manufacturing resource planning can be divided into three parts which are composed of:

(*i*) Product planning functions which take place at the top management level.

(*ii*) Operations planning handled by staff units.

(*iii*) Operations control functions conducted by manufacturing line and staff supervisors.

Check points among the three divisions provide feedback regarding the adequacy of overall resources, completeness of resource commitments and the quality of performance in carrying out the plans. Feedback based on these checks permits a quick response to changing conditions using the latest operating data.

MRP-II integrated system for planning and control is shown in Fig. 18.5.

The process (as shown in figure) involves developing a production plan from the business plan

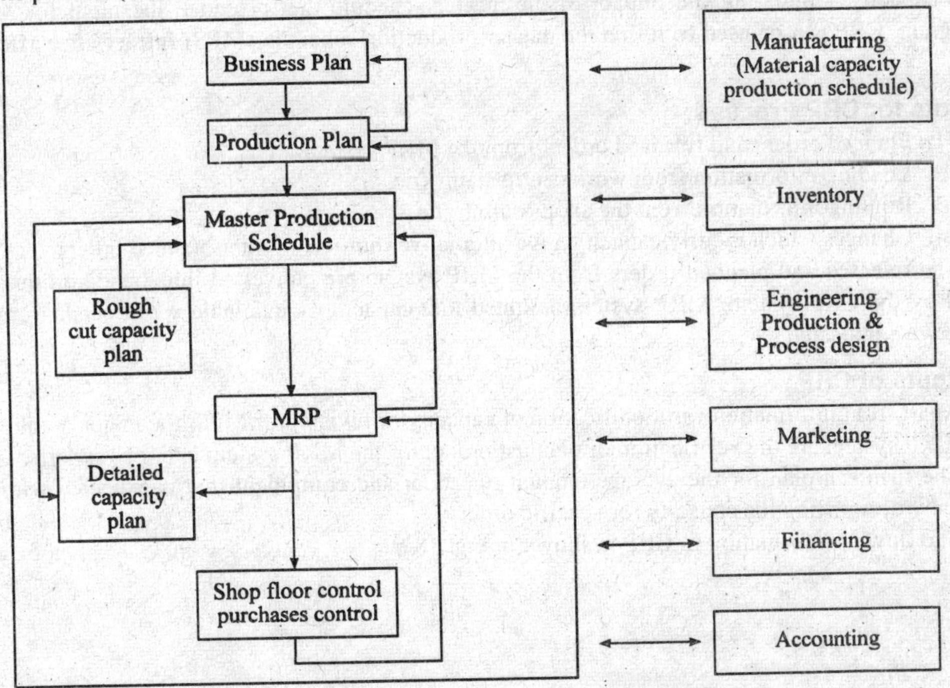

Fig. 18.5: MRP II—An integrated system for planning and control.

to specify monthly levels of production for each product line over the next five years. The production department then is expected to produce at the committed levels, the sales department to sell at these levels, and the finance department to ensure adequate financial resources for these level. Then the production plan guides the master schedule and gives the weekly quantities of specific products to be built. Then the capacity available is roughly adequate. If not, either master schedule or capacity is changed. Once settled, this master schedule is used in MRP to create material requirement and priority schedules for production. Then an analysis of detailed capacity requirement determines whether capacity is sufficient at each work centre during scheduled time periods. Then the execution and control activities are determined to ensure that the master schedule is met.

18.11. CAPACITY REQUIREMENTS PLANNING (CRP)

Capacity is a measure of the productive capability of a facility per unit of time. Capacity decisions begin with the initial facility layout and extend to aggregate planning, master scheduling, capacity requirements planning.

"CRP is a technique for determining what personnel and equipment capacities are needed to meet the production objectives embodied in the master schedule and the material requirements plan."

CRP is an effort to develop a match between the MRP schedule and the production capacity of the company. Determination of the capacity of the work centre and the capacity requirements imposed on those work centres by a particular product mix enables a company to known what level of sales, its production system can support. Thus, company will be able to make realistic sales commitments. Capacity planning helps to avoid under utilisation of capacity and also CRP enables the company to anticipate production bottlenecks in some work centres in time to take corrective actions.

To be effective, capacity requirements planning must be coordinated with MRP. Working together MRP and CRP programmes translate the master schedule to requirements for components and capacity, simulating the impact of the master schedule that provided the input for MRP program. CRP can be used to refine the master production schedule (MPS) further after MRP is run.

Inputs for CRP Process

1. Planned orders and released orders from the MRP system.
2. Loading information from work centre status file.
3. Routing information from the shop routing file.
4. Changes which modify capacity, give alternative routings or alter planned orders.

The released and planned orders from the MRP system are converted into standard hours of load by the CRP system. MRP system assumed that capacity is available when needed unless otherwise indicated.

Outputs of CRP

Apart from information for modification of capacity or revision of MPS, the major outputs of the CRP system are the verification of planned orders for the MRP system and load reports.

The firm can plan for the average amount of labour and equipment that is expected without actually designating the capacity for specific orders.

The flow of information in CRP is shown in Fig. 18.6.

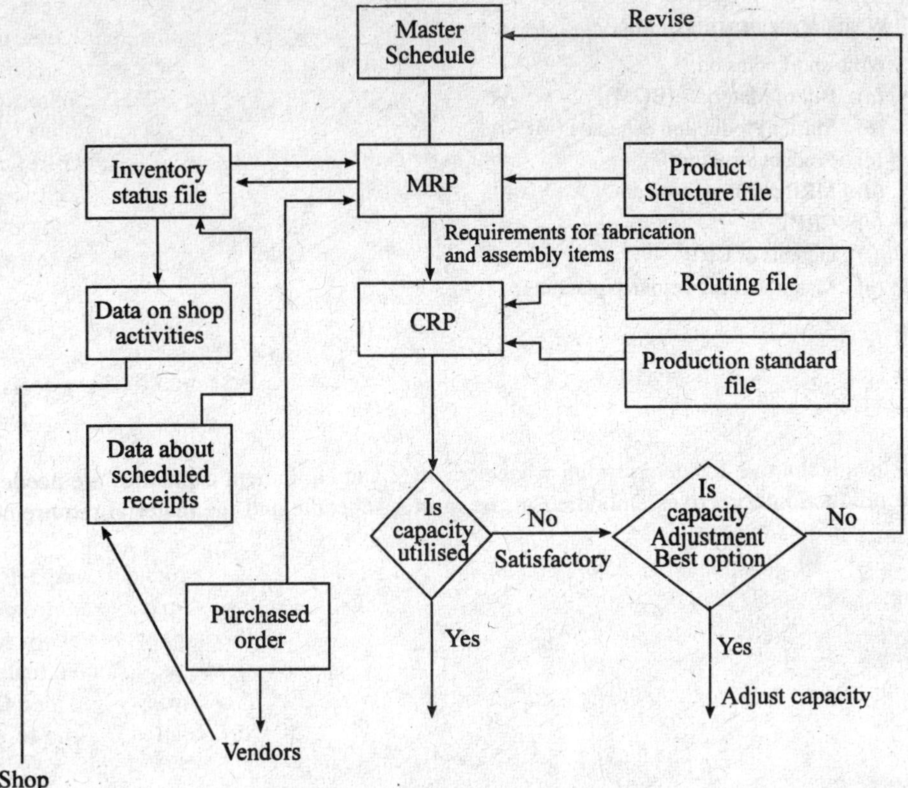

Fig. 18.6: Flow of information in CRP.

References for Further Reading

1. Joseph Monks, *Operations Management*, Third Edition, McGraw Hill Book Company, New York, (1987).
2. James Riggs, *Production System—Planning Analysis and Control*, 4th Edition, John Wiley & Sons, New York, (1987).
3. James Dilworth, *Production and Operations Management*, 5th Edition, McGraw Hill Company, New York, (1993).
4. Orlicky Joseph A., *Materials Requirement Planning*, McGraw Hill, New York, (1975).
5. Vollman Thomas *et al.*, *Manufacturing Planning and Control Systems*, 3rd Edition, Homewood III, Irwin, (1992).

REVIEW QUESTIONS

1. Define MRP and CRP and BOM.
2. How MRP differs from inventory control system?
3. Explain the various terms associated with MRP.
4. Distinguish between planned receipt and scheduled receipt.
5. What are the inputs to MRP?
6. Explain MRP logic and what do you mean by time phasing.
7. What are various lot sizes used for MRP?

8. What are the benefits of MRP?
9. Write short notes on:
   (*a*) Bill of Materials (BOM)
   (*b*) Master Production Schedule (MPS)
   (*c*) Product Structure
   (*d*) MRP planning horizon
   (*e*) CRP
   (*f*) Outputs of CRP
   (*g*) Manufacturing resource planning.

19

PROCESS PLANNING

• Introduction • Framework for Process Engineering • Process and Equipment Selection • Application of BEA for selecting the machine • Machine requirements • Machine output • Manpower Planning • Assembly Line Balancing • Process Planning—Definition; Uses of Process Planning • Information Required for Process Planning • Factors Affecting Process Planning • Steps in Process Planning.

19.1. INTRODUCTION

A conversion process can be defined as a set of operations that are performed at workstations for the purpose of transforming inputs into outputs.

Process engineering (process design) is concerned with successfully transforming the design into a physical product.

Process engineering is concerned with determining the method of manufacture of a product, establishing the sequence and type of operations involved, tools and equipment's required and analysing how the manufacturing of a product will fit into the facilities.

In transformation of raw materials into finished products, several questions need to be answered; such as:

1. What will be the production quantity?
2. What are characteristics of the products to be manufactured?
3. The availability of equipment and what kinds of equipment are to be purchased and what will be the investment?
4. What kinds of labour is required?
5. What should be the level of automation?
6. Make or buy the components required?

Once these questions are answered, the process planning activity can be carried out with minute detail as to how each component can be manufactured.

19.2. FRAMEWORK FOR PROCESS ENGINEERING

The needs of the customers are translated into technical specifications and the product is designed with due consideration to functional, operational, quality and reliability and cost considerations. The output of the design will be in the form of assembly and part drawings and Bill of Materials (BOM).

This serves as an input to process design.

The information from process R&D and product final design are the inputs to the process design stage.

The process design stage is composed of two levels;

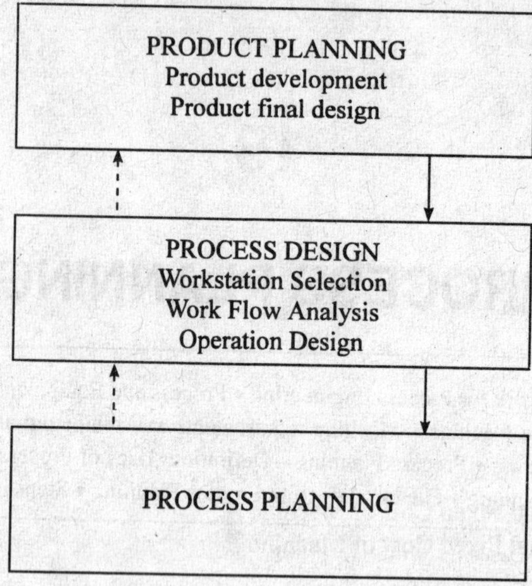

Fig. 19.1: Framework for process engineering.

- **Selection of the workstations:** This involves the specification of the process, selection of plant and equipment to be included in the process.
- **The flow analysis of materials between the various departments (or workstations):** This stage decide the type of manufacturing whether it is a continuous (mass production) or intermittent production.
- The Flow patterns for materials, type of layout and material handling systems are specified at this stage. The selection of type and number of equipment's is decided at this stage. The decision to make or buy is taken during this level.
- The lower level process design is concerned with job design or workstation design. The focus here is on the various aspects like balancing production line. Steps that should be included in a particular workstation and the method executing, analysis of operator variability and efficiency, labour and time standards, job specialisation and job enlargement needs to be analysed in detail.

Process Planning Defined

It is defined as the systematic determination of methods by which a product is to be manufactured economically and competitively. It consists of selecting the proper machines, determining the sequence of operations, specifying the inspection stages, and tools, jigs and fixtures such that the product can be manufactured as per the required specification. The detailed process planning is done at each component level.

The framework for process engineering is shown in the Fig. 19.1.

19.3. PROCESS AND EQUIPMENT SELECTION

This is an important step. Specification and selection of a process or equipment is not an easy job. There are many factors which influence the selection. The following considerations are to be given while selecting a process or machine.

1. **Economic considerations:** Due analysis should be made with respect to the initial cost, maintenance and running cost. An alternative which results in lower total cost should be selected.

2. **Production rate and unit cost of production.**
3. **Durability and dependability, *i.e.*, quality and reliability aspects.**
4. **Lower process rejection**
5. **Minimum set-up and put away times.**
6. **Longer productive life of machines or equipment.**
7. **Functional versatility—Should be able to perform more than one functions.**

19.4. APPLICATION OF BEA IN THE CHOICE OF MACHINES OR PROCESS

This analysis is the most convenient method for selecting the optimum method of manufacture or machine amongst the competing ones. The cost estimates of the competing methods (both fixed and variable costs) is prepared and a particular quantity N is determined at which the alternatives give the same cost.

If the quantity to be manufactured is less than N the process with lower fixed cost is selected and if the quantity to be produced is more than N the process with lower variable cost is selected.

Let F_A the Annual Fixed Cost of Machine A

F_B the Annual Fixed Cost of Machine B

V_A = Variable Cost per unit for Machine A

V_B = Variable Cost per unit for Machine B

N = Quantity at which costs on both machines will be equal.

∴ Total cost on machine A = Total cost on Machine B for Quantity N

i.e., $F_A + V_A \cdot N = F_B + V_B \cdot N$

$$N(V_A - V_B) = F_B - F_A$$

$$N = \frac{F_B - F_A}{V_A - V_B}$$

The alternative with lower fixed cost will be more economical for manufacturing up to N and once the quantity exceeds N, it is economical to select an alternative with lower variable cost.

Problem 1: A component can be manufactured either on centre lathe or on a turret lathe. The cost and time information to process a component is given below.

Particulars	*Centre lathe*	*Turret lathe*
• Set up time	30 minutes	120 minutes
• Processing time	10 minutes	5 minutes
• Tooling up cost (Rs.)	200	500
• Labour cost/hr	Rs. 2	Rs. 2
• Depreciation and other cost per hour	Rs.10	Rs. 20

The tooling costs are to be recovered within a year. There are no repeat orders. The requirements are to be met in two lots.

(*i*) Find the quantity at which both alternatives results in equal cost. (BEP)

(*ii*) Give the decision rule regarding the choice of lathes

(*iii*) If the quantity required is 800 Nos./year, which of the machine do you propose.

Solution: Let F_1 = Fixed cost for the centre lathe.

Fixed cost consists of set-up and tooling up costs.

∴ Fixed cost for centre lathe (F_1) = Set-up cost + Tooling up cost

∴ F_1 = No. of set-ups/year × set-up time /set-up (hrs) × [(set-up labour rate) + (Depreciation and other expense/hr)] + tooling up costs.

Output/day = 440/12 = 36.4 = 37 pieces

Percentage increase in output = $\frac{37-34}{34} = 9\%$

So the redesign of jigs is not justified.

(*c*) **Suggestion if the number of pieces is large:** The critical operation (bottleneck) is processing on machine *B* which requires 10 minutes. If we introduce one more machine of kind *B*, then the cycle time will be reduced to 8 min.

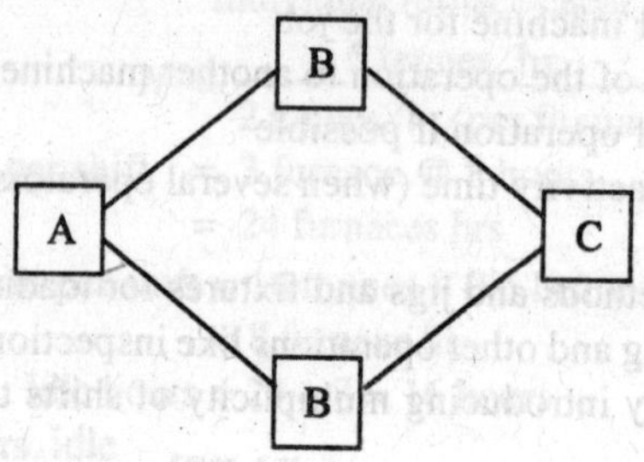

The new output = $\frac{440}{8} = 55$ pieces

Problem: A component is to be processed on two machines Lathe and milling machine. The sequence of operation is first turning and then milling.

The machine times are given below:

Turning 12 minutes.
Milling 20 minutes.

1. Estimate the number of machines required to machine 2500 components per week if available machine hours per week are 48.
2. What are the steps that you propose to reduce number of machines.

Solution: No. of components required to be machined = 2500/week

The available time (in hr.) per week = 48

Assuming 50 weeks in a year, no. of components to be machined/annum = 2500 × 50 = 1,25,000 annum.

The time required to machine 1,25,000 components = $1{,}25{,}000 \times \frac{12}{60} = 25{,}000$ hrs

Time available per annum = 48 × 50 = 2400 hrs

$$\text{No. of lathes required} = \frac{25{,}000}{2400} = 10.41 \approx 11$$

$$\text{No. of milling machines required} = \frac{1{,}25{,}000 \times \frac{20}{60}}{2400} = 17.36 \approx 18$$

2. Steps to reduce the number of machines. Increase the number of shifts. Addition of second shift is going to reduce the number of machines by 50 per cent.

19.7. MANPOWER PLANNING

Manpower planning involves the forecasting of human resource needs in the organisations and designing appropriate course of action such as recruitment, training, career development based on these needs. Planning for human resource is major managerial responsibility in today's industrial economics.

Planning for people becomes important when job requirements specify scarce skills and capabilities. Qualified and skilled people have become scarce, and human resource planning has become a necessity for long-term survival in industrial economics.

Definitions: "Manpower planning is a process by which management determines how the organisation should move from its current manpower position to its desired manpower position."

Through manpower planning, management strives to have right number and right kinds of people at the right place at right time doing things that result in long-term benefits to organisations.

According to Geisler: "Manapower planning is the process including forecasting developing and controlling by which the firm ensures that it has the right kind of people of right number and at right place doing work for which they are economically most suitable."

Reasons for Manpower Planning

1. Job and job requirements are changing faster than in earlier periods because of pace of change of technology, change in ways the goods and services are produced.
2. The occupational structure of the workforce in industrialised economies has shifted to meet changes in job.
3. Within existing occupations rising job requirements make retaining a must for current job holder.
4. National concern about levels of employment and effective unionisation of human has resulted in National manpower programmes.
5. The skill shortages have become the major problem for progress towards industrialisation for less developed countries.
6. Increased mobility of human resource has worked both to assist the organisations and nations in meeting new job requirements and to complicate the efforts to retain qualified employees.
7. Rising interest and activity in the total process of management planning has stimulated attention to the need for human resource planning.

The definition of manpower planning should include the functions and process of,

- Effective utilisation.
- Forecasting the needs.
- Developing appropriate policies and programmes to meet the requirements.
- Reviewing and controlling the total process.

Objectives of Manpower Planning

1. **Utility as a planning and control technique:** A manpower plan because it is systematically done, enables a manager to predict the manpower needs and requirements, controls the deployment of manpower, a more precise matching of manpower needs to the firms business plans.

2. **Manpower planning** is the necessary for management to get information about manner in which existing personnel are deployed, the kinds of skills required for the various categories of jobs and manpower requirements over a specified period of time in relation to the organisational goals.

Process of Manpower Planning

The various steps involved in manpower planning:

1. **Manpower demand forecasting:** Manpower demand forecasting at the micro level (organisational level) can be done in two ways,

– by ascertaining the total manpower requirements for the entire organisation for a given period and then estimating requirements of each unit, division or department.

OR

– First determine the manpower requirement of each department and subsequently make a total projection.

Many forecasting methods are available like simple and multiple regression models. It should be clearly mentioned as to up to what future period the forecasting is done.

2. **Manpower supply forecast:** The supply of manpower should include both, internal supply (effected by promotion and transfer), and external supply (study of labour market). There are many activities which give significant information on which manpower planner builds up his plan.

- Manpower inventory sets out what the firm has in stock or can expect to have in stock in future. Comparison of this data, against the requirements gives an immediate picture of the shortfall.
- Appraisal of the existing performance level tells us the present level of manpower utilisation.
- Assessment of labour market situation tells us the availability of the required manpower from which planner may like to source the requirements.

3. **Manpower inventory:** If the manpower planning to be realistic then it should be based on sound foundation of factual information. Thus, the planner must have as clear picture as possible of the existing staff. A manpower inventory provides the information about existing manpower. The manpower inventory gives the information about existing employees with regard to number, skill, age group and many other details.

4. **Manpower audit:** Manpower audit requires systematic analysis of data and it describes the collected data together with its analysis:

Manpower Audit answers the following questions:

(*i*) What is the position of starts and termination?
(*ii*) What is the position of absenteeism?
(*iii*) What type of labour is difficult to recruit?
(*iv*) Salary and age distribution.
(*v*) Trends in labour market with respect to needed skills?
(*vi*) Reasons for employee turnover?
(*vii*) From where do our recruits come from?

5. **Assessment of market supply situation.**

6. **Estimating manpower supply:** Manpower supply can be from both internal sources and external sources. The manpower supply from internal sources depends on two factors they are—(*a*) The extent to which the present employees survive in the organisation, and (*b*) The rate of internal turn over (*i.e.*, transfers and promotion) in the organisation.

7. **Analysing the internal movements.**

8. **Manpower supply from external sources:** Four determinants of final manpower plan are

(*i*) Manpower utilisation.
(*ii*) Manpower supply.
(*iii*) Training and development.
(*iv*) Personnel policies.

Fig. 19.2 shows the manpower planning process.

Advantages of Manpower Planning

1. By anticipating the need for various types of skill requirements and levels of personnel, well in advance, a manpower planning will be able to give adequate lead time for recruitment, selection and training of such persons.

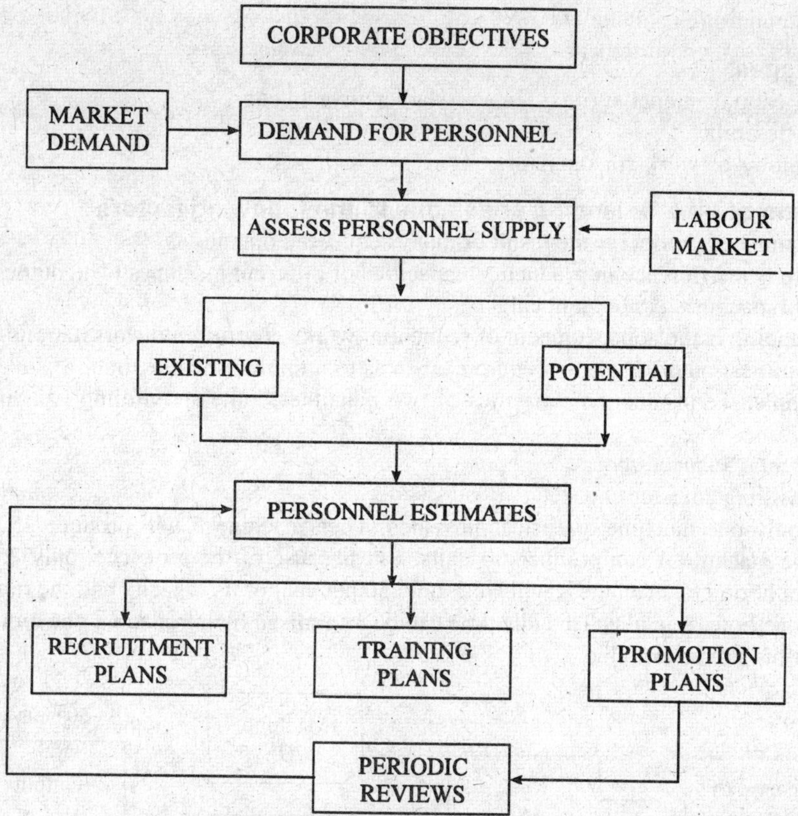

Fig. 19.2: Manpower planning process.

2. A manpower plan could give an overall picture for identification of surplus or shortage of personnel.
3. An effective labour cost control and manpower development.
4. In the absence of availability of required skills in the labour market, the steps are taken to promote the personnel from internal sources with training.
5. Manpower inventory can provide information to management for internal succession of managerial personnel if there is an unanticipated turnover.
6. Manpower planning will help managers to firm upon their long-term supply and demand expectations.

19.8. LINE BALANCING

Assembly line balancing is associated with a product layout in which products are processed as they pass through a line of work centres. An assembly line can be considered as a "PRODUCTION SEQUENCE" where parts are assembled together to form an end product. The operations are carried out at different workstations situated along the line.

Advantages of Assembly Line (or flow line)

1. Uniform rate of production.
2. Less material handling.
3. Less work-in-process.
4. Easy production control.
5. Effective use of facilities/ labour.

6. Less congestion.

Disadvantages

- More capital intensive (*i.e.*, demands larger investments).
- Low flexibility.
- Monotony of work for operators.

The Problem of Line Balancing arises due to the following factors

1. The finished product is the result of many sequential operations.
2. There is a difference in production capacities of different machines (The output from different machines is not identical).

Line balancing is the apportionment of sequential work activities into workstations in order to gain a high utilisation of labour and equipment so as to minimise the idle time.

For example, the production capacities of two machines, lathe and milling is as under for a particular job.

Lathe 50 pieces/hour
Milling 25 pieces/hour

Now if only one machine of each is provided, Then machine *B* will produce 25 units/hour where as the machine *A* can produce 50 units. But because of the sequence, only 25 units are produced per hour, *i.e.*, machine *A* will work only 50 per cent of its capacity and the remaining 30 minutes in one hour, it is idle. This idle time can be minimised by introducing one more machine of kind *B* in the production line.

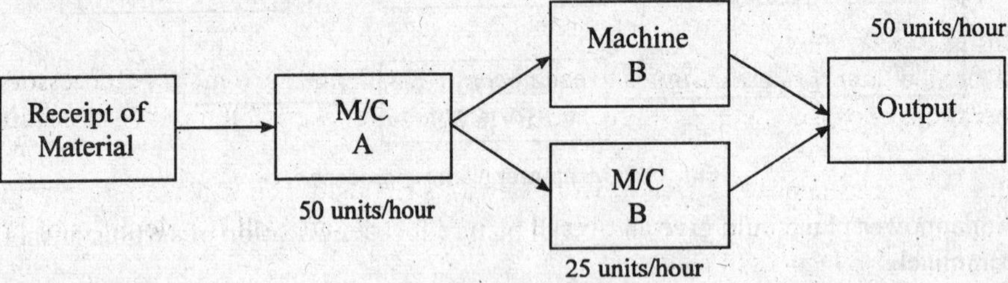

Some Definitions

1. **Workstation:** A work station is a location on assembly line where given amount of work is performed.
2. **Cycle time:** It is the amount of time for which a unit that is assembled is available to any operator on the line or it is the time the product spends at each work station.

$$\text{Cycle Time } (CT) = \frac{\text{Available time period}}{\text{Output units required/ period}} = \frac{AT}{\text{Output}}$$

3. **Task:** The smallest grouping of work that can be assigned to a workstation.
4. **Predecessor task:** A task that must be performed before performing another (successor) task.
5. **Task time (ti):** Standard time to perform element task.
6. **Station time (sk):** Total standard work content of specific workstation.
7. **Balance Delay (BD):** Percentage of total idle time on the line to total time spent by the product from beginning to end of line.

$$B.D. = \frac{n.CT - \sum_{k=1}^{n} sk \times 100}{n \times CT}$$

$B.D.$ = Balance delay

n = number of work stations
CT = Cycle time
sk = Station time

Steps in Solving Line Balancing Problems

1. Define task.
2. Identify precedence requirements.
3. Calculate minimum number of workstations required to produce desired output.
4. Apply heuristics to assign task to each station.
5. Evaluate effectiveness and efficiency.
6. Seek further improvement.

Three important parameters in line balancing

1. Line efficiency $(LE) = \dfrac{\text{Total station time}}{\text{Cycle time} \times \text{no. of workstations}} \times 100$

2. Balance delay $(BD) = \dfrac{\text{Total idle time for all workstations}}{\text{Total available working time on all stations}} \times 100$

   $BD = (1 - LE)$

3. Smoothness Index $(SI) = \Sigma_{i=1}^{k}$ (Max. station time – station times of station $i)^2$

   SI = 0, means a perfect balance
   K = total number of workstations < total number of elements
   Also, $CT \geq$ maximum time of any work element n

Heuristic Method

In this method, numbers are assigned to each operation to denote how many predecessors it has. Those operations showing the lowest predecessor number are taken first on the workstations. .

Steps

1. Draw the precedence diagram of work elements first and then succeeding elements. Elements within the columns are assigned to workstation after all the elements of previous columns have been assigned.
2. Select Cvcle time (CT) which is feasible.

   i.e. $T_{max} \leq CT \leq \Sigma_{i=1}^{n} T_i$

   where T_i = Time for work element.
   n = number of work elements
   T_{max} = Maximum work element time
   CT = Cycle time
3. Assign work elements to workstations sum of elemental times should not exceed cycle time (CT) while doing so proceed from column I to column II and so on. Break intra column tie by using minimum number of precedence.
4. Deduct assigned work elements from total elements. Repeat step (3).
5. If workstations (ws) time is more than CT, identify work element due to which this happens and carry it forward to next workstation.
6. Repeat steps (3) to (5) till all elements are fully assigned.

***Illustratan 1*:** The precedence diagram is shown below for six workstations. Assign the work elements to workstations. Calculate line efficiency and balance delay.

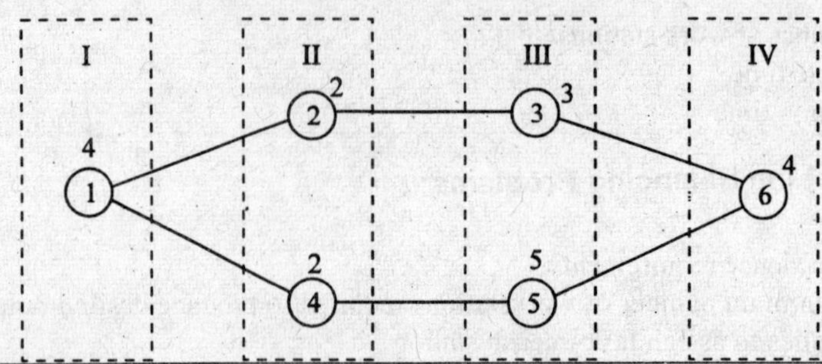

Work elements	*No. of precedence*	*Ti*
1	0	4
2	1	2
3	2	3
4	1	2
5	2	5
6	5	4

***Solution*:** Let the cycle time be (CT) = 8 min.

Assignment of work elements to work stations

Station	*Element*	*Ti*	*Stations sum*	*Idle time*
I	1 2	4 2	6	2
II	4 5	2 5	7	1
III	3 6	3 5	8	0

$$\text{Line efficiency} = \frac{\text{Total stationtime}}{\text{CT} \times \text{no. of workstations.}}$$

$$= \frac{21}{3 \times 8} \times 100 = 87.5\%$$

$$\text{Balance delay } (BD) = \frac{\text{Idle time of all wor kstations}}{\text{Available working time for all stations}} \times 100$$

$$= \frac{03}{21} \times 100 = 14.29\%$$

Heuristic method will not give optimal solution.

***Illustraton 2*:** For the given precedence diagram, carry out the line balancing and improve the solution using heuristic method.

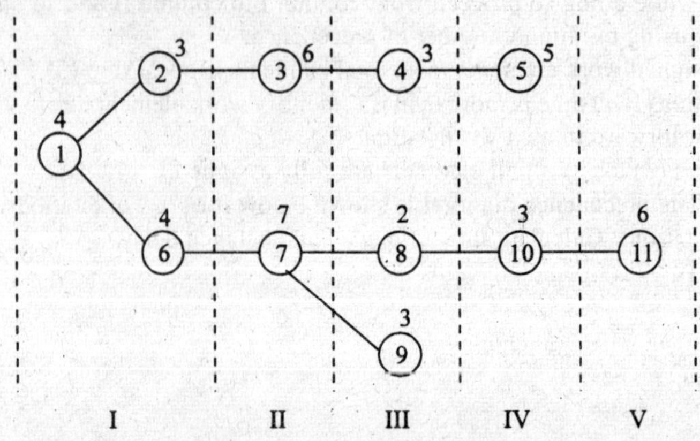

Assignment of work elements to workstations
The cycle time is 13

Work station	I	II	III	IV	V
Time (min)	11	13	8	8	6

$$\text{Idle time} = 2 + 5 + 5 + 7 = 19$$

$$\text{Idle time \%} = \frac{19}{13} \times 100 = 150\%$$

Rearranging the workstations with a cycle time of 13 minutes.

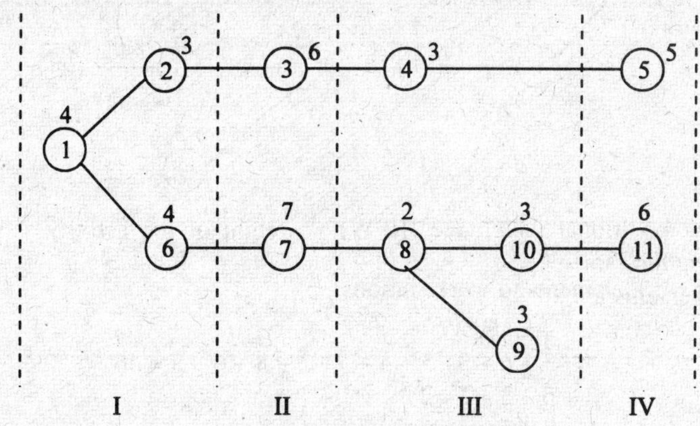

Workstation	I	II	III	IV
Time in (min)	11	13	11	11

Here the idle time is 6 min *i.e.*, 46%.

Step 3. Line balancing by rearranging the workstations,

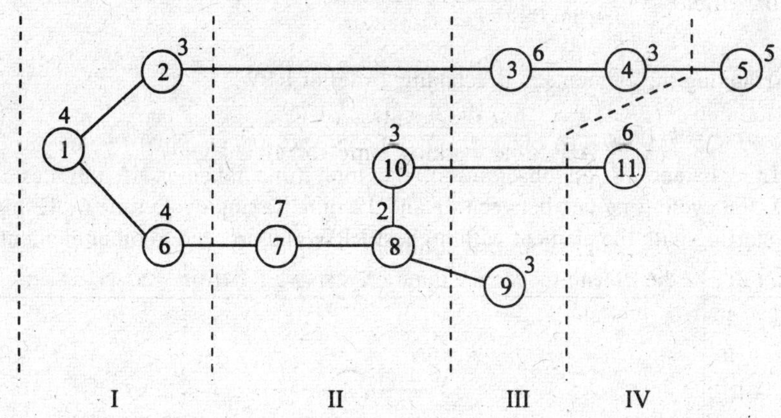

Workstation	I	II	III	IV
Time in (min)	11	12	12	11

Idle time is 2 minutes cycle time is 12.
Idle time % = 16.6% only.
An alround improvement.

2. RANK POSITION WEIGHTAGE (RPW) METHOD

(Helgeson and Burnie)

The method is illustrated through the following example:

***Illustration 3*: Consider the problem**

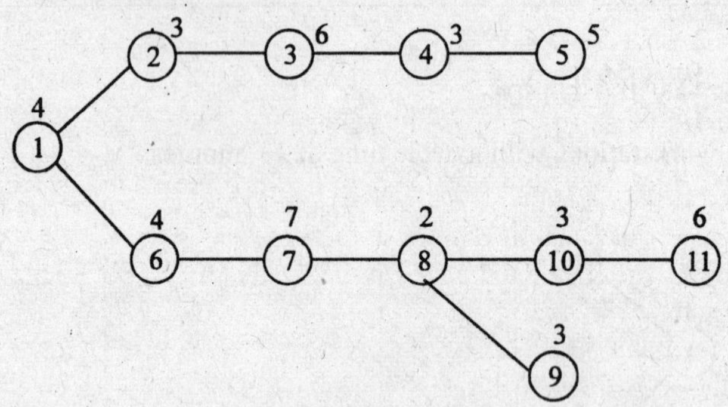

***Solution*:** Rank positional weightage (RPW) = weight of the activity + weightage of all activities following the same.

RPW for the elements are,

Element	**RPW**
1	4 + 3 + 6 + 3+ 5 + 4 + 7+ 2 + 3 + 6 + 3 = 46
2	3 + 6 + 3 + 5 = 17
3	6 + 3 + 5 = 14
4	3 + 5 = 08
6	4 + 7 + 2 + 3 + 3 + 6 =25
7	7 + 2 + 3 + 3 + 6 = 21
8	2 + 3 + 6 = 11
9	3
10	3 + 6 = 09
11	6

Step 2. Arranging the elements in descending order of RPW.

Element	1	6	7	2	3	8	10	4	11	5	9
RPW	46	25	21	17	14	11	9	8	6	5	3

Step 3. Line balancing. on observation, the total time taken is 46 minutes assuming 4 workstations, the cycle time lies between 11 and 12 min. Taking cycle time as 12 min, fill each workstation starting with the element with highest RPW and proceed to other elements.

Elements		1	6	7	2	3	8	10	4	11	5	9
Workstations	I	4	4	–	3	–	–	–	–	–	–	–
	II	–	–	7	–	–	2	3	–	–	–	–
	III	–	–	–	–	6	–	–	3	–	–	3
	IV	–	–	–	–	–	–	–	–	6	5	–

Workstation	I	II	III	IV
Idle time	1	0	0	1

As the workstation I concerned,

Element 1 has a time of 4 minutes.

We are left with 12 – 4 = 8 minutes. Look for next element in the same order, *i.e.*, element No. 6 has the time of 4 minutes. Enter this in workstation I. Now there is a scope of accommodating elements for 4 minutes (as cycle time is 12 minutes). Element number 2 having time of 3 minutes is assigned which makes the time as 11 minutes. So idle time is 1 minute.

Similarly proceed further for all the remaining workstations

Total idle time is 2 minutes.

% of idle time = 16.67 %.

19.9. PROCESS PLANNING

Process planning establishes the shortest route that is followed from raw material stage till it leaves as a finished part or product.

The activities that are associated with process planning are:

- List of operations to be performed and their sequence.
- Specifications of the machines and equipment required.
- Necessary toolings jigs and fixtures.
- Gives the manufacturing details with respect to feed, speed, and depth of cut for each operation to be performed.
- It gives the estimated or processing times of operations

All the above information is represented in the form of a document called process sheet or route sheet.

The information given in the process sheet can be used for variety of activities.

- It becomes the important document for costing and provides the information on the various details like set-up and operation times for each job.
- The machine and manpower requirements can be computed from the set-up and operational times.
- Helps to carryout scheduling.
- The material movement can be traced.
- It helps in cost reduction and cost control.
- It helps in cost reduction and cost control.
- It helps to determine the efficiency of a work centre.

The information required for process planning

1. **Assembly and component drawings and bill of materials (part list)**

These details give the information regarding the general description of part to be manufactured, raw material specification, dimensions and tolerances required, the surface finish and treatment required.

2. **Machine or equipment details with respect to—**

The various possible operations that can be performed.

The dimensions (maximum and minimum) that can be machined on the machines.

The accuracy of the dimensions that can be obtained.

Available feeds and speeds on the machines.

3. **The standard times for operation and details of set-up time for each job:** This helps to compute the standard time of the operation and hence the production rate.
4. **Availability of toolings (both standard and special purpose toolings):**

Factors affecting process planning

(*i*) Volume (quantity) of production.

(*ii*) Delivery dates for components or products.

(*iii*) Accuracy and process capability of machines.

(*iv*) The skill and expertise of manpower.

(*v*) Material specifications.

(*vi*) Accuracy requirements of components or parts.

Steps in Process Planning

1. Detailed study of the component drawings to identify the salient features that influence process selection, machine selection, inspection stages and toolings required.
2. List the surfaces to be machined.
3. The surfaces to be machined are combined into basic operations. This step helps in selection of machines for operation.
4. Determine the work centre, tools, cutting tools, jigs and fixtures and inspection stages and equipment.
5. Determine the speed, feed and depth of cut for each operation.
6. Estimate the operation time.
7. Find the total time to complete the job taking into account the loading and unloading times, handling times, and other allowances.
8. Represent the details on the process sheet.

References for Further Reading

1. Buffa Elwood S., *Modern Production/Operations Management,* 7th Edn., Willey, New Delhi, (1984).
2. Barry Shore, *Operations Management*, Tata McGraw Hill Company, New Delhi, (1980).
3. Joseph Monks, *Operations Management Theory and Problems*, Third edition, McGraw Hill International, New York, (1987).
4. L.C., Jhamb, *Production Planning and Control,* Aditya Publishing House, Pune.

REVIEW QUESTIONS

1. Define process design and explain the framework of process design by means of a block diagram?
2. What are the considerations in selection of a equipment or process?
3. Illustrate with an example how B.E.A. can be used in choice of machines or process amongst alternatives.
4. How do you find out the machine requirements?
5. What are the various methods to reduce the idle time of machines?
6. What is manpower planning? Explain the reasons for manpower planning?
7. What are the various steps in manpower planning?
8. What are advantages of manpower planning?
9. What is flow line production? What are its advantages and limitations?
10. What is line balancing? Illustrate with an example.
11. Explain the various terms associated with line balancing.
12. Explain the following parameters w.r.t. line balancing:
    (*i*) Balance delay
    (*ii*) Line efficiency
    (*iii*) Smoothness index
13. Explain heuristic methods of line balancing.
14. Define process planning. What are activities associated with it?
15. List the information required for process planning.
16. What are the factors that influence process planning?
17. Explain the steps in process planning.

PROBLEMS

1. A product requires three processes which are given below. The requirement of the product is 500 units/week. Assume 48 hours per week and determine the number of machines of each kind:

Process	*Time (min)*	*Efficiency (%)*
Turning	50	90
Grinding	40	80
Polishing	06	90

2. A gear manufacturer has gear shaper and gear hobbers. The gear can be processed on gear shaper as well as on gear hobber.

   The following information is given.

	Gear shaper	*Gear hobber*
1. Machining time per piece	12 min	4 min
2. Machine cost per hour	Rs. 45	Rs. 120
3. Set-up time (in minutes)	60	90
4. Tooling up cost	400	2000

   Which of the two machines will you choose to do the job if the order quantity is,

   (*i*) 1000 number and order is unlikely to repeat

   (*ii*) 1000 numbers and the order is likely to repeat for 3 years

3. The manufacturer of electronic equipment wants to add a component subassembly operation that can produce 80 units per shift of 8 hours. There are 3 activities which are shown below:

Operation	*Activity*	*Std. time (min)*
A	Assembly	12
B	Electric wiring	16
C	Testing	03

   (*a*) How many workstations are required for each activity?

   (*b*) Find the idle time percentage at each workstation.

4. A toy manufacturer produces dolls on a product line geared to an output of one unit per unit. The assembly precedence relationships and activity times are shown.

   (*a*) Group the activities into most efficient arrangement.

   (*b*) Determine the line efficiency

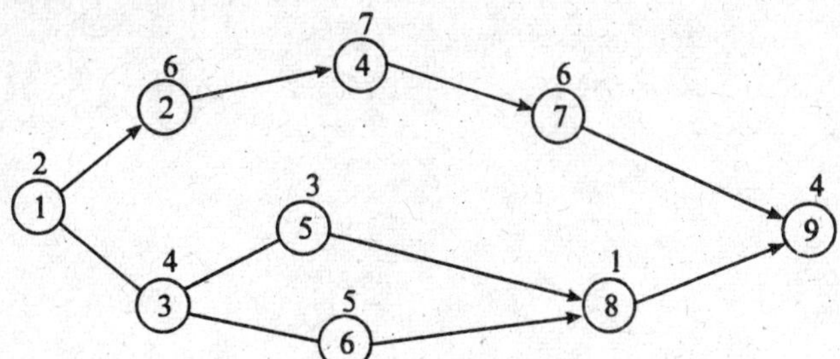

5. The task timings and precedence relationships are given below.

Task	*Time (min)*	*Preceding task*
A	10	None
B	24	None
C	17	*A*
D	49	*A*

E	12	*C*
F	14	*C*
G	27	*B*
H	9	*E*
I	20	*F, G*
J	23	*D, H, I*
K	36	*I*
L	18	*J, K*

(*a*) Draw the precedence diagram.

(*b*) Group the tasks into 5 station assembly line.

(*c*) What is cycle time?

(*d*) What is line efficiency?

20

PROJECT SCHEDULING WITH CPM AND PERT

• Introduction • Objectives of network analysis • Applications of network analysis • Basic concepts in network analysis • Numbering of events—Fulkerson's Rule • Critical path method (CPM)—Programme Evaluation and review technique (PERT) — Time cost trade off (Crashing) the network.

20.1. INTRODUCTION

A project is any task which has definable beginning and definable end, and requires investment and expenditure of one or more resources in each of the separate but interrelated and interdependent activities which must be accomplished to achieve the objectives for which the task or project was instituted.

It is essential to manage effectively the projects through proper planning, scheduling and control as project requires a heavy investment, and is associated with risk and uncertainties.

Network scheduling is a technique used for planning and scheduling large projects in the fields of construction, maintenance, fabrication and any other areas. This technique is the method of minimising the bottlenecks, delays and interruptions by determining the critical factors and coordinating various activities. There are two basic planning and control techniques. They are Critical Path Method (CPM) and Programme Evaluation and Review Techniques (PERT).

20.2. OBJECTIVES OF NETWORK ANALYSIS

1. A powerful coordinating tool for planning, scheduling and controlling of projects.
2. Minimisation of total project cost and time.
3. Effective utilisation of resources and minimisation of effective resources.
4. Minimisation of delays and interruption during implementation of the project.

20.3. APPLICATIONS OF NETWORK ANALYSIS (PERT AND CPM)

1. Research and development projects
2. Equipment maintenance and overhauling
3. Construction projects (building, bridges, dams)
4. Setting up new industries
5. Planning and launching of new products
6. Design of plants, machines and systems
7. Shifting the manufacturing location from one location to another
8. Control of production in large job shops
9. Market penetration programmes
10. Organisation of big programmes, conferences, etc.

20.4. BASIC CONCEPTS IN NETWORK

(*a*) **Network:** It is a graphical representation of the project and it consists of series of activities arranged in a logical sequence and show the interrelationship between the activities.

(*b*) **Activities:** An activity is a physically identifiable part of the project, which consumes time and resources. Each activity has a definite start and end. Activity is represented by an arrow (→).

(*c*) **Event:** An event represents the start or the completion of an activity. The beginning and end points of an activity are events.

Example: Machining a component is an activity
Start Machining is an event
Machining completed is an event

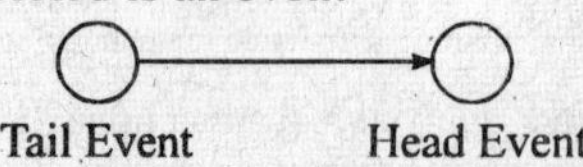

In a network a number of activities may terminate into single node, called merge node and a number of activities may emanate from a single node called burst node.
The merge event, and burst events are shown in Fig. 20.1.

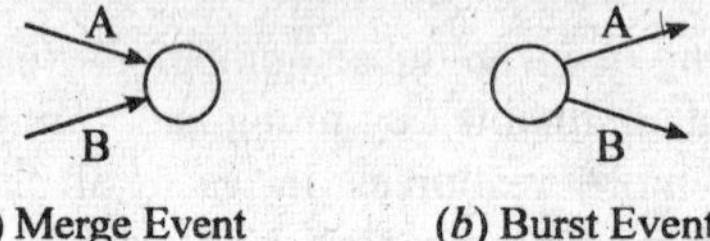

(*a*) Merge Event (*b*) Burst Event

Fig. 20.1: Merge and burst events.

(*d*) **Predecessor and successor activities:** All those activities, which must be completed before starting the activity under consideration are called its predecessor activities. All the activities which have to follow the activity under consideration are called its successor activities.

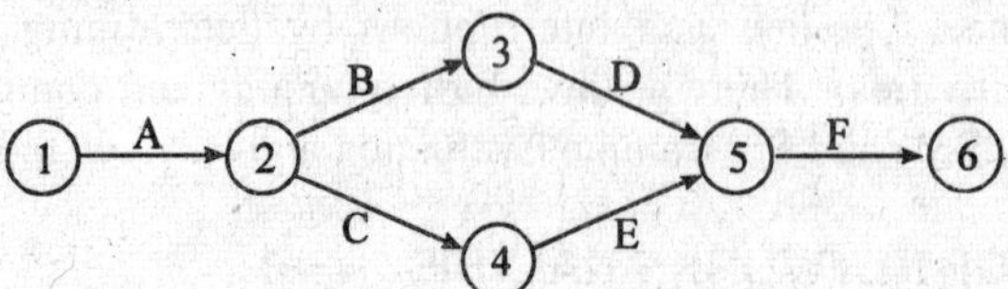

Fig. 20.2: Network diagram.

In Fig. 20.2., Activities 2-3 and 2-4 are immediate successors to activity 1-2.
Activities 2-3 and 2-4 . 3-5, 4-5 and 5-6 are its successor activities.
Activities 1-2, 2-3, are predecessors to activity 3-5 and 2-3 is the immediate predecessor.
The precedence relationship between the activities can be expressed as —

Activity	*Immediate Predecessor*
A	–
B	*A*
C	*A*
D	*B*
E	*C*
F	*D, E*

(*e*) **Path:** An unbroken chain of activities between two events is called a path. For example, in Fig. 20.2. *A-B-D-F* is a path connecting events (1) and (6). There are number of paths traced through the network.

(*f*) **Dummy activity:** An activity which depicts the dependency or relationship over the other but does not consume time or resources. It is used to maintain the logical sequence. It is indicated by a dotted line. A dummy activity shown in Fig. 20.3.

Consider the network of activities *P, Q, R* and *S*. activity *R* is preceded by activities *P* and *Q*. While activity *S* is preceded by *Q* only. The wrong and correct representation of these activities are shown in Fig. 20.3.

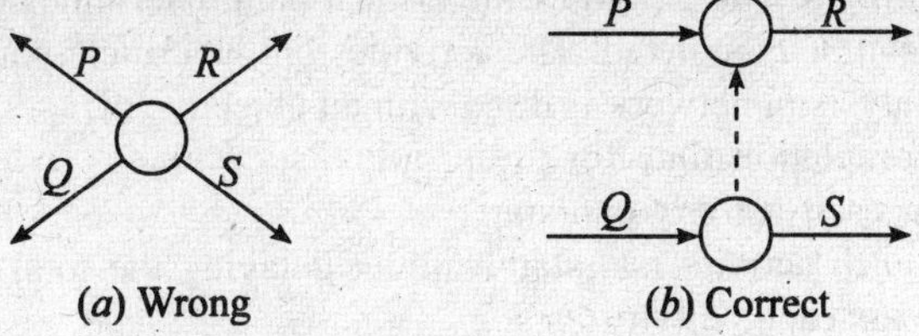

Fig. 20.3: Representation of dummy activity.

20.5. NUMBERING OF EVENTS (FULKERSON'S RULE)

The steps involved in numbering the events as per the Fulkerson's rule are:

1. The initial event which has all outgoing arrows with no incoming arrow is numbered "1".
2. Delete all the arrows coming out from node "1". This will convert some more nodes (at least one) into initial events number these events 2, 3 .... etc.
3. Delete all the arrows going out from these numbered events to create more initial events. Assign next numbers to these events.
4. Continue until the final or terminal node which has all arrows coming in, with no arrow going out is numbered.

Figure 20.4 shows numbering of the network.

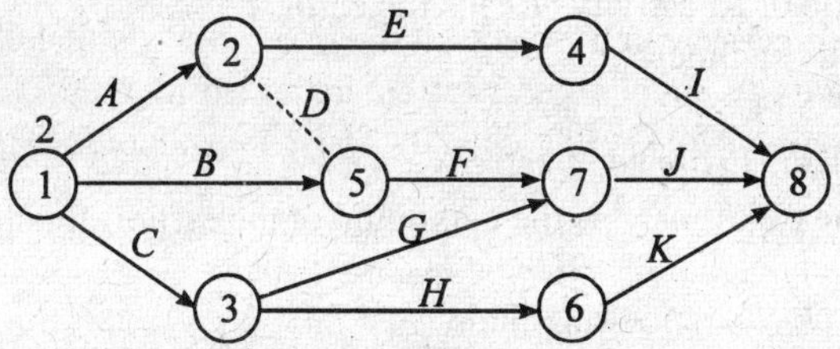

Fig. 20.4: Numbering the network.

Problem 1: Construct the network from the information given.

Activity	*Immediate Predecessor*	*Activity Time (Weeks)*
A	–	6
B	–	10
C	–	14
D	*C*	6
E	*A, B*	14
F	*E, D*	6
G	*D*	4
H	*F, G*	4

Solution: Construction of the network is shown in Fig. 20.5.

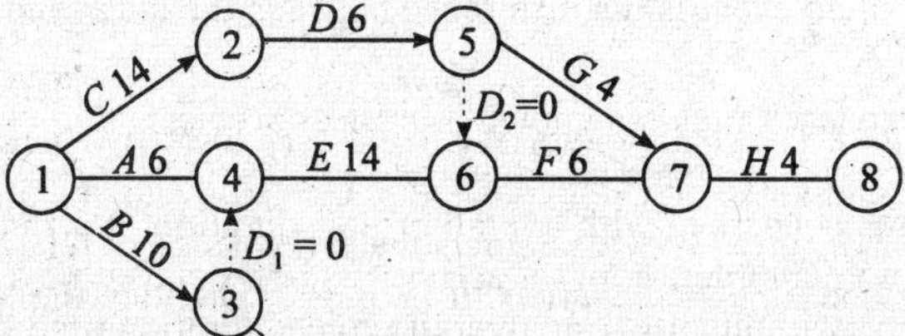

Fig. 20.5: Network diagram.

20.6. CRITICAL PATH METHOD

In critical path method (CPM) the activity times are known with certainty. For each activity earliest start time (EST) and latest start times (LST) are computed. "The path with the longest time sequence is called critical path. The length of the critical path determines the minimum time in which the entire project can be completed. The activities on the critical path are called 'Critical activities'." The time analysis in network is done with an objective of:

- Determining the completion time for the project
- Earliest time when each activity can start
- Latest time when each activity can start without delaying the total project
- Determining float for each activity
- Identification of the critical activities and critical path.

20.6.1. FORWARD PASS COMPUTATIONS (EARLIEST EVENT TIME)

Forward pass computation gives the earliest expected start and finish times for each activity (indirectly earliest occurrence time for each event). The computations start from the initial node and move to the end node. Forward pass computation, starts with, an assumed earliest occurrence of zero for the initial event.

(*a*) **Earliest start time (ES):** It is the earliest event time of the tail end event—

$$ES_{IJ} = E_i$$

where, ES_{IJ} = Earliest start for an activity (i, j)

E_i = Earliest event occurrence time of event i.

(*b*) **Earliest finish time** of an activity (*EF*) is the earliest Starting time + Activity time

$$EF_{ij} = ES_{ij} + t_{ij}$$

t_{ij} = Time Estimate of activity

(*c*) **Earliest event time** for event '*j*' is the maximum of the earliest finish times of all activities ending into that event.

E_j = Maximum of $(ES_{ij} + t_{ij})$ = Maximum $(E_i + t_{ij})$

20.6.2. BACKWARD PASS COMPUTATIONS (LATEST ALLOWABLE TIME)

The latest event times specifies the time by which all activities entering into that event must be completed without delaying the total project.

(*a*) Latest finish time for an activity (i, j) equals the latest event time of event $j - LF_{ij} = L_{ij}$.

(*b*) Latest starting time of activity (i, j) is the latest completion time of (i, j) minus the activity time.

i.e., $LS_{ij} = LF_{ij} - t_{ij}$

(*c*) Latest event time for event i is the minimum of the latest start time of all activities originating from that event.

L_i = Maximum $j\,(LS_{ij})$

= Minimum $(LF_{ij} - t_{ij})$ = Min $j\,(L_j - t_{ij})$

(*d*) **Slack** – The slack of an event is the difference between the latest and earliest event times

Slack $(i) = L_i - E_i$

The events with zero slack time are known as critical events.

(*e*) **Floats**

(*i*) **Total float:** It is concerned with overall project duration. It is defined as "the amount of time by which completion of an activity can be delayed beyond earliest expected completion time without affecting overall project duration time."

Total float for an activity (i, j) is the difference between the latest start time and earliest start time for that activity.

$$TF_{ij} = LS_{ij} - ES_{ij}$$
$$= (L_j - E_i) - t_{ij}$$

where E_i = earliest expected completion time of tail event
= Earliest starting time for an activity (i, j)

L_i = Latest allowable completion time of head event
= latest finish time for activity (i, j)

(*ii*) **Free float:** It is the time by which the completion of an activity can be delayed beyond the earliest finish time without affecting the earliest start of a subsequent activity.
Free float for an activity

$$= (E_j - E_i) - E_{ij}$$

(*iii*) **Independent float:** It is the amount of time by which the start of an activity can be delayed without affecting earliest start time of any immediately following activities assuming that the preceding activity has finished at its latest finish time.
Independent float is given by:

$$(E_i - L_i) - E_{ij}$$

***Problem 2*:** The activity details and their predecessors are given below along with their activity times.

Construct the network diagram

Activity	*Predecessors*	*Activity time (Weeks)*
A	–	4
B	*A*	3
C	*A*	2
D	*B*	5
E	*B*	3
F	*C, D*	4
G	*E, F*	3

***Solution*:** The network diagram is shown in the Fig. 20.6.

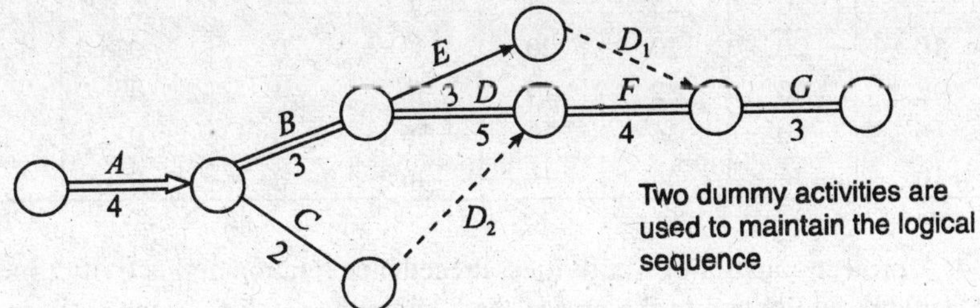

Fig. 20.6: Network diagram.

The slack for the activities is calculated as shown in Table 20.1.

Table 20.1: Computation of slack.

Activity	*Activity Time*	*ES*	*EF*	*LS*	*LFT*	*Slack*
A	4	0	4	0	4	0
B	3	4	7	4	7	0
C	2	4	6	8	12	2
D	5	7	12	7	12	0
E	3	7	10	9	12	3

Activity	*Activity Time*	*ES*	*EF*	*LS*	*LFT*	*Slack*
F	4	12	16	12	16	0
G	3	16	19	16	19	0

The critical path is one that connects the activities with zero slack. The critical paths

A— B — D — F — G

***Problem 3*:** The activities involved in a small project are given below along with relevant information. Construct the network and find the critical path

Find the floats for each activity.

Activity	1 – 2	1 – 3	2 – 3	2 – 4	3 – 4	4 – 5
Duration	20	25	10	12	6	10

***Solution*:** The network diagram along with timing is shown in the Fig. 20.7.

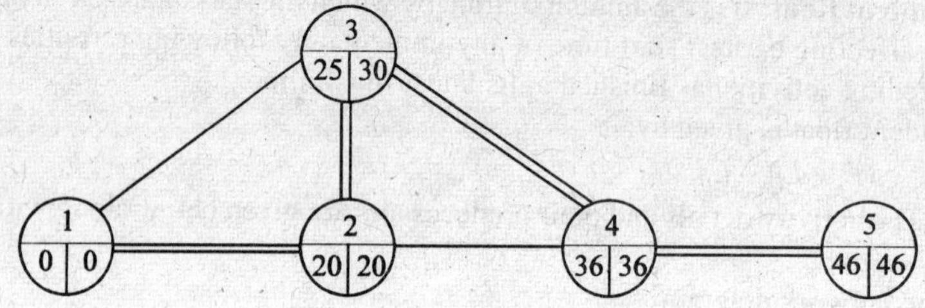

Fig. 20.7: Network diagram.

Critical path is 1 – 2 – 3 – 4 – 5

The floats are calculated as shown in Table 20.2.

Table 20.2: Calculation of floats

Activity	*Duration*	*Earliest*		*Letest*		*Float*		
		Start	*Finish*	*Start*	*Finish*	*Total*	*Free*	*Independent*
1 – 2	20	0	20	0	20	0	0	0
1 – 3	25	0	25	5	30	5	5	5
2 – 3	10	20	30	20	30	0	0	0
2 – 4	12	20	32	24	36	4	4	4
3 – 4	6	30	36	30	36	0	0	0
4 – 5	10	36	46	36	46	0	0	0

***Problem 4*.** A project consists of 8 activities. Precedence relation and activity times are given.

Draw the network and compute the critical path show the slack for each activity in a tabular form.

Activity	*Immediate Predecessor*	*Activity Time (Weeks)*
P	–	12
Q	–	20
R	–	28
S	*R*	12
T	*P, Q*	28
U	*T, S*	12
V	*S*	8
W	*U, V*	8

***Solution*:** The network is shown in Fig. 20.8.

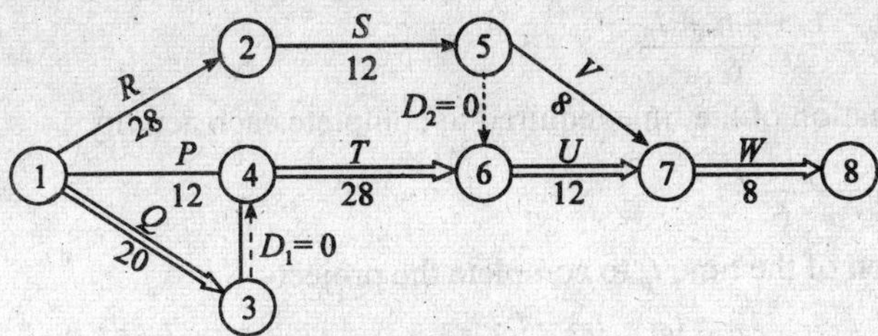

Fig. 20.8: Network diagram.

Table 20.3 shows the activities times.

Table 20.3: Computation of slack

Activity	*Duration*	*ES*	*EF*	*LS*	*LF*	*Slack*
P	12	0	12	8	20	8
Q	20	0	20	0	20	0
R	28	0	28	8	36	8
S	12	28	40	36	48	8
T	28	20	48	20	48	0
U	12	48	60	48	60	0
V	8	40	48	52	60	12
W	8	60	68	60	68	0
D_1	0	20	20	20	20	0
D_2	0	40	40	48	48	8

The critical path is: $Q - T - U - W$

Project duration = 20 + 0 + 28 + 12 + 8 = 68 weeks.

20.7. PROGRAMME EVALUATION AND REVIEW TECHNIQUES (PERT)

Critical path method has only one time estimate. So it does not consider uncertainty in time. In reality the duration of activities may not be deterministic (certain) in all cause.

PERT takes into account the uncertainty of activity times. It is a probabilistic model with uncertainty in activity duration.

PERT makes use of three estimates of time:

(*i*) Optimistic time (t_o)

(*ii*) Most likely time (t_m)

(*iii*) Pessimistic time (t_p)

Optimistic time (t_o) is the shortest possible time, if everything goes perfectly without any complications. It is an estimate of minimum possible time to complete the activity under ideal condition.

Pessimistic time (t_p) is the longest time taking into consideration all odds. This is the time estimate if everything goes wrong.

Most likely time (t_m) is the best estimate of the activity time. This lies between the optimistic and pessimistic time estimates.

The three time estimates t_o, t_p and t_m are combined to develop expected time (t_e) for an activity. The fundamental assumption in PERT is that the three estimates of time follow a β (Beta) distribution.

The expected time (t_e) is given by

$$t_e = \frac{t_o + 4\,t_m + t_p}{6}$$

The standard deviation of the time required to complete each activity

$$\text{Std. deviation } (\sigma) = \frac{t_p - t_o}{6}$$

Standard deviation of the time t_p to complete the project—

$$= \frac{t_{p1} - t_{o1}}{6} + \frac{t_{p2} - t_{o2}}{6} + \ldots\ldots\ldots\ldots + \frac{t_{pn} - t_{o1}}{6}$$

Probability of Completion of the Project within a Scheduled Time

The probability of completing a project is given by the probability of occurrence of the end event of the network. The probability distribution is assumed to be the normal distribution.

The probability of completion of the project within scheduled is computed as:

1. Calculate the mean of the event time (t_e) by adding the times of the activities along the critical path leading to the event.
2. Calculate the variance of the event time by adding up the variances of the activities on the critical path. Take the squire root of this variance to get T (standard deviation).
3. Compute standard normal variate,

$$\therefore \quad Z = \frac{T_s - T_e}{\sigma T}$$

From tables of normal curve, the value corresponding to Z gives the required probability.

Problem 5: The PERT time estimates of the project are given

Construct the PERT Network—Find the critical path and variance of each event. Find the project duration at 95 per cent probability.

Activity	*Optimistic* (t_a)	*Most expected* (t_b)	*Pessimistic* (t_m)
1 – 2	1	5	1.5
2 – 3	1	3	2
2 – 4	1	5	3
3 – 5	3	5	4
4 – 5	2	4	3
4 – 6	3	7	5
5 – 7	4	6	5
6 – 7	6	8	7
7 – 8	2	6	4
7 – 9	5	8	6
8 – 10	1	3	2
9 – 10	3	7	3

Solution: Computation of expected time and variance of each activity is shown in Table 20.4.

Table 20.4: Expected time and variance of activities

Activity	*Optimistic* (*a*)	*Pessimistic* (*b*)	*Most likely* 4*m*	*Expected*	
				Time a + 4m + b	*Variance [(b – a/6)]*
1 – 2	1	5	6	2	4/9
2 – 3	1	3	8	2	1/9

Activity	Optimistic (a)	Pessimistic (b)	Most likely 4m	Expected	
				Time a + 4m + b	Variance [(b − a/6)]
2 – 4	1	5	12	3	4/9
3 – 5	3	5	16	4	4/9
4 – 5	2	4	12	3	1/9
4 – 6	3	7	20	5	4/9
5 – 7	4	6	20	5	1/9
6 – 7	6	8	28	7	4/9
7 – 8	2	6	16	4	4/9
7 – 9	5	8	24	6.1/6	1/4
8 – 10	1	3	8	2	1/9
9 – 10	3	7	20	5	4/9

The construction of the network is shown in the Fig. 20.9.

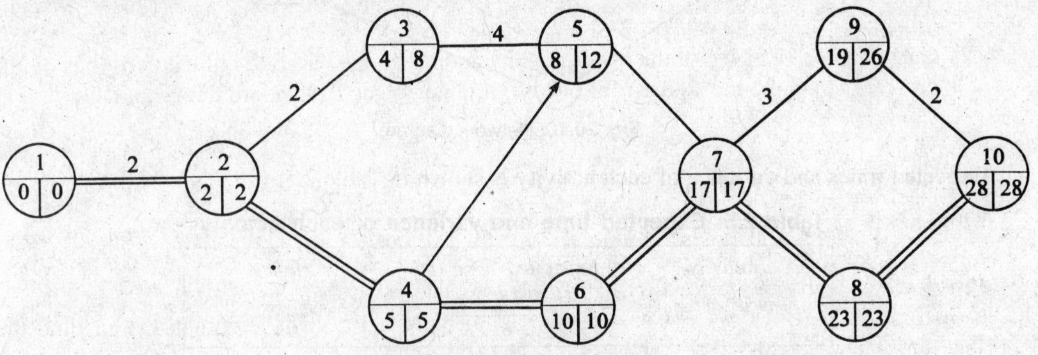

Fig. 20.9: Network diagram.

The Critical path is 1 – 2, 2 – 4, 4 – 6, 6 – 7, 7 – 9, and 9 – 10

Expected the duration of the project = 2 + 3 + 5 + 7 + 6 + 6.1/6 + 5 = 28.1/6 days

$$\text{Project variance} = \frac{4}{9}+\frac{4}{9}+\frac{4}{9}+\frac{1}{9}+\frac{1}{4}+\frac{4}{9}=\frac{77}{36}$$

$$Z = \frac{\text{Due date} - \text{Expected date of completion}}{\sigma T}$$

$$\frac{X - 28.1/6}{77136} = 0.8229 \qquad X = 29.6 \text{ days} \quad \textbf{(Ans)}$$

Problem 6: A small project is composed of the following activities whose time estimates are given below:

(Estimated activity duration in weeks)

Activity	Optimistic	Most likely	Pessimistic
1 – 2	1	1	7
1 – 3	1	4	7
1 – 4	2	2	8
2 – 5	1	1	1
3 – 5	2	5	14
4 – 6	2	5	8
5 – 6	3	6	15

(*a*) Draw the network and find the critical path
(*b*) Find the expected duration and variance for each activity
(*c*) Calculate the standard deviation of project length
(*d*) What is the probability that the project will be completed
  (*i*) At least 4 weeks earlier than expected time?
  (*ii*) No more than 4 weeks later than 4 weeks.
  (*iii*) If the project due date is 19 weeks what is the probability of not meeting the due date.
  (*iv*) Find the project duration at 95 per cent probability?

***Solution*:** The network diagram is shown in Fig. 20.10.

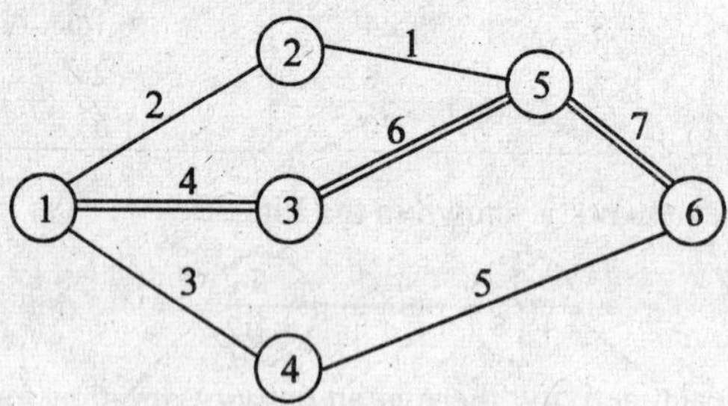

Fig. 20.10: Network diagram.

Expected times and variance of each activity is shown in Table 2.5.

Table 2.5: Expected time and variance of each activity

Activity	*Expectied Time* (t_e) $\frac{t_o + 4\,t_m + t_p}{6}$	*Variance* $\frac{t_p - t_o}{6}$
1 – 2	2	1
1 – 3	4	1
1 – 4	3	1
2 – 5	1	0
3 – 5	6	4
4 – 6	5	1
5 – 6	6	4

Critical path is: 1 – 3 – 5 – 6

∴ Expected time of the project = 4 + 6 + 7 = 17 weeks

Standard deviation of the project = Variance = 9 = 3 weeks

(*a*) Probability that the project will be completed at least 4 weeks earlier than expected time *i.e.*, 17– 4 = 13 weeks

$$Z = \frac{\text{Due date} - \text{Expected time}}{\text{Std. deviation}} = \frac{13 - 17}{3} = -1.33$$

(From the normal table value of Z = – 1.33 is 0.0918)

∴ Probability that the project will be completed 4 weeks earlier than expected time is 9.18 per cent.

(*b*) Probability that the projected will be completed no more than 4 weeks later than expected time,

$$Z = \frac{21 - 17}{3} = \frac{4}{3} = 1.33$$

Ref. Normal table the value of Z = 90.82 per cent

Probability is 90.82 per cent

(*c*) If the project due date = 19 weeks probability of not meeting the schedule

$$Z = \frac{19-17}{3} = 0.67$$

$$Z = 0.2514$$

Probability of not meeting the schedule = 1 – 0.2514 = 74.56 per cent.

(*d*) If probability is 95 per cent, the project duration

$$Z = \text{value for } 0.95 = 1.65$$

$$1.65 = \frac{X-17}{3} \qquad X = 22 \text{ weeks}$$

20.8. TIME-COST TRADE OFF (CRASHING)

The total overall cost will consist of direct and indirect costs. The indirect cost consists of overheads, office expenditures, administrative expenses, cost penalty for delay in work, etc., the indirect time varies with time. Any reduction in project time means reduction in indirect costs.

Direct costs includes cost of materials, machinery used in the project and payments towards labour and subcontracting.

It is assumed that for each activity, there is an activity duration for which the direct cost is minimum. If activity results in more than this time, more resources and, hence, more funds are required. For instance a point will be reached beyond which no further reduction in time will be possible irrespective of resources spent.

The time for the activity at minimum cost is called normal time and the minimum time for the activity is called crash time. The costs associated with these times is called normal and crash costs. A linear relationship is assumed between time and cost as shown in Fig. 20.11.

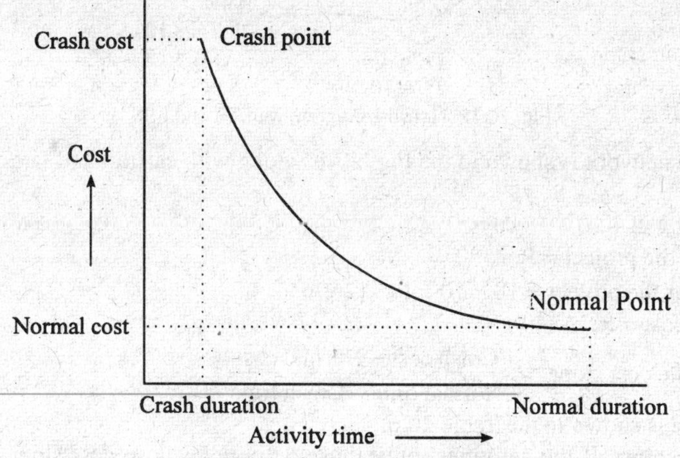

Fig. 20.11: Time-cost relationship.

$$\text{Cost slope } (CS) = \frac{\text{Crash cost} - \text{Normal cost}}{\text{Normal time} - \text{Crash time}}$$

Procedure for crashing —

The process of reducing the project duration is called crashing. The crashing is at the cost of extra resources, *i.e.*, at an extra cost.

The procedure for project crashing are:

1. Construct the network diagram and for a given network diagram find the critical path.
2. Calculate the cost slope for different activities.

3. Crashing the network — crashing the activities in critical paths as per the ranking, *i.e.*, the activity having lower cost slope would be crashed first to the maximum possible extent.
4. Determine the total cost of the project.

***Problem 7*:** From the activity details given below. Determine the optimal project duration.

Table 20.6: Cost time and cost slopes

	Normal		*Crash*		*Cost Slope*
Activity	*Time*	*Cost*	*Time*	*Cost*	
1 – 2	8	100	6	200	50
1 – 3	4	150	2	350	100
2 – 4	2	50	1	90	40
2 – 5	10	100	5	400	60
3 – 4	5	100	1	200	25
4 – 5	3	80	1	100	10

Indirect cost = Rs. 70 per day

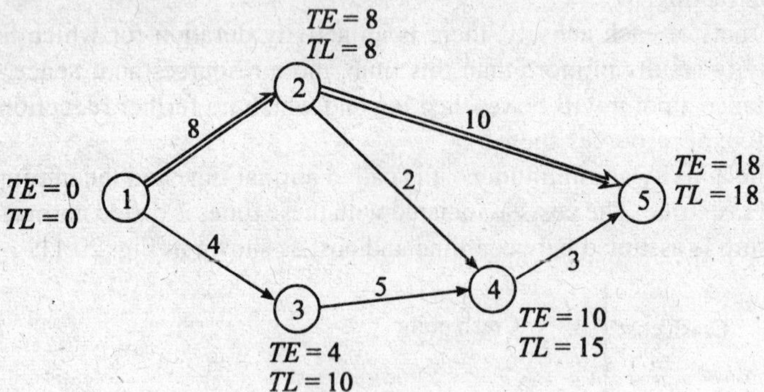

Fig. 20.12: Network diagram with ES and LS.

Solution: The network is shown in the Fig. 20.12 along with earliest and latest starts.

Critical path = 1 – 2 – 5

= 8 + 10 = 18 days

Direct cost of the project = Rs. 580

Indirect cost of the project = 18×70 = Rs. 1260

Total project cost = Rs. 1840

$$\text{The cost slope} = \frac{\text{Crash cost} - \text{Normal cost}}{\text{Normal time} - \text{Crash time}}$$

The cost-slope is shown in the Table 20.6.

Instead of crashing all the activities, only those activities whose cost slope is less than the indirect cost.

Along the critical path activities 1 – 2 – 5, both activities 1 – 2 and 2 – 5 have cost-slopes less than indirect cost.

The minimum cost slope is associated with 1 – 2, and it can be crashed by two days.

∴ Project duration is reduced to 16 days while critical path remains unchanged

Direct cost = $580 + 2 \times 50 = 680$

Indirect cost = $70 \times 16 = 1120$

∴ Total cost = $680 + 1120 = 1800$

Network after crashing is shown in Fig. 20.13.

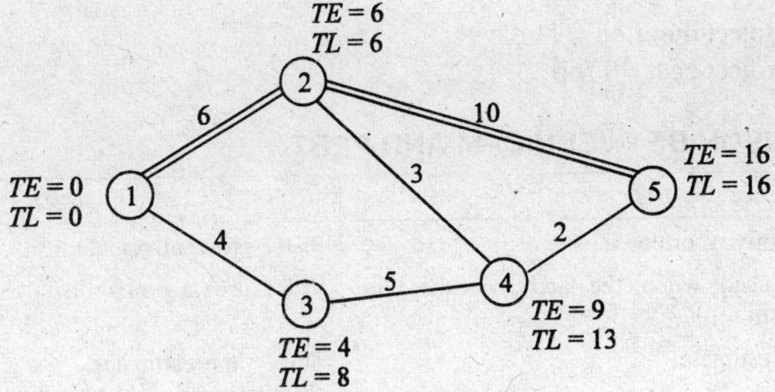

Fig. 20.13: Network after crashing.

Next, crash the activity 2-5 by 5 days

The critical path is changed to 1-3-4-5. This is the shown in the Fig. 20.14.

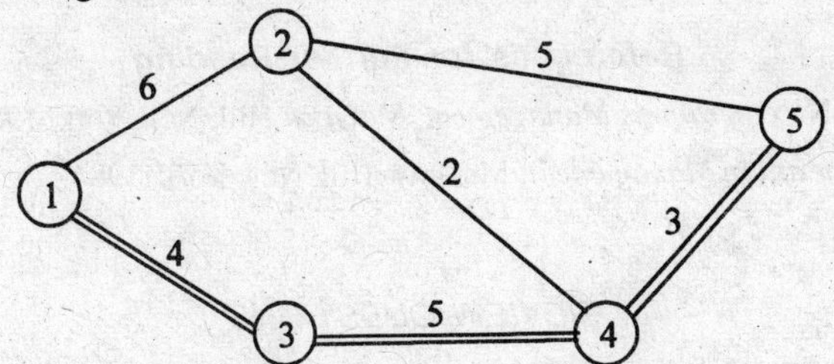

Fig. 20.14: Network after crashing.

Direct cost $= 680 + 5 \times 60 = 980$

Indirect cost $= 12 \times 70 = 840$

Total cost= $980 + 840 = 1820$

Activity on new critical path (Slope less than 70 are 3-4 and 4-5) crash on new critical the activity 4-5 by days.

Critical path again chances to 1– 2–5

Total cost $= 980 + 2 \times 10 + 11 \times 70 = 1770$

This is shown in Fig. 20.15.

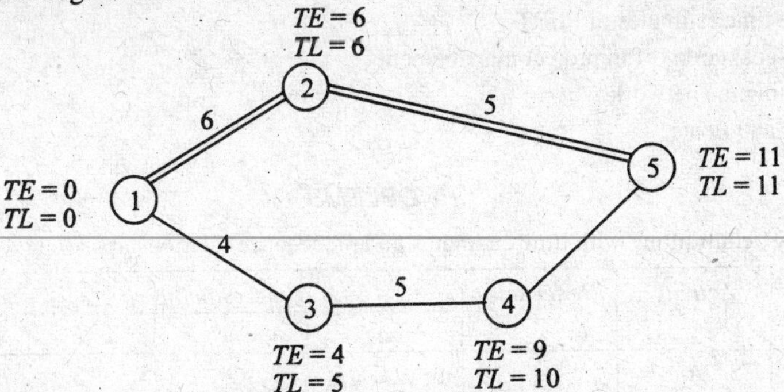

Fig. 20.15: Network after crashing.

The project cannot be crashed any further as the activities on the critical path are already crashed. From the network diagram, the activity 4–5 can be expended by one day without affecting the project duration.

This will save crashing cost of activity 4–5 for one day

Total cost = 1760

∴ Optimal project duration = 11 days
Optimal project cost = 1760

20.9. COMPARISON BETWEEN CPM AND PERT

	CPM	PERT
1.	CPM is activity oriented.	PERT is event oriented.
2.	CPM is used when the activity times are deterministic.	PERT uses a probabilistic times.
3.	One time estimate.	Three time estimates. (*a*) Optimistic, (*b*) Most likely, (*c*) Pessimistic.
4.	CPM directly introduces cost concept analysis.	PERT indirectly currents for costs.
5.	CPM is a planning device.	PERT is a control device.

References for Further Reading

1. Devisiotis Kostas N., *Operations Management*, McGraw Hill, New York, 1981.
2. Monks Joseph, *Operation Management*, McGraw Hill, New York, 1984.
3. Srinath L.S., *PERT and CPM.*

REVIEW QUESTIONS

1. Explain the areas of applications of network techniques.
2. Mention the objectives of network scheduling.
3. CPM uses a deterministic time estimate and PERT uses probabilistic time estimates. Explain.
4. What is a critical path? What is its importance?
5. Explain the significant information revealed by PERT.
6. Bring out the difference between PERT and CPM.
7. Write short notes on
   (*a*) Fulkerson's rule numbering events
   (*b*) Dummy activities
   (*c*) Three time estimates of PERT
   (*d*) Time-cost tarde off in project management
   (*e*) Crashing the network
   (*f*) Slack and floats.

PROBLEMS

1. The activity details along with time estimates and precedence relationships are given below:

Activity	*Predecessor*	*Time estimate*		
		a	*m*	*b*
A	–	1	2	3
B	*A*	1	2	3
C	*A*	2	4	6
D	*A*	2	5	14
E	*C, D*	6	12	18
F	*D*	1	3	5

Activity	Predecessor	Time estimate		
		a	*m*	*b*
G	*E*	10	12	30
H	*G*	3	5	7
I	*H*	1	2	3
J	*B* and *I*	5	10	15

(*a*) Construct the network

(*b*) Calculate the expected time and variance for each activity

(*c*) Critical path and project duration

(*d*) Probability of the completion of the project within the expected project time.

2. Draw the network, determine the earliest and latest start and finish times of the activities. Identify the critical activities.

Activity	1 – 2	1 – 4	1 – 7	2 – 3	3 – 5	4 – 6	4 – 8	5 – 6	6 – 9	7 – 8	8 – 9
Duration	2	2	1	4	1	5	8	4	3	3	3

3. A small project is composed of time activities whose time estimates are given below.

Activity	*A*	*B*	*C*	*D*	*E*	*F*	*G*	*H*	*I*
Optimistic time	2	2	4	2	2	3	2	5	3
Most likely time	2	5	4	2	5	6	5	8	6
Pessimistic time	8	8	10	2	14	15	8	11	15

Activities *A*, *B* and *C* can start simultaneously., Activity *D* follows activity *A* while *E* follows *B*, Activities *D* and *E* are followed by activity *G* while *F* is dependent on *C*, *H* depends on *D* and *E* while *I* depends on *F* and *G*.

(*a*) Find the expected duration and variance of each activity.

(*b*) Calculate the slack for each event.

(*c*) What is the expected project duration? Determine the variance and standard deviation of the project.

(*d*) If the project due date is 28 days, what is the probability of not meeting the due date.

(*e*) What should be the project duration, for the probability of completion of 95 per cent.

4. The following data gives details of normal and crash costs and times of a project:

Activity	*Normal*		*Crash Time (Weeks)*	*Cost (Rs.)*
	Time (Week)	*Cost (Rs.)*		
1 – 2	3	300	2	400
2 – 3	3	30	3	30
2 – 4	7	420	5	580
2 – 5	9	720	7	810
3 – 5	5	250	4	300
4 – 5	0	0	0	0
5 – 6	6	320	4	410
6 – 7	4	400	3	470
6 – 8	13	780	10	900
7 – 8	10	1000	9	1200

(*i*) Draw the network and identify the critical path.

(*ii*) What is the normal project duration and associated cost?

(*iii*) Find out the total float associated with each activity.

(*iv*) Crash the relevant activities and determine the optimal project duration and cost.

21

PRODUCTION CONTROL

21.1. INTRODUCTION

Production control provides the foundation on which most of the other controls are based. Control is described as the constraining the activities to follow the plans. **"Production control is the function of management which plans, directs and controls the material supply and processing activities of an enterprise, so that specified products are produced by specified methods to meet an approved sales programme."** These activities being carried out in such a manner that the labour, plant and capital available are used to the best advantage." The production control specifies three levels—programming, ordering and dispatching.

- Programming plans the production output of products.
- Ordering plans the output of components from the suppliers and departments which is necessary to meet the programme.
- Dispatching considers each department in turn and plans the output from machines and work centres necessary to carry out the orders.

21.2. OUTLINE OF PRODUCTION CONTROL

Loading and scheduling: The sales department will issue works order which will authorise the manufacture of a product or group of products. This order is the starting point for all activities of the production control department concerned with the manufacturing of products. The master production schedule (MPS) is prepared which involves assessing labour, and material requirements and availability and determining the dates by which major functions must be completed. The loading of various work centres is carried out. A copy of the Master Schedule will be passed to the material control which will check material availability.

Material control: The function of material control section of production control is that of assessing the need for material, and then taking appropriate steps (actions) to meet these requirements.

Dispatch and progress: A manufacturing is actually initiated at an appropriate time which collects together all relevant documents, verifies availability of each of the factors of production and authorises the start of production activities by issue of authorising documents. The progress section will monitor performance and verifying that the requirements of the master schedule are being fulfilled. Any deviation from this schedule are brought to the notice of the concerned persons and corrective actions are devised to keep the deviation at minimum. Outline of functions of production control are shown in Fig. 21.1.

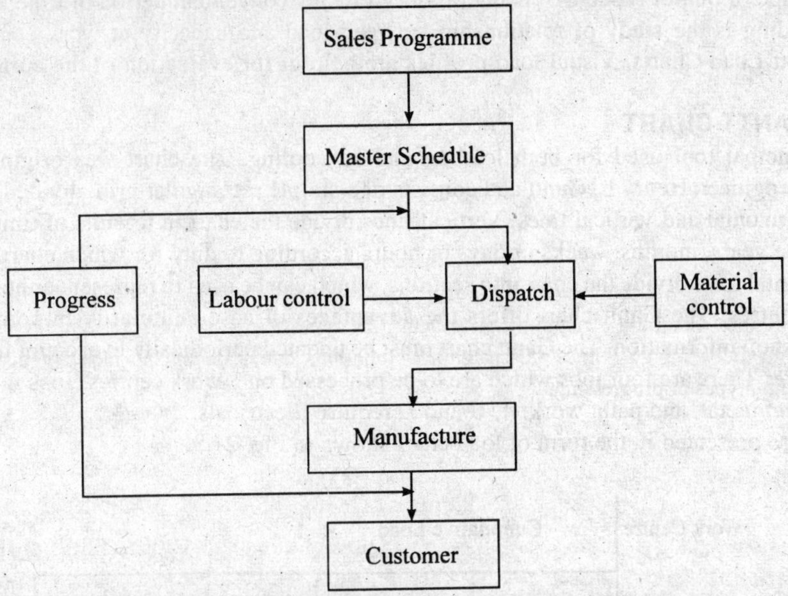

Fig. 21.1: Outline of functions of production control.

21.3. LOADING, SEQUENCING AND SCHEUDLING

Output plans specify when products are needed but these specifications must be transformed into operational terms to be implemented on the shop floor. The operations scheduling system is shown in Fig. 21.2.

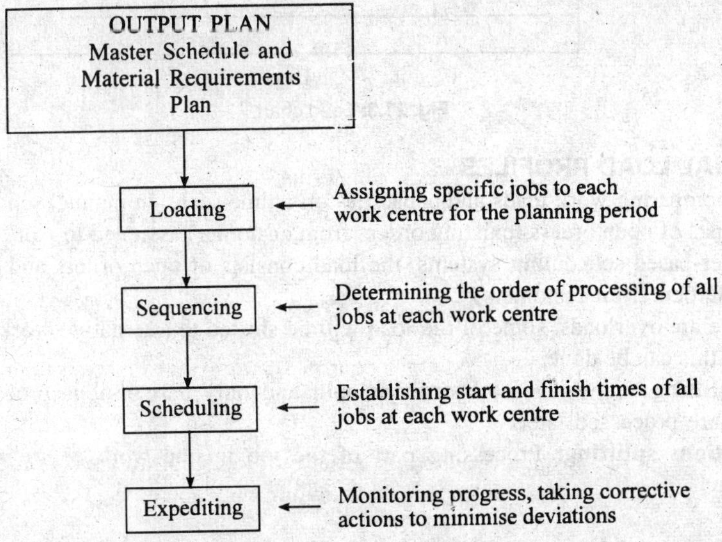

Fig. 21.2: Operations scheduling system.

21.4. LOADING

Each job may have a unique product specification and has a unique routing through various work centres, when the job orders are released, they are allocated to the work centres thus establishing the quantity of load each work centre should carry during the specific planned period. This assignment is called loading. Load is the work assigned to a machine or an operator and capacity is the volume of output capable of being produced in any convenient period of time.

- Loading is the study of relationship between load and capacity at work centres.
- Gantt Load Charts, Visual load profiles are helpful for evaluation of the current loading.

21.4.1. GANTT CHART

It is a principal tool used for both loading and scheduling. The chart was originated by the American engineer, Henry L. Gantt and consists of a simple rectangular grid, divided by series of parallel horizontal and vertical lines. Vertical lines divide the chart in to units of time. The scale units can be years, months, weeks or days or hours according to duty for which chart is required. The horizontal lines divide the chart into sections, which can be used to represent either work tasks or work centres. The Gantt chart offers the advantages of ease and clarity in communicating important shop information. The Gantt chart must be updated periodically to account for new jobs.

Example: There are four jobs which are to be processed on 3 work centres. Jobs *A, B, C* and *D* require sheet metal, and paint work. *A, C* and *D* require Electricals.

These are presented in the form of load chart shown in Fig. 21.3.

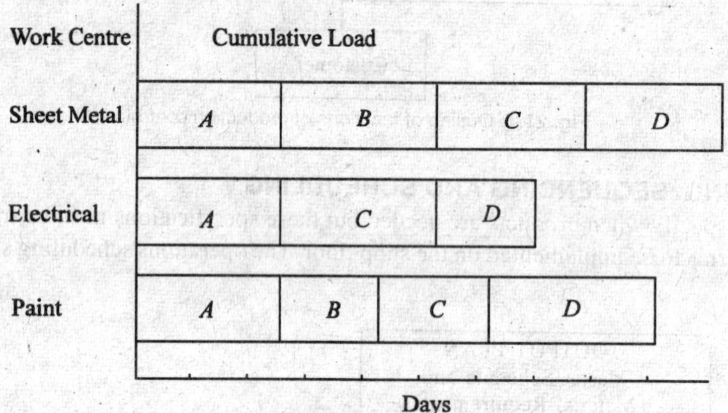

Fig. 21.3: Load chart.

21.4.2. VISUAL LOAD PROFILES

It is a graph comparing work loads and capacities on a time scale. In manual scheduling system, the load consists of open orders (existing orders from customer) assigned to work centre.

In computer-based scheduling systems, the load consists of open orders and planned orders (Prospective Orders from customers).

When there are overloads, some of the load will be shifted to alternative work centres. There are two ways this can be done:

- **Lot splitting:** In which a job order is split and only part of it is processed now and others are processed later.
- **Operations splitting:** Processing part of the job in one work centre and the rest at another.

21.5. PRIORITY SEQUENCING

When jobs compete for work centres capacity, which job should be done next? Priority sequence rules are applied to all jobs waiting in the queue. Then when the work centre becomes open for the job, the one with the highest priority is assigned. "Priority Sequencing" is a systematic procedure for assigning priorities to waiting jobs thereby determining the sequence in which the jobs will be performed.

Choosing criteria for sequencing

1. Set up cost.
2. In process inventory.
3. Idle times.
4. Average time to complete jobs.
5. Average number of jobs waiting in the queue.
6. Average time the jobs are late.

The criteria like set up cost, inventory cost and idle times are primarily concerned with internal facility efficiency and other criteria reflect both customer service and internal efficiency.

Priority sequencing rules

1. **First come first served (FCFS):** Gives top priority to the waiting job that arrived earliest in the production system.
2. **Earliest due date (EDD):** Gives top priority to the waiting job whose due date is earliest.
3. **Shortest processing time (SPT):** Gives top priority to the waiting job whose operation time at work centre is shortest.
4. **Least slack (LS):** Slack is calculated as the difference of length of time remaining until the job is due and the length of its operation time.
5. **Longest processing time (LPT).**
6. **Preferred customer order (PCO).**

Illustration **(*SPT Rule*):** There are five jobs which are to be processed on work centre sheet metal shop. The processing times are given below:

Job	*A*	*B*	*C*	*D*	*E*
Processing time (days)	4	17	14	9	11

Determine the sequencing using SPT rule.

The data resulting from SPT rule is shown in table below

Waiting Job (SPT Seq)	*Processing Time (Days)*	*Flow Time (Days)*	*Due Date (Days)*
A	4	4	6
D	9	13	12
E	11	24	12
C	14	38	18
B	17	55	20

Sequencing data for shortest processing time priority.

Performance Using SPT Sequence

1. Total competition time = 55 days

2. Mean flow time = $\dfrac{4 + 13 + 24 + 38 + 55}{55} = 2.8$ days

3. Average number of jobs in the system each day $= \dfrac{5(4)+4(9)+2(14)+1(17)}{55} = 2.44$ jobs.

4. Average lateness – Lateness of job in this sequence are 0, 1, 12, 20 and 35 days.

$$\text{Average lateness} = \frac{0+1+12+20+35}{5} = 13.6 \text{ days.}$$

21.6. SEQUENCING PROBLEMS

The sequencing problem arises whenever there is a need for determining an optimum order of performing a number of jobs by number of facilities according to some pre-assigned order so as to optimise the output in terms of cost, time or profit.

Production of finished goods from raw materials consists of several operations to be performed in a given sequence. Frequently, similar operations required for several products are performed at the same work stations particularly in intermittent or batch production. Under such situations, it is required to select a preferred order for products passing through a work station. The problem becomes complicated when the several work stations serve many products. In such a problem, the criteria is minimum total processing time.

The general sequencing problem is stated as: There are n jobs (1, 2, 3 .... n) each of which must be processed through each of M machines (m_1, m_2, m_3 ...... mn) one at a time. The order of processing each job through the machines is given and also the time taken to process each job on each machine is known.

The problem is to determine the order of processing of n jobs so that the total elapsed time for all the jobs will be minimum. The general sequencing problem is to determine the optimal sequence from amongst $(n\,!)^M$ sequences that minimises the total elapsed time.

Assumptions

1. No machine can process more than one job at a time.
2. Processing times are independent of processing of jobs.
3. Each job once started on one machine is continued till completion on it.
4. Time involved in moving a job from one machine to another is negligibly small.

The satisfactory solutions are available for the following problems

1. n jobs and 2 machines.
2. n jobs and 3 machines.
3. 2 jobs and m machines.
4. n jobs and m machines.

(1) *n* Jobs and 2 Machine Problem (Johnsons Algorithm)

The problem is defined as,

There are only two machines A and B

Each job is processed in the order A and B

The processing times of n jobs (1, 2, 3 ..... n) on each of the two machines is known. Let A_1, A_2 .... A_n are processing times on A and B_1, B_2 ...... B_n are processing times on B.

The problem is to find the order in which the n jobs are to be processed to minimise the total elapsed time (T) to complete all the n jobs.

Procedure

Step 1: Select the smallest processing time from the given list of processing times A_1, A_2 .... A_n and B_1, B_2 ......B_n.

Step 2: If the minimum processing time is A_r (*i.e.*, Job number r on Machine A), do the rth job first in the sequence. If the minimum processing time is B_s (*i.e.*, job number *s* on Machine *B*, do the *s*th job last in the sequence.

Step 3: After doing this step, (*n*–1) jobs are left to be sequenced. Repeat step (1) and step (2) till all the jobs are ordered.

Step 4: Find the total processing time as per the sequence determined and also determine idle time associated with machines.

Problem 1: Five jobs are to be processed on two machines M_1 and M_2 in the order M_1M_2. Processing time in hours are given below:

Job	*Processing times (hrs)*	
	Machine M_1	*Machine* M_2
1	5	2
2	(1)	6
3	9	7
4	3	8
5	10	4

Determine the sequence that minimises total elapsed time. Find out the total elapsed time and idle time (if any) on M_2.

Solution: (1) The minimum processing time is 1 for job 2, machine M_1. So, job 2 should be processed first in the sequence.

2				

(2) Once the job 2 is over, it is excluded from the list. The reduced list of processing times for remaining jobs are:

Job	*Processing times (hrs)*	
	Machine M_1	*Machine* M_2
1	5	(2)
3	9	7
4	3	8
5	10	4

Now, the minimum processing time is 2 for job 1, machine M_2. As per the algorithm, process job 1 last in the sequence.

2				1

(3) The processing times for remaining jobs are:

Job	*Processing times (hrs)*	
	Machine M_1	*Machine* M_2
3	9	7
4	(3)	8
5	10	4

The minimum processing time is 3 for job 4 machine M_1. So process job 4 next in the sequence.

2	4			1

The processing times for remaining two jobs are:

Job	*Processing times (hrs)*	
	Machine M_1	*Machine M_2*
3	9	7
5	10	(4)

The minimum processing time is 4 for job 5 machine M_2. So process job 5 last in the sequence. The optimal sequence is:

2	4	3	5	1

Total elapsed time

Optimal Sequence	*Machine*		*Machine M_2*		
	In	*Out*	*In*	*Out*	*Idle time M_2*
2	0	1	1	7	1
4	1	4	7	15	–
3	4	13	15	22	–
5	13	23	23	27	1
1	23	28	8	(30)	1

To compute the elapsed time,

- The starting time onmMachine M_1 is assumed as 0.00.
- The machine M_2 starts processing job 2 only when it comes out of machine M_1 after completion. So it is idle for 1 hour in the start till job comes to it from machine M_1.
- The minimum total elapsed time is 30 hrs. to process all the 5 jobs through two machines M_1 and M_2.

The idle time on Machine $M_1 = 1 + 1 + 1 =$ **3 hrs** **(Ans)**

Problem 2: There are seven jobs which are to be pressed first on Machine I and then on Machine II. Processing times in hours are given below:

Job	*A*	*B*	*C*	*D*	*E*	*F*	*G*
Machine I	6	24	30	12	20	22	18
Machine II	16	20	20	13	24	2	6

Find the optimal sequence and total elapsed time. Compute the idle time on Machine II. The optimal sequence (as per the Johnson's algorithm) is:

A	*D*	*E*	*C*	*B*	*G*	*F*

Total elapsed time is calculated as follows:

Optimal Sequence	*Machine I*		*Machine II*	
	In	*Out*	*In*	*Out*
A	0	6	6	22
B	6	18	22	35
E	18	38	38	62
C	38	68	68	88
B	68	92	92	112
G	92	110	112	118
F	110	132	132	134

Minimum total elapsed time = 134 hrs
Idle time on Machine II = 6 + 3 + 6 + 4 + 14 = 33 hours.

n jobs 3 machines problem (Johnson's algorithm)

Conditions to be satisfied to solve the above problem by Johnson's method.
There are three machines M_1, M_2 and M_3
Each job has to go through three machines in the order M_1, M_2 and M_3.

1. The smallest processing time on machine M_1 $\geq$ largest processing time on machine M_2.
2. The smallest processing time on machine M_3 $\geq$ largest processing time on machine M_2.

If either or both of the above stated conditions are satisfied, the given problem can be solved by Johnson's algorithm.

Procedure

Step I: Convert the three machine problem into two machine problem by introducing two fictitious machines *G* and *H*

Such that

$$G_i = M_1\,i + M_2\,i$$
$$H_i = M_2\,i + M_3\,i \qquad i = 1, 2, 3, \ldots\ldots\ldots\ldots n$$

Step II: Once the problem is converted to *n* job 2 machine the sequence is determined using Johnson's algorithm for *n* Jobs and 2 machines.

Step III: For the optimal sequence determined, find out the minimum total elapsed time and idle times associated with machines.

Tie breaking Rules

1. If there are equal smallest processing times one for each machine, place the job on machine 1, first in the sequence and one in machine 2 last in the sequence.
2. If the equal smallest times are both for machine 1, select the job with lower processing time in machine 2 for placing first in the sequence.
3. If the equal smallest times are both formMachine 2, select the one with lower processing time in machine 1, for placing last in the sequence.

***Problem 3*:** Seven jobs are to be processed through 3 machines M_1, M_2 and M_3 in the order M_1, M_2, M_3. The processing times are given in hrs to process each one of the 3 jobs through all the machines find the optimal sequence of the jobs. Also find the minimum total elapsed times and idle times on M_2 and M_3.

Jobs	M_1	M_2	M_3
A	3	4	6
B	8	3	7
C	7	2	5
D	4	5	11
E	9	1	5
F	8	4	6
G	7	3	12

***Solution*:**

Step I: Check for the conditions to be satisfied.

Minimum processing time on M_1 = 3

Minimum processing time on M_3 = 5

Maximum processing time on M_2 = 5

Condition I: Minimum time on $M_1 \geq$ Max. time on M_2

$$3 \geq 5$$

condition I is not satisfied.

Condition II: Minimum time on $M_3 \geq$ Max. time on M_2.

$$5 = 5$$

condition II is satisfied

Hence, the problem can be solved by using Johnson's *n* job 3–Machine algorithm

Step II: Convert *n* job 3 machine problem into *n* job 2 machine problem by assuming two fictitious machines *G* and *H* such that,

$G = M_1 i + M_2 i$

$H = M_2 i + M_3 i$ (where $i = 1, 2, 3 .... n$)

For the given problem

Jobs	G_i	H_i
A	7	10
B	11	10
C	9	7
D	9	16
E	10	6
F	12	10
G	10	15

Step III: Determine the sequence using *n* job 2 machine procedure.

On observation, the minimum time is 6 and is associated with job *E* and machine H_i. So process job *E* last in the sequence.

The sequence is

						E

The problem is reduced to,

Jobs	G_i	H_i
A	7	10
B	11	10
C	9	7
D	9	16
E	10	6
F	12	10
G	10	15

The minimum time is 7. It is associated with job *A* machine (G_i) and job *C* machine (H_i)
So, process job *A* first and job *C* last as per tie breaking rule.
The sequence after this step is,

A					*C*	*E*

The remaining jobs are represented as

Jobs	*Machine*	
	G_i	H_i
B	11	10
D	9	16
F	12	10
G	10	15

The minimum time is 9 associated with machine (G_i) and job *D* .
Process the job next in the sequence is:

A	*D*				*C*	*E*

The remaining jobs are

Jobs	*Machine*	
	G_i	H_i
B	11	10
F	12	10
G	10	15

The minimum time is 10 associated with

(*i*) Machine G_i job (*G*). So do the job *G* next in the sequence.

(*ii*) There is a tie both job (*B*) and job (*F*) have same minimum time on machine (H_i).

As per the tie breaking rule, job *B* should be processed last.

The final sequence is

A	*D*	*G*	*F*	*B*	*C*	*E*

(*iii*) Minimum total processing time to complete all the 7 jobs through all the three.

Job Sequence (Optimal)	*Machines*					
	M_1		M_2		M_3	
	In	*Out*	*In*	*Out*	*In*	*Out*
A	0	3	3	7	7	13
D	3	7	7	12	13	24
G	7	14	14	17	24	36
F	14	22	22	26	36	42
B	22	30	30	33	42	49
C	30	37	37	39	49	54
E	37	46	46	47	54	(59)

Minimum total elapsed time = 59 hrs
Idle time on machine M_2 = 3 + 2 + 5 + 4 + 4 + 7 = 25 hrs
Idle time on machine M_3 = 7 hrs

Processing *n* jobs through *m* machines

Let there *b n* jobs which are to be processed through *m* machines $M_1, M_2 \ldots M_n$ in the order $M_1, M_2 \text{----} M_m$.

Let T_{ij} denote the time taken by *j*th job on the *i*th machine.

Procedure

Step I: Find (*i*) Min T_{ij} (Minimum time for the first machine)

(*ii*) Min T_{mj} (Minimum time on the last machine)

(*iii*) Maximum time on intermediate machines.

Max ($T_{2j}, T_{3j} \ldots\ldots T_{m\text{-}1j}$)

For $j = 1, 2, 3 \ldots\ldots n$)

Step II: Check for the following conditions.

(*i*) Min $T_{ij} \geq$ Max T_{ij} (For $i - 2, 3, \ldots\ldots m-1$)

(*ii*) Min $T_{mj} \geq$ Max T_{ij} (For $i - 2, 3, \ldots\ldots m-1$)

i.e., the minimum processing time on the machines M_1 and M_m (First and last machines) should be ≥ maximum time on any of the 2 to *m*-1 machines.

Step III: If the conditions in step II are not satisfied, the problem cannot be solved by this method, and otherwise go to next step.

Step IV: Convert the *n* job *m* machine problem into *n* job 2 machine problem by considering two fictitious machines *G* and *H*.

Such that,

$$TG_j = T_{ij} + T_{2j} + \ldots\ldots + T_{m-1j}$$
$$TH_j = T_{2j} + T_{3j} + \ldots\ldots + T_{mj}$$

Step V: Now obtain the sequence for *n* jobs using Johnson's Algorithm.

Step VI: Determine the minimum total elapsed time and idle times associated with machines.

Problem 4: Jobs each of which must be processed on the machine $M_1, M_2, \ldots\ldots M_6$. The processing times in hrs are given

(*i*) Find the optimal sequence.

(*ii*) Minimum total elapsed time.

(*iii*) Idle times associated with machines.

Jobs	*Processing times*					
	M_1	M_2	M_3	M_4	M_5	M_6
A	18	8	7	2	10	25
B	17	6	9	6	8	19
C	11	5	8	5	7	15
D	20	4	3	4	8	12

***Solution*:** Minimum time on first Machine $M_{ij} = 11$
Minimum time on last Machine $M_{6j} = 12$
Maximum times on intermediate machines
$(M_{2j}, M_{3j}, M_{4j}, M_{5j})$ are (8, 9, 6, 10)

Conditions: (*i*) Min $T_{ij} \geq$ Max T_{ij} $(i - 2, 3, 4, 5)$

i.e. $11 \geq (8, 9, 6, 10)$

The condition is satisfied

(*ii*) Min $T_{mj} \geq$ Max T_{ij} $(i = 2, 3, 4, 5)$

$12 \geq (8, 9, 6, 10)$

As both conditions are satisfied this problem can be solved using Johnson's algorithm.
Given problem can be converted in to 4 job 2 m/c problem by

$$TG_i = \sum_{i=1}^{m-1} T_{ij} \qquad TH_i = \sum_{j=2}^{m} T_{ij}$$

Convert the problem into 2 Machine Problem.

Jobs	G_i	H_i
A	45	52
B	46	48
C	36	40
D	39	31

The optimal sequence is

C	*A*	*B*	*D*

(*ii*) Minimum total elapsed time.

Jobs	*Processing times*					
	M_1	M_2	M_3	M_4	M_5	M_6
C	0 – 11	11 – 16	16 – 24	24 – 29	29 – 36	36 – 51
A	11 – 29	29 – 37	37 – 44	44 – 46	46 – 56	56 – 81
B	29 – 46	46 – 52	52 – 61	61 – 67	67 – 75	80 – 100
D	46 – 66	66 – 70	70 – 73	73 – 77	77 – 85	100 – 112

Minimum Total elapsed time = 112 hrs.

21.7. ASSIGNMENT MODEL

In assignment problems the objective is to assign the number of resources (men, machines, etc.) to the equal number of jobs at a minimum cost. The assignment is to be made on one to one basis.

Assignment Problem

There are n jobs and these n jobs are to be processed on n available machines. C_{ij} represents the cost of assigning job i to the machine j. The assignment of the jobs should be made in such a way that each job should be allocated to only one machine. The problem is to determine the best assignment of jobs to machines that result in minimum total cost of processing.

Assumptions

1. Only one job should be allocated to each machine.
2. Jobs differ in their work contents and machines differ in their capabilities.
3. The cost of processing each job on each of the machines (C_{ij}) is known.

Mathematical statement of the problem.

$$\text{Minimise } Z = \sum_{i=1}^{n} \sum_{j=1}^{n} C_{ij} X_{ij}$$

$$i = 1, 2, \ldots\ldots\ldots n$$
$$j = 1, 2, \ldots\ldots\ldots n$$

Subjected to the following restrictions

$X_{ij} = 0$ if the ith job is not assigned to jth machine.
$\quad\quad\; 1$ if the ith job is assigned to jth machine.

$$\sum_{i=1}^{n} X_{ij} = 1 \qquad j = 1, 2, 3 \ldots\ldots\ldots n$$

Only one person should be assigned to jth job.

$$\sum_{j=1}^{n} X_{ij} = 1 \qquad j = 1, 2, \ldots\ldots\ldots n$$

(Only one job is done by jth person)

Hungarian Method for Solving Assignment Problem

Step 1: Subtract the minimum cost of each row of the cost matrix from all the other elements of respective rows.

Step 2: In reduced matrix of step 1 subtract minimum cost of each column of the cost matrix from all the other elements of the respective column.

Step 1 and 2 will result in minimum one zero in each row and column.

Step 3: Draw the minimum number of horizontal and vertical lines to cover all zero in the matrix.

Let N be the minimum number of lines to cover all zero.

(*i*) If $N = n$, the order of the matrix (number of rows or columns) make the zero assignments to get the solution.

(ii) If $N < n$, then proceed to step 4.

Step 4: Determine the smallest cost element in the matrix not covered by N lines. Subtract this element from all the elements which are not covered by the lines and add this elements at the junction of (Intersection) horizontal and vertical lines. The second modified matrix is obtained.

Step 5: Repeat step (3) and (4) until minimum number of lines to cover all zero will be equal n (Number of rows or column).

Step 6: Procedure to make zero assignment.

- Examine the rows until a row which has single zero is found. Mark the row by a square □ to indicate the assignment.
- Mark a cross (×) over all zero lying in the column of the marked row which indicates that no other zero is considered for future assignment.
- Continue in this manner until all rows have been examined.

- Repeat the same procedure for columns also.

Step 7: Repeat step 6 until the following situation occurs:

(*i*) No unmarked zero is left.

(*ii*) There are more than one unmarked zero in the one column or row.

In case (2), mark □ one of the unmarked zero arbitrarily and marked (×) in cells containing remaining zero in its row or column.

Repeat this step until no unmarked zero is left in the matrix.

Step 8: The exactly one marked (□) zero in each column and each row is obtained.

The assignment corresponding to marked (□) zero will give the optimal assignment.

Procedure to draw minimum number of lines to cover all zero

(*i*) Mark (√) the rows for which no assignment is made.

(*ii*) Mark (√) the columns (not ready marked) which have zero in the marked rows.

(*iii*) Mark (√) rows (not already marked) which have assignments in marked columns

(*iv*) Repeat steps (*ii*) and (*iii*) until no more markings are possible.

(*v*) Draw the lines through all unmarked rows and through marked columns.

This gives the minimum number of lines to cover all zero.

***Problem 5*:** 4 different jobs are to be done on 4 different machines. The matrix below gives the cost (in Rs.) of producing each job *i* on each one of the machines *j*. How should the jobs be assigned to the machine so that the total cost is minimum.

Jobs	*Machines*			
	A	*B*	*C*	*D*
J_1	5	7	11	6
J_2	8	5	9	6
J_3	4	7	10	7
J_4	10	4	8	3

***Solution*:**

1. **Row Operation:** Select the smallest element from each row and subtract it from all other elements from that row. Resulting matrix is

	A	*B*	*C*	*D*
J_1	0	2	6	1
J_2	3	0	4	1
J_3	0	3	6	3
J_4	7	1	5	0

2. **Column Operation:** Select the smallest element in each column and subtract it from all other elements of the column. The reduced matrix after column operation is

	A	*B*	*C*	*D*
J_1	[0]	2	2	1
J_2	3	[0]	~~0~~ (×)	1
J_3	~~0~~ (×)	3	2	3
J_4	7	1	1	[0]

3. Make the assignments and draw the minimum number of lines to connect all zero. As the number of lines required to connect all zero (N) is less than number of rows or columns. Then proceed further to step 4.
4. Select the smallest element not covered by lines and subtract it from all the other elements which are not covered by lines and add this element at the intersection of horizontal and vertical lines. Other elements will remain unchanged. The modified matrix is shown below:

Jobs	*A*	*B*	*C*	*D*
J_1	0	1	1	0
J_2	4	0	0	1
J_3	0	2	1	2
J_4	8	1	1	0

Since ($N = 3$) no. of rows or columns.
Repeat step (4)
The resulting matrix after the repetition of step (4) is below

	A	*B*	*C*	*D*
J_1	[0]	~~0~~	~~0~~	~~0~~
J_2	5	[0]	~~0~~	2
J_3	0	1	[0]	1
J_4	8	0	~~0~~	[0]

Minimum 4 lines are required to connect all zero
Therefore $N = n$ (no. of rows or columns)
Hence, optimality has reached.
The optimal assignment is

Assignment	*Cost*
$J_1 - A$	5
$J_2 - B$	5
$J_3 - C$	10
$J_4 - D$	3
	23

Total cost of assignment = Rs. 23.

Maximisation of Assignment Problem

Maximisation problem can be converted into a minimisation problem by multiplying each element of the matrix by -1 (or subtracting all the elements of the matrix from the largest elements in the matrix).

Then the problem is solved using Hungarian method.

***Problem 6*:** A company wishes to assign 4 salesmen to 4 districts. The volume of sales matrix is given below. Make the optimal assignment which results in maximum volume of sales.

Salesman	*District*			
	A	*B*	*C*	*D*
1	250	300	420	400
2	350	400	200	250
3	500	375	400	350
4	400	350	420	300

Solution: Convert the problem into minimisation problem by subtracting all the elements of the problem by the largest element in the matrix. The resulting matrix is,

	A	*B*	*C*	*D*
1	250	200	80	100
2	150	100	300	250
3	0	125	100	150
4	100	150	80	200

Performing row operations, the resulting matrix is,

	A	*B*	*C*	*D*
1	170	120	0	20
2	50	0	200	150
3	0	125	100	150
4	20	70	0	120

Performing column operation, the resulting matrix is,

	A	*B*	*C*	*D*
1	170	120	~~0~~	[0]
2	50	[0]	200	130
3	[0]	125	100	130
4	20	70	[0]	100

Optimal assignment and total sales are given below

Salesmen	*Districts*	*Sales volume*
1	D	400
2	B	400
3	A	500
4	C	420
		1720 units

Unbalanced Assignment Problem

The Hungarian method for solving assignment problem requires that number of rows and number of columns must be equal (Square ($N \times N$)) matrix. If the number of rows and number of columns are not equal, the problem becomes an unbalanced assignment problem. To solve an unbalanced assignment problem, the number of rows or number of columns are added as per the requirements to make it a square matrix. If the number of columns are more than number of rows, add dummy rows with zero elements to make it a square matrix and *vice versa*. Once it becomes a square matrix, solve the problem by Hungarian method to get the optimal assignment.

Problem 7: There are four machines W, X, Y and Z. Three jobs A, B and C are to be assigned to the 3 machines out of total 4 machines. The cost of assignment is given below. Find out the optimal assignment.

	W	X	Y	Z
A	18	24	28	32
B	8	13	17	18
C	10	15	19	22

Solution: 1. No. of rows = 3
No. of colomns = 4

This is an unbalanced assignment problem as the number of rows are not equal to number of columns. Therefore, in order to make it a square matrix, add a dummy row (D) with zero cost elements. Thus, the modified matrix becomes

	W	X	Y	Z
A	18	24	28	32
B	8	13	17	18
C	10	15	19	22
D	0	0	0	0

2. The reduced matrix after row operation is; (Column operation is not necessary as there is already one and in each column).

	W	X	Y	Z
A	0	6	10	14
B	0	5	9	10
C	0	5	9	12
D	0	0	0	0

3. Minimum number of lines to connect all zeros is two (2) which is no. of rows or columns. Therefore, proceed to next step.
4. Select the smallest element which is not covered by the lines and subtract it from all other elements which are not covered by the lines and add it at the intersection of two lines.

The resulting matrix is:

	W	X	Y	Z
A	0	1	5	9
B	0	0	4	5
C	0	0	4	7
D	5	0	0	0

Minimum number of lines = 3 (less than the order of the matrix)
Therefore, repeat step 4. The result matrix is:

	W	X	Y	Z
A	[0]	1	1	5
B	⊠	[0]	⊠	1
C	⊠	⊠	[0]	2
D	9	4	⊠	[0]

Optinal assignment is
A – W – 18
B – X – 17
C – Y – 15
Total – 50

21.8. SCHEDULING

Scheduling can be defined as "Prescribing of when and where each operation necessary to manufacture the product is to be performed." It is also defined as "establishing of times at which to begin and complete each event or operation comprising a procedure." The principle aim of scheduling is to plan the sequence of work so that production can be systematically arranged towards the end of completion of all products by due date.

21.8.1. PRINCIPLES OF SCHEDULING

1. **The principle of optimum task size:** Scheduling tends to achieve maximum efficiency when the task sizes are small, and all tasks are of same order of magnitude.

2. **Principle of optimum production plan:** The planning should be such that it imposes an equal load on all plants.

3. **Principle of optimum sequence:** Scheduling tends to achieve the maximum efficiency when the work is planned so that work hours are normally used in the same sequence.

21.8.2. INPUTS TO SCHEDULING

1. Performance standards: The information regarding the performance standards (standard times for operations) helps to know the capacity in order to assign required machine hours to the facility.
2. Units in which loading and scheduling is to be expressed.
3. Effective capacity of the work centre.
4. Demand pattern and extent of flexibility to be provided for rush orders.
5. Overlapping of operations.
6. Individual job schedules.

21.8.3. SCHEDULING STRATEGIES

Scheduling strategies vary widely among firms and range from "no scheduling" to very sophisticated approaches.

The strategies are grouped into four classes:

(*i*) Detailed scheduling.
(*ii*) Cumulative.
(*iii*) Cumulative detailed.
(*iv*) Priority decision rules.

1. Detailed scheduling for specific jobs that are arrived from customers is impracticable in actual manufacturing situation. Changes in orders, equipment breakdown, unforeseen events deviate the plans.

2. Cumulative scheduling of total work load is useful especially for long range planning of capacity needs. This may load the current period excessively and under load future periods. It has some means to control the jobs.

3. Cumulative detailed combination is both feasible and practical approach. If master schedule has fixed and flexible portions.

Capacities are planned on a broad basis first in terms of total labour and machine hour requirements per week at key work centres. As changes occur during the weeks prior to manufacturing, the computer updates material and capacity requirements automatically. Capacity may then be allocated to specific jobs later a few days before the commencement of job. The shortest scheduling unit for job shop is one day.

4. Priority decision rules are scheduling guides that are used independently and in conjunction with one of the above strategies, *e.g.*, First Come First Serve. These are useful in reducing work-in-process (WIP) inventory.

21.8.4. FORWARD SCHEDULING AND BACKWARD SCHEDULING

Forward scheduling (set forward) is commonly used in job shops where customers place their orders on "needed as soon as possible" basis. Forward scheduling determines start and finish times of next priority job by assigning it the earliest available time slot and from that time, determines when the job will be finished in that work centre. Since the job and its components start as early as possible, they will typically be completed before they are due at the subsequent work centres in the routing. The set forward method generates in process inventory that are needed at subsequent work centres and higher inventory cost.

Forward scheduling is simple to use and it gets jobs done in shorter lead times, compared to backward scheduling.

Backward scheduling: (Set backward) is often used in assembly type industries and commit in advance to specific delivery dates. Backward scheduling determines the start and finish times for waiting jobs by assigning them to the latest available time slot that will enable each job to be completed just when it is due, but not before. By assigning jobs as late as possible, backward scheduling minimises inventories since a job is not completed until it must go directly to the next work centre on its routing.

Bill of materials (BOM) and lead time estimates are maintained for all work centres otherwise the system breaks down and due dates are violated. Fig. 21.4 shows forward and backward scheduling.

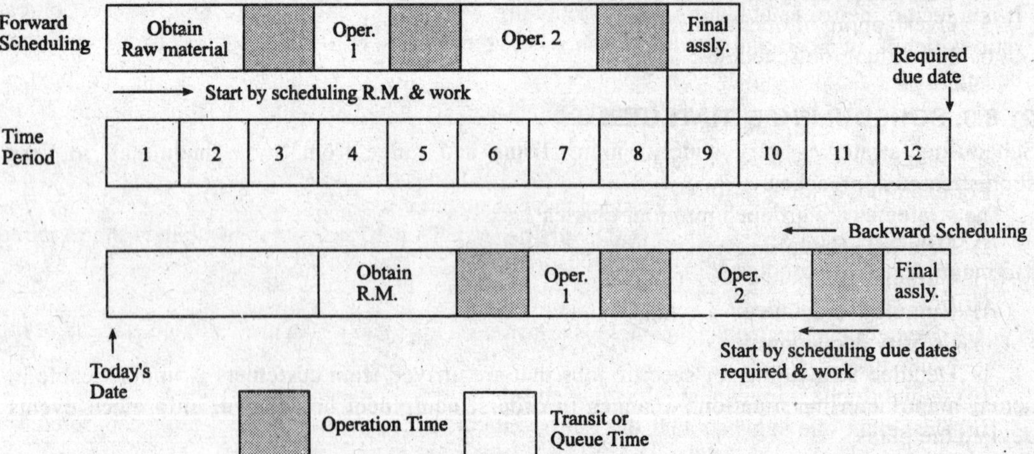

Fig. 21.4: Forward and backward scheduling.

21.8.5. FINITE LOADING

It is a scheduling procedure that assigns jobs into work centres and determines their starting and completion dates by considering work centre capacities.

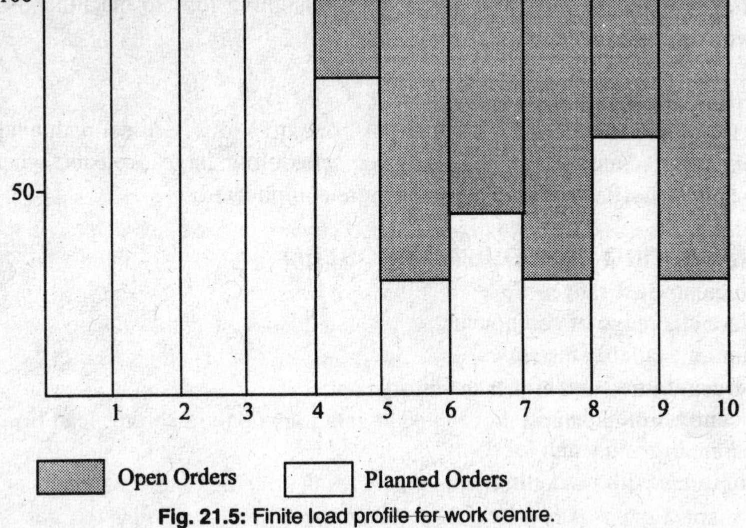

Fig. 21.5: Finite load profile for work centre.

This is an alternative scheduling technique that combines into a single system the loading, sequencing and scheduling. Finite loading systems start with a specified capacity for each work centre and list of jobs. The work centres capacity is then allocated unit by unit (*e.g.*, labour hours, machine hours, etc.) to the jobs by simulating job starting and completion dates. Thus, the system creates a detailed schedule for each job and each work centre based on centres capacities.

The finite capacity load profile is shown in Fig. 21.5.

21.8.6. CRITICAL RATIO SCHEDULING

It is a technique for establishing and maintaining priorities among the jobs in a factory. Critical ratio which is an index and sets relationship between when a product is required and when it can be supplied.

$$\text{Critical ratio} = \frac{\text{Time the job is needed (Demand time)}}{\text{Supply lead time.}}$$

$$\text{For example, Critical ratio} = \frac{\text{Product is required in 20 days}}{\text{Product can can be made available in 10 days}} = 2$$

A critical ratio of 2 means that time available to deliver the product is twice the time required to manufacture the product.

Critical ratio of

(*i*) greater than one means that there is a sufficient time and job can be completed ahead of the schedule.

(*ii*) equal to one implies that job is just on schedule and requires close watch.

(*iii*) less than one suggests that the job is critical and needs to be expedited to complete on schedule.

Advantages of Critical Ratio Scheduling

- To fix up relative job priorities for scheduling.
- Status of each job is determined.
- Schedule can be adjusted automatically when there are changes in job demand and progress.

21.8.7. INDEX METHOD

It is a subjective method of evaluation. It is simple and easy to operate and provides considerable improvements over conventional scheduling techniques. Normally jobs are assigned to the best machine till it is loaded to its full capacity and remaining jobs to the next best method. This does not result in optimum loading. The best way of assigning jobs to machines is on the basis of relative effectiveness of machines.

21.8.8. CRITICAL PATH METHOD

It is useful technique to determine schedule of "one off" jobs, Critical path analysis overcomes the deficiencies of Gantt Chart and used for scheduling large projects, where relationships between various activities of the project are more complicated.

21.8.9. SIMPLIFYING SCHEDULING PROBLEM

1. Reduce product range.
2. Reduce the range of component.
3. Examine available resources.
4. Enlarge all jobs wherever possible.
5. Subcontract the jobs.
6. Separate big and small jobs.
7. Be in touch with marketing.

21.8.10. SCHEDULING AND LOADING GUIDELINES

A realistic schedule is essential for getting the job done and maintaining credibility in the system. Guidelines for scheduling and loading are given below:

1. Provide a realistic schedule.
2. Allow adequate time for operations, before, after and between operations.
3. Don't release all the available jobs to the shop.
4. Do not load all the available capacity in the shop.
5. Load only selected work centres.
6. Allow for necessary changes.

21.9. DISPATCHING

Dispatching is the routine of setting productive activities in motion through the release of orders and instruction in accordance with previously planned times and sequences in the route sheets.

The Functions of Dispatching

1. Primary function of dispatching is to prepare manufacturing orders which consists of shop orders, move orders, tool orders, etc. These are to be issued at the right time to the concerned persons.
2. Release of necessary order and production forms so that the operations can be started.
3. Withdraw required quantity of material from the stores and deliver to the work centre where first operation is to be completed through stores issue order.
4. Issue of tools required for production.
5. Inter-departmental transport (move order).
6. Stage inspection.
7. Coordination with scheduling.
8. Forwarding materials to dispatch or to finished parts stores.

Documents Raised by the Dispatcher

A. Material requisitions.

B. **Job cards:** Which authorise the workman to start the work on certain material, indicate what to do and also serves as a means of production progress.

C. **Labour cards:** Which are used to report labour time utilised and quantity of work performed and to supply other information which are required in the preparation of production reports and payrolls.

D **Move cards:** Which authorise the movement of materials as per requirements of the job and used in production progress reports.

Duties of a Dispatcher

1. The receipt and filing of all shop orders and associated documents.
2. Selection of jobs for issue, in the most favourable sequence.
3. The issue of job cards or other forms of instructions to the operations.
4. Issue of instructions to setters regarding what machines are to be set up, for which jobs and when.
5. Issue of instructions concerning the movement of materials between work centres.
6. Issue of instructions concerning the issue and return to stores of special tooling.
7. The maintenance of records of production.

21.10. PROGRESSING

Progressing is that part of the production control function which is responsible for making routine comparisons between production performance and production plans and for reporting exceptional variances to the line staff so that they can be corrected. Progressing in the production control completes the loop and which by giving early warning when actual production deviates from planned production makes it possible for corrective action to be taken in order to regain the desired course.

Progressing can be divided into four main tasks:

1. Recording of actual production.
2. Comparison of actual production with planned production.
3. Measurement of deviation.
4. Reporting of all excessive deviations to the authorities responsible for executing the plans.

Types of Progressing

1. **Programme control:** It is the job of comparing the actual production output with the production programme and reporting deviations from plan to the line management for consideration and correction. There are different ways in which product output can be recorded and compared with a programme control. Gantt charts, tabulated records and Z-charts are used for recording. The Z-chart is a programme control method which shows the actual outputs at the end of each chosen interval and also show cumulative output since the beginning of the financial year and the moving annual totals at the same intervals. The Z-chart can be used to show both the plan and the performance and deviation from the plan can be easily indicated by knowing the gap between two curves.

2. **Order progressing:** It is concerned with the control of internal orders and purchase requisitions. There are four main progress record systems in use for this purpose.

(a) *Due date filing*: It is simplest of all order progressing systems. It involves the filing of copies of all orders in a box file in due date sequence. Order copies are removed from the file only when they are completed. Feedback is though orders overdue list normally prepared once in a week and circulated to all the concerned.

(b) *Order delivery records*: The typical order delivery record shows the deliveries and schedules requirements for purchased part.

(c) *Operation progress records*: It is a record concerned with the type of order normally completed and delivered all at the same time as one complete batch. It is a record used to show the state of completion of a batch.

The advantages include the following:

(*i*) It shows the position of all work in the shop.

(*ii*) It shows operations which are behind schedule.

(*iii*) It shows how much work has been scrapped at each operation.

(*d*) *List order progress*: It is a type of order processing associated with standard batch control and occasionally with base stock control, job loading and period batch control. A measure of the progress achieved is given by the number of items on the list which still await completion and for full control of the orders it is necessary to maintain records showing which items still have to be finalised.

3. **Shortage chasing:** It is the job of comparing the actual availability of materials and parts with the quantities required for production and of reporting any shortages so that they can be produced quickly.

4. **Daily plan progressing:** It is the control used at the third level of production control to see that the daily plans made during dispatching are achieved.

The most effective and general method used is to have a departmental meeting every morning to review the progress during the previous day. A weekly reporting by the departmental manager to his superior is usually sufficient to maintain the efficiency of progressing.

5. **Departmental progress control:** It is a method used to assess the efficiency of the different production departments in a factory by recording number of failures to complete orders by due-date in each department at regular intervals and by comparing these quantities with the prescribed limits of performance. Feedback system determines the efficiency of the control systems.

References for Further Reading

1. John L. Burbidge, *Principles of Production Control*, 3rd Edition , ELBS, London.
2. Alan Mukhemann, *Production and Operations Management*, 6th Edition, ELBS, London.
3. Monks J., *Operations Management. Theory and Practice*, 3rd Edition, McGraw Hill, New York.
4. Lockyer, *Factory and Production Management*, 3rd Edition, ELBS, London.
5. L.C. Jhamb, *Production Planning and Control*.

REVIEW QUESTIONS

1. Define production control. How production control forms the basis for other control?
2. Explain the three levels of production control.
3. Explain with the block diagram the outline of functions of production control.
4. Explain the framework of operations scheduling system.
5. What is loading? What are various techniques of loading?
6. What is Gantt chart? How it is constructed? How it is used to represent the loading and scheduling?
7. How assignment model is used for loading?
8. What is priority sequencing? What are the criteria for priority sequencing?
9. What are the priority sequencing rules?
10. Define scheduling. What are principles of scheduling?
11. What are inputs to scheduling?
12. Explain the various scheduling strategies.

13. Explain the following:
    (*i*) Forward and backward scheduling.
    (*ii*) Finite loading.
    (*iii*) Critical ratio scheduling.
14. What are the ways of reducing the scheduling problems?
15. Write note on guidelines for loading and scheduling.
16. What are the functions of dispatching?
17. Explain the various documents raised by the dispatcher.
18. What are the duties of the dispatcher?
19. What is progressing? What are the tasks of progressing?
20. Explain the following:
    (*i*) Programme control.
    (*ii*) Order progressing.
    (*iii*) Shortage chasing.
    (*iv*) Daily plan progressing.
    (*v*) Departmental progress control.

22

INVENTORY CONTROL

• Introduction • Meaning of inventory • Types of inventories • Reasons for keeping inventories • Inventory control • Objectives of inventory control • Benefits of inventory control • Costs associated with inventories • Terminology • Inventory cost relationships • Inventory models • EOQ with instant stock replenishment • Production Model (EOQ when stock replenishment is gradual) • Inventory model with price discounts • Safety stock and determination of safety stock • Inventory control systems • Fixed order quantity system • Periodic review system • Selective control of inventory • ABC analysis • VDE analysis • SDE analysis • HML analysis • FSN analysis • SOS analysis • XYZ analysis.

22.1. INTRODUCTION

In majority of the organisations, the cost of material forms a substantial part of the selling price of the product. The interval between the receiving the purchased parts and transforming them in to final products varies from industries to industries depending upon the cycle time of manufacture. Materials are procured and held in the form of inventories. It is therefore necessary to hold inventories of various kinds to act as a buffer between supply and demand for efficient operation of the system. Thus, an effective control on inventory becomes a must for smooth and efficient running of the production cycle with least interruptions. The stocking of anything that is tangible in order to meet the future demand is the subject-matter of inventory theory.

22.2. MEANING OF INVENTORY

Inventory generally refers to the materials in stock. It is also called the idle resource of an enterprise. Inventories represent those items which are either stocked for sale or they are in the process of manufacturing or they are in the form of materials which are yet to be utilised.

22.3. TYPES OF INVENTORIES

A manufacturing firm generally carries the following types of inventories:

1. **Raw materials:** Raw materials are those basic unfabricated materials which have not undergone any operation since they are received from the suppliers, *e.g.*, round bars, angles, channels, pipes, etc.
2. **Bought out parts:** These parts refer to those finished parts, subassemblies which are purchased from outside as per the company's specifications.
3. **Work-in-process inventories (WIP):** These refer to the items or materials in partially completed condition of manufacturing, *e.g.*, semi-finished products at the various stages of manufacture.
4. **Finished goods inventories:** These refer to the completed products ready for dispatch.
5. **Maintenance, repair and operating stores:** Normally these inventories refer to those

items which do not form the part of the final product but are consumed in the production process, e.g., machine spares, oil, grease.

6. **Tools inventory:** Includes both standard tools and special tools.
7. **Miscellaneous inventories:** Miscellaneous inventories—office stationaries and other consumable stores.

Inventories can also be classified as: (*i*) Fluctuation inventories, (*ii*) Anticipation inventories, (*iii*) Lot size inventories, and (*iv*) Transportation inventories.

Fluctuation inventories have to be carried for the reason that sales and production times for the product cannot be always predicted with accuracy. There are variations in demand and lead times required to manufacture items. Thus, there is a need for reserve stock or safety stock to account for the fluctuations in demand and lead time.

Anticipation inventories are built up in advance for big selling season, promotion programme or anticipation of likely change in demand suddenly and in case of plant shutdown period. It is the inventory for the future need.

Lot size inventory refers to producing and storing at the rate higher than the current consumption rate. The production in lots is going to help the advantage of price discounts for quantities purchased in bulk and fewer set-ups and, hence, the lower set-up cost.

The transportation inventories exist because materials must be moved from one place to another. When transportation requires a long time, the items in transport represent the inventory. Thus, transportation inventory is a result of extended or longer transportation time.

22.4. REASONS FOR KEEPING INVENTORIES

1. **To stabilise production:** The demand for an item fluctuates because of the number of factors, *e.g.*, seasonality, production schedule etc. The inventories (raw materials and components) should be made available to the production as per the demand failing which results in stock out and the production stoppage takes place for want of materials. Hence, the inventory is kept to take care of this fluctuation so that the production is smooth.

2. **To take advantage of price discounts:** Usually the manufacturers offer discount for bulk buying and to gain this price advantage the materials are bought in bulk even though it is not required immediately. Thus, inventory is maintained to gain economy in purchasing.

3. **To meet the demand during the replenishment period:** The lead time for procurement of materials depends upon many factors like location of the source, demand supply condition, etc. So inventory is maintained to meet the demand during the procurement (replenishment) period.

4. **To prevent loss of orders (sales):** In this competitive scenario, one has to meet the delivery schedules at 100 per cent service level, means they cannot afford to miss the delivery schedule which may result in loss of sales. To avoid this the organisations have to maintain inventory.

5. **To keep pace with changing market conditions:** The organisations have to anticipate the changing market sentiments and they have to stock materials in anticipation of non-availability of materials or sudden increase in prices.

6. Sometimes the organisations have to stock materials due to other reasons like suppliers minimum quantity condition, seasonal availability of materials or sudden increase in prices.

22.5. INVENTORY CONTROL

Inventory control is a planned approach of determining what to order, when to order and how much to order and how much to stock so that costs associated with buying and storing are optimal without interrupting production and sales. Inventory control basically deals with two problems, (*i*) When should an order be placed? (Order level), and (*ii*) How much should be ordered? (Order quantity)

These questions are answered by the use of inventory models. The scientific inventory control system strikes the balance between the loss due to non-availability of an item and cost of carrying the stock of an item. Scientific inventory control aims at maintaining optimum level of stock of goods required by the company at minimum cost to the company.

22.6. OBJECTIVES OF INVENTORY CONTROL

1. To ensure adequate supply of products to customer and avoid shortages as far as possible.
2. To make sure that the financial investment in inventories is minimum (*i.e.*, to see that the working capital is blocked to the minimum possible extent).
3. Efficient purchasing, storing, consumption and accounting for materials is an important objective.
4. To maintain timely record of inventories of all the items and to maintain the stock within the desired limits.
5. To ensure timely action for replenishment.
6. To provide a reserve stock for variations in lead times of delivery of materials.
7. To provide a scientific base for both short-term and long-term planning of materials.

22.7. BENEFITS OF INVENTORY CONTROL

It is an established fact that through the practice of scientific inventory control, the stocks can be reduced anywhere between 10 per cent to 40 per cent. The benefits of inventory control are:

1. Improvement in customers relationship because of the timely delivery of goods and services.
2. Smooth and uninterrupted production and, hence, no stock out.
3. Efficient utilisation of working capital.
4. Helps in minimising loss due to deterioration, obsolescence damage and preliferage.
5. Economy in purchasing.
6. Eliminates the possibility of duplicate ordering.

22.8. COSTS ASSOCIATED WITH INVENTORY

1. **Purchase (or production) cost:** The value of an item is its unit purchasing (production) cost. This cost becomes significant when availing the price discounts. This cost is expressed as Rs./unit.

2. **Capital cost:** The amount invested in an item, (capital cost) is an amount of capital not available for other purchases. If the money were invested somewhere else, a return on the investment is expected. A charge to inventory expenses is made to account for this unreceived return. The amount of the charge reflects the percentage return expected from other investment.

3. **Ordering cost:** It is also known by the name procurement cost or replenishment cost or acquisition cost. Cost of ordering is the amount of money expended to get an item into inventory. This takes into account all the costs incurred from calling the quotations to the point at which the items are taken to stock.

There are two types of costs—Fixed costs and variable costs.

Fixed costs do not depend on the number of orders whereas variable costs change with respect to the number of orders placed. The salaries and wages of permanent employees involved in purchase function and control of inventory, purchasing, incoming inspection, accounting for purchase orders constitute the major part of the fixed costs. The cost of placing an order varies from one organisation to another. They are generally classified under the following heads:

(*i*) **Purchasing:** The clerical and administrative cost associated with the purchasing, the cost of requisitioning material, placing the order, follow-up, receiving and evaluating quotations.

(*ii*) **Inspection:** The cost of checking material after they are received by the supplier for quantity and quality and maintaining records of the receipts.

(*iii*) **Accounting:** The cost of checking supply against each order, making payments and maintaining records of purchases.

(*iv*) **Transportation costs.**

4. **Inventory carrying costs (holding costs):** These are the costs associated with holding a given level of inventory on hand and this cost vary in direct proportion to the amount of holding and period of holding the stock in stores. The holding costs include.

(*i*) Storage costs (rent, heating, lighting, etc.).

(*ii*) Handling costs: Costs associated with moving the items such as cost of labour, equipment for handling.

(*iii*) Depreciation, taxes and insurance.

(*iv*) Costs on record keeping.

(*v*) Product deterioration and obsolescence.

(*vi*) Spoilage, breakage, pilferage and loss due to perishable nature.

5. **Shortage cost:** When there is a demand for the product and the item needed is not in stock then we incur a shortage cost or cost associated with stock out. The shortage costs include:

(*i*) Backorder costs.

(*ii*) Loss of future sales.

(*iii*) Loss of customer goodwill.

(*iv*) Extra cost associated with urgent, small quantity ordering costs.

(*v*) Loss of profit contribution by lost sales revenue.

The unsatisfied demand can be satisfied at a later stage (by means of back orders) or unfulfilled demand is lost completely (no back ordering, the shortage costs become proportional to only the shortage quantity).

22.9. INVENTORY CONTROL—TERMINOLOGY

1. **Demand:** It is the number of items (products) required per unit of time. The demand may be either deterministic or probabilistic in nature.
2. **Order cycle:** The time period between two successive orders is called order cycle.
3. **Lead time:** The length of time between placing an order and receipt of items is called lead time.
4. **Safety stock:** It is also called buffer stock or minimum stock. It is the stock or inventory needed to account for delays in materials supply and to account for sudden increase in demand due to rush orders.
5. **Inventory turnover:** If the company maintains inventories equal to 3 months consumption. It means that inventory turnover is 4 times a year, *i.e.*, the entire inventory is used up and replaced 4 times a year.

15. **Re-order level (ROL):** It is the point at which the replenishment action is initiated. When the stock level reaches R.O.L., the order is placed for the item.
16. **Re-order quantity:** This is the quantity of material (items) to be ordered at the re-order level. Normally this quantity equals the economic order quantity.

22.10. INVENTORY COST RELATIONSHIPS

There are two major costs associated with inventory. Procurement cost (ordering cost) and inventory carrying cost. Annual procurement cost varies with the number of orders. This implies that the procurement cost will be high, if the item is procured frequently in small lots. The procurement cost is expressed as Rs./Order.

The annual inventory carrying cost (Product of average inventory × Carrying cost) is directly proportional to the quantity in stock. The inventory carrying cost decreases, if the quantity ordered per order is small. The two costs are diametrically opposite to each other. The right quantity to be ordered is one that strikes a balance between the two opposing costs. This quantity is referred to as "Economic order quantity" (EOQ). The cost relationships are shown in the Fig. 22.1.

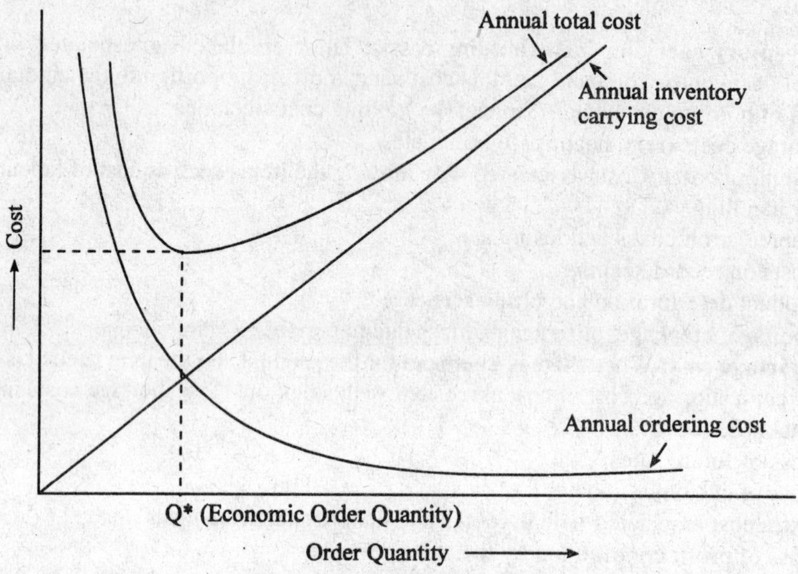

Fig. 22.1: Inventory carrying cost.

22.11. INVENTORY MODELS

One of the basic problem of inventory management is to find out the order quantity so that it is most economical from overall operational point of view. Here the problem lies in minimising the two conflicting costs, *i.e.*, ordering cost and inventory carrying cost. Inventory models helps to find out the order quantity which minimises the total costs (sum of ordering costs and inventory carrying costs). Inventory models are classified as shown in Fig. 22.2.

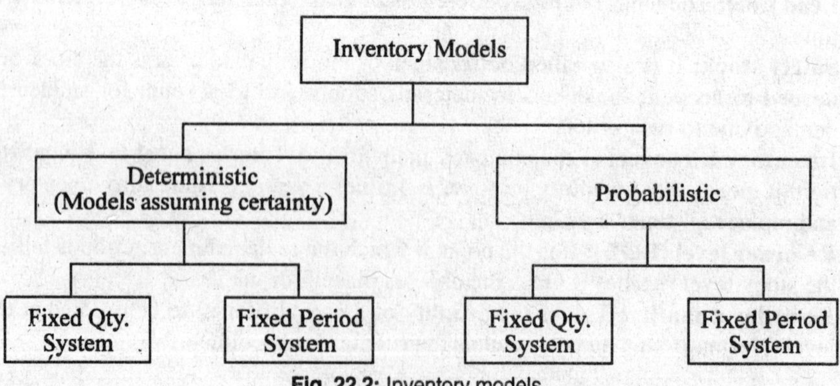

Fig. 22.2: Inventory models.

Model-I. ECONOMIC ORDER QUANTITY WITH INSTANTANEOUS STOCK REPLENISHMENT (BASIC INVENTORY MODEL)

Assumptions

1. Demand is deterministic, constant and it is known.
2. Stock replenishment is instantaneous (lead time is zero)
3. Price of the materials is fixed (quantity discounts are not allowed)
4. Ordering cost does not vary with order quantity.

Graphical representation of the model is shown in Fig. 22.3.

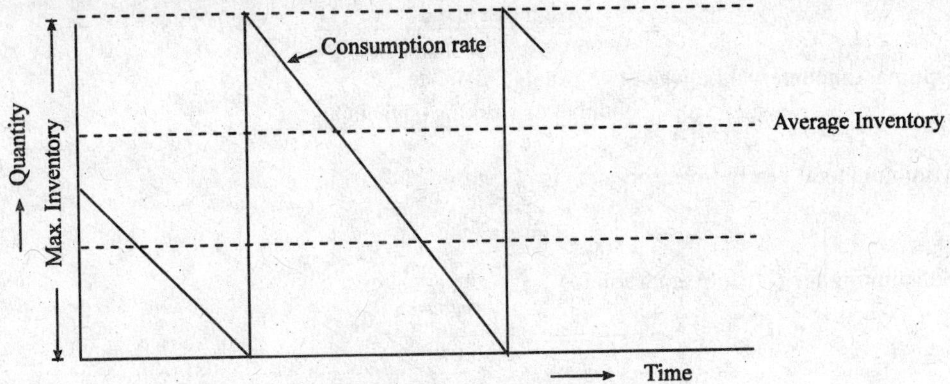

Fig. 22.3: Basic inventory model.

Let D be the annual demand (units per year)

C_o = Ordering costs (Rs./order)

C_h = Inventory carrying costs (Rs./unit/unit time)

C_p = Price per unit

Q = Order quantity

Q^* = Economic order quantity

N = Number of orders placed per annum

T_c = Total cost per annum

$$\text{Annual ordering cost} = \text{No. of orders} \times \text{Ordering cost/order}$$

$$= \frac{\text{Annual demand}}{\text{Order Quantity}} \times \text{Ordering cost/order}$$

$$= \frac{D}{Q} \times C_o \qquad \text{...(1)}$$

$$\text{Annual inventory carrying cost} = \text{Average Inventory investment} \times \text{inventory carrying cost.}$$

$$= \left(\frac{\text{Max. Inventory} - \text{Min. Inventory}}{2}\right) \times \text{inventory carrying cost}$$

$$= \frac{Q\,C_h}{2} \qquad \text{... (2)}$$

$$\text{Annual total cost } (T_c) = \text{Annual ordering cost} + \text{Annual inventory cost.}$$

$$= \frac{D\,C_o}{Q} + \frac{Q\,C_h}{2} \qquad \text{... (3)}$$

To determine Economic order quantity (EOQ), differentiate, Annual Total cost. Equation (3) w.r.t. Q and set the first derivative equal to zero.

$$\frac{dT_C}{dQ} = \frac{-D\,.\,C_o}{Q^2} + \frac{C_h}{2} = 0$$

$$Q^* = \sqrt{\frac{2DC_o}{C_h}} \qquad \text{... (4)}$$

Q^* is economic order quantity.

Note: If the inventory carrying cost is expressed as a percentage of annual average inventory investment, then

$$Q^* = \sqrt{\frac{2DC_o}{CpI}}$$

where, it is expressed as a percentage of annual inventory investment.

The optimal number of orders placed per annum

$$N^* = \frac{\text{Annual demand}}{\text{Economic order quantity}} = \frac{D}{Q^*}$$

Optimal time interval between two orders

$$T^* = \frac{\text{Number of working days in a year}}{N^*}$$

Minimum total yearly inventory cost is given by

$$T_{cm} = \frac{D\,C_o}{Q_o} + \frac{Q^*\,C_h}{2}$$

Substituting for Q^* from equation (4)

$$T_{cm} = \frac{D.C_o}{\sqrt{\frac{2DC_o}{C_h}}} \times \sqrt{\frac{2D \cdot C_o}{Ch}} \times \frac{C_h}{2}$$

$$\boxed{T_{cm} = \sqrt{2\,D\,C_o \cdot C_h}}$$

Model-II. ECONOMIC ORDER QUANTITY WHEN STOCK REPLENISHMENT IS NON-INSTANTANEOUS (PRODUCTION MODEL)

This model is applicable when inventory continuously builds up over a period of time after placing an order or when the units are manufactured and used (or sold) at a constant rate. Because this model is specially suitable for the manufacturing environment where there is a simultaneous production and consumption, it is called **"Production Model"**.

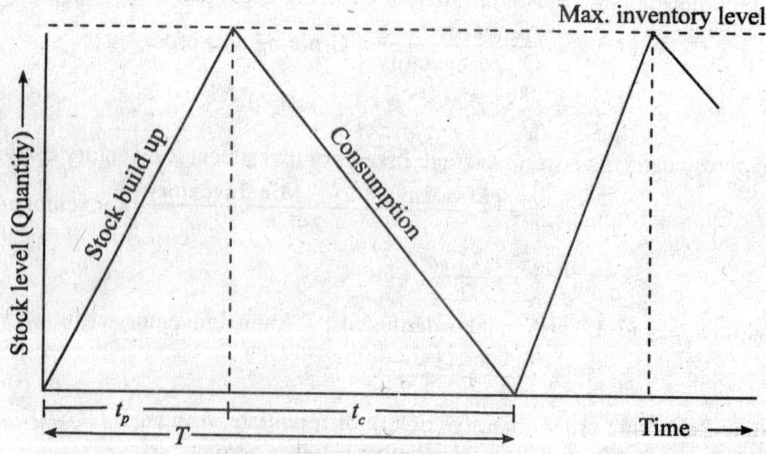

Fig. 22.4: Production inventory model.

Assumptions

1. The item is sold or consumed at the constant demand rate which is known.
2. Set up cost is fixed and it does not change with lot size.
3. The increase in inventory is not instantaneous but it is gradual.

Production inventory model is represented in the Fig. 22.4.

Let p be the production rate.

d is the demand or consumption rate.

Replenishment of inventory under this system build-up during the period t_p and consumption takes place during the entire cycle T.

Note: All other symbols are same as in Model-I.

Inventory under this situation, builds at the rate of $(p - d)$ units and inventory is maximum at the end of production period t_p.

Maximum inventory at the end of production run $= (p - d) \times t_p$

$$\text{Average inventory} = \frac{(p-d)\,tp}{2}$$

Now, the quantity produced during production period $(Q) = p \times t_p$

$$\therefore \quad t_p = \frac{Q}{p}$$

Substituting for t_p,

$$\text{Average inventory} = (p-d)\,.\,\frac{Q}{2p}$$

$$= \frac{Q}{2}\left(1 - \frac{d}{p}\right)$$

$$\text{Annual inventory carrying cost} = \text{Average inventory} \times \text{Inventory carrying cost}$$

$$= \frac{Q}{2}\left(1 - \frac{d}{p}\right) C_h$$

$$\text{Annual set-up cost} = \text{No. of set-ups} \times \text{Set-up cost/set-up}$$

$$= \frac{D}{Q} \times C_o$$

$$\text{Total annual cost } (T_c) = \frac{D.C_o}{Q} + \frac{Q}{2}\left(1 - \frac{d}{p}\right) C_h \quad \text{... (1)}$$

To determinc, the economic Batch (Lot) size (EBQ) differentiate equation (3) with respect to Q, set the first derivative equal to zero.

$$\frac{dT_c}{dQ} = \frac{1}{2}\left(1 - \frac{d}{p}\right) C_h - \frac{D.C_o}{Q^2} = 0$$

$$\therefore \quad Q^* = \sqrt{\frac{2DC_o}{(1 - d/p)\,C_h}}$$

where Q^* is economic batch quantity.

Note: If the inventory carrying cost is expressed as a percentage of annual inventory investment, then

$$\therefore \quad Q^* = \sqrt{\frac{2DC_o}{\left(1 - \frac{d}{p}\right) C_h I}} \qquad \text{(where } I \text{ is expressed in \%)}$$

Optimal total cost is given by

$$T_{cm} = \sqrt{2\,D\,C_o{\cdot}C_h\,(1 - d/p)}$$

Optimal number of production runs

$$N^* = \frac{\text{Annual demand}}{\text{Economic Batch Qty (EBQ)}} = \frac{D}{Q^*}$$

Model-III: A INVENTORY MODEL WHEN SHORTAGES ARE PERMITTED

In many practical situations, shortages or stock outs are not permitted. So, it is must that stock out situations are to be avoided. There are occasions where stock out are economically justifiable. This situation is observed normally when cost per unit is very high.

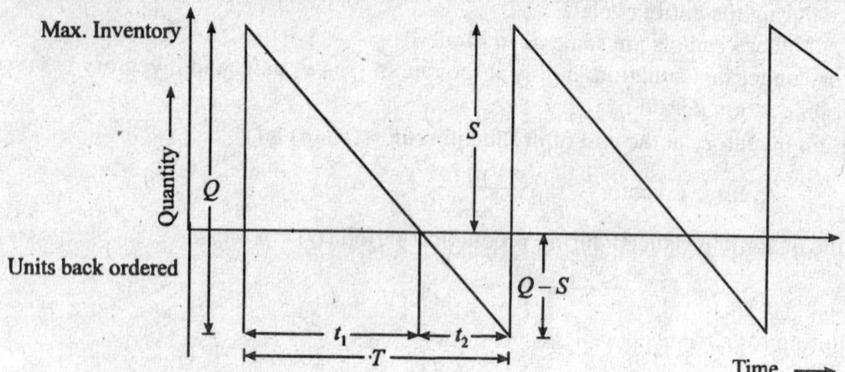

Fig. 22.5: Inventory model with shortages permitted.

C_s = Shortage Cost (Stock out cost) per unit per period.
S = Balance units after back orders are satisfied.
$Q - S$ = Number of shortages per order.
t_1 = Time period during which inventory is positive.
t_2 = Time during which shortage exists.
T = Time between the receipt of orders.

The basic assumption is that there is no loss of sales due to stock out or shortages.

The economic order quantity is expressed as follows:

1. $Q^* = \sqrt{\dfrac{2DC_o}{C_h}\;\dfrac{(C_s + C_h)}{C_s}}$
2. Optimal remaining units after back ordering.

$$S^* = \sqrt{\frac{2DC_o}{C_h}\;\frac{C_s}{(C_s + C_h)}}$$

3. Optimal amount back ordered

$$Q^* - S^* = Q^*\left(1 - \frac{C_s}{(C_h + C_s)}\right)$$

4. Total optimal inventory cost

$$T_{cm} = \sqrt{2DC_o \cdot C_h\left(\frac{C_s}{C_h + C_s}\right)}$$

Model-IV: INVENTORY MODEL WITH PRICE DISCOUNTS

When items are bought in large quantities, the supplier often gives discounts. However, if the material is purchased to take advantage of discount, the average inventory level and so the inventory carrying costs will increase.

Benefits for the purchaser from large orders are, lower cost per unit, lower shipping and transportation cost, reduced handling cost and reduction in ordering costs due to less number of orders.

These benefits are to be compared with the increase in carrying costs. As the order size increases, more space should be provided to stock the items.

A decision is, therefore, to be taken whether the buyer should stick to economic order quantity or increase the same to take advantage that, at large quantities, the production costs per piece are lower (economics of scale) and, hence, part of the savings can be passed on to the customer.

Price-discount Model

Let D be the annual consumption.(Demand)

C_1 is the price per unit. (Basic price)

C_2 is the discounted price per unit.

C_o is the ordering cost.

I is inventory carrying cost expressed as a percentage of average inventory investment.

Q_B be the price break quantity. (Quantity at which the price changes)

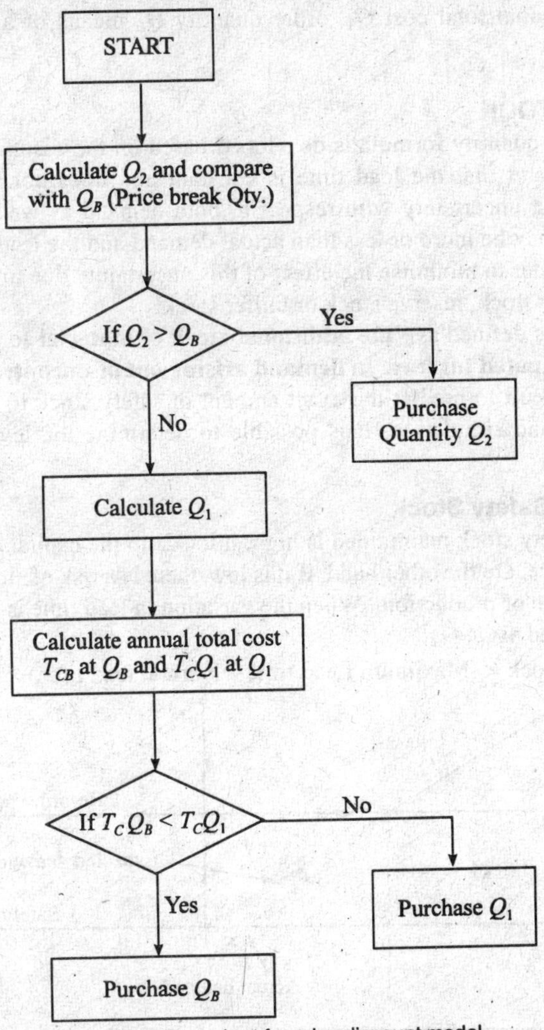

Fig. 22.6: Flow-chart for price discount model.

Procedure (Decision Rules)

1. Calculate Q_2 (Economic order Qty. at discounted price (C_2)

$$Q_2 = \sqrt{\frac{2\,D\,C_o}{C_2 I}}$$

Compare Q_2 with Q_B (Price break quantity)

If $Q_2 \geq Q_B$, Order quantity Q_2.

If $Q_2 < Q_B$.

Compute Q_1 (Economic order Qty. at basic price C_1) and calculate Annual Total cost at Q_1 and Q_B

$$\text{Annual Total cost at } Q_1\ (T_C\,Q_1) = D \cdot C_1 + \frac{D.C_o}{Q_1} + \frac{QC_1 I}{2}$$

$$\text{Annual Total cost at } Q_1\ (T_C\,Q_B) = D \cdot C_2 + \frac{D.C_o}{Q_B} + \frac{Q_B C_2 I}{2}$$

If $T_C Q_B < T_C Q_1$ select Q_B, otherwise purchase Q_1. (If Annual Total cost at price break quantity (Q_B) is less than annual total cost Q_1, order quantity Q_B means of a flow chart as shown in Fig. 22.5.

22.12. SAFETY STOCK

The economic order quantity formula is developed based on the assumption that the demand is known and certain and that the lead time is constant and does not vary. In actual practical situations, there is an uncertainty with respect to both demand as well as lead time. The total forecasted demand may be more or less than actual demand and the lead time may vary from the estimated time. In order to minimise the effect of this uncertainty due to demand and lead time, a firm maintains safety stock, reserve stock or buffer stock.

The safety stock is defined as **"the additional stock of material to be maintained in order to meet the unanticipated increase in demand arising out of uncontrollable factors."**

Because it is difficult to predict the exact amount of safety stock to be maintained, by using statistical methods and simulation, it is possible to determine the level of safety stock to be maintained.

Determination of Safety Stock

If the level of safety stock maintained is high, it locks up the capital and there is a possibility of risk of obsolescence. On the other hand, if it is low, there is a risk of stock out because of which there may be stoppage of production. When the variation in lead time is predominant. The safety stock can be computed as:

$$\text{Safety stock} = (\text{Maximum Lead time} - \text{Normal lead time}) \times \text{Consumption rate.}$$

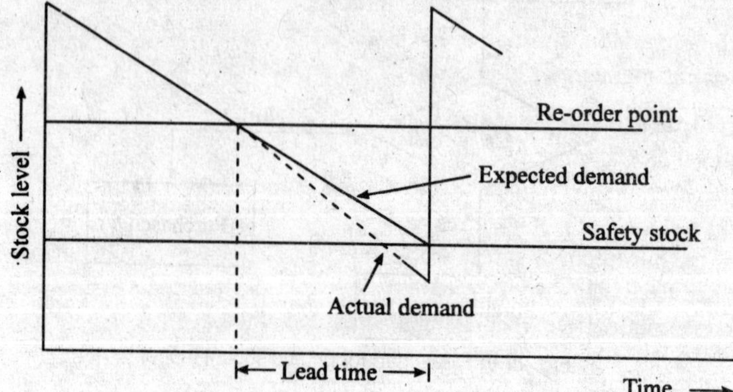

Fig. 22.7: Inventory pattern with safety stock and actual demand exceeds the expected demand.

The service level of inventory thus depends upon the level of safety stocks. Larger the safety stocks, there is lesser risk of stock out and, hence, higher service level. Sometimes higher service levels are not desirable as they result in increase of costs, thus, fixing up a safety stock level is critical. Using the past date regarding the demand and lead time data, reliability of suppliers and service level desired by the management, safety stock can be determined with accuracy.

The inventory pattern with safety stock and variation in demand and lead time is shown in Fig. 22.7. and Fig. 22.8.

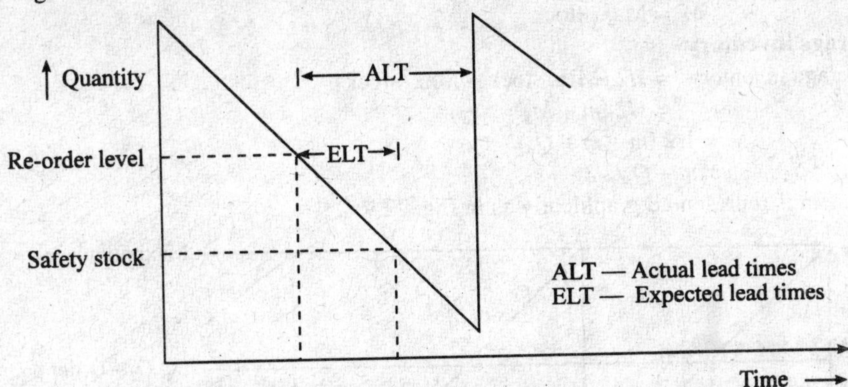

Fig. 22.8: Inventory pattern with safety stock and lead time.

22.13. INVENTORY CONTROL SYSTEM

The inventory systems are developed to cope with the situations where the demand or lead time or both will fluctuate. The basic approach to all stock control methods is to establish a re-order level which, when reached would indicate the signal for the replenishment action. Thus the replenishment of the inventory means determining the quantity to be ordered and the time of ordering.

Basically, there are two types of replenishment systems.

(*i*) Fixed quantity system (Q-system)

(*ii*) Fixed period system (P-system)

Fixed Order Quantity System

This is also called perpetual inventory system or Q-system. In this system, the order quantity is fixed and ordering time varies according to the fluctuation in demand.

The characteristics of this system are:

(*i*) Re-order quantity is fixed and normally it equals Economic order quantity (EOQ).

(*ii*) Depending upon the demand, the time interval of ordering varies.

(*iii*) Replenishment action is initiated when stock level falls to Re-order level (ROL).

(*iv*) Safety stock is maintained to account for increase in demand during lead time.

Parameters to Operate the System

1. Re-order level (ROL)

This equals the sum of safety stock and lead time consumption.

$$R.O.L. = m + L \times C$$

where m – is the minimum or safety stock.

L – Lead time (days/weeks/months)

C – consumption rate. (per day/per week/per month)

2. Re-order quantity (Q)

This normally equals Economic order quantity (*EOQ*)

3. Maximum stock level (M)

It equals the safety stock + order quantity

$$M = m + Q_o$$

where Q_o is order quantity

m = Safety stock

M = Max. stock

4. Average inventory

$$\begin{aligned}\text{Average inventory} &= 1/2\ (\text{Min. stock} + \text{Max. stock})\\ &= 1/2\ (m + M)\\ &= 1/2\ (m + m + Q_o)\\ &= m + Q_o/2\end{aligned}$$

The system is represented graphically as in Fig. 22.9.

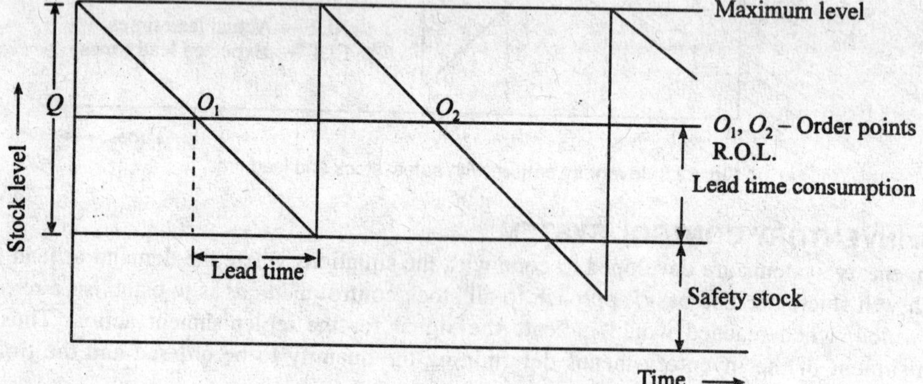

Fig. 22.9: Fixed order quantity system.

Advantages

1. Simple and cheaper to operate.
2. Stock control will be accurate as the replenishment action is initiated soon after the stock reaches R.O.L.
3. Suitable for low value items.
4. Appropriate for variety of inventory maintained within the organisation.

Limitations

(*i*) In this inventory system, there will be a load on the re-ordering system if many items reach R.O.L. at the same time.

(*ii*) The stock leves records and usage rate data are to be maintained.

Application of Q-System—TWO BIN SYSTEM

Two bin system operates on R.O.L. system and it physically segregates the stock of entire items into two bins.

The second bin contains quantity equal to R.O.L.

i.e., $(m + LC)$ and it is means to satisfy demand during the replenishment period.

The first bin contains the quantity (Order quantity) = ($Q - LC$) to satisfy demand between the receipt of materials and placing the next order. LC is the lead time consumption.

The working of the system

To begin with, the stock from the first bin is consumed. The emptying of first bin indicates that the stock has reached R.O.L. and the replenishment action is initiated.

The Quantity in the second bin are thus consumed during the replenishment period.

This system reduces the work involved in record keeping and entering (clerical) errors.

Periodic Review System

It is also called fixed period system or *P*-system.

This system has a fixed ordering interval but the size of the order quantity may vary with changes in demand.

In this system,

The inventory position is verified at prefixed interval, (weekly/monthly/quarterly), then depending upon the situation replenishment action is initiated.

The characteristics of the system are:

1. Order interval is fixed for individual item or group of items.
2. Stock is reviewed at periodic interval and the quantity (Q) which will bring the inventory to maximum level is ordered.

The periodic review system is shown in Fig. 22.10.

Parameters to Operate the System

1. Maximum level (M)

It is sufficient to satisfy demand during review period and lead time.

Max. level (M) = Min. stock + Consumption during review period and lead time.

$$M = m + C(R + L)$$

2. Re-Order Quantity.

(*i*) when lead time is less than review period,

Q = Maximum stock – stock actually held at the time of review.

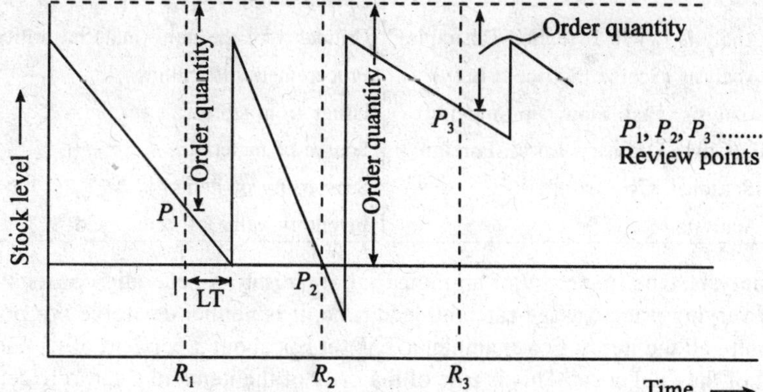

Fig. 22.10: Periodic review inventory system.

(*ii*) when lead time is more than review period.

Q = Max. Stock – (Stock on hand + Stock on order)

This system is suitable for high value items which require strict control on stock levels.

Comparison between Fixed Quantity (Q) and Fixed Period (P) System

Q-system	*P-System*
1. The Quantity to be ordered each time is fixed and normally it is equal to *EOQ*.	The period of ordering the inventory is fixed and the order quantity depends on the stock on hand.
2. It is suitable for low unit cost, high volume items.	Suitable for high unit cost and less in number items.
3. Normally preferred when supplier puts minimum quantity restriction.	Preferred when supplier delivers at fixed periods.

22.14. SELECTIVE CONTROL OF INVENTORY

Selective control refers to the variation in method of control from item to item on some selective basis. There are many criteria used for this purpose. They are:

- based on the cost of the product/item.
- Lead time.
- Usage rate.
- Procurement difficulties, criticality, frequency of usage.

The selective control is more effective and is directed to more significant groups of items. In this system, the items are categorised in a few groups depending upon the selected criteria such as value, usage and frequency of consumption. Such grouping helps the organisation for scientific inventory control. The various types of classifications are shown in Table 22.1.

Table. 22.1: Types of Classifications

Classification	*Criteria*
1. A.B.C. Analysis	Annual usage value of items
2. H.M.L. Analysis (High, Low, Medium)	Unit price of the material (it does not depend upon consumption)
3. V.E.D. Analysis (Vital, Essential, Desirable)	Criticality of the item (material criticality)
4. S.D.E. Analysis (Scarce, Difficult, Easy)	Procurement difficulties
5. F.S.N. Analysis (Fast, Slow, Non-moving)	Issues from stores
6. G.O.L.F. (Govt., Ordinary, Local, Foreign)	Source of material
7. S.O.S. (Seasonal, Off Seasonal)	Seasonality of items
8. X.Y.Z. Analysis	Inventory value of items used

1. ABC analysis: The inventory of an industrial organisation generally consists of thousands of items with varying prices, usage rate and lead time. It is neither desirable nor possible to pay equal attention to all the items. For example, a T.V. set has about 5 per cent of its parts contribute to 80 per cent of the total costs. This is true of majority of the items like car, refrigerator, etc.

ABC analysis is a basic analytical tool which enables management to concentrate its efforts where results will be greater.

The Pareto principle (20/80) of cause and effect is a useful concept in business where it can be used to solve majority of production, quality and inventory problems.

The concept applied to inventory control is called as *ABC* analysis.

Statistics reveal that just a few items account for bulk of the annual consumption of the materials. These few items are called *A* class items which hold the key to business. The other items known as *B* and *C* which are numerous in number but their contribution is less significant. *ABC* analysis thus tends to segregate the items into three categories *A*, *B* and *C* on the basis of their annual usage. The categorisation is made to pay right attention and control demanded by items.

A class items: These items hardly constitute 5-10 per cent of the total items and account for 70–75per cent of the total money spent on inventories. These items require rigid and strict control and need to be stocked in smaller quantities. These items are to be procured frequently and each time less quantity is procured. The inventory of *A* class items is kept at minimum.

B class items: These items are generally 10-15 per cent of total items and represent 10-15 per cent of the total expenditure on materials. These are intermediate items. The control on these items should be intermediate between A and C items.

C class items: These are about 70-80 per cent in number and constitute only 5-10 per cent total expenditure on materials.

These items being less expensive does not require strict control. These are ordered in bulk as against infrequent ordering of *A* class items.

Advantages of ABC Analysis

This approach helps the manager to exercise selective control and focus his attention only on a few items.

By exercising strict control on A class items, the materials manager is able to show the results within a short period of time. It results in reduced clerical costs, saves time and effort and results in better planning and control and increased inventory turnover. ABC analysis, thus, tries to focus and direct the effort based on the merit of the items and, thus, becomes an effective management control tool.

Limitations of ABC Analysis

ABC analysis is a fundamental tool for exercising selective control over numerous inventory items but in present form do not permit precise consideration of all relevant problems of inventory management.

ABC analysis is not one time exercise and items are to be reviewed and recategorised periodically.

Features and Policy Guidelines for ABC Analysis is shown in Table 22.2.

Table 22.2. Features and Policy Guidelines for ABC Analysis

	A Class (High Value)	*B Class (Moderate Value)*	*C Class (Low Value)*
1.	Tight control on stock levels	Moderate control	Less control
2.	Low safety stock	Medium safety stock	Large safety stock
3.	Ordered frequently	Less frequently	Bulk ordering
4.	Individual posting in stores	Individual posting	Collective postings
5.	Continuous check on schedules and revision when called for	Broad check on schedule revisions	Hardly any check required
6.	Weekly control statements	Monthly control reports	Quarterly control reports
7.	Procured from multiple sources	Two or more reliable sources	Two reliable sources for each item
8.	Minimise waste, obsolete and surplus	Quarterly control over waste	Annual review regarding waste
9.	Continuous effort to reduce lead time	Moderate efforts	Minimum efforts

Procedure for making ABC Analysis

1. Calculate the total inventory value for each item held in inventory by multiplying the number of units used in a year by its unit price.
2. Tabulate these items in descending order of their values placing first the item having the highest total value and so on.
3. Prepare a table showing item No., unit cost, annual units consumed and annual rupee value of units used.
4. Compute the running total item by item for the items and also for rupee value of consumption.

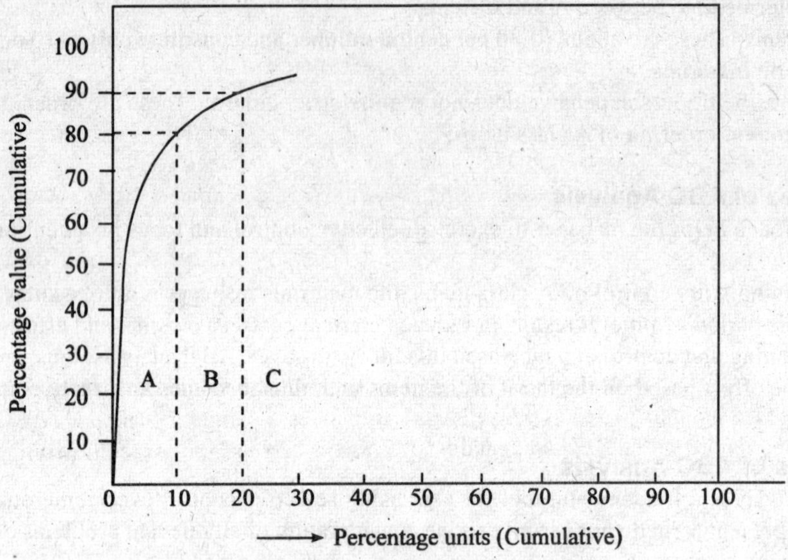

Fig. 22.11: ABC analysis.

5. Compute the cumulative percentage for the item count and cumulative annual usage value.
6. Classify the items as per the norms for ABC items.
7. The cumulative percentages are represented graphically as in Fig. 22.11.

2. V.E.D. analysis: This analysis represents classification of items based on criticality. The analysis classifies the items into three groups called vital, essential and desirable.

Vital items are those items the unavailability of which will stop the production.

Essential items are those items whose stock out costs are very high.

Desirable items will not cause any immediate production stoppages and their stock out costs are nominal.

This analysis is mainly carried out to identify critical items. An item which may belong to *C*-category may be critical from production point of view.

The service level for each item may be determined and the inventory can be planned accordingly.

3. S.D.E. analysis: S.D.E. analysis classifies the items into three groups namely—Scarce, Difficult and Easy. These are based on the problems of procurement.

Scarce, classification includes items which are in short supply, imported items. Such items are procured once in a year because of effort and expenditure involved in its import.

Difficult items are those items which are available indigenously but are difficult to procure.

Easy classification covers those items which are readily available and easy to procure. These are mostly commercially available standard items.

This SDE analysis is applied by purchased department.

4. HML analysis: This analysis is similar to ABC analysis but here the criteria is 'price' instead of usage value.

The items in this analysis are classified into three groups, *i.e.*, high, low and medium. The management decides the cut off lines or prices for the three categories. This analysis helps to keep control over consumption as per the price and helps to assess storage and security requirements, *i.e.*, the high priced items are to be stored in the cupboards. It helps to outline the buying policies and to delegate authorities to buyers.

5. FSN analysis: All the items in the inventory are not required at the same frequency. Some are required regularly, some occasionally and some very rarely.

FSN analysis classifies items into fast moving, slow moving and non-moving.

6. SOS analysis: This classification is based on the seasonality of the items as seasonal and off seasonal. Seasonal items are available only for a limited period and, hence, they are procured to meet the demand till the next season.

7. XYZ analysis: This analysis is based on the value of the stocks on hand (*i.e.*, capital employed to procure inventory). Items whose inventory values are high are called *X* category and whose values are low are called *Z* items. Usually XYZ analysis is used in association with A.B.C. analysis.

***Problem 1*:** ABC corporation has got a demand for particular part at 10,000 units per year. The cost per unit is Rs. 2 and it costs Rs. 36 to place an order and to process the delivery. The inventory carrying cost is estimated at 9 per cent of average inventory investment. Determine

(*i*) Economic order quantity.

(*ii*) Optimum number of orders to be placed per annum.

(*iii*) Minimum total cost of inventory per annum.

***Solution*:** Annual demand for parts (D) = 10,000 units/annum

Ordering cost (C_o) = Rs. 36/order

Cost per unit (C_p) = Rs. 2 /unit

Inventory carrying cost (I) = 0.09

(*i*) Economic order quantity (*EOQ*)

$$Q^* = \sqrt{\frac{2DC_o}{C_p I}}$$

$$= \sqrt{\frac{2 \times 10{,}000 \times 36}{2 \times 0.09}}$$

$$= \mathbf{2000} \text{ units}$$

(*ii*) Optimum Number of orders $= \dfrac{D}{Q^*}$

$$= \frac{10{,}000}{2000} = \mathbf{5}$$

(*iii*) Total Annual inventory cost

T_{cm} = Ordering cost + inventory carrying cost

= 2 × Ordering cost (at *EOQ*, order cost equals carrying cost)
= **360** Rs./annum.

***Problem 2*:** A manufacturer has to supply his customers 3600 units of his product per year. Shortages are not permitted. Inventory carrying cost amounts Rs. 1.2 per unit per annum. The set-up cost per run is Rs. 80. Find:

(*i*) Economic order quantity.
(*ii*) Optimum number of orders per annum.
(*iii*) Average annual inventory cost (minimum).
(*iv*) Optimum period of supply per optimum order.

***Solution*:** Annual demand (D) = 3600 units
Inventory carrying cost (C_h) = 1.2 Rs./unit /annum.
Ordering cost (C_o) = 80 Rs./order.

(*i*) Economic Order Quantity.

$$Q = \sqrt{\frac{2DC_o}{C_h}} = \sqrt{\frac{2 \times 3600 \times 80}{1.2}}$$

$$= \sqrt{\frac{2 \times 3600 \times 80}{1.2}}$$

$$= 692.82 = (\mathbf{693}) \text{ units}$$

(*ii*) Optimum number of orders per annum.

$$N = \frac{\text{Annual demand}}{E.O.Q.} = \frac{D}{Q} + \frac{3600}{693} = 5.19 = \mathbf{5} \text{ orders.}$$

(*iii*) Minimum annual inventory cost

$$T_{cm} = \sqrt{2\,D\,C_o \cdot C_h}$$

$$= \sqrt{2 \times 3600 \times 80 \times 1.2} = \mathbf{831.38} \text{ Rs.}$$

(iv) Optimum period of supply per optimum order

$$= \frac{1}{N} = \mathbf{0.2} \text{ yrs.}$$

***Problem 3*:** Usha corporation currently practices the following system for the procurement of an item.

No. of orders placed in a year = 8, Ordering cost = Rs. 750/order
Each time order quantity = 250, Carrying cost = 40 per cent
Comment on the ordering policy of the company and estimate the loss to the company in not practising scientific inventory policy

***Solution*:** Annual Demand (D) = 250 × 8
Ordering cost (C_o) = 750 Rs./order
Inventory carrying cost (I) = 0.4
Cost per unit (C_p) = 40

Economic order quantity

$$Q = \sqrt{\frac{2\,D.C_o}{C_p}} = \sqrt{\frac{2 \times 250 \times 8 \times 750}{40 \times 0.4}}$$

$$= \mathbf{433} \text{ units}$$

Minimum total cost (annual) corresponding to *EOQ*

$$T_{cm} = \sqrt{2 \times D \times C_o \times (C_p \times I)}$$

$$= \sqrt{2 \times (250 \times 8) \times 750 \times (40 \times 0.4)}$$

$$= \mathbf{6928.28}$$

Total cost of inventory under present system.

T_c = Annual ordering cost + Annual inventory carrying cost
= No. of orders/annum × ordering cost/order + Annual inventory carrying cost

$$= 8 \times 750 + \frac{250}{2} \times 40 \times 0.4$$

$$= 6000 + 2000 = \mathbf{8000} \text{ Rs.}$$

Loss to the company in not operating scientific inventory

Policy $= T_c - T_{cm}$
$= 8000 - 6928.28 = \mathbf{1071.72}$ Rs.

Problem 4: Indian Telecoms entered into contract with precision instruments for the purchase of 12,500 instruments at the rate of Rs. 250 per instrument during the year. The deliveries of the instruments will be made each time half a month after the order is placed. Indian telecoms estimates its carrying cost at Rs. 48 per instrument per annum. The cost of paper work, follow up transport and receipt work out to be Rs. 2000. How frequently should Indian Telecom place orders with precision instruments? What is the re-order point?

Solution: Annual Requirement (D) = 12,500 units
Ordering Cost (C_o) = 2000 Rs./order
Inventory carrying cost (C_h) = 48 Rs./unit/annum

Economic order quantity

$$Q^* = \sqrt{\frac{2\,D.C_o}{C_h}} = \sqrt{\frac{21 \times 12{,}500 \times 2000}{48}} = 1020.62$$

$= \mathbf{1021}$ numbers

$$\text{No. of orders to be placed in a year} = \frac{D}{Q^*} = \frac{12{,}500}{1021} = 12.24$$

$= \mathbf{12}$

The Re-order point (ROP) = Lead time consumption
Lead time is given as 1/2 month.

$$\text{Re-order point } (ROP) = \frac{12{,}500}{24} = 520.83$$

$= \mathbf{521}$ units (without considering safety stock)

Problem 5: ABC company produces a cable at the rate of 5000 metres per hour. The cable is used at the rate of 2500 metres/hour. The cost of the cable is Rs. 5 per metre. The inventory carrying cost is 25 per cent and set-up costs are Rs. 4050 per set-up. Determine the optimal number of cycles required in a year for the manufacture of this cable.

Solution: Production rate (p) = 5000 metres/hr.
Demand rate (d) = 2500 metres/hr.
Price per unit (C_p) = 5 Rs./metre.
Inventory carrying cost (I) = 0.25
Set up Cost (C_o) = 4050 per order
$D = 2500 \times 8 \times 365$
Metres/hrs per day × days

Economic Batch Quantity

$$EBQ = \sqrt{\frac{2\,DC_o}{1 - d/p)\,C_p \cdot I}}$$

$$= \sqrt{\frac{2 \times (20{,}000 \times 365)}{\frac{(1 - 2500) \times 5 \times 0.25}{5000}}} \times 4050 = 307584.13$$

$= \mathbf{307585}$

Optimal number of cycles per annum

$$= \frac{\text{Annual demand}}{EBQ} = \frac{D}{EBQ} = \frac{2500 \times 8 \times 365}{307585} = 23.75$$

$$= \mathbf{24}$$

***Problem 6*: A contractor undertakes to supply diesel engines to a truck manufacturer at the rate of 25 per day. He finds that the cost of holding a completed engine in stock is Rs. 16 per month. Production of engines is in batches and each time a new batch is started, there are set-up costs of 10.000 Rs. How frequently should the batches be started and what will be the minimum average inventory cost and production time if production rate is 40 engines/day**

Assume 300 working days in a year.

***Solution*:** Production rate (p) = 40 per day

Demand rate (d) = 25 per day

Annual demand (D) = 25 × 300 = 7500 per annum

(Assuming 300 working days)

Set-up Cost (C_o) = 10,000

Inventory carrying cost (C_h) = Rs. 1.6 × 12 per year.

Economic manufacturing quantity

$$EBQ = \sqrt{\frac{2\,D.C_o}{C_h\,(1 - d/p)}}$$

$$= \sqrt{\frac{2 \times 7500 \times 10{,}000}{19.2\,(1 - 25/40)}}$$

$$= 4564.35 = \mathbf{4565}$$

No. of production runs per annum

$$N = \frac{D}{Q*} = \frac{7500}{4585} = 1.642 = \mathbf{2} \text{ orders}$$

Frequency of production run (Time between two runs)

$$= \frac{1}{N} = \frac{1}{2} = \mathbf{0.5} \text{ yrs.}$$

Total annual inventory cost

$$T_{cm} = \sqrt{2\,D\,C_o\,C_h\,(1 - d/p)}$$

$$= \mathbf{32863.35} \text{ Rs./annum.}$$

***Problem 7*:** An automobile manufacturing company is purchasing an item from outside suppliers. Demand is 10,000 units per annum. Cost of the item is Rs. 5 per unit and procurement cost is estimated to be Rs. 100 per order. Cost of carrying inventory is 25 per cent. If the consumption rate is constant determine EOQ.

In the above problem, if the company decides to manufacture the above item with an equipment which produce 100 units per day. The cost of units thus produced is Rs. 3.5 per unit setup cost is Rs. 150. How your answer is changed in the second case.

***Solution*:** Annual Demand (consumption) (D)= 10,000 units/annum

Cost per unit (C_p) = Rs. 5/unit

Ordering cost (C_o) = 100 Rs./order

Inventory carrying cost (I) = 0.25

Economic order quantity

$$EOQ = \sqrt{\frac{2\,D.C_o}{C_p.I}} = \sqrt{\frac{2 \times 10{,}000 \times 100}{0.25 \times 5}}$$

$= \mathbf{1265}$ units

(*b*) If the company decides to manufacture the above item,

Production rate (p) = 100 per day

Cost per unit produced (C_p) = 3.5 Rs./unit

Setup cost (C_o) = Rs. 150/order.

Inventory carrying cost remaining the same.

Economic manufacturing quantity.

$$EBQ = \sqrt{\frac{2\,DC_o}{(1 - d/p)\,C_pI}}$$

$$= \sqrt{\frac{2 \times 10{,}000 \times 150}{(1 - 10{,}000/30{,}000) \times 0.25 \times 3.5}}$$

$= \mathbf{2262}$

The company has to manufacture 2262 unit in a batch for minimum cost.

***Problem 8*:** A company consumes 12000 units of a particular item. The company has a production capacity of 60 units/day. The cost of each unit produced by the company is Rs. 8. The setup and tooling up cost is Rs. 96 per setup. The carrying charges are 15 per cent of cost per unit.

Determine

(*i*) Economic quantity to be manufactured in each batch.

(*ii*) How frequently should the production runs be made.

(*iii*) Determine the production period.

Assume 300 working days per annum.

***Solution*:** Annual consumption (D) = 12000 units per annum

Production rate (p) = 60 per day

Setup cost (C_o) = Rs. 96 per order

Consumption rate (d) = $\frac{12{,}000}{300}$ = 40 per day

Inventory charges/unit/annum (C_h) = $8 \times 0.15 = 1.2$ Rs./unit/annum.

Economic manufacturing quantity.

(*i*) $$EBQ = \sqrt{\frac{2DC_o}{C_h(1 - d/p)}}$$

$$= \sqrt{\frac{2 \times 12{,}000 \times 96}{1.2\,(1 - 40/60)}} = \mathbf{2400}$$

(*ii*) Frequency of production run $= \frac{\text{Annual consumption}}{EBQ} = \frac{D}{Q*} = \frac{12{,}000}{2400} = \mathbf{5}$

(*iii*) Production time

$$T_p = \frac{Q*}{p} = \frac{2400}{60} = \mathbf{40} \text{ days.}$$

***Problem 9*:** Universal toolings has a requirement for 1,50,000 metal bushings per annum. The company orders the metal bushing in lots of 40,000 units from a supplier. The ordering cost is Rs. 40 and the carrying charges are expressed 20 per cent of the unit cost. The bush costs Rs.15 each. The company wants to know what percentage of their order quantity differs from economic order quantity and how the cost varies for the two. Find the optimal order quantity.

***Solution*:** Annual demand (D) = 1,50,000

Order quantity (Q) = 40,000 (presently ordered)

Ordering cost (C_o) = 40

Cost per unit (C_p) = Rs.1.5 /unit

Inventory cost (I) = 0.20

(*i*) Economic order quantity

$$EOQ = \sqrt{\frac{2\,DC_o}{1\,C_p}} = \sqrt{\frac{2 \times 1{,}50{,}000 \times 40}{0.2\,(1.5)}}$$

Q^* = **6325** (optimal order quantity)

(*ii*) Comparison between EOQ (Q^*) and present order quantity (Q)

$$\frac{Q}{Q^*} = \frac{40{,}000}{6325} = \mathbf{6.3}$$

(*iii*) Comparing total costs. (T_cQ and T_cQ^*)

$$T_cQ^* = \sqrt{2\,DC_oC_h}$$

$$= \sqrt{2 \times 15000 \times 40 \times 0.2 \times 1.5}$$

= **1897.36** Rs./annum

Annaul total cost T_cQ

$$T_cQ = \frac{1{,}50{,}000}{40{,}000} \times 40 + \frac{40{,}000}{2} \times 1.5 \times 0.2$$

= 150 + 6000 = **6150** Rs./annum.

The optimal order quantity = EOQ = **6325**.

***Problem 10*:** A pharma company has a demand for 10,00,000 bottles. Each empty bottle costs the company Rs. 1. Empty bottles are supplied by M/s Rupa Glass Ltd. The R.O.L. system of stock replenishment is followed. Ordering cost is Rs. 12.5 /order and inventory carrying cost is 25 per cent of cost per bottle. The demand is constant throughout the year. The lead time is 15 days. Determine

1. Economic order quantity.
2. Lead time consumption.
3. Re-order level.
4. Average inventory.

***Solution*:** 1. Economic order quantity (EOQ)

$$Q^* = \sqrt{\frac{2\,DC_o}{I \cdot C_p}}$$

$$= \sqrt{\frac{2 \times 10{,}00{,}000 \times 12.5}{2.25 \times 1}}$$

= **10,000** bottles

2. Lead time consumption = Lead time (in months) × monthly consumption.

$$= \frac{15}{30} \times \frac{10{,}00{,}000}{12}$$

= **41667** units

3. Safety stock is assumed equal to Lead time consumption.

Re-order level = Safety stock + Lead time consumption

= 41667 + 41667

= **83334** units

4. Average inventory = $\dfrac{\text{Max. inventory} + \text{Min. inventory}}{2}$

$$= \frac{(41667 + 10,000) + (41667)}{2}$$

$= \mathbf{46667}$ bottles.

Problem 11: A contractor has to supply 10,000 bearings per day to an automobile manufacturer. He finds that when he starts the production run, he can produce 25,000 bearings per day. The cost of holding the bearings in stock for one day is 2 paise and setup cost of production run is Rs. 18.

How frequently should the production runs be made.

Solution: Demand Rate (d) = 10,000 bearings/day

Production Rate (p) = 25,000 bearings/day

Holding cost (C_h) = 0.02 per unit/day

Setup cost (C_o) = Rs. 18 per setup

$$\text{Economic Lot Size } (Q^*) = \sqrt{\frac{2\,DC_o}{C_h\,(1 - d/p)}}$$

$$= \sqrt{\frac{2 \times 10,000 \times 18}{0.02\,(1 - 10,000/25,000)}} = \mathbf{5478} \text{ bearings}$$

Time interval between two consecutive production runs

$$t = \frac{2\,C_o}{d.C_h} \times \frac{p}{p - d}$$

$$= \frac{2 \times 18}{10,000 \times 0.02} \times \frac{25000}{25,000 - 10000}$$

$= \mathbf{10.5}$ days

Problem 12: The demand for an item is uniform at the rate of 25 units/month. Ordering cost is Rs. 30 and cost per unit is Rs. 2 per unit. Inventory carrying cost is Rs. 0.5 per unit per month and if shortage cost is Rs. 0.3 per unit per month. Determine Economic order, quantity, and how often to make production run.

Solution: This is *EOQ* model with shortages.

(*i*) *EOQ* with shortage is computed as

$$Q^* = \sqrt{\frac{2DC_o}{C_h} \cdot \frac{C_h + C_s}{C_s}}$$

In this problem, C_h = 0.5 Rs./unit/month

C_s = 0.3 Rs./unit/month

$$Q^* = \sqrt{\frac{2 \times 25 \times 30}{0.5} \times \frac{(0.5 + 0.3)}{0.3}}$$

$= \mathbf{90}$ units.

(*ii*) Optimal scheduling between two consecutive orders

$$= \frac{Q}{D} = \frac{90}{30} = \mathbf{3.6} \text{ months}$$

Problem 13: The following information on production inventory system of a manufacturing company is given below.

Determine (*i*) Optimal lot size.

(*iii*) Manufacturing time and time between set-ups

Demand per annum: 6000 units

Unit cost : Rs.40

Set-up cost : Rs. 500
Production rate : 3600 units
Holding cost : Rs. 8/unit/annum
Shortage cost/unit : Rs. 20/unit/annum

***Solution*:** The economic lot size

$$(i)\ Q^* = \sqrt{\frac{2DC_o}{C_h\,(1 - d/p)}} \times \sqrt{\frac{(C_h + C_s)}{C_s}}$$

$$= \sqrt{\frac{2 \times 6000 \times 500}{8\,(1 - 6000/3600)}} \times \sqrt{\frac{(8 + 20)}{20}}$$

$$= 1122.7$$

$$= \mathbf{1123} \text{ units}$$

$$(ii)\ \text{Manufacturing time} = \frac{Q^*}{P} = \frac{1123}{36000}$$

$$= \mathbf{0.031} \text{ years}$$

$$(iv)\ \text{Time between set-ups} = \frac{Q^*}{D} = 0.187 \text{ years}$$

***Problem 14*:** Annual requirement of an item is 2400 units. Each item costs the company Rs. 6. The manufacturer offers discount of 5 per cent if 500 or more quantities are purchased. The ordering cost is Rs. 32 per order and inventory cost is 16 per cent.

Whether it is a advisable to accept the discount? Comment.

***Solution*:** Annual demand (D) = 2400 units.
Ordering cost (C_o) = 32 per order
Inventory carrying cost (I) = 0.16
Price per unit (Basic) (C_1) = 6 Rs./unit
Discounted price (C_2) = 5 per cent discount on Rs. 6 = 5.7 Rs./unit
Price break quantity (Q_B) = 500 units

Step 1: Calculate Q_2 (EOQ at price C_2)

$$Q = \sqrt{\frac{2\,D.C_o}{C_2 I}} = \sqrt{\frac{2 \times 2400 \times 32}{5.7 \times 0.16}}$$

$$= \mathbf{410} \text{ units.}$$

Step 2: Comparing Q_2 and Q_B

$$Q_2 < Q_B\ (410 < 500)$$

So the condition is not satisfied Q_2 cannot be ordered.

Step 3: Compute Q_1 (Economic order quantity at C_1)

$$Q_1 = \sqrt{\frac{2\,D.C_o}{C_1 I}} = \sqrt{\frac{2 \times 2400 \times 32}{6 \times 0.16}}$$

$$= \mathbf{400} \text{ units.}$$

Step 4: Compute the Annual total costs for Quantity Q_1 and Q_B

Annual total cost for Quantity Q_1 at price C_1

$$A\,T.C_i = D.C_I + \frac{D}{Q_1}.C_o + \frac{Q_1}{2} \cdot C_1.I$$

$$= 2400 \times 6 + \frac{2400}{400} \times 32 + 6 \times 0.16$$

$$= \mathbf{14784} \text{ Rs./annum}$$

Annual Total cost for price break quantity Q_B at price C_2

$$ATC = D.C_2 + \frac{D}{Q_B} C_o + \frac{Q_B}{2} \cdot C_2.I$$

$$= 2400 \times 4.7 + \frac{2400}{500} \times 32 + \frac{500}{2} \times 4.7 \times 0.16$$

$$= \mathbf{140061.6} \text{ Rs. /annum.}$$

Annual Total cost at Q_B is less than Annual Total cost at Q_1

$$ATCB < ATC_1$$

$$14006.6 < 14784$$

Order quantity equal to Q_B = **500** units.

Thus, discount should be availed and order quantity should be **500** units.

Problem 15: A materials manager adopts the policy to place an order for a minimum quantity of 500 of a particular item in order to avail a discount of 10 per cent. It was found from the company records that for last year 8 order were placed each of size 200 Nos. ordering cost is Rs. 500 per order. Inventory carrying charges at 40 per cent cost per unit = Rs. 400.

Is the purchase manager justified in his decision .

What is the effect of this decision on the company.

Solution: Economic order quantity (*EOQ*)

$$Q = \sqrt{\frac{2\,DC_o}{C_p.I}} = \sqrt{\frac{2 \times 1600 \times 500}{400 \times 0.4}}$$

$$= 100 \text{ units}$$

(*i*) Annual Total inventory costs at $Q = 100$

$$ATCQ = D \times C_1 + \frac{D}{Q^*} C_o + \frac{Q}{2} \times C_1 \times I$$

$$= 1600 \times 400 + \frac{1600}{100} \times 500 + \frac{100}{2} \times 400 \times 0.4$$

$$= \mathbf{6{,}56{,}000} \qquad \text{... (1)}$$

(*ii*) Annual Total cost of present policy (*i.e.*, order size Q = 200 units)

$$ATC_Q = D \times C_2 + \frac{D}{Q} \times C_o \times \frac{Q}{2} \times C_1 \times I$$

$$= 1600 \times 400 + \frac{1600}{200} \times 500 + \frac{200}{2} \times 400 \times 0.4$$

$$= \mathbf{660{,}000} \text{ units} \qquad \text{... (2)}$$

(*iii*) Proposed inventory costs

On an order 500 units, a discount of 10 per cent is offered.

Discounted price (C_2) = 460 Rs./unit

Total annum cost at order Quantity of 500

$$ATC = 1600 \times 360 + \frac{1600}{500} \times 500 + \frac{500}{2} \times 360 \times 0.4$$

$$ATQ = 500$$

$$= \mathbf{613600} \qquad \text{... (3)}$$

From the cost consideration, the managers decision is justified as it is going result in savings of Rs. 46,400 /annum over present system.

Problem 16: Monthly consumption of an item is 300 units. The price per unit is Rs.10. Inventory carrying cost is 18 per cent and ordering cost is Rs. 36 per order. Lead time of 1 months stock. Assuming *ROL* system.

Determine

1. Re-order quantity.
2. Re-order level.
3. Minimum level.
4. Maximum level.
5. Average inventory.

***Solution*:** (*i*) In *ROL* system, usually the order quantity is *EOQ*.

$$Q = \sqrt{\frac{2\,D.C_o}{C_p}.I} = \sqrt{\frac{2 \times 3600 \times 36}{10 \times 0.18}}$$

$$= \mathbf{120} \text{ units}$$

(*ii*) Reorder level = Lead time consumption + safety stock
= 300 + 300 = **600** units

(*iii*) Minimum level = Safety stock = **300** units

(*iv*) Maximum level = Safety stock + order quantity
= 300 + 120 = **420** units

(*v*) Average inventory $= \frac{420 + 300}{2}$ = **360** units

***Problem 17*:** Illustration on *ABC* analysis.

Ten items are kept in the inventory. The details regarding the number of items used per annum and price per unit are given below.

Classify the items into *A*, *B* and *C* class.

Item No.	*Annual usage*	*Price*
101	200	40.00
102	100	360.00
103	2000	0.20
104	400	20.00
105	6000	0.04
106	1200	0.80
107	120	100.00
108	2000	0.70
109	1000	1.00
110	80	400.00

***Solution*:** Compute the annual usage value

Item No.	*Annual usage value*		*Rank*
101	200 × 40	= 8000	4
102	100 × 360	= 36000	1
103	2000 × 0.2	= 400	9
104	400 × 20	= 8000	5
105	6000 × 0.04	= 8000	10
106	1200 × 0.8	= 960	8
107	120 × 100	= 12000	3
108	2000 × 0.7	= 1400	6
109	1000 × 1	= 1000	7
110	80 × 40	= 32000	2

The annual usage value in descending order is written

Item No.	Class	Annual Usage	Cum Annual	Cum usage value %	% of item in class	
102		36000	36000	36%	2/10 × 100 = 20%	*A*
110		32000	68000	68%		
107		12000	80000	80%		
101		8000	88000	88%	3/10 × 100 = 30%	*B*
104		8000	96000	96%		
108		1400	97400	97.4%		
109		1000	98400	98.4%		
106		960	99360	99.36%	5/10 × 100 = 50%	*C*
103		400	99760	99.76%		
105		240	100000	100%		

The classification is

	Item Nos.	% of Annual value	% of Item in class
A Class	102, 110	68	20
B Class	107, 101, 104	28	30
C Class	108, 109, 106, 103, 105	4	50

References for Further Reading

1. Ammer Deans, *Materials Management,* Irwin Homewood, Illinois, (1980).
2. Buffa E.S. and W.H., *Taubert Production Inventory System Planning and Control*, Irwin Homewood, Illinois.
3. Starr and Miller, *Inventory Control Theory and Practice*, Prentice Hall of India, New Delhi, (1997).
4. Mayor Reymond, *Production and Operations Management*, McGraw Hill Co. Ltd., International Editt.
5. Everette Adams *et al.*, *Production and Operations Management*, Prentice Hall of India.

REVIEW QUESTIONS

1. What are inventories? Why does it is essential to keep inventories?
2. What are the different types of inventories a manufacturing organisation keeps?
3. What are the objectives of inventory control?
4. What are the various costs associated with inventory.
5. By means of a graph, explain the quantity cost relationship in inventory.
6. What is *E.Q.O.*? Derive an expression for the economic order quantity when the stock replenishment is instantaneous giving the assumptions made.
7. What is economic manufacture quantity? Derive the expression for the same.
8. What is the set up cost in a manufacturing situation? What elements does it comprise of ?
9. What are the limitations of *EOQ* model?
10. Define the following:
    (*i*) Re-order point
    (*ii*) Lead time.
    (*iii*) Minimum level.
    (*iv*) Maximum level.

(*v*) Average inventory.

(*vi*) Safety stock.

11. What is the purpose of keeping safety stocks?
12. Explain *Q* system (Fixed order Qty.) and *P* system (Periodic Review system).
13. Describe the general procedure for arriving at the value of an economic lot size when the quantity discount is available.
14. Differentiate between fixed order system and fixed period system.
15. What is selective control of inventory and explain various selective control techniques.
16. Describe the procedure for *ABC* analysis. Bring out the merits and demerits of *ABC* analysis.
17. What is the effect of storage costs on *EOQ* model?

PROBLEMS

1. The demand for a product is 3600 units per annum. Inventory carrying cost amount to Rs. 1.6 per unit per year. The setup cost per run is Rs. 800.

   Find (*i*) Economic order quantity.

   (*ii*) Minimum average inventory control.

   (*iii*) Optimum number of orders per year.

   (*iv*) Optimum period of supply per optimum order.

2. The annual demand for an item is 3200 units. The unit cost is Rs. 6. Inventory charges are 25 per cent procurement cost is Rs.150.

   Determine (*i*) Economic order quantity.

   (*ii*) No. of orders per year.

   (*iii*) Optimum number of orders per year.

   (*iv*) The optimal cost.

   [**Answer:** *EOQ* = 800 units. No. of orders = *A*, Time between orders = 3 months,

   Optimal cost – Rs. 775 per annum

3. An item is produced at the rate of 50 units/day. The demand is 25 units per day. Setup cost is Rs.100 per setup holding cost is Rs. 0.01 per unit per day.

   Find (*i*) Cycle time.

   (*ii*) Minimum total cost per run.

   [**Answer:** (*i*) *EOQ* = 1000 units, (*ii*) Cycle time 40 days, (*iii*) Total cost Rs. 200/annum]

4. An aircraft company rivets at a rate of 2500 kg per annum. The rivets cost Rs. 301 kg and it cost Rs. 130 to place an order and carrying cost is 10 per cent how frequently should the orders be placed for rivets and what quantities should be ordered.

   [**Answer:** *EOQ* = 466 kgs. No of orders 5 / year]

5. An Illustration on ABC analysis

The following information is available about the group of 10 items classify the items in to *A*, *B* and *C* class items.

Item No.	*Annual consumption (units)*	*Units price (Rs.)*
1001	3000	100
1002	28000	150
1003	300	100
1004	11000	50
1005	400	50
1006	22000	100
1007	1500	50
1008	8000	50
1009	6000	150
1010	800	100

6. A contractor undertakes to supply diesel engines to a truck manufacturer at the rate of 25 per day. He finds that the cost of holding a complete engine in stock is Rs. 16 per month and there is a clause in the contract penalising him Rs. 10 per engine per day late for missing the scheduled delivery date. Production of engine is in batches and each time a new batch is started the setup cost incurred is Rs. 10,000. How frequently should batches be started and what should be the initial inventory level at the time each batch is completed.

7. The annual demand for a product is 10,000 units. Each units costs Rs.100. If the order is placed in quantities below 200 units but for the orders of 200 or above the price is Rs. 95. The annual inventory holding cost is 10 per cent of value of the item and ordering cost is Rs. 32 per order. Determine the optimal order quantity.

8. A factory uses Rs. 32,000 worth of raw material per year. The ordering cost per order is Rs. 50 and carrying cost is 20 per cent of the average inventory investment. The company follows E.O.Q. purchasing policy. Calculate the re-order point, maximum and minimum inventory levels and average inventory. It is given that the factory works for 300 days a year. Replenishment time is 9 days and safety stock is worth Rs. 300.

9. One item is manufactured at the rate of 4800 units per week and is consumed at the rate of 3000 per week. If the setup cost of manufacturing is Rs. 450 per setup and cost of holding the stock is Rs. 0.4 per unit per week.

   Find (*i*) Economic batch quantity to be produced and max. inventory

   (*ii*) Manufacturing period

   (*iii*) Cycle time of operation.

10. Following data was obtained for EOQ system of ordering

    Annual demand — 10,000 units at price of Rs./per unit
    Ordering cost is Rs. 12 /order
    Inventory carrying cost is 24 per cent
    Max. lead time = 30 days. Normal lead time = 15 days
    Calculate (*i*) EOQ, (*ii*) safety stock, (*iii*) Re-order level.

11. A company has a demand of 12,000 units per year and it can produce 2000 units per month. Setup cost is Rs. 400 and holding charges are Re. 0.15 per unit per month.

(*i*) Find the optimal lot size and total cost of operating inventory per year.

(*ii*) Find maximum inventory and manufacturing cycle time.

(*iii*) If the cost of shortage is Rs. 20 per unit per annum. Find ordering quantity, manufacturing and total time.

12. An item is produced at the rate of 128 per day. Annual demand is 6400 units. Setup cost of production run is Rs. 24 and holding cost is Rs. 3/unit/annum. There are 250 working days for production each year.

    Develop an inventory policy for this item.

13. A certain trailer manufacturing company uses 2400 wheels per year in the manufacture of trailers. Lead time is 30 days. Minimum stock maintained is equal to one month's consumption. Inventory carrying cost is 20 per cent ordering cost is Rs.100 cost of one wheel is Rs. 50.

    Determine.

    1. Optimal order quantity.
    2. Max. inventory level.
    3. Avg. inventory level.
    4. Re-order point.

14. ABC company buys a particular item at a cost of Rs. 5 per unit.

    The company is offered the discount as follows:

    Order quantity less than 500 units — Rs. 5.00 per unit

    500 to 1250 units — Rs. 4.80 per unit

    1250 units or more — Rs. 4.80 per unit

    The cost of placing an order is Rs. 36. Annual demand is 2800.

    Inventory carrying cost is 20 per cent of average inventory investment

    Give your decision regarding the quantity to be ordered.

23

PRODUCTION COST CONCEPTS AND BREAK-EVEN ANALYSIS

• Introduction • Costs of production • Concept of cost • Cost centre — Cost unit • Classification of costs • Analysis of production costs • Break-even Analysis • Assumptions • Break-even point • Margin of safety • Angle of incidence • Methods of lowering BEP • Applications of BEP • Cost-Volume-Profit analysis.

23.1. INTRODUCTION

The production manager must maintain a close watch over the costs that are incurred in production department. The type and nature of the costs incurred must be known before appropriate measures can be implemented.

Two broad heads of costs are associated with production, *i.e.*, costs of capital assets and costs of production. The classification and analysis of costs are important to enable the manager to exercise proper control on costs so as to manufacture the products at the pre-established cost level.

23.2. COSTS OF PRODUCTION

The costs of production include:

1. Purchase costs of raw materials, bought out components and subassemblies, procurement and transportation costs.
2. Purchase costs of supplies such as oils, lubricants, tools of small value, fuel oil, machinery spares, cotton waste, etc.
3. Wages and salaries paid to direct production workers, maintenance inspection, stores staff, supervisors and other staff.
4. Costs paid to subcontractors for the orders placed on them.
5. Cost of production line rejections, wastage, spoilage and rework.
6. Expenses towards rent and insurance of factory buildings, insurance on plant and machinery, stores, etc.
7. Interest on working capital to the extent it relates to inventory.
8. Cost of procurement of capital assets like buildings, machinery, tooling, inspection equipment, furniture, etc., and the depreciation of these capital assets.

23.3. CONCEPT OF COST

Cost is the amount of resources sacrificed or given up to achieve a specific objective which may be the acquisition of goods or services. Costs are always expressed in money terms, *e.g.*, a manufacturer incurs costs in buying materials and in hiring labour, etc.

23.4. COST CENTRE

Cost centre is defined as "a location or item of equipment, (or group of these) for which costs may be ascertained and used for the purpose of cost control".

Usually a department or responsibility centre or area of responsibility under the control of manager who is responsible for costs incurred is referred to as cost centre. There are some cost centres which do not play a direct role in production, e.g., stores, finance, etc., and they are called service cost centres. They provide services to other cost centres.

23.5. COST UNIT

It is the unit of output in relation to which costs are determined.

Examples

Industry	*Costy Unit*
Cement	Tonnes
Steel	Tonnes
Automobile	Number
Power	KW-hr
Wood	Cubic Feet
Transport	Tonnes-Km or Seat-Km

23.6. CLASSIFICATION OF COSTS

Classification of costs are based on the following:

1. Natural characteristics (material, labour and overhead)
2. Changes in activity or volume (fixed, variable, mixed)
3. Degree of tracebility to the product (direct cost, indirect cost)
4. Costs for analytical and decision-making (sunk costs, opportunity costs, controllable and non-controllable, differential, imputed costs)
5. Other classifications (product cost, period cost).

1. Natural Classification of Costs

This classification refers to the basic physical characteristics of the cost. In a manufacturing company, the total cost of a product includes the following four elements:

(A) Direct material: Direct material refers to the cost of materials which become a major part of the finished product. They are the raw materials that become an integral part of the finished product and are traceable to specific units of output.

Examples of direct materials are: Raw cotton in textiles, crude oil to make diesel, steel to make automobile parts. The following groups of materials come under direct material:

(*i*) All materials purchased for a particular job, process or product.

(*ii*) All materials acquired from stores for production.

(*iii*) Components or parts purchased or produced.

(*iv*) Materials passing from one process to another process.

(B) Direct labour: Direct labour is defined as the labour associated with workers who are engaged in the production process. It is the labour costs for specific work performed on products that is traceable to end products.

Example: Labour of machine operators, assembly operators.

(C) Direct expenses: The expenditure incurred (other than direct material and direct labour) on a specific job or product are included in direct expenses. These are also called chargeable expenses.

Examples: Cost of special layout, design or drawings, hiring special machines for specific product manufacture, etc.

(D) Factory overheads: These are also called manufacturing costs. These include the costs of indirect materials, indirect labour and indirect expenses.

(*i*) Indirect material refers to materials that are needed for the completion of the product but it is not possible to trace or identify it with end product, *e.g.*, cutting oil, lubricants cannot be charged to specific product.

(*ii*) Indirect labour refers to the labour hours expended which will not directly affect the composition or construction of the finished product.

Examples: Foreman, shop clerks, material handlers, maintenance employees. Their labour is considered indirect because it is not economically possible to trace their with specific product.

(*iii*) Indirect expenses are the expenditure incurred by the manufacturing company from the beginning (start) of production to its completion and transfer to finished goods store.

Direct costs and factory overheads together are called conversion costs.

(E) Distribution and administrative overheads: Distribution overheads are also called marketing or selling overheads. These costs include advertising, salesmen salaries, and commission, packaging, storage, transportation and sales administrative costs. Administrative overhead includes costs of planning and controlling of general business operations. All costs which are not charged to production and sales are included in administrative overheads, *e.g.*, Chairman's salary, fees of board of directors, rent of administrative office.

These costs are represented in the Table 23.1.

Table 23.1: Costs of Manufacturing Company

1. Direct Material + Direct Labour + Direct Expenses = Prime Cost
2. Indirect Material + Indirect Labour + Indirect Expenses = Factory Overhead
3. Prime Cost + Factory Overhead = Factory Cost
4. Factory Cost + Distribution And Administrative Overhead = Total Cost

2. CLassification Based on Activity or Volume

(A). Fixed cost: The costs which do not change for a given period in spite of change in volume of production. This cost is independent of volume of production.

Examples of fixed costs are rent, taxes, salaries of supervisors, depreciation, insurance, etc. Fixed costs are normally expressed in terms of time period.

- *i.e.*, per day, per annum, etc.

Fixed costs are represented as shown in Fig. 23.1.

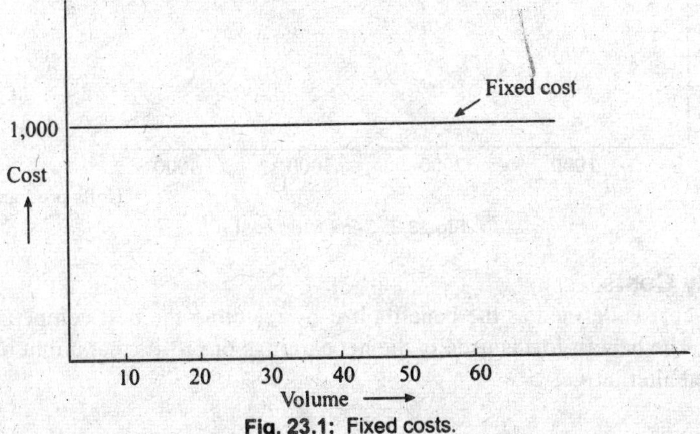

Fig. 23.1: Fixed costs.

- Fixed cost does not mean that they never change. They are constant up to specific volume or range of volume

(B) Variable costs: These vary directly and proportionately with output. There is a constant ratio between the change in the cost and change in the level of output. Direct material cost and direct labour costs are generally variable costs. Variable costs results from the utilisation of raw materials and direct labour in production departments.

Variable cost is represented in Fig. 23.2.

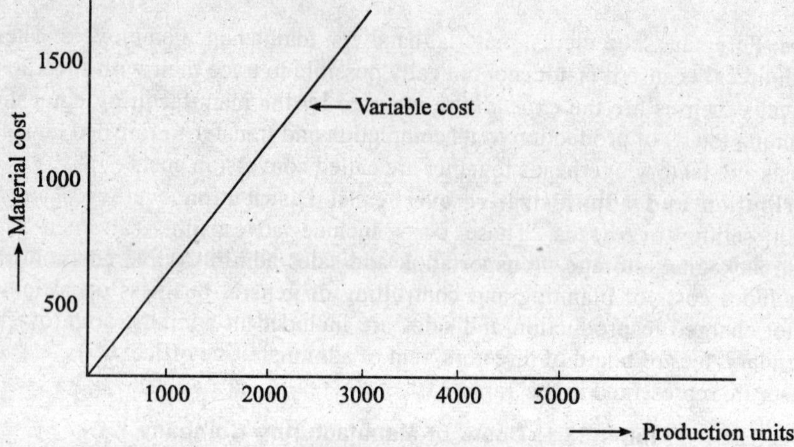

Fig. 23.2: Variable cost

(C) Mixed costs: Mixed costs are made up of fixed and variable costs. They are combination of semi-variable and semi-fixed costs. Because of variable component, they fluctuate with volume, because of fixed component, they will not change in direct proportion to output. Semi-fixed costs are those costs which remain constant up to a certain level of output after which they become variable as represented in Fig. 23.3. Semi-variable cost is the cost which is basically variable but whose slope may change abruptly when a certain output level is reached.

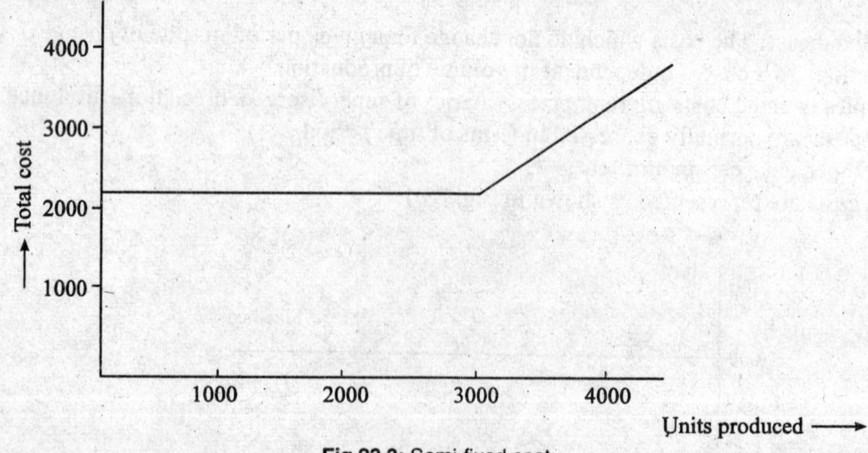

Fig 23.3: Semi-fixed cost.

3. Opportunity Costs

Opportunity cost is defined as the benefits lost by rejecting the best competing alternative to the one chosen. The benefit lost is usually the net earnings or profits that might have been earned from the rejected alternative.

4. Sunk Costs

It is an expenditure for equipment or productive resources which has no economic relevance to the present decision-making process. It is a cost that has either already been incurred or is yet be incurred but will be same no matter which alternative course of action is selected. Generally, it is known as unavoidable cost.

5. Controllable and Non-contrallable Costs

A controllable cost is the cost over which a manager has direct and complete decision authority, *i.e.*, the manager has complete control over these costs, *e.g.*, indirect labour, cutting tool, lubricants.

A cost which cannot be influenced by the action of the specified member of an organisation is referred to as uncontrollable cost.

6. Imputed Costs

Imputed costs are costs not actually incurred in some transaction but which are relevant to the decision as they pertain to a particular situation. These costs do not enter into traditional accounting systems, *e.g.*, interest on internally generated funds, rental value of company owned property, etc.

7. Out of Pocket Costs

Out of pocket costs signify the cash cost associated with an activity. Non-cash costs such as depreciation are not included in out of pocket costs.

23.7. ANALYSIS OF PRODUCTION COSTS

1. Direct Material Costs

The materials that can be directly identified with the production can be classified as direct materials. It is essential to know what items make up the direct material cost.

Direct materials costs are analysed with regards to purchase, purchase price of materials and storage and issue.

(*i*) **Direct material purchase:** An effective procurement procedure would greatly help in minimising costs of purchase of direct materials.

- Materials should be standardised and classification and codification is necessary.
- A sound purchase policies and procedures.
- Procurements should be based on authorised material requisitions.
- Purchases should be made preferably after floating tenders.
- The company should develop a continual vendor rating system.

(*ii*) **Direct material storage and issue:** After the material is received and incoming inspection is carried out, all materials would be moved to stores for future uses. Record of these receipts will be made by the storekeeper on the bin cards.

The organisation should follow a procedure for issues and implement measures for issue control. A lot of costs can be saved and unnecessary wastage of materials avoided if proper control over issues is exercised.

2. Direct Labour Costs

In organisations it is normally easy to distinguish between the direct workers and indirect workers. In case of machine operators and tools setters a particular machine or group of machines may be considered as the cost centre and wages of these machine operators and tool setters may be allocated to the cost centres.

Labour cost control is very much essential. A part from fixing proper time standards, measuring the actual time spent and comparing actual with standard, a continuous monitoring of attendance time with actual time booked to the various jobs and isolation of idle time costs and overtime costs will greatly help in control of labour costs.

Any idle time involves payment of wages for not working and, hence, idle time should not be allowed to occur. The idle times which are unavoidable, the costs should be minimised and other idle times are to be avoided.

Overtime is going to cost the company double the regular time. Since it is costly, a close control over sanction of overtime and production during overtime should be maintained. The normal salaries and wages during overtime are considered as direct labour costs and additional overtime premium is considered as indirect expense.

3. Production Overhead

The costs which cannot be traced with the product like indirect material, indirect labour and indirect expenses incurred at the cost centre are called overheads and that portion of the overhead relating to the production function is called the production overhead.

Allocation and Apportionment of Overhead Costs

(*i*) **Primary overheads:** The overhead costs can be directly identified with a particular department or cost centre as having been incurred for that cost centre. These items of overhead cannot be traced to products or jobs but can be allocated specifically to departments.

Examples are — repairs and maintenance expenses, overtime, indirect materials, supplies, etc.

Expenses such as power, light, rent, depreciation of factory and buildings, expenses shared by all departments cannot be charged directly to the department.

These expenses do not originate in any department. Therefore, these costs should be apportioned to any or all departments using such items. Cost apportionment is the process of charging expenses in an equitable proportion to the various cost centres or departments.

The following are the basis of apportionment.

1. Direct labour hours/or machine hours.
2. Number of workers employed.
3. Floor area occupied.

(*ii*) **Secondary distribution:** The reassignment of service departments overhead to producing departments or centres is termed as secondary distribution.

The basis of secondary distribution are:

1. Benefits obtained.
2. Ability to pay.
3. Efficiency or incentives.

(*iii*) **Methods of absorption of factory overheads:** The allotment of overhead to cost units (output) is called overhead absorption.

Methods

1. Percentage on direct labour.
2. Percentage on direct wages.
3. Prime cost percentage.
4. Unit of production basis.
5. Machine hour rate.

23.8. BREAK-EVEN ANALYSIS

Break-even analysis establishes the relationship among the factors affecting profit. It indicates at what level cost and revenue are in equilibrium. It is a simple method of presenting to management the effect of changes in volume on profit. The detailed analysis of break-even data will help the management to understand the effect of alternative decisions that convert costs from variable to fixed, the costs which increase sales volume and revenue. It is a powerful tool in evaluating alternative course of action.

23.8.1. ASSUMPTIONS IN BREAK-EVEN ANALYSIS

1. Selling prices will remain constant at all sales levels (Quantity aiscounts are not available)
2. There is a linear relationship between sales volume and costs.
3. The costs are divided into two categories—Fixed costs, those costs which does not vary with volume (quantity) and variable costs will be varying in direct proportion to quantity.
4. Production and sales quantities are equal. (There is no inventory)
5. No other factors will influence the cost except the quantity.

23.8.2. BREAK-EVEN POINT

Break-even point refers to the level of sales (sales volume) at which the sales income (revenues) equal the total costs. It is a point at which the profit is zero. The quantities produced (sold) above break-even point result in profits and quantity below break-even point result in losses. The break-even point is reached when the fixed costs are completely recovered.

The break-even point is represented as shown in Fig. 23.4.

Let F represents the fixed cost.

Q is quantity produced and sold.

b is the sales price per unit.

a is variable cost per unit.

bQ total income (revenue).

aQ total variable cost.

Total costs = Fixed cost + variable cost = $F + aQ$

At BEP, Total costs equal total income

Therefore, Total cost = Total income.

$$F + aQ = bQ$$

$$Q = F/b–a$$

$$\text{Contribution} = \text{Sales} - \text{Variable cost} = b - a$$

$$\text{Break-even Quantity (units)} = \frac{\text{Fixed cost}}{\text{Contribution}}$$

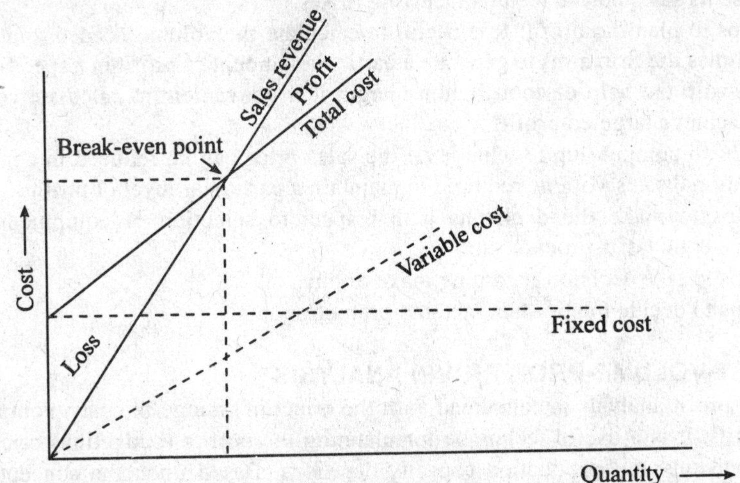

Fig. 23.4. Break-even chart.

Margin of Safety

Margin of safety is the difference between the existing level of output and the level of output at BEP.

$$\text{Margin of safety} = \frac{\text{Sales at BEP}}{\text{Sales}} \times 100$$

Angle of Incidence

This is an angle at which the sales line cuts the total cost line. The management aims at large angle of incidence because large angle of incidence indicates a high profit rate. A narrow angle will show that even fixed overheads are absorbed and relatively low rate of return.

23.8.3. METHODS OF LOWERING BEP

Every organisation aims at lower break-even point so that their fixed costs are recovered and soon the profit begins after the sales starts exceeding BEP. Lower BEP increases the safety margin.

BEP can be lowered by the following methods:

1. Reduce fixed costs

Fixed cost is reduced from F to F'

The reduced BEP $(Q') = Q \,.\, F'/F$

Where Q is Break-even Quantity at Fixed Cost F.

2. Reduce the variable cost

Variable cost is reduced from a to a'.

Reduced BEP $(Q') = Q \,.\, (b - a/b - a')$

$(b - a)$ is contribution at variable cost a

$(b - a')$ is contribution at variable cost a'

3. Increase the slope of the income line from *b* to *b*.

Reduced BEP $Q' = Q \,.\, b - a/b' - a$

b' is the increased price per unit.

b is the current price per unit.

23.8.4. APPLICATIONS OF BEP

1. Safety margin: Break-even chart helps the management to know at a glance the profits at different levels of activity and the safety margin refers to the extent to which an organisation can afford to loose its sales before it starts incurring losses.
2. It helps to plan the profit: It is useful to calculate the volume needed to attain the target profit. Sometimes the firm aims to generate a particular amount of profit in a specified period. For this purpose, with the help of contribution margin it is convenient to calculate volume of sales necessary to achieve targeted profit.
3. It helps to compute up to what level the sales price can be reduced in competition or to compute additional sales volume required to maintain a particular level of profit.
4. It helps to make the decisions with respect to selection of equipment amongst the alternatives, selection of a process, etc.
5. It helps to take decision regarding make or buy.
6. It helps to decide the product mix and promotion mix.

23.8.5. COST-VOLUME-PROFIT (CVP) ANALYSIS

Cost-volume-profit analysis is concerned with the effect of change in costs, volume and selling price on profits. It is a useful technique for planning the profits (budgeting) pricing decisions, sales mix decisions and production capacity decisions. Based upon the concept of fixed and variable costs, it is possible to determine break-even sales volume to calculate the sales level necessary to generate desired profits and answer many questions that arise during management planning.

Profit-Volume (P/V) Graph

A *P/V* graph is used along with break-even chart. Profit (or loss) is plotted on a *y*-axis, and units of products (quantity) Sales revenue or percentage of activity is plotted on *x*-axis. A horizontal line is drawn on the graph to separate profits from losses.

Profits and losses at various sales level are plotted and are connected by a straight line.

Break-even point is measured at the point where the profit line intersects the horizontal line.

P/V graph focuses on the relationship between volume and profit.

P/V graph is shown in Fig. 23.5.

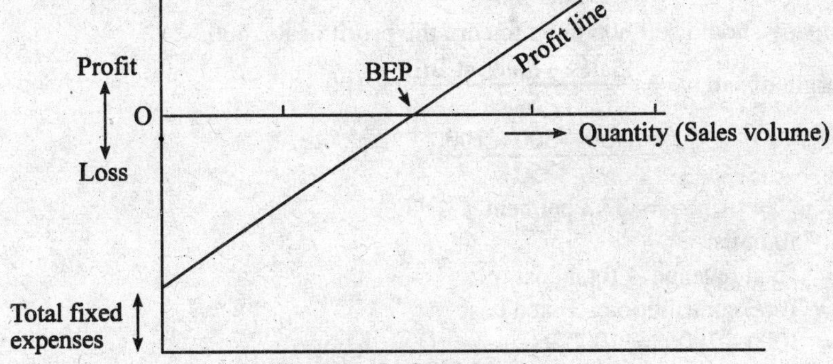

Fig. 23.5: Profit volume chart.

The important relationships are:

$$\text{BEP (units)} = \frac{\text{Fixed Cost}}{\text{Contribution}}$$

$$\text{Contribution} = \text{Sales} - \text{Variable Cost}$$

$$P/V\ \text{Ratio} = \frac{\text{Contribution}}{\text{Sales}}$$

$$\text{Contribution} = \text{Sales} \times P/V\ \text{Ratio}$$

$$\text{BEP} = \frac{\text{Fixed Cost}}{P/V\ \text{Ratio}}$$

$$\text{Sales(units)} = \frac{\text{Fixed Cost} + \text{Profit}}{\text{Contribution}}$$

***Problem 1*:** A manufacturing firm incurs a fixed cost of Rs. 18000. The variable costs accounts Rs. 8 per unit and selling price is Rs. 13.

Find the number of pieces to be produced to break-even.

***Solution*:** Fixed Cost (F) = 18000

Variable cost (a) = 8 Rs./unit

Selling price (b) = 13 Rs./unit

$$\text{BEP} = \frac{\text{Fixed Cost}}{\text{Contribution}} = \frac{F}{b-a} = \frac{18000}{13-8}$$

$$= 3600 \text{ pieces}$$

***Problem 2*:** ABC company plans to sell an article at a local market. The articles are purchased at Rs. 5 on the condition that all unsold articles shall be returned. The rent for the space is Rs. 2000. The articles will be sold at Rs. 9. Determine the number of articles which must be sold (*a*) To break-even, (*b*) To earn Rs. 400 as profit, (*c*) If the company sells 750 articles. Calculate margin of safety and profit.

Solution: (*a*) $\text{BEP} = \dfrac{\text{Fixed cost}}{\text{Contribution}} = \dfrac{F}{b-a} = \dfrac{2000}{9-5}$

$= \mathbf{500}$ units

(*b*) To earn a profit of Rs. 400

$$\text{No. of articles to be sold} = \frac{\text{Fixed cost} + \text{Profit}}{\text{Contribution.}} = \frac{F + \text{Profit}}{b-a}$$

$$= \frac{2000 + 400}{9-5}$$

$$= \mathbf{600} \text{ units}$$

The company should sell 600 units to earn the profit of Rs. 400.

(*c*) $$\text{Margin of safety} = \frac{\text{Sales} - \text{Sales at BEP}}{\text{Sales}} \times 100$$

$$= \frac{750 - 500 \times 100}{750}$$

$$= \mathbf{33.3} \text{ per cent}$$

Profit at 750 units.

Profit = Total revenue – Total cost

= Total contribution – Fixed cost

= 3000 – 2000 = 1000 Rs.

Problem 3: Determine the amount of fixed cost from the following information:

Sales = 24,000

Direct material = 80,000

Direct labour = 50,000

Variable overheads = 20,000

Profit = 50,000

Solution:

Variable cost = Direct Material + Direct labour + Variable Overhead

= 80,000 + 50,000 + 20,000 = 1,50,000

We have the relation,

Sales – variable cost = Fixed cost + profit.

$$2{,}40{,}000 - 1{,}50{,}000 = F + 50{,}000$$

$$F = \mathbf{40{,}000}$$

Problem 4: PQR Limited company has been offered a chance to buy between Machine *A* and Machine *B*. You are required to compute.

(*i*) BEP of each machine.

(*ii*) The level of sales at which both machines earn equal profits?

The following data is given

	Machine A	*Machine B*
Annual output (in units)	10,000	10,000
Fixed costs	30,000	16,000
Profit at above level	30,000	24,000

The market price of the product is expected to be Rs. 10/unit.

***Solution*:**

Particulars	*Machine*	
	A	*B*
Sales Rs. (10,000 × 10)	1,00,000	1,00,000
Contribution (Fixed cost + Profit)	60,000	40,000
Variable cost	40,000	60,000
P/V Ratio (Contribution/Sales)	60%	40%
BEP (Fixed cost / *P/V* Ratio)	50,000	40,000
Contribution per unit	6	4
Variable cost/unit	4	6

(*b*) Since selling prices of *A* and *B* are equal, the machines will earn equal profit when total cost of operations on both machines is equal.

Let *X* be the output when the total costs on both are equal.

Machine *A* Total cost = $4x + 30{,}000$ ... (1)

Machine *B* Total cost = $6x + 16{,}000$

$4x + 30000 = 6x + 16000$

Solving the equation for *x*

$x =$ **7000** units.

At production level of 7000 units both machine give equal profits.

***Problem 5*:** An analysis of the company reveals the following information

Cost element	*Variable cost*	*Fixed cost*
Direct material	32.8	—
Direct labour	28.4	—
Factory overheads	12.6	1,89,900
Distribution overheads	4.1	58,400
General administrative overheads	1.1	66,700
Budgeted sales are		18,50,000

Determine

1. Break-even sales volume.
2. The profit at the budgeted sales volume.
3. The profit if the actual sales (*a*) drop by 10 per cent, (*b*) increase by 5 per cent from budgeted sales.

***Solution*:** Total variable costs as percentage of sales = 32.8 + 28.4 + 12.6 + 4.1 + 1.1

= 79 per cent

$$P/V \text{ Ratio} = \frac{\text{Contribution}}{\text{Sales}} = \frac{100-79}{100} = 21 \text{ per cent}$$

Total fixed costs = 3,15,000

1. Break-even Sales Volume

$$\text{B.E.P.} = \frac{\text{Fixed Cost}}{P/V\text{ Ratio}} = \frac{315000}{0.21}$$

$= \mathbf{1{,}500{,}000}$

2. Profit at budgeted Sales volume

Profit = Sales × *P/V* Ratio – Fixed cost

= 18,50,000 × 0.21 – 31,5000 = 3,88,500 – 3,15,000

= **73,500**

3. Profit when actual sales drop by 10 per cent

Actual sales = 18,50,000 – 1,85,000 = 16,65,000

Profit = Sales × *P/V* Ratio – Fixed cost

= 16,65,000 × 0.21 – 3,15,000

= **34,650**

4. Profit when actual sales are increased by 5 per cent

Actual sales = 18,50,000 + 92,500 = 19,42,500

Profit = 19,42,500 × 0.21 – 3,15,000

= 4,07,925 - 3,15,000

= **92,925**

Problem 5: ASMIT corporation has given the following information on its capacity, sales and cost as follows:

1. Current capacity = 1,00,000 units.
2. At current level of operations, its margın of safety is 5 per cent of its break-even point.
3. Contribution Margin P/V Ratio = 2 5 per cent.
4. The unutilised capacity at present is 10,000 units
5. Sales price Rs. 40 per unit.

(*a*) **Find**

(*i*) Break-even point in sales volume.

(*ii*) Fixed costs.

(*iii*) Variable costs per unit.

(*iv*) Margin of safety in units.

(*b*) If the fixed costs are decreased by Rs. 1,80,000 to what extent can the price be reduced to maintain the total profit at current level.

***Solution*: (*a*) (*i*) Break-even point in sales volume**

Current capacity	1,00,000 units
Less unutilised capacity	10,000 units
Utilised capacity	90,000 units

Utilised capacity – BEP = Margin of safety

Utilised capacity – BEP = 50 per cent of BEP

Utilised capacity = BEP + 50 per cent BEP

= 1.1/2 (BEP)

BEP = 90,000/1.5

= 60,000 units

BEP (Value) = 40 × 60,000

= **2,40,000**

(*ii*) Fixed costs

$$\text{BEP} = \frac{\text{Fixed cost}}{\text{Contribution per unit}}$$

$$60{,}000 = \frac{\text{Fixed cost}}{\text{Sales price} \times \text{contribution margin}}$$

$$60{,}000 = \frac{\text{Fixed costs}}{40 \times 0.25}$$

Fixed cost = 6,00,000

(*iii*) Variable cost per unit

Sales price	Rs. 40
Contribution (Based on contribution margin ratio of 25%)	Rs. 40
Variable cost	Rs. 30

(*iv*) **Margin of safety** = 50% of BEP

= $.50 \times 60{,}000$

= 30,000 units.

(*b*) Currently the contribution is Rs. 9,00,000 (90,000 × 10). After reducing the fixed costs the profit is Rs. 300000. If the fixed costs were to decline by Rs. 1,80,000 to maintain same profit level, the contribution can come down by the same amount and even the sales value. The reduction in sales price would, therefore, to be equal to reduction in sales value/sales quantity

$$= \frac{1{,}80{,}000}{90{,}000} = \text{Rs. } \mathbf{2\ per\ unit}$$

***Problem*:** For a particular product, the following information is given:

Selling price per unit	Rs. 10
Variable cost per unit	Rs. 6
Fixed Costs	Rs. 1,00,000

Due to inflation variable costs increase by 10% while fixed costs increase by 5%. If the break-even quantity is to remain constant by what percentage should the sales price to be raised.

***Solution*:**

Selling price per unit	Rs. 10
Less: Variable cost per unit	Rs. 6
Contribution per unit	4

Break-even quantity $= \frac{1{,}00{,}000}{4} = 25{,}000$ units

New variable cost = 6.6 Rs./unit

New fixed cost = 1,05,000 Rs.

At, BEP, Sales = Total costs

As, BEP is to be unchanged at 25,000 units

Total costs = $25000 \times 6.6 + 1{,}05{,}000 = 2{,}70{,}000$ Rs.

Sales = Total cost = 2,70,000 at BEP

Revised Sales Price $= \frac{270000}{25000} = 10.8$ Rs./unit.

Increase in sales price to maintain same BEP.

$\frac{10.8}{10} - 10 =$ **8** per cent.

REVIEW QUESTIONS

1. Define cost and give the basis of classification of costs?
2. What are direct and indirect costs example? Give an example of each?
3. Give the classification of costs based on behaviour?
4. What do you mean by factory costs and factory overheads?
5. Explain the concept of cost centre and cost unit.
6. Explain the various costs associated with production. Why it is necessary to analyse the production costs?
7. What is BEP? What is its significance?
8. What are the basic assumptions made in BEP?
9. How BEP can be lowered ?
10. Write a note on cost volume profit analysis.

PROBLEMS

1. Asian industries specialise in the manufacture of small capacity motors. The cost structure of the motor is as under

Material	Rs. 50
Labour	Rs. 80
Variable overhead	75% of labour cost.

Fixed overheads of the company amount to 2,40,000 Rs./annum.

The sales price of the motor is Rs. 230 each.

(*a*) Determine the number of motors to be manufactured to break-even.

(*b*) How many motors are to be sold to make a profit of Rs. 1 lakh/unit.

(*c*) If the sales price is reduced by Rs. 15, how many motors are to be sold to break-even.

[**Ans.** (*a*) BEP– 6000 units, (*b*) 8,500 units (*c*) 9,600 units]

2. Menon and Company, makers of quality hand made pipes has experienced steady growth in the sales for the past five years. However, the increased competition has made the company to take up aggressive Advertising. To prepare for next years advertising campaign, the company's accountant has prepared and presented the data for the current year 1996.

Variable expenses	
Direct Material	Rs. 3.25 per piece
Direct Labour	Rs. 8.00 per piece
Variable Overhead	Rs. 2.50 per piece
	Rs. 13.75 per piece
Fixed Expenses	
Manufacturing	Rs. 25,000
Selling	Rs. 40,000
Administrative	Rs. 70,000
Total fixed expenses	Rs. 1,35,000
Selling price per pipe	Rs. 25
Expected sales 1996-97 (2000 units)	Rs. 50,0000

The company has set the target sales of Rs. 5,50,000 for 1997.

(*a*) What is projected profit for 1986?

(*b*) What is the break-even point for 1996?

(*c*) If the additional advertising expense of Rs. 11,250 in 1997 is necessary to attain 1997 sales target with all other expenses remaining constant. What will be profit for 1993 if additional Rs. 11,250 is spent and 22,000 pipes are sold?

(*d*) What will be BEP in sales for 1997 if the additional Rs. 11,250 is spent for advertising?

(*e*) What is the required sales level in Rs. sales to equal 1996 profit if Rs. 11,250 is spent on advertising.

(*f*) A sales level of 22,000 units what is the maximum amount which can be spent on advertising of a profit of Rs. 1,00,000 is desired.

[**Answer:**(*a*) 90,000 (*b*) 12,000 units (*c*) 1,01,250

(*d*) 3,25,000 (*e*) 3,25,000 (*f*) 12,500]

3. From the following data calculate:

(*i*) P/V Ratio

(*ii*) Profit when sales is Rs. 20,000

(*iii*) New BEP if selling price is reduced by 20 per cent

Given: Fixed expenses Rs. 4,000

Beak-even point Rs. 10,000

[**Answer:** (*i*) 40 per cent, (*ii*) 4000 Rs., (*iii*) 16,000 Rs.]

24

MAINTENANCE MANAGEMENT

• Introduction • Maintenance Objectives • Maintenance Costs • Benefits and Limitations of Failure Statistics • Types of Maintenance • Preventive Maintenance System • Preventive versus Breakdown Maintenance • Condition Based Maintenance (CBM) System • Basic Maintenance Decisions • Maintenance Performance Evaluation.

24.1. INTRODUCTION

Maintenance of facilities and equipment in good working condition is essential to achieve specified level of quality and reliability and efficient working. Plant maintenance is an important service function of an efficient production system. It helps in maintaining and increasing the operational efficiency of plant facilities and, thus, contributes to revenue by reducing the operating costs and increasing the effectiveness of production.

24.2. MAINTENANCE OBJECTIVES

Maintenance in any activity is designed to keep the resources in good working condition or restore them to operating status.

The objectives of plant maintenance are :

1. To increase functional reliability of production facilities.
2. To enable product or service quality to be achieved through correctly adjusted, serviced and operated equipment.
3. To maximise the useful life of the equipment.
4. To minimise the total production or operating costs directly attributed to equipment service and repair.
5. To minimise the frequency of interruptions to production by reducing breakdowns.
6. To maximise the production capacity from the given equipment resources.
7. To enhance the safety of manpower.

24.3. MAINTENANCE COSTS

Breakdown of equipment makes the workers and the machines idle resulting in loss of production, delay in schedules and expensive emergency repairs. These downtime costs usually exceed the preventive maintenance costs of inspection, service and scheduled repairs up to the point M shown in Fig. 24.1. Beyond this optimal point an increasingly higher level of preventive maintenance is not economically justified and it is economical to adopt breakdown maintenance policy. The optimal level of maintenance activity M, is easily identified on a theoretical basis, to do this the details of the costs associated with breakdown and preventive maintenance must be known.

Costs associated with maintenance are :

1. Downtime (Idle time cost) cost due to equipment breakdown.
2. Cost of spares or other material used for repairs.
3. Cost of maintenance labour and overheads of maintenance departments.
4. Losses due to inefficient operations of machines.
5. Capital requirements required for replacement of machines.

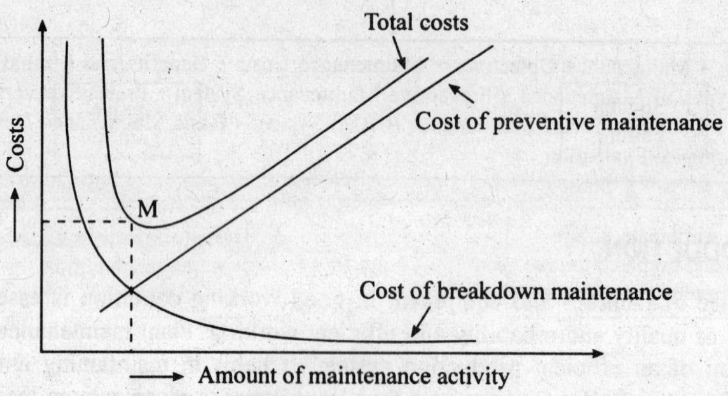

Fig. 24.1. Maintenance costs.

24.4. BENEFITS AND LIMITATIONS OF FAILURE STATISTICS

Benefits :

1. Helps to diagnose failures when they occur

   Depending upon the time to failure, one can immediately identify whether the cause of failure is due to design defect, or a wear out failure or any other reason and helps to take maintenance decisions whether to repair or replace.

2. Provides a valuable information regarding the life and reliability of the equipment to the design engineer.

3. Failure statistics helps to take maintenance policy decisions as to whether to opt for repair or preventive or replacement options.

4. It is important information for spare parts management.

Limitations :

The information or failure data is to be collected over a large population and also for extensive period of time. If the equipment is unique, then the failure statistics is limited. By the time we get a collection of data for meaningful analysis the life of equipment would have been over. This difficulty can be overcome, if the data on similar equipments elsewhere is available.

The failure analysis is represented in the figure 24.2.

The wear out failures are due to,

1. Failures based on inherent quality and reliability characteristics in design, manufac-turing.
2. Manufacturing defects
3. Poorly installed equipment or machinery
4. The chance failures will occur due to operation included failures, damage of equipments due to improper handling, accidents etc.

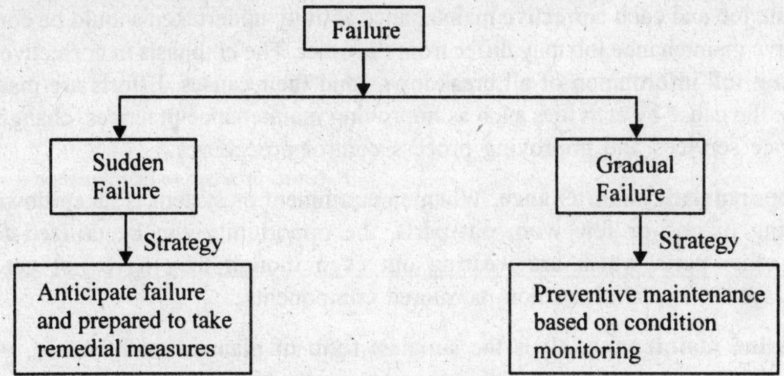

Fig. 24.2. Failure analysis.

24.5. TYPES OF MAINTENANCE

The various forms of maintenance and their relationships are represented in figure 24.3.

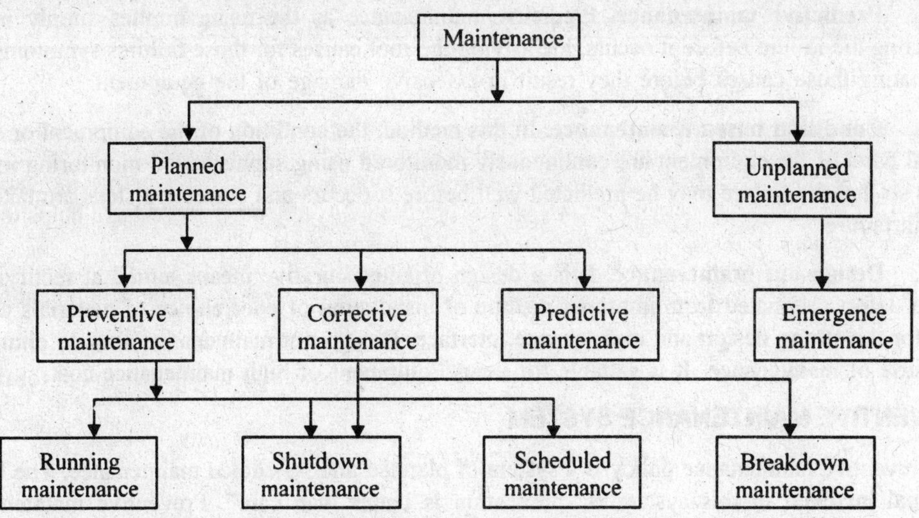

Fig. 24.3. Types of maintenance.

1. Planned maintenance. It is an organized maintenance work carried out as per recorded procedures having control.

2. Breakdown maintenance. It is an emergency based policy in which the plant or equipment is operated until it fails and then it is brought back into running condition by repair. The maintenance staff locate any mechanical, electrical or any other fault to correct it imme-diately.

The beak down maintenance policy may work good and feasible for small factories where,

(*a*) There are few types of equipment.

(*b*) Machines/equipments are simple and does not require any specialist.

(*c*) Where sudden failure does not cause any serious financial loss.

3. Corrective maintenance. This is an organized maintenance work intended to restore a failed unit. It includes different types of actions like typical adjustments to redesign of equipments. It is a one-time job and each corrective maintenance activity undertaken should be completed fully. Each corrective maintenance job may differ from the other. The emphasis in corrective maintenance is on obtaining full information of all breakdowns and their causes. Efforts are made to identify and eliminate the cause by activities such as improving maintenance practices, changing frequency of maintenance services and improving process control procedures.

4. **Opportunistic maintenance.** When an equipment or system is taken down for maintenance/changing of one or few worn out parts, the opportunity can be utilized for changing/maintaining other parts which are wearing out even though they have not yet failed. This maintenance system is useful for non-monitored components.

5. Routine maintenance. It is the simplest form of planned maintenance, which is very essential. Routine maintenance means carrying out minor maintenance jobs at regular intervals. It involves minor jobs such as cleaning, lubrication, inspection and minor adjustments. Routine maintenance needs very little investment in time and money.

6. Preventive maintenance. Preventive maintenance is a planned maintenance of plants and equipments in order to prevent or minimize the breakdown.

7. Predictive maintenance. Predictive maintenance as the name implies simply means predicting the failure before it occurs, identifying the root causes for those failures symptoms and eliminating those causes before they result in extensive damage of the equipment.

8. Condition based maintenance. In this method, the condition of the equipment or some critical parts of the equipment are continuously monitored using sophisticated monitoring instruments so that the failure may be predicted well before it occurs and corrective steps are taken to prevent failure.

9. Design-out maintenance. It is a design oriented curative means aimed at rectifying a design defect originated from improper method of installation or poor choice of materials etc. It calls for the strong design and maintenance interface. Design out maintenance aims to eliminate the cause of maintenance. It is suitable for items/Equipment of high maintenance cost.

PREVENTIVE MAINTENANCE SYSTEM

The preventive maintenance policy is a system of planned and scheduled maintenance. The basic principal involved in this system is "prevention is batter than care". Preventive maintenance includes :

1. Proper identification of all items, their documentation and coding
2. Inspection of plant and equipments at regular intervai (Periodic inspection)
3. Proper cleaning, lubrication of equipments.
4. To upkeep the machine through minor repairs, major overhauls etc.
5. Failure analysis and planning for their elimination.

Preventive maintenance schedules are normally of two categories:

1. Fixed time maintenance (firm) schedule.
2. Condition based maintenance.

Majority of P.M. schedules are of fixed time maintenance. Simple routines or cycles of workloads are established for different days of work, fortnight month or quarter and are regularly followed. Crew carrying out Preventive. Maintenance jobs should normally be separate from crew attending breakdowns. Frequency of. Schedule is not same throughout the life cycle of the equipment. Failure rates are high during initial stage is just after commissioning, failure rates are less during normal working period and failures are again high towards the end of the working cycle that is just before the discard of equipment or taken for major overhauling Preventive. Maintenance. Frequency between these three phases is decided accordingly during the chance failure phase, the inspection, cleaning, lubrication etc. .May have predetermined fixed frequency and interval but the major overhauls and capital repair etc .may same time follow slightly differing frequency and intervals.

24.6. PREVENTIVE VERSUS BREAKDOWN MAINTANANCE

Preventive maintenance is the routine inspection and service activities designed to detect potential failure conditions and make major adjustments or repairs that will be help prevent major operating problems where as, breakdown maintenance is the emergency repair and it involves higher cost of facilities and equipments that have been used until they fail to operate

An effective preventive maintenance programme for equipments requires properly trained personnel, regular inspection and service and should maintain regular records. It is planned in such a way that it will not disturb the normal operations, hence no down time cost of equipment. Breakdown maintenance stops the normal activities and the machine as well as the operators is rendered ideal till the equipment is brought back to normal condition of working

24.7. CONDITION BASED MAINTENANCE (CBM) SYSTEM

Predictive maintenance is more feasible today because of technology that available for equipment surveillance and diagnosis of problems while the machines are still operating. The condition of a machine can be monitored by several means. Sensors may be installed, or periodic readings may be taken with portable units to measure vibration or temperature. Vibration sensors and ultrasonic sensors are used to feed data in to a computer for analysis. Tthe deviation from the normal vibration pattern are recorded when the machine is running properly are analyzed to determine where the problem is developing and when it will become serious. A problem of this type prevents unplanned downtime that disrupts production schedules.

The objectives of condition-based maintenance are :

To detect the failures before they occurs.

To carry out maintenance only when required.

To reduce the maintenance costs and downtime costs.

The methodology of C.B.M. consists of following steps :

1. Proper identification and location of machines/equipments by codification.
2. Selection of critical machines and systems.
3. Identifying components/elements.

4. Fixing condition parameters.
5. Monitoring techniques.
6. Monitoring schedule and frequency.
7. Trend monitoring.
8. Repair schedule and execution.
9. Follow up.

The conditioning monitoring techniques are broadly classified as

- Visual
- Temperature
- Vibration
- Lubricant monitoring
- Leakages monitoring
- Noise and sound monitoring.

Benefits of condition monitoring

- Increased system availability
- Improved plant operation and safety
- Improved maintenance
- Improved product quality

2. Maintenance planning and scheduling

Once the list of the defect is known and the type of the maintenance is decided, the next step is to plan and schedule the maintenance jobs in order to execute the jobs properly and to get the desired results. Planning and scheduling of maintenance jobs/activities are the important components of maintenance functions.

3. Job planning

Maintenance planning basically deals with answering two questions "what " jobs/activities to be done and "How" these activities are done. The various steps in job planning includes –

(*i*) Knowledge about equipment, job, available technique and facilities.

(*ii*) Job investigation at the site – to ascertain the physical access and space limitations, accessing if the available lifting and handling tackles/ facilities are enough or special facilities are required etc.

(*iii*) Development of the repair plan – preparation of a step-by-step procedure, which would accomplish the work with the most economical use of time, manpower and material.

(*iv*) Preparation of list of materials required.

(*v*) Preparation of list of special tools and equipments.

(*vi*) Estimation of the time required to accomplish the job or work.

(*vii*) Provision of necessary safety devices and safety instructions.

3.1. Job manuals

Job manuals are also permanent records about methodology, spares, tools and facilities etc. for all maintenance jobs, which may have to be done for the future.

The following steps are generally involved in preparing the job manuals :

(*i*) List of all major and medium maintenance jobs of the plant are prepared and they are properly coded for identification.

(*ii*) For each coded job, a separate job manual is to be prepared which includes the following :

- Sequence-wise break of the job in to activities with instructions to carry out those jobs
- List of tools, tackles, spares and consumables needed to perform the activity
- List of jigs and fixtures and handling facilities for each activity
- Safety instructions
- Time estimate for the job or job activities

(*iii*) Make available the manuals to all the users and update the manuals as and when needed.

4. Job scheduling

Scheduling of maintenance job basically deals with answering the two questions "who" and "when" of the job that is who is to do the job and when the job is to be started.

Realistic schedules are the functions of realistic thinking, availability of the data and records. The scheduler should obtain knowledge about the following facts before starting his job.

(*i*) Availability of competent manpower- by trade, location, shifts etc.

(*ii*) Availability of equipments, spares and proper tools where the work is to be carried out.

(*iii*) Starting data for the job and past schedules and charts.

Scheduling techniques like bar charts and network techniques are used.

Systematic maintenance. The various elements of systematic maintenance are :

- Codification and Cataloguing
- Equipment/machine history cards and sheets
- Instructions manuals and operating manuals.
- Standard operating practices and work instructions.
- Maintenance time standards (MTS)
- Maintenance — operation liaison
- Maintenance work order and work permit
- Job cards and job cards procedures
- Job execution, monitoring, feed back and control
- Maintenance records and documentation.

24.9. BASIC MAINTENANCE DECISIONS

1. Centralized versus decentralized maintenance
2. Contract versus in-house maintenance.
3. Repair verses replacement

4. Individuals versus group replacement
5. Standby equipments
6. Amount of maintenance capacity
7. Preventive versus predictive maintenance.

24.10. MAINTENANCE PERFORMANCE EVALUATION

The following criteria can be used for measuring the effectiveness of maintenance function.

1. Total plant maintenance productivity (TPMP)

$$\text{TPMP} = \frac{\text{Plant output in the given period}}{\text{Total maintenance cost in the given period}}$$

2. $$\text{Down time index} = \frac{\text{Down time hours}}{\text{Production hours}} \times 100$$

3. $$\text{Maintenance cost index} = \frac{\text{Maintenance cost}}{\text{Capital cost}} \times 100$$

4. $$\text{Equipment utilization} = \frac{\text{Hours the equipment ran at capacity}}{\text{Hours the equipment was available to run at capacity during evaluation period}}$$

5. $$\text{Man power efficiency} = \frac{\text{Total man hour actually worked on the job}}{\text{Total man hour scheduled for these jobs}}$$

References for Further Reading

1. **B. N. Saha,** *'Integrated Maintenance Management — Concept to computerization.'* SBA Publications, New Delhi.
2. **James Riggs,** *'Production Systems Planning, Analysis and Control',* 4th Edition, Willey Eastern Publishing, New Delhi.
3. **Mayer,** *'Production Management',* 4th Edition, McGraw Hill, New York.

25

MAKE OR BUY DECISIONS

• Introduction • Make or buy decision when? • Factors influencing make or buy decisions • Functional aspects of make or buy decision • Economic and non-economic factors influencing make or buy decision.

25.1. INTRODUCTION

The make or buy decision refers to the problem encountered by an organisation when deciding whether a product or service should be purchased from outside sources or manufactured internally. Theoretically, every item which is currently purchased from an outside supplier is always a candidate for internal manufacture and every item currently manufactured in house is a potential candidate for purchase.

The majority of the make or buy decisions are made on the basis of price. But this is only one of the criteria which is to be evaluated in this strategic decision. Many non-cost factors encourage long-term contracts with the suppliers to aid in the achievement of production and quality levels and encourage investments in appropriate resources and new ideas. This results in excellent, mutually beneficial customer-supplier relationships developed over long periods based on trust and achievement of common objectives. Most of the make or buy decisions are complex, time consuming and affect many parts of the organisation. Senior management involvement is required in a number of the stages of this strategic decision.

25.2. MAKE OR BUY DECISION WHEN?

The following situations demand for the evaluation of make or buy decisions:

1. When the organisation introduces new products.
2. The fluctuating demand for the company's products.
3. When the organisation carries out value analysis or cost reduction programs.
4. Deteriorating quality and delivery commitment of the supplier if presently the item is bought.
5. The scarcity of funds for investment in additional plant and equipment.

25.3. FACTORS INFLUENCING MAKE OR BUY DECISION

1. **Volume of production:** The quantity or volume of production affects the make or buy decision to the greater extent. If the volume of production is high, it favours the make decision and low volume favours buy decisions.

2. **Cost analysis:** The cost analysis refers to the determination of costs to make an item as well as the cost to buy it. The cost to make include—the material cost, direct labour cost, setup and tooling up costs, depreciation, administrative overheads, interest, insurance, taxes and inventory carrying costs of raw materials and work-in-process. The cost to make also includes the

appropriate allowances, spoilage of work or scrap, and the risk associated with doing business. The cost to buy an item should include—purchase price of the item or component, transportation cost, sales tax and octroi, procurement cost, carrying cost, receiving and incoming inspection costs. The analysis of these two costs helps take decision whether to make or buy.

3. **Utilisation of production capacity:** The organisation which has created large production capacity favors the decision to make.

4. **Integration of production system:** The vertical integration favours the make decision where as horizontal integration favours buy decision.

5. **Availability of manpower:** Availability of skilled and competent manpower favours make decision whereas scarce manpower prefers buy decision.

6. **Secrecy or protection of patent, right:** This condition favours the make decision.

7. **Fixed cost:** A lower fixed cost favours the decision to make and higher fixed cost the buy decision.

8. **Availability of competent suppliers or vendors.**

9. **Quality and reliability of vendors.**

25.4. FUNCTIONAL ASPECTS OF MAKE OR BUY DECISION

Make or buy decision should be viewed with both long-term and short-term perspectives in mind. Some of the effects are tangible and other are intangible.

These are classified as follows:

- Financial Aspects
- Technological Aspects
- Marketing Aspects
- Purchasing Aspects
- Strategic Aspects

1. **Financial aspects:** The make decision always demands an investment in plant, machineries and equipment. The investments can be categorised into fixed cost and variable cost. The buy decision is associated with only variable cost. A thorough and comparative analysis is carried out by expressing all factors into money terms. Then the decision is to be taken based on which one is more economical, to make or to buy.

2. **Technological aspects:** The make or buy decision is influenced by

   (*a*) The access to the latest technology to the organisation.

   (*b*) Feasibility and terms and conditions of technology transfer.

   (*c*) Outdating of technology.

   (*d*) Product life cycle.

3. **Marketing aspects:** The marketing aspects have the influence on make or buy decision. When there is a fierce competition, an organisation tries to enhance the quality and cut down the costs. The make decision assures the quality and reliability of the parts. Under the situation of increasing market share and a good future sales potential company can have an additional investment potential and, hence, can opt for make decision. When there is a doubt about the market potential, the company should opt for buy decision. The large organisations pay greater attention to quality which favours the make decision to maintain quality and reliability of items.

4. **Purchasing aspects:** The decision is influenced by.

   - The availability of items or components in sufficient quantities.
   - The delivery commitments must be reliably met.
   - The acceptable quality and price level of the product.
   - Economy in transportation from the source to the organisation.
   - Competence and reliability of vendors.

5. **Strategic aspects:** Any decision that is to be taken including make or buy decision should

be taken with the consideration to overall objective of the organisation. Due importance should be given to the economy, secrecy and flexibility in taking decision regarding make or buy.

6. **Intangible aspects:** Intangible aspects like environmental factors, labour union acceptance, goodwill, support to ancillarisation and growth of SSI units, technical assistance to vendors also influence the make or buy decision.

25.5. ECONOMIC AND NON-ECONOMIC FACTORS INFLUENCING MAKE OR BUY DECISIONS

Decisions regarding whether to make or buy the components involve both economic and non-economic factors. Economically, an item or component is a candidate for in house production, if the company has sufficient capacity and if the components value is high enough to cover the variable costs of production and make some contribution towards the fixed cost. Low volumes favour buying which incurs very little or no fixed costs. Fig. 25.1 shows the relationship between cost factors.

The non-economic factors are:

- Availability of infrastructure and skilled personnel
- Desire for alternate sources of supply
- Employee preferences and stability concerns
- Need to maintain trade secrets
- Desire to expand into new product line
- Forward or backward integration
- Long lasting and mutually rewarding relationships with vendors

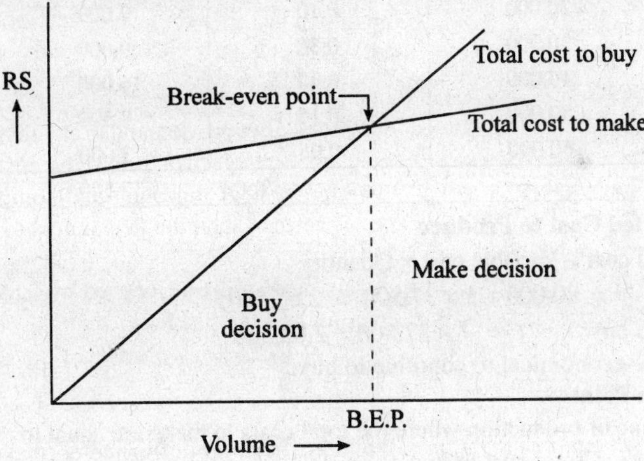

Fig. 25.1.

Problem 1: Demand for the component is at the rate of 6000 per year and this demand is going to continue for next three years. The company has two options. It can get the component manufactured from outside or it can manufacture in house. It costs the company Rs. 2.8 per unit to buy the component. The in-house manufacture will incur a fixed cost to the extent of Rs. 10,000 and variable cost of Rs. 1.5 per unit. Give the decision rule for make or buy.

Solution: Let x represent the number of units.

The total cost incurred in buying the component is

$$\text{Total cost } (Y) = 2.8\,x \quad \text{... (1)}$$

Total cost incurred in making the component in-house

$$\text{Total cost } (Y) = 10{,}000 + 1.5\,x \quad \text{... (2)}$$

At BEP, both alternatives result in equal total cost

∴ At BEP, equation (1) = equation (2)

$$2.8\,x = 10{,}000 + 1.5\,x$$
$$1.3\,x = 10{,}000$$
$$x = 7692.3 \text{ say } 7693$$

The decision rules are:

1. If the quantity is 7693, both make and buy are equally economical (results in same costs)
2. Quantity less than 7693, it is economical to buy
3. Quantity more than 7693, it is economical to manufacture

***Problem 2*:** The ABC company is investigating the decision whether to make or buy a plastic packaging which is currently being purchased at Rs. 7 each. The demand estimates are shown below.

Demand (Units)	20,000	30,000	40,000	50,000	60,000
Chance (%)	10	30	40	15	5

The decision to manufacture in-house costs the company an annual fixed cost of Rs. 80,000 towards renovation and conditioning variable costs are estimated at Rs. 5 per unit.

Give your decision whether to make or buy. At what quantity it is profitable to produce rather than buy.

***Solution*:** The expected demand (or volume) is determined treating the percentage change as a probability.

Demand (D)	*Probability (P)*	*D.P.(D)*
20,000	0.10	2,000
30,000	0.30	9,000
40,000	0.40	16,000
50,000	0.15	7,500
60,000	0.05	3,000
	Total	37,500

(*i*) The Expected Cost to Produce

$$TC = \text{Fixed cost} + \text{Variable cost} \times \text{Quantity}$$
$$= 80{,}000 + 5 \times 37{,}500 = 2{,}67{,}500$$

Expected cost to buy $TC = 7 \times 37{,}500 = 2{,}62{,}500$

The decision is—economical to continue to buy.

(*ii*) Break-Even Point

BEP is the volume of production where the total costs to make are equal to the total costs to buy.

Total cost to make = Total cost to buy

$$\therefore \quad 80{,}000 + 5\,x = 7\,x$$
$$\therefore \quad x = 40{,}000$$

It is economical to produce, if the quantity is > 40,000 units.

***Problem 3*:** An item which is required by the company can be manufactured on any of the three following machines and also it can be purchased at a price or Rs. 1.2 per component:

Machine	*Fixed Cost (Rs.)*	*Variable Cost (Rs./Unit)*
M_1	9,000	0.75
M_2	3,500	0.50
M_3	92,000	0.05

suggest the best option if the requirement is 2,000 units and give the decision rules.

***Solution*:** Let x be the quantity required. The cost equations for 4 options are:

Buy	$TC_B = 1.2\,x$	...(1)
M_1	$TC_{M1} = 9{,}000 + 0.75\,x$	...(2)
M_2	$TC_{M2} = 3{,}500 + 0.50\,x$	...(3)
M_3	$TC_{M3} = 92{,}000 + 0.05\,x$	...(4)

The relationship between cost and volume is established. The total costs for all the four alternatives are determined at various quantities (volume).

The cost are tabulated below

Alternative	*Buy*	*Machine*		
Volume (x)	*(TC_B)*	*TCM_1*	*TCM_2*	*TCM_3*
10,000	12,000	16,500	40,000	97,000
20,000	24,000	24,000	45,000	93,000
50,000	60,000	46,500	60,000	94,500
60,000	72,000	53,000	65,000	95,000
80,000	96,000	69,000	75,000	96,000
1,00,000	1,20,000	84,000	85,000	97,000
1,10,000	1,32,000	91,000	90,000	97,500
1,20,000	1,44,000	99,000	95,000	98,000
1,30,000	1,56,000	1,06,500	1,00,000	98,500
1,50,000	1,80,000	1,21,000	1,10,000	99,500

Plot the cost *Vs* quantity on a graph

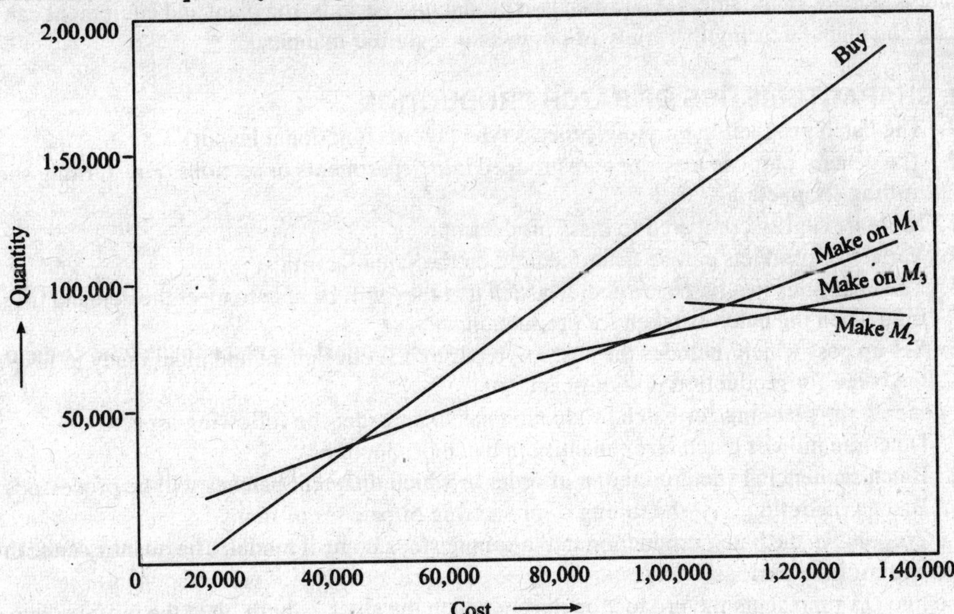

In the particular case

The requirement is to the extent of 1,20,000.

The best option is to make on M_2 which results in lowest cost.

26

PLANNING AND CONTROL OF BATCH PRODUCTION

• Introduction • Characteristics of batch production • Determination of batch size • Criteria for selection of batch size • Production range • Minimum cost batch size • Maximum profit batch size • Sequencing and scheduling of batches • Line of Balance (LOB) technique.

26.1. INTRODUCTION

The Batch production is recommended when the variety of products are to be manufactured and the quantity is not high enough to demand the dedicated product line. Batch production is necessary when the demand for the product is limited or when the rate of production is high compared to rate of consumption. The product is, therefore, manufactured periodically in a quantity that will be sufficient to satisfy the demand for some time until manufacture of this product is started. In the interval between two production periods, the plant and equipment can be used for the manufacturing of variety of products in a similar manner.

26.2. CHARACTERISTICS OF BATCH PRODUCTION

1. The batch production employs process type layout (functional layout).
2. The general purpose machines are grouped into departments or sections (*e.g.*, turning shop, milling shop, etc.).
3. More flexibility compared to mass production.
4. Variety of products can be manufactured on the same facilities.
5. The quantities in a batch are decided such that they will be able to meet the demand till the time when the batch is taken for production.
6. Set-up cost which includes the cost of preparing the equipment and plant ready to take up the batch for production is significant.

In general, the planning for batch production has to consider the following aspects:

1. Determination of batch size (quantity in batch production).
2. Batch sequencing (determination of order in which different batches will be processed).
3. Batch scheduling, *i.e.*, the timing of processing of batches of items.

Fig. 26.1 shows the batch production in the simple stock control model. The quantity (stock) is plotted against time periods.

The line *OA* represents the production during which the stock is built up at the uniform rate of p units per unit time. After the production period P, the production stops when the stock level reaches Q_1. The line *AB* represents the variation in stock level due to consumption at the uniform rate of d units per unit time. At point B, the stock level becomes zero and production starts. This cycle will be repeated. During the consumption period the plants and machineries can be utilised for manufacture of other products.

From the $\Delta\, OAB$,

$$Q_1 = P.\, T_p$$

and

$$Q_1 = d.T_c$$

Hence

$$\frac{T_p}{T_c} = \frac{d}{p} = \gamma$$

γ represents the ratio of the production time to consumption time. Higher value of γ represents higher production time.

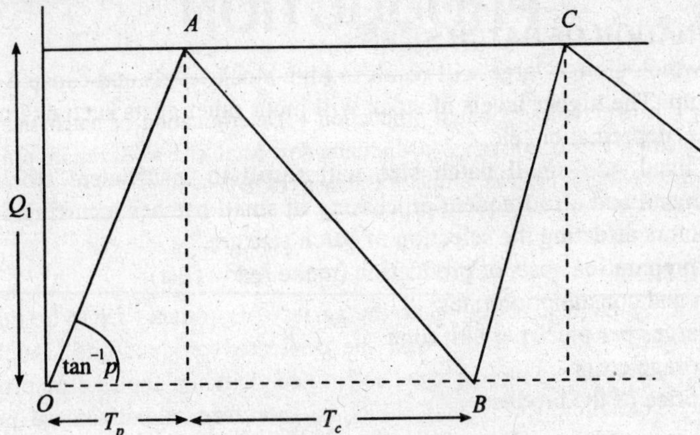

Fig. 26.1: Simple stock control model.

This is an ideal condition. An unrealistic assumption is made namely, that there is no consumption during production period. But the consumption will be continuous and consumption takes place during production period also. Secondly, in order to ensure that the stock level will never become zero, an indication should be given well in advance to start the production that will allow for building up of stock to a predetermined level at the completion of production period. The Fig, 26.2 shows the change in stock pattern when these facts are considered.

The production should be started at point B'' in order to accumulate the stock requirement at point C. Q_o represents the safety stock. The actual preparation for production will start before the point B'', such that the set-up of machines will be ready at point B''. The net increase in stock level is given by the difference between the production and consumption rates.

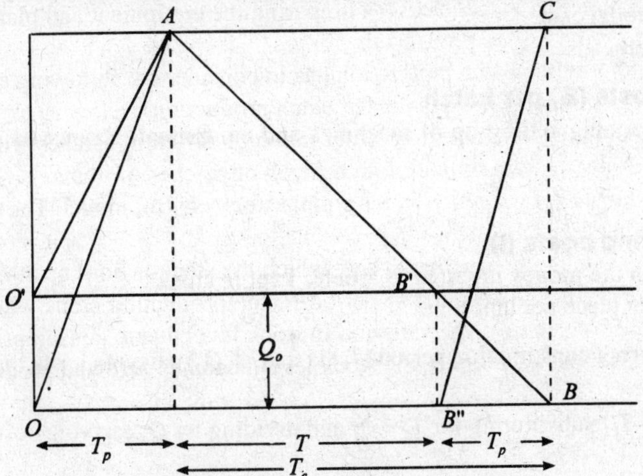

Fig 26.2: Stock control with safety stock.

i.e., the rate of stock built up $= p - d$

The time interval T can be utilised for production of other batches of products

i.e., $$T = T_c - T_p - T_s \quad (1)$$

where T_s is the set-up time required for preparing machines for production.

The safety stock Q_o can be determined from $\Delta B' B'' B$ as,

$$Q_o = T_p . d \quad (2)$$

or $$\frac{Q_o}{Q} = \frac{d}{p} = \gamma \quad (3)$$

Equation (3) is obtained by dividing equation (2) by the equation $Q = p.\ T_p$

26.3. DETERMINATION OF BATCH SIZES

Batch quantities which are too large will result in high stock levels and cause a large amount of capital to be tied up. The higher levels of stock will incur other costs such as inventory or stock keeping, insurance, depreciation, etc.

On the other hand, too small batch size will result in insufficient stock level to meet fluctuations in demand and also frequent processing of small batches incurring set up costs each time. The main factors affecting the selection of batch size are,

1. Setup and preparation costs of production (make ready cost)
2. Production and consumption rates
3. Interest charges per piece per unit time
4. Average storage costs
5. Unit sales price of the product.

26.4. CRITERIA FOR SELECTION OF BATCH SIZES

- Minimum costs per piece
- Maximum profits per batch
- Maximum return batch (Maximum ratio of profit to cost of production)
- Maximum rate of return per unit time

26.5. MINIMUM COST BATCH SIZE

The production cost per piece is expressed as

1. Constant cost per piece (C)

$$C = L + M + O$$

where L = Labour cost

M = Material cost

O = Overheads

2. Preparation costs (S) per batch

This includes planning setting up of machines and equipment. Expressed as set-up cost per piece $\frac{S}{Q}$.

3. Interest Carrying costs (I)

Interest is paid on the money invested in articles kept in stock

$I = i\,C$ per piece per unit time.

Average stock during consumption period (T_c) is $\frac{1}{2}(Q + Q_o)$ carrying costs for the stock for this period are; $I \frac{Q + Q_o}{2} T_c$. Substituting for $T_c = \frac{Q}{d}$ and dividing by Q, carrying cost per piece are,

$$\frac{I}{2d}Q(1+\gamma)$$

2. Storage, carrying costs

If the average storage costs per piece are B per unit time. For the consumption period T_c, they are,

$$BT_c \quad \text{or} \quad \frac{BQ}{d}$$

These costs are represented as in Fig. 26.3.

The total cost per piece (T_c)

$$T_c = C + \frac{S}{Q} + \frac{Q}{2d}[I(1+\gamma) + 2B]$$

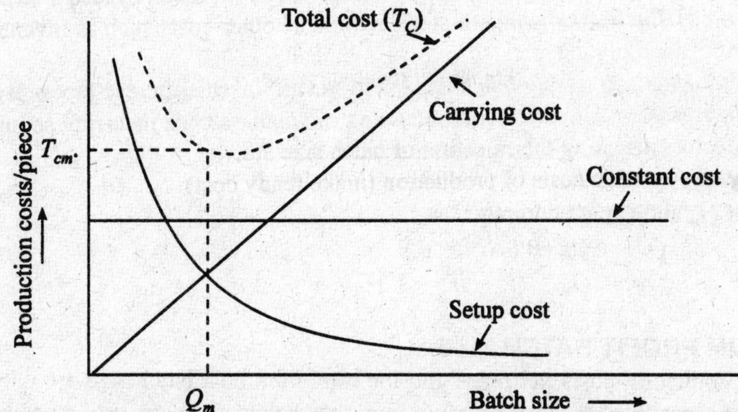

Fig 26.3: Relationship between production costs *vs* quantity.

Differentiating total cost (T_c) equation (1) with respect to Q and putting $\frac{dT_c}{dQ} = 0$,

we get $$Q_m = \sqrt{\frac{2dS}{I(1+\gamma)+2B}} \qquad Q_m = \text{Economic batch size.}$$

at Q_m, carrying costs per piece and preparation cost per piece are equal, *i.e.*,

$$\frac{S}{Q_m} = \frac{I(1+\gamma)+2B}{2d} \cdot Q_m$$

Minimum total cost per piece (T_{cm})

$$T_{cm} = C + \left[\frac{I(1+\gamma)+2B}{2d}\right] Q_m$$

26.6. PRODUCTION RANGE

An increase in total production costs, above minimum total costs result into two batches on the either side of Q_m. If the deviation in cost is allowed, then any batch size which lies between two limits of Q_m without causing an increase in production costs above a predetermined value is determined. The range between the two limits is called "Production range". This is shown in Fig. 26.4.

To determine the value of the two limits of production range QA and QB, a non-dimensional ratio P is defined as

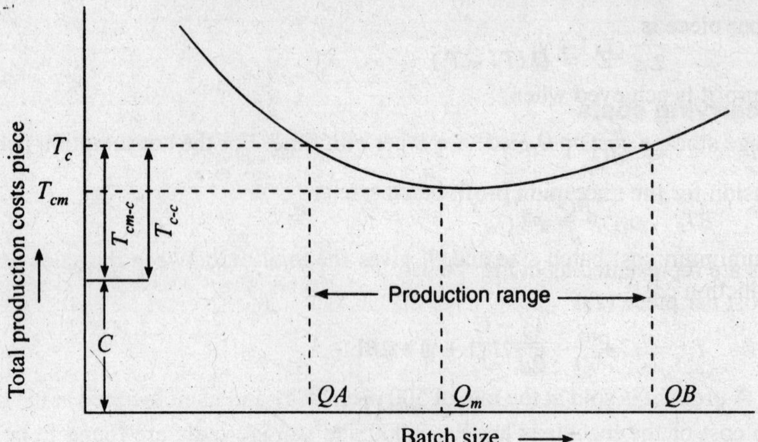

Fig. 26.4: Production range.

$$P = \frac{T_c - C}{T_{cm} - C} = \frac{\text{Variable cost}}{\text{Minimum variable cost}}$$

The values of Q_A and Q_B are computed as

$$Q_A = Q_m (P - \sqrt{P^2 - 1})$$

$$Q_B = Q_m (P + \sqrt{P^2 - 1})$$

26.7. MAXIMUM PROFIT BATCH SIZE

The minimum production costs per piece and the minimum cost batch size are unrelated to the sales price of the product. In case of minimum cost batch size, the profit per piece will be maximum but the profit for the whole batch is dependent upon number of pieces in a batch. It is clear that, when the batch size is larger than Q_m is produced, the total profit will increase. There is a limit on increase in batch size as an increase in batch size corresponds to rise in production costs, higher carrying costs. The maximum profit batch is shown in Fig. 26.5.

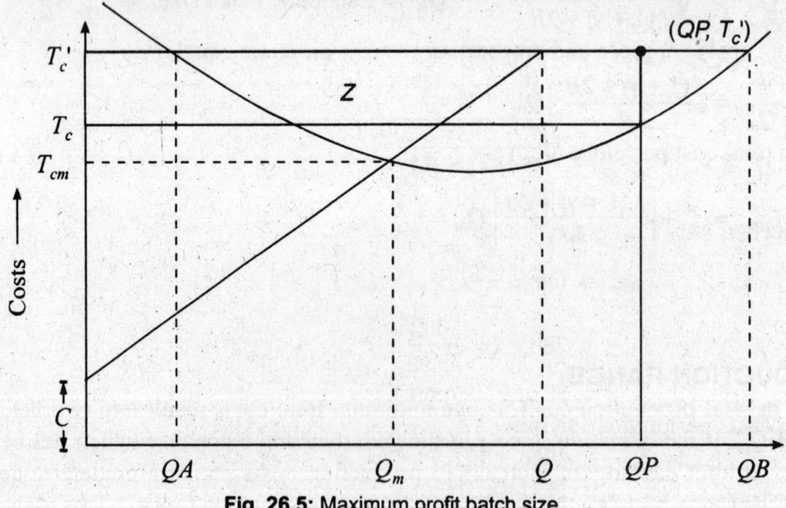

Fig. 26.5: Maximum profit batch size.

Profit can be expressed in terms of batch size. If Q pieces are produced at the cost of T_c per piece and if sales price is T_c'.

The profit per piece is

$$Z = Q\,(T_c' - T_c)$$

Maximum profit is achieved when,

$$\frac{dZ}{dQ} = 0$$

The expression for the maximum profit batch size is

$$QP = P'\,Q_m$$

where Q_m is minimum cost batch size and P' gives the relation between the sales price and the minimum production costs.

$$P' = \frac{T_c' - C}{T_{cm} - C}$$

***Problem 1*:** A product is sold at the rate of 500 pieces/day and manufactured at the rate of 2500 per day. Setup cost of the machines are Rs. 3,000 and storage costs are found to be 4.5×10^{-3} Rs./unit/day. Labour, material and overhead charges are Rs. 95, Rs. 65 and Rs. 120 respectively. If the interest charges are 16 per sent. Find the minimum cost batch size and cost of production run.

***Solution*:** Set-up cost (S) = Rs. 3,000

Inventory carrying cost (B) = 0.0045 Rs./unit/day

Demand rate (d) = 500 pieces/day

Production rate (p) = 2,500 pieces/day

$$\gamma = \frac{\text{Production rate } (d)}{\text{Demand rate } (p)} = \frac{500}{2500} = 0.2$$

Constant costs per piece (C) $= L + M + O$

$= 95 + 65 + 120$

$=$ Rs. 280

Interest charges per day (assuming 300 working days)

$$i = \frac{16}{100} \times \frac{1}{3000} = 5.3 \times 10^{-4}/\text{day}$$

$$I = i\,C = 5.3 \times 10^{-4} \times 280$$

$$= 0.149 \approx 0.15 \text{ Rs./piece/day}$$

Minimum cost batch size

$$Q_m = \sqrt{\frac{2\,d\,S}{I\,(1+\gamma) + 2\,B}}$$

$$= \sqrt{\frac{2 \times 500 \times 300}{0.15\,(1+0.2) + 2 \times 0.0045}}$$

$$= 3984.$$

Production costs per (T_{cm}) piece

$$T_{cm} = C + 2\frac{S}{Q_m}$$

$$= 280 + 2 \times \frac{3000}{3984}$$

$$= 281.50 \text{ Rs./unit}$$

Production costs per production run

$$T_{cm} \times Q_m = 3984 \times 281.5$$

$$= 1121496 \text{ Rs.}$$

***Problem 2*:** Machine components supplied to the assembly shop are produced in a plant at the rate of 100 pieces/day. A cost analysis showed that the constant costs are 150 Rs./unit and storage costs are 0.015 Rs./piece/day. If the preparation and machine set-up costs are Rs. 1,500 and assembly requires 40 such pieces per day. Find the minimum cost batch size and length of the production run. Estimate the production cost per production run. Assume interest rate of 18 per cent.

***Solution*:** Constant costs per piece (C) = Rs. 150
Set-up cost (S) = Rs. 1500
Storage cost (B) = Rs. 0.015 /unit/day
Demand rate (d) = 40 pieces/day
Production rate (p) = 100 pieces/day

Interest charges per day @ 18 per cent assuming 300 working days in a year

$$Q_m = \frac{18}{100} \times \frac{1}{300} = 6 \times 10^{-4} \text{ Rs./day}$$

$$I = i\,C = 6 \times 10^{-4} \times 150$$
$$= 0.09 \text{ Rs./piece/day}$$

(*I*) Minimum cost batch size (Q_m)

$$Q_m = \sqrt{\frac{2\,d\,S}{I\,(1+\gamma) + 2B}}$$
$$= \sqrt{\frac{2 \times 40 \times 1500}{0.09\,(1+0.4) + 2 \times 0.15}}$$
$$= 877$$

(*II*) Production period (T_p)

$$T_p = \frac{Q_m}{P} = \frac{877}{100} = 8.77 \text{ days}$$

(*III*) Production costs per run

$$\text{Total minimum cost per piece } (T_{cm}) = C + 2\frac{S}{Q_m}$$
$$= 150 + 2\frac{1500}{877}$$
$$= 153.42 \text{ Rs./unit}$$

$$\text{Production cost per run} = Q_m \times T_{cm}$$
$$= 877 \times 15.42$$
$$= 134550 \text{ Rs.}$$

***Problem 3*:** The following information regarding the batch production is given below, using the data determine the production range.

Set-up cost (S) = Rs. 3000
Carrying cost factor including interest factor (K) = 7.5×10^{-3} per unit per day
Constant costs per piece (C) = Rs. 50
Allowable increase in total costs per piece = 2.5 per cent

***Solution*:** The minimum cost batch size $= \sqrt{\frac{S}{K}}$

$$= \sqrt{\frac{3000}{7.5 \times 10^{-3}}}$$
$$= 632.4$$
$$\approx 633 \text{ units}$$

The ratio $U = \frac{C}{S/Q_m} = \frac{50}{3000/633} = 10.55$

The factor P can be found from the equation.

The increase in costs due to deviation from Q_m is given by

$$Z = \frac{\Delta T_c}{T_{cm}} = \frac{P-1}{\frac{1}{2}U+1}$$

Z is given as 2.5% substituting

$$0.025 = \frac{P-1}{\frac{1}{2} \times 10.5 + 1}$$

$$P = 1.156$$

$$\left.\begin{cases} Q_1 = 633\,(1.156 - \sqrt{(1.156)^2 - 1}) = 365 \\ Q_2 = 633\,(1.156 + \sqrt{(1.156)^2 - 1}) = 1262 \end{cases}\right\} \text{Production range}$$

26.8. SEQUENCING AND SCHEDULING FOR BATCH PRODUCTION

Once the optimal size of the batch for each item which is to be processed is determined the next step is to consider the sequencing. Sequencing can be done using priority sequencing rules and algorithms.

Consider a situation in which two items (A and B) are to be processed successively on the same equipment. The processing of batches is shown in Fig. 26.6.

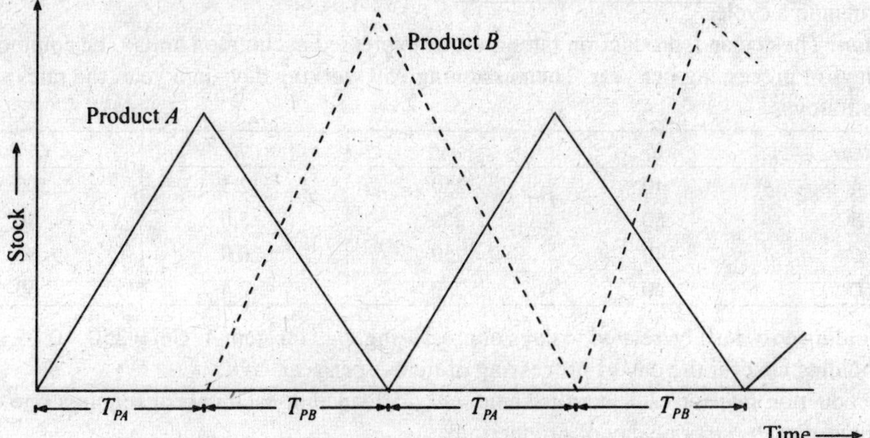

Fig. 26.6: Successive processing of batches of product A and B.

The order in which (sequence) different products A and B are processed will be determined by either process itself or set-up or preparation requirements.

The problem is to find out the most economical cycle, *i.e.*, the one that minimises the set-up or inventory cost. The setup cost increase and inventory cost decreases as the number of cycles increases.

When a number of items are to be processed successively the total number of cycles per unit time is given by

$$N = \sqrt{\frac{\sum\limits_i Ch_i \times d_i \left(1 - \frac{d_i}{p_i}\right)}{2\sum\limits_i CS_i}}$$

where N = Number of complete processing cycles, each consisting of the processing of a batch item.

d_i = Consumption rate for item i

p_i = Production rate for item i

Ch_i = Holding cost per unit of time for item i
CS_i = Set-up for batch item i
Since, $Q_i = d_i / N$

$$Q_i = \sqrt{\frac{2\, d_i \sum_i CS_i}{\sum_i Ch_i\, d_i \left(1 - \frac{d_i}{p_i}\right)}}$$

Problem 4: Four items A, B, C and D are to be processed successively in batches on the same facility. The demand and production rates, set-up costs and holding costs for each product are given below:

Item	*Demand per year*	*Production rate per day*	*Holding cost/unit/annum*	*Set-up cost per batch*
A	10,000	250	0.05	100
B	5,000	100	0.05	50
C	8,000	200	0.10	80
D	12,000	300	0.05	60

Determine the number of cycles per annum and the quantity in batches and the production times for each item in a cycle.

Solution: The demand, production rates must be expressed in common units. The common unit used is days of processing per year. Then assuming 250 working days in a year, the rates will be shown as follows:

Item	d_i	p_i	Ch_i	C_s
A	40	250	12.5	100
B	50	250	5.0	50
C	40	250	20.0	80
D	40	250	2.4	60

The holding cost must be related to days of processing, *i.e.*, for item A, $Ch_i = 250 \times 0.05 = 12.5$, *i.e.*, the holding cost for the day of processing of item A per year.

The production rate for each item is same, *i.e.*, 250 since in each case it requires one day to process unit of each (*i.e.*, one day of production)

Item (1)	d_i/p_i (2)	$1 - d_i/p_i$ (3)	$Ch_i.d_i$ (4)	$Ch_i.d_i\ (\ - d_i/p_i)$ (5)	CS_i (6)
A	0.160	0.84	500	420	100
B	0.200	0.80	250	200	50
C	0.160	0.84	800	672	80
D	0.160	0.84	900	806	60
				Σ (5) = 2098	Σ (6) = 290

$$N = \sqrt{\frac{2098}{2 \times 290}} \text{ cycles per year}$$

For minimum cost, 1.9 complete runs per year should be made. Each complete cycle will consist of 4 batches as follows:

$$QA = \frac{\text{Consumption}}{N} = \frac{10{,}000}{1.9} = 5{,}250 \text{ items}$$

$$QB = \frac{5{,}000}{1.9} = 2{,}630 \text{ items}$$

$$QC = \frac{8,000}{1.9} = 4,220 \text{ items}$$

$$QD = \frac{12,000}{1.9} + 6,300 \text{ items}$$

Each complete run lasting

$$TA = \frac{5,250}{250} = 21 \text{ days}$$

$$TB = \frac{2,630}{100} = 26.3 \text{ days}$$

$$TC = \frac{4,220}{200} + 21.1 \text{ day}$$

$$TD = \frac{6,300}{300} = 21 \text{ days}$$

$$\text{Total} = 89.4 \text{ days}$$

The production rate for each product is that total annual consumption or, demand can be satisfied by substantially less than one year's processing. Unless consumption is increased or additional items are manufactured, the equipment will spend sometime idle in a year.

26.9. LINE OF BALANCE (L-O-B) TECHNIQUE

Line of balance technique (L-O-B) was developed before the net work technique. This method was developed by the Navy during World War II. It is a method of production control which combines the features from a Gantt chart and CPM time chart with graphs of material requirement. It is most appropriate for assembly operations involving a number of distinct components. LOB can be used to schedule and control of a single batch (of the same job). The following requirements should be satisfied,

1. There must be an identifiable stages in production at which managerial control can be exerted.
2. The manufacturing times between these stages must be known.
3. A delivery schedule must be available.
4. Resources can be varied as required.

The application of line of balance technique consist of the following stages:

1. A graphic representation of the delivery objective.
2. Activity on arrow (AOA) diagram to show the logical sequence and timing of activities of production.
3. Progress chart of the current status of component completion.
4. A line of balance drawn to show the relationship of component progress to the output needed.

LOB diagram is a representation of the state of the production process at a specified time. For a particular review period it represents the progress of all the activities involved in the production and indicates if the delivery schedule can be met. For each individual activity, it shows the current level of production against the planned production. LOB helps the management to identify the activities which are out of balance and to take the corrective action.

Steps in Line of Balance Technique

1. Construct an AOA (Activity On Arrow) diagram to represent the logic in the tasks which make up the manufacture of the product.
2. Carry out a reverse forward pass starting at time zero with the final node, to calculate the equivalent week number.

3. Rank the activities in descending order of equivalent week number, prepare an LOB diagram – a time scaled network.
4. Prepare a calendar and accumulated delivery quantity table.
5. From the calendar and accumulated delivery quantity table.
6. From the calendar and LOB diagram, prepare an LOB table showing the quantity of each activity which should be completed by each week number. For each week number this gives an LOB chart, and over the life of the project at LOB table.
7. The progress at any point in time can be recorded on the LOB chart and LOB table.

References for Further Reading

1. James L. Riggs, *Production Systems, Planning, Analysis and Control*, 4th edition John Wiley and Sons, New York (1987).
2. Samuel Eilon, *Production Planning and Control.*
3. Dervisiotis Kostas N., *Operations Management*, McGraw Hill, New York (1981).

PART III
ADVANCED TOPICS IN PRODUCTION MANAGEMENT

27

APPLICATION OF LINEAR PROGRAMMING TECHNIQUE IN PRODUCTION MANAGEMENT

• Introduction • Product mix decisions • Standard form of L.P.P. • Formulation of L.P. problem • Graphical method for solving L.P.P. • Simplex procedure • Big M. method.

27.1. INTRODUCTION

In practice, linear programming is one of the powerful technique for managerial decision-making. The application of this technique has helped to solve many complex problems which otherwise are more difficult to solve. The specific problems where this technique can be successfully applied are:

- Production scheduling
- Product mix decisions
- Capital budgeting
- Plant location
- Resource allocation and optimal utilisation of resources

27.2. PRODUCT MIX DECISIONS

Product mix decision is an important planning activity.

The optimal product mix selection has far reaching implications and it contributes to the survival and growth of an industry. There are many constraints under which the product mix decision is to be made, *e.g.*, cost, capacity and other constraints.

Linear programming problem (LPP) is a technique useful to select the optimum product mix.

Linear programming is a quantitative technique and the relationships between variables in the problem is linear, hence, the name linear programming.

27.3. STANDARD FORM OF LINEAR PROGRAMMING PROBLEM

Let $x_1\ x_2\ x_3\ ....\ x_n$ are the decision variables. Optimise (maximise or minimise)

$$Z = c_1x_1 + c_2x_2 + ... + c_nx_n \qquad \text{(objective function)}$$

Subject to the constraints:

$$a_{11} \cdot x_1 + a_{12} \cdot x_2 + ... + a_{1n} \cdot x_n \le b_1$$
$$a_{21} \cdot x_1 + a_{22} \cdot x_2 + ... + a_{2n} \cdot x_n \le b_2$$
$$am_1 \cdot x_1 + am_2 \cdot x_2 + ... + am_n \cdot x_n \le b_n$$

$x_1, x_2\ ......\ x_n \ge 0$ (non-negative restriction)

where, c_1, c_2 ..... c_n are cost or profit coefficients.

a_{ij} ($i = 1, 2, ... n, j = 1, 2 ..... n$) are structural coefficients b_1, b_2 ..... b_n are called requirements or availability.

The LPP can be solved by two methods

(*i*) **Graphical method:** It is used only when two decision variables are involved. This is more simple.

(*ii*) **Simplex method:** This is useful for any number of decision variables in the problem and there are number of constraints on the problem.

27.4. FORMULATION OF L.P. PROBLEM

Steps involved in formulations:

1. From the given problem, identify the key decision to be made.
2. Identify the decision variables . . . whose values give the solution to the problem.
3. Write the objective in the quantitative terms, and express it as a function of linear variables.
4. Study the constraints and express them as a linear equations.

Problem 1: A firm can produce three types of cloth say *A*, *B*, and *C*. Three kinds wool are required for it say red wool, green wool and blue wool. One unit length of type *A* cloth needs 2 yards of red wool and 3 yards of red wool. One unit length of type *B* cloth needs 3 yards of red wool, 2 yards of green wool and 2 yards of blue wool and one unit length of type *C* cloth needs 5 yards of green and 4 yards of blue wool. The company has a stock of only 8 yards of red, 10 yards green and 15 yards blue wool. The profit from sale of one unit length of type *A* is Rs. 10 type *B* is Rs. 8 and type *C* is Rs. 5.

Determine how the firm should use the available material so as to maximise the profit.

Formulate this as a L.P. problem.

Solution: (*i*) The key decision here is to find the units of three type of clothes *A, B* and *C* to be produced by the company.

Let x_1, x_2 and x_3 be the number of units of cloth of type *A*, type *B* and type *C* to be produced respectively.

(*ii*) The objective is to maximise the profit by selling three types of clothes.

The profit equation is written as $z = 10\,x_1 + 8\,x_2 + 5\,x_3$

Here the co-efficient represent the contributions towards the profit equations.

(*iii*) Requirements and availability of wool.

Requirement of wool	*Clothes* *A*	*B*	*C*	*Availability of wool*
(*a*) Red wool	2	3	—	8
(*b*) Green wool	—	2	5	10
(*c*) Blue wool	3	2	4	15

These can be expressed as a linear equations

$$2x_1 + 3x_2 \le 8 \quad \text{(availability of red wool)}$$

$$2x_2 + 5x_3 \le 10 \quad \text{(availability of green wool)}$$

$$3x_1 + 2x_2 + 4x_3 \le 15 \quad \text{(availability of blue wool)}$$

These are represented in standard form as:

Maximise

$$Z = 10x_1 + 8x_2 + 5x_3$$

Subjected to

$$2x_1 + 3x_2 \le 8$$

$2x_2 + 5x_3 \leq 10$

$3x_1 + 2x_2 + 4x_3 \leq 15$

$x_1, x_2, x_3 \geq 0$

Problem 2: A paper company produces rolls of papers used in cash registers each roll of paper is 200 metres in length and can be produced in the width of 2.5, 5, 7.5 and 12.5 cms. The company's production process results in 200 metres rolls that are 30 cms in width. The company must cut its 30 cm roll to the desired width. It has six basic cutting alternatives.

Cutting Alternative	*Number of Rolls*				*Waste (cms)*
	2.5	5	7.5	12.5	
1	6	3	0	0	0
2	0	3	2	0	1
3	1	1	1	1	1
4	0	0	2	1	1
5	0	4	1	0	1
6	4	2	1	0	1

The maximum demand requirements for the 4 rolls are as follows:

Roll width (cms)	*Demand (Rolls)*
2.5	3000
5	2000
7.5	1500
12.5	1000

The company wishes to minimise the waste created by its production process, while meeting its requirements.

Formulate the problem as a L.P. model.

Solution: Let x_j be the number of cutting alternatives

where $j = 1, 2, 3, 4, 5$ and 6

Let x_3, x_4, x_5 and x_6 are the waste produced in alternatives 3, 4, 5 and 6 respectively.

Minimise $Z = x_3 + x_4 + x_5 + x_6$

Subjected to

$6x_1 + x_3 + 4x_6 \leq 3000$ ... (1) (width of roll 2.5 cm)

$3x_1 + 3x_2 + x_3 + 4x_5 + 2x_6 \leq 2000$ ... (2) (width of roll 5 cm)

$2x_2 + x_3 + 2x_4 + x_5 + x_6 \leq 1500$ ... (3) (width of roll 7.5 cm)

$x_3 + x_4 \leq 1000$ ... (4) (width of roll 12.5 cm)

$x_1, x_2 \ldots x_6 \geq 0$

Problem 3: A manager of an oil refinery has to decide upon the optimal mix of two blending processes to which inputs and outputs are given per production run.

Process	*Input*		*Output*	
	Crude A	*Crude B*	*Gasolene X*	*Gasolene Y*
1	5	3	5	8
2	4	5	4	4

The maximum amount of crude A and B are 200 units and 150 units respectively. Market and requirements show that at least 100 units of Gasoline X and 80 units of Gasoline Y must be produced. The profit per production run for process 1 and 2 are Rs. 3 and Rs. 4 per unit respectively. Formulate the problem as L.P.P.

***Solution*:** Let x_1 and x_2 be the number of units produced per production run by process (1) and process (2) respectively.

Then objective function

Maximise $Z = 3x_1 + 4x_2$

Subjected to:

$5x_1 + 4x_2 \le 200$ (Maximum availability of crude A)

$3x_1 + 5x_2 \le 150$ (Maximum availability of crude B)

$5x_1 + 4x_2 \ge 100$ (Market requirement of Gasoline X)

$8x_1 + 4x_2 \ge 80$ (Market requirement of Gasoline Y)

$x_1, x_2 \ge 0$

***Problem 4*:** An advertisement planning is to be carried out for a company which uses three media T.V., radio and print to get maximum exposure of its products to large number of customers. The following information is obtained from the market survey report.

	T.V.	*Radio*	*Print media*
Cost of Advt./unit	30,000	20,000	18,000
No. of customers reached	2,00,000	6,00,000	1,50,000
No. of targeted customer (female) reached	1,50,000	40,000	70,000

The company intends to spend maximum Rs. 5,00,000 on advertisement. The requirements of the company are:

(*i*) At least 1 million exposures take place among female customers.
(*ii*) The maximum unit on advertising in print media is Rs. 1,50,000
(*iii*) At least 3 advertisements must be brought in print media.
(*iv*) No. of advertisement units on TV and radio should each be minimum 7 and respectively.

Formulate as L.P.P.

***Solution*:** The key decision to be made here is to determine the number of advertisement units to be brought on TV, radio and print media.

Let x_1, x_2 and x_3 be the number of advertisement to be brought on TV, radio and print media respectively.

Objective is to maximise the number of customers.

Objective function is

$$Z = 10^5 [2x_1 + 5x_2 + 1.5x_3]$$

Subjected to constraints

$30{,}000\, x_1 + 20{,}000\, x_2 + 18{,}000\, x_3 \le 5{,}00{,}00$ (Advt. budget)

$1{,}50{,}000\, x_1 + 4{,}00{,}000\, x_2 + 70{,}000\, x_3 \le 10{,}00{,}000$ (Female exposure)

$x_3 \le 1{,}50{,}000$(expenses on print media)

$x_3 \ge 3$ (at least 3 advt. in print media)

$x_1 \ge 7$ (at least 7 advt. on TV)

$x_2 \ge 12$ (at least 12 advt. on radio)

$x_1, x_2, x_3 \ge 0$

27.5. GRAPHICAL METHOD FOR SOLVING L.P. PROBLEMS

Step I: State the problem in a mathematical form.

Step II: Plot on a graph the problem constraints by temporarily ignoring the inequality sign and decide upon the area of feasibility solution as per the inequality sign of the constraints. Indicate the area of the feasible solution by the shaded area which forms a convex polyhydron.

Step III: Determine the coordinates of all points at the corner of the feasible solution.

Step IV: Find out the value of objective function corresponding to all the solution points determined in step (II).

Step V: Determine the feasible solution which optimises the value of the objective function.

***Problem 5*:** A company produces two types of dolls *A* and *B*. Doll *A* is of superior quality and *B* is of lower quality. Profit on doll *A* and *B* is Rs. 5 and Rs. 3 respectively. Raw material required for each doll *A* is twice that is required for doll *B*. The supply of raw material is only 1000 per day of doll *B*. Doll *A* requires a special crown and only 400 such clips are available per day. For doll *B*, only 700 crowns are available per day. Find graphically the product mix so that the company makes maximum profit.

***Solution*:** The formulation of L.P.P. is

Maximise $Z = 5x_1 + 3x_2$

Subjected to constraints

$$2x_1 + x_2 \leq 1000$$
$$x_1 \leq 400$$
$$x_2 \leq 700$$
$$x_1, x_2 \leq 0$$

The straight lines corresponding to the constraint equation are plotted as shown in Fig. 27.1.

Equation (1) $2x_1 + x_2 = 1000$

$x_1 = 0$, gives $x_2 = 1000$

$x_2 = 0$, gives $x_1 = 500$

(0, 1000) (500, 0) are the coordinates.

Equation (2) $x_1 = 400$

The coordinates are (0, 4000)

Equation (3) $x_2 = 700$

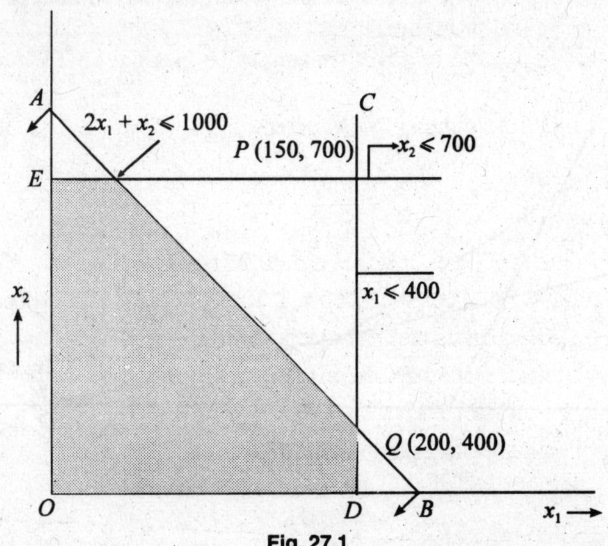

Fig. 27.1

The co-ordinates are (0, 700)

These are plotted on a graph as shown in Fig. 27.1.

The coordinates of the corner point are:

0 (0, 0), $D = (400. 0)$ $Q = (200, 400)$, $P = (150, 700,)$ $E = (0, 700)$

Then substituting the values off these points in objective function the maximum value is got for point P (150,700).

The value of $Z = 2050$.

150 number of doll A and 700 number of doll B are to be produced to maximise the profit.

Problem 2: Find the optimal solution to the L.P. problem given using graphical method.

Minimise $Z = 600x_1 + 500x_2$

Subjected to

$$3000x_1 + 1000x_2 \geq 24,000 \quad \text{... (1)}$$

$$1000x_1 + 1000x_2 \geq 16,000 \quad \text{... (2)}$$

$$2000x_1 + 6000x_2 \geq 48,000 \quad \text{... (3)}$$

$$x_1, x_2 \geq 0$$

Solution: This is a minimisation problem.

The constraint equation are (making them equalities)

Equation (1) $3x_1 + x_2 = 24$

$x_1 = 0$, gives $x_2 = 24$ and $x_2 = 0$, $x_1 = 8$

co-ordinates are (0, 24), (8, 0)

Equation (2) $x_1 + x_2 = 16$

co-ordinates are (0, 16) (16, 0)

Equation (3) $2x_1 + 6x_2 = 48$

Then $x_1 = 0$ gives $x_2 = 8$, $x_2 = 0$ gives $x_1 = 24$

co-ordinates are (0, 8), (24, 0)

These are plotted on the graph as shown in Fig. 27.2.

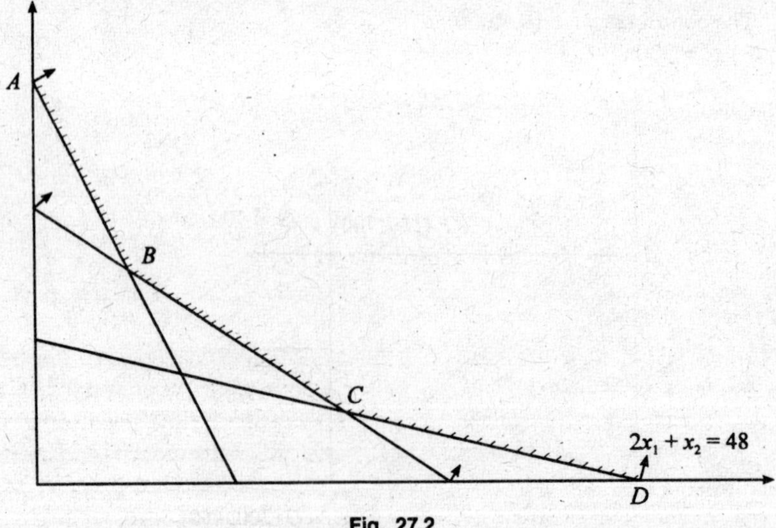

Fig. 27.2

The values of objective function

Point	Coordinate	Objective function value ($z = 600x_1 + 400x_2$)
A	(0, 24)	9600
B	(4, 12)	**7200 Minimum**
C	(12, 4)	8800
D	(24, 0)	14600

The optimal value of objective function 14,600
The optimal solution is $x_1 = 4, x_2 = 12$

27.6. SIMPLEX PROCEDURE

1. Formulate the problem as the L.P.P. problem in standard form.

2. Convert the inequalities into equalities by introducing slack or surplus and/or artificial variables as required by the problem.

3. Initial solution—obtain the initial or starting solution by setting *n-m* variables in the problem equal to zero and *m* represents the number of constraint equations *n* represents number of variables. The variables set equal to zero are called non-basic variables. Those (*n-m*) variables should be set equal to zero, which give unique solution to the problem.

4. The initial solution is represented in the tabular form and it is called initial simplex table.

5. The next step is to improve the solution by the number of iterations till optimal solution is obtained. This step is explained as follows:

(*i*) **Determine the entering variable:** The variable with the most negative value in the z equation of the initial table corresponds to the entering variable. This is called entering variable as it enters the basis.

(*ii*) **Leaving variable:** This is the variable which leaves the basis to give place for new entering variable. To identify the leaving variable—the ratio of entering column co-efficient is calculated for the rows except z (objective) row only +ve values are considered. The variable associated with smallest ratio is called the leaving variable.

(*iii*) **The column corresponding to the entering variable** is called entering or pivot column. The row corresponding to the leaving variable is called pivot row. The element at intersection of pivot row and pivot column is called the pivot element or key element.

(*iv*) **For computing the improvement**, the pivot element should be 1 and there should be zeros in the pivot column in all other places. This is done using the Gauss-Jordan elimination method.

The new values are calculated using the following equations:

(*a*) New pivot equation $= \dfrac{\text{Old pivot equation}}{\text{pivot element}}$

(*b*) New Z equation = (Old Z equation) – (its entering column coefficient) × New pivot equation.

(*c*) For all other equations,

New Equation = (Old equation) – (its entering column coefficient) × New pivot equation.

The values obtained as represented in the table.

6. Test for optimality—

If the values of the non-basic variable in Z row are all + ve, then optimal solution is reached. Get the value of Z and decision variables from the table otherwise.

Repeat the step 5 to get the improved solution till optimal solution is obtained.

Simplex procedure for maximisation problem is shown in fig. 27.3.

FLOWCHART OF SIMPLEX METHOD

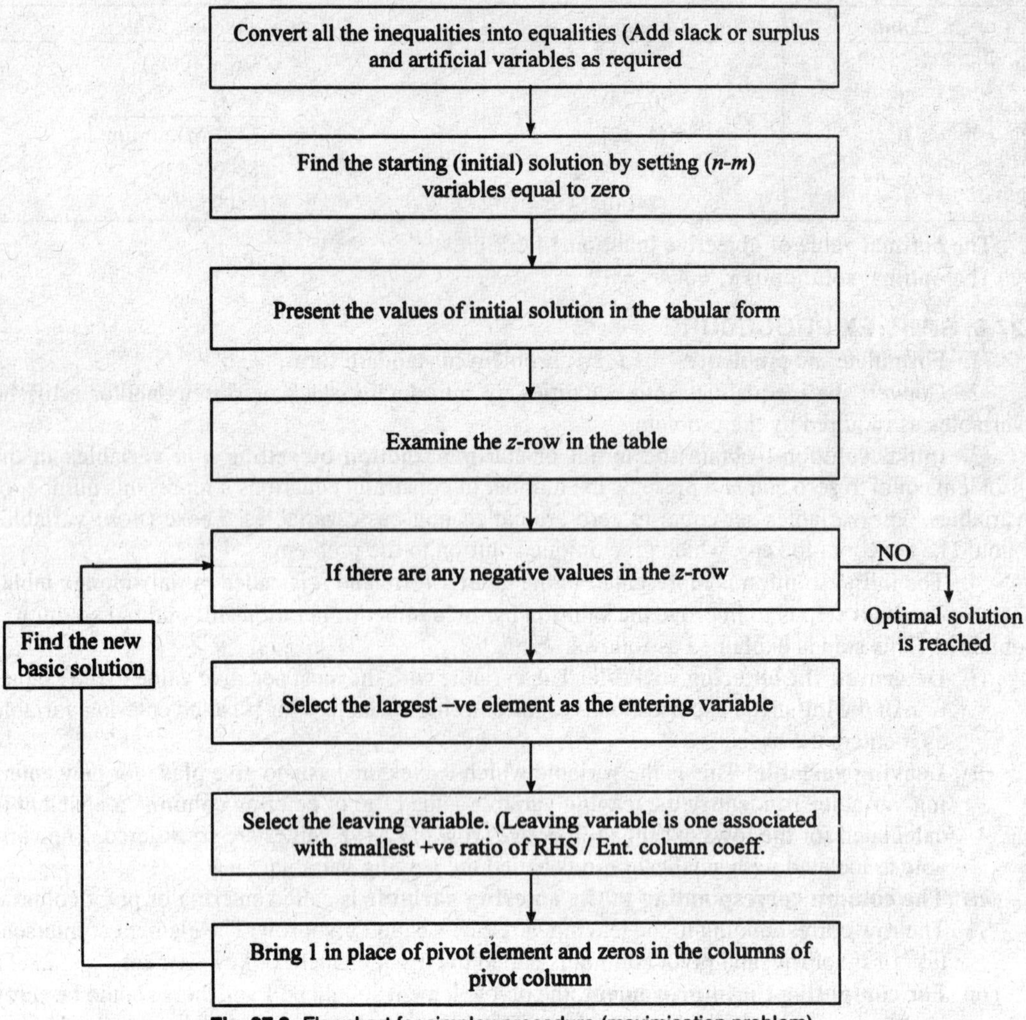

Convert all the inequalities into equalities (Add slack or surplus and artificial variables as required

↓

Find the starting (initial) solution by setting (n-m) variables equal to zero

↓

Present the values of initial solution in the tabular form

↓

Examine the z-row in the table

↓

If there are any negative values in the z-row → NO → Optimal solution is reached

↓

Select the largest –ve element as the entering variable

↓

Select the leaving variable. (Leaving variable is one associated with smallest +ve ratio of RHS / Ent. column coeff.

↓

Bring 1 in place of pivot element and zeros in the columns of pivot column → Find the new basic solution → (back to "If there are any negative values in the z-row")

Fig. 27.3: Flowchart for simplex procedure (maximisation problem).

A firm engaged in a manufacture of two products A and B performs only three operations—painting, assembly and testing. The relevant data regarding the products is given bellow:

Product	*Sales price (Rs. per unit)*	*Hours required per unit*		
		Assembly	*Painting*	*Testing*
A	50	1.0	0.2	–
B	80	1.5	0.2	0.1

Total number of hours available each week are:

Assembly – 600 hrs
Painting – 100 hrs
Testing – 30 hrs

The firm wishes to determine its weekly production so as to maximise its revenue. Formulate the model and solve by simplex method.

Solution. Let x_1 and x_2 be the number of units of of product *A* and *B* respectively to be produced in order to maximise revenue.

The problem can be represented as a L.P.P., model as shown below:

Maximise $Z = 50x_1 + 80x_2$ (objective function)

Subjected to

$x_1 + 1.5x_2 \leq 600$ ... (1) – assembly hours constraint

$0.2x_1 + 0.2x_2 \leq 100$ ... (2) – painting hours constraints

$0.1x_2 \leq 30$ ... (3) – testing hours constraints

$x_1, x_2 \geq 0$ – Non-negativity restrictions

Simplex procedure.

1. Initial or starting solution

The problem can be rewritten by introducing slack variables to convert inequalities into equalities.

The problem after introducing slack can be written as,

Maximise $Z = 50x_1 + 80x_2$

Subjected to

$$x_1 + 1.5x_2 + x_3 = 600 \quad \text{... (1)}$$

$$0.2x_1 + 0.2x_2 + x_4 = 100 \quad \text{... (2)}$$

$$0.1x_2 + x_5 = 30 \quad \text{... (3)}$$

x_3, x_4, and x_5 are slack variables

$x_1, x_2, x_3, x_4, x_5 > 0$

Number of variables $(n) = 5$

Number of equations $(m) = 3$

To get the starting solution, $(n - m)$ variables should be set equal to zero. The variables set equal to zero should be such that the should give an unique solution.

$(n - m) = 5 - 3 = 2$ variables

Two variables x_1, x_2 should be set equal to zero.

Substituting which gets

$$x_3 = 600 \quad \text{... (1)}$$

$$x_4 = 100 \quad \text{... (2)}$$

$$x_5 = 30 \quad \text{... (3)}$$

$$Z = 0$$

The starting solution is represented in the form of an initial table (Table 1).

The variables which are set equal to zero (*i.e.*, x_1 and x_2) are called non-basic variables and variables which are set equal to zero are called basic variables.

Basic variables are represented in the basic column of the initial table.

Table 1: Initial table

			Pivot Column					
Basic	*Z*	x_1	x_2	x_3	x_4	x_5	*Solution*	*Ratio*
Z	1	– 50	– 80	0	0	0	0	–
x_3	0	1	1.5	1	0	0	600	400
x_4	0	0.2	0.2	0	1	0	100	500
x_5	0	0	0.1	0	0	1	30	300

↑ Pivot row ↑ Pivot element

***Iteration 1*:** The highest negative value in the z-row is associated with the variable x_2. So, x_2 is the entering variable.

To find out the leaving variable, the ratio of $\frac{\text{RHS}}{\text{Entering column co-efficient}}$

Only for +ve coefficients of the entering column the ratio is to be calculated. The ratio is to be calculated only for constraint equation.

The minimum positive ratio is 300 associated with variable x_5 so x_5 is the leaving variable.

The pivot element should be 1 and the other co-efficient in pivot column should be zero.

To get this, Gauss-Jordan elimination method is used or the following equations are used to calculated new equation.

$$\text{New pivot equation (NPE)} = \frac{\text{old pivot equation}}{\text{pivot element}}$$

New z-equation = old z-equation – (its entering column co-efficient) × NPE

Other equations can be calculated as,

New equation = old equation – (its entering column co-efficient) × NPE

The values are calculated are represented as shown in Table 2.

Table 2: Iteration No. 1

Basic	Z	x_1	x_2	x_3	x_4	x_5	*RHS*	*Ratio*
Z	1	– 50	0	0	0	800	2400	–
x_3	0	1	0	1	0	– 15	150	150
x_4	0	2.0	0	0	1	– 2	40	200
x_2	0	0	1	0	0	10	300	–

***Iteration 2*:** The variable x_1 (which was the co-efficient of –50 in z-row) is the entering element To determine the leaving variable, the ratio of

$$\frac{\text{R.H.S.}}{\text{Co-efficient of entering column is taken}}$$

The ratio corresponding to the variable x_3 is minimum (150)

$\therefore$ x_3 is the leaving variable.

Now, is there is already 1 in the position of pivot element there is no need to compute the new pivot equation.

The z-equation and x_4 and x_2 equations are calculated and the value are tabulated as shown in Table 3.

Table 3: Iteration No. 2

Basic	Z	x_1	x_2	x_3	x_4	x_5	*Solution*
Z	1	0	0	50	0	50	31,500
x_1	0	1	0	1	0	–15	150
x_4	0	0	0	– 0.2	1	1	10
x_2	0	0	1	0	0	10	300

Since all the non-basic variables in the z-equation (row) are non-negative, optimally has reached.

The Solution is

$$x_1 = 150$$
$$x_2 = 300$$
$$x_3, x_4 \text{ and } x_5 = 0$$

Substituting these values in objective function.

$$Z = 50x_1 + 80x_2$$
$$= 50 \times 50 + 80 \times 300$$
$$= 31500$$

27.7. BIG M. METHOD (ILLUSTRATION)

Maximise

$$Z = 4x_1 + x_2$$

subjected to

$$3x_1 + x_2 = 3$$
$$4x_1 + 3x_2 \geq 3$$
$$x_1 + 2x_2 \leq 4$$
$$x_1, x_2 \geq 0 \quad \text{(Non-negativity restriction)}$$

Solution.

The inequalities can be converted into equalities by introducing the appropriate variables like slack, surplus and artificial variables and after augmenting the artificial variables, the objective function and the constraint equations can be expressed as :

Maximise

$$Z = 4x_1 + x_2 - MR_1 - MR_2$$

subjected to

$$3x_1 + x_2 + R_1 = 3 \quad ...(1)$$
$$4x_1 + x_2 - x_3 + R_2 = 6 \quad ..,(2)$$
$$x_1 + 2x_2 + x_4 = 4 \quad ...(3)$$
$$x_1,\ x_2,\ x_3,\ x_4\ R_1,\ R_2 \geq 0$$

where R_1 and R_2 are artificial variables.

From equation (1), $R_1 = 3 - 3x_1 - 3x_2$

From equation (2), $R_2 = 6 - 4x_1 - 3x_2 + x_3$

Substituting the values of R_1 and R_2 in objective function.

$$Z = 4x_1 + x_2 + M(3 - 3x_1 - 3x_2) + M(6 - 4x_1 - 3x_2 + x_3)$$
$$= (4 - 7M)\,x_1 + (1 - 4M)\,x_2 + Mx_3 + 9M$$

This can be written as

$$Z - (4 - 7M)\,x_1 - (1 - 4M)\,x_2 - Mx_3 = 9M.$$

Starting solution.

Number of variables $(n) = 6$

Number of constraint equation = 3

By setting $(n - m)$ variables equal to zero, we get the starting solution.

$x_1,\ x_2,\ x_3\ =\ 0$ gives, $R_1 = 3$ and $R_2 = 6$, $Z = 9M$.

This initial solution is represented in Table 1.

Table 1: Initial solution

Basic	*Z*	x_1	x_2	x_3	R_1	R_2	R_4	*RHS*
Z	1	$-4 + 7M$	$-1 + 4M$	$-M$	0	0	0	$9M$
R_1	0	3	1	0	1	0	0	3
R_2	0	4	3	-1	0	1	0	6
x_4	0	1	2	0	0	0	1	4

Table 2: Iteration No. 1

Basic	Z	x_1	x_2	x_3	R_1	R_2	R_4	RHS
Z	1	$\frac{1+5M}{3}$	$-M$	$\frac{4-7M}{3}$	0	0	0	$4+2M$
x_1	0	1	⅓	0	⅓	0	0	1
R_2	0	0	5/3	– 1	– 4/3	1	0	2
x_4	0	0	5/3	0	– ⅓	0	1	3

Table 3: Iteration No. 2

x_2 is entering variable and R_2 is leaving variable.

Basic	Z	x_1	x_2	x_3	R_1	R_2	R_4	RHS
Z	1	$1+5M/3$	0	$20-49M/3$	$-4M/5$	$3M/5$	0	$20+16M$
x_1	0	1	0	3/15	2/9	4/15	– 1/15	3/5
x_2	0	0	1	– 3/15	– 4/5	3/5	0	6/5
x_4	0	0	0	– 1	– 5/3	1	1	1

Table 4: Optimum Solution

Basic	Z	x_1	x_2	x_3	R_1	R_2	R_4	RHS
Z	1	0	0	0	$7/5-M$	$-M$	– 1/5	17/5
x_1	0	1	0	0	2/5	0	– 1/5	2/5
x_2	0	0	1	0	0 – 1/5	0	3/5	9/5
x_3	0	0	0	1	1	– 1	1	1

The optimal solution is,

$$x_1 = 2/5$$

$$x_2 = 9/5 \text{ and } x_3 = 1$$

Substituting these values in objective function,

$$Z = 4x_1 + x_2 = 4 \times \frac{2}{5} + \frac{9}{5}$$

$$= 17/5$$

Maximum value of objective function = 17/5.

REVIEW QUESTIONS

1. Discuss the application of L.P. in production decisions ?
2. "L.P.P. is a powerful tool for decision-making in the area of production management ?" Comment.
3. Explain the graphical method for solving L.P. problems ? Discuss the advantages and limitations.
4. Describe the simplex procedure for solving L.P. problems.
5. Explain the following terms with respect to linear programming:
   (*a*) Slack and surplus variables.
   (*b*) Decision variables.
   (*c*) Basic and non-basic variables.
   (*d*) Degeneracy in linear programming.

(*e*) Basic solution, feasible solution, basic feasible solution and optimal solution.
(*f*) Big M method.

PROBLEMS

1. Solve the following problems using simplex method.

(*a*) Maximise $Z = 3x_1 + 2x_2 + 5x_3$

Subjected to

$2x_1 + 3x_2 + 5x_3 \leq 30$

$x_1 + x_2 + x_3 \leq 11$

$2x_1 - x_2 - x_3 \leq 8$

$x_1, x_2, x_3 \geq 0$

(**Ans.** $x_1 = 5, x_3 = 4, x_2 = 0, Z = 35$)

(*b*) Maximise $Z = 3x_1 + x_2$

Subject to

$2x_1 + x_2 \leq 1$

$3x_1 + 4x_2 \geq 4$

$x_1, x_2 \geq 0$

[**Ans.** $x_1 = 0, x_2 = 1, Z = 2$]

2. A manufacturer of leather belts makes three types of belts *A*, *B* and *C* which are processed on three machines M_1, M_2 and M_3 melt *A* requires 2 hours on machine M_1 and 3 hrours on machine M_3. Belt *B* requires 3 hours on machine M_1, 2 hours on machine M_2 and 2 hours on machine M_3 and belt *C* requires 5 hours on machine M_2 and 4 hours on machine M_3. There are 8, 10 and 15 hours available per day on machines M_1, M_2 and M_3 respectively. The profit gained in Rs. 3, Rs. 4 and Rs. 5 from belts *A*, *B* and *C* respectively. What should be daily production of each type of belt so that the profit is maximum?

3. A firm manufactures a product each unit of which consists of 7 units of part P_1 and 6 units of part P_2. The two part P_1 and P_2 require different raw materials of which 320 units and 450 and units respectively are available. These parts can be manufactured by three different methods and details are given below:

Method	*Input per urn (units)*		*Output per urn (units)*	
	Raw Mat-I	*Raw Mat-II*	*Part P_1*	*Part P_2*
1	7	5	6	4
2	4	7	5	8
3	2	9	7	3

Formulate this as a linear programming problem to determine the number of production runs for each method so as to maximise the total number of complete units of final product.

28

TQM – CONCEPT AND PHILOSOPHY

• Review of Quality Management • Evolution of Quality Management • Total Quality Management Approach • Stages of Implementation of TQM • List of Techniques for TQM. TQM Model TQM Critical Success Factors • TQM Dimensions.

28.1 REVIEW OF QUALITY MANAGEMENT

The quality idea has been around for hundreds of years. It has progressed from inspection to today's Total Quality Management. Inspection and protection was established as a management idea. Products were inspected and the quality image of the company protected by the removal of poor quality products before the customer applied their own inspection and reaction. Inspection and protection is however little more than reactive management, reacting when poor quality has already entered the product. Now a days, quality has encompassed an entire organisation including all the processes and functions. Mere inspection of products has become a primitive idea instead quality management has become proactive; making plans to bring about continuous quality improvement and to achieve a more desirable future. The objective here is to get rid of poor quality from the product rather than get rid of poor quality product. Quality management has progressed, establishing proactive rather than reactive organisations. A step ahead towards quality management is Total Quality Management (TQM). TQM is a whole system concept recognising the need to manage sets of interacting issues; technical, cultural and political nature.

28.2 EVOLUTION OF QUALITY MANAGEMENT

Traditionally, building quality into product was the aim of skilled craftsman. Tradesman gained the reputation for quality products through craftsmanship that was maintained. Industrial revolution, which led to the establishment of factories and mass production, led to inspection, which was the sole guarantee of quality. The First World War demanded yet large-scale production and demanded reliable products. This in part led to the formation of associations and institutions and to the publication of formalized ideas. The Second World War led to the formation of American Society for Quality Control (ASQC) by the thousands of quality specialists who have been trained mostly by the war production board. However, the real success story for quality thinking ironically emerged in one of the defeated nations. The Japanese launched a new nationalistic drive for expansion, pursuing economic rather than Military goals. One famous guru who played a dominant role in the process of quality improvement in Japan was W. Edward Deming, but there were others from United States such as J.M. Juran. They have the benefit of an intimate involvement in working out sound techniques during the war and in the post war period. By the 1970s, the Japanese have become "Masters" at achieving quality in their manufacturing sector. But they never stopped on this achievement and their quest for superior production by continuous improvement in knowledge, methods and techniques is still continuing. The Japanese were prime in switching commercial interests from competition in productivity to competitiveness in quality. The Japanese success story has, however urged some managers in western and other countries to wake up to the quality issue. People have recognized that Japanese success was not only due to national, cultural and social differences but reflected strongly a new attitude and desire of Japanese management to ensure that consumers receive what is promised. the 1980's therefore became an era of competitive challenge with increasing number of companies adopting qualtiy Management System. the development of International Quality Assurance Management system (ISO 9000) standards in the 1980's -1990's has also acted as a catalyst in many countries. During 1990's and beyond, the quality management has become International Management Philosophy. The evolution of quality Management is represented in table (28.1).[15]

28.3 TOTAL QUALITY MANAGEMENT

Total Quality Management incorporates the features like: products that meet customer needs, control of processes to ensure their ability to meet design requirements and quality improvements for the continued enhancement of quality.

Table 28.1 Evolution of Quality Management

1. **Quality Inspection**
   Salvaging, sorting, investigating, corrective actions, identifying sources of non conformance and dealing with them.
2. **Quality Planning**
   Developing quality manuals, producing process performance data, planning for quality.
3. **Quality Management**
   Statistical process control, third party approval, quality system audit, use of quality costs, involvement of non-production operations.
4. **Total Quality Management (TQM)**
   Continuous improvement system perspective involves all operations and at all levels (company wise), undertakes performmance measurement , focus on leadership, teamwork and participation of every one, employee empowerment.

Customer is the driving force behind quality of design. Customer satisfaction is based on the subjective comparison between the expectations and the actual quality received. The sales of the product clearly generate the hard currency, one must also recognise that customer satsifaction is derived from ancillary services associated with product and the sensitivity and timeliness with which the problems are handled. Quality system should possess a sound behavioural as well as technical perspective. To develop such a quality system, the management should research the customer preferences, train employees to be sensitive to customer needs and reward employees for making customer satisfaction a primary objective.

Total quality management is based on the premise that any production and/or service can be improved and that successful organisation must consciously seek out and exploit improvement. The essence of TQM is continuous improvement through collaborative efforts across functional boundaries and between organisational levels with the ultimate goal of providing customer satisfaction. Each work in TQM has a special significance. Total means here comprehensive ways of dealing with complex sets of interacting issues – involving every one at all elvels, addressing all major issues. It is also referred to as performance encompassing both the quantitative and qualitative aspects of product/service. "Quality" can be defined in variety of ways. The following definition takes into account several ideas expressed by quality Gurus. Quality means meeting customers (agreed) requirements, formal and informal at lowest cost, first time and every time.

Total quality means that everyone should be involved in quality at all levels and across all functions ensuring that quality is achieved according to the requirements in everything they do. The word "Total" injects a systematic meaning idealness into quality. "Management" in TQM denotes the system supporting the achievement of quality and performance on a continuously improving path. The management responsibility refers to the need for every one to be responsible for managing their own jobs, which incorporates managers with workers and all others concerned. Thus, TQM portrays a whole systems view for quality management.

An Organisation that endures itself to the ultimate customer, also fulfills commitments in terms of performance levels demanded by a number of related sets of customers. Such customers could be shareholders who expect a sizable return on their investment, employee's serving such an institution naturally respects it for providing a livelihood, the suppliers to the company as well as the dealers who distribute products are very similar to employees of the company, they enjoy considerable confidence in being associated with its long term planning. In turn, they are motivated to make the best use of their services available. All these sets of people are referred to as stakeholders, whether they are users of the product, employees, shareholders, vendors or dealers. They all have the expectation of qualitative and quantitative performance being fulfilled by the organisation they patronise and serve.

At the product level, a customer may consider performance reflected in material factors such as safety, reliability and value for money. In terms of service, performance can mean the delivery of the product on time at committed schedules without any hidden costs. However, a customer can equally evaluate organisational performance from the standpoint of qualitative measures e.g. prestige associated with use of the product and pleasure or ease of use associated with the product and the quality of the customer support. If the customer feels confident in dealing with the product, it generates brand loyalty on an ongoing process. This in turn generates a feeling where the customer confidently recommends the product and the company to friends and associates.

The relevance of TQM to business is world-class productivity. Basically, the essence of TQM is value addition. A business unit draws on its resources and adds value in order to create an output that delivers customer delight. TQM perspective of productivity recognise both the qualitative and quantitative aspects of relationship between inputs and outputs. It recognises the qualitative aspects of input other than considering organisation as a mechanical system transforming inputs into outputs. It considers creative talent as well s the motivation with which people engage themselves in the creation of the final output. Value addition is not merely reflected in physical transformation in shape, size, structure but the organisational learning that occurs in the process and the patented know how. The output from the system does not confine itself only to physical goods but includes the added dimensions of prestige, pride of ownership, warmth and pleasure in long lasting relationships with customer.

28.4 TQM APPROACH

Quality is a continuous process that can be broken anywhere in the system of supply and customer service. By litting every person know how their activities help fulfill customer's requirements, the organisation can motivate their employees and supplies to provide quality consistently. They must also realise that throughout the organisation they will have both internal customers and supplies to those outside the organisation. In general, a process helps to change a set of inputs into desired output in the form of products or services. Proper investigation of the inputs and outputs of the organisation help to determine the action to be taken for the improvement of the quality. The quest for continuous improvement of quality is a continuous cycle. The process on which continuous improvement is based is generally known as "Deming Wheel". The wheel represented in Fig (28.1) shows a continuous movement in a certain direction. The idea behind this that the input, which generates activities with measurable output, is process and the perfection of the process is the ultimate objective.

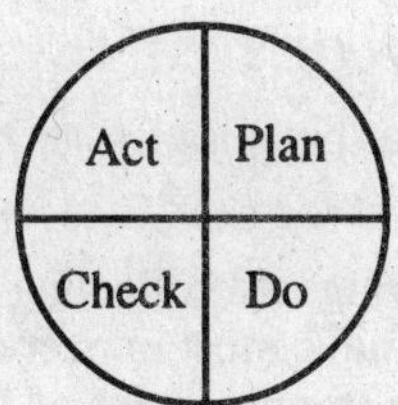

Fig (28.1) PDCA Cycle

In a Deming's wheel, the plan defines the process, which ensures documentation and sets measurable objectives against it. The "Do" executes the process and collects the information required. The CHECK analyses the information in suitable format. The ACT obtains corrective action using TQM techniques and methods and assesses future plans. At the end of each cycle the process is either standardised or targets are adjusted based on the analysis and the cycle continuous. The link between customer/supplier with process improvement is shown in Fig (28.2).

The TQM approach is both a practical working process and a quality philosophy for the organisations committed to growth and survival. TQM approach starts with a vision that a concentrated management action can improve the quality of service and products of the organisation at a very competitive cost satisfying customer's need and increasing the market share. This increased market share will be stable because it has been earned with the help of solid customers goodwill and not by gimmickry advertising.

Table 28.2 Principles and Actions of TQM

Principles	Actions
The approach	Management led
The scope	Company wide
The scale	Every one is responsible for quality
The philosophy	Prevention not detection
The standard	Right first time
The control	Cost of quality
The theme	Continuous improvement
The dimensions	Human, technical and cultural

To develop TQM process the organisation has to be guided through the following basic rules of action and is given by the following principles and actions as represented in table (28.2).

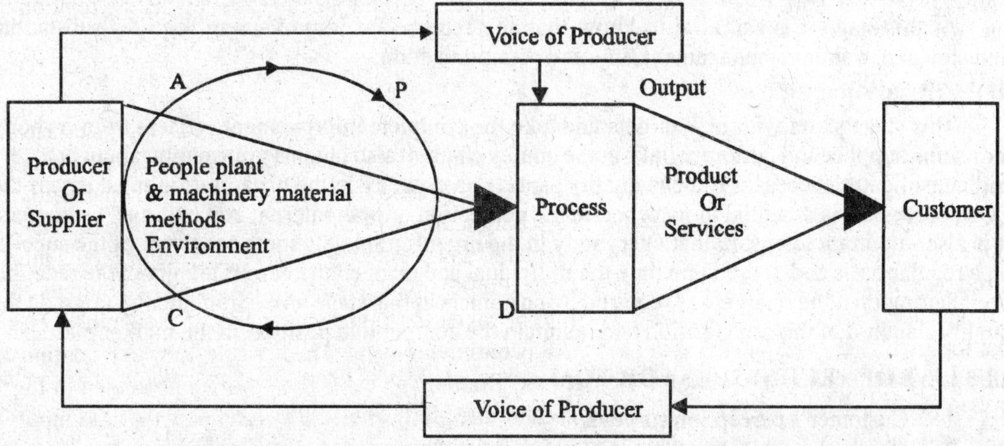

Fig. (28.2): Supplier, Customer continuous improvement interface

28.5 STAGES OF IMPLEMENTATION OF TQM

The process of implementing TQM in an organisation can be organised in the following four stages:

(*i*) Identification and Preparation

This stage is concerned with identifying and collecting information about the organisation in the prime areas where improvement will have most impact on the organisation's perormance and preparing the detailed basic work for the improvement of the organisation's activities. It is also important to find out the cost of quality, which incorporates the total cost of waste, error correction, failure appraisal and prevention in the organisation. It is also required to understand the views and opinions of the customers, suppliers the managers and the employees. The differences between their views and opinions will provide an idea of the scale of the problem and task ahead. The measurements of the cost of quality made at the beginning of the TQM process can be compared with measurement at a later stage to establish the achieved improvements. The initial measurements of the costs will also indicate the potential areas for improvement and direct efforts towards the areas where they are most needed. All data and information must therefore be identified, prepared and summarized in a manner to ensure that the managers get the correct information to make their decision.

(*ii*) Management Understanding

This step is concerned with making sure that the management understands the objective and methodology of TQM and is ready to adopt them all the time. For many companies, TQM means a major change in the management practice and it is difficult to implement over a short period of time. However, to make a significant change in management practice, it is necessary to educate the managers in their understanding and approach to TQM. Once they have mastered the principle and practice of TQM the managers can then demonstrate their total commitment and take the lead in the quality improvement process.

(*iii*) Scheme for Improvement

This stage is concerned with identifying quality issues and affects a resolution of them by management led improvement activities. To develop quality improvement scheme, it is necessary to identify the quality problems in each division, in each department and throughout the whole organisation. A scheme of training for improvement can be established after the realization of the following aspects of the organisation. They are:

- Purpose of the department
- Customer's and suppliers relationship
- Meeting customer needs
- Problem causes and best solutions
- Prevention of recurring problems
- Customer satisfaction
- Priorities for improving efficiency

At this stage it is essential to know that any scheme for improvement requires substantial investment in training, management time and communication.

(*iv*) Critical Analysis

This stage starts with new targets and take the complete improvement process to everybody indicating supplier and customer links in the quality chain. It also obtains information about progress and consolidates success. To focus quality aspects, everybody in the organisation must assess the TQM process. It is essential to incorporate the perception of both internal and external customers. It is also important to ensure that everybody in the organisation gets some feedback of the success on a regular basis and at the same time the individual and team contributions are given the recognition. Setting up of new targets as required by customers at this stage will automatically upgrade the quality standard of the organisation and maintain the competitive position in the market place.

28.6 LIST OF TECHNIQUES FOR TQM

1. Customer's perception surveys
2. Quality function deployment
3. Cost of quality statement

4. Top team workshops
5. Total quality seminars
6. Departmental purpose analysis
7. Quality training
8. Improvement action team
9. Quality circles
10. Suggestion schemes
11. Help calls
12. Visible data
13. Process management
14. Statistical process analysis
15. Process capability analysis
16. Fool proofing
17. Just in Time Manufacturing (JIT)
18. Business Process Reengineering (BPR)
19. Quality Improvement Team (QIT)

28.7 TQM Model

Customer satisfaction is the focus of TQM. The model shown in Fig (28.3) highlights how the implementation of TQM benefits the company in both long term and short term and in turn achieves the customer satisfaction.

Bascially, the customer satisfaction depends upon the gap between the expected and actual quality of products offered to the customer. When the customer's expectations of product/service quality balance the actual product quality offered by the company, the customer satisfaction results. If the customer's exoectatuibs exceed the actual results in customer delight. TQM aims at customer delight going one step ahead of mere satisfaction of customers. The delighted customer will become the loyal customer and have a complete trust in the offering of the company's products and services. The quality of the product results in higher reliability of which in turn helps to attain the retention of loyal customer base.

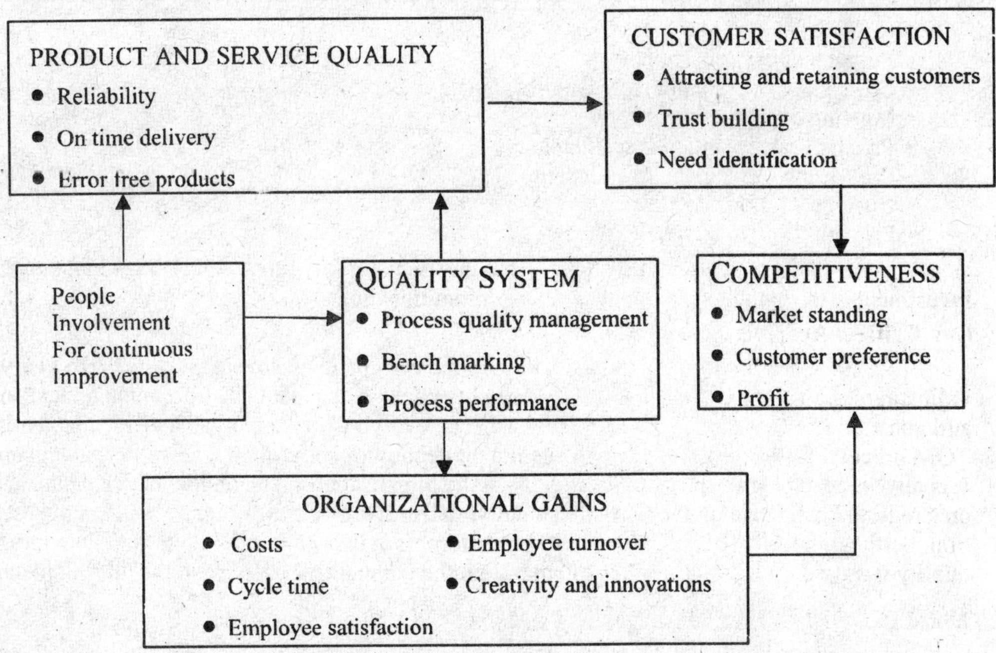

Fig. 28.3. Total Quality Management Model

The quality of the product depends on the ability of the company to identify both stated and unstated needs, translation of these needs into design specifications, and designing and managing the process to keep quality level as per design specifications and ensuring performance. This inturn is possible through a well-designed quality system and involvement of each and every employee at levels. The continuous improvement in quality is the result of empowered employees and the leadership of the management.

Thus, higher quality levels of products/services accompanies by loyal and satisfied customer base results in enhancing compeititive position of the company. The organisational benefits of implementing TQM include – reduces cost and cycle time, job satisfaction and reduced turnover of employees, increase in productivity and a good reward for all the stakeholders.

28.8 TQM CRITICAL SUCCESS FACTORS

The successful implementation of TQM depends upon the following key factors.

1. Training
2. Bench marking
3. Customer satisfaction surveys
4. Recognition and rewards
5. Management commitment

28.9 TQM DIMENSIONS

Total quality management has basically three dimensions: Technological dimension, Human (people) dimension and cultural dimension. The technological dimension is concerned with the process of designing and building quality into the product/service, human dimension is concerned with empowering people to demonstrate mastery over the tasks performed and the cultural dimension encompasses the organisational environment to foster quality mindedness. The three dimensions together create an organisational climate where continuous improvement results because of the innovations and creativity, technology and the channeling potential of the people.

29

BUSINESS PROCESS REENGINEERING

• Definition • Characteristics of BPR • Need for reengineering • Methodology of BPR • Framework for reengineering • Process of BPR • Industrial Engineering and reengineering • Advantages of reengineering.

29.1. DEFINITION

Business process reengineering (BPR) today is the latest, most radical, revolutionary and extremely powerful management tool.

BPR as defined bv Michael Hammer as —

"The fundamental rethinking and radical redesign of the business process to achieve dramatic improvements in critical contemporary measures of performance such as cost, quality, service and speed."

29.2. CHARACTERISTICS OF BPR

1. **It is process centred:** BPR offers process based approach to strategy rather than market based approach. The tool concentrates on process activities that convert inputs to output for the customer.

2. **It is redesign governed:** It will not believe in minor improvements through modifications. It is going to strike the very aspect of design.

3. **It is radical:** The entire form or structure of the process may change during the reengineering and the reengineered process may not resemble the old one.

4. **It is dramatic:** The tool is not much useful for marginal improvements of business performance.

5. **It is customer oriented:** The entire process of reengineering revolves around how to give the customer what he wants at the right time and in the most cost effective manner.

29.3. NEED FOR REENGINEERING

BPR is a powerful tool that promises the higher order magnitudes of improvements in revenues, profits, productivity, cycle time and efficiency.

There is an universal need to subject all the processes to reengineering.

At the macro-level, because of the environmental demands and internationalisation (or globalisation) of the economies the following factors become crucial for survival and development of economies.

They are:

1. Crisis of energy, 2. Crisis of conflict, 3. Crisis of confusion, 4. Crisis of stress, and 5. Crisis of culture.

Therefore, it is evident that these environmental demands calling for an action, clearly need, a new conceptual approach and requires to re-examine and create our personal, organisational and global systems and practices.

At the organisational level, the factors that favour the reengineering of the process are:

1. **Competition**—The present economic scenario is marked by competition in almost all sectors of economy. The competition is becoming tougher and satisfaction of customers expectations have become key element in competitive strategies for the 1990s and beyond.

Companies which delight the customer with quality, variety and fast cycle times at competitive prices will gain a competitive edge and it takes away the major share of the business.

Thus because of competitive pressure for survival in the global economy the organisations have to reengineer.

2. **Environmental demands.**
3. **Technological advancement.**
4. **Sinking profitability and market share.**
5. **Declining share price.**

The need for reengineering is shown in Fig. 29.1.

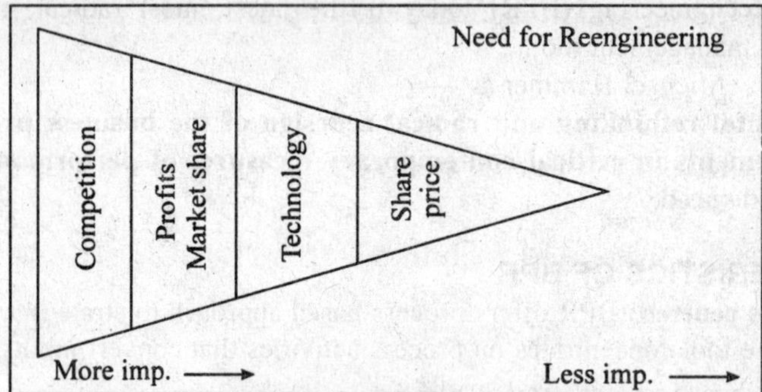

Source. Reengineering hand book by Raymond Mananelli and Klein.

Fig. 29.1: Need for reengineering.

29.4. STEPS IN REENGINEERING (METHODOLOGY)

The flowchart for implementing BPR is shown in Fig. 29.2.

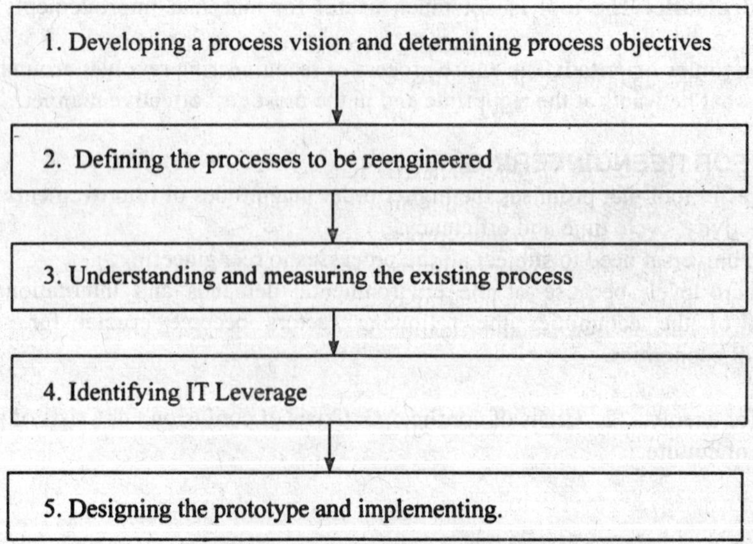

Fig. 29.2: Steps in implementing BPR.

Step 1: Developing process vision and determining process objectives

The objective of reengineering is to provide competitive advantage to the organisation. So the starting point for BPR is organisations business vision and from the vision the processes that have to be the best in the world are derived in order to realise the vision.

This is a focus phase which demands the top management involvement to arrive at an objective. The process selected for reengineering are those which provide value to the customer. *e.g.*, the vision of an organisation may be to become number one company in its business in next 10 years.

Thus, the vision or focus phase has to take place at a senior executive level where the business is perceived as an integrated whole.

Step 2: Defining process to be reengineered

This phase starts with identification of critical activities within the processes in consultation with top management, process attributes (cost, quality and time) and process measures (results) to be achieved.

The setting up of bench marks (standards) for processes and sub-processes is essential to control and ensure process outcome. Identification of processes with clarity and defining the boundaries and measures is the difficult task.

Step 3: Understanding and measuring existing process

The detailed information on how the existing process works and what are the weaknesses of the present system and about the process measures are sought.

This step makes the concerned people clear about the terms like delay, response, waiting time and non-value addition, etc. This analysis automatically creates a dissatisfaction about the existing process.

Step 4: Identifying Information Technology (IT) leverage

The reengineering achieves the improvement through effective and innovative deployment of information technology. Thus, the assistance of IT is sought to simplify the processes.

Step 5: Designing, building and implementing a prototype of new process

This is a crucial step in reengineering. The top management should provide enough resources in terms of personal skills and technology, thus, creating a conducive climate for implementation. The prototypes of the suggested processes are made and then the users try out the new system to give their comments. The practicability of the design, its working limitations are addressed at this stage.

To implement, the reengineered process should be owned by the people manning them for success. Sometimes, the facilitator does this job of linking people concerned and the reengineered process.

29.5. FRAMEWORK FOR REENGINEERING

The conventional practice of dividing and subdividing the process to perform round the structural process is taken over by the "complete process focus" in BPR. This makes employees focus on how work is done rather than trying to improve what is being correctly done.

This clearly gives way for reengineering case for action to all of the employees with emphasis on factors like context, problem identification, market, demands, diagnostics, cost of inaction and a clear vision.

The framework for reengineering is shown in Fig. 29.3.

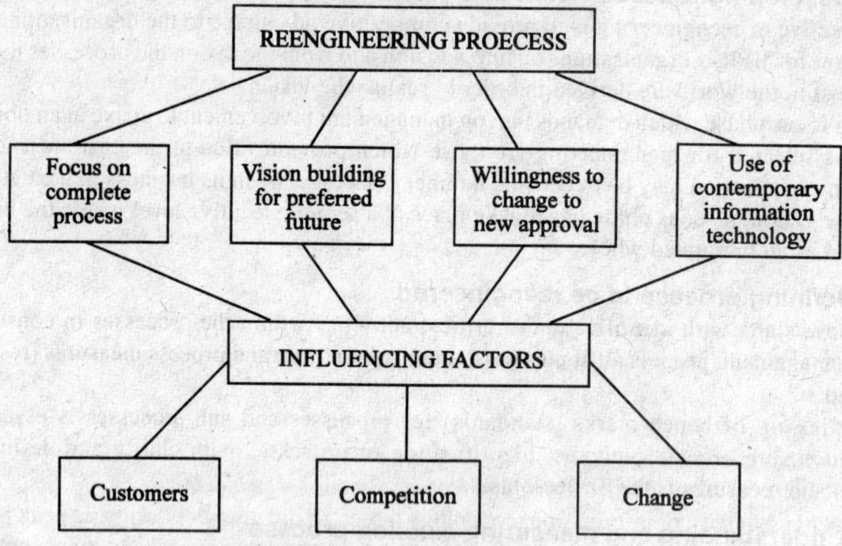

Fig. 29.3: Framework of reengineering.

29.6. PROCESS OF REENGINEERING

Reengineering should focus on top down, structured reintegration across several areas of activity.

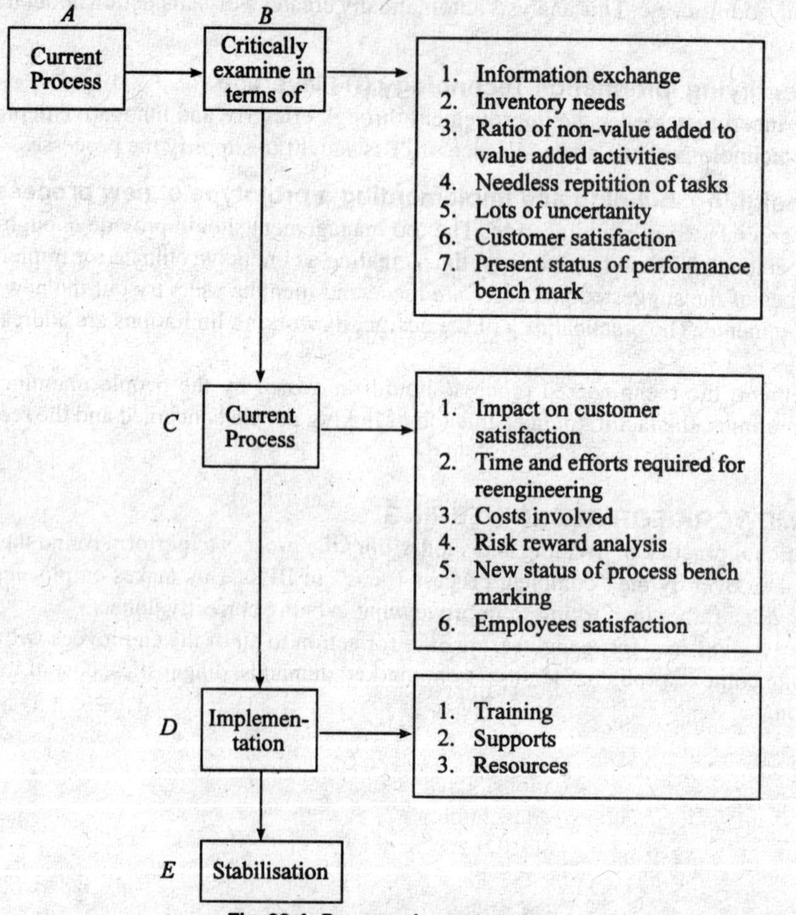

Fig. 29.4: Process of reengineering.

29.7. INDUSTRIAL ENGINEERING AND REENGINEERING

A. Similarities between I.E. and Reengineering

1. Both are concerned with reducing costs, waste and cycle time.
2. Both aim at reducing non-value adding activities.
3. Both believe in working smarter not harder.
4. Questions what, when, where, why and how and who are common to both.
5. Reducing paper work, evolving performance standards are common to both.
6. Both are useful in repetitive activities.
7. In short and medium term, both may reduce employment.
8. Both stress on multi-skilling and combining of jobs.

B. Differences between I.E. and Reengineering

1. BPR aims at radical, not marginal improvement, IE aims at whatever improvements is immediately possible.
2. BPR deals with process whereas IE usually takes up activities or procedures for streamlining within each functional area.
4. BPR believes in total redesigning and a cross-functional perspective.
5. BPR has a benefit of information technology which IE did not have during its peak periods.
6. BPR is customer focused whereas IE directly concentrates on benefits to management.
7. BPR emphasises external bench marking and beating competition. IE usually develops standards by internal work study.
8. BPR can be applied to select few business processes whereas IE has universal application.

29.8 INFORMAITON TECHNOLOGY LEVERAGE IN REENGINEERING

Information is extremely powertul torm of energy and nowadays it is becoming readily available because of the databases and powerful communication networks. IT is bringing a total transformation in the business arena.

Total integration of business is made possible because of IT and computers from the first, customer to order acceptance, and to the internal factory operations, suppliers control and delivery processing is carried out entirely through IT today.

IT has also significantly transformed the marketing operations of organisations.

29.9. SUCCESS FACTORS IN REENGINEERING

- Setting an aggressive reengineering performance target.
- Review of customer needs, economic leverages and market trends.
- Top management commitment.
- Precise vision.
- Picking up of right process for reengineering.
- Empowering people and getting them to take initiative.
- In calculating a feeling of ownership amongst as many people as possible.

29.10. ADVANTAGES OF REENGINEERING

1. Improvement in entire organisation as a whole.
2. Better systems and management improvements in the areas of—
   - Products and services.
   - Design and operations.
   - Improved system operations.
3. Takes advantage of improved technology.
4. Improved application of industrial engineering in the areas of—

- Organisational strategies.
- Management functions.
- Plant utilisation.
- Quality improvement
- Creativity and innovation.
- Confidence in competition.

5. Improvement in customer satisfaction.

References for Further Reading

1. Michael Hammer and Steven A. Staton, *Reengineering Revolution*.
2. Raymond Manganelli and Klein, *Reengineering Handbook*.
3. Peter Drucker, *Managing for the Future*.
4. M.S. Jayaraman *et al.*, *Business Process Reengineering*.
5. *Business Today* (7-21 Dec., 1995).
6. *Analyst* (Dec. 1994).
7. *Hindu Newspaper* (24 Jan., 1996).

30

GROUP TECHNOLOGY

• Introduction • Definition of GT • Group Layout • Desirable characteristics of groups • Benefits of groups • Stages of GT manufacture.

30.1. INTRODUCTION

Group technology is an approach to organising manufacturing which can be applied in any industry where a small batch variety production is used. The basic concept of group technology (GT) is relatively simple. It is the identification and bringing together related or similar parts and processed, to make advantage of the similarities which exist, during all stages of design and manufacture. GT is based on the fact that small lots of different parts can be produced more economically, if they are grouped and scheduled for production according to the common characteristics.

30.2. DEFINITION OF GROUP TECHNOLOGY

1. It is an approach to the organisation of work in which organisational units are relatively independent groups, each responsible for the production of a given family of products.
2. A group is a combination of a set of workers and a set of machines and/or other facilities laid out in one reserved area, which is designed to complete a specified set of products.
3. A family is the set of products produced by a group.

30.3. GROUP LAYOUT

With the line layout, machines or workstations are arranged in a line in their sequence of operation. It is preferred when in all the components made on the line use the same workstations in the same sequence.

When line layout cannot be used, the machines and workstations are used to process the parts in batches and the facilities are arranged with respect to functions.

New method of facilities layout called group technology is based on product specialisation in which group of machines chosen for each family are situated together in a group layout in such a way that parts flow from one machines to the next in the sequence of operation. It is not necessary for every part to visit each machines but machines in a group or cell should ideally be capable of carrying out all the operations required in the family. The group layout is shown in the Fig. 30.1.

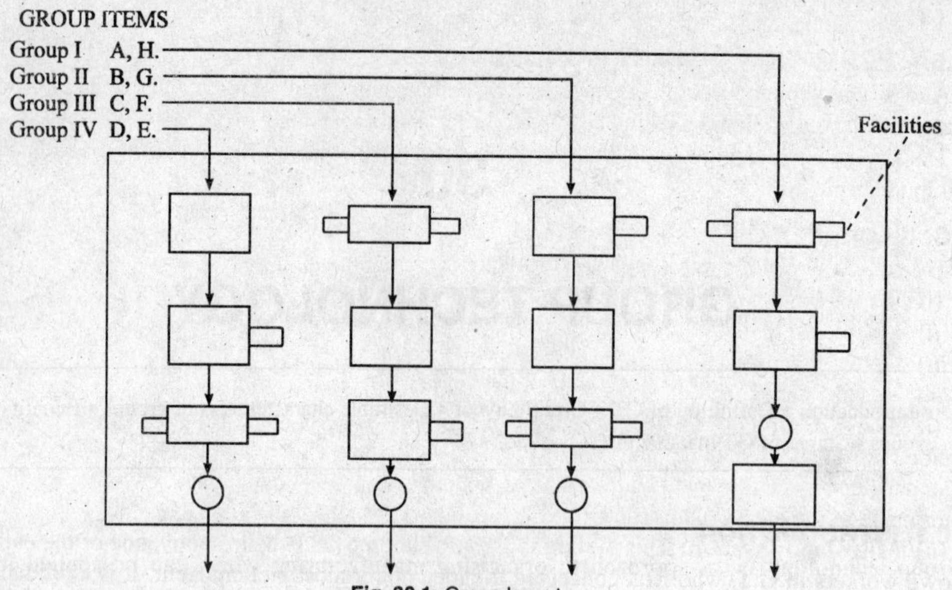

Fig. 30.1: Group layout.

30.4. DESIRABLE CHARACTERISTICS OF GROUP

1. **There should be group of workers special to the group:** The group of workers associated with a specific group make a good team and they learn to understand the special problems associated with machines and equipment and with the family of components which they make. Workers should not be allowed to move or transferred out of the group.

2. **There should be group of machines and equipment special to each group.** The group should be supplied with machines that are needed to process all the components in the their families. The parts should not move out of the group for operations on machines belonging to other groups. This helps to reduce throughput time, WIP inventory and handling time.

3. **Specially allocated area to house machines and equipment:** This is an essential characteristic if full benefits of GT is to be obtained.

4. Group should have its own special products to be completed.

5. A common output target for the group.

6. This characteristic is essential if full advantage is to be taken of the use of planned sequence of loading to reduce set-up times and throughput times and to increase machine capacity.

7. The number of workers in the group should be small:

30.5. BENEFITS OF GROUPS

1. Reduction in throughput time as all operations on a part are done inside the group.
2. Decrease in work-in-process inventory (WIP).
3. Improvement in quality levels.
4. Completion of work as per due date or schedule.
5. Reduced set-up time hence increase in productive capacity.

30.6. STAGES IN GROUP TECHNOLOGY MANUFACTURE

1. The parts of each of the items processed are examined and placed into logical classes or families and the operations sequence for each class of parts is determined and specified.

2. Groups of facilities suitable for processing of these classes of parts are specified using the operations planning details and forecasted demand for the items and hence the components.

3. The sequencing of each class of parts for each group of facilities.

30.6.1. FORMATION OF PART FAMILIES

A family is a collection of parts which are similar for purpose of manufacture. Usually they are related by size geometry shape and required similar manufacturing operations.

The key factor in using Group Technology successfully is the ability to identify readily items within the same family.

1. Coding and Classification System

The four systems are commonly used for classification:

(*i*) British System
(*ii*) Mitrofanov System
(*iii*) Optiz System
(*iv*) Vuoso System

(*i*) **The British system:** When applied to GT, this requires an eight-digit primary code or monocode which effectively sets down the design characteristics of the parts, followed by a secondary code or polycode, identifying the manufacturing characteristics. All other digits are tailor made to the each organisation.

(*ii*) **Mitrofanov system:** This system is a contribution by S.P. Mitrofanov, one of the earliest known workers in G.T. who first conceived the idea of composite component. It is a production oriented code of seven digits—

First digit (0-9)	Section	
Second digit (0-9)	Class	Parts characterised by common functions and structural shape.
Third digit (0-9)	Sub-class	Parts characterised by common shapes and similar processing methods.
Fourth digit (0-9)	Group	Characterised by similar shapes and number of manufacturing operations.
Fifth digit (0-9)	Type	Operation type
Sixth and Seventh digit (0-9)	Size	

(*iii*) **Optiz System:** This system is applicable to only machined components. It is uses 5 digit to define shape, followed by four "Optional extra" digits specifying size, material, original shape of raw materials and accuracy.

(*iv*) **Vuoso system:** This was developed by Vuoso Research Institute for machine tools and metal cutting in Czechoslovakia in 1959. It is a simple 4 digit code. A family is a collection of parts, which are similar for purpose of manufacture. Usually they are related by size and geometric shape and require similar manufacturing operations.

2. Production Flow Analysis

This technique was derived by Burbidge (1975) to enable groups of parts to be defined in terms of the operations that they will require for their manufacture. Major groups are formed using factory flow analysis on the basis of similar operation route sequence.

Production flow analysis has three stages :

(*a*) **Factory flow analysis:** This is concerned with the first major division into large groups of department size and into very large families of parts to be made in these departments.

(*b*) **Group analysis:** This concerned with the division of the plant assigned to each department on to groups and the division of parts into associated families. The primary aim is to achieve the simplest possible material flow system inside each department and, to help to achieve this, the three secondary aims are adopted. As far as possible, each part should be

processed in one group only, each machine type should exist in one group only and incompatible processes should be in different groups.

(c) **Line analysis:** This is concerned with the flow of materials between machines inside the groups and with planning the best layout for machines. The aim is to find the sequence of machines which will give the nearest approximation to line flow.

3. Cluster Analysis

Clustering is the science of classification of objects based upon their possession or lack of defined characteristics. This approach provides a way to study the similarities between a diverse population of objects in a quantitative manner.

The clustering approach provides us with algorithms for the study of similarities between objects in a quantitative manner. This involves three stages:

1. **Preparation of part operation matrix.** This indicates whether the features are either present or absent.
2. **Compute a similarity co-efficient matrix.** This is based on the extent to which the parts share common characteristics.
3. **Perform a cluster analysis.**

The similarity between each pair of objects is examined and groups of objects formed such that within each group the objects are similar to each other according to the set of rules.

30.6.2. CHOICE OF FAMILY

The composition of family of parts which are "housed" in a group is largely determined by the equipment available within the organisation. Four aspects of the group likely to result from family should be examined:

1. What load will the family generate?
2. What capacities and capabilities would be needed?
3. Is it possible to set up the group for the family?
4. Are the necessary machines available or to be obtained?

30.6.3. COMPOSITE COMPONENT

Once a family has been identified, a composite component may be conceptualised. It is the one which contains all of the features of all the members of the family. Composite cum part family

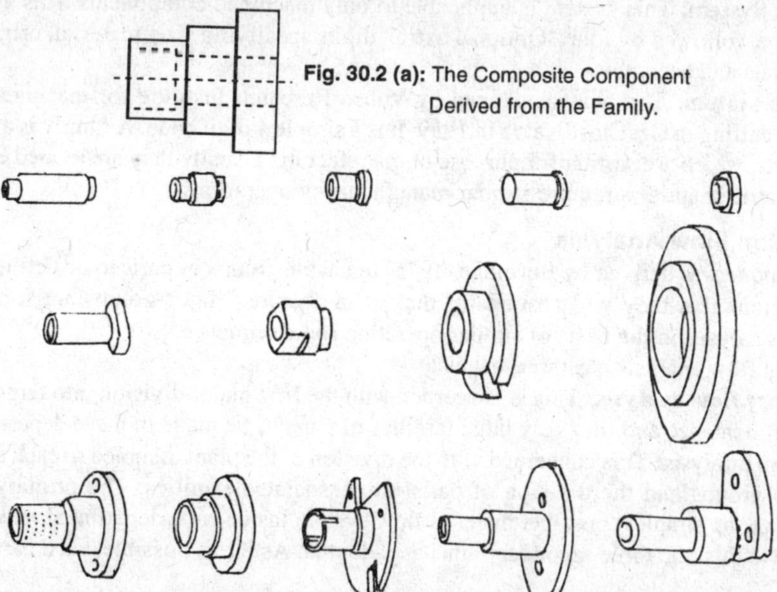

Fig. 30.2 (a): The Composite Component Derived from the Family.

Fig. 30.2 (b): A typical GT family (*The Production Engineer*, February 1970).

shown in Fig. 30.2 (*a*) and (*b*). The available machines are then surveyed to find which group can be best put together to produce the family.

30.6.4. FACILITIES SEQUENCING

The facilities necessary to perform all operations on the part family and the expected load on each equipment can be listed for each family identified.

"Singleton" outlined a simple method for determining a layout sequence for a number of operations or machines through which a variety of parts is processed, each part having a particular route through the operation. Travel or cross charts are sometimes used in developing layouts. Such charts show the nature of the inter-operation movements for all parts for a given period of time.

30.6.5. SEQUENCING

The determination of sequence in which batches of parts are loaded on to the group technology cell or 'Line' will be influenced by the desire to minimise set-up cost and minimise throughput time.

30.7. INTRODUCING GROUP TECHNOLOGY

Introduction of group technology may affect wide areas of the organisation depending on the extent to which it is applied. Considerable preparatory effort is required, including the training of the staff to get acceptance of the accepted change over.

New procedures in production planning and control are needed if the advantages of G.T. are to realised fully.

30.8. BENEFITS OF GROUP TECHNOLOGY

1. Reduced unit set-up time and increased capacity.
2. Improved labour efficiency resulting from standardisation and simplification.
3. Simplification of the material flow system.
4. Simplifies production control, material handling.
5. Reduction in throughput time.
6. Reduce-work-in process.
7. Simplification in planning procedures.

References for Further Readings

1. Ivanov E.K., *Group Production Organisation and Technology*, Business Publication Ltd. (1968).
2. Gallagher C.C. and Knight W.A., *Group Technology Production Methods in Manufacture*, Elis Horwood Ltd., New York (1986).
3. J.L. Burbidge, *Introduction to Group Technology*, Heinemann, London (1975).

31

JUST IN TIME (JIT) MANUFACTURING

• Introduction • Seven wastes • Basic elements of JIT • Benefits of JIT • JIT Philosophy • Kanban System • JIT implementation.

31.1. INTRODUCTION

Most successful companies develop and implement strategies that will give them a competitive advantage. A company that improves performance on a regular and continuous basis certainly will gain the competitive edge. Companies seek competitive advantage by emphasising on performance factors such as flexibility, quick responsiveness, cost, efficiency, quality and reliability and service.

Just in time manufacturing is a philosophy rather than a technique. By eliminating all wastes and seeking continuous improvement, it aims at creating a manufacturing system that is responsive to the market needs.

The phrase just in time is used because this system operates with very low WIP (work in process) inventory and often with very low finished goods inventory. Products are assembled just before they are sold, subassemblies are made just before they are assembled and components are made and fabricated just before sub-assemblies are made. This leads to lower WIP and reduced lead times. To achieve this organisations have to be excellent in other areas, *e.g.*, quality.

JIT is a manufacturing system whose goal is to optimise processes and procedures by continuously pursuing waste reduction.

According to Voss, JIT is viewed as a **"Production methodology which aims to improve overall productivity through elimination of waste and which leads to improved quality."** JIT provides for the cost efficient production in an organisation and delivery of only the necessary parts in the right quantity at the right time and place while using the minimum of facilities. JIT enables one to conceive, design, implement and operate a manufacturing and supporting systems, as an integrated whole based on the principles of continuous improvements and elimination of all kinds of waste.

31.2. SEVEN WASTES

Shigeo Shingo, a Japanese JIT authority and engineer at the Toyota Motor Company identifies seven wastes as being the targets of continuous improvement in production processes. By attending to these wastes, the improvement is achieved.

1. **Waste of overproduction :** Eliminate by reducing set-up times, synchronising quantities and timing between processes, layout problems. Make only what is needed now.
2. **Waste of waiting :** Eliminate bottle necks and balance uneven loads by flexible workforce and equipment.

3. **Waste of transportation :** Establish layouts and locations to make handling and transport unnecessary if possible. Minimise transportation and handling if not possible to eliminate.
4. **Waste of processing itself :** Question regarding the reasons for existence of the product and then why each process is necessary.
5. **Waste of stocks :** Reducing all other wastes reduces stocks.
6. **Waste of motion :** Study motion for economy and consistency. Economy improves productivity and consistency improves quality. First improve the motions, then mechanise or automate. Otherwise, there is a danger of automating the waste.
7. **Waste of making defective products :** Develop the production process to be prevent defects from being produced, so as to eliminate inspection. At each process, do not accept defects and make no defects. Make the process fail safe. A quantify process always yields quality product.

31.3. BASIC ELEMENTS OF JIT

1. **Flow layout :** The physical layout of production facilities is arranged, so that the process flow is streamlined, *i.e.*, for each component, the proportion of value-added time should be more, there should be minimum queuing and non-value-added times.

Use of dedicated lines, U-shaped or parallel lines, use of small machines is preferred. Flexibility of equipment is essential to adjust quickly to changing market demand.

2. **Smoothed build up rate :** The build up rate should be smooth over a monthly cycle. To achieve this, under capacity scheduling is resorted to so that they can respond to demand changes.

3. **Mixed model scheduling :** JIT objective is to match the production rate to demand as closely as possible. One way of doing this is to increase the flexibility of production lines to allow concurrent assembly of different models on the same line.

4. **Small lots and minimum set-up time :** The objective of minimising set-up times is to reduce the batch sizes to the minimum possible. This reduces the manufacturing cycle time and inventory. Use of SMED technique (single minute exchange of dies) is recommended.

5. **Buffer stock removal :** Constant elimination of buffer stocks is emphasised to highlight production problems scheduled by high inventory levels.

6. **Kanban card :** It is a pull system of managing material movement comprising of "Kanban card" based on information system. It helps to trigger the movements of material from one operation to another (next). Merely by alternating the frequency of the circulating Kanban, the production system can be made to adjust to demand fluctuations within limits. The number of cards in the system determine the total inventory. Hence, the objective is to minimise the number of kanbans.

7. **Quality :** The achievement of high quality levels is a prerequisite of successful JIT. Zero defect, statistical process control, process data collection and worker centred quality are commonly used quality programmes.

8. **Product and process simplification :** This is achieved through

(*i*) Rationalisation of product range

(*ii*) Simplification of methods of manufacture

(*iii*) Simplification through component item standardisation.

9. **Standard container :** JIT emphasize small standardised containers. This simplifies materials movement and use of M.H. equipment.

10. **Preventive maintenance :** JIT requires removal of causes of uncertainty and waste. Breakdown is a major cause of the uncertainty. Rigorous preventive maintenance attempts to remove the uncertainty.

11. **Flexible workforce :** This is the critical requirement of JIT. Flexible workforce is developed through cross functional training. It is necessary to match production rate and demand rate as closely as possible.

12. **Organisation in modules or cells :** Many JIT factories are organised in small autonomous modules or cells, each cell being totally responsible for its own production and supply of adjacent module. The cells are designed so that material flow between the cells is minimised.

13. **Continuous improvement :** JIT is not one time effort. It is a philosophy of continuous improvement. It seeks the involvement of everyone in the continuous improvement.

14. **JIT purchasing :** Materials and components are purchased in accordance with well defined requirements in terms of quality, quantity and delivery. JIT purchasing vendor development, long-term buyer-seller relationship, vendor involvement in design of products high quality if purchased material, frequent part delivery, etc. Supplier JIT is a prerequisite in JIT manufacture. The key elements are represented in a Table 31.1.

Table 31.1 : Key Elements of JIT

- Have a flexible workforce capable of using multiple skills.
- Strive for reduced set-up times and small lot sizes.
- Work for a constant master schedule.
- Insist for defect free materials and supplies be delivered when needed.
- Use a kanban or comparable system to pull needed inventory through the system
- Necessary support system to be developed
  (reliable vendors, employee involvement and cooperation, total productive maintenance).

31.4. BENEFITS OF JIT

The most significant benefit of JIT is to improve the responsiveness of the form to the changes in the marketplace thus providing an advantage in competition. The benefits are :

1. **Product Cost :** is greatly reduced due to reduction of manufacturing cycle time, reduction of waste and inventories and elimination of non-value-added operations.
2. **Quality :** is improved because of continuous quality improvement programmes.
3. **Design :** Due to fast response to engineering change, alternative designs can be quickly brought on the shop floor.
4. **Productivity improvement.**
5. **Higher production system flexibility.**
6. **Administrative ease and simplicity.**

31.5. JIT PHILOSOPHY

The roots of the JIT system can probably be traced to the Japanese environment. Japan has inherent limitations of lack of space and lack of natural resources. Japanese have developed an aversion towards all kinds of wastes. They view scrap and rework as waste and hence strive for perfect quality. They strongly believe that inventory storage wastes space and results in locking up of valuable materials and capital. Any thing that does not contribute value to the product is viewed as waste. Thus, it is quite natural for the JIT philosophy to develop in Japan. Apart from eliminating wastes JIT has another important feature utilizing the full capability of the worker. Workers in JIT system are charged with responsibility for producing quality parts just in time to support the next production process. The objective of JIT system is to improve profits and return on investment through cost reductions, inventory reduction and quality improvement. Involvement of workers and elimination of waste are the means of achieving these objectives.

So, JIT manufacturing is a broad philosophy of continuous improvement that includes three mutually supportive components such as,

1. People participation and involvement
2. Total quality control
3. Just in time flow

The People Involvement

The stock less production or zero inventories have created an impression that JIT is only an inventory program. JIT has a strong human resource management component that must be recognized in order to exploit the full potential of technology component. The success of JIT depends upon how the companies train their human resource to have an appropriate skill, responsibility and co-ordinate and motivate people JIT seeks to fully utilize the creative talents of employees, suppliers, subcontractors and others who may contribute to the company's improvement. Teamwork, discipline and supplier involvement are the important components that contribute to the success of JIT.

Total Quality Control (TQC)

Total Quality Control refers to the achievement and improvement in quality in a JIT company, which involve every department and each employee in the company. All employees should seek ways to serve the final customer better so that the company can remain competitive.

Internal customer Concept — JIT companies believe in broad definition of a customer. The traditional organization define that customer is a person outside the company who buys and uses the products and services. JIT companies adds to this definition the concept of immediate customer (or internal customer) who is the next person or department or process who uses or further processes them. If each worker sends only defect free items to his immediate customer. no defective final products will be produced.

Quality at Source — Each employee is given the responsibility for quality at his work-station. Employees are trained in quality principles and testing procedures. They inspect their own work to ensure that the defectives are not passed onto the next process. The defective element is more easily detected by the immediate customer than by the person who is responsible for it. *e.g.,* a part may not fit in to the assembly if it is not properly made. A procedure called "JIDOKA" is brought in to effect. Any employee who senses that a process is producing defects or is about to go out of proper specification has the authority and the responsibility to stop the process. The concept behind this is that it is better to stop the production rather than producing defects.

TQC is a culture not a program

The TQC philosophy aims at the culture of continuous improvement in which people always strive to do better. Companies continue to look for incremental product improvements and process refinements. The objective here is to reduce variability in processes and in parts because it is the variation, which makes the product deviation from quality. Total quality efforts extends to suppliers. When suppliers quality reaches a consistently high quality, there is no need for the supplier to go through incoming inspection.

JIT Flow

A queue in front of the work center represents the WIP. Any form of inventory is a waste as per the JIT philosophy when the queues are long, the cost of holding the WIP becomes high and the time required for a job to flow through the required work centers becomes excessive. The major objective of JIT is to have only the right item at the right place at the right time. This practice reduces the WIP and hence the working capital requirement but also the floor space and the flow through time. Thus the important aspects that support the JIT flow are

- Uniform production rate and mixed model assembly.
- Pull method of co-coordinating work centers.
- Quick and inexpensive set ups.
- Multi skilled work force and flexible facilities.
- High quality levels with no rejects or reworks.

The JIT system is represented in fig (31.1).

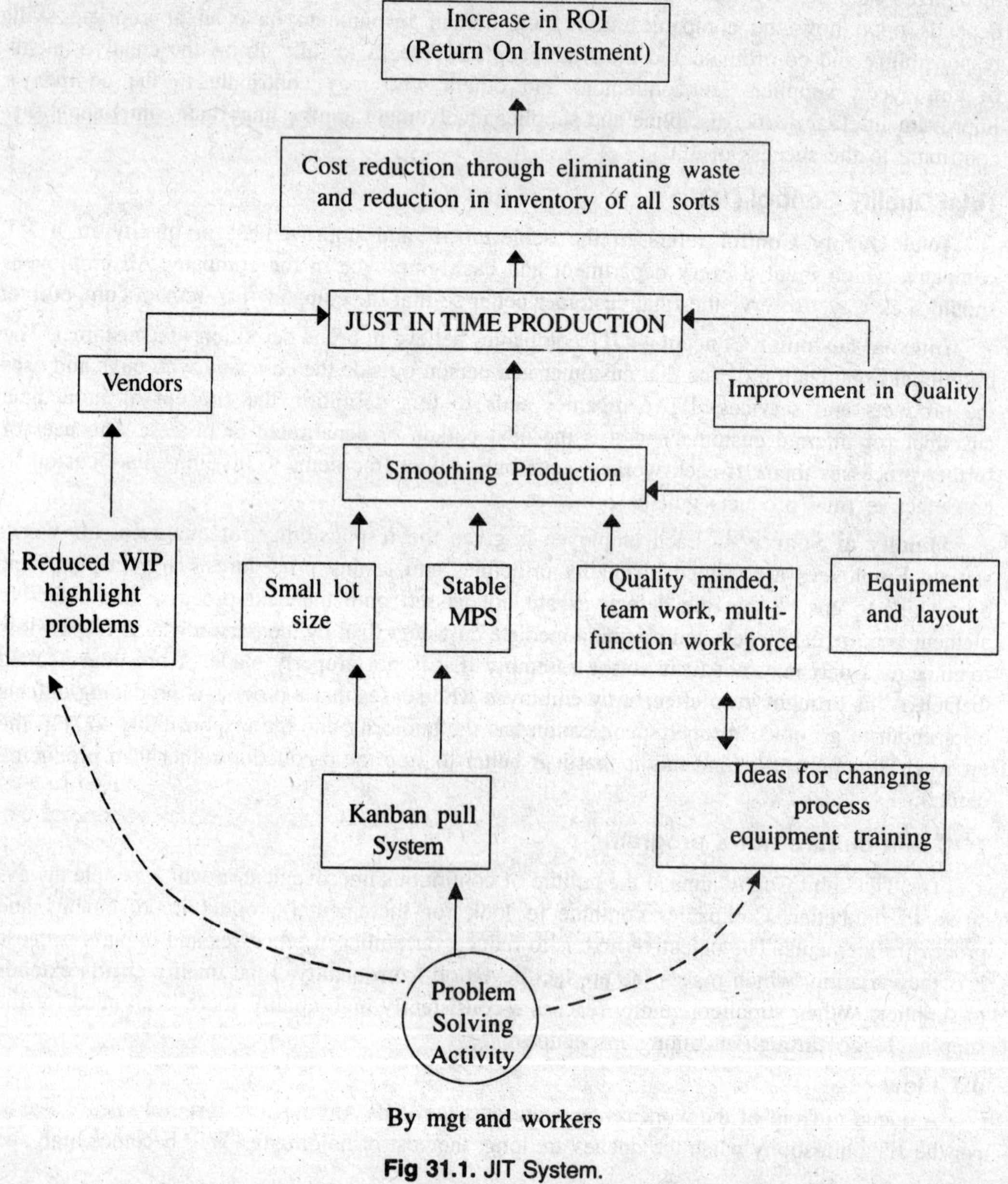

Fig 31.1. JIT System.

31.6. KANBAN SYSTEM

Kanban system is a simple information system used by a work center to signal its supplier work center to request a replacement container and to authorize production of another container of that particular item. The name comes from a Japanese word "Kanban" which means card or sign. Originally a card was used to signal the supplying work center. A work center can use a variety e.g. a flashing light, the empty container or a computer communication or a message. The purpose of the Kanban system is to signal the need for more parts and to ensure that those parts are produced in time to support subsequent fabrication or assembly. This is achieved by pulling parts through the assembly line. Only final assembly line receives a schedule from the dispatch office. All other operators and suppliers receive production orders (Kanban cards) from the subsequent (using) work centers.

The Kanban system is a physical control system consisting of cards and containers. In the two cards Kanban system, two types of cards are used. The production card (P-card) authorizes the work centers to make on standard container of a particular part specified on the card. The second type of one container of the specified part from a particular work center to another work, center as specified on the card. Since these cards are continually reused, they are issued only when the production of an item is to be started or changed significantly.

A working of a Kanban system is shown in fig (31.2).

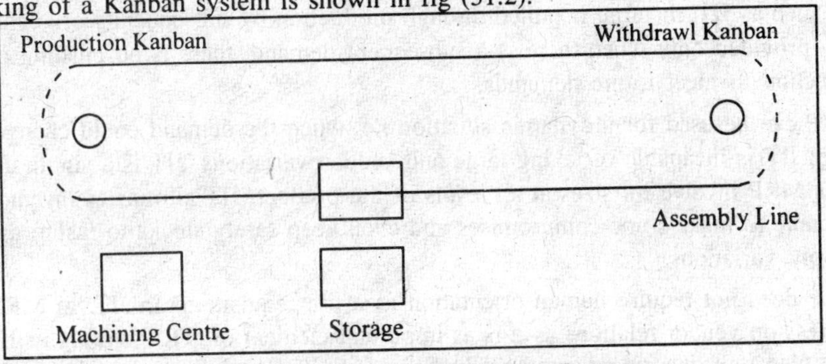

Fig. 31.2. Working of Kanban System.

The working of the system is as follows :

Assume that the containers are moved one at a time. When the containers of the parts is emptied at worker center B, the empty container and the withdrawal card are taken back to work center A. The production card from a full container of parts is removed from its container and replaced by the withdrawal card. The production card is then placed in the Kanban receiving post at work center A, there by authorizing production of another container of parts. The empty container is left at work center A. The full container and its withdrawal card are moved to work center B and placed in the input area. When this container of parts is used, its withdrawal card is empty container are taken back to work center A and the cycle is repeated.

The number of containers needed to operate a work center is a function of the demand rate, container rate and the circulating time for a container.

The number of containers is computed using the formula;

$$n = \frac{DT}{C}$$

where n = total number of containers.

D = Demand rate of using work center.

C = Container size in number of parts.

T = Time for the container to complete entire cycle. (Referred to as lead time)

Inventory can be decreased by reducing the size of the containers or the number of containers used. This is done by reducing the time required to circulate the container including its machine setup time operation time, waiting time or move time.

Kanban system can be a very simple and effective method of co-coordinating work centers and vendors. The organization must be well disciplined so that there is always an authorizing Kanban with every container, ensuring that only appropriate items are produced and excessive inventory does not build up.

Comparison Between JIT and MRP

In comparing the relationship between JIT and MRP, it is important to understand the distinction between push and pull system of production control. A push system like MRP, pushes material into production to meet future needs, Master schedule is based on future forecasts/orders which determines what components are to be ordered and pushed through production. In pull system, such as JIT, material is pulled through production by subsequent work centers. Materials are only provided only when there is a subsequent demand, there is no pushing of materials in to production to meet future demands.

MRP can be used for a dynamic situation i.e. when the demand could change significantly in future; JIT is incapable of taking large and sudden variations. JIT is a single unit production where as MRP involve lot sizes at all levels of the product. JIT eliminates inventories where as MRP is able to make some compromises and even keep safety stocks to insure against demand and supply variations.

MRP does not require human orientation as a pre requisite as in JIT. In MRP there is not much stress on vendor relations as it is as important element of JIT. Planning in JIT is easy one needs to plan only the smoothened production of the finished products, where as, in MRP one needs to plan for every intermediate product and process as well.

JIT is preferred in case of repetitive (mass) production where as MRP is suited for job or batch production.

31.8. IMPLEMENTATION OF JIT

To facilitate the implementation of JIT, Hall suggests the following approach is suggested by Hall.

1. Obtain commitment from top management.
2. Prepare an implementation plan.
3. Gain the co-operation of the work force.
4. Create a strong leadership on the shop floor.
5. Guarantee stable employment, engage training and encourage participation and teamwork.
6. Level the production and smoothen the flow.
7. Reduce set up times of machines/equipments.

8. Balance fabrication rates with final assembly rates.
9. Provide spare capacities in all areas.
10. Extend JIT to suppliers.
11. Remove the bottlenecks and stabilize delivery schedules.

References For Further Reading

1. Korgaonkar M.G., *Just in Time manufacturing*, Macmillan India Ltd., Delhi (1992).
2. Monks Joseph, *Operations Management*, 3rd edition, McGraw Hill International Edition, New York (1984).
3. Lubben R.T., *Just in Time Manufacturing*, McGraw Hill, New York (1988).
4. Shingo S., *Study of Toyota Production System from Industrial Engineering viewpoint*, J.M.A., Tokyo (1981).

32

OPERATIONS STRATEGY

• Introduction • Operations Strategy • Operations Decisions • Operations Priorities • Components of Production Strategy • Frame work for manufacturing/operations Strategy • Developing and Implementing Production Strategy — Steps in developing a strategic plan for an S. B. U. • Implementation of strategy • Focused Operations • Types of Operation Strategy — Differentiation and Cost Leadership Strategy • Interface between Operations and Production function • Porters Five Forces model.

32.1. INTRODUCTION

Operations management may be defined as the design, operations and improvement of the production systems that create firm's primary products and services. The word operations here include both the manufacturing and non-manufacturing operations (services). As a basic business function, along with marketing and finance, the operations functions play a vital role in achieving a company's strategic plan. It has become all the important that operations must help the firm to achieve a competitive advantage in the market place. Operations should not be considered as only a place to make organization's products and services but should be able to provide some competitive strength to the business as well. Gaining a competitive advantage through improved operations performance requires a strategic response on the part of the operations function.

32.2. OPERATIONS STRATEGY

Productivity improvement is an imperative in operation strategy for all firms. In order to remain competitive, a firm must continually seek ways of continually reducing costs and must seek new ways of improving the operations. For firms operating in global markets, productivity improvement needs to become a managerial religion. Productivity improvement get implemented through production/operations strategy.

Definition of Operation Strategy

Operation Strategy is concerned with setting broad policies and plans for using the resources of the firm to support in a best possible way firms long-term competitive strategy.

Schroeder, Anderson and Cleveland (1986) have defined operations strategy as consisting of four components — mission, distinctive competence objectives and policies. These four components help define what goals operations should accomplish and how it should achieve these goals.

The resulting strategy should guide decision-making in all phases of manufacturing.

Thus, Operations strategy is a vision for the operations functions that sets an overall directions or thrust for decision making.

32.3. OPERATIONS DECISIONS

Operations strategy is a functional strategy and it should get its direction from business or corporate strategy. A simple operations strategy model is represented in the fig (32.1).

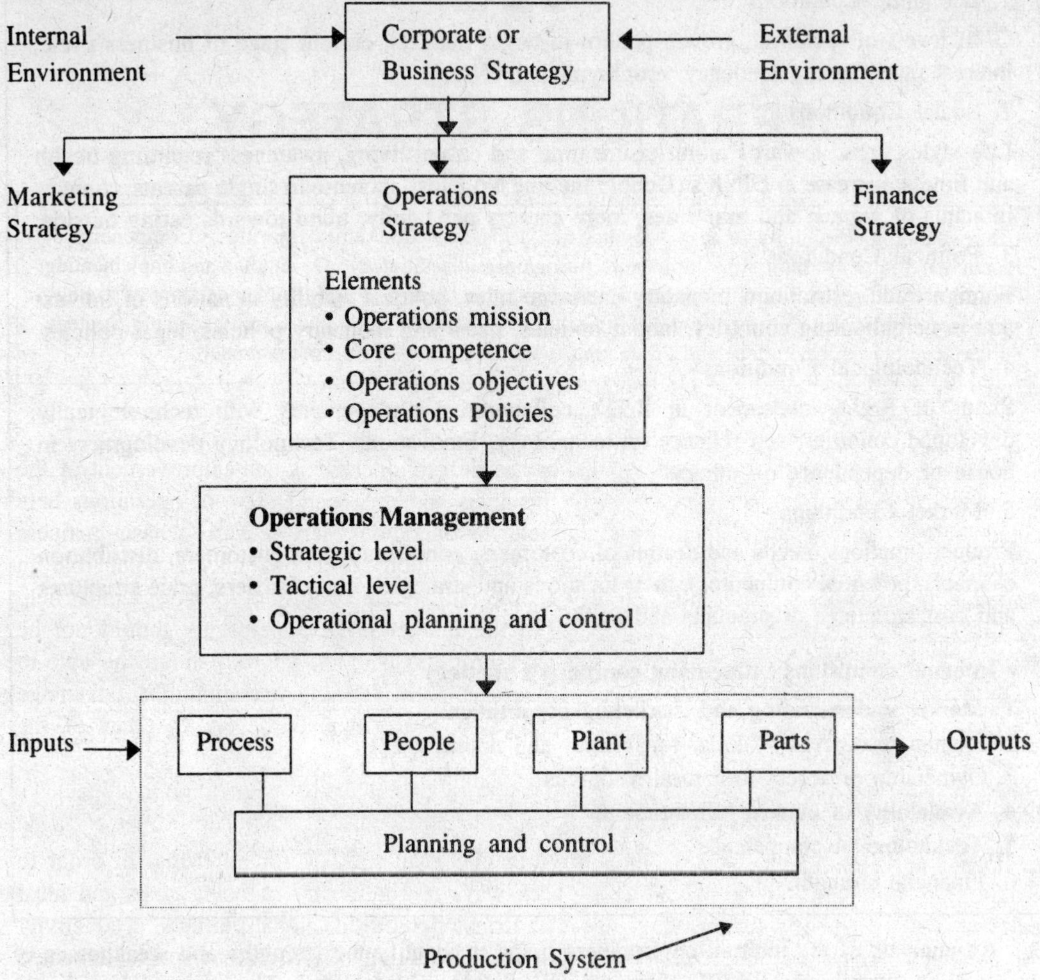

Fig. 32.1. Operations Strategy Model.

Operations decisions are made in the context of the firm as a whole. To start with, at the top we have the corporate or business strategy. This strategy defines how a particular business will compete and it is based on the corporate mission. In particular, the corporate strategy reflects how the firm plans to use all its resources and functions (marketing, finance and operations) to gain competitive advantage.

Many organisations consist of a group of related businesses, each of which is identified as a separate division or Strategic Business Unit (SBU). Each SBU is required to find its own basis of competition based on the products it offers and the selected market segment. Therefore, it is extremely important that the business or corporate strategy be defined before operations strategy is formulated.

In order to formulate corporate strategy, as well as the operations strategy, a thorough analysis should be made of the external and internal environment. The factors pertaining to internal and external conditions are represented in the table 32.1.

Table 32.1

EXTERNAL AND INTERNAL CONDITIONS INFLUENCING STRATEGY

- **External conditions influencing company's strategy**

1. Economic Conditions

GNP, levels of spending, growth pattern in target markets, current stage of business cycle, interest rates, saving tendency, employment levels.

2. Social Conditions

Life styles, trend towards more leisure time and casual living, awareness regarding health and fitness, increase in DINK's (Double Income No Kids), increase in single parents, changes in status of women and minorities, more earners per family, trend towards eating outside.

3. Political Conditions

Foreign trade restrictions, monetary exchange rates, political stability in nations of interest and in neighbouring countries, labour policies, fiscal and monetary policies, legal policies.

4. Technological Conditions

Status of R&D, investment in R&D, collaborative arrangements with technologically developed countries, self-reliance on technology, Innovations, Technology development in house or dependence on others.

5. Market Conditions

Product functions, needs and desires of customers, concentration of customers, distribution channels, potential competitors, their locations and strategies, entry barriers, price structures and cost structures of products and PLC.

- **Internal conditions influencing company's strategy**

1. Market understanding and marketing capabilities
2. Human resources — skills, capabilities and management
3. Ownership or access to natural resources
4. Availability of current infrastructure
5. Technological competence
6. Financial strength.

An analysis of the internal environment helps to identify the strengths and weaknesses of the existing operations. SWOT (Strength, Weakness, Opportunity, Threat) analysis should indicate a "Proper Fit" between the proposed strategic response of the operations function and the strengths of the company. Strengths and weaknesses of the company need not be taken as static or as constants. A company should always introspect, "What should be our strengths? What weaknesses should be overcome or how best to convert the weaknesses into strengths to cash on potential opportunities." The SWOT analysis is represented in the fig (32.2). This SWOT analysis forms the basis for strategy formulation. Operations strategy attempts to overcome weaknesses and build on existing strengths.

Operations strategy specifies how the firms will employ its production capabilities to support its corporate strategy. Mainly there are four important elements that define the purpose of operations functions in relation to corporate strategies.

(*i*) Operations Mission

It defines the purpose of operations function. It should decide the priority among the operations objectives of cost, quality delivery and flexibility.

(*ii*) Core (Distinctive) Competence

Core competence refers to those capabilities that differentiate the manufacturing (or service) firm from its competitors. The core competence should support the mission of operations and it leads to competitive advantage of the firm. Distinctive advantages can take on a variety of forms. *e.g.*, Operations can be distinctive in terms of operations objectives such as lowest cost, highest quality, best delivery or greatest flexibility or they may be distinctive in terms of best technology compared to competitors, most people oriented, by its resources etc.

(*iii*) Operations Objectives

Objectives in operations should be stated in specific quantitative and measurable terms. They should reflect the actual results to be achieved in the short and long run. The operations objectives are to be stated in terms of cost, quality, delivery and flexibility.

(*iv*) Operations Policies

The operations policies define broadly how the objectives of operations will be achieved. Operations policies should be developed for each of the decision categories such as process, quality, capacity, inventory and work force.

Within the operations function, management decisions can be categorized into three broad areas.

- Strategic (Long term) decisions
- Tactical (Intermediate term) decisions
- Operational planning and control (Short term) decisions

The strategic issues are usually broad based, addressing the issues such as — How to make a product ? (Process selection). Where to locate the facilities? (Plant location). How much capacity to be created ? (Capacity planning) etc.

Operations management decisions at the strategic level have impact on the company's long-range effectiveness in terms of how best it addresses the customer needs. Thus, for success of the organisation, these decisions should be in line with the corporate strategy. The strategic level operations decisions become operating constraints under which the firm must operate in both the intermediate and short term.

Tactical planning normally addresses how to efficiently schedule material and labour within the constraints prefixed by strategic decisions. The decisions that normally come under the purview of tactical level are — How many workers we need? When do we need? Should we work overtime? What should be the stock level? These decisions in turn will become the operational constraints under which planning and control decisions are made.

Table 32.2 : SWOT ANALYSIS OF COMPANY

STRENGTHS (What the firm has)	WEAKNESSES (What the company has to overcome)
• Resource availability	• Industrial relations
• Technology	• Employee age
• Capacity	• Equipment/plant age
• Flexibility	• Management style
• Skills and Competency	• Work culture
• Motivation	• Competitive position
• Quality mindedness	• Job satisfaction
• Financial healthiness	
• Cost structure	
• Loyal customer base	

OPPORTUNITIES (The company has to cash on or utilize)	THREATS (Anticipate and Combat)
• Population characteristics • Government policy changes • New products, new markets • Interest rates • Exchange rates • Improved communication • Globalization of business	• Competition and competitors strategies • New competitors • New technologies • Consumption trends — supplementary and complementary products

32.4 OPERATIONS PRIORITIES

According to skinner of Harward Business School and Hill Terry of London Business School, a few basic operations priorities are determined. The priorities include -

1. Cost

Every industry has a price sensitive segment, which makes the buying decision on the basis of low cost. To successfully compete in this segment, a firm must be a low cost producer. The products strictly sold on the basis of cost are typically - low valued commodity in nature, *i.e.*, it is difficult to distinguish products of one producer from another. Competition is fierce in this segment and the low cost producer will determine the price of the product.

2. Quality

Quality refers to both the product quality and process quality. The level of quality offered in the products design will vary with the market segment to which the product is offered. The objective of establishing the proper level of product quality is to focus on the requirements of the customer. So, level of quality is a balanced with the cost, which is a compromise.

Process quality is critical as it relates directly to the reliability of the product. The goal of process quality is error free production of products. Reliability and quality are built in to the product during design and subsequently during manufacturing phase.

3. Delivery

The speed of delivery i.e. the company's ability to deliver goods and services more quickly than its competitors is critical for some products and services i.e. specially for repairs. Also along with the delivery speed, delivery reliability i.e. the ability of the firm to supply the product or service on or before the scheduled delivery due date is also critical. Delivery reliability is an indicator of dependability of the vendor and is one of the criteria for vendor evaluation or rating.

4. Changes in Demand

The ability of the company to respond to increase or decrease in demand is an important factor in their ability to compete. The ability to effectively deal with dynamic market demand over the long term is an essential element of operations strategy.

5. Flexibility

Flexibility here refers to the ability of a company to offer a wide variety of products to meet the needs of the customers. The prime element of this ability to offer different products is the time required for a company to develop new products.

The other product specific criteria include;

• The degree of customization or the variety and range of products.

- Supplier after sales support.
- Meeting the product launch dates.
- Technical liaison and support.

32.5 COMPONENTS OF PRODUCTION STRATEGY

All the activities carried out in the firm from material flow by suppliers through fabrication and assembly, physical distribution must be integrated for sensible production strategy formulation. Apart from the materials, the other crucial inputs of labour, job design and technology must be the parts of the integrated strategy.

The six basic components of production strategy are

- Positioning the production system
- Capacity/location decisions
- Product and process technology
- Work force and job design
- Suppliers and vertical integration.

These components are basic to production strategy because there is a wide managerial choice available within each and each affects the long-term competitive position of the firm by making influence on cost, quality, production availability and flexibility/service.

Positioning the Production System

If production is not made a part of the corporate strategy, there are all the chances that a mismatch between system and markets is high, resulting in conflict between production and marketing functions. A firm without a unified strategy that includes the production function is likely to anticipate obtaining low cost, high quality, product availability and flexibility from its production system all at the same time or realizing that one cannot optimise all these dimensions simultaneously, that there are trade-offs between them. A firm that attempts to be "all things" in its production system is likely to compromise all four dimensions of production competence and end up. **"Stuck in the middle"** with low margins.

It is useful to think of product volume as independent variable and productive system type as the dependent variable. As the product develops through its life cycle, the productive system goes through a life cycle of its own from job shop system (Process focused made to order) when the product is in its initial stages through intermediate stages to a continuous system (Product focused, made to stock) when the product is demanded in large volumes. These stages of product system and process development are interdependent. The production system is thus dependent upon the volume of product sold and the volume of product sold is dependent upon the costs of production and the price quality competitive position. Wrong positioning of production system will make the production system ineffective. Thus, production strategy should be the integral part of corporate strategy and positioning of production system should be in line with market requirements (demand) and in accordance with the life cycle of the product.

Capacity/Location Decisions and Product — Process Strategies

A large part of the problem is in managing the potential or real over capacity in the company and in industry. The appropriate capacity moves can be rather obvious when new products are in their rapid growth phases. Since the potential costs of lost sales are so great in these instances, capacity expansion is clearly justified. Over capacity is helpful in cases where there is a certainty of explosive increase in demand, otherwise over capacity carries with it a higher overhead cost. Thus, over capacity is justified when the new products are in their growth stage.

Advanced technology will make production capacity more flexible and are subjected to the effects of product and schedule changes. The firms that develop the capacity with greatest flexibility will have competitive advantages in terms of fast responsiveness to major shifts in product design and demand. Though operations decisions do not have a strategic impact, the Japanese have been particularly successful in creating effective operations systems that have a strategic significance in cost reduction and quality control. Japanese firms put their maximum effort towards reducing the setup time and costs through efficient tool design, quick clamping devices carefully worked out procedures etc. They developed a technique, which is popularly known as "SMED" (Single Minute Exchange of Dies) to reduce the setup time to the minimum possible. The ultimate objective here is to reduce setup costs to EOQ equal to one unit. This reduces the in process inventory and the change over time from one product to another is minimized thus maximizing the flexibility.

It is needless to make a distinction between the long-term strategic issues and short term operating decisions. Though capacity, process technology and labour costs have strategic significance, but inventory quality and other shop floor issues tend to be neglected as if "production" has no long term importance. Yet, quality, cost and timeliness of delivery of products can be extremely important in the basic strategy of the firm.

Strategies Regarding Suppliers and Vertical Integration

Purchasing and relationship with suppliers must be carefully formulated to be a part of the production strategy. In dealing with suppliers, there are many issues concerning production that need to be addressed. The focus of attention is on strategic issue of choosing among alternative suppliers, judging the strength of our own position in dealing with suppliers and using the strength effectively in choosing the suppliers. There should be a compatibility of the suppliers objectives with company's own objectives to achieve overall strategic goals.

Since supplier processes are really an extension of manufacturing processes of firm, the question of whether or not to integrate backwards and develop internal capacity to produce the component must be considered. The firm has to think under what conditions vertical integration is an appropriate strategy.

Purchasing and supplier relationships should be carefully formulated to be the part of the production strategy. In examining supplier relationships and purchasing strategy, the major issue is whether or not the vertical integration is the logical step. In vertical integration decisions, the emphasis is on the logic of change within the strategy of the firm to produce some thing of economic value.

32.6. FRAMEWORK FOR MANUFACTURING/OPERATIONS STRATEGY

Operations strategy cannot be formulated in isolation from other functional areas of business; Operations strategy must be linked vertically to the customer needs and desires and horizontally to other parts of the organisation like marketing, finance etc. The operations strategy framework for manufacturing is shown in fig (32.2).

Fig (32.2) shows the linkages between customers needs and desires, performance parameters for manufacturing operations, operations and other related enterprise resource capabilities to satisfy the needs. Strategic vision of the top/senior management identifies the target market, firm's product line, the operations capabilities and core competence.

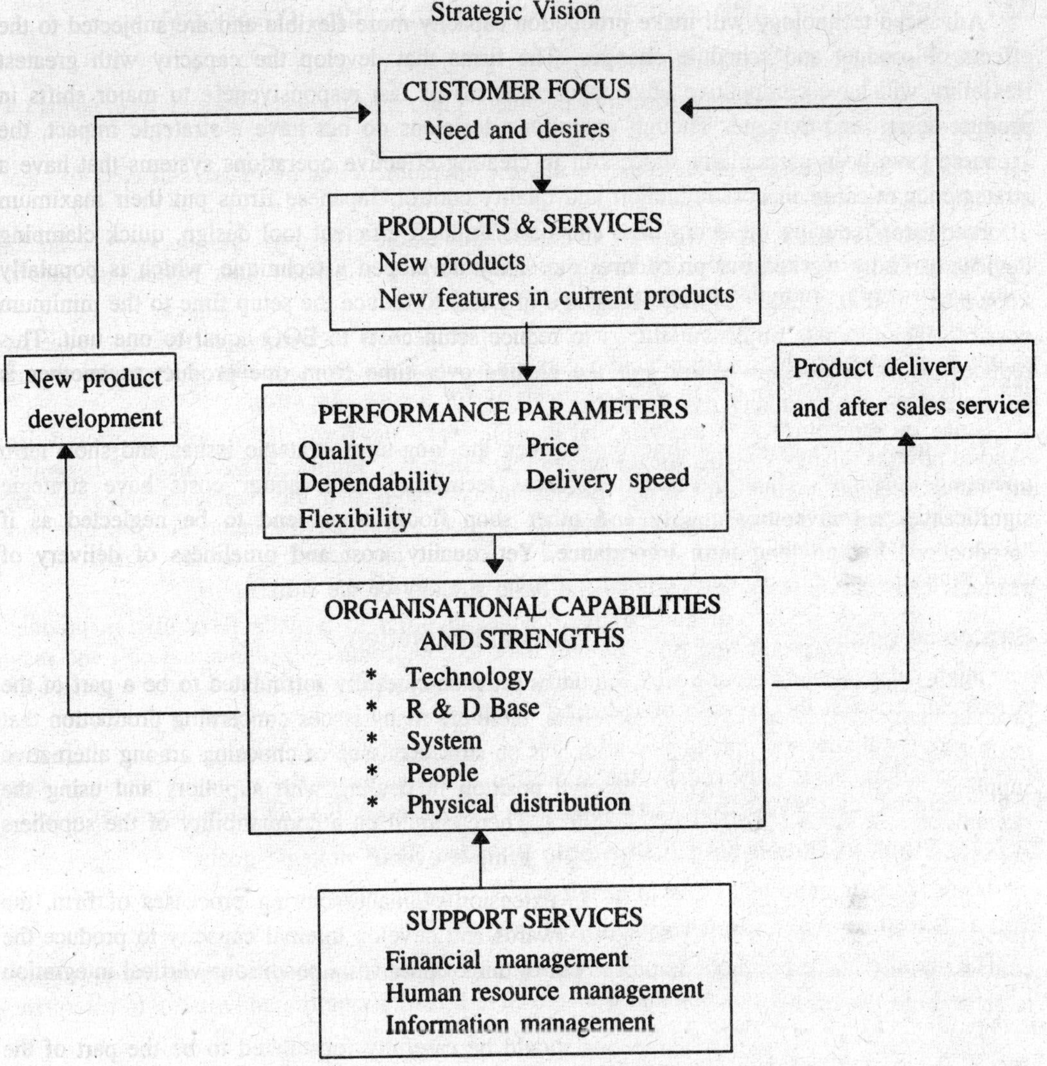

Fig. 32.2. Operations Strategy Framework

The customer focused new products or current product requirements give rise to performance parameters, which become the required priorities for operations. These performance requirements are linked to organizational strengths and capabilities. A close coordination of support services like financial management, information management, human resource management etc. is essential for the operations to satisfy customer needs. Based on the operations performance requirements an operations division uses its capabilities to win the orders. These capabilities include technology, systems and people. Computer Integrated Manufacturing (CIM), Just In Time (JIT) and Total Quality Management (TQM) represent fundamental concepts and tools used in each of the three areas. The supplier's capabilities are also equally important to get the supply of materials and parts as per the specifications. Each and every part is now subjected to an interrogation as to whether to make in house or buy from outside. Thus, as per the argument of Wickham Skinner, what companies need now in this world of intense global competition is not mere technique but a way to structure a whole new product realization system differently and better than any competitor.

32.7. DEVELOPING AND IMPLEMENTING PRODUCTION STRATEGY

The main objectives of production strategy developments are :

(*i*) Translation of required performance parameters into specific performance requirements for operations.

(*ii*) To make necessary plans to assure that the operations capabilities are sufficient to meet performance requirements.

Production strategy must fit into overall organizational strategy. A thorough strategic analysis involves an examination of major competitors in the industry and its markets, identifying its structural strengths and weaknesses.

A novel approach for formulating a consistent strategy is to involve different managerial groups in prioritizing the four production performance criteria, cost, quality, dependability and flexibility. These criteria must be defined precisely for a particular firm.

Based on the competitive analysis and capabilities of the firm, the managerial groups are asked to provide relative weights for the above four criteria. A significant advantage of assigning importance weights to the appropriate production criteria is that the managers are given an opportunity to reveal their assumptions about the product market environment. The rationale and discussion of their view points accompanied by deeper understanding of the other's views, application of trade offs and compromises results in a consensus.

The marketing group will have to realize that in order to provide flexibility in product designs and features, production function will have to compromise on production cost and may incur higher capital and labour costs. Similarly production group will have to realize emphasis on reducing costs at the expense of reducing flexibility is not in the interest of the firm.

A successful strategic planning requires the involvement and commitment of top managers and the commitment is indicated by the allocation of resources to the strategic planning. The involvement of key persons from marketing, personnel, finance, R & D is an important aspect.

32.7.1. Steps in Developing a Strategic Plan for SBU

There are four important steps in developing a strategic plan for an SBU. They are :

Step 1: Situation Analysis (Production Audit)

The purpose of the production audit to take stock of production to obtain a realistic estimate of the status of production function and how effective it is in giving life and vitality to enterprise strategy.

Step 2: Develop Strategic Alternatives

The audit provides insight in to the existing problem areas and highlights some possible directions. In developing strategic alternatives the following questions are to be answered.

- What we are trying to achieve?
- What are the problems with current strategies and their implementation?
- Whether the available capacity is sufficient?
- Whether the technology that is used currently is appropriate?
- If the firm is not cost competitive, how can we change this situation?
- What control techniques are available to maintain the performance level?

These questions should be considered for developing strategic alternatives for productions.

Step 3: Select a strategy in harmony with the enterprise strategy

Step 4: Formulate detailed plans and budgets to implement the strategy.

32.7.2. Implementation of Strategy

The strategy developed must be communicated to every one who has a role in its implementation. Policies and procedures must be established, the roles are assigned to people and resources are allocated in the form of budgets to achieve results.

Technology and production environments constantly change in most of the industries. Implementation of the strategy is an ongoing process rather than onetime allocation of resources. During the period of implementation and subsequent phases, several decisions have to be made. These decisions must be evaluated using the defined criteria consistent with the production strategy.

To achieve the desired goals and successful implementation of the production strategy, the internalization of production priorities and corporate mission should take place at all levels.

The strategic decisions in production are shown in table (32.3).

Table 32.3 : STRATEGIC DECISIONS IN PRODUCTION

1. Capacity decisions	Amount, timing and type
2. Facilities decisions	Size, location and specialization
3. Technology decisions	Process choice, equipments, automation direction,
4. Vertical integration decisions	extent and balance.
5. Work force decisions	Skill level, wage policy, extent of security.
6. Quality decisions	Quality policy, defect prevention, monitoring and control.
7. Production, planning and materials control	Sourcing policies, centralization, and decision rules.

32.8. FOCUSED OPERATIONS

It is required to identify for each product type what performance capabilities operations would have to exhibit for what product to compete successfully in the market. Each business has to identify the relative importance of four major performance capabilities (shown in fig (32.3)) that the operations functions can have. One or more of the performance capabilities are used to describe the critical success factors for company's success.

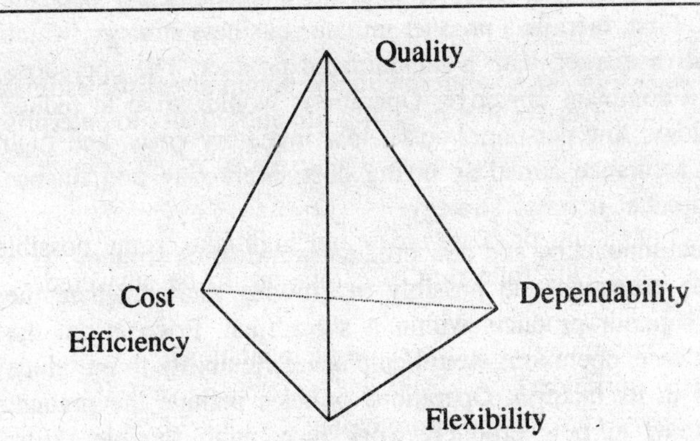

- Cost Efficiency: A firm that emphasizes on cost efficiency will keep all the costs (capital and operation) low compared to all other competing firms.
- Quality: Do all characteristics of the product make it suitable and reliable for intended use
- Dependability refers to the reliability of the service.
- Flexibility refers to the ability of the company to offer a wide variety of products to its customers.

Fig 32.3. Performance Capabilities of Operations.

The concept of operations focus and trade offs came along with operations strategy during 1960s and 1970s onwards. The logic behind the operation focus was that an operation could not simultaneously out perform on all parameters of performance measures. Consequently, management has to decide which parameters of performance were critical to the firm's success and then concentrate or focus the resources of the firm on those particular characteristics.

For example, if a company wanted to focus on speed of delivery, then it could not be very flexible in terms of its ability to offer wide range of products. Similarly, a low cost strategy is not compatible with either speed of delivery or flexibility. High quality is also viewed as a trade off to low cost.

Skinner has suggested the creation of "plant within plant" (PWP) concept for large organisations. Skinner argues that some plants have been made too large and complex under the concept of economics of scale. In this case, the plant should be divided and focused on a specific product or product group, which demands a consistent set of manufacturing tasks. Each smaller plant or PWP could then concentrate on the particular set of cost, quality, delivery and flexibility objectives, which are more appropriate to do it.

Under the PWP concept, even the workers would be separated to minimise the confusion associated with shifting of one type of strategy to other. Thus, a concept of focused factory and PWP are in wider use today. Focused operations can be developed as part of the strategy process several focus dimensions are combined when focusing on operations.

Types of focus dimensions are :

1. Product focus
2. Process type — job or project, batch, continuous
3. Technology
4. Volume of sales
5. New and matured products.

32.9. TYPES OF OPERATIONS STRATEGY

Operations/manufacturing strategy should be a part of overall business strategy and it should be linked to marketing and financial strategies. Basically, there are two diametrically opposite business strategies that can be selected. First, there is a product imitator business strategy, which is characterized by mature, price sensitive market, with a standardized product. The operations mission would emphasize on cost as a dominant objective. Operations would strive to reduce costs through superior process technology, low personnel costs, low inventory costs, and high degree of vertical integration, quality assurance aimed at saving cost. Marketing and finance would also practice and support the product imitator strategy.

The second strategy is called product innovation and new product introduction strategy. This strategy would typically be used in an emerging and possibly in growing market where the advantage is gained by bringing out superior products within a short span. Price is not the dominant factor for competition. In this case, operations would emphasize flexibility to introduce new products as a dominant objective in its mission. Operations policies include the product teams, flexible automation to be adapted to new products work force with flexible skills, outsourcing some key services and parts to retain flexibility. Japanese strategy believed and is successful in using imitator strategy.

The company after deciding the mission, objectives and the target market segment in which the company will compete, the company has to choose between the fundamental strategic competitive options like

(*i*) Differentiation strategy
(*ii*) Overall cost leadership strategy
(*iii*) Focus strategy

Differentiation Strategy

Differentiation strategy aims to create a new product or service, which is unique in that industry. The uniqueness may be achieved through design or brand image, technology, and customer service or dealer network. Differentiation is a strategy to win customers and retaining them for a longer period. There are many differentiation strategies of which flexibility is important one.

The flexibility may be in terms of

(*i*) Product — range, design or product mix
(*ii*) Product volumes
(*iii*) Quick deliveries
(*iv*) Minimum lead time between the product concept and introduction
(*v*) Quicker response to changing needs of the customer

Flexibility is being part of fast response to changing needs of the customers and includes the flexibility of both manufacturing and service, and design functions. The manufacturing or operations part consist of flexibility in machines, process and man power to process alternative product designs (variety) in small batches with minimum wastage of time and at optimal cost. An operations system should offer market flexibilities through flexibilities in machinery, processes technology implementation, men, systems and utilization of capacity and time. The flexibility should improve the capacity and capability to adapt to the changing external and internal environment.

Cost Leadership Strategy

The overall cost leadership aims to produce and deliver the product or service with specified quality, at a lower cost relative to the competitors. The firm practicing this strategy will be distinguished as a lowest cost producer and seller. The lowest cost should be achieved through cost reduction. Approaches to cost leadership as a strategy are represented in the table (32.4).

Focus Strategy

Organisations using focus strategy aim to concentrate on a particular group of customers (nicthe market), geographical market or product lines in order to serve a well defined but narrow market better than its competitors.

Table 32.4 : COST LEADERSHIP STRATEGY —
TRADITIONAL VS MODERN APPROACH

Traditional approach	*Modern approach*
• Cost control and cost reduction are primary focus.	• Focus on elimination of costs due to non-value adding activities.
• Stress is on reduction in budgetary allocations such as training, HRD etc.	• Stress is on improvement in quality if design and thus reduce costs of input materials and costs of processing and packaging etc.
• Defer investment in new machines and replacement of new machines.	• Process improvement and increase productivity reduce scrap and rework.
• Reduce inventories of raw materials, WIP and finished goods.	• Reduce procurement lead times and need to keep inventory
• Focus on reduction in cost of material, labour and overheads.	• Focus on efficient utilization of productive resources
• Techniques used include – Method study – Conventional cost reduction techniques – Product/process audit – Design, modification.	• Techniques used include – Value engineering and analysis – Business process reengineering – Process mapping

32.10. INTERFACES BETWEEN OPERATIONS AND MARKETING FUNCTION

Operations function has to work in synergy with all the other functional areas of business. A company's competitive advantage stems from such holistic approach. There should be a support approach between various functional areas specially operations and marketing.

Thus, an interface between marketing and operations is necessary to provide a business with an understanding of its markets from both perspectives. Terry Hill has introduced the words "Order winners" and "order qualifiers" to describe market oriented priorities that are key to competitive success.

An order winner is a criterion that differentiates the products and services of one firm from another. Depending on the situation, the order winning criteria may be cost of the product (price), product quality and reliability or any other priority developed by the firm. An order qualifier is a evaluation criterion that permits firm's products to be considered for purchase by customers.

The order qualifying criteria may include conformance to specification, quality, on time delivery, product reliability. Low price is emerging as the order winner. It is important to note that the order winning and order qualifying criteria may change over time.

Marketing strategy based on any of the P's such as price, positioning, place should be congruent to manufacturing strategy and vice versa. For example, if "Place" is important in the marketing strategy, then the manufacturing function must gear up to the challenge of multilocational operations or providing a product on time at different locations. Marketing strategy may lead to company's technology strategy and operations strategy should provide perfect back up for the customer service goals of marketing function. Marketing function should provide market feed back for operations.

32.11 PORTER'S FIVE FORCES MODEL

Porter's five forces analysis is useful in formulating an appropriate operations strategy. The model is represented in fig (32.5).

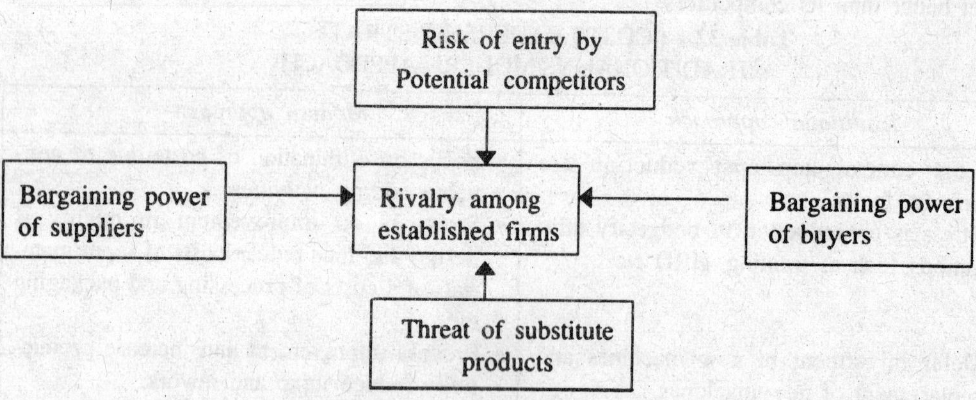

Fig 32.5. Porter's Five Forces Model.

Porter's model is one of the structured ways in which the industry environment is analyzed. He argues that the stronger each of these forces, it will be more difficult for the company to raise prices and make more profits. A strong force is equivalent to a threat; a weak force is equivalent to an opportunity. Those who are involved in operations strategy formulations have to recognize the opportunities and threats as they arise and design an appropriate strategic response from the operations function. In a dynamic business environment, each of the forces can change over time due to factors beyond the firm's control. The relative strengths of these forces can also be decided by the company through appropriate choice of the strategy and turn the forces to its advantage. *e.g.*, the threats of substitute products can be minimized by providing variety of products. This requires an appropriate operation's strategic response through a cost conscious yet flexible system of manufacturing. Risk of entry of potential competitors can be reduced through providing a long lasting relationship between the two companies through an atmosphere of transparency and through dedicated service orientation. Operations strategy and consequent actions can be built around these organizational requirements.

References For Further Reading

1. Ansoff Igor, "Corporate Strategy", Penguin Books London, U.K., 1968.
2. Aaker David A., "Developing Business Strategies", John-wiley* Sons, New York, USA, 1995.
3. Hill Terry, "Manufacturing Strategy", Macmillan Education, London, 1985.
4. Ohmae Kenichi, "The Mind of the Strategist, the Art of Japanese Business", Mcgraw Hill, New York, USA, 1982.
5. Peter Thomas J and Waterman Jr. Rober H., "In Search of Excellence", Harper and Row, New York, 1982.
6. Michael Porter E., "Competitive Advantage", The Free Press, New York, 1985.
7. Skinner Wickham, "Manufacturing — A formidable Competitive Weapon", John Willey and Sons, New York, 1985.

33

MATERIALS MANAGEMENT

• Introduction • Objectives of Materials Management • Scope • Integrated Materials Management • Flow of Materials in Manufacturing • Purchasing — Purchasing Cycle • Methods of Buying • Source Selection • Vendor Rating • Vendor Relations • Supply Chain Management.

33.1. INTRODUCTION

The basic objective of business is to satisfy the customer. Managers have to ensure that there is a continuous flow and optimum utilization of materials. Manufacturing organisations must schedule the simultaneous availability of appropriate materials and work capacity so that the materials will be transformed into goods/service that will provide the necessary functions required by the customer. Materials management is concerned with management functions supporting the complete cycle of material flow, from the purchase and internal control of production materials to planning and control of work in process, to warehousing, shipping and distribution of the finished product. An effective materials management process ensures that the right kinds of materials are at the right place whenever needed.

Materials management is concerned with planning, directing and controlling the kind, amount, location, movement and timing of various flow of materials used in and produced by the process.

33.2. OBJECTIVES OF MATERIALS MANAGEMENT

Materials management objectives are categorized into

• Primary objectives

• Secondary objectives

Primary objectives are stated as

"Making available (supply) of materials in specified quantity and quality at economic cost and maintaining the continuity of supply. Minimisation of investments in materials and inventory costs. And assuring high inventory turnover."

Secondary objectives

Secondary objectives help to achieve the primary objectives. The secondary objectives can be stated as

1. Purchasing the items from a reliable source at economic price.
2. Reduction of costs by using various cost reduction techniques such as variety reduction, standardization and simplification, value analysis, inventory control, purchase research etc.
3. Co-ordination of the functions such as planning, scheduling, storage and maintenance of materials.

33.3. SCOPE OF MATERIALS MANAGEMENT

Materials management encompasses all the aspects of the materials, *i.e.*, material costs, material supply and material utilization. Materials management is concerned with material planning and materials control activities. The details of planning and control activities are represented in table (33.1).

Table 33.1: Planning and Control Activities of Materials Management

• Planning	– Right type, quantity of purchasing of materials.
• Sourcing	– Selecting a right source of material supply and procurement from these sources.
• Controlling	– Follow up and monitoring, inventory control.
• Storing	– Receiving, inspection, storing and issue of materials.
• Handling	– Flow or movement, packaging, transportation.
• Distribution	– Distribution of finished goods, warehousing and shipping.

33.4. INTEGRATED MATERIALS MANAGEMENT

Materials required for production purpose are normally procured and stored in the plant and issued to manufacturing when there is a requisition. Materials are to be purchased in advance and stored to ensure uninterrupted supply.

There should be a proper co-ordination and co-operation among different functional heads of materials department to optimise the operations of materials management. The materials function to be effective, the objective must be to maximise materials productivity. An integrated approach to materials management i.e. materials planning and control must look in to the problem areas in a co-coordinated manner in order to maximise the effectiveness of materials management.

An integrated materials management will result in the following advantages :

- Better accountability for materials and material concerned costs.
- Better co-ordination within the materials functions and also other functional areas of business.
- Better performance and effectiveness.
- Adaptability to automated and computerised systems.

The important areas to improve materials planning and control are :

1. Value analysis and purchase price analysis.
2. Materials planning and control (Inventory Control).
3. Stores control.
4. Waste management.
5. Materials handling.

Various elements of integrated materials management are represented in table (33.2). The inventory control is covered in chapter (17).

Table 33.2: Elements of Integrated Materials Management

1. **Materials Planning**
   - Forecasting materials and parts requirements.
   - Preparation of material budgets.
   - Forecasting material, inventory levels.
   - Scheduling orders and monitoring of performance.
2. **Inventory Control**
   - Selective control of materials.
   - Determining economic order quantity (EOQ).
   - Fixing level of safety stock and reorder level.
   - Lead time analysis.
3. **Purchasing**
   - Selection of source and supplier evaluation.
   - Finalization of terms and conditions of supply (negotiation).
   - Planning orders and follows up.
4. **Stores Management**
   - Receipts, issue and storage of materials.
   - Preservation of stores.
   - Efficient handling and disposal.
   - Maintenance of stores records.

33.5 FLOW OF MATERIALS IN MANUFACTURING

In any manufacturing organisation, there is a flow of materials at various stages of manufacturing, *i.e.,* from input to output. The flow of tangible materials from input through manufacturing to output of a manufacturing operation is represented in the fig (33.1). The flow with reference to inputs involves such activities as purchasing, traffic control and receiving. The activities associated with flow within the factory includes the material handling activities at different work stations as per the order of processing. Output related activities include packaging, shipping and distribution.

Irrespective of the type of the organisation, the basic material related functions performed at the organisation are

Purchasing
Inbound traffic from suppliers to company
Receiving
Inventory control
Production control
In plant storage
Material handling
Packaging and shipping
Outbound traffic
Warehousing and distribution.

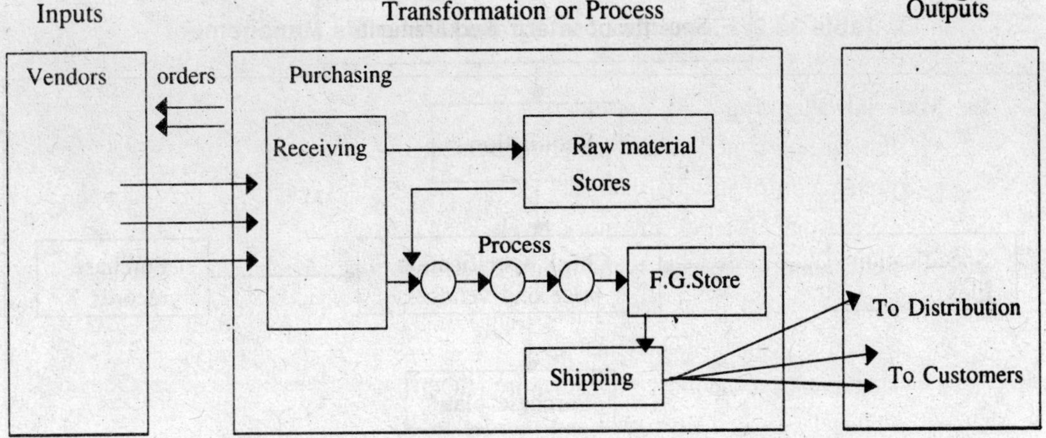

Fig 33.1. Flow of Materials through a company.

33.6. PURCHASING

Purchasing plays a crucial role in the materials management because it is concerned from input stage up to the consumption in manufacturing. Purchasing functions as a monitor, clearing house and a pipe line to supply materials needed for production.

Dr. Walters defines scientific purchasing as

"Procurement by purchase of the proper materials, machinery and equipment and supplies of stores used in manufacturing of the product, adopted to marketing in the proper quantity and quality at the proper time and at the lowest price consistent with quality desired."

Purchasing Cycle

The purchase procedure followed varies from company to company and also from one industry to other. The purchasing cycle is represented as shown in fig (33.2).

The basic elements in purchasing are :

- The origin of demand for materials and components based upon the requisitions made to purchase department by user departments with all the details like descriptions, quantity and quality specifications.
- Specifications are checked and verified and purchase plan is made for items demanded.
- Selection of source of supply.
- Preparation of purchase order by supplier (order acceptance) and acceptance of terms and conditions.
- Follow up to ensure prompt delivery of right quality and quantity of materials.
- Incoming inspection of materials (both to check quality and quantity) to ensure correct material as per specification.
- Checking supply invoice against purchase order and goods received and payments are made.

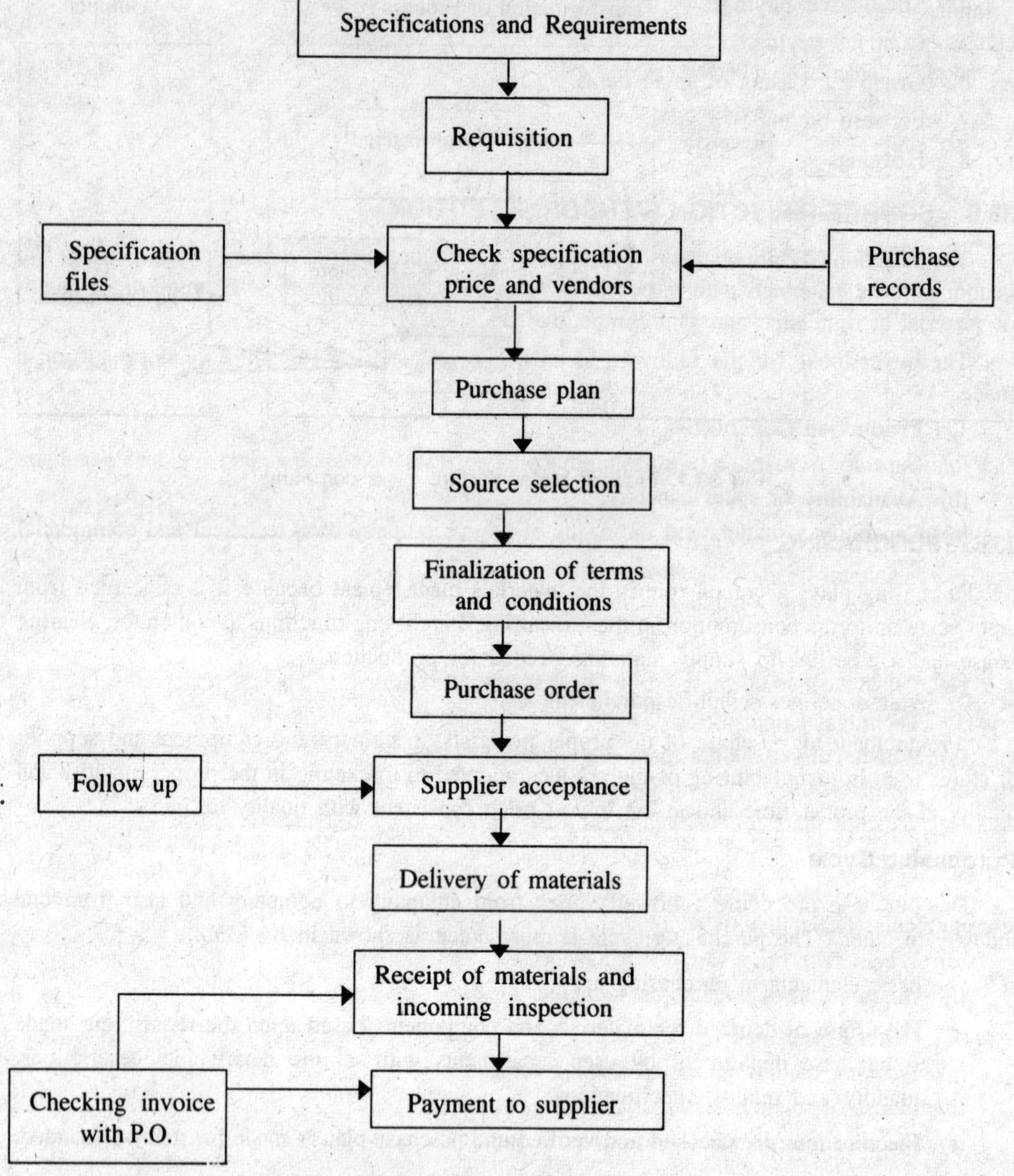

Fig. 33.2. Purchasing cycle

33.7. METHODS OF BUYING

Methods of purchasing will vary according to the nature of demand and the market conditions.

The principle methods of buying are

1. Purchasing by requirement (Hand to mouth buying).
2. Purchasing for a specified future period.
3. Market purchasing.

4. Speculative buying.
5. Contract buying.
6. Group purchasing of small items.
7. Forward buying.
8. Hedging.

33.8. SOURCE SELECTION (VENDOR SELECTION)

The selection of right source of supply is an important aspect in materials management. The vendor is to be examined with respect to his capability and competency to supply right quality of material at right time and at a competitive price.

The buyer looks for the following details before a decision regarding vendor selection is made.

1. **Production Capabilities**

(a) Capacity to manufacture the products as per the specifications and required quantities.

(b) Availability of spare capacity.

(c) Capability to understand the needs of buyer company both technical and commercial.

2. **Financial Position of the Vendor**

(a) The type of the company - Private limited, Partnership or Sole proprietorship.

(b) Company's capital structure.

(c) Financial position and profitability of the company since last 3-4 years.

3. **Technical Capabilities**

(a) Whether the available plant and equipments are in a position to meet the quality and quantity specifications of the customer.

(b) Whether there are enough technically skilled and trained people.

(c) Whether R & D facilities are available.

(d) What is the market standing of the vendor with respect to quality and delivery commitments (Reliability of supply).

(e) Whether he has enough storing and warehousing facilities.

(f) Quality control procedures - whether an ISO-9000 certified supplier.

4. **Other Conditions**

(a) Working conditions in the vendor company.

(b) Industrial relations and bargaining power of unions.

(c) Possible reasons for interruptions in supply.

33.9. VENDOR RELATIONS

Purchase department should establish a sound relationship with vendors based on mutual trust and benefit to ensure smooth supply of materials/parts as per the quality and quantity required. So, apart from a formal commercial relationship as a customer, a long lasting and mutually rewarding relationship is to be established.

A strategic partnership between the buyer and supplier is defined as a continuing relationship involving a commitment over an extended time period, an exchange of information and acknowledgment of risks and rewards of the partnership.

A sound relationship emerges from the proper help and co-ordination on the part of both buyer and supplier. The expectations from buyers include :

- Promptness in delivery.
- Meets the quality and quantity requirements.
- Entertains occasional rush orders.
- Flexibility in quantity to be supplied.
- Can wait for the bills.
- Competitive pricing.

The expectations from the vendors

- Consistent orders
- Prompt payment of the bills.
- Extend help during difficulty both financial and technical.
- Continuity of orders.

33.10. VENDOR RATING (EVALUATION OF SUPPLIERS)

The appraisal of the vendor performance is a continuous process. The vendor can be rated on various characteristics such as

1. Delivery (To deliver as on time as per the order).
2. Quality (To delivery as per specifications).
3. Price.
4. Other factors such as capability to meet urgent/rush orders, readiness to try out new designs or new methods etc.

In vendor rating, one usually gives weightage to these various characteristics and measures the performance of the vendors periodically on the basis of certain norms and procedures. Every company should have a formal vendor rating system. It is not only beneficial to the buyer company but also for the supplier company. Vendors will get the feed back based on objective evaluation and can compare his own performance with that of competitors. It is a fair evaluation since the rating is based on facts and not on opinions or prejudices. Vendor company's can know their shortcomings and improve upon them.

Generally, there are three methods of vendor rating

1. Categorical plan
2. Weighted point method
3. Cost ratio method.

33.11. JUST IN TIME (JIT) PURCHASING

Just in time purchasing is a major component of JIT manufacturing system. The basic concept of JIT purchasing is to establish agreement with vendors to deliver small quantities of materials/parts just in time for production. This approach is quite contrast to traditional approach of bulk buying.

The features of JIT purchasing are :

- Reduced lot sizes
- Frequent and reliable delivery schedules.
- Reduced and highly reliable lead times.
- High quality level of purchased parts.

The JIT purchasing aims at a single reliable source for each item and consolidation of several items from each supplier. The reduction of number of suppliers in JIT purchasing have the following advantages.

1. Consistent quality
2. Minimum investment and resources such as buyer's time, travel and engineering
3. Focused attention on vendors
4. Savings on toolings
5. Establishment of long-term relationship.

For making JIT work, the following conditions are put on purchasing department.

- Reduction in number of suppliers.
- Locating the suppliers who are nearby.

The success of JIT purchasing depends on how well the firm establishes the strategy of single sourcing. The suppliers should be seen as "Outside" partners who can contribute to the long run wellfare of buying firm.

33.12. SUPPLY CHAIN MANAGEMENT

33.12.1. Introduction

The term supply chain management is used to describe the management of material, suppliers, production facilities, distribution services and customers linked together through the forward flow of information and the backward flow of materials [Evans-1995]. The idea here is to apply a total systems approach to managing an entire flow of information, materials and services from raw materials supplier through factories and warehouses to the end user. The focus is in those core activities that a business must operate efficiently each day to meet the demand.

Supply chains are now common not only within the country where the business is operating, but they are extended as international supply chain in both low and high technology products ranging from textiles to sophisticated electronics and automobiles where raw materials are processed in one country, intermediate processing or assembly in another country and final processing in the third country. If distribution is included in the supply chain, there could be many stages in that like warehousing third party freight forwarders, wholesalers, retailers and service providers.

The term supply chain forms the picture of how organisations are linked together. If we start with a purchasing department and move towards the supply side, its has number of suppliers, each of which in turn, has its own set of suppliers and so on. The result is a supply network or series of chains.

This supplier network is represented in the fig (33.3) for one purchasing department and three of its suppliers.

33.12.2. Objectives of Supply Chain Management

The primary objective of supply chain management is to reduce risks and uncertainties in to supply chain, thereby positively affecting inventory levels, operation and product cycle time, processes and ultimately end user's service levels. The focus is on system optimization and enhancement of performance effectiveness. The various tools that are useful in system optimization are forecasting, aggregate planning inventory systems and operations scheduling.

A forecast is developed from the common database, which in turn becomes the input for aggregate planning. The aggregate plan sets the guidelines and constraints for the inventory plan and from this requirement of work force and equipment schedule is prepared.

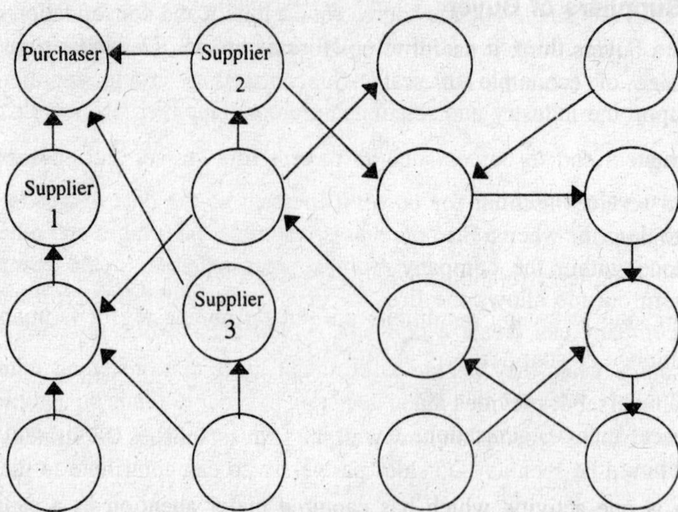

Fig. 33.3. Complexity of Supplier Network.

33.12.3. Supply Chain Management Strategy

The strategic goals of supply chains have to match those of the constituent firms. These goals have to set a scene for medium and long term and operational management for those firms. The interface between the buyer and suppliers, buyers and other buyers of suppliers, suppliers and final customer is equally important.

Buyer-Supplier Relationship

Buyers today have greater concerns for their suppliers and they are involved in the operations of their suppliers directly such as operational audits, open costing procedures etc. or indirectly through training of workers of suppliers etc.

Audits are useful means of monitoring supplier performance and the audit focuses on

- Evaluation of organizational effectiveness
- Planning systems and documentation
- Cost monitoring and reduction
- Scheduling and delivery capabilities
- Technology and Research & Development Capabilities

Audits can be used to evaluate the need for improvement or corrective actions.

Relationship Between Buyers and Other Buyers of Suppliers

The sophisticated buyer can provide a standard (or benchmark) regarding the performance of suppliers with the best practices available. Large firms are very much effective in disseminating good manufacturing practices to suppliers and this is often arrived at in collaboration with other large firms. The potential risk involved here is that of confidentiality of technical and commercial information.

Supplier Relationship with Customer

Each supplier in the supply chain should be aware of the final customer who determines the value or worth of product/service offered in the product market. This can be achieved through various ways like warranty, Quality Function Deployment (QFD) and concurrent engineering and logistics.

Supplier and Other Suppliers of Buyer

Several suppliers to a buyers think it useful to collaborate in such areas like transportation. They also take advantage of economic of scale. Subcontracting among suppliers is also encouraged depending upon the industry and relationship among supplier firms with the buyer.

33.12.4. Out Sourcing

Purchased items and services account for 60 to 70 percent of the cost of goods sold. Out sourcing is a term used to describe when a firm purchases materials, assemblies and other services that were previously done within the company from sources external to the company. Out sourcing, *i.e.*, buying from outside allows the firm to focus on activities that represent its core competencies. Thus, a company can create a competitive advantage while reducing cost.

Organisations out source when they decide to buy something that are being manufactured in house currently. It is a reversal of previous make decision. In order to focus on activities, which constitute core competence, many organisations out source the non-value adding activities like cleaning and house keeping.

Information system is one activity, which has captured major attention as a candidate for outsourcing. An entire function may be out sourced or some activities may be outsourced and others are kept in house. Depending upon the importance of the function, its nature as to whether it is a strategic or routine the decision regarding outsourcing will be taken.

Reasons for Outsourcing

- Cost reduction
- Focus on core competencies
- Reduction in man power
- Reduce development and production cycle
- Improve efficiency
- Minimizes inventory, material handling and non-value added costs.

Risks and Disadvantages of Outsourcing

- Higher exit barriers
- Loss of control
- Exposure to supplier risks
- Difficulty in quantification
- Possibility of bearing tied to obsolete technology
- Attention required by top management
- Supply restrictions

33.12.5. Value Density

Value density is expressed as a value per unit of weight. An important decision in purchasing is to take decision regarding the mode of shipping, *i.e.m* whether they should be shipped by air, or by ground transportation. Value density the value of item per Kg of weight is an important measure when taking decision regarding where the items should be stocked geographically and how they should be shipped ?

Strategic Partnership — A Close Buyer-supplier Relationship

A close relationship is both desirable and possible with all suppliers. The extent of the effort at this relationship building is determined by number of considerations like (*i*) Based on value of suppliers of goods and services. This will establish a potential cost and other benefits of developing a close relationship.

(*ii*) Percentage of supplies coming from one supplier.

33.12.6. Stages of Development of Buyer-Supplier Relationship

Once the organisation makes the decision to buy, and then the next logical step is identification of sources of supply. The basic stages of development of Buyer-Supplier relationship :

1. Specification of basic contract between the two parties. Suppliers are provided with drawings and basis specifications [Quantities to be supplied, rates and period of contract].
2. Specifications of critical parameters for incoming parts. Aggregate production plans known to suppliers.
3. Standardization of quantity procedure at suppliers dispatch. Buyer is informed about upstream development and testing procedures of suppliers, die-design, patterns and toolings, supplier product constraints are taken into account.
4. Initiation of improvements by buyer after supplier audit of supplier operations.
5. Initiation of improvement by supplier with sharing of gains. This could conclude value engineering and design improvements.
6. Common strategy planning of long-term relationship spanning many contracts.

34

PROJECT MANAGEMENT

• Introduction to Project Management • Project Concept and Definition • Characteristics of Project • Project Identification • Sources of Project Ideas • Project Identification for Existing Industries • Project Feasibility • Marketing Feasibility • Technical Feasibility • Financial and Economic Feasibility • Cost of Project and Means of Finance • Cost of Production • Working Results • Project Planning • Project Organisation • Project Plan • Tools and Techniques of PM • Monitoring and Control of Projects • Project Information System.

34.1. INTRODUCTION TO PROJECT MANAGEMENT

Project management is an emerging profession. The concept of project management encompasses a set of principles, methods and techniques that helps in the effective planning and completion of tasks under the given constraints imposed on a project. The basic characteristics of a capital expenditure (project) is that it generally involves a current outlay (investment) of funds that generate benefits for the future period. The formulation of sound projects is of significance in industrial development of any planned economy. A systematic evaluation of proposed projects based on thorough investigation of their economic and technical feasibility is a pre-requisite for selecting viable projects and providing financial and technical resources to them. Project formulation and evaluation are particularly important in a developing country like India because of limited resources in capital and skills. In order to make most rational distribution and select those financial resources, it is essential to identify and select those projects which are to be given priority over other projects competing for the same resources. This leads to systematic analysis and planning using suitable criteria which may form the basis for evaluating projects before an investment decision is taken.

34.2. PROJECT CONCEPT AND DEFINITION

A project in an economic sense is an activity which is undertaken to generate additional production and resources of an economy.

A project is defined as, "one shot time bound major activity demanding the commitment of varied skills and resources to achieve specific goal".

Project includes both setting up of a new unit or substantial expansion of existing units of production. Project is thus a combination of human and non-human resources integrated together in a time bound temporary organisation to achieve a specified objective.

34.3. CHARACTERISTICS OF PROJECT

1. Project is one time activity which will never be repeated exactly the same manner.

2. A project has a definite start and finish, *i.e.*, a project is executed in a definite time bound schedule.

3. A project uses a cross-functional relationships because it needs diversified skills and talents from different professions.

4. A project has definable goals or end results that can be defined in terms of cost, schedule and performance requirements.

5. Project demands the investment (current outlay) and the benefits are spread for number of future periods.

6. Once the project goals are achieved, the project team will be either disbanded or reconstituted for another new project.

7. Project passes through several distinct activities which constitute a project life cycle.

34.4. PROJECT IDENTIFICATION

An entrepreneur has a wide choice of projects. The various aspects of the choice of the products include — the product, market, technology, equipment, scale of production, time phasing and location. The selection of the project involves a detailed analysis of the environment in which it operates. Both objective as well as subjective analysis of various factors have an influence on the selection of a project.

Government policies and regulatory framework are going to influence the identification of the project to a greater extent. The regulatory framework seeks to define various avenues available to business firms in different categories and channalise the investments in prior direction by providing incentives, concessions etc. and also specify the kinds of operational control that are exercised by the government over the functioning of business enterprise.

The important elements of the regulatory framework are given as shown in the table 34.1.

Table 34.1: Governmental Regulatory Framework

- Industrial policy
- Industrial development and regulation act
- Foreign exchange regulation act (FERA)
- Monopolies and restrictive trade practices act (MRTP)
- Export and import policy (EXIM)
- Incentives for export oriented units
- Incentives for units situated in industrial backward areas
- Incentives for small scale industrial units
- Liberalisation and Globalisation of markets

34.5. SOURCES OF PROJECT IDEAS

The success of a project depends upon the sound selection and planned execution of the project work. A variety of sources are to be tapped to stimulate the generation of project ideas.

The various sources of project ideas are :

(*a*) Thorough analysis of performance of existing industries

A thorough analysis of existing industries with respect to the profitability, capacity utilisation and other critical facts helps to locate and select the project.

(*b*) Examination of inputs and outputs

A study of inputs required for various industries may aid the selection process. Opportunities for investment in projects exist when (*i*) the materials purchased from distant sources with higher transportation cost and higher lead time, (*ii*) several industries produce internally some components which can be out sourced. The examination of output of existing industries may reveal the opportunity for further processing of outputs.

(*c*) Examination of imports and exports

Import strategies for a longer specified period (5 to 6 years) is helpful in understanding the trend of imports of various goods and the potential for import substitution. Likewise, an examination of exports statistics is useful in exploring the possibilities of exports of various products.

(*d*) Have a knowledge of guidelines and government development plan outlays

Governments proposed outlays and priority in different sectors of economy acts as a pointer for project selection. These indicate the potential demand for goods and services required by different sectors.

(*e*) Guidelines and suggestions given by the financial institutions and other promotional agencies.

(*f*) Investigate local materials and resources.

(*g*) Analyse economic and social trends.

(*h*) Exploring the possibilities and reviving sick units.

(*i*) Study of changes in technology and keeping pace with technological developments.

(*j*) Trends analysis and study of consumption pattern of foreign markets.

(*k*) Identification of unfulfilled psychological needs.

(*l*) Analysis of demand-supply gap.

(*m*) Participate in seminars/conferences and attend trade fairs, industrial exhibitions to provide an access for new products and new technology.

(*n*) Stimulate creativity for generating new product ideas.

34.6. PROJECT IDENTIFICATION FOR EXISTING INDUSTRIES

An existing company which seeks to identify new project opportunities should undertake a SWOT analysis to evaluate its strengths and weaknesses, opportunities and threats in the external environment. The SWOT analysis will provide the basis for the corporate strategy to be followed and indicate major areas of thrust. These include :

(*i*) Expansion of existing capacity.

(*ii*) Enhancing the existing product range.

(*iii*) Vertical and horizontal integration.

(*iv*) Related diversification.

(*v*) Mergers and acquisitions.

34.7. PROJECT FEASIBILITY

A project involves a large amount of investment and it is a one-time activity and the investment cannot be reversed. Once the resources are committed and expanded. So, it is essential to evaluate the process thoroughly in all aspects before the investment decision is made. The project is said to be feasible, if the project is sound from marketing, technical and economic point of view. The various aspects of feasibility is represented in the fig. 34.1.

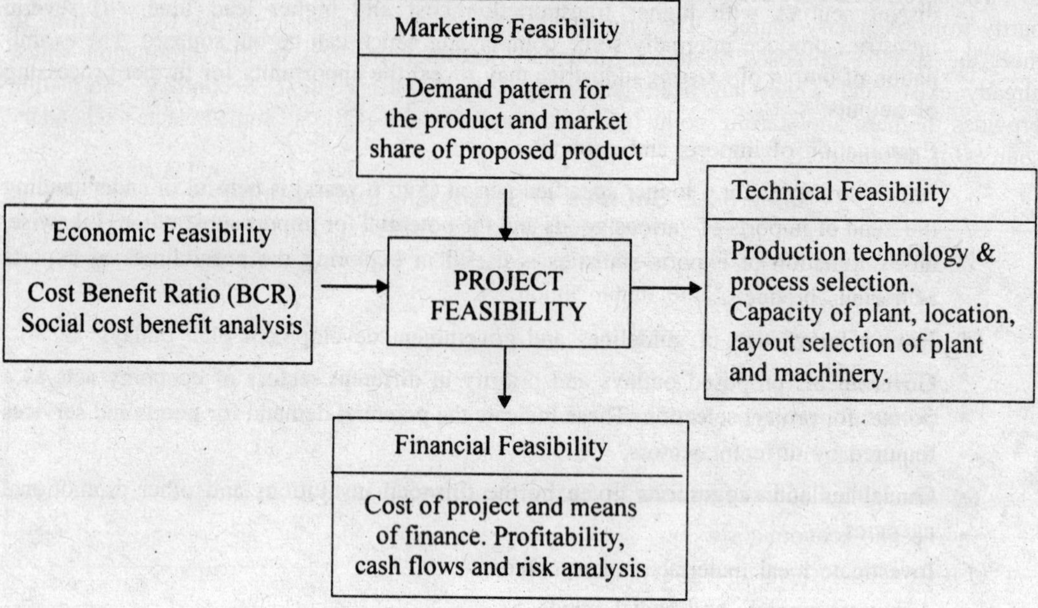

Fig. 34.1. Project Feasibility.

34.8. MARKETING FEASIBILITY (Appraisal)

Marketing appraisal is concerned with determination of the aggregate demand for the proposed product/service and the market share of the project under appraisal it is going to capture. If there, is no sufficient demand exists and there is no positive sign of increasing trend, then the projects market feasibility does not exist. Hence, there is no point in continuation of the project.

To carryout the market appraisal, the analyst requires variety of information and he should seek answers to the following questions :

Who is the customer for the proposed product ?

Where he is located ?

How can I reach them ?

What is the expected sales for the next 5 or 10 years ?

To find the answers to the above questions, information on various aspects of marketing and demand are required.

The information requirements are :

- Past and present consumption and demand pattern and likely projected trend
- Demand supply gap
- Production capacity and price structure
- Methods of distribution
- Consumer behaviour, intentions, motivations, tastes and preferences
- Price elasticity of demand
- Government policy and restrictions
- Competitors
- Supplementary and complementary goods.

The information required for demanu market analysis is partly through market survey and partly from secondary sources. The primary information which is collected for the first time to meet the specific purposes on hand. Secondary information refers to the information which already exists and which has been gathered in some other context. Secondary information provides the base and starting point for market and demand analysis. Some selected secondary sources of information :

Table 34.2: Sources of Secondary Information

- Indian year book
- Statistical abstract of the Indian union
- National sample survey reports
- Statistical year book
- Economic survey
- Annual survey of Industries
- Annual bulletion of statistics of export and import
- Techno-economic survey
- Monthly bulletin of Reserve Bank of India

Industry specific information

- Annual reports of association of Indian automobile manufacturers
- Build Machine — Build India — Directorate general of Technical Development
- Annual reports of Indian Electrical Manufacturers Association
- Iron-Steel Control Bulletin (Quarterly)
- Materials and Metals Review
- Textile, Cement, Glass, Pharmaceutical, Rubber, Packaging Industry Association Publications

34.9. TECHNICAL FEASIBILITY (Appraisal)

Technical and engineering analysis of the proposed project is done continually when the project is formulated. Technical appraisal seeks to determine whether the pre-requisite for the successful commissioning of the project have been considered and reasonably good choices have been made with respect to location, size and process etc. It also helps to locate source of suppliers of raw material and fixing up the final cost of the project.

Important questions that are raised in technical appraisal are :

- Whether the technology for the proposed product is available commercially or should be developed in house.
- What are the arrangements made to acquire the technology ?
- Whether the conversion process chosen is appropriate and suitable for the proposed product.
- Whether the selected capacity is optimum ?
- Whether the equipment and machines chosen are appropriate ?
- Whether the location, proposed layout, building and plant are sound ?
- What are the provisions made for treatment of waste and affluents (discharge) ?

Thus, the major aspects of technical appraisal should include

(a) Choice of technology which is influenced by various factors such as — plant capacity, principle inputs, investment outlay and cost of population, product mix, ease of absorption.

(b) Means of acquiring technology — This is the options available are — Technology, Licensing, Purchase of technology, Joint venture or collaboration arrangements.

(c) Product mix design.

(d) Plant capacity decision — The factors like demand, technological requirement, input constraints, investment outlay, government policy and resources of the firm.

(e) Plant capacity decision.

(f) Choice of equipment's and machines.

(g) Decisions regarding structure and civil works.

(h) Considerations of alternatives — The technical analysis is a pre-requisite establishing a feasibility for the project.

34.10. FINANCIAL AND ECONOMIC FEASIBILITY

Financial appraisal analyses whether proposed project will be financially viable in the sense that it is able to meet the burden of servicing the debts and whether the project will satisfy the return expectations of those who have invested capital.

The various aspects that are considered in financial appraisal are :

1. Cost of the project
2. Means of finance
3. Profitability and break-even point
4. Financial statements
5. Criteria for evaluation of projects
6. Risk analysis

Economic appraisal is considered to be social cost benefit analysis. It is concerned with judging the project from the larger social point of view.

34.11. COST OF THE PROJECT

Funds are required to be arranged before the actual commencement of the project because of the following purposes :

- Purchase of land
- Purchase of machinery and equipment
- Initial working capital, preliminary and preoperative expenses. Though the whole capital is not required at the beginning of the project, however, adequate arrangements should be made in advance to avoid bottlenecks in the way of targets. Money will be required in the phased manner as per the plan of the project and schedule.

The cost of project includes the following :

- Land and site development.
- Buildings and other civil construction works.
- Technical know how fees and other fees payable to collaborations.
- Expenses on foreign technicians and training of Indian technician's abroad.
- Miscellaneous fixed assets. (furniture, office equipment, transportation etc.).

- Preliminary expenses.
- Pre-operative expenses.
- Provision for contingencies.
- Margin money for working capital.

The table 34.3. shows the items which are normally included as per the planning commission's guidelines for correctly estimating the fund requirements. This helps to estimate the total requirement of capital as well as the phased requirement.

Table 34.3: Cost of the Project

S. No.	*Particulars of cost*	*Total cost (Rs)*	*Phases of capital expenditure*			
			I	*II*	*III*	*IV*
1.	Land and site development					
2	Buildings and civil work					
3.	Plant and machinery					
4.	Technical know how fees					
5.	Misc. fixed assets					
6.	Preliminary expenses					
7.	Pre-operative expenses					
8.	Provision for contingencies					
9.	Margin money for working capital					

34.12. MEANS OF FINANCE

Means of finance refers to the various sources of finance for the project. These sources include both debt and equity. A typical proportion of debt and equity is reflected in the capital structure represented in table 34.4 and various sources of financing are represented in table 34.5.

Table 34.4: Capital Structure and Means of Finance

	ABC Industries Ltd.		
			Rs (Lakhs)
1.	Equity Share Capital		1,300
2.	Loans from financial institutions		2,800
	IDBI	1,500	
	IFCI	650	
	ICICI	650	
3.	Loans from Bank		1,000
	State Bank of India	400	
	Canara Bank	600	
4.	HDFC Loan		150
5.	Unsecured loan from promoters (TIDCO)		250
	Total		5500

Table 34.5: Sources of Finance

Lending period	*Sources*	*Used to finance*
Short term financing	• Commercial paper (Hundi) • Trade credit • Loan from commercial banks • Accounts receivable financing	Working Capital
Median term financing	• Hire purchase • Leasing • F.D. from public • Subsidy by Govt.	Capital Cost
Long term financing	• Equity — issue of shares • Debentures • Preference stock	Capital Cost

34.13. COST OF PRODUCTION AND WORKING RESULTS

The cost of production includes — the material cost, cost of utilities, labour cost and factory overheads. The statement of estimated cost should be prepared for 8 to 10 years.

The table 34.6 shows the estimation of cost of production.

Table 34.6: Estimation of Cost of Production

Particulars	*Year ending*			
	1	2	3	___ 10
A. Total material cost (Raw materials, components, chemicals, consumable stores)				
B. Total utilities cost (Power, fuel, water)				
C. Total Labour Cost (Wages, factory supervisory salaries, labour overheads, other facilities to labour like P.F., Bonus, etc.)				
D. Total Factory Overheads (Repairs and maintenance, lighting, rent and taxes, insurance on factory overheads, miscellaneous, factory expenses and contingency)				
E. Estimate of cost of production [A + B + C + D]				

Working Results :

This represents the profitability of the proposed project. The profitability projections can be made based upon the expected sales and cost of production and other details of cost like administrative overheads, total financial expenses, sales overheads, taxes and dividends proposed etc.

Table 34.7 shows the estimates of working results.

Table 34.7: Estimates of Working Results

Particulars	*Year ending*				
	1	2	3	______	10
A. Cost of production As per table (34.6)					
B. Total Administrative expenses [Administrative salaries and expenses, professional fees, directors fees, lighting, postage and telephone, office supplies, insurance and taxes on property]					
C. Total Sales expenses					
D. Royalty and know how payable					
E. Total cost of production [A + B + C + D]					
F. Total expected sales					
G. Gross Profit before interest [F – E]					
H. Total financial charges					
I. Depreciation					
J. Operating profit [G – H – I]					
K. Other income (if any)					
L. Preliminary expenses written off					
M. Profit/Loss before tax [J + K + L]					
N. Provision for taxation					
O. Profit after tax [M – N]					
P. Retained profit					
Q. Dividend					

34.14. PROJECT PLANNING

A project involves a large number of activities and also a large amount of resources are to be expended both human and non-human which are scarce. Thus, the management must make effective utilisation of resources which is possible only through proper planning, monitoring and controlling of projects.

Project planning can be defined as the process of stating the project objectives and then determining the most effective activities necessary to reach the objectives. The planning defines the activities and actions, time and cost targets and performance milestones which will result in successful implementation and achievement of project objectives. The plan must indicate the materials, equipment's, facilities, human resources and other resources that are necessary to complete the project.

The steps in project planning are represented in Fig. 34.2.

Step I Determination of project objectives

This step involves setting project and results to be obtained, time, cost and performance standards to achieve the objectives.

Step II Identification of specific work activities

This refers to breaking the entire project into definable and executable jobs using work break down structure (WBS). Thus, this step defines and lists various tasks need to achieve the project objectives.

Step III Creating project organisation

This is done by creating suitable organisation structure, departments and managers responsible for work activities.

Step IV Preparation of project schedule

This is done by indicating the timing of each activities, dead lines and milestones.

Step V Preparation of Budget and Resource Plan

Estimation of amount and timing of resources and expenditure for the work activities and related activities.

Step VI Forecasting of time, cost and performance

This step is concerned with determining in advance the time, cost and performance projections for the completion of the project.

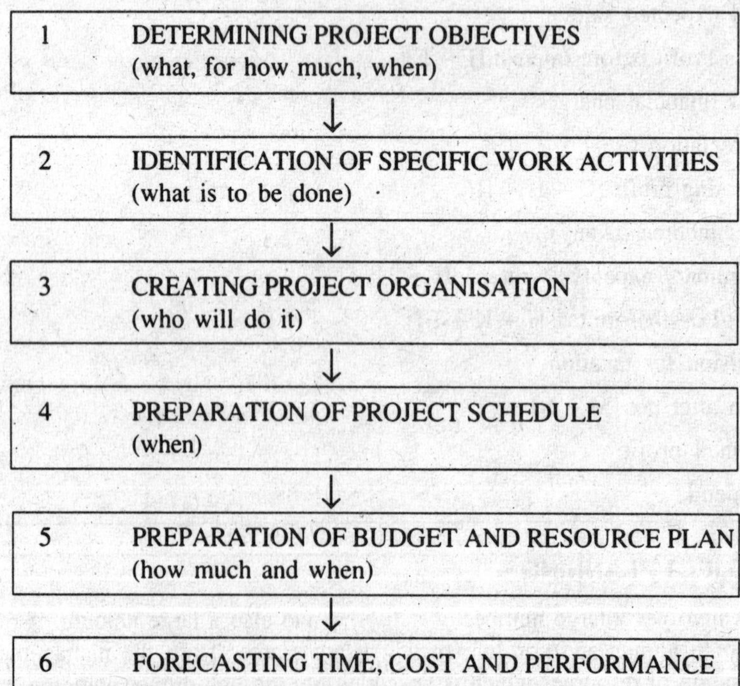

Fig. 34.2. Steps in Project Planning.

34.15. PROJECT ORGANISATION

A project requires a special organisation. There is no one unique project organisation for all situations, there are certain common requirements of all project organisations.

The desirable properties of a project organisation are :

(*i*) Commitment of the organisation towards the overall goals and objectives of the project.

(*ii*) It must be a temporary organisation which will be in operation till the implementation of the project.

(*iii*) It should be dynamic and sensitive to handle the uncertainties.

(*iv*) It should be able to integrate diversified activities of the project.

(*v*) It should be capable enough to accommodate large variety of specialised tasks.

(*vi*) It should be flexible to accept the changes in the structure when the scope of the project changes.

The responsibilities of project manager are :

1. What is to be done ?
2. When will the task be done ?
3. How much money is available to do the task ?
4. How will has the project been done ?

Functional manager's responsibilities are :

1. How will the task be done ?
2. Where will the task be done ?
3. Who will do the task ?
4. How well the functional inputs are integrated with project ?

The project manager is responsible for the overall integration of the total project system, while the functional manager is responsible for technical direction in his discipline.

Merits and demerits of matrix organisation structure.

Merits	*Demerits*
1. Efficient utilisation of the organisation's human and physical resources. 2. Clear and executable mechanism for achieving project integration. 3. Sensitive to changes in goals and tasks. 4. Effective dissemination of information. 5. Team of experts work together. 6. Morale of the team members are high.	1. Needs information systems and human behaviour to support multi-dimensional working. 2. Unclear authority and responsibility. 3. Conflicts may be develop between project managers and functional managers. 4. Decision-making may be delayed. 5. Project personnel have to be report to more than one base.

Pure Project Organisation Structure

If a highly complex project exists, it requires major resources commitments and involves heavy stakes in results (outcome). In such a situation, an organisation considers a pure project organisation. Project organisation is a separate organisational entity headed by a project manager. A typical pure project organisation structure is shown in Fig. 34.3.

By creating a separate dedicated organisation for the project. The seminar management of the organisation has offered it the highest priority and resources both internal and external.

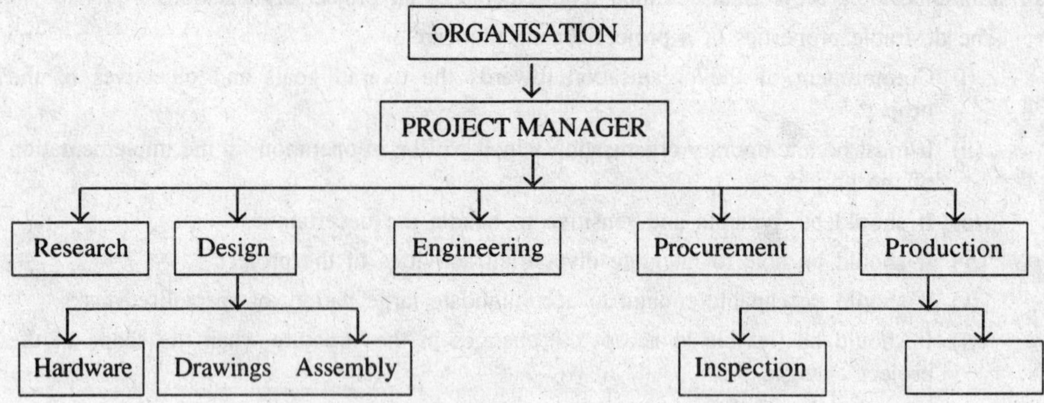

Fig. 34.3. Pure Project Organisation.

Merits and demerits of pure project organisation :

Merits	*Demerits*
1. Project manager has complete line authority over the project	1. High cost of maintaining the separate organisation.
2. Strong communication lines and perfect reporting system.	2. The project personnel will have no home at all once the project is completed.
3. Fast response time and reduced lead time on project activities.	3. Resources dedicated to project may have to be acquired in advance of work schedule.
4. Flexibility in time, cost and performance trade off.	4. Lowered motivation and morale as the project comes to an end.
5. Retains qualified and specialised personnel for dedicated work.	

34.16. A DETAILED PROJECT PLAN

The contents of the project plan are as follows :

1. Project Summary

   It should essentially executive summary that can be easily absorbed by the concerned people in shortest possible time. It should identify the objectives, the goals and the constraints of the project.

2. Specifications

   This aspect of the project plan should define the characteristics and the performance goals for the final end product.

3. Work statement

   This is the part of the proposal request that tells the contractor what is desired or the customer exactly what the contractor proposes to do. This is the basic document on which the project plan is built.

4. Master schedule

The project master schedule interrelates all tasks on a common time scale. For simple projects, the master schedule may consist of a bar or Gantt chart, a complex project will necessitate use of network analysis techniques like PERT and CPM. A master schedule should contain the following information.

(*a*) Name of work packages listed in WBS.

(*b*) Name of persons having responsibilities for work packages.

(*c*) Starting date and expected duration of each work package.

(*d*) The completion date for each work package.

5. Procedures Guide
6. Budgets and control systems.
7. Activity or event network plan.
8. Materials and equipment forecast.
9. Responsibility matrix.
10. Project organisation and project personnel plan.
11. Reporting and review procedure.

34.17. TOOLS AND TECHNIQUES OF PROJECT MANAGEMENT

1. PROJECT SELECTION TECHNIQUES
   - Cost — benefit analysis
   - Risk and sensitivity analysis
2. PROJECT PLANNING TECHNIQUES
   - Work break down structure (WBS)
   - Project responsibility matrix
   - Project management manual
3. PROJECT SCHEDULING AND COORDINATING TECHNIQUES
   - Bar charts
   - Net work techniques (PERT – CPM)
   - Line of balance (LOB) technique
4. PROJECT COST AND PRODUCTIVITY CONTROL
   - Progress Measurement Technique (PROMPT)
   - Performance Measurement System
   - Updating Reviewing and Reporting Techniques (URT)
5. PROJECT COST AND PRODUCTIVITY CONTROL TECHNIQUES
   - Productivity budgeting technique
   - Value engineering
6. PROJECT COMMUNICATION TECHNIQUE
   - Project information system

34.18. MONITORING AND CONTROL OF PROJECTS

Effective management of a project during its entire life cycle requires a well-organised control system be designed, developed and implemented so that effective and efficient feedback on the project's progress can be attained.

The monitoring and control stage starts as soon as the execution of the projects. In fact, the execution process could be considered to start as soon as a project is conceived. The main objective of monitoring is to ensure that various time and cost targets are met and the network as well as its operational plans prepared for execution of the projects are adhered to.

Monitoring process ensures some positive action and see that there is no gap between the desired and actual achievements and targets. What is to be measured, reviewed and reported will depend on who is to take action and what action he is likely to take. The project organisation has various levels and each level has different duties and take different types of actions which fall under their jurisdiction.

Steps in monitoring :

1. Setting a monitoring environment

   The role of monitor is well-defined, known and accepted by all the participating agencies whose activities are to be monitored. Project manager sets up an environment in which he exercises his authority and responsibilities.

2. Setting performance standards

   The project monitor on behalf of the project manager, sets systems and procedures for monitoring the projects. He also reviews performance in terms of time schedule, budget and quality of the project against given standards.

3. Measuring the progress of the project

   The project monitor keeps a close watch on the progress of activities by the way of collecting information regarding "what has been done with respect to". He then quantifies them for comparison with targets. Then the corrective actions can be devised based on feedback.

4. Reviewing

   In case of non-permissible activities, he takes the corrective decision as to how these should be handled.

5. Reporting

   The project monitor report to project manager if it is not possible to take early action.

6. Action

   The project manager acts promptly on all unresolved issues.

Monitoring and control system starts from the point when the planning phase is over, *i.e.*, when time plans, schedules, as well as the operational plans are ready and are to be implemented.

The monitoring areas are as follows :

1. Monitoring of time
2. Monitoring of manpower
3. Monitoring of material resources
4. Monitoring of costs in relation to work done
5. Monitoring of funds

Project Control

Project monitoring ensures adherence to schedules and target control is a process of keeping the project on target and as close to plan as possible. Thus, the project control like any other control system should perform tracking, monitoring goals of time, cost and performance.

The scope of control activities are represented in the table 34.8.

Table 34.8: Project Control

1. Scope/Progress control
(*a*) List of activities
(*b*) Progress measurement
(*c*) Expediting and follow up
2. Performance control
• Performance indices
3. Schedule control
• Schedule analysis
4. Cost control
• Cost analysis
• Cost reduction
• Cost status reports

- **Scope and progress control**

The process of control should begin at the last level of WBS. The completion of these activities in time within the allocated budget and as per the desired quality will lead to the completion of activities at the next higher level of WBS and ultimately the total project.

The detailed list of activities to be performed by various agencies involved in the project helps to exercise functional control over activities.

The measurement of progress of project stresses up on the efforts and money spent.

To calculate the overall progress, the weightage to each phase of the project is to be assigned. The progress of the complete project is given by,

$$\text{Total project progress } (P) = \frac{P_1W_1 + P_2W_2 + P_3W_3}{W_1 \times W_2 \times W_3} \times 100$$

where P_1, P_2 and P_3 are the percentage progress made in design and engineering phase, procurement phase and construction phase responsibility and W_1, W_2 and W_3 are the respective weightage for the stages of the project.

- **Expediting and control is a process which ensures the progress.**

Expediting acts as an external force to do work on a project towards its completion.

- **Performance Control**

Before exercising the performance control, it is essential to identify performance parameters and ensure right specifications, right selection of source of suppliers/contractors etc. The plant design and specifications must be made in accordance with performance requirements.

Control charts can be used to keep the quality of the various activities under control.

The performance analysis can be done by the following methods.

1. Cost and Schedule can be assessed by three parameters.
   (*a*) Budgeted cost of work scheduled (BCWS)
   (*b*) Actual cost of work performed (ACWP)
   (*c*) Budgeted cost of work performed (BCWP)

With the cost and schedule analysis, the following performance indices can be determined.

1. Schedule performance index (SPI)

$$\text{SPI} = \frac{\text{Budgeted cost of work pertormed (BCWP)}}{\text{Budgeted cost of work scheduled (BCWS)}}$$

2. Cost performance index

$$\text{CPI} = \frac{\text{Budgeted cost of work performed (BCWP)}}{\text{Actual cost of work performed (ACWP)}}$$

• **Schedule Control**

Schedule control is exercised to ensure compliance with time schedule of the project. Schedules are prepared to provide a basic for direction, communication, co-ordination and progress control. To ensure the progress of the project of each agency work which helps to stick to the project schedule.

• **Cost Control**

The objective of cost control should be to monitor all cost components and ensure completion of the project at any optimal cost without lowering the performance.

• Variance Analysis of Project Performance

The various variances of project performance are computed as follows :

1. Accounting Variance (AV)

$$\text{AV} = \text{ACWP} - \text{BCWS}$$

The accounting variance (AV) is the difference between the actual cost of work performed to date and the value of the work scheduled to have been completed.

2. Schedule Variance (SV)

$$\text{SV} = \text{BCWS} - \text{BCWP}$$

Schedule variance is the difference between the value of the work scheduled to have been completed and that which has actually been completed.

3. Time Variance (TV)

$$\text{TV} = \text{SD} - \text{BCSP}$$

where SD = status data or date of review.

BCSP = The date on which BCWP equals BCWS.

Time variance (TV) in a project gives the time lag, at the date of review, by indicating time equivalent of the value of the work which has actually to be completed to stay on schedule.

34.19. PROJECT INFORMATION SYSTEM (PIMS)

The technical specifications for a project are described in sufficient detail in the definition phase. Based on these specifications, dates and costs are planned at the end of this phase in order to supply the various participating agencies in the projects with standards in the form of partial target dates and cost estimates.

A project management information system (PIS) is an instrument for the project manager which is employed in the implementation on phase of the project. This instrument assists project manager in his task namely to implement successfully the project under him within the limits of schedules and budgets.

Concepts of PIS

The project information must cover the complete information cycle during the entire lifetime of the system, *i.e.*, the project duration. The information cycle for the project is shown in the fig. 34.4.

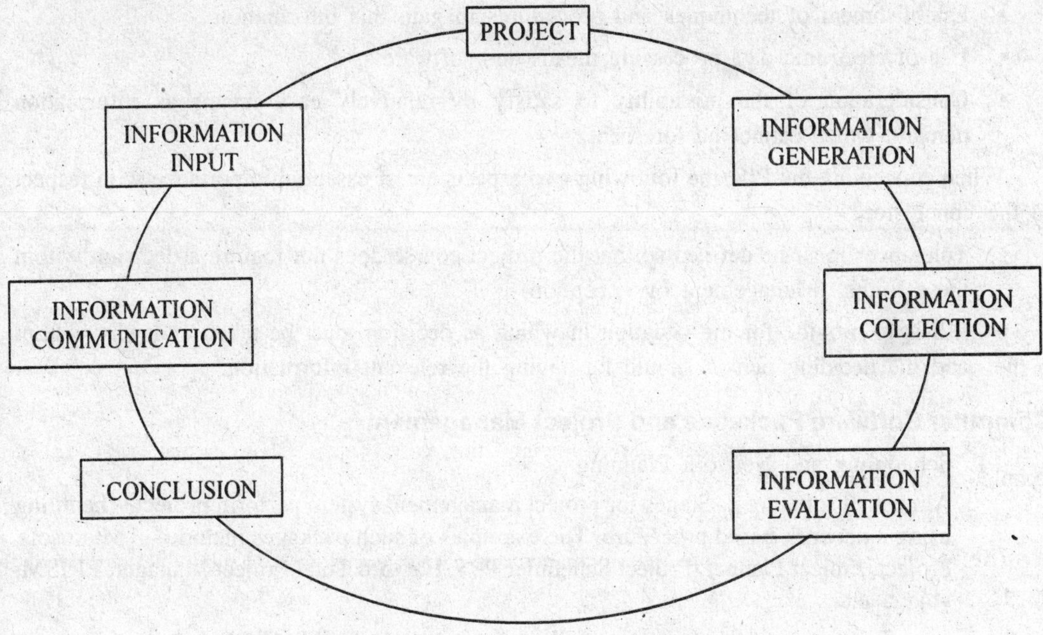

Fig. 34.4. Project Information System.

Objectives of PIS

1. To provide the relevant, timely and useful information to all the participants concerned with the project.

2. To systematise the control to a large extent by the reduction of all the activities to those absolutely necessary and by their simplification.

3. If possible to objectively control by quantification of all the decision findings.

4. To automate control to a high degree by the restriction of human action and their formalisation.

5. To facilitate a timely finding for all stages of hierarchy in practice to avoid the information explosion.

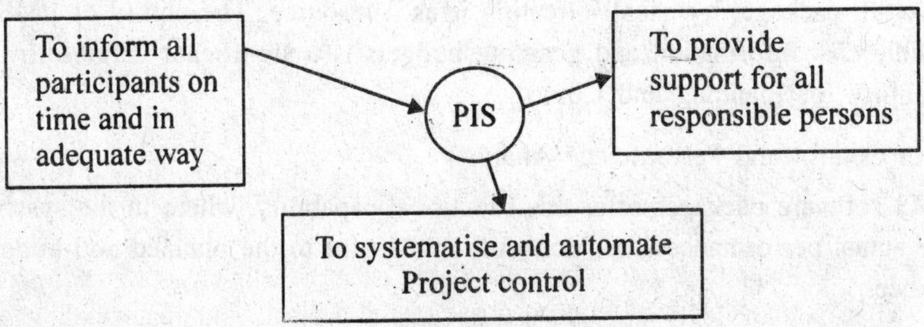

Thus, the project information system (PIS) includes rules and policies for all project activities and the procedures and techniques necessary to implement these activities. The objectives of project information system.

The focal point of any PIS is the information analysis, *viz.* Obtaining of new information by combination of source data. The following aspects play an important role.

- Which data to be obtained, *i.e.*, what kind of new information is to be obtained.
- Establishment of techniques and procedures to gain this information.
- Use of electronic data processing means and software's.
- Consideration of the possibility to satisfy by relatively easy means as information demand which cannot be foreseen.

When conceiving the PIS, the following two aspects are of essential importance with respect to the conclusion.

- Tolerances must be defined so that the project course does not require a decision within these limits. (Management by exception)
- PIS is to provide for the situation in which as decision must be taken and why, whom and the deciding person should be having the relevant information.

Computer Software Packages and Project Management

1. Scheduling and Network Planning

   Almost all software packages for project management system perform project scheduling using a network-based procedure. The examples of such packages include — Microsoft, Project, Project Planner, Project Scheduler PS4, Harvard Total Project Manager, PRISM, yojana etc.

   (*a*) Type of network procedure followed — namely PERT, CPM etc.

   (*b*) Maximum number of allowable activities.

   (*c*) The coding scheme of activities.

   (*d*) The use of probabilities and time estimates.

   (*e*) Quality and clarity of output format.

2. Resource Management

   Some project management software packages perform resource management functions such as resource loading, levelling and allocations. However, the analytical capabilities and the quality of the project of the reports may vary tremendously between the system.

3. Budgeting

   In many PMS packages, it is possible to associate cost information with each activity or work package, by usually treating it as a resource. The ability of PMS systems to handle cost information and generate budgets is a significant variable in the systems usability for Planning and Control.

4. Cost Control and Performance Analysis

   PMS software packages offer this functional capability, where in the system compares the actual performance (cost, schedule and work) to the planned and budgeted performance.

Review Questions

1. Define project and project management. What are the characteristics of projects ?
2. Discuss the various stages in the Life Cycle of a project.
3. What are the sources of project ideas ? From where you get information regarding the sources ?
4. What are the guidelines of identification of the projects for existing industries ?
5. What do you mean by project appraisal ? Why it is necessary to appraise the project ?
6. What do you mean by marketing appraisal ? Why it is essential to carryout the market appraisal ?
7. How do you estimate the cost of the project ? How is the capital structure decided ?
8. Discuss the means of finance.
9. What are the sources of Long term and working capital ?
10. How the projected cost of production and working results are computed ?
11. Briefly given an account of project appraisal criteria.
12. How the feasibility report for the project is prepared ?
13. What is project planning ? What are the steps involved in project planning ?
14. Write the note on Project Organisation.
15. Briefly describe matrix organisation and pure project organisation. Mention their advantages and limitations.
16. What is responsibility matrix and responsibility charts ?
17. Discuss the various tools and techniques of Project Planning.
18. What are the contents of the Project Plan ?
19. What are the objectives of monitoring and control of projects ? What are the steps in monitoring ?
20. What are the various activities of Project Control ?
21. Write notes on
    (*a*) Performance and Schedule Control
    (*b*) Project Information System
    (*c*) Project Cost Control

35

SERVICE MANAGEMENT

• Introduction • Definition of service • Service Characteristics • Intangibility aspects of service • Reasons for growth of service sector • Product levels • Moments of truth and manpower implications • Service matrix • Service model • Service strategy • Service product • Service delivery system and service quality.

35.1. INTRODUCTION

Economies are becoming more service oriented. Especially in the developed economies like America, Europe etc. services account for about 80% of the workforce and over 70% of the GNP. Service sector's contribution to Indian economy is on the rise. The last two decades have seen a boom in consumerism in India. Services related to the consumer goods and consumer services are increasing significantly in our country. The social characteristics of India and life styles of people of India are changing very fast Indians have become increasingly pleasure seeking and the purchasing power of middle class, so also the size of the middle class have increased substantially. The focus of the world is towards India as an emerging market and more number of multinationals have started their operations in India. Indian consumer is also changing fast. The trend is towards the value for money, quality of services and the consumer has become more demanding. Thus, to survive and grow in these competitive market conditions, a substantial service component is to be added to company's product offerings.

35.2. SERVICE DEFINED

Most definitions of service stress the intangibility aspect of service as contrasted with tangibility of products. A service is an activity which has some element of intangibility associated with it, which involves some interaction with customers or with property in their possession and does not result in a transfer of ownership. A change in condition may occur and production of service may or may not be closely associated with the physical product. A product is an overall package of objects or processes which provide some value to customers, while goods and services are subcategories, which describe two types of products.

Philip Kotler has distinguished four categories of offers varying from pure tangible good to a pure intangible service.

(*i*) A pure tangible good such as soap, toothpaste, salt etc. No service accompanies the product.

(*ii*) A tangible good with accompanying services to enhance its consumer appeal, *e.g.*, computers.

(*iii*) A major service with accompanying minor goods and services, *e.g.*, first class air travel.

(*iv*) A pure service, *e.g.*, psychotherapy.

35.3. CHARACTERISTICS OF SERVICES

- Intangibility — Services to a large extent are abstract and intangible. Services cannot be touched and they should be experienced.
- Heterogeneity — Services are non standard and highly variable. It is difficult to achieve standardization in services.
- Inseparability — Services are typically produced and consumed at the same time with customer participation in the process.
- Perishability — It is not possible to store the services (non inventorable).
- Patent protection is not possible.
- Services do not involve transfer of ownership.

35.4. INTANGIBILITY ASPECT OF SERVICES

Services are activities performed by the provider. They are essentially intangible in nature. When goods are bought, the purchase enhances the physical possession of the buyer and he has some thing tangible to show for the money he has spent. On the other hand, in the purchase of a service, the buyer does not get any thing tangible "A service is a deed, performance, an effect." Therefore, it is consumed and possessed. In airline service, the buyer does not get anything in tangible; he consumes the services to meet the transportation needs.

Commonly used services are repairs, hair dressing, tuitions, car renting, taxi services, motels and hotels, theaters, banking and insurance, health care, couriers, legal services and advertising etc. Intangibility is the important feature of services that distinguishes services from goods. Intangibility means palpable, the customer cannot touch a service product.

35.5. REASONS FOR GROWTH OF SERVICE SECTOR

Following are the changes that account for growth in service sector.

1. **Demographic Changes**
   - Life expectancy has risen producing an expanding retired population. This sector has created new demands for leisure and travel as well as for health care and nursing.
   - Structural shift in communities have affected where and how people live.
   - Increased urbanization.
2. **Social Changes**
   - Increased number of women in the work force.
   - Domestic functions being carried outside the home. Rapid rise in fast food industry, childcare facilities and other personal services.
   - Double income family has created a greater demand for consumer services including retailing, real estate and personal financial services.
   - Quality of life has improved. Smaller families with double income have higher disposable income to spend on entertainment, travel and hospitality services.
   - Communication and travel have increased the aspiration levels.
3. **Economic Changes**
   - Globalization has increased the demand for communication travel and information services (IT).
   - Increased specialization within the economy has led to great reliance on specialist service providers.

4. **Political and Legal Changes**
   - Government has grown in size creating huge infrastructure
   - Internationalization of business gives way to legal specialists.
   - Trend to subcontract the services.

The above changes account for the rapid growth of service sector in Indian context.

35.6. PRODUCT LEVELS

A product is anything that can be offered to a market for attention, use or consumption that might satisfy a want or a need. It includes physical objects, services and ideas etc.

A service product refers to an activity or activities that a marketer offers to perform, which results in satisfaction of a need or want of predetermined target customers. It is the offering of the firm in the form of activities that satisfy the need such as a surgery performed etc.

Basically, there are four levels of products.

(*i*) Generic product is the rudimentary substantive thing. It is the product at its basic level.

(*ii*) Expected product is the customer's minimum set of expectations from a product or a service.

(*iii*) Augmented product is an offering (product benefits) in addition to what the customer expects.

(*iv*) A potential product is doing everything potentially feasible to hold and attract the customer.

Most organisations stagnate around the generic product level. At this level of a product delivery, the customer is not satisfied enough to be a hard-core loyalist. Therefore, the competitor who opts to operate at higher product level will be able to break in to service company's market share. While average companies hover around the generic products successful company enhances their customer satisfaction by adding pleasant surprise in to their services packages. When a product exceeds its expected benefits, it comes as a pleasant surprise to the customer leading to his loyalty.

The three levels beyond the core or generic product represents opportunities to provide added values to the customers. The added value may be solely at the emotional level and it is never real to the customers. The differentiation of the brand is achieved through adding value to the core service product. The product surround is shown in fig (35.1).

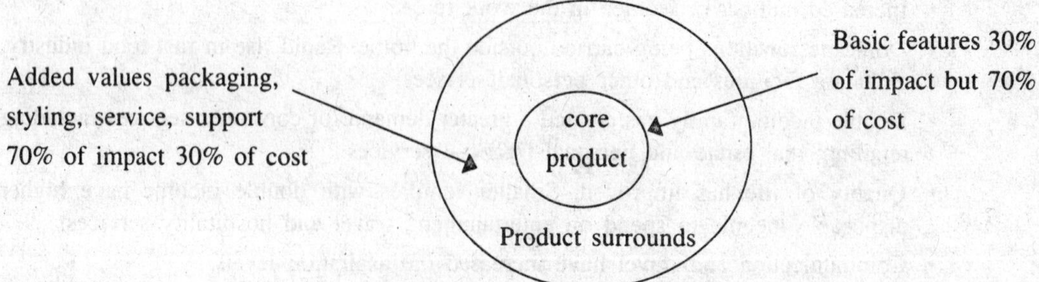

Fig. 35.1. Product surround concept.

35.7. MOMENTS OF TRUTH AND MANPOWER IMPLICATIONS

Albercht and Zemke have termed the point in time when the customer comes in contact with the service/delivery system as "moments of truth" during the service delivery cycle. At that instant when customer avails the service, he forms an impression about the service received. The cumulative effects of these moments of truth are important. A bad moment of truth can nullify many good moments of truth. The customer's perception of service is a function of all the moments of truth he has experienced.

The distinguishing feature about services is that most of the moments of truth are involved with the lower or the lowest level of staff. *e.g.*, in a hotel, the room service boy, reception clerk, waiter, rarely, the customer comes in contact with the service managerial or supervisory staff. Certain problems are to be sorted out as and when they arise instantly. Three things are expected from service provider to offer the satisfactory service.

1. The routine part of the operations should be planned to the minute level.

2. Personnel providing the service have to undergo adequate training to perfect these micro tasks .along with verbal and facial expressions.

3. They should be empowered to sort out any variations themselves because; in services the reaction of the customer is spontaneous and could be varied. This empowerment of people is an essential ingredient of the service management specially where the customer contact is high.

Roles of people in service are based on the degree of frequency of customer contact and the extent to which staff are involved with conventional service marketing activities. This categorization results in four groups.

(*i*) Conactors — who have regular or frequent contact with customers?

(*ii*) Modifiers — receptionist, credit department, switch board operators etc.

(*iii*) Influencers — market research, product development staff.

(*iv*) Isolated — perform various support functions and have neither regular customer contact nor major decision making, *e.g.*, purchasing, data processing.

	Involved with conventional Marketing mix	Not directly involved in marketing mix
Frequency or Periodic customer Contact	CONTACTORS	MODIFIERS
Very low or no Customer contact	INFLUENCERS	ISOLATED

Fig. 35.2. Roles of people in service marketing.

35.8. SERVICE MATRIX

A 2 × 2 matrix is used to represent some of the characteristics and requirements of a service delivery system. This matrix divides services in to four groups or quadrants, with the complexity of the service as one dimension and the degree of service customization as the other. The fig (35.3) shows the matrix with several services listed in each of the four quadrants. The complexity of service here refers to the degree of knowledge and skill or capital investment that the service provider must posses in relation to the amount that a typical consumer might have.

Services on the left side of the matrix require a great deal of training and/or capital investment to perform. The services with upper half of the matrix are more personalized (or customized) in the sense that they are designed specifically for the individual needs of particular consumer. Services in the lower half of the matrix are more standardized in the way that more number of customers will receive the same service.

Since the services are people oriented, businesses in all the four segments select persons who are courteous and have good interpersonal skills for the jobs that entail significant contact with the customers.

Many of the services in II and IV quadrants can train their own employees in the necessary skills. Companies in fourth quadrant are more likely to develop procedures and to train employees for reliable, consistent service encounters. Workers in II quadrant need to have broader skills and are to be more flexible in response to customer requirements. The workers involved in businesses that fall in I quadrant require extensive professional skills and they should highly trained outside the organisation. Investment capabilities in equipment and facilities is often a significant characteristic of services in quadrant III. These business must interpret the needs of people and offer appropriate mix of services if they want to attract large number of people.

Complexity of service provided

Degree of Customization	High	Low
Customized	I Physician, Attorney, Dentist, Auto repair, Appliance repair, Air charter	II Beauty and hair care, Lawn care, House painting, Wall papering, Restaurant, Taxi
Standard	III Radio and TV, Movie Zoo, Museum, School Long distance call, Routine auto tune up Air line	IV Fast food, Car wash, Car rental, Dry cleaning storage, Retail transport

Fig. 35.3. Matrix of service characteristics.

The service matrix can be used to illustrate how the operations management task varies with different types of services. The highly automated services require familiarity with technology correct capital decisions, management of demand to avoid peaks and careful scheduling to utilize expensive capacity. Highly labour-intensive services require more attention to managing large work force associated with these services.

35.9. SERVICE MODEL

A service model includes service strategy, service product design, service delivery system and the integration of the elements like technology, workforce and the service strategy is shown in fig (35.4).

Services are customer oriented and customer centered and strategy, system and people in the operations of the services also focus on the customer. Customer's expectations are central to the design of service strategy of the firm. People are extremely important in producing and delivering the service to the customer. The strategy to system link means that the systems and procedures should follow from the service strategy.

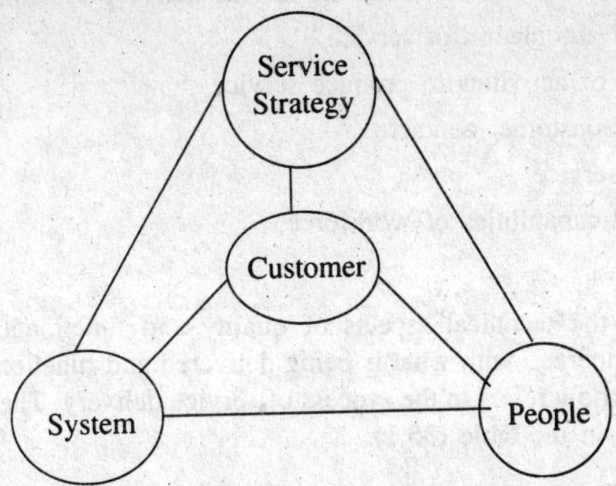

Fig. 35.4. Integration of service strategy, system and people.

Service strategy defines what business the company is carrying on (which is service mission). It provides guidance for designing products, delivery systems and measurements. It is should provide the vision for the business and describe how the business is perceived by the customers and employees.

Service strategy should consider the international scope of services offered. Many firms will adopt global service strategy where the service is standardized around the world, the basis of competition is international and the scale of service operations is global.

The Service Product

The service product comes as a bundle with goods as well. Sasser and Others (1990) have defined service product as consisting of following bundle of goods and services.

(*i*) The physical items or facilitating goods.

(*ii*) The sensual benefits or explicit services.

(*iii*) The psychological benefits or implicit services.

For example, in case of a restaurant,

The physical items consist of the facilities, food, drinks etc.

Sensual benefits include taste, service level, and smell of food.

Psychological benefits include comfort, status etc.

Key to the design of service products is how best the company defines the mix of these three ingredients. It is essential to define the quality standards for services. In designing services, management should carefully read customer expectations. The management should make the service guarantee to back up their good service bundle.

Customer contact is the key step in service process. Key element in choosing a process is the amount of customer contact. In service where customer contact is high, he can disrupt the production process of services or special treatment. Thus, a high customer contact may lead to inefficient processes. The degree of customer contact can be an important ingredient in designing processes or service delivery systems. Higher customer contact leads to higher price and customization and lower customer contact leads to lower price and higher customization.

Service Delivery System

This consists of the work force providing the service and the other physical elements used to produce service. The important elements in service delivery systems are

1. The degree of automation of service
2. The sequence of activities to produce service
3. The degree of customer contact
4. The place of service
5. The skills and capabilities of workforce

Service Quality

For services both the technical aspects of quality and functional quality are applicable. Technical quality is concerned with what is being delivered and functional quality refers to how it is developed, *i.e.*, the how refers to the process of service delivery. The determinants of service quality are represented in the table (35.1).

Table 35.1: Determinants of service quality

- Reliability

Should provide consistent service, which is dependable.

- Responsiveness

Prompt/timely response to customer need.

- Access

Ease to approach

- Communication.
- Image and credibility, courtesy.
- Understanding the needs of customer.
- Total service package — core plus product surrounds.

36

PRODUCT AND SYSTEM RELIABILITY

• Introduction • Definition • Failure — Causes of Failure, Nature of Failures • Phases of Failures (Bath tube curve) • Measures of Reliability • Reliability Function derivation • Hazard Rate • Hazard and Reliability Functions for well known distributions — exponential, normal, log-normal, Weibull and Gamma • System reliability — Systems with components in series, parallel, series — Parallel Combination, Reliability of K out of m system, Reliability Improvement, Redundancy Maintainability, Availability • Reliability Life Testing.

36.1. INTRODUCTION

Customers always want that the products that are purchased should have a long service life and during this life it should give intended service and utility with few failures. As product becomes more complex, the problems of failures will increase over time. The improvement in effectiveness of such complex systems has therefore acquired special importance in recent years. The effectiveness of a system is its suitability for the fulfilment of the intended function and the efficiency of utilizing the means put in to it. The suitability of performing definite task is primarily determined by the reliability and quality of the system. An evaluation of a system reliability becomes essential to decide whether a system will accomplish its mission successfully.

During the past, very high safety factors were introduced which added tremendously to its weight as well as the cost and another approach was to learn from the previous mistakes (failures) of previous designs. These approaches become impractical for the design of new products and systems where each design is different from other and each design has to be right first time and all the time. There is no much time for trial and error. Reliability is to be built in to the product right at the design stage and quality and hence the reliability should be built during subsequent stages of manufacture.

Reliability engineering is a emerging area, which is a collection of many tools. Reliability engineering is a combination of management and technical disciplines, which has a specific objective — that is the assurance of intended performance for a specified time span of products. It is a formalized approach to achieve optimum product reliability using management, engineering, mathematical and statistical elements and concepts. A high level of reliability can be achieved only through an integrated effort, which ensures that there are no weak links in the total process of design, development and manufacturing.

36.2. DEFINITION OF RELIABILITY

"Reliability is the ability of an item to perform its intended function under stated operation conditions for a given period of time."

The definition stresses on four significant elements

(*i*) Probability

(*ii*) Intended Function

(*iii*) Time

(*iv*) Operating Conditions.

- **Probability**

Consideration of variation makes reliability a probability. It is possible to identify the frequency distribution of an item, which permits prediction of life of the item. *e.g.*, the probability of an item functioning is 0.85 for 60 hours indicates that only 85 times out of 100, we would expect the item to be functioning for a period of 60 hours.

- **Intended Function**

For an item to be reliable, it must perform a certain functions satisfactory when called upon to do, while considering the reliability of an item, the criteria of what is considered as the required function have to be exactly spelled out in advance. Thus, criteria must be established in all cases, which clearly specify and define what is considered as intended function.

- **Time**

Time is the most important factor in the assessment of reliability, since it represents a measure of the period during which one can expect a certain degree of performance from an item.

- **Stated Conditions**

The application and operating circumstances under which an item is put to use is an important component of reliability. As the operating conditions, change the reliability of an item also changes. Operating conditions such as temperature, humidity, torque, and corrosive atmosphere all have a definite effect on performance.

Thus, reliability can be stated as follows :

"The reliability of a 60 watts incandescent bulb has been estimated to be 0.95 for 1200 hours providing 20 candles output under 180-230 volts and at normal environmental conditions."

36.3. FAILURE

Failure of an item represents unreliability. Thus, to compute the reliability of an item, it is necessary to understand the concept of failure. A deviation in the properties of an item from the prescribed conditions is considered as fault. A state of the fault is denoted as "Failure".

An item is considered to have failed under one of the following conditions

1. When it becomes completely inoperable.
2. When it is still operable, but no longer able to perform a required function.
3. When a serious deterioration makes the item unsafe for its continued use.

36.3.1. Causes of Failures

There are many specific causes of failures of components and systems. Due to the complexity of the system, some are known and some are unknown. Some of the causes of failures are

- Deficiencies in design.
- Improper selection of process and manufacturing technique.
- Lack of knowledge and experience.
- Errors of assembly.
- Improper service conditions.

- Inadequate maintenance.
- Variation in environmental and operating conditions.
- Human errors.

36.3.2. Nature of Failures

An item may fail in many ways. An understanding of these failures help in taking appropriate corrective measures for achieving better reliability. The different modes of failure are :

1. **Catastrophic Failures**

In this case, a normally operating item suddenly becomes inoperative. Example : Blowing of a fuse or electric bulb.

2. **Degradation (Creeping) Failures**

These failures occur gradually because of change in some parameter with time. Example, change in resistance will affect the performance of a resistor.

3. **Independent Failures**

These are the failures, which occur independently and does not depend on failure of the other.

4. **Secondary Failures**

A secondary failure occurs as a result of some primary failure.

5. **Failure due to Improper Handling and Misuse**

e.g., overloading (stressing beyond the capacity).

36.3.3. Phases of Failures (Bath Tub Curve)

Analysis of failure data has shown that failures in general can be grouped in to different modes depending upon the nature of the failure. When a large number of units are put into operation, it is likely that there is large number of failures initially. These failures are called Initial Failures or Infant Mortality. After the initial failures, for a long period of time of operation fewer failures are reported but it is difficult to determine their cause. The failures during this period are often called random failures or catastrophic failures. This is the period of normal operation. As the time passes, the units get worn out due to wear and tear and begin to deteriorate. Here in this period, the failures are due to wear and tear and due to ageing. This region is called the wear out region.

These three phases of failures are represented in fig. (36.1) and the characteristics of each phase are shown in table (36.1).

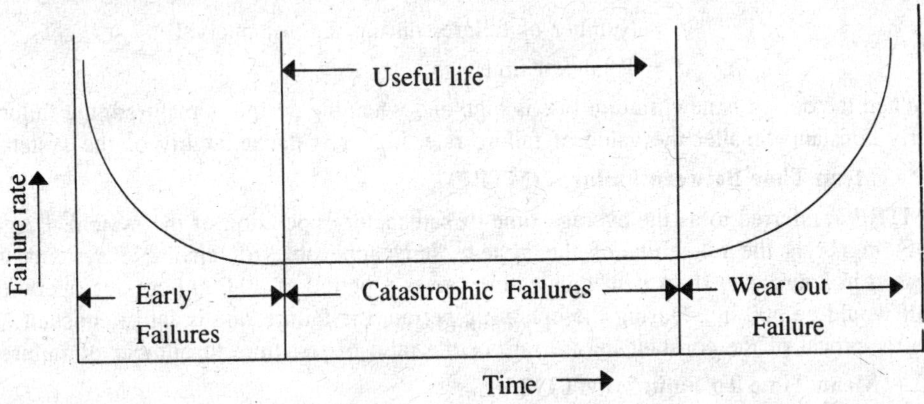

Fig. 36.1. Bath Tub Curve.

Table 36.1: Characteristics of Various Phases of Bath Tub Curve

1. **Early Failures**
   - These failures occur at the beginning due to the probability of defective design, manufacturing or assembly and quality control techniques during manufacturing.
   - These are eliminated by debugging or burn in process. The weak and substandard products/components that fail during early hours of system operation are replaced by good or tested components. Debugging is a method of accelerating the completion of early failures by operating the system continuously for number of hours, correcting them and then releasing the system for actual use.
   - Debugging is done generally prior to dispatch to the user to ensure the detection and elimination of early failures.
   - Warranty is based on the concept of early failures.

2. **Catastrophic (Chance) Failures**

These failures are predominant during actual working of the system. They occur randomly and unexpectedly. The failure rate is fairly constant. These are caused due to sudden stress accumulation beyond the design strength of the material. This phase is called the useful life of the component. The failures at this stage can be minimised by introducing redundancy in the system.

3. **Wear Out Failures**

The item is more likely to fail due to wear and tear and the number of failures will be high. This is a typical ageing problem. Proper care and maintenance will reduce the failures at this stage.

36.4. MEASURES OF RELIABILITY

1. **Failure Rate**

Failure rate is expressed in terms of failures per unit time i.e. as failures per hour, or failures per 100 or 1000 hours. Failure rate is the ratio of number of failures (f) during a specified test interval to the total test time of items undergoing test.

$$\lambda = \frac{f}{T}$$

λ = Failure rate

f = Number of failures during the test interval.

T = Total test time.

When the design is new, failure rate is high and when the design is matured, the failure rate is fairly constant. Smaller the value of failure rate, higher is the reliability of the system.

2. **Mean Time Between Failures (MTBF)**

MTBF is referred to as the average time of satisfactory operation of the system. Larger the MTBF, higher is the reliability of the system. It is applicable to repairable systems and is expressed in hours, *e.g.* If an item fails 8 times over a period of 40,000 hours of operation the MTBF would be 500 hrs. During the operating period, the failure rate is fairly constant MTBF is the reciprocal of the constant failure rate or the ratio of test time to number of failures.

3. **Mean Time To Failure (MTTF)**

This is applicable to non-repair systems. The mean time to failure is expressed as the average time an item is expected to function before failure. If we have the life test information on '*n*' items with failure times t_1, t_2, ... t_n, then the mean time to failure is defined as

$$\text{MTTF} = \frac{1}{n} \sum_{i=1}^{n} t_i$$

36.5. RELIABILITY FUNCTION DERIVATION

Reliability is defined as the probability that a system (component) will function over some period of time 't'. To express this relationship mathematically, we define the continous randon. variable 'T' to be the time to failure of the system (or component), $T \geq 0$.

Then, reliability can be expressed as

$$R(t) = \text{Prob}\{T \geq t\} \qquad \text{...(1)}$$

Where $R(T) \geq 0$, $R(0) = 1$ and $\lim_{t \to \infty} R(t) = 0$

For a given value of t, $R(t)$ is the probability that the time to failure is greater than or equal to 't'.

We define,

$$F(t) = 1 - R(t) = \text{Prob}\{T < t\} \qquad \text{...(2)}$$

Where $F(0) = 0$,

and $\lim_{t \to \infty} F(t) = 1$

then, $F(t)$ is the probability that a failure occurs before time 't'. Thus, $R(t)$ is the reliability function and $F(t)$ is the cumulative distribution function (CDF) of the failure distribution. A third function called probability density function (PDF) is defined as

$$f(t) = \frac{d\,F(t)}{dt} = -\frac{d\,R(t)}{dt} \qquad \text{...(3)}$$

This function describes the shape of the failure distribution. The three functions are illustrated in the fig (36.2).

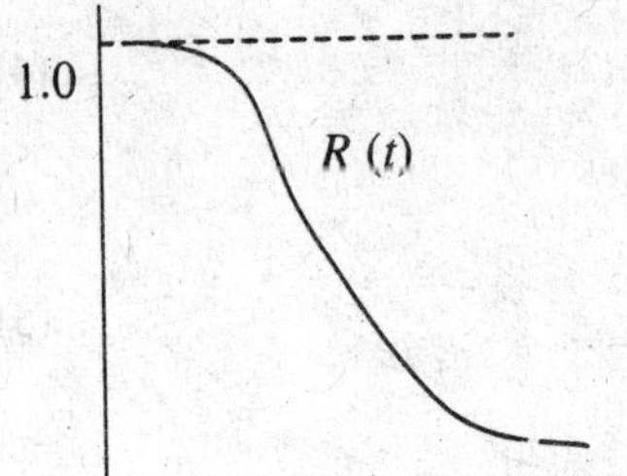

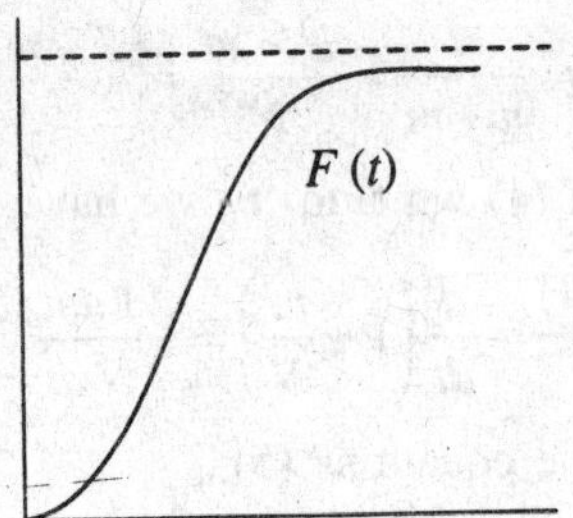

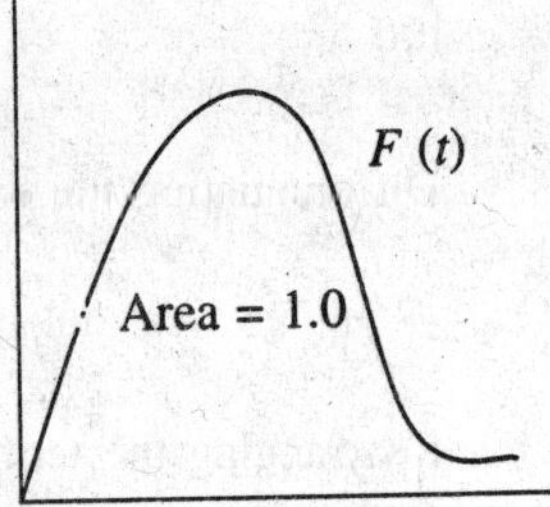

Fig. 36.2. (a) Reliability function (b) Cumulative distribution function (c) Prob. density function.

The probability density function (pdf), has the following two properties

$$f(t) \geq 0 \text{ and } \int_0^{\infty} f(t)\,dt = 1$$

Given $Pdf.\, f(t)$, then

$$F(t) = \int_0^t f(t')\,dt'$$

$$R(t) = \int_t^{\infty} f(t')\,dt'$$

Both reliability function $R(t)$, and the cumulative density function represent areas under the curve defined by $F(t)$. Since the area under the curve is equal to one, both the reliability and failure probability will be defined so that,

$$O \le R(t) \le 1 \text{ and } O \le F(t) \le 1$$

The function $R(t)$ is normally used when reliabilities are being computed and the function $F(t)$ is normally used when the failure probabilities are being computed. The graphical representation of *pdf* $[f(t)]$ provides a visual representation of the failure distribution.

Component Reliability From Test Data

Consider a set of 'N' components in operation from time $t = 0$. With the progress of the time, the components fail. After a certain period t, consider that number of components surviving are n and the number of components that have failed are m.

The components fail independently with the probability of failure.

$\therefore$ $N = (n + m)$ is constant through out the test. This is because, as the test proceeds, the number of failed components 'm' increases as the number of surviving components (n) decreases.

Probability of survival (reliability) is expressed as the fraction at any time 't' during the test and is given by

$$R(t) = \frac{n}{N} = \frac{n}{m+n} \quad \text{...(1)}$$

The probability of failure or unreliability is expressed at any time 't' can be expressed as

$$F(t) = \frac{m}{m+n} \quad \text{...(2)}$$

At any time t,

$$R(t) + F(t) = 1 \quad \text{...(3)}$$

Because $R(t)$ and $F(t)$ are mutually exclusive events equation (1) can be rewritten as

$$R(t) = \frac{n}{n+m} = \frac{N-m}{N} = \left(1 - \frac{m}{N}\right) \quad \text{...(4)}$$

Differentiating the equation (4) w.r.t. to 't', we have,

$$\frac{d\,R(t)}{dt} = \frac{d}{dt}\left(1 - \frac{m}{N}\right) = -\frac{1\,dm}{N\,dt} \quad \text{...(5)}$$

Rearranging the terms in the equations (5),

$$\frac{dm}{dt} = -N\frac{d\,R(t)}{dt} \quad \text{...(6)}$$

This equation (6) represents the rate at which the components fail since $m = (N - n)$, on differentiation.

$$\frac{dm}{dt} = \frac{d}{dt}(N-n)$$

$$= -\frac{dn}{dt} \quad \text{...(7)}$$

This is the negative rate at which the components survive. The term $\frac{dm}{dt}$ can be interpreted as — number of components failing in the time interval 'dt' between the times t and $t + dt$ which is equivalent to the rate at which the components still in the test at time 't' is failing.

Dividing equation (6) by 'n' on both sides

$$\frac{1}{n}\frac{dm}{dt} = -\frac{N}{n}\frac{d\,R(t)}{dt} \qquad \text{...(8)}$$

The L.H.S. of equation is the failure rate (λt) and $\frac{n}{N} = R(t)$.

Thus, $$\lambda(t) = -\frac{1}{R(t)}\frac{d\,R(t)}{dt} \qquad \text{...(9)}$$

Relation Between f (t), R (t) and λ (t)

1. **Probability Density Function [f (t)]**

It is the probability that a random trial yields the value of 't' within the interval from t_1 to t_2 and expressed as

$$\int_{t_1}^{t_2} f(t)\,dt$$

$f(t)$ is the density function for a continuous random variable.

2. **Distribution Function [F(t)]**

It is the probability that in a random trial, the random variable is not greater than 't'

$$\therefore \qquad F(t) = \int_{-\infty}^{t} f(t)\,dt$$

$F(t)$ is recognized as unreliability function.

3. **Reliability [R(t)]**

It expresses the probability that the variable is at least as large as 't'.

$$R(t) = \int_{t}^{\infty} f(t)\,dt$$

and $$R(t) = 1 - F(t).$$

4. **Failure Rate [λ (t)]**

The rate at which failures occur in the interval t_1 and t_2 is called the failure rate. It is expressed as the conditional probability that failures occur in the interval t_1 and t_2, given that failures have not occurred prior to t_1, *i.e.*, the start of the interval.

Failure rate is given by

$$\lambda(t) = \frac{\int_{t_1}^{t_2} f(t)\,dt}{(t_2 - t_1)\int_{t_1}^{\infty} f(t)\,dt}$$

$$= \frac{\int_{t_1}^{\infty} f(t)\,dt - \int_{t_2}^{\infty} f(t)\,dt}{(t_2 - t_1)\int_{t_1}^{\infty} f(t)\,dt}$$

$$= \frac{R(t_1) - R(t_2)}{R(t_1)(t_2 - t_1)}$$

If we substitute $t_1 = t$ and $t_2 = t + \Delta t$

The failure rate can be expressed as

$$\lambda(t) = \frac{R(t) - R(t + \Delta t)}{\Delta t \, R(t)}$$

As a special case,

where $f(t)$, the probability density function is exponential the various functions can be computed as

Probability density function $f(t)$ is

$$f(t) = \left(\lambda e^{-\lambda t}\right)$$

where λ is the constant failure rate.

$$f(t) = \int_0^t \lambda e^{-\lambda t} \, dt$$

$$= 1 - e^{-\lambda t}$$

Reliability function can be calculated as

$$R(t) = \int_t^{\infty} \lambda e^{-\lambda t} \, dt$$

$$= e^{-\lambda t}$$

$$R(t) = 1 - F(t)$$

36.6. HAZARD RATE Z (t)

Hazard rate or instantaneous failure rate is defined as the limit of failure rate as the time interval length approaches to zero. It is a measure of instantaneous speed of failure.

It is expressed as

$$Z(t) = \lim_{\Delta t \to 0} \left[\frac{R(t) - R(t + \Delta t)}{\Delta t \, R(t)}\right]$$

$$= \frac{-1}{R(t)} \frac{d(Rt)}{dt}$$

$$\therefore \quad Z(t) = \frac{f(t)}{R(t)}$$

as $$f(t) = \frac{-d\,R(t)}{dt}$$

Relationship between R (t) and Z (t)

$$Z(t) = \frac{F(t)}{R(t)} = \frac{d\,F(t)}{dt} \cdot \frac{1}{R(t)}$$

Integrating the above equation between 0 to t

$$\int_0^t Z(t)\,dt = \int_0^t \frac{d\,F(t)}{dt} \frac{1}{R(t)}$$

$$= \int_0^t \frac{d\,F(t)}{1 - F(t)} = -\log\left[1 - F(t)\right]_0^t$$

$$= -\left[\log R(t)\right]_0^t$$

$$= -\log R(t) \text{ as } R(0) \text{ and } \log R(0) = 0$$

$$\therefore \quad \boxed{R(t) = e^{-\int_0^t Z(t).dt}}$$

Cumulative Distribution Function

$$F(t) = 1 - R(t) = 1 - e^{-\int_0^t Z(t).dt}$$

Probability density function

$$F(t) = Z(t).e^{-\int_0^t Z(t).dt}$$

Rearranging the equation and integrating with proper limits, we have,

$$\lambda(t).dt = \frac{-d\,R(t)}{R(t)}$$

or
$$\int_0^t \lambda(t)\,dt = -\int_0^R \frac{d\,R(t)}{dt} = -\ln R(t) \qquad ...(10)$$

$$\therefore \quad \ln R(t) = -\int_0^t \lambda(t).dt \qquad ...(11)$$

Initially, at $t = 0$, $R(t) = 1$, we obtain

$$R(t) = \exp\left[-\int_0^t \lambda(t)\,dt\right] \qquad ...(12)$$

This is (equation 12) is a general formula for computing reliability. $\lambda(t)$ can be any variable and integratable function of time. But, if we specify that $\lambda(t)$ is constant overtime and $\lambda(t) = \lambda$ (say).

The reliability formula becomes

$$R(t) = e^{-\lambda t}$$

Properties

The probability of survival or reliability $R(t)$ at time 't' has the following properties.

1. $0 \le R(t) \le 1$
2. $R(0) = 1$ and $R(\infty) = 0$

As special case, when the probability function is exponential and failure rate is constant.

Then
$$Z(t) = \frac{f(t)}{R(t)} = \frac{\lambda e^{-\lambda t}}{e^{-\lambda t}}$$
$$= \lambda$$

When the failure rate is constant, hazard rate is also constant and is equal to the failure rate.

Mean Time To Failure (MTTF)

The Mean Time To Failure will assume to be the same for all the components, which are identical in the design and operate under the conditions, which are identical.

MTTF is given by the mathematical expectation of the random variable '*T*' describing the MTTF of the component.

Therefore, $$\text{MTTF} = E(T) = \int_0^{\infty} t \, . \, f(t) \, dt$$

This is the mean or expected value of the probability distribution defined by *F* (*t*).

We have the probability density function (PDF)

$$F(t) = \frac{d \, . \, f(t)}{dt} = \frac{-d\,R(t)}{dt}$$

$$\text{MTTF} = \int_0^{\infty} \frac{-d\,R(t)}{dt} t \, . \, dt$$

Using integration by parts

$$\text{MTTF} = -t \, . \, R(t) \Big|_0^{\infty} + \int_0^{\infty} R(t) \, dt$$

$$= \int_0^{\infty} R(t) \, dt$$

Since, $$\lim_{t \to \infty} t \, . \, R(t) = \lim_{t \to \infty} t \, . \exp\left[-\int_0^t \lambda(t^1) \, . \, dt^1 = 0 \right]$$

and *R* (0) = 0.

Constant Hazard Model

The constant hazard model can be expressed as

$$Z(t) = \lambda$$

where λ is constant and independent of time.

An item with constant hazard rate will have the following reliability and associated functions.

$$f(t) = \lambda \, . \, e^{-\lambda t}$$

$$R(t) = e^{-\lambda t}$$

$$F(t) = 1 - e^{-\lambda t}$$

The mean time to failure of the item is

$$\text{MTTF} = \int_0^{\infty} e^{-\lambda t} dt = \frac{1}{\lambda}$$

36.7. RELIABILITY AND HAZARD FUNCTIONS FOR WELL KNOWN DISTRIBUTIONS

36.7.1. Exponential Distribution

Exponential distribution is widely used in reliability. A constant failure rate model for continuously operating system leads to an exponential distribution. Replacing a time dependent failure rate λ (*t*) by a constant λ in the *pdf* equation.

We have

$$F(t) = \lambda . e^{-\lambda t}$$

Similarly, CDF becomes

$$F(t) = 1 - e^{-\lambda t}$$

and the reliability can be written as

$$R(t) = e^{-\lambda t}$$

$$\text{MTTF} = \frac{1}{\lambda}$$

The exponential failure density function is represented in fig 36.3.

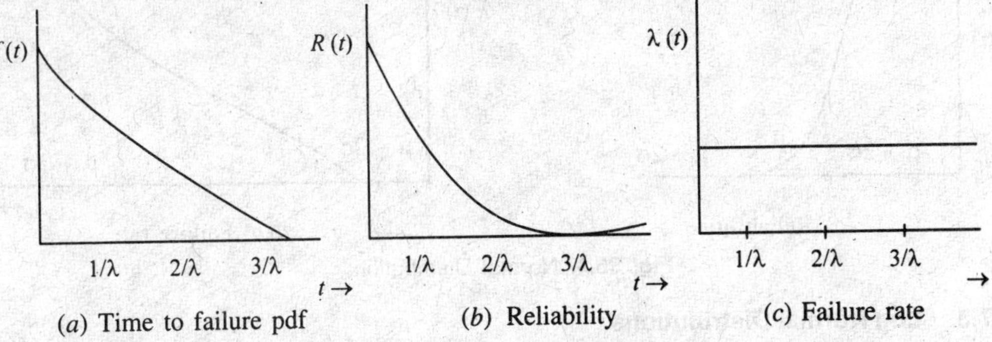

(*a*) Time to failure pdf (*b*) Reliability (*c*) Failure rate

Fig. 36.3. The exponential Distribution.

A device described by a constant failure rate and therefore by an exponential distribution of times to failure has the following property of "Memorylessness". The probability that it will fail during same period in the future is independent of its age.

36.7.2. Normal Distribution

The normal distribution takes the well-known bell-shape and to describe the time dependence of reliability problems. The pdf for normal distribution is given by the following equation with '*t*' as a random variable.

$$F(t) = \frac{1}{\sqrt{2\pi}\sigma} \exp\left[-\frac{(t-\mu)^2}{2\sigma^2}\right]$$

where μ is here is the MTTF.

$$F(t) = \int_{-\infty}^{t} \frac{1}{\sqrt{2\pi}\sigma} \exp\left[-\frac{(t-\mu)^2}{2\sigma^2}\right] dt$$

In the standardized normal form,

$$F(t) = \phi\left[\frac{t-\mu}{\sigma}\right]$$

and Reliability for normal distribution is given by

$$R(t) = 1 - \phi\left[\frac{t-\mu}{\sigma}\right]$$

and the failure rate is obtained by the equation,

$$\lambda(t) = \frac{1}{\sqrt{2\pi}\sigma} \exp\left[-\frac{1}{2}\left(\frac{t-\mu}{\sigma}\right)^2 \left(1 - \phi\frac{t-\mu}{\sigma}\right)\right]^{-1}$$

The reliability and *pdf* are plotted for times to failure as shown in fig 36.4.

As indicated by the behaviour of failure rate, normal distributions are used to describe the reliability of the equipment to situations other than to which constant failure rates are applicable. It is useful in describing the reliability in situations in which there is a reasonably well-defined wear out time μ.

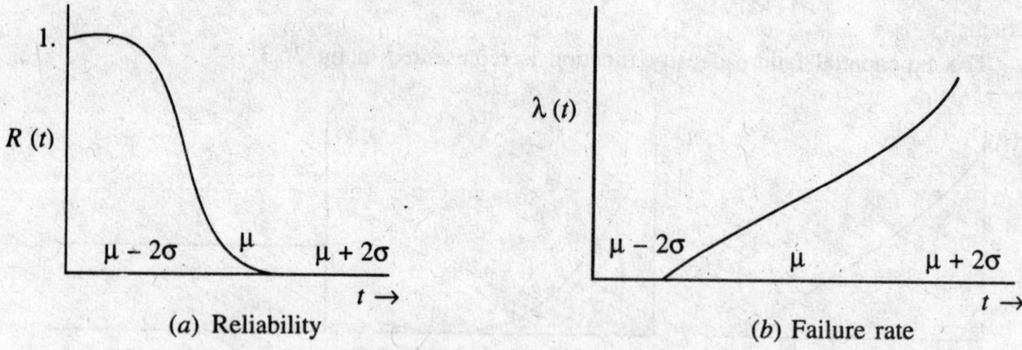

(*a*) Reliability (*b*) Failure rate

Fig. 36.4. Normal Distribution.

36.7.3. Log Normal Distributions

The log normal is related distribution that has been found to be useful in describing failure distributions for a variety of situations especially when the time to failure is associated with large uncertainty. The pdf for the time to failure is given by

$$f(t) = \frac{1}{\sqrt{2\pi}\, St} \exp\left\{-\frac{1}{2S^2}\left[\ln\frac{t}{t_0}\right]^2\right\}$$

corresponding CDF is expressed as

$$F(t) = \phi\left[\frac{1}{S}\ln\frac{t}{t_0}\right]$$

however to is not the MTTF.

$$\text{MTTF} = \mu = \text{to} \,.\, \exp\left(S^2/2\right)$$

The log normal distribution is frequently used to describe fatigue and other phenomenon that are caused by ageing or wear and that result in failure rates that increase with time.

The log normal reliability function and hazard functions are shown in fig (36.5). The failure can be increasing or decreasing depending on value of '*S*'.

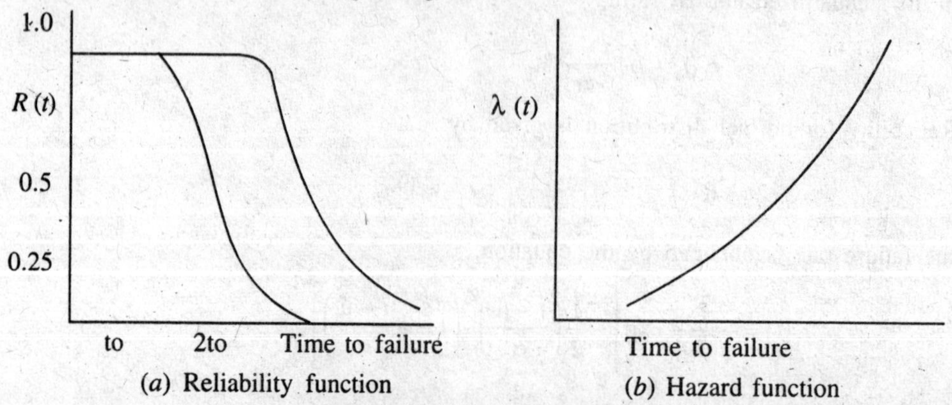

(*a*) Reliability function (*b*) Hazard function

Fig. 36.5. Log normal distribution.

36.7.4. Weibull Distribution

One of the most useful probability distribution i.. reality is the weibull. Weibull failure distribution may be used to model both increasing and decreasing failure rates. It is characterised by the hazard rat: function of the form.

$$\lambda(t) = a \, . \, t^b$$

which is a power function.

The function $\lambda(t)$ is increasing for $a > 0$, $b > 0$ and is decreasing for $a > 0$, $b < 0$. For convenience, $\lambda(t)$ can be expressed as

$$\lambda(t) = \frac{\beta}{\theta}\left(\frac{t}{\theta}\right)^{\beta - 1} \qquad \theta > 0,\ \beta > 0, t \geq 0$$

$R(t)$ is expressed as

$$R(t) = \exp\left[-\int_0^t \frac{\beta}{\theta}\left(\frac{t}{\theta}\right)^{\beta-1} . dt\right]$$

and

$$f(t) = -\frac{d\,R(t)}{dt} = \frac{\beta}{\theta}\left[\frac{t}{\theta}\right]^{\beta-1} . e^{-(t/\theta)^{\beta}}$$

Beta (β) is referred to as the shape parameter. Its effect on the distribution can be represented in fig (36.6) for several different values β.

For $\beta < 1$, the pdf is similar in shape to the exponential and $\beta \geq 3$ [large values], the *pdf* is somewhat symmetrical like normal distribution.

For $1 < \beta < 3$. The density function is skewed. When $\beta = 1$, $\lambda(t)$ is constant and the distribution is identical to exponential.

Theta (θ) is a scale parameter that influences both the mean and dispersion of the distribution. As 'θ' increases the reliability increases at a given point in time. The parameter is also called the characteristic life.

MTTF and variance of the Weibull distribution is given by

$$\text{MTTF} = \theta\,\Gamma\left(1 + \frac{1}{\beta}\right) \quad \text{and}$$

$$\sigma^2 = \theta^2 \left\{\Gamma\left(1 + \frac{2}{\beta}\right) - \left[T\left(1 + \frac{1}{\beta}\right)\right]^2\right\}$$

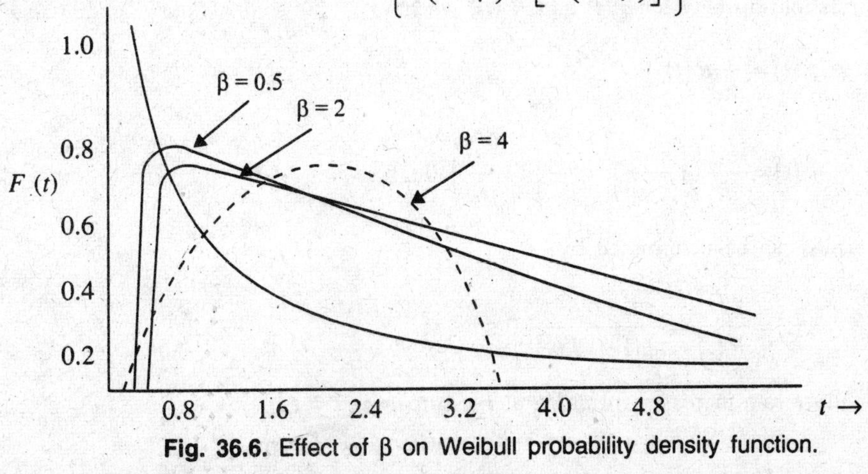

Fig. 36.6. Effect of β on Weibull probability density function.

where $\Gamma(x)$ is Gamma function and is given by

$$\Gamma(x) = \int_0^{\infty} y^{x-1} . e^{-y} \, dy \quad \text{where} \quad y = \left(\frac{t}{\theta}\right)^{\beta}$$

36.7.5. Gamma Distribution

The failure density function for gamma distribution is

$$f(t) = \frac{\lambda^{\eta}}{\Gamma(\eta)} t^{\eta-1} \, e^{-\lambda t} \qquad t \geq 0, \ \eta > 0, \ \lambda > 0$$

where η is a shape parameter and λ is the scale parameter

$$F(t) = \int_0^t \frac{\lambda^n}{\Gamma(n)} \tau^{n-1} \, e^{-\lambda t} d\tau$$

If η is an integer, it can be shown by successive integration by parts that

$$F(t) = \sum_{k=\eta}^{\infty} \frac{\left((\lambda t)^k \exp[-\lambda t]\right)}{k!}$$

Reliability function

$$R(t) = \sum_{k=0}^{\eta-1} \frac{\left((\lambda t)^k \exp[-\lambda t]\right)}{k!}$$

The gamma distribution can be used to model the time the nth failure of the system, if the underlying failure distribution is exponential.

Problem 1. *The reliability of a cutting assembly is given by*

$$R(t) = \begin{cases} (t - t/t_0)^2 & 0 \leq t \leq t_0 \\ 0 & t \geq t_0 \end{cases}$$

Determine (*i*) *the failure rate*

(*ii*) *does failure rate increase or decrease with time*

(*iii*) *determine the MTTF.*

Solution.

From the relationship between $f(t)$ and $R(t)$, we have

(*i*) $$f(t) = \frac{d}{dt} R(t)$$

$$\therefore \qquad f(t) = \frac{-d}{dt}\left(1 - \frac{t}{t_0}\right)^2 = \frac{2}{t_0}\left(1 - \frac{t}{t_0}\right) \quad 0 \leq t \leq t_0,$$

and the failure rate can be determined by

$$\lambda(t) = \frac{f(t)}{R(t)} = \frac{2}{t_0 (1 - t/t_0)} \quad 0 \leq t \leq t_0$$

(*ii*) The failure rate increases from $2/t_0$ at $t = 0$ to ∞ at $t = t_0$.

(*iii*) Mean time to failure

$$\text{MTTF} = \int_0^\infty R(t)\, dt$$

$$= \int_0^\infty dt\,(1 - t/t_0)^2 = t_0/3$$

Problem 2. *Pdf for a random variable T, the time in operating hours to failure of an engine is given. What is the reliability for 100 hr operating life.*

$$f(t) = \begin{cases} \dfrac{0.001}{(0.001\,t+1)^2} & t \geq 0 \\ 0 & \text{otherwise} \end{cases}$$

Solution.

$$R(t) = \int_t^\infty f(t)\, dt = \int_t^\infty \frac{0.001}{(0.001\,t+1)^2}\, dt$$

$$= \left.\frac{-1}{(0.001\,t^1 + 1)}\right|_t^\infty = \frac{1}{0.001\,t+1}$$

and

$$f(t) = 1 - R(t) = 1 - \frac{1}{0.001\,t+1} = \frac{0.001\,t}{0.001\,t+1}$$

then,

$$R(100) = \frac{1}{0.1+1} = 0.909.$$

Problem 3. *The probability density function is given by*

$$f(t) = \begin{cases} 0.002\, e^{-0.002\,t} & t \geq 0 \\ 0 & \text{otherwise} \end{cases}$$

with 't' hours. Determine R (t) and MTTF and also find the median time to failure.

Solution.

$$R(t) = \int_t^\infty 0.002\, e^{-0.002\,t}\, dt$$

$$= e^{-0.002\,t}$$

and

$$\text{MTTF} = \int_0^\infty e^{-0.002\,t}\, dt$$

$$= \left.\frac{-e^{0.002\,t}}{-0.002}\right|_0^\infty = \frac{1}{0.002} = 500 \text{ hrs.}$$

To find the median to failure

Put $R\,(t_{med}) = e^{-0.002\,t}$ med = 0.5

Solving for t_{med}, $t_{med} = \dfrac{\ln 0.5}{-0.002} = 346.6$ hrs.

Problem 4. *A particular machine has a constant failure rate of* $\lambda = 0.02$ *hrs.*

(*a*) *What is the probability that it will fail within first 10 hours.*

(b) Suppose that the machine has operated successfully operated for 100 hrs, what is the probability that it will fail during the next 10 hours of operation.

Solution. (*a*) Probability of failure for first 10 hours is

$$P\{t \le 10\} = \int_0^{10} f(t)\,dt = F(t)$$

$$= 1 - e^{-0.02 \times 10} = 0.181$$

(*b*) The conditional probability

$$P\{t \le 110 \mid,\ t > 100\} = P\left\{\frac{(t \le 100) \cap (t > 100)}{P(t > 100)}\right\}$$

$$= P\left\{\frac{(100 \le t > 100)}{P\{t > 100\}}\right\}$$

$$P\{t \le 100 \mid,\ t > 100\} = \int_{100}^{110} \frac{f(t)\,dt}{1 - F(100)}$$

$$= \int_{100}^{110} \frac{0.02\, e^{-0.02t}\,dt}{1 - 1t \exp(-0.02 \times 100)}$$

$$= \frac{\exp(-0.02 \times 100) - \exp(-0.02 \times 100)}{\exp(-0.02 \times 100)}$$

$$= 1 - \exp(-0.02 \times 100) = 0.181$$

Problem 5. *If the two reliability functions have the same mean, show that their reliabilities may be different for the same operating time.*

Solution. Let $R_1(t) = e^{-0.002t}$ $\quad t \ge 0$

with $\quad \text{MTTF}_1 = 500$ hrs. $\quad$ [from problem 3]

and $\quad R_2(t) = \dfrac{1000 - t}{1000} \quad 0 \le t \le 1000$

where $\quad \text{MTTF}_2 = \displaystyle\int_0^{1000} \left(1 - \frac{t}{1000}\right) dt = t - \frac{t^2}{2000}\Bigg|_0^{1000} = 500 \text{ hrs.}$

By computing reliabilities for an operating time of 400 hrs. we obtain,

$$R_1(400) = e^{-0.002(400)} = 0.449$$

and $\quad R_2(400) = \dfrac{1000 - 400}{1000} = 0.60$

Thus, though the mean for the two functions is same, reliabilities are different.

Problem 6. *A linear hazard function $\lambda(t) = 5 \times 10^{-6}\, t$, where 't' is measured in operating hours. If the reliability of 0.98 is desired, what is the design life ?*

Solution. $\quad R(t) = \exp\left[-\int_0^{t} 5 \times 10^{-6}\, t^1\, dt^1\right]$

$$= \exp\left[-2.5 \times 10^{-6}\, t^2\right]$$

$$= 0.98$$

$$t_{0.98} = \sqrt{\frac{\ln 0.98}{-2.5 \times 10^{-6}}}$$

$$= 89.89 \approx 90 \text{ hrs.}$$

Problem 7. *A device has a decreasing failure rate characterized by two parameter Weibull distribution with a wear out linear hazard function*

$$\lambda(t) = \frac{2}{1000}\left(\frac{t}{1000}\right) = 2 \times 10^{-6}\, t$$

The shape parameter $\beta = 2$ and scale parameter, $\theta = 1000$. The device is required to have a design life reliability of 0.99. Determine the design life and MTTF.

Solution. $R(t) = e^{-(t/1000)^2} = 0.99$

The design life is given by

$$t_{\text{Design}} = 1000\sqrt{-\ln 0.99} = 100.25 \text{ hrs.}$$

$$\text{MTTF} = 1000\,\Gamma\left(1 + \frac{1}{2}\right) = 886.23 \text{ hrs.}$$

and

$$\sigma^2 = 10^6 \left\{\Gamma(1+1) - \left[\Gamma\left(1 + \frac{1}{2}\right)\right]^2\right\}$$

$$= 214601.7$$

or $\sigma = 463.25$ hrs.

where $\Gamma\left(1 + \frac{1}{2}\right) = 0.886227$ from the table of Gamma function.

36.8. SYSTEM RELIABILITY

It is always difficult to estimate the reliability of the system comprising of many elements. One approach for analyzing such systems is to decompose the system in to subsystems who's individual reliability factors can be estimated or determined. Depending upon the manner in which these subsystems are connected to constitute the given system, the combinational rules of probability are applied to obtain the system reliability.

The following types of systems are analyzed

1. Systems with components in series
2. Systems with components in parallel
3. Combination of series and parallel systems.

36.8.1. Systems with Components in Series

A series system is represented in the fig (36.7). System generally consists of large number of components connected in series. The successful operation of the system depends on the proper operation of all the components i.e. if one of these components fails, the system fails.

In terms of survival, the system can be no better than a component with the lowest probability of survival.

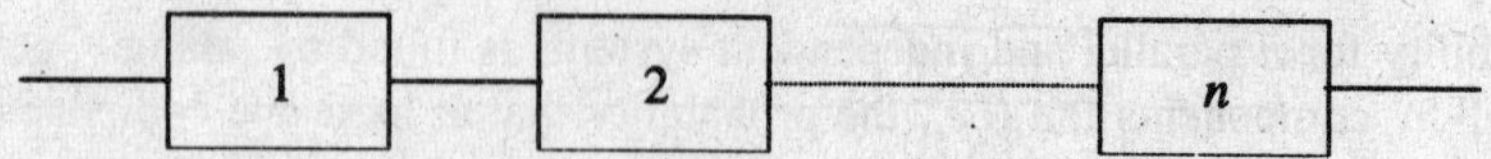

Fig. 36.7. Reliability of System in Series.

The reliability of the system (Rs) can be determined in the following manner.

Let E_1 = the event that component 1 does not fail

E_2 = the event that component 2 does not fail

Then, $P(E_1) = R_1$ and $P(E_2) = R_2$

$\therefore$ $Rs = P(E_1 \cap E_2) = P(E_1) \cdot P(E_2) = R_1 \cdot R_2$ assuming that the two components are independent, *i.e.*, failure of one component does not change the reliability of other component, *i.e.*, in order for the system to function, both components 1 and 2 must function.

For '*n*' mutually independent components in series.

$$Rs(t) = R_1(t) \cdot R_2(t) \cdot R_3(t) \text{ ....} R_n(t)$$

In this case, the system reliability decreases rapidly as the number of series components increases and the reliability will always be less than or equal to the least reliable component.

Thus, for a series system,

$$Rs \leq \min \{R_i\}$$

For example, a system having three subsystems with reliability of 0.6, 0.9 and 0.8 will have reliability.

$$Rs = 0.6 \times 0.9 \times 0.8 = 0.432$$

This system has a reliability Rs = 0.432 which is much less.

Necessary Level of Subsystem Reliability

Let q = the probability that a subsystem will fail.

Assuming 'q' to be identical for all subsystems

$$Rs = (1 - q)^n$$

For example, if we want Rs = 0.9999 in a given system with twenty components, Then we have,

$$Rs = (1 - q)^n$$

Application of binomial series gives,

$$Rs = 1 + n(-q)^1 + \frac{n(n-1)}{2}(-q)^2 + \ldots + (-q)^n$$

By ignoring higher order terms assuming that 'q' is small, we have,

$$Rs = 1 - nq$$

$$= 0.9999 \quad = 1 - 20 \times q$$

$$q = 0.0000005.$$

This is a subsystem reliability to meet the requirements of system.

36.8.2. Reliability of a System in Parallel

If two or more components are in parallel (or redundant), the successful operation of any one of the components leads to the successful operation of the system, *i.e.*, any one element fails the system will continue to operate (to function). The block diagram of parallel system is represented in the fig (36.8).

System reliability for n parallel and independent systems is found by taking. One minus the probability that all 'n' components fail (*i.e.*, the probability that at least one component does not fail).

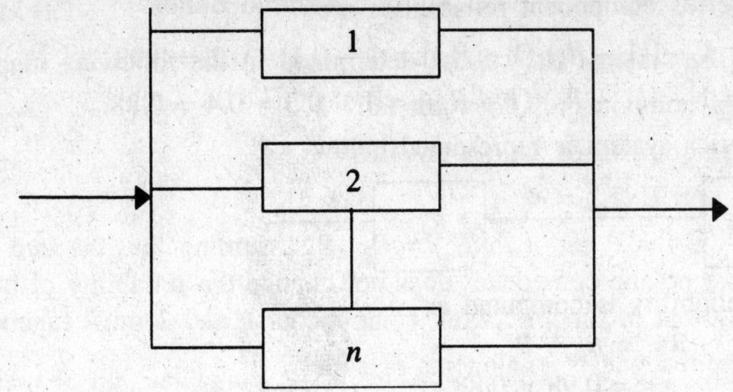

Fig. 36.8. Reliability Block Diagram of Parallel System.

For the two components in parallel,

Consider $$\text{Rs} = P\left[E_1 \cup E_2\right] = 1 - \left[E_1 \cup E_2\right]^C = 1 - P\left[E_1^{\,C} \cap E_2^{\,C}\right]$$

$$= 1 - P\left[E_1^{\,C}\right] \cdot P\left[E_2^{\,C}\right]$$

$$= 1 - (1 - R_1)(1 - R_2)$$

Generalizing,

$$\text{Rs}(t) = 1 - \prod_{i=1}^{n}\left[1 - R_i(t)\right]$$

It is always true that,

$$\text{Rs}(t) \geq \text{Max}\left\{R_1(t),\ R_2(t), \ldots, R_n(t)\right\}$$

For example, a system consisting of two elements connected in parallel have their probability of functioning as 0.9 and 0.8. Then, the reliability of the system is,

$$\text{Rs} = 1 - [(1 - 0.9) \times (1 - 0.8)]$$

$$= 1 - 0.02 = 0.98.$$

36.8.3. Combined Series Parallel System

Simple combinations of parallel and series subsystems can be easily analyzed by successfully representing subsystems in to equivalent parallel or series components.

Consider a system shown in fig (36.9)

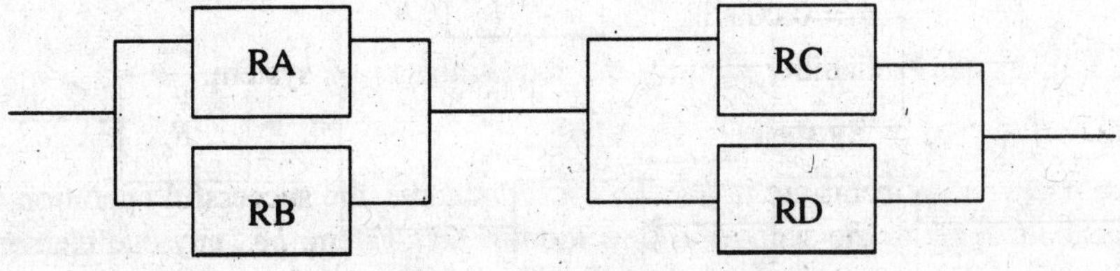

Fig. 36.9. Series Parallel System.

To calculate the reliability of this system, first the system is converted into equivalent series components.

Let $R_A = 0.9$, $R_B = 0.8$, $R_C = 0.7$ and $R_D = 0.6$

Then equivalent series component reliabilities are

$$R_{AB} = 1 - [(1 - R_A)(1 - R_B)] = 1 - 0.1 \times 0.2 = 0.98$$

$$R_{CD} = 1 - [(1 - R_C)(1 - R_D)] = 1 - 0.3 \times 0.4 = 0.88$$

This equivalent series system is represented below

∴ The system reliability is computed as

$$\begin{aligned} Rs &= R_{AB} \cdot R_{CD} \\ &= 0.98 \times 0.88 \\ &= 0.8624 \end{aligned}$$

Consider a parallel series system as shown in fig. (36.10)

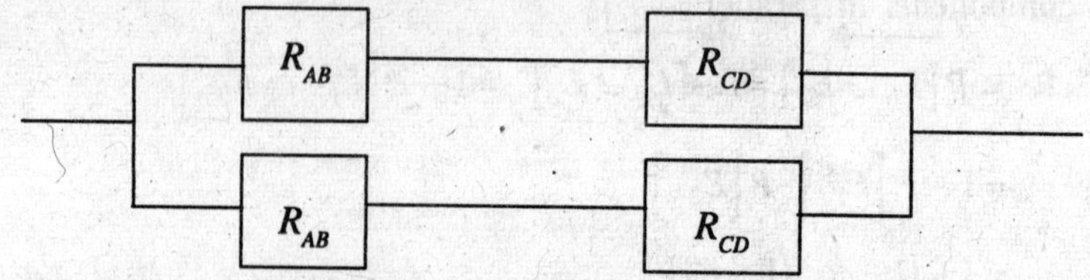

Fig. 36.10. Parallel Series System.

First convert the series subsystems in to equivalent parallel components.

Let $R_A = 0.9$, $R_B = 0.8$, $R_C = 0.7$ and $R_D = 0.6$

Equivalent parallel components are

$$R_{AC} = R_A \times R_C = 0.9 \times 0.7 = 0.63$$

$$R_{BD} = R_B \times R_D = 0.8 \times 0.6 = 0.48$$

System reliability,

$$\begin{aligned} Rs &= 1 - [1 - R_{AC}][1 - R_{BD}] \\ &= 1 - (0.37) \times (0.52) \\ &= 0.8076. \end{aligned}$$

Now, Consider a system as represented in fig. (36.11).

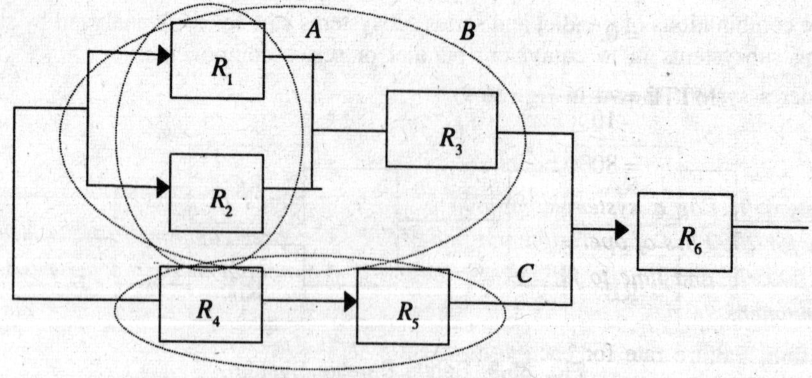

Fig. 36.11. Series Parallel Combination System.

The system reliability is obtained on the basis of relationship among the subsystems. The reliabilities of subsystems are computed as follows.

For a subsystem A,

$$R_A = [1 - (1 - R_1)(1 - R_2)]$$

For a subsystem B,

$$R_B = R_A R_3$$

$$= [1 - (1 - R_1)(1 - R_2)] \cdot R_3$$

and for a subsystem C,

$$R_C = R_4 R_5$$

The equivalent system is represented as

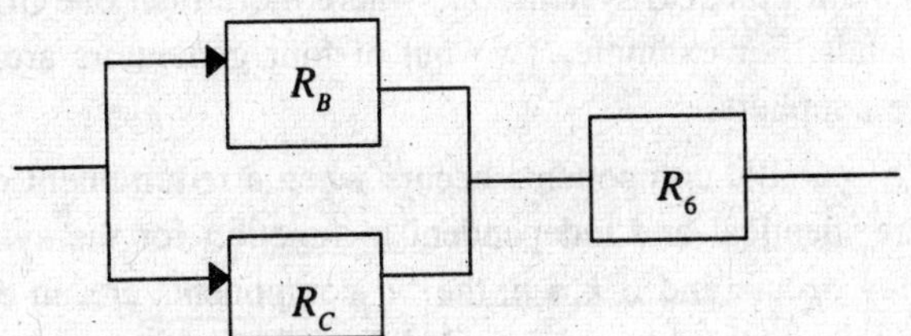

Since R_B and R_C are in parallel with each other and in series with R_6,

$$\text{Rs} = [1 - (1 - R_B)(1 - R_C)] \times R_6$$

Problem 8. *A system is composed of 10 components connected in series. Each component has an exponential time to failure distribution with a constant failure rate of 0.5 per 4000 hours. Compute the reliability of the system for 2000 hours of operation and find MTTF.*

Solution. Component failure rate $(\lambda) = \dfrac{0.5}{4000} = 12.5 \times 10^{-6}$ per hour

Reliability of each component after 2000 hours of operation is

$$R = e^{-\left[(12.5 \times 10^{-6}) \times 2000\right]}$$

$$= 0.975$$

Reliability of the system after 2000 hours of operation

$$\text{Rs} = \exp\left[-\left(10 \times 12.5 \times 10^{-6}\right) \times 2000\right]$$

$$= 0.779$$

$$\text{MTTF} = \frac{1}{10 \times 12.5 \times 10^{-6}}$$

$$= 8000 \text{ hours.}$$

Problem 9. *For a system composed of three elements in parallel, determine system of reliability for 2000 hrs of operation and find MTTF. The three components have identical failure rate of 0.0005/hr and time to failure distribution is exponential in each case. What is MTTF of each component.*

Solution. Failure rate for each component = 0.0005/hr

System reliability for 2000 hrs of operation is

$$Rs = 1 - [1 - e^{-0.0005 \times 2000}]$$
$$= 1 - (0.63212)^3$$
$$= 0.7474.$$

$$MTTF = \frac{1}{0.0005}\left[1 + \frac{1}{2} + \frac{1}{3}\right] = 3666.67 \text{ hrs.}$$

MTTF for each component is

$$MTTF = \frac{1}{\lambda} = \frac{1}{0.0005} = 2000 \text{ hrs.}$$

36.8.4. Reliability of k-out of 'm' System

There is another important practical system, one where more than one of its components are required to meet the demands. For example, Two out of four generators are required to supply the required power to the suppliers.

A generalization of 'n' parallel components occurs when a requirement exists for 'k' out of 'n' components, which are identical and independent to function for the system to function. If $k = 1$, complete redundancy occurs and if $k = n$, the 'n' components are, in effect in series. The reliability can be obtained from the binomial probability distribution.

If 'R' is the reliability of each component which are independent. If p is the probability of success of each component, then we have

$$P(x) = \begin{bmatrix} n \\ x \end{bmatrix} R^x (1 - R)^{n-x}$$

This is the probability of exactly x components operating. This is true since,

$$\binom{n}{x} \frac{n!}{x!(n-x)!}$$

is the number of ways (arrangements) in which successes can be obtained from 'n' components.

$R^x (1 - R)^{n-x}$ is the probability of x successes and $n - x$ failures for a single arrangement of successes and failures.

$\therefore \quad Rs = \sum_{x=k}^{n} P(x)$ is the probability of k or successes from among the 'n' components.

36.8.4. Complex System (Non Series Parallel System)

In practice, the systems are not always simple series, parallel systems. A complex system on simplification may produce a non series parallel configuration. A simple structure of structure of non-series parallel structure is represented in the fig. (36.12).

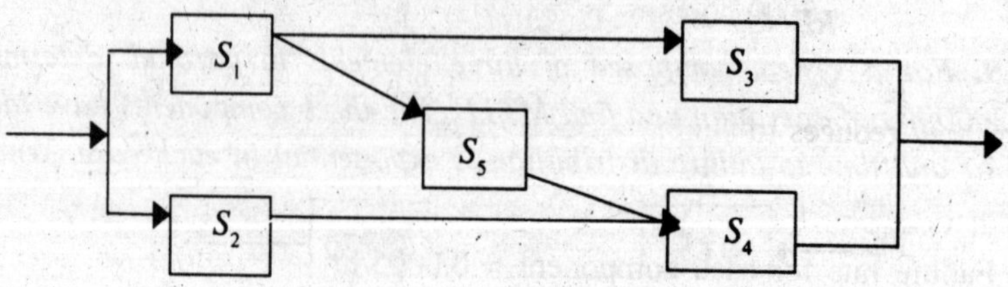

Fig. 36.12. A Bridge Configuration.

The logic diagram approach converts the system diagram into a logic diagram, which consists of simple parallel paths between input and output terminals. Each path contains elements whose successful operation can lead to the success of the system. For successful operation of the system, there should be at least one continuous path between in and out terminals.

The logic diagram for fig. (36.12) is shown in fig. (36.13).

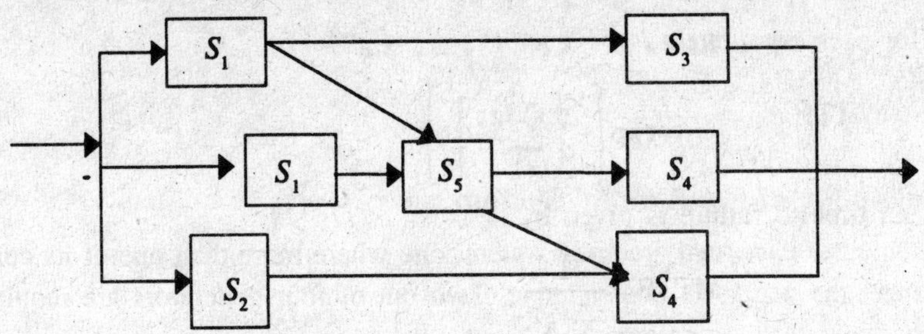

Fig. 36.13. Logic Diagram for fig. 36.12.

It may be noted that the subsystem S_5 is unidirectional like other subsystem.

Problem 10. *Calculate the system reliability for units connected as shown below.*

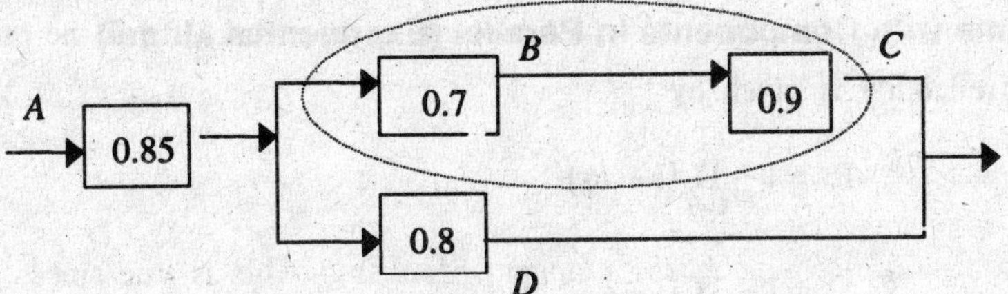

Solution. First, the reliability of system *B* and *C* is determined.

$$R_{BC} = R_B \times R_C = 0.7 \times 0.9 = 0.63$$

This is represented in the block diagram below.

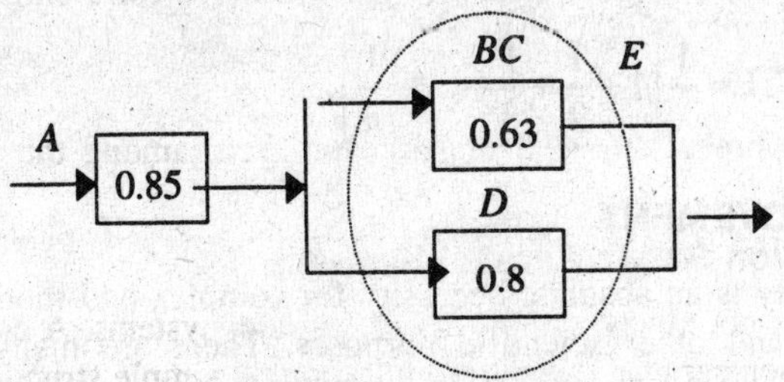

Secondly, *BC* and *D* are then replaced by *E* as *B* & *D* are parallel.

$$RE = 1 - [(1 - 0.63)\ (1 - 0.8)]$$
$$= 0.926.$$

The system now reduces to simple series system,

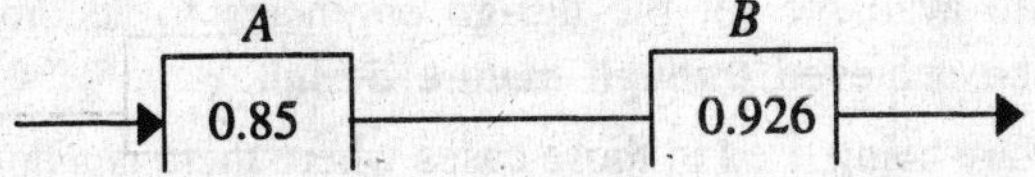

System reliability $Rs = R_A \times R_E$

$= 0.85 \times 0.926 = 0.7870.$

36.8.6. Systems with Components in Series (Exponential Model)

Suppose that the system has 'n' components in series, each with exponentially distributed time to failure with failure rates $\lambda_1, \lambda_2, ..., \lambda_n$. The system reliability is given by

$$Rs = e^{-\lambda_1 t} \times e^{-\lambda_2 t}, ..., \times e^{-\lambda_n t}$$

$$= \exp\left[-\left(\sum_{i=1}^{n} \lambda_i\right) t\right]$$

The mean time to failure is given by

$$\text{MTTF} = \frac{1}{\sum_{i=1}^{n} \lambda_i}$$

If all the components in series have identical failure rates (λ).

$$\text{MTTF} = \frac{1}{n\lambda}$$

36.8.7. Systems with Components in Parallel (Exponential Model)

The system reliability is given by

$$\text{Rs} = 1 - \prod_{i=1}^{n} (1 - R_i)$$

$$= 1 - \prod_{i=1}^{n} \left(1 - e^{-\lambda_i t}\right)$$

As a special case, if all the components have the same failure rates (λ), system reliability is given by

$$\text{Rs} = 1 - \left(1 - e^{-\lambda_i t}\right)^n$$

$$\text{MTTF} = \frac{1}{\lambda}\left[1 + \frac{1}{2} + \frac{1}{3} + ... + \frac{1}{n}\right].$$

36.9. RELIABILITY IMPROVEMENT

A high degree of reliability is an absolute necessity for complex and modern systems to be used for industrial, military and other scientific purposes. There are many ways by which reliability of a component or system can be enhanced. These methods are discussed below.

1. **Design and Safety Factor**

In order to design reliability into products, reasons for product failures must be analyzed thoroughly. Generally, a product fails prematurely because of inadequate design features, manufacturing and part defects, abnormal stresses induced, environmental condition and human error. Several methods are available for the design engineers to accomplish the requirements. Higher reliability could be achieved through mature design.

Higher safety factors are being used in those cases where there is a doubt regarding the ability of the certain structure/component to withstand a particular load. Nowadays, exhaustive testing methods are available to achieve required degree of reliability.

2. **Parts and Material Selection**

Designer has to choose between selecting standard parts and manufactured specialized parts with higher reliability and greater tolerances. The trade off is usually in cost but ease of parts availability, ease of repair, energy requirements, weight and size may also be considerations. The historical databases can assist in determining relative reliabilities among competing parts. Knowledge of material properties and the external stresses the system will experience is important. The material properties of materials such as metals, polymers, ceramics and composites include tensile strength, hardness, impact strength, fatigue life and creep.

3. **Redundancy**

When it is not possible either to manufacture a highly reliable component or the cost associated with such manufacturing is too high, the system reliability can be improved by the techniques of introducing redundancy. This involves the creation of additional parallel paths in the system. Generally, there are two types of redundancies - parallel and stand by to improve system reliability.

In a system of complex nature, redundancy can be applied at various levels.

The various approaches for introducing redundancy in the system are :

(*a*) To provide a duplicate or an additional path for the entire system itself. This is known as system or unit redundancy.

(*b*) To provide redundant path for each component individually which is called component redundancy.

(*c*) Weak component should be identified and strengthened by reliability.

(*d*) Use a combination of the above methods depending upon the configuration called mixed redundancy.

Component Versus Unit Redundancy

It is easier to introduce redundancy at unit level rather than at component level. However, component level redundancy provides higher reliability than unit redundancy.

The two ways of applying redundancy is shown in fig (36.14).

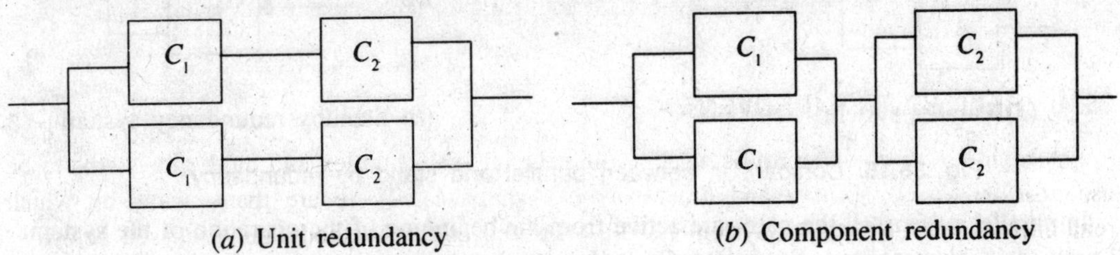

(*a*) Unit redundancy (*b*) Component redundancy

Fig. 36.14. Redundancy in Systems.

For unit redundancy, the reliability of the system.

$$\text{Rsu} = 1 - \left[(1 - R_1 R_2)(1 - R_1 R_2)\right]$$

where R_1 and R_2 are the reliabilities of components of components C_1 and C_2.

For the case of component redundancy, the reliability is

$$\text{Rsc} = 1 - \left[(1 - R_1)^2\right]\left[1 - (1 - R_2)^2\right]$$

Suppose, if $R_1 = R_2$, then

$$\text{Rs}_U = 2R^2 - R^4$$

$$\text{Rs}_C = R^2\,(2 - R)^2$$

The difference between Rs_C and Rs_U is

$$\text{Rs}_C - \text{Rs}_U = 2R^2(1 - R)^2$$

i.e., $$\text{Rs}_C - \text{Rs}_U > 0$$

i.e., the redundancy at the component level is better than redundancy at the unit level as far as reliability is concerned.

Parallel redundancy also referred to as "hot redundancy" is generally resorted to when the individual unit doesn't have the required reliability. If the unit is highly critical or strategic, then more than one unit can be put in parallel redundancy to compensate for unreliability. A compromise is to be struck between the improvement in reliability and the cost of each additional unit.

Stand by Redundancy

In stand by redundancy the units are duplicated but one or more units more remain idle (called secondary unit) until the primary unit fails. In standby redundancy the failed equipment or unit is replaced manually or automatically by its equivalent and in such cases the reliability of the operator or sensing and switching mechanism is to be taken into consideration.

Comparison between parallel (redundancy) system and stand by system is shown in the fig (36.15).

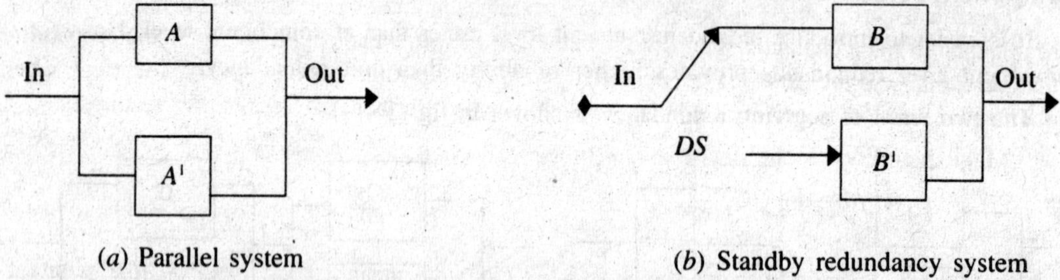

(*a*) Parallel system (*b*) Standby redundancy system

Fig. 36.15. Comparison between parallel and stand by redundancy.

In parallel system, all the paths are active from the beginning of the operation of the system till both of its elements fail. But in standby redundancy, all the paths are not in operation since starting, but will be operating when the first system fails.

In fig (36.15), when the system starts operating, decision switch (*DS*) connects the input to the element *B*, where as the other element B^{I} is in reserve in a "Off" condition. In case element '*B*' fails, the decision switch senses this by a built in mechanism and the connection is made to the stand by element B^{I}.

As per the fig. ( ),

The reliability of the parallel system,

$$Rp = 1 - [(1 - p)(1 - p)^1$$

Reliability for the stand by system.

$$Rs = 1 - p(B) \times p\left(\frac{B^1}{B}\right)$$

If both the elements are identical, and reliability of each element is 'p'. Then reliability of the system is

$$Rs = p\,(1 - \log_e p)$$

4. **Marginal Testing**

Marginal testing is prescribed by the designer as a method of predicting probability of failure due to degradation. Marginal testing involves periodic testing on a programmed basis. This helps to isolate degraded parts or components and are replaced before the actual failure occurs.

5. **Derating**

Derating consists of using a component under stress significantly below its rated value. This is proved to be more beneficial when applied to electronic components, in which case the designed voltage or current strength of the part is well above the normal operating level. Derating curves are provided in military handbook — reliability prediction of electronic components.

6. **Quality Control and Z-D Programme**

7. **Maintainability**

Maintainability is used to provide high effective reliability. If parts are readily interchangeable and replaceable, failures can be repaired at a faster rate by replacing defective parts with operating spares. This increases the available of the system.

36.10. MAINTAINABILITY

Goldman and Slattery (1967) has given the quantitative definition of maintainability. Maintainability is a characteristic of design and installation, which is expressed as the probability that an item will be restored to specified conditions within a given period of time when maintenance action is performed in accordance with prescribed procedures and resources.

Maintainability can be expressed as

$$M = 1 - e^{-t/\mathrm{MTTR}}$$

$$= 1 - \exp\,[-t/\mathrm{MTTR}]$$

where 't' is the specified time to repair and MTTR is the mean time to repair. Thus, maintainability refers to the ease with which preventive and corrective maintainance on a product can be achieved.

Specifications of Maintainability

1. Mean time to repair (MTTR).
2. Median time to repair.
3. Maximum time (Tp) in which a certain percentage of failures must be repaired.
4. Mean System Down Time — It is the average down time including scheduled maintenance but without maintenance delay times.

5. Mean Time to Restore (MTR) — It is the average unscheduled system down time including delays for maintenance and supply resources.

6. Maintenance work hours per operating hour (MH/OH).

Methods to Increase Maintainability

1. Fault isolation and self-diagnostics.
2. Parts standardization and interchangeability.
3. Modularization and accessibility.

36.11. AVAILABILITY

Availability depends on both reliability and maintainability. The availability can be expressed as

$$\text{Availability} = \frac{\text{Up time}}{\text{Up time} + \text{Down time}}$$

"Availability is the probability that a system or component is performing its required functions at a given point in time or over a stated period of time when operated and maintained in a prescribed manner."

Forms of Availability

1. **Inherent Availability**

The probability that a system, when used under conditions, without consideration of any preventive action in an ideal support situations shall operate satisfactory at a given point of time.

$$A_i = \frac{\text{MTBF}}{\text{MTBF} + \text{MTTR}}$$

Mean time between failures is a factor of reliability and is given by

$$\text{MTBF} = \frac{1}{\lambda} = \frac{1}{\text{Failure rate}}$$

2. **Operational Availability**

$$A_o = \frac{\text{Mean time between failures (MTBF)}}{\text{MTBF} + \text{Mean time waiting for spares} + \text{Mean administrative time} + \text{Mean time for repairs}}$$

3. **Use Availability**

$$\text{Au} = \frac{\text{Operation time} + \text{Off time}}{\text{Operation time} + \text{Off time} + \text{Total down time}}$$

36.12. RELIABILITY LIFE TESTING

The basic objective of reliability life testing is to obtain information concerning failures in order to express reliability quantitatively and to ascertain whether reliability and safety goals are met and to improve product reliability.

The factors that are to be considered before any reliability test is being carried out are

- Objective of the test.
- Types of the test to be performed.
- Operating and environmental conditions.

- Number of units to be tested (Sample size).
- Duration of the test.

The various types of tests are described below :

1. **Burn-in Testing**

This test is carried out to eliminate or reduce infant morality failures by accumulating initial equipment hours and resulting failure before user acceptance.

Primary objective of burn in testing is to increase the mean residual life of components as a result of being survival this test. Items that have failed during the burn in may be discarded and replaced or be repaired. The burn-in testing requires testing of all the units produce for the designated time, so it increases the production lead time and costs.

2. **Acceptance Testing and Qualification Testing**

This testing demonstrates through life testing that the reliability goals or specifications have been met or determines whether parts or components are within acceptable standards. It demonstrates that the system design meets performance and reliability requirements under specified operating and environmental conditions. Acceptance testing is based on predetermined sample size.

3. **Sequential Testing**

Sequential testing provides an efficient method for a accepting or rejecting a statistical hypothesis when the sample is highly favourable to one of the two decisions. This test is based on the sequential probability ratio test developed by Wald and is used for reliability and maintainability demonstration or in acceptance or qualification testing.

4. **Accelerated Life Testing**

This comprises of techniques for reducing the length of the test period by accelerating failures of highly reliable products.

5. **Experimental design** involves statistical methods that are useful in isolating causes of failures in order to eliminate them.

6. **Reliability Growth Testing**

The objective of reliability growth testing is to improve reliability over time through changes in product design, in manufacturing processes and procedures. Reliability test and assessments are conducted on prototypes to determine whether reliability goals are being met, if not a failure analysis will determine the high failure modes and the corresponding fixes. The failure modes are eliminated through engineering redesign and the cycle is repeated.

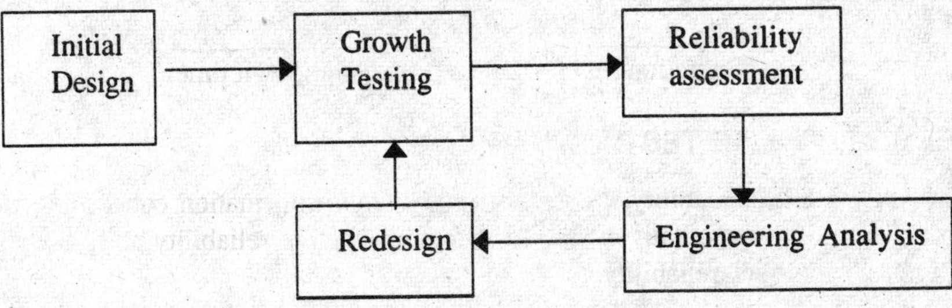

Fig. 36.16. Reliability Growth Cycle.

References For The Further Reading

1. E. Balgurusamy, "Reliability Engineering", Tata McGraw Hill Company Ltd., New Delhi (1984).
2. A.K. Govil, "Reliability", TMH, New Delhi.
3. A. K. Gupta, "Reliability Engineering and Terotechnology", Mcmillan India Ltd., New Delhi (1986).
4. Charles Ebeling, "An Introduction to Reliability and Maintainability Engineering", McGraw Hill International, New York (1997).
5. Carter, "Mechanical Reliability", John Wiley and Sons, New York (1972).
6. Smith C. O., "Introduction to Reliability in Design", McGraw Hill, New York.
7. Lewis E. E., "Introduction to Reliability Engineering", John Wiley and Sons, New York.

37

THEORY OF CONSTRAINTS (TOC)

• Introduction • Synchronous Manufacturing • Performance Measurements • Bottlenecks and Unbalanced Capacity • Managing Bottlenecks • Drum Buffer — Rope • Components of Production Cycle Time • Goldratts Theory of Constraints • Cost Accounting System for T.O.C. • Comparison of T.O.C. with J.I.T. and M.R.P. • VAT Classification of Firms.

37.1. INTRODUCTION

To achieve excellence in manufacturing and to compete in a global level, we have to excel in all aspects of manufacturing within the limited time and resources. In the last few years, new powerful techniques have emerged that have helped enhance manufacturing productivity. Amongst the most important and powerful of these is the Theory of Constraint (TOC). Formerly known as optimum Production Technology is a management philosophy that is about logistics, performance measurement and logical thinking. Logistics include drum buffer scheduling buffer management and VAT analysis.

37.2. SYNCHRONOUS MANUFACTURING

Synchronous manufacturing refers to the entire production process working together in harmony to achieve the goals of the organization. The concept of synchronous manufacturing refers to logic of co-ordination of all resources so that they work together and are in harmony or are synchronized. In such a situation, emphasis is on total system performance not on any localized performance measures such as labour and machine utilization.

In a typical manufacturing situation, in which the raw material is processed at various work centers before it reaches the final product stage, usually, the production departments attempt to maximize production at each work center to achieve maximum utilization and efficiency. They implement various measures to increase production by introducing quota systems and incentives. Buffers are established between work centers to take care of uncertainties like breakdown, unauthorized absenteeism etc. To some extent these help to achieve the goals of the production. But, a careful analysis will reveal that in increasing finished goods production, costs have also increased. Thus, in order to meet delivery and quality commitment often leaves behind chaos and confusion, highly stressed employees and an undesirable compromise on quality.

Manufacturing system is like a chain. When the performance of the plant is restricted by the capacity of the work center, it becomes the constraint or bottleneck. As the system performance is determined by the capacity of this constraint, the possibility of enhancing its capacity can be explored by working overtime or hiring additional capacity. This is the basic tenet of theory of constraints.

Taking the analogy of a chain further, the pulling of the chain breaks at its weakest link; so the weakest link determines the strength of the chain. By strengthening the weakest link, we can increase the strength of the chain. Similarly, by strengthening the weakest work centers, we can increase the performance of the overall manufacturing system.

In a typical manufacturing organizations, we have the various departments like manufacturing, purchase, finance and marketing, maintenance etc. Each functional department is trying to maximize its output or performance. Most of the time, this optimization results from a direct contradiction from other departments objectives. For example, the purchase department opt for high volume purchase, which may be consumed in a number of months to avail quantity discount. But this becomes detrimental to the performance of the inventory management and finance department. So the optimization of individual department do not result in achieving the organizational goal. Theory of constraints (TOC) considers an organization as one system and looks at ways and means to optimize its holistic performance.

37.3. PERFORMANCE MEASUREMENTS

The objective of all organizations is to make profits in the present and in the future through better customer service, larger market share, lower cost, high quality etc., *i.e.*, the company needs to make profits in relation to the investment committed which is monitored as "ROI" (Return On Investment).

To adequately measure a firms performance, two sets of measurements are needed, one from financial point of view and the other from operations points of view.

- **Financial Measurements**

The three measures of ability to make money are :

1. Net Profit: An absolute measurement in rupees.
2. Return on Investment : A relative measurement based on investment.
3. Cash Flow: a survival measurement.

The three measurements must be used together. It is observed that many companies are profitable as per their balance sheet but their cash flow is so poor that at times it is difficult to meet day-to-day cash requirements.

- **Operational Requirements**

At the operational level, we have the three measurements that guide.

1. **Throughput**

It is the rate at which the system generates money through sales. It is specifically defined as "Good Sold". An inventory in finished goods is not throughput. Actual sales must occur. This definition is specifically intended to prevent the system from continuing to produce under the illusion that goods might be sold. Such a decision simply increases costs, builds inventory and consumes cash. Normally inventory is valued (both WIP and Finished goods) only at the material cost it contains. Labour cost and machine hours are ignored. To calculate the throughput per unit of each product, we need to subtract Total Variable Cost (TVC) from its selling price. TVC is the cost that varies for the every extra unit produced. This will tell us how much money the company generates with sale of one unit or production.

2. **Inventory**

All the money that the system has invested in purchasing things it intends to sell.

3. **Operating Expenses**

All the money that a company spends to transform inventory in to throughput. Operating expenses include production costs (such as direct and indirect labour, inventory carrying costs, equipment depreciation and materials and supplies) and administrative costs.

The objective of the firm is to treat all three measurements simultaneously and continually and this achieves the goal of the firm, *i.e.,* making the money. The operational goal and the operational performance parameters are represented in the fig (37.1).

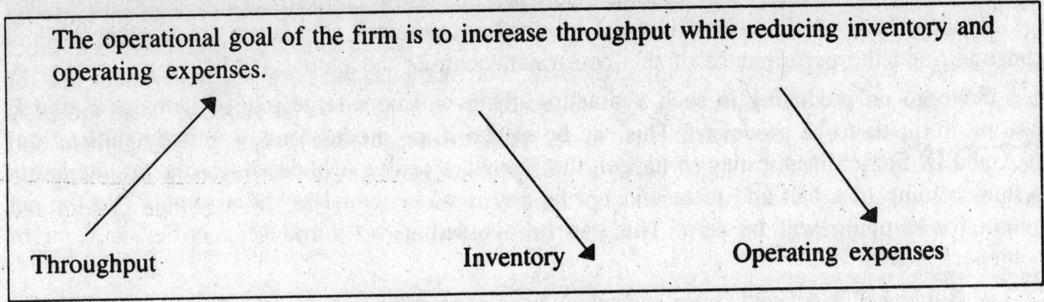

Fig. 37.1. Operational Goal of the Firm.

37.4. BOTTLENECKS AND UNBALANCED CAPACITY

Majority of the manufactures still try to balance capacity across a sequence of processes in an attempt to match capacity with market demand.

Consider simple manufacturing systems where a raw material is to be converted in to finished goods through different operations as shown in the fig (37.2).

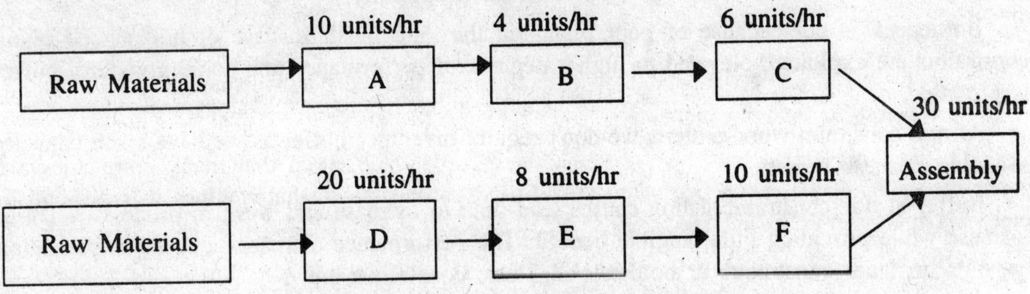

Fig. 37.2. A Manufacturing System.

The different operations A, B, ..., F are performed and then after assembly the products are shipped to the customer. Once the output rate of the line has been established, production people try to make the capacities of all stations the same. This is done by adjusting machines or equipments used, workload skills and types of labour assigned and so on.

In synchronous manufacturing, making all the capacities same is considered as a bad decision. Such a balance would be possible only if the output time of all workstations were constant or had a very narrow distribution. A normal variation in output times causes downstream stations to have an idle time, when upstream stations take longer to process. If the upstream stations process in a shorter time, inventory builds up between work stations only way to smoothen the variations is to increase WIP inventory to absorb variations (a bad decision) or increasing capacities downstream to be able to make up for the longer upstream times. Here, the capacities within the process sequence should not be balanced to the same extent, but attempts should be made to balance the flow of product through the system. When flow is balanced, capacities are unbalanced.

Referring to the fig (37.2), different operations have different capacity. Assuming that, there is a continuous flow of materials, based on the operations capacities only 4 units will be ready for shipping every hour. Operation A processes 10 units and transfers it to B. Operation B can process maximum 4 units/hr leaving 6 units/hr to be added to queue in front of B. Operation B will deliver 4 units/hr to C even though C is capable of processing 6 units/hr. So only 4-units/hr come out of C. Assembly will be completed when one unit each from C and F are joined. Thus, the output of the whole system is only 4 units/hr. Operation C is then the bottleneck or a constraint and the performance of this constraint, controls the output.

If we go on producing in such a situation, there will be a large pile of items at B and E waiting in queue to be processed. This can be solved if we process only 4 units/hr (utilization) at A and D. Some times it may so happen that the work center A break down leaving the entire system coming to a halt and there will not be any item for line 1 to the assembly. So the net output for shipping will be zero. This can be avoided if we introduce a buffer in front of bottleneck center B.

- Bottleneck is defined as any resource whose capacity is less than the demand placed upon it. A bottleneck is a constraint within the system that limits throughput. It is that point in the manufacturing process where flow is to a narrow stream. A bottleneck may be a machine, scarce or highly skilled labour, or a specialized tool.
- A non-bottleneck is any resource whose capacity is greater than the demand placed on it. A non-bottleneck should not be working constantly because it can produce more than is needed. A non-bottleneck contains idle time.
- A capacity constrained resources (CCR) is one whose utilization is close to capacity and it could be bottleneck, if it is not scheduled carefully.

Bottleneck is not because of poor plant but they are must. In case of unbalanced plant, constraints are exploited; elevated to higher degrees of performance and some inventory buffer is provided.

At non-constraint work centers, we don't require inventory buffer as we have spare capacity available at these centers.

Buffer at non-constraint station causes lead-time to increase and work in process (WIP) to increase while providing little tangible benefit. The performance of manufacturing organization depends on these constraints or bottlenecks. These constraints are

- Internal resource constraint.
- Market constraint.
- Policy constraint.

Internal resource constraint exists if the resource can produce less output than market demand when it is working for the 24 hours per day. If we have a surplus capacity in the existing plant but if there is no sufficient market demand, in such a situation, market is a constraint. Some times, the policy guidelines laid out will become a constraint like no work in night shifts, not off loading certain jobs.

37.5. MANAGING BOTTLENECKS

The loading pattern on each resource changes as the product mix changes. It is impracticable to restructure the shop floor with every product mix change. In TOC, the unbalanced plant is accepted as such, and increases the performance of the system by exploiting the bottlenecks and then elevating bottlenecks to the higher degree of performance. This is done by saving time at bottleneck resources by eliminating tasks, which can be carried out at other work centers, or working all the three shifts.

The various ways of managing the bottlenecks are :

A non-bottleneck resources are scheduled with larger batch sizes, which will create a bottleneck, which is to be avoided.

Drum, Buffer, Rope

The production system needs some control point i.e. to control the flow of products through the system. If the system contains the bottleneck, the bottleneck is the best place of control. This control point is called the drum.

The bottleneck is working throughout and the reason for using it, as a control point is to make sure that the operations upstream do not over produce and build excess WIP inventory that a bottleneck cannot handle.

If there is no bottleneck, the next best place to set the drum would be a capacity constrained resource (CCR). CCR is one that is operating near capacity but on an average has adequate capability as long as it is not incorrectly scheduled.

There are two things that must be done to the bottleneck :

1. Keep a buffer inventory infront of the bottleneck operation to make it sure that it always has something to work on. Because it is a bottleneck, its output determines the throughout of the system.
2. Communicate back upstream what the bottleneck operation has produced, so that the first operation produces and provides only that amount. This avoids inventory building up. This communication is called a "Rope".

The buffer inventory in front of the bottleneck operation is time buffer :

The drum, buffer and rope are represented in the fig (37.3).

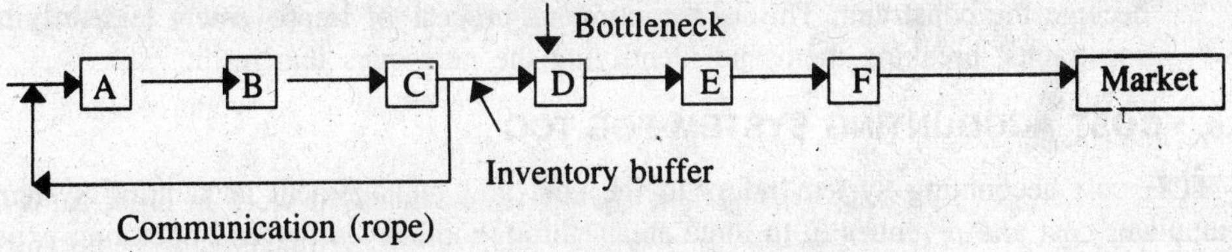

Fig. 37.3. Flow of Product with a Bottleneck.

37.6. COMPONENTS OF PRODUCTION CYCLE TIME

The component of the production cycle time are :

1. Set up time : The time that a part spends waiting for a resource to be set up to work on this same part.
2. Process time : The time that the part is being produced.
3. Queue time : The time that a part waits for a resource while the resource is busy with something else.
4. Waiting time : The time that a part waits not for a resource but for another part so that they can be assembled together.
5. Idle time : The unused time, *i.e.*, the cycle time is less than the sum of set up times, processing time, and waiting time and queue time.

Queue time is more for the parts waiting to go through the bottleneck as the bottleneck has fairly large amount of work assigned. For non-bottleneck, wait time is the greatest. Schedule always try to save the setup times. The batch sizes are doubled to reduce the setup times to half. With the increase in batch size, all the other times are doubled, where as setup time reduced. The net result is that the work in process is approximately doubled, as in the investment in inventory.

Saving Time

As the bottleneck is a resource whose capacity is less than the demand placed on it. Bottlenecks are restricting throughputs. There are number of ways we can save time on bottlenecks like better tooling, higher quality labour, large batch sizes, reducing setup etc.

It is to be noted that :

An hour saved at the bottleneck adds an extra hour to the entire production system. Where as, an hour saved at non-bottleneck only adds to an hour of its idle time.

37.7. GOLDRRATS THEORY OF CONSTRAINTS

1. Identify the system constraints

   i.e., it is essential to find out the weakest link for the improvements to happen.

2. Decide how to exploit the system constraints

   i.e., make the constraints as effective as possible.

3. Align every other part of the system to support the constraints even if this reduces the efficiency of non-constraint resources.
4. Elevate the system constraints

   If the output is still inadequate, acquire more of this resource so that no longer it is a constraint.

5. After step 4, there is a possibility that some other work center or operation might have become the constraint. This is a continuous process of improvement : identifying the constraints, breaking them and identifying the new ones that result.

37.8. COST ACCOUNTING SYSTEM FOR TOC

TOC cost accounting system refers to the cost and management accounting system that accumulates cost and revenues in to three areas- throughout, inventory and operating expenses. This system is considered to provide a truer reflection of actual revenues and cost than traditional costing system. TOC accounts provide a simplified and more accurate method considering the costs, which vary with throughput.

The primary focus of TOC costing is on aggressively exploring the constraints to make more money for the firm. In TOC, all the measures are in terms of money — measurement for incoming money (throughput), inventory for money blocked inside and one for money going out (operational expenses). These three measurements are sufficient to make a bridge between Net Profit (NP) and ROI.

$$NP = T - OE$$

$$ROI = \frac{T - OE}{I}$$

With these three measures (T, I and OE), we know the impact a decision has on company's profits. The ideal decision will be one, which increases 'T' and decreases I and OE.

In TOC, the focus is on increasing throughput and not on inventory reduction. The entire bottleneck concept is not geared to decrease operating expenses but instead, focused on increasing throughput. Throughput is more important than inventory. We need to shift the focus from cost mode to throughput mode. TOC, suggests two performance measures - Inventory Rupee days as a measure of things done ahead of schedule and the second is throughput rupee days as a measure of things done behind the schedule.

37.9. COMPARISON OF TOC (SYNCHRONOUS MANUFACTURING) WITH JIT AND MRP

The comparison is represented in the table.

MRP	JIT	TOC
1. Balance capacity and then balance the flow.	Balance capacity and then balance the flow.	Flow is balanced and not the capacity.
2. An hour lost at bottleneck is an hour lost at that resource.	–	An hour lost at the bottleneck is an hour lost for the entire system.
3. Bottleneck temperorily limits throughput.	Predetermined inventory buffers regulate rate of production for assembly.	An hour saved at non-bottle-neck counts for nothing.
4. Schedules should be deter-mined by sequence.	Schedule is determined by market demand.	Capacity constraint resource dictate the schedule based on market demand and its own potential.
5. Ensures global optimum by ensuring local optimum.	–	Sum of local optimum does not equal global optimum.
6. MRP uses backward sche-duling.	–	Approach is forward sche-duling.

37.10. VAT Classification of firms

Depending upon the products and processes, all manufacturing firms can be classified in to one or combination of three types, designated as V, A and T.

V Plant

In a V-plant there are very few raw materials and they are transformed through a relatively standard processes in to a large number of end products. *e.g.*, a steel plant, a few raw materials are converted in to a large number of types of steel sheets, rods, wires etc.

Characteristics

- There are large number of end items.
- Product uses the same sequence or processes.
- The number of routes are limited.
- Usually capital intensive and specialized.
- Needs high investment.
- High inventory levels of finished goods.
- Too large batch sizes.

A Plant

In A-plant, many raw materials, components and parts are converted in to one or few end products. *e.g.*, Aeroscope making jet engines, planes, automotive, machine tools.

Characteristics

- Dominant assembly feature.
- General-purpose machines to a larger extent compared to SPM's.
- Resources are shared across and within the routings.
- Less utilization of resources and efficiency.
- Completed parts/sub assemblies are maximum in inventory.
- Frequent use of bottleneck.

T Plant

In this T-plant, the final product is assembled in many different ways out of similar parts and components. Normally, there are two stages in the production process — First, the basic parts and components are manufactured and stored. Second, assembly takes place, combining these common parts into many possible options to create the final product.

Example, consumer products and appliances.

Characteristics

- Two distinctive stages of manufacturing - fabrication and assembly.
- Commonality of parts is maximum.
- Fabrication is done in large batches.
- There is a large inventory between the two stages.

References For Further Reading

1. **Goldratt Eliyahu M. and Jeff Cox,** "The Goal-excellence in Manufacturing Croton on Hudson", Ny North River Press, 1992.
2. **Richard B. Chase E.** "Production and Operations Management", TMH Company Ltd., New Delhi, 1999.
3. **Rajeev Nambiar,** "Theory of constraints and Organisational Productivity", M.&M. March, 2000, pp 40-46.

38

ADVANCED MANUFACTURING TECHNOLOGIES AND SYSTEMS

• Introduction • Challenges Facing Organizations • Growth of Technology • Advanced Manufacturing Philosophies and Technologies • Integrated Manufacturing Activities • Computer Aided Design (CAD) • Computer Aided Manufacturing (CAM) • Computer Aided Process Planning • Computer Integrated Manufacturing (CIM) • Lean Production and Agile Manufacturing • Manufacturing Excellence.

38.1. INTRODUCTION

Technology is a resource of maximum importance not any to production and operations but also to the organizations profitability. It is technology that drives the change in this world. Technology influences both the competitive (position) capabilities of the organization in particular and the nations in general. Technology is changing at a faster rate and these rapid changes in technology are posing many problems and challenges to the organizations. The rate of obsolescence of technology is on the rise. Firms that have made technology a competitive weapon have integrated the company's technology strategy and business strategy. Information technology (IT) has changed significantly the way the business is being run with a wide adoption of the Internet and web based information systems. The cost of gathering the information and processing the information has dropped quickly. This technological advancement is owing to the innovations in information technology, electronics and other fields.

Technological advancement is not limited to computer technology. New types of materials, methods and concepts have evolved and these advances in technologies are always associated with higher standards of living. In order to successfully cope up with the changes occurring in the manufacturing environment and achieve manufacturing excellence, an organization must proactively and intelligently incorporate advanced manufacturing technologies (AMT) and philosophies into its strategic plan.

38.2. CHALLENGES FACING THE ORGANISATIONS

For their survival and growth, organizations have to be responsive to the changes that are taking place. This makes the company to face challenges on all fronts, *i.e.*, changing demands of customers, quality revolution, reducing life cycle of products etc.

The main challenges that organizations have to face are -

1. Quality expectations of customers about products/services are sharply increasing with a demand for higher reliability and safety.
2. Rapid change in technology makes the existing manufacturing infrastructure outdated.

3. Emphasis is on quality control at source.
4. New ways of thinking about value chain improvements and use of process focused production concepts.

Advanced manufacturing technologies are the powerful tools to meet the above challenges.

38.3. GROWTH OF TECHNOLOGY

The early production activities were dependent on human effort for controlling the processes and to provide energy. As the days went on, the machines were used to supply power to the processes but still the processes were manually controlled. Then, it was automation which provided automatic control of processes, *i.e.,* the machine, could sense its output, compare with its preset target value and make necessary adjustments if required. Advent of computers and information technology (IT) has changed the face of manufacturing system and revolutionized the manufacturing and business processes. There is a trend to redesign the existing processes using the IT leverage (BPR). The trend is towards automation with no or minimum human interference.

Automated manufacturing is becoming more flexible. Today, the material handling systems are totally automated and they can move objects from one location to another in response to the signals they receive from the computers e.g. automated storage and retrieval systems (ASRS), Automated Guided Vessels (AGV). Computer controlled robots can execute various manipulations of work object and tools. There are production systems in which there is no human interference i.e. unmanned machining centers. Combinations of automatic machines and automatic material handling systems co-coordinated under the control of a computer are being operated without human interference. Designers can use computer graphics and powerful simulation programmes to develop and test programme.

38.4. ADVANCED MANUFACTURING PHILOSOPHIES AND TECHNOLOGIES

1. Materials Requirements Planning (MRP) and Manufacturing Resources Planning (MRP-II).
2. Just In Time manufacturing (JIT).
3. Computer Aided Design and Computer Aided Manufacturing (CAD/CAM).
4. Robotics.
5. Computer Integrated Manufacturing Systems (CIMS).
6. Flexible Manufacturing Systems (FMS).
7. Optimized Production Technology (OPT).
8. Agile Manufacturing.
9. Lean Manufacturing.

MRP, Just in time (JIT) manufacturing are discussed in detail chapters 18 and 31 respectively.

38.5. INTEGRATED MANUFACTURING ACTIVITIES

The four basic functions to be performed by an integrated manufacturing system are :

1. Production Functions : converts the resources in to products
2. Design Functions : plans and draws the products/parts.
3. Management Functions : plans and controls design and production activities.
4. Strategic Planning Functions : consists of decision making of strategic issues.

The framework of the above four fundamental functions in manufacturing industries is represented in the fig 38.1.

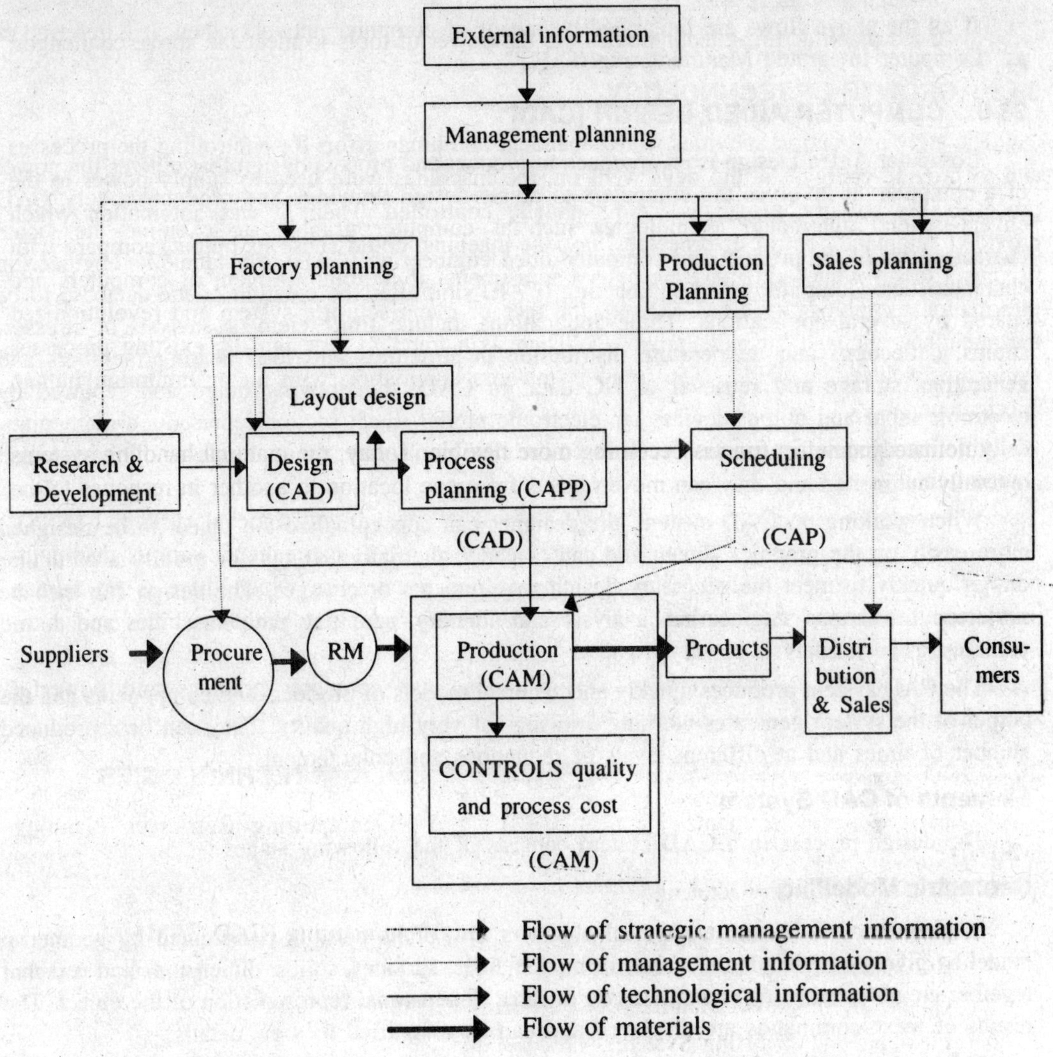

Fig. 38.1. Frame work of integrated manufacturing activities.

- **Production function** — is the key activity in the manufacturing industries. Production consists of fabrication of parts and assembly of products in engineering industries or a series of unit operations in process industries. The raw materials are changed in shape as they pass through various stages and are eventually stored in warehouse and become final products. This is referred to as flow of materials.

- **Design function** — designs and prepares assembly and part drawings of products to be manufactured and decides the procedure by which to produce them (process planning). Some times, it is required to prepare the plant and facilities (layout planning).

- **Management functions** — establishes a master plan to determine the product mix and amounts to be manufactured (production planning) according to the long term forecast and marketing (sales) planning. A schedule is prepared based upon this plan (scheduling). The control aspects include process control, quality control and cost control. The series of all these information is referred to as 'flow of managerial information'.

- **The strategic planning function** — establishes a strategic plan in both short and long term based upon the external information. This broadly includes factory planning, research and development planning.

If all the above flows are integrated by means of computer networks then, it is referred to as "Computer Integrated Manufacturing (CIM)".

38.6. COMPUTER AIDED DESIGN [CAD]

Computer Aided Design is an approach to product and process design that utilizes the power of a computer. It involves the computers to create design drawings and product models. CAD covers several automated technologies such as computer graphics to examine the visual characteristics of the product and computer aided engineering (CAE) to evaluate its engineering characteristics. Computer aided engineering (CAE) simplifies the creation of the database to be shared by several applications. These applications include finite element analysis of stresses, strains, deflections and temperature distribution in structures and load bearing members and generation, storage and retrieval of NC data. In CAD, the drawing board and replaced by electronic input and output devices, an electronic plotter. Each section represents a mathematically defined geometric function such as co-ordinate point, line, plane, circle or cylinder called menu items.

When working on CAD system, the designer can conceptualize the object to be designed more easily on the graphics screen and can consider alternative designs or modify a particular design quickly to meet the necessary design requirements or changes. The design can then be subjected to various engineering analysis and identify potential problems. The speed and accuracy of such analysis is very high.

The CAD system produces quickly and accurate models of products and components and the output of the system generates working drawings of very high quality. They can be reproduced number of times and at different levels of reductions and enlargements.

Elements of CAD System

The design process in a CAD system consists of the following stages :

Geometric Modelling

In geometric modelling, a physical object or any of its parts is represented by geometric model by giving commands that create or modify lines, surfaces, solids, dimensions and text that together are an accurate and complete two or three dimensional representation of the object. The results of these commands are displayed and can be magnified to view details.

The models can be represented in three different ways :

In line representation (wire frame), all edges are visible as solid lines. This image can be ambiguous particularly for complex shapes. Different colours are generally used for different parts of the object, thus making the object easier to visualise. In the surface model, all the surfaces are shown but the data describe the interior volume.

Design Analysis and Optimization

Once the graphic model is created, the design is subjected to engineering analysis. This phase may consist of analysing stresses, strains, deflections and other parameters. Various sophisticated packages having capabilities to compute these quantities accurately are now available. Because of the relative ease with which such analysis can be made designers are increasingly willing to thoroughly analyse a design before it moves on to production. Experiments and field measurements may be necessary to determine the effects of loads, temperature and other variables.

Design Review and Evaluation

An important design stage is review and evaluation to check for any interference between various components in order to avoid difficulties during assembly or use of the part and whether the moving members such as linkages are going to operate as intended. The designer can 'zoom' in on part design details and magnify the image on the graphics screen for close scrutiny. Layering is one of the important methods used in CAD design review. During design review, the part is precisely dimensioned and tolerenced as required for manufacturing it.

Documentation and Drafting

After analysis and review, the design is reproduced by automated drafting machines for documentation and reference. Detailed and working drawing are developed and printed. The automated drafting features include automated dimensioning, scaling of the drawing, development of sectional views and enlarged views of particular part details.

The comparison between CAD and conventional design procedure is represented in fig (38.2).

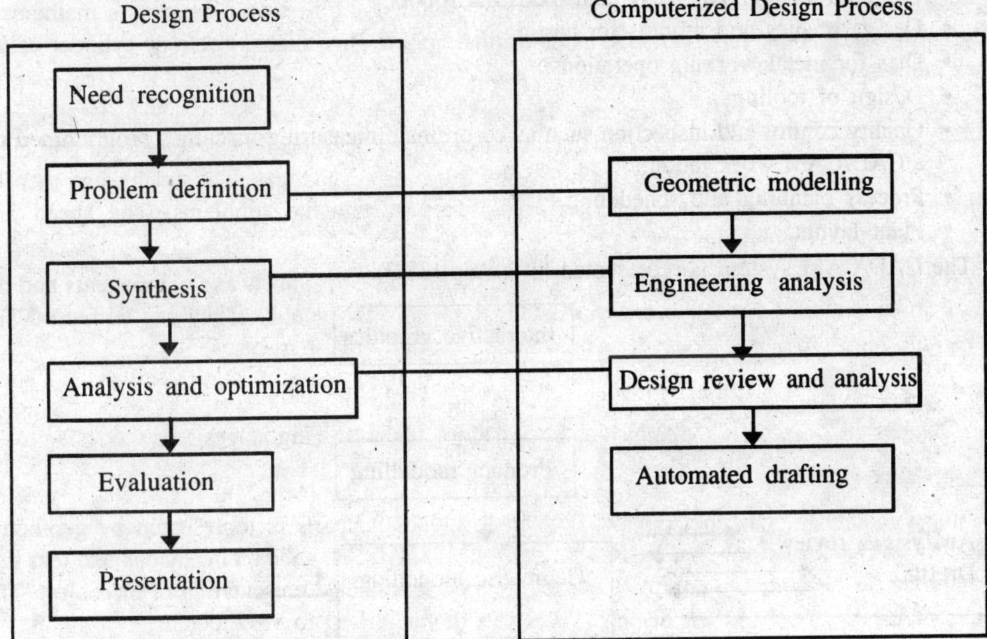

Fig. 38.2. Comparison CAD and Conventional Design Process.

38.7. COMPUTER AIDED MANUFACTURING (CAM)

Computer aided manufacturing is the use of computers and computer technology to assist in all phases of manufacturing a product including process and production planning, machining, scheduling management and quality control. Because of the benefits CAD and CAM are often combined into CAD/CAM systems. This combination allows the transfer of information from design into the planning for manufacturer of a product without the need to manually reenter the geometric model. The database developed in CAD is stored and processed further by CAM in to the necessary data instructions for operating and controlling production machinery, material handling equipment and automated testing and inspection for product quality.

The manufacturing planning applications of CAM are those in which computers are used indirectly to support the production function but there is no direct connection between the computer and the process. The computer is used off-line to provide information for the effective planning and management of production activities.

The important applications include :

- Cost estimating.
- Computer aided process planning.
- Computerized machinability data systems.
- Computer assisted NC part programming.
- Computer aided line balancing.

The second category of CAM applications are concerned with developing computer systems for implementing the manufacturing control function. Manufacturing control is concerned with managing and controlling the physical operations in the factory. Process control, quality control, shop floor control and process monitoring are the scope of this function.

Thus, the scope of CAD/CAM includes design, manufacturing planning and manufacturing control. Typical applications of CAD/CAM includes :

- Programming for NC, CNC and industrial robots.
- Design of dies and moulds for casting.
- Dies for metal working operations.
- Design of toolings.
- Quality control and inspection such as co-ordinate measuring machines programmed on a CAD/CAM workstation.
- Process planning and scheduling.
- Plant layout.

The CAD/CAM system is represented in the fig (38.3).

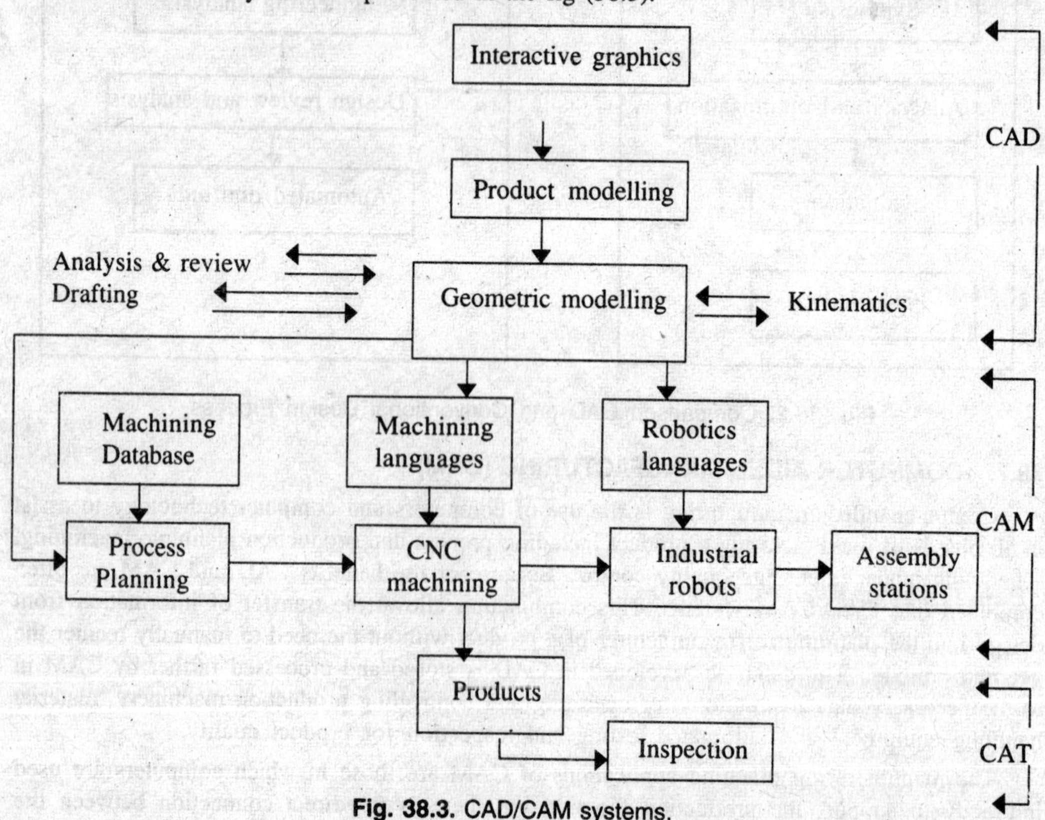

Fig. 38.3. CAD/CAM systems.

38.8. COMPUTER AIDED PROCESS PLANNING (CAPP)

Process planning is concerned with selecting methods of production, toolings, fixtures and machinery, sequence of operations and assembly. It involves the preparation and documentation of the plans for manufacturing the products. Computer aided process planning (CAPP) is a means of implementing the planning function by computer. CAPP helps in determining the processing steps required to make a part after CAD has been used to define what is to be made. CAPP accomplish these complex steps by viewing the total operation as an integrated system, so that the individual operations and steps involved in making each parts are co-coordinated with others and are performed efficiently and reliably.

There are two types of CAPP systems : Variant and Generative process planning.

Variant CAAP System

In this system, the computer files contain a standard process plan for the part to be manufactured. The search for a standard plan is then retrieved, displayed for review and printed as a routing sheet. The process plan includes operations such as type of tools, and machines to be used, the sequence of manufacturing operations to be performed, speeds, feeds, time required for each sequence and so on. If the standard plan for a particular part is not in file of the computer, one that is close to it with similar code number and an existing routing sheet is retrieved. If a routing sheet does not exist, it is prepared for the new part and stored in the computer memory. The general procedure for using the retrieval computer aided process planning is shown in fig (38.4).

Generative CAPP Systems

It is an alternative approach to automated process planning. In this system, a process plan is automatically generated on the basis of the same logical procedures that would be followed by a traditional process planner in making that particular part. Such a system is complex because it must contain comprehensive and detailed knowledge of part shape, dimensions process capabilities. Selection of manufacturing methods and machinery and tools required and sequence of operations to be performed. These capabilities of computers are referred to as expert systems.

An expert system is a computer program that is capable of solving complex problems that normally require a human being with experience and knowledge.

There are several ingredients required in a fully generative process planning system.

1. The technical knowledge of manufacturing and the logic that is used by successful process planners must be captured and coded in to a computer program.
2. Computer compatible description of the part to be produced. This description contains all of the pertinent data and information needed to plan the process sequence.
3. Capability to apply the process knowledge and planning logic contained in the knowledge base to a given part description.

Benefits of CAPP

1. Standardisation of productivity of process plan.
2. Standardisation of productivity of process planners resulting in reduced lead times, reduced planning costs.
3. Increases efficiency, consistency of product quality and reliability.
4. Process plan and route can be prepared more quickly.

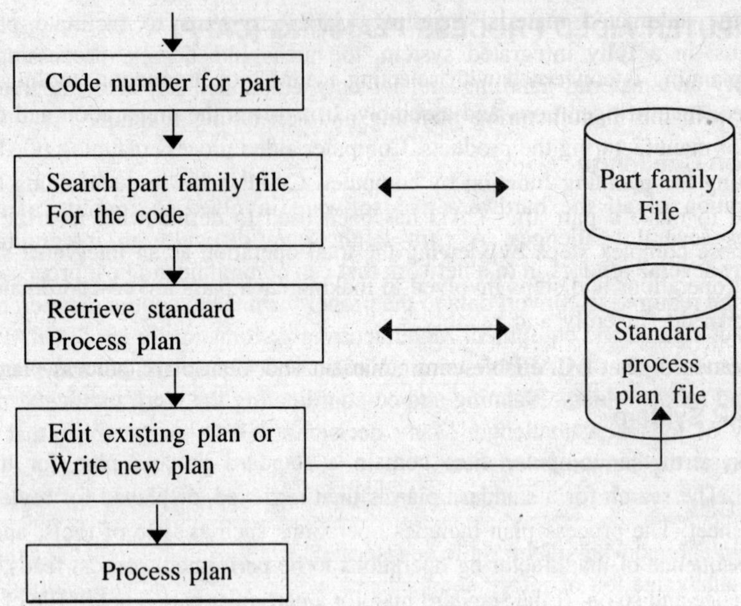

Fig. 38.4. The general procedure for using the retrieval computer aided process planning.

38.9. COMPUTER INTEGRATED MANUFACTURING (CIM)

Computer integrated manufacturing includes all the engineering functions of CAD/CAM, but also includes the business functions of the firm as well. The ideal CIM system applies computer technology to all of the operational functions and information processing functions in manufacturing from order receipt, through design and production to product delivery. At the broader level. CIM can be viewed as, an integration of

- Product and process design.
- Production planning and control.
- Production process.

Computer integrated manufacturing is the automated version of the manufacturing process where the three major manufacturing functions product and process design, planning and control and production process are replaced by the automated technologies CAD/CAM, CAPP and automated material handling systems (Robots, AS/RS, AGV) and computerized business systems like order entry, payroll and billing. This is also called "Completely Automated Factory with no human interference and factory of the future."

CIM calls for the co-coordinated participation in all phases of manufacturing enterprises for the purpose of integration and supervision. Thus, it includes — design of parts and components, Planning of its manufacture, automatic production of parts, automatic assembly, automatic testing and the computer controlled flow of materials. The purpose of CIM is to enable the company to transform ideas into a high quality products in the minimum time and at minimum cost and the CIM goes beyond the scope of FMs or CAD/CAM system. The concept is to integrate information from marketing, order entry maintenance, accounting, shipping. The components of CIM are shown in the fig (38.5).

All the CIM technologies are tied together using a network and integrated databases, *i.e.*, the data integration allows CAD systems to be linked to computer aided manufacturing (CAM), numerical control (NC) part programmes and manufacturing planning and control systems can be linked to the automated material handling (AMH) systems to facilitate parts pick list generation. Thus, in a fully integrated system, the areas like design, processing, inspection, testing, assembly and material handling are not only automated but also integrated with each other and also with the manufacturing planning and scheduling function.

The Integration Challenge

The integration of all the hardware and software involved in running a manufacturing operation poses several challenges. A firm might have difficulty in integration equipment manufactured by several vendors in to a network that can communicate. Post processors or special translators may be required to convert data to the proper form. Data communication requires some standardization of signals and equipment Manufacturing Automation Protocol (MAP) is evolving as one of the standards that will enable communication between manufacturing equipment such as conveyors and robot welders. Planning and co-coordinating the work of several machines that perform variety of jobs is a challenge. Many decisions will be so complex that they require decision support at the human interfaces.

Definition of CIM

Computer Integrated Manufacturing is an interdisciplinary science applied to manufacturing. It involves the amalgamation of information science with automated manufacturing. Information science in CIM may have direct application or indirect application. Japan adopted the concept of flexible manufacturing where as USA strongly believed in rigid automation.

Direct Application

In this application, computer or network of computers monitor and control the hardware in the manufacturing plant such as machine tools, inspection machines, and robots and automated guided vehicles, which physically transform or transport or assemble components.

Computer Aided Design (CAD)

Geometric modelling engineering analysis. Design review and evaluation automated drafting.

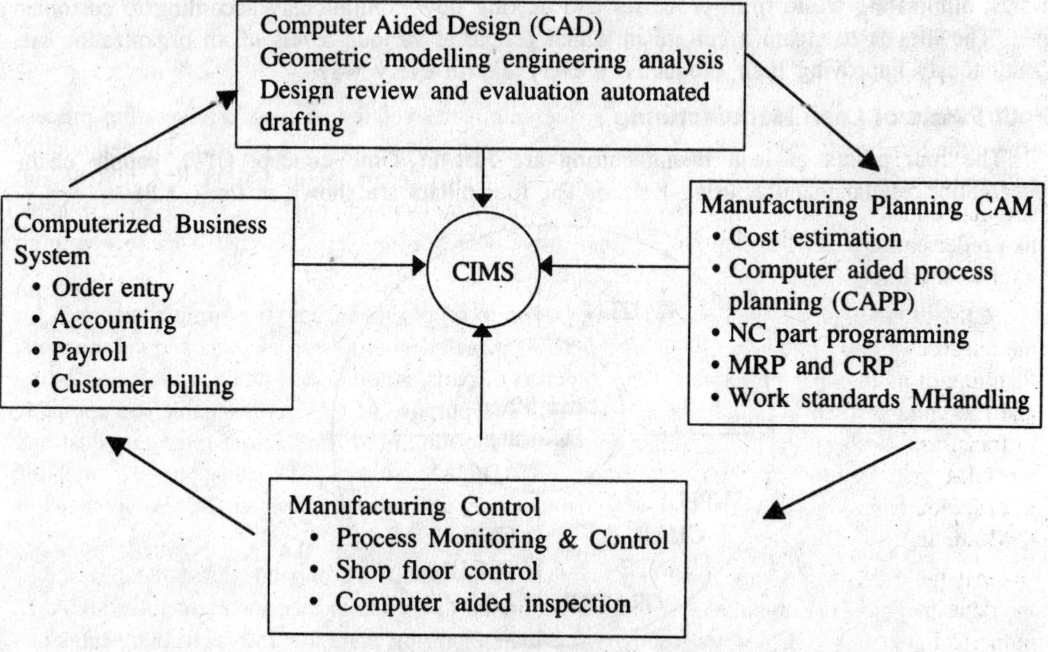

Fig. 38.5. Elements of CIM system.

Indirect Applications

The following applications are included in indirect category.

1. Computer aided design (CAD)
2. Production Planning and Control (PPC)
3. Inventory Control
4. Marketing/sales, factory management
5. Business softwares such as,
   - Materials requirement planning (MRP)
   - Manufacturing Resource Planning (MRP II)
   - Just In Time (JIT) manufacturing technique
   - Optimized Production Technology (OPT)

CIM includes utilization of both direct and indirect applications.

38.10. LEAN AND AGILE MANUFACTURING

38.10.1. Lean Manufacturing

Lean manufacturing basically involves the assessment of each of the company's activities — the efficiency and effectiveness of its various operations, the reason for retaining specific operations and manpower, the efficiency of equipment and machinery in the operation, number of people involved in particular operation and detailed analysis of costs associated with each activity including both productive and non productive operations.

Lean manufacturing is a production strategy that aims at high levels of production using lesser effort, time and materials. It is an integrated business approach adopted to eliminate non-value added activities from the customer delivery cycle in the operations. This approach enables companies to respond quickly and profitability to changes in customer demands. It is not only restricted to shop floor. Actually lean is a way of thinking, an attitude. The technique of lean manufacturing can be applied to every situation in a company by finding out what the customer wants, eliminating waste from processes and making flow continuously according to customer pull. The idea is to create a culture in which people at various levels of an organization are continuously improving their productivity every day in every way.

Four Pillars of Lean Manufacturing

The four pillars of lean manufacturing are Just in Time concept (JIT), supply chain integration, cellular manufacturing Kaizen. The four pillars are shown in fig. (38.6).

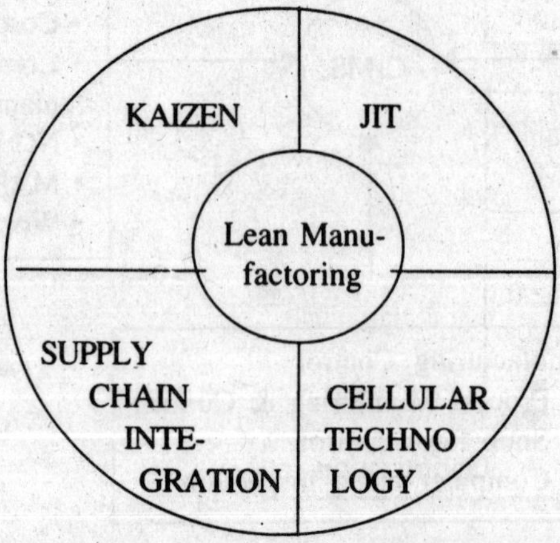

Fig. 38.6. Pillars of Lean Manufacturing.

JIT concept is utilized to procure raw materials in the exact amounts at the exact time required, thus avoiding inventory costs. JIT can be implemented effectively only through an optimum supply chain integration which maintains the smooth flow of materials from the suppliers to customer. JIT leads to reduced inventory cost and is achieved through standardization of processes, TPM and Kanban.

The lean system can be effective only if it is executed all along the production chain, *i.e.*, from the supplier's supplier to the customer's customer. Every link along this chain is affected if a single member does not deliver. Thus, supplier base management becomes all the important.

The cellular technology is an important aspect of lean manufacturing. Manufacturing cells are designed to process parts of the product in separate fixed areas, thus eliminating non-value added activities. Ultimately, the layout creates a single piece flow. This reduces the order flow time, WIP, material handling costs and thus elevating customer satisfaction.

Lean manufacturing produces optimum results only, if it is implemented as an ongoing improvement process involving every one at each level. Kaizen demands prioritization of demands standardization, and continuous improvement from the members of the organization.

Special Features of Lean Manufacturing

- Lean manufacturing produces goods with few people, little inventory and little waste.
- Lean ensures that each production stage processes exactly what, how much and exactly when next stage wants it.

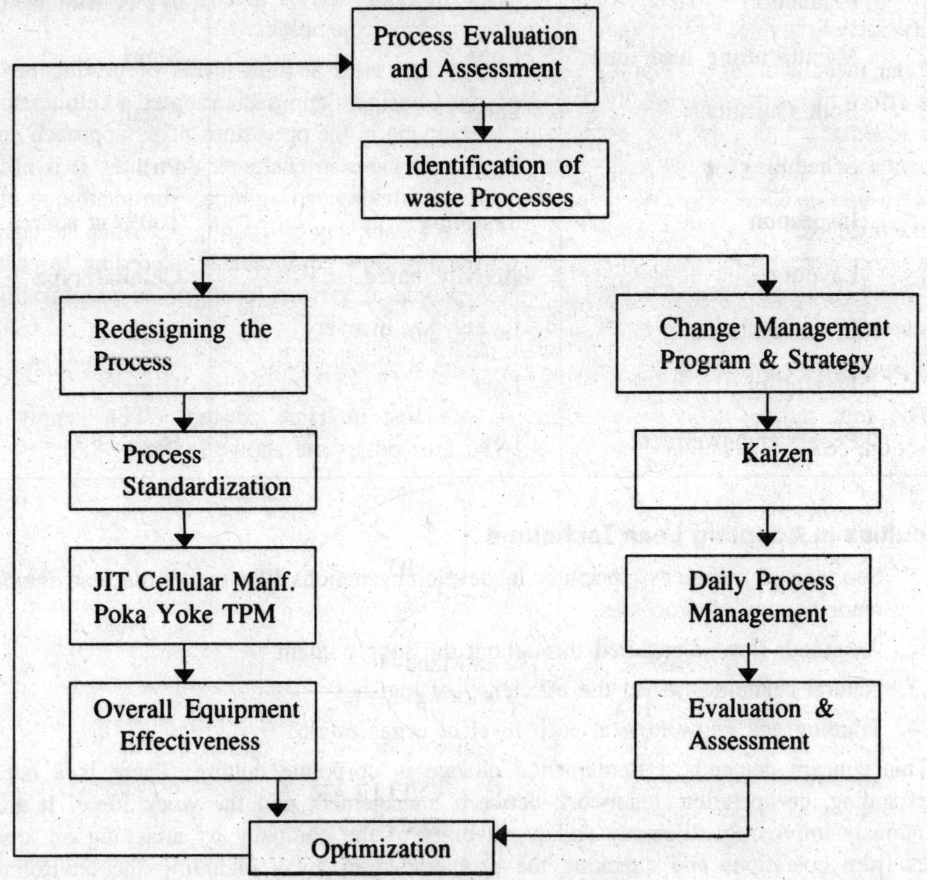

Fig. 38.6. Lean Manufacturing Process.

- Lean allows large variety in products without the kind of changeover costs that customized mass manufacturing involves.
- Lean embraces every facet of the organization sourcing, manufacturing, marketing, accounting and strategic planning.

The lean process is represented in the fig (38.6) and comparison between traditional and lean manufacturing is shown in table (38.1).

Advantages and Limitations

1. Inventory levels are very low because of use of JIT.
2. Simple and multitask machinery which aids variety products manufacturing.
3. More value to the customer.
4. Reduction in manufacturing lead-time.
5. Integrates people and techniques to improve the work place and strengthens company's competitive position.

Table 38.1: Traditional Vs Lean Manufacturing

Sr. No.	*Parameters*	*Traditional Manufacturing*	*Lean Manufacturing*
1.	Production	Made to stock	JIT (Customized)
2.	Manufacturing lead time	Long	Short
3.	Both Quantity	Large	Small
4.	Scheduling	Push	Pull
5.	Inspection	Sampling	100% at source
6.	Layout	Activity based	Cellular type
7.	Inventory twins	Low	High
8.	Flexibility	Low	High
9.	People empowerment	Low	High

Difficulties in Adopting Lean Technique

1. Shortage of resources (specially in developing nations like India) that can inhibit the reengineering of processes.
2. A smooth flow is required throughout the supply chain.
3. Natural calamities affect the efficiency of logistics.
4. Training and know how at each level of organization.

This concept demands a fundamental change in corporate culture. There is a need for understanding, co-operation, teamwork between management and the work force. It aims at continuously improving efficiency and profitability of the company by attacking all kinds of wastes from operations and attacking the problems right away including the environmental issues.

38.10.2. Agile Manufacturing

Agile manufacturing is a word, which has been recently coined to indicate the use of principles of lean production on a broader scale. The principle underlying agile manufacturing is ensuring agility and hence flexibility in the manufacturing organization so that it can quickly respond to changes in product demand and customer needs. This is to achieved through people, equipment, computer hardware and software and sophisticated communication systems.

This manufacturing approach requires that manufactures benchmark their operations. This means understanding the competitive position of other manufactures with respect to theirs and setting ambitious still realistic goals for the future. Thus, bench marking is a reference from where different measurements can be made and compared.

Agile manufacturing approach discovers various guiding principles for the manufacturing and infuse the important principles and technologies in to manufacturing practice. AMA recognize the need for integration of technology, management, information system and the work force. It integrates key functions and disciplines involved in creating, designing, making, selling/ servicing products. It encompasses not only the critical operations like technology, product and process engineering, administration and marketing /sales/services within an organization but also venders/suppliers, Community and Government. This is represented in fig (38.8).

Thus, Agile manufacturing (AM) is the science of business system that integrates management, technology and workforce making the system flexible enough for manufacturing to switch over from one product (that is being produced) to another product (desired to be produced) in a cost effective manner within the framework of the system.

Agile manufacturing encompasses entire business process commencing from planning, finance and process design, toolings, layout, materials and inventory, cost specifications, pricing, marketing, sales and services.

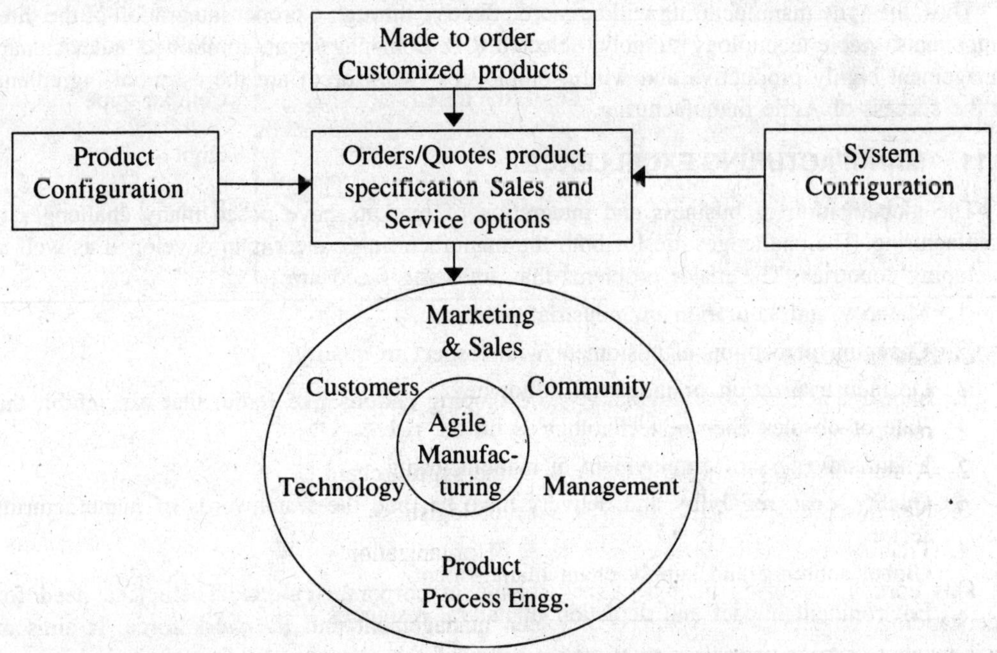

Fig. 38.8. Agile manufacturing.

Some of the functions of Agile Manufacturing can be made available as special modules. The industry can select the right modules they desire to have and then they can add other modules as they provide the desired interfaces. Finally each industry can emerge as a unit that has adopted agile manufacturing in its entity.

The integration of system parameters in agile manufacturing can be divided in to three categories.

(1) Management with Technology and Workforce

In order to provide flexibility to agile manufacturing, the traditional management accounting should be replaced by Activity Based Costing (ABC). As the technology that is adopted is to be user oriented, it is essential that the user must have control over the technologies adopted. This can be brought about effectively by organizing the interactions between the user and the workforce in an industry. The training of entire workforce is essential. Organizational pattern must be flexible enough to provide perfect co-operation between the three components of agile manufacturing — manufacturing, technology and work force.

(2) Technology with Management and Workforce

In order to achieve a balanced integration of management, technology and workforce, an interdisciplinary design methodology must be adopted to bring the concept of total enterprise design. Strategies must be directed to enhance skill and knowledge of work force and not on replacing the old technology by new one. Research in technology must involve all the three components to arrive at an appropriate technology most suitable for the system.

(3) Workforce with Management and Technology

The workforce must be built up to produce products that would satisfy the customers. Innovative skills should be rewarded and encouraged. The communication, which is vital for success of the organization must be given the proper importance.

Thus, the agile manufacturing will be more effective through a proper integration of the three components, viable technology suitably, Selected R & D management, committed and visionary management highly productive and willful employees (workforce) are the essential ingredients for the success of Agile manufacturing.

38.11. MANUFACTURING EXCELLENCE

The globalization of business and integration of markets have posed many challenges to manufacturing. The challenges are for both the manufacturing concerns in developed as well as developing countries. The major problems that are being faced are

1. Maturity and saturation of industrial products.
2. Changing perception of customers with respect to quality.
3. De industrialization or industrial hollowness.
4 Rate of obsolescence of technology is on the rise.
5. Unattractiveness of employment in manufacturing.
6. Quality, cost, reliability and delivery have become the watchwords of manufacturing sector.
7. Global sourcing and supply chain management.
8. Environment impact and depletion of natural resources.

Manufacturing Excellence — Definition

Manufacturing excellence is a key word expressing the enhancement of the conditions of goods production for pursuing human happiness while eliminating the "hollowing" of production. [Hollowing refers to the decline of domestic manufacturing labour force and production activities.] Thus, manufacturing excellence consists of just in time, total quality and total people involvement.

Manufacturing Culture

Manufacturing culture implies a combination of intellectual, conceptual, philosophical activities and ideas with human relationships, motivation and involvement.

Approaches to manufacturing Excellence

The following methods are effective in moving towards excellence in manufacturing.

- Automated production [Computer integrated manufacturing (CIM)]
- Flexible/human centered production [Mass customization]
- High value added production
- Do it right first time and all the time [deflect free production]
- Manufacturing for customer satisfactions.

The trends in manufacturing are represented in Table (38.2).

Table 38.2: Trends in Manufacturing

Trends in Manufacturing
• Shorter product life cycle.
• Increased emphasis on quality and reliability.
• More customized products.
• New and advanced materials.
• Growing use of electronics.
• Pressure to reduce inventories.
• Faster response time.
• Global Sourcing.
• Out Sourcing.
• Just in Time production.
• Point of use manufacture
• Greater use of computers in manufacture.

Strategies for Achieving Excellence in Manufacturing

The excellence in manufacturing is an integrated outcome of appropriate technology committed and visionary management and dedicated and productive workforce. This is represented in the fig (38.9).

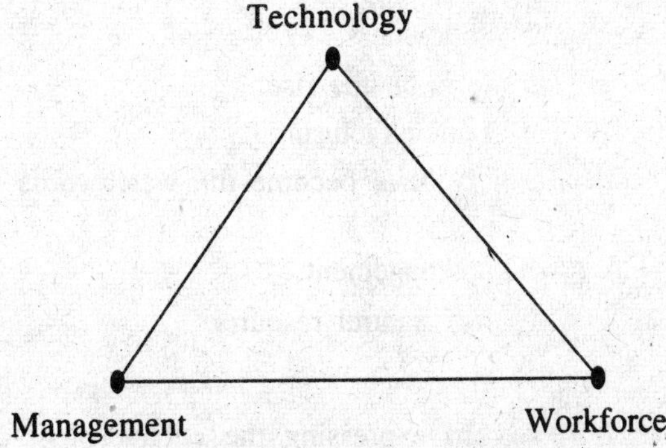

Fig 38.9. Integration of Manufacturing Elements.

Technology

Technology is a broad based activity ranging from product design through production, product use and maintenance. Technology understanding and its integration into an overall system is a must. Technology can be developed with a proper understanding of product design, production technologies process management and control, production planning, material selection and management, logistics, marketing, vendor support, economics, CAD, TQM, manufacturing automation and control, and information technologies.

Management

Management in a manufacturing programme should always aim at newer and innovative ideas and newer methodologies so as to enable them to remain as leaders. There should be a methodology for creating values and methods for life long learning and collaboration. There should be a continuous and rewarding effort to disseminate and implement programme knowledge base. There should be a methodology through which managers can conceive and effect positive changes in the manufacturing protocol, rapidly exchange, concepts and ideas.

Workforce

Commitment of workers to career in manufacturing is essential to achieve excellence in manufacturing. Workers should have aptitude to learn but also have strong talents for working in teams. There should be an active encouragement for the work force to cut across traditional disciplinary boundaries to address manufacturing issues.

The following policies for workforce will help to achieve excellence in manufacturing :

- Development of integrated and market driven system of life long learning.
- Focus on job evaluation and job rotation, job enlargement and enrichment.
- Participation of work force in decision-making.
- Focus in continuous improvement (Kaizen).
- Promote team work.

SUPPLEMENTS

Supplement – One

INVESTMENT DECISIONS FOR CAPITAL PROJECTS

Investment refers to an economic activity of committing various resources like land, labuor and capital with an expectation of receiving a stream of benefits in future periods. If the time horizon over which the benefits are accrued is longer than one year, the resources committed are called "Investment" and the money spent is called the "Expenditure".

Evaluation Criteria

The essence of capital investment analysis is in computing and comparing benefits. (Cash inflows) obtainable over a period of time with costs (Cash outflows).

Important criteria used for evaluation are :

1. Pay back period
2. Average return on investment
3. Net present value (NPV)
4. Internal rate of return (IRR)
5. Benefit cost ratio (BCR)
6. Capital Rationing (CR)
7. Profitability Index.

1. Pay Back Period

Pay back period is the length of time required to recover the investment made on the project, which is recovered by the net returns from the project.

For example, a project involves an initial investment of Rs. 6.0 lakhs and generates the following cash flows

Year	1	2	3	4
Cash inflows (Rs.Lakhs)	1.0	1.5	1.5	2.0

In this case, the pay back period is 4 years because, the sum of cash inflows during the first four years equal the initial investment of 6 Lakhs.

If the cash inflows are same in each year, the payback period is expressed as

$$n = \frac{I}{B}$$

where n is the pay back period in years,

I is the initial investment,

B is the net returns (benefits) per annum.

This in general, payback period (n) is expressed as

$$n = \frac{I}{\sum_{i=0}^{n} R_i - D_i}$$

where R_i is the income during the period i [$i = 1, 2, 3, ..., n$]

D_i is the operating expenses during period i.

I is the initial investment.

According to pay back period criteria, shorter the pay back period the more desirable is the project. Firms using this criteria generally specify the maximum acceptable pay back period. If the maximum acceptable period is 'n', then only projects with a pay back period of 'n' years or less are worth considering.

Advantages

- Simple both in concept and application.
- Does not involve any complicated and tedious computations.
- It is a ready method for dealing with risk as it emphasizes on the earliest return of invested capital.

Disadvantages

- This method ignores the cash flows beyond the pay back period.
- It does not consider the time value of money.
- It measures the projects capital recovery and not the profitability of the firm.

2. **Average Return on Investment**

This is also called "Return On Investment" (ROI). The average return on investment is defined as. "A ratio of net average income from the project to the investment made in the project". The net income is defined as the difference between the net cash inflows and cash outflows. This is also called "Accounting Rate of Return". Average rate of return is expressed as :

$$\text{Average rate of return} = \frac{\text{Profit after tax (PAT)}}{\text{Book value of the investment}}$$

$$= \frac{\sum_{t=1}^{n} P_t / n - 1}{\sum_{t=0}^{n} B_{vt} / n}$$

where P_t is profit after tax,

B_{vt} is the book value of the investment at t[th] year.

ROI criteria is traditionally popular, computation is simple and based on easily available information.

Advantages

- It is simple to calculate without any complicated procedure.
- It is based on accounting information, which is readily availabie.
- It considers benefits over the entire period of project life.

Disadvantages

- It is based on accounting profit not on net cash flow.
- Does not take in to account the time value of money.
- Average rate of return measure is internally little consistent.
- This method also does not take in to account the time value of money.

3. Net Present Worth (NPW) Method

This method of project evaluation takes into account the time value of money. The NPW method converts the entire streams of cash flows at a time period zero. In the context of NPW method of evaluation the discount rate is also called "Minimum Attractive Rate of Return (MARR)". The net present (NPV) of a project is equal to the sum of the present values of all the cash flows associated with the project. Symbolically, it is represented as

$$\text{NPV} = \frac{F_0}{\left(1+\frac{i}{100}\right)^0} + \frac{F_1}{\left(1+\frac{i}{100}\right)^1} + \frac{F_2}{\left(1+\frac{i}{100}\right)^2} + \ldots + \frac{F_n}{\left(1+\frac{i}{100}\right)^n}$$

Decision Rule Associated with NPV

Accept the project if the net present value (NPV) is positive and reject the project if it is negative. A net present value zero represents indifference.

When there are more than one projects compete for limited resources, then the proposals are to be selected such that the total amount to be invested would not exceed the funds available. Under such circumstances, the procedure for selection will be to rank the competing projects in the descending order of the rank till the available fund is exhausted.

Advantages

- It takes in to account the time value of money.
- It considers cash flows throughout the life of the project.
- NPV contributes to the wealth maximizations of shareholders.

Illustration

The cash flows of a project are given below :

Year	0	1	2	3	4	5
Cash flow	-10	2	2	3	3	3.5

(Rate in lakhs)

Discount 10%, Net present value of the proposal is computed as,

$$\text{NPV} = \frac{-10}{(1+0.1)} + \frac{2}{(1.1)} + \frac{2}{(1.1)^2} + \frac{3}{(1.1)^3} + \frac{3}{(1.1)^4} + \frac{3}{(1.1)^5}$$

$$= -5.273$$

Note. Since the NPV is negative, the project is not favourable.

Problem 1. *Two bids are offered to install an elevator for the newly constructed building. The following information is available.*

Alternative Bids	*Initial Cost Rs*	*Estimated service Life (Rs)*	*Annual operation and maintenance costs*
Hilton Elevators	*4,50,000*	*10*	*27,000*
Batton Elevators	*5,40,000*	*10*	*28,500*

The applicable interest rate is 15%. Determine NPV of both projects and suggest the alternative to be chosen.

Solution. Alternative I – Hilton elevators

The cash flow diagram is represented as follows :

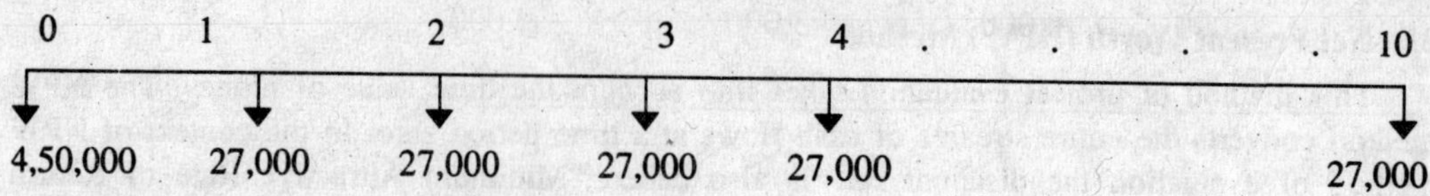

Present worth of alternative 1

Present worth @ 15% = 4,50,000 + 27,000 (P/A, 15%, 10) from tables.

= 4,50,000 + 27,000 × 5.019

= Rs. 5,85,513.0

Present worth of alternative 2

The cash flow diagram is shown in the fig below :

0 1 2 3 4 . . . 10

5,40,000 28,500 28,500 28,500 28,500 28,500

Present worth @ 15% = 5,40,000 + 28,500 (P/A, 15%, 10)

From tables (P/A, 15%, 10) = 5.019

= 5,40,000 + 28,500 × 5.019

= Rs. 6,83,042.00

The present worth of alternative 1 is less than that of alternative 2,

Hence, alternative 1 is to be selected.

Problem 2. *The initial investment and net cash flows for two project proposals A and B are given in the table below :*

Proposals	*End of years*				
	0	*1*	*2*	*3*	*4*
A	*– 10000*	*3000*	*3000*	*7000*	*6000*
B	*– 10000*	*6000*	*6000*	*3000*	*3000*

Select the project based on NPV assuming the discount factor of 15%.

Solution. The present worth of proposal A at I = 15%

The cash flow diagram is represented as follows

3,000 3,000 7,000 6,000

0 1 2 3 4

10,000

$$\text{NPV} = -10000 + \frac{3000}{(1+0.15)} + \frac{3000}{(1+0.15)^2} + \frac{7000}{(1+0.15)^3} + \frac{7000}{(1+0.15)^4}$$

$$= -2908.60$$

The present worth of project *B*

The cash flow diagram is represented as follows :

	6,000	6,000	3,000	3,000
0	1	2	3	4
10,000				

$$NPV = -10000 + \frac{6000}{(1+0.15)} + \frac{6000}{(1+0.15)^2} + \frac{3000}{(1+0.15)^3} + \frac{3000}{(1+0.15)^4}$$

$$= -3442.70$$

The present worth of proposal B is higher than that of A, hence select the proposal B.

4. **Internal Rate of Return (IRR)**

One of the important requirements of NPW method is discount rate. Prior choice of discount rate will make or break the project. The Internal Rate of Return (IRR) is that rate of discount rate that equates the initial investment in the project with the future net income stream, *i.e.*, it is the discount rate which makes the net present value of the project equal to zero.

This can expressed as

$$-I_0 + \frac{B_1 - C_1}{\left(1+\frac{i}{100}\right)} + \frac{B_2 - C_2}{\left(1+\frac{i}{100}\right)^2} + \ldots + \frac{B_n - C_n}{\left(1+\frac{i}{100}\right)^n}$$

where, I_0 is the initial investment in Rs. [cash out flow, hence – ve sign]

$B_1, B_2, \ldots, B_n$ are the cash inflows for the n years [$n = 1, 2, \ldots, n$]

$C_1, C_2, \ldots, C_n$ are the cash inflows for the n years [$n = 1, 2, \ldots, n$]

and i is the discount rate.

Advantages

- It takes into account the time value of money.
- It considers cash flow in its entirety.
- It makes sense to businessmen who are talking in terms of IRR.

Limitations

- There may be multiple rates of return.
- It may be misleading when the choice is for mutually_exclusive projects that have different outlays.
- It is cumbersome in calculation.
- IRR is only a figure reflecting the rate of return or the efficiency in the use of capital. It does not give an absolute value of the benefits or net incomes generated by the project.

Problem 3. *Compute the internal rate of return (IRR) for a project given below with the following cash flows :*

Year	*0*	*1*	2	*3*	*4*
Cash flows (Rs)	*– 100000*	*30000*	*30000*	*40000*	*45000*

Solution. Now, we have the expression

$$-I_0 + \frac{B_1 - C_1}{\left(1+\frac{i}{100}\right)} + \frac{B_2 - C_2}{\left(1+\frac{i}{100}\right)^2} + \frac{B_3 - C_3}{\left(1+\frac{i}{100}\right)^3} + \frac{B_4 - C_4}{\left(1+\frac{i}{100}\right)^4} = 0$$

In this case, C_1, C_2, etc. are zero as there are no cash outflows other than initial investment.

$$\therefore \quad -10,000+\frac{30000}{\left(1+\frac{i}{100}\right)}+\frac{30000}{\left(1+\frac{i}{100}\right)^2}+\frac{40000}{\left(1+\frac{i}{100}\right)^3}+\frac{45000}{\left(1+\frac{i}{100}\right)^4}=0$$

Substituting, the value of I in the L.H.S. of the above equation, by trial and error such that the value of I would make L.H.S. zero.

By trial and error, find the value of I that makes LHS = 0.

i = 12%, gives LHS = 7733.

I = 15%, gives LHS = 802.

i = 16%, gives LHS = – 1359.

Thus, the desired value of I lies between 15% and 16%. If the estimated NPV were plotted against corresponding discount rates, a curve as shown below would be obtained. The IRR of the project proposal is 15.37% i.e. the point where the curve intersects the x-axis i.e. discount rate axis or alternatively IRR can be found by interpolation.

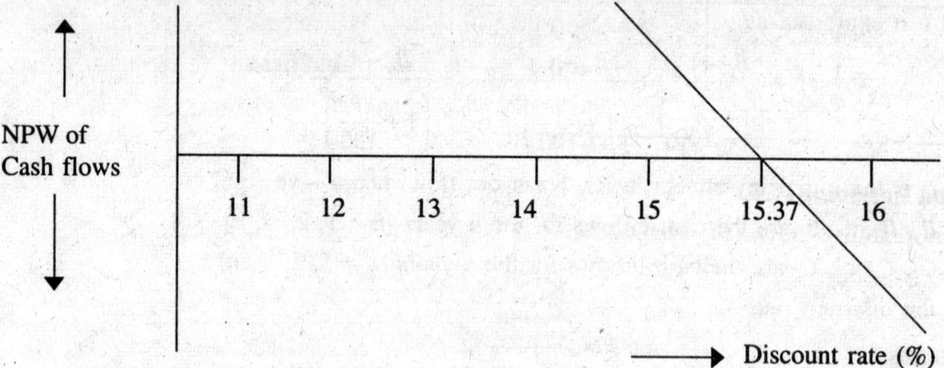

5. **Benefit Cost Ratio (BCR)**

The cost benefit ratio can be defined in two ways.

The first definition is given by

$$BCR=\frac{P_{VB}}{I}$$

where BCR = Benefit cost ratio

P_{VB} = Present values of cash inflows

I = Initial investment.

It relates the present values of cash inflows (benefits) to the initial investment. Second definition is a net measure and is called "Net benefit cost ratio" (NBCR).

$$NCBR = BCR - 1$$

Advantages

1. It takes in to account the time value of money.
2. It is able to discriminate between the small and large projects as it measures NPV rupee of investment.

Illustration

Initial investment Rs. 1,00,000, cost of capital is 12%.

Year	1	2	3	4
Benefits	25000	40000	40000	50000

$$BCR = \frac{\frac{25000}{(1.12)} + \frac{40000}{(1.12)^2} + \frac{40000}{(1.12)^3} + \frac{50000}{(1.12)^4}}{100000}$$

$$= 1.145$$

NBCR = BCR – 1 = 1.145 – 1 = 0.145.

Decision rules associated with BCR are

BCR	*NBCR*	*Decision rule*
> 1	> 0	Accept
= 1	= 0	Indifferent
< 1	< 0	Reject

6. **Capital Rationing (CR)**

The profitability index (*PI*) of the project is given by

$$PI = \frac{\sum_{t=1}^{n} \frac{B_t - C_t}{(1+i)^t}}{I}$$

where B_t is the cash inflows for different years [t = 1, 2, ..., n]
C_t is the cash outflows for different years.

The projector is accepted/rejected based on the decision rules.

- Project is accepted if the profitability index (*PI*) > unity and the funds are limited.
- Project is rejected if the profitability index (*PI*) < unity.

This criterion is used for mutually exclusive projects with limited resources (capital rationing).

There may be number of project proposals which have NPV greater than zero, IRR greater than cost of capital and *PI* greater than one. But because of budget constraints, all of these projects cannot be undertaken.

The objective of the firm is to maximise profits. With a limited finance and number of project proposals, one can maximise the profit by maximizing the NPV of combination or package of project (proposals) investments.

The procedure for selection will be

1. Identify the different packages, which can be considered under fund constraints
2. Select the package of projects, which has the maximum of NPV.

Illustration

The available project opportunities along with their investment outlays. NPV and IRR are represented in the table (S-1).

Table S-1: Capital Rationing Situation

Project	Investment (Rs.Lakhs)	NPV	IRR (%)
1	100	40	16
2	180	30	12
3	160	30	17
4	140	25	13.5
5	120	20	12
6	100	20	16
7	80	20	18

Budget constraint is of Rs. 600 lakhs.

The package (combinations), which can be accommodated within a budget of Rs.600 lakhs and their respective NPV's, are given.

Package of Projects	Total Investment (Rs.Lakhs)	Total NPV
1+2+3	540	100
2+3+4+5	600	105
3+4+5+6+7	600	115
1+4+5+6	560	105

Since the combination of projects 3, 4, 5, 6 and 7 yields the maximum net present value (NPV), this package of investments will be selected.

Supplement – Two

QUALITY FUNCTION DEPLOYMENT (QFD)

With the changing techno-economic scenario around the entire world, the market has turned from seller to buyer type. The main objective of every organisation is to satisfy the stated and implied needs of the customer. It has been observed and demonstrated that fast growing companies are those, which take care of their customer's satisfaction. To ensure continued business success every company should have process to constantly monitor and update its knowledge of customer. Thus, voice of the customer has become the important factor in designing products and services to satisfy the needs.

This concept of utilizing the voice of the customer for designing the products was first introduced at Mitsubishi's Kobe shipyard in 1972. This method of structuring customer requirements and translating in to technical specifications as a basis for new product development is called "HIN SHITU KI NOTEN KAI" which has been translated in to English as "Quality Function Deployment (QFD)".

QFD Concept

QFD has been developed to help translate requirements as seen by the customer into technical specifications that can be used by designer and production. It is a method for linking customer requirements to technical specifications. QFD is not a tool, it is a planning process. It helps the organisation plan for the effective use of the other technical tools to support and complement each other and address issues on priority.

QFD helps to focus on needs of the customers by using matrices and charts to help organisation hear and understand the customer voice, develop the definition of quality for themselves and then deploy that definition to the development and production of all services and components of the system. It is not a design or problem solving tool but its purpose is to inform and educate the managers and employees in the product characteristics that will interest the customer.

The QFD concept is represented in the table (S2.1)

Table S2.1: QFD Concept

• QFD is a planning process. • Inputs — Customers wants and needs. • Matrix format is used for recording vital information. • It permits analysis and determination of priority issues. • Output — Key action issues for improved customer satisfaction based on customer inputs.

Thus, QFD is a systematic approach of translating customer requirements into appropriate requirements at each stage right from.

Research and development to engineering

Engineering to manufacturing

Manufacturing to distribution and sales

The complete journey of the product is covered right from conception till it reaches the customer.

Features of QFD

Some of the features of QFD methodology are as follows

(*i*) It is a planning process as opposed to a tool for problem solving.

(*ii*) It uses matrix to display the information.

(*iii*) The customer's desires and needs are the inputs to the matrix.

(*iv*) The collection of information in the matrix format facilitates analysis, examination and cross checking.

Steps in QFD Process

The QFD process has three requirements

- Development of customer requirements.
- Development of technical information.
- Analysis of results.

The steps in QFD Process are

1. Determine the voice of the customer.
2. Survey the customers for importance rating and competitive evaluations.
3. Develop customer portion of the matrix.
4. Develop technical portion of the matrix.
5. Analyze the matrix, choose priority items.
6. Compare the proposed design concept, synthesize the best.
7. Develop a part characteristic matrix for priority design requirement.
8. Develop a process-planning matrix for priority process requirement.
9. Develop a manufacturing-planning chart.

The steps involved in QFD are described as follows

- **Voice of the Customer**

The voice of the customer is the input for QFD process. An analysis carried out depends upon this voice of the customer.

- **Survey**

Survey is the best method to determine target market. First one should have to determine which people to survey. Normally the procedure is

— Determine the target market.

— Determine the demographics.

— Determine the geographical condition and distribution.

— Survey people external to the organisation.

— Survey with or without samples of the current product.

How to Obtain the Voice of the Customer

The ways to obtain customer voice is represented in table (S2.2).

1. **Focus Group**

Focus group usually involves 12-15 people. A number of discussion and issues are agreed in advance. A facilitator works with the group to develop conversation on attitude, wants and needs of the participations. The facilitator should take care that activity must not be dominated by the indivisual.

2. **Interviews**

Interviewing customer by telephone or one to one is a good method to obtain customers voice. The one to one contact gives the detailed information, which is very much useful.

3. **Mail Questionnaires**

This can be sent to many people with minimum cost. The response rate of mail questionnaire is around 15% to 30%. Customer has sufficient time to respond to questions and hence the customer ratings can be analyzed with objectivity.

4. **Product Clinics**

Product clinics are excellent ways to develop perspective on how people feel about a specific proposal or concept. Product clinic provides an organisation the opportunity to get customers opinions on a variety of proposed projects.

5. **Personal Observation**

It is the best approach to judge customer's reaction Japanese use the term "Murmur" to describe this simple process of listing and observing.

6. **Root Wants**

All the above ways are to know the needs and wants of customer. The process of questioning and interviewing people to understand their root wants. The root wants are the key to achieve the customer satisfaction.

Table S2.2: Ways to Obtain Customer's Voice

- Focus Group
- Interviews
- Telephone
- One to One
- Mail Questionnaire
- Product Clinics
- Murmurs Observations
- Root Wants

Quality Function Deployment (QFD) — Matrices

QFD matrix is the main tool to bridge the customer requirement and process evaluation. This matrix helps in understanding customer voice and response and transformation of this voice in to technical specification. The cascading of the information is achieved by a series of matrices (called houses) in which the target values or prioritized values developed at each previous house are utilized as inputs for succeeding house. This process is continued until each objective is refined to an actionable level. QFD is known by the first house — House of quality shown in fig (S2.1). It is called a house of quality because of its triangular roof structure. The house of quality is a kind of conceptual matrix that provides the means for interfunctional planning and communication.

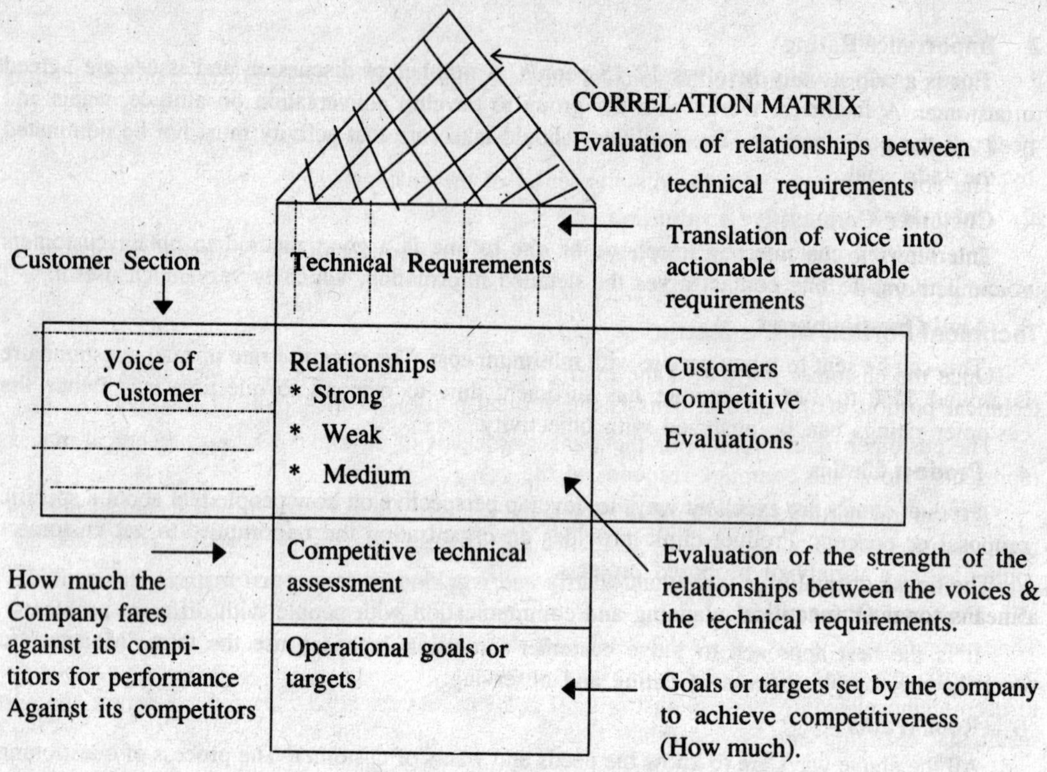

CORRELATION MATRIX
Evaluation of relationships between technical requirements

Customer Section

Technical Requirements

Translation of voice into actionable measurable requirements

Voice of Customer

Relationships
* Strong
* Weak
* Medium

Customers Competitive Evaluations

Competitive technical assessment

Evaluation of the strength of the relationships between the voices & the technical requirements.

How much the Company fares against its compititors for performance Against its competitors

Operational goals or targets

Goals or targets set by the company to achieve competitiveness (How much).

Fig. 2.1. Houses of Quality.

The QFD matrix consists of two portions.

1. Horizontal portion or customer matrix
2. Vertical portion or technical matrix

Customer Matrix

This is the horizontal part in the house of quality. This matrix evaluates the voice and response of customer on particular product. There are several problems in understanding the exact need of the customer. *e.g.*, normally the customers express their opinion in a vague and ambiguous manner, *i.e.*, very pleasant to hear but difficult to understand.

The customer portion of the matrix is shown in fig (2.2).

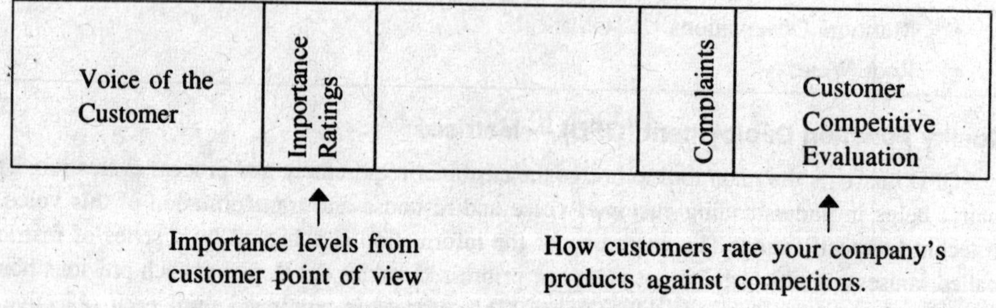

Fig. 2.2. Customer Matrix.

The customer matrix is divided into four sections/compartments.

1. **Voice of the Customer**

It is an input to QFD. This implies actual needs and wants of the customer and it is referred to as the heart of the matrix as further evaluations depend on this matrix.

2. **Importance Rating**

This is a priority setting to the voice of the customer based on the importance of the function to customer.

3. **Complaints**

The complaints represent the dissatisfaction of the customer.

4. **Customer Competitive Evaluation**

This refers to the fact that how customers rates the product of the company in relation to its competitors.

Technical Portion of the Matrix

Once the customer portion of the matrix is created and then the next step is to create the technical portion of the matrix. This is the vertical portion of the house of quality.

The customer matrix represents the wants and needs of customer where as technical matrix represents "How" the company responds to the voice of customer.

The central portion of house of quality is the portion where customer and technical portion interacts and provide an opportunity to record the relationship between input items and actions.

The house of quality is a resultant matrix and is a kind of conceptual matrix. This provides a means for inter functional planning and communication with people with different problems. The triangular roof is the extension of the main building and is known as correlation matrix. In this matrix, the trade off can be achieved comparing and calculating the degree of correlation of each technical requirement against other. Each requirement is examined to determine the net result that changing one has on other.

QFD Approach

The product-planning matrix of house of quality contains the most critical information that one needs regarding relationship to customer and comparative position in the market.

QFD is a 5 step, 4 phase technique to integrate the

- Needs
- Engineering
- R & D
- Manufacturing management

The customer is cascaded into design, process, and production as shown in the fig (S2.3). This is achieved by creating new matrices.

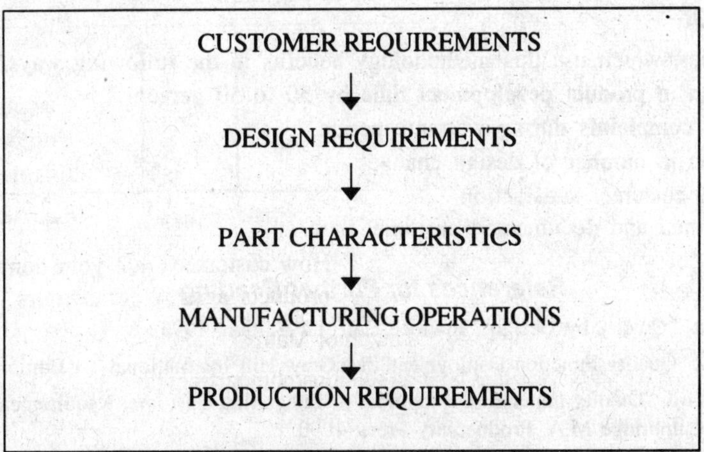

Fig. S2.3. QFD Approach.

In this new matrices "Hows" of previous matrix become the "Whats" for this matrix. The process continues until each objective is redefined to an actionable level. In the product development process this means

1. Taking customer requirement and defining design requirement.
2. Considering design requirement and defining part characteristics.
3. Considering part characteristics, defining manufacturing operations.
4. Considering manufacturing operations, defining production planning.

The phase of QFD and their analysis is represented in the fig (S2.4).

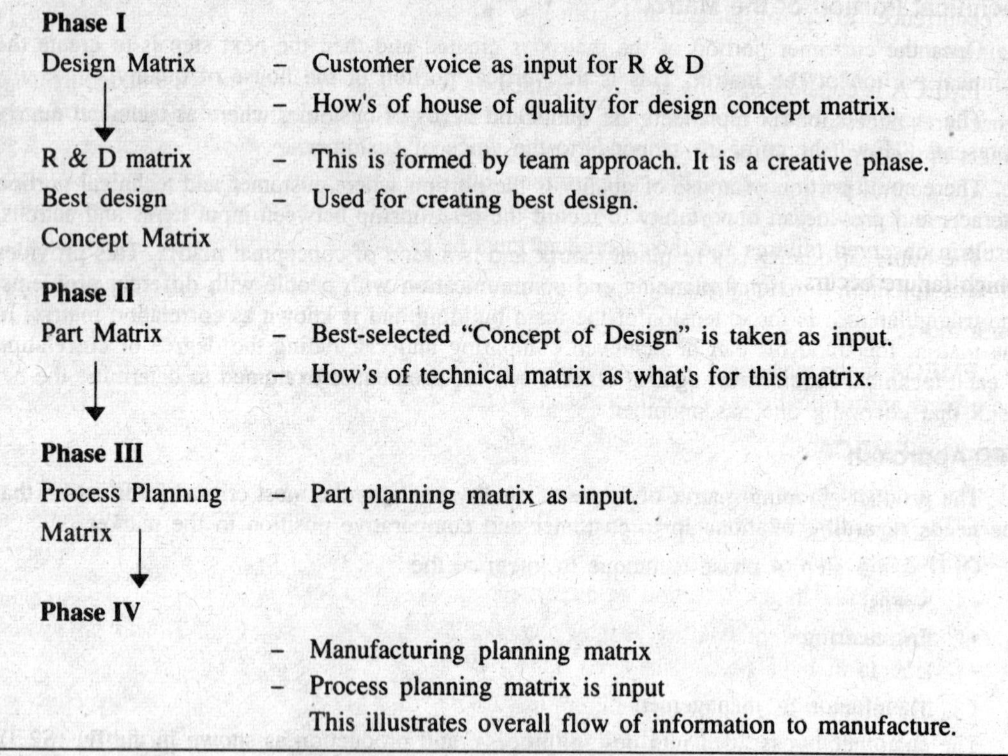

Fig. S2.4. Analysis of Phases of QFD.

Benefits of QFD

The companies, which use this methodology benefits in the following ways.

1. Reduction in product development time by 30 to 50 percent.
2. Reduced complaints during warranty period.
3. Reduction in number of design changes.
4. Increased customer satisfaction.
5. Well-defined and documented customer base.

References for Further Reading

1. **Balvendra,** "Quality by Design", Prentice Hall, 1st Edition, 1995.
2. **Day R.G.,** "Quality Function Deployment", McGraw Hill International 1st Edition, 1995.
3. **Akao & Yoji,** "Quality Function Deployment — Integrating Customer Requirements into Product Design", Cambridge M.A. Productivity Press, 1990.
4. **Choen L.,** "Quality Function Deployment", Addison Wisley, 1995.

Supplement – Three

FAILURE MODE EFFECT AND CRITICALLY ANALYSIS (FMECA)

The Failure Mode, Effect and Critically Analysis (FMECA) is a systematic process whereby each mode of failure of every component or function within a system is assessed for probability of occurrence, effect of failure and criticality in terms of successful operation, safety and maintenance.

FMECA is considered to be an effective method for establishing the formal and documented procedures necessary for manufacturing and to predict, eliminate and reduce the loss producing events in the product. It identifies loss producing events at the product design and process stage i.e. where could be the potential for going wrong with a product during its manufacture or during its end use at the customers end. Most failure events have several causes, which combine together result in observed failures and thus a product must be examined for all possible causes through which failure occurs.

Definition

FMECA is the systematic process to evaluate failure modes, causes associated with failures, the effects and criticality of effect associated with design and manufacturing of new product.

Need for FMECA

For a design to be successful, one must be able to predict the problems and try to prevent them. The need for FMECA can be analyzed in a way, in which it strengthens and supports the design process.

1. Facilitates in the selection of alternatives during the design process.
2. Increases the probability that a potential failure mode and their effects on system operation have been considered during design.
3. Develops a list of potential failure modes ranked according to their probable effect on the customer/criticality. These failure modes can then be minimized or eliminated by design efforts.
4. Helps to detect primary or often minor failures, which may cause serious secondary failures.
5. Helps to detect areas where fail safe/fail safe features are needed.

SCOPE OF FMECA

FMECA is used as a pre-design step to decide the probable life and reliability of the product, which is going to be designed. It helps in deciding the alternatives for designing a product. It helps to decide what might be the probable failure modes, their effects, causes and how to overcome them.

FMECA Process

The steps involved in FMECA are represented in the flow chart in fig (S3.1) and table (S3.1).

The process of FMECA starts with identifying the parts, their areas of failure and then focusing on the function of each element or process and its effects on the total system FMECA provides the possible failure mechanism and permits the assessment of risk level based on likelihood of occurrence, severity and ease of detection. The three terms are then multiplied to provide the "Risk Priority Number (RPN)".

$$\text{RPN} = O \times S \times D$$

where O = Occurrency probability

S = Severity of the process

D = Ease of detection.

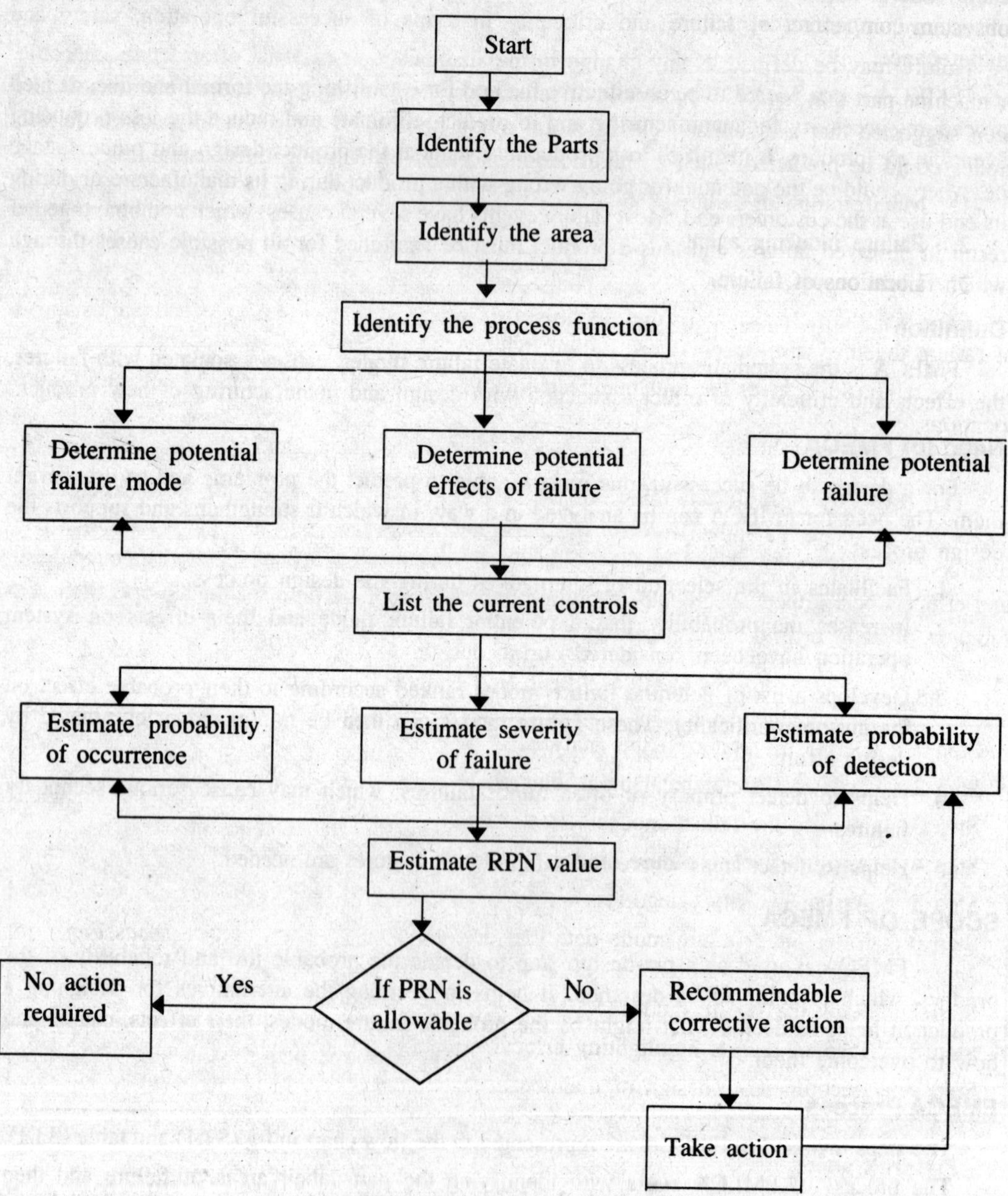

Fig. S3.1. Flow Chart for FMECA.

This is followed by the selection of an appropriate allowable RPN, which is generally done by heuristic approach. If the calculated RPN for the producing event is less than the allowable RPN, then no corrective action is required, otherwise the design or process is attempted again by the same procedure.

ELEMENTS OF FMECA

1. **Failure Mode**

Failure mode define the ways in which failures occur and help to determine causes and effects of the failure that has occurred. Basically, it identifies the manner of failure of the function, subsystem component or part.

Failure may be defined as any change in the size, shape or material of structure, machine or machine part that renders it incapable of satisfactory performing its intended functions. It has been suggested that a systematic classification might be devised by which all possible failure modes could be predicted. Such a classification is based on defining three categories.

1. Manifestation of failure
2. Failure inducing agent
3. Locations of failure.

Each specific failure mode is then identified as a combination of one or more manifestations of failure together with one or more failure inducing agents and a failure location.

The table (S3.2) gives the four manifestations of failure, failure inducing agents and failure locations.

2. **Failure Effect**

This shows the result of any failure on the system i.e. small crack may lead to breakage. The consequences of each failure mode on the items operation should be carefully examined and recorded. The effects can be distinguished at three levels — local, not higher abstraction level and effect. Local effects specifically show the impact of the failure mode on the operation and function of an item under consideration.

Table S3.1: Steps in FMECA Process

Step 1 :	Define the system to be analyzed.
Step 2 :	Construct hierarchical block diagram.
Step 3 :	Identify the failure modes
Step 4 :	Assign effect to failure modes
Step 5 :	Assign severity categories
Step 6 :	Enter other failure mode data like detection, failure rates and compensating provisions.
Step 7 :	Rank the failure modes
Step 8 :	Prepare reports highlighting critical failures
Step 9 :	Recommend redesign or maintenance actions to reduce critical failure.

3. **Failure Causes**

These are the reasons due to which the failure has taken place. In depth analysis of these along with their remedies helps to overcome the mode of failure.

4. **Environment**

The overall design must consider how the environment influences the system during testing, installing and operation. Environment includes conditions, circumstances, stresses, influences, and surroundings affecting the system. Environment also includes the program and people selected to perform the work.

5. **Failure Detection**

Failure detection features for each failure mode should be described. The described symptoms can cover the operations of a component under consideration or can be both the component and the overall system. A detected failure should be corrected so as to eliminate its propagation to the whole system thus maximizing its reliability.

6. **Severity Classification**

Severity ranking for FMECA is an estimate of the serious potential failure on the performance of the system or measure of seriousness of the consequences of the failure.

The severity levels are classified as follows :

1. Catastrophic : A failure mode that may cause death or complete mission.
2. Critical : A failure mode that may cause severe injury or major system degradation, damage or reduction in system performance.
3. Marginal : A failure mode that may cause minor injury or degradation in the system performance.
4. Minor : A failure that does not cause injury or system degradation but may result in system failure and unscheduled maintenance/repair.

7. **Criticality Analysis**

Criticality analysis is the combination of a probabilistic determination of the occurrence of failure modes and determination of the impact of the failure mode on the reliability of the system.

Requirements for FMECA

Following are the requirements for FMECA :

- A team of people with a commitment to improve the ability of the design to meet the customer needs.
- Block diagram of each level of the system from subassembly to the complete system.
- Components specification, part lists and design data.
- Functional specification of modules, sub assemblies.
- FMECA form (paper or electronic) and a list of any specific considerations.

Table S3.2
Failure Manifestations, Failure inducing agents and Locations of failure

- **Manifestations of failure (Category I)**

(*i*) Elastic deformation

(*ii*) Plastic definition

(*iii*) Rupture or Fracture

(*iv*) Material changes — metallurgical, chemical, and nuclear.

- **Failure inducing agents (Category II)**
  - (*i*) Force
    - * Steady * Transient * Cyclic * Random
  - (*ii*) Time
    - * Very short * Short * Long
  - (*iii*) Temperature
    - * Low * Elevated * Steady * Transient * Cyclic * Random
  - (*iv*) Reactive environment
    - * Chemical * Nuclear
- **Failure Locations (Category III)**
  - (*i*) Body type
  - (*ii*) Surface type

Note. To describe a specific mode of failure, it is necessary to select appropriate categories. *e.g.*, plastic deformation from category I steady state and room temperaure from category II. Body type from category III.

Benefits of FMECA

Following are the benefits of FMECA :

1. FMECA is a useful technique for evaluating the safety and reliability of a design and highlights potential problems early in the development cycle.
2. The short-term benefits are more prominent and they represent savings in cost of repair, retest and down time.
3. The long-term benefit is much more difficult to measure since it relates to the customer satisfaction with the products.

References for further Reading

1. **Arvindan Sundarajan, Sivanandam,** "FMEA", National Conference on Intelligence Manufacturing System, Feb., 6,7, 1998.
2. **Jack Collins,** "Failure of Material Mechanical Design", Wiley Eastern Publication Ltd., 1992.
3. **Sushil Kumar Srivastava,** "Industrial Maintenance Management", S. Chand and Co. Ltd., New Delhi, 1998.

Supplement – Four
TOTAL PRODUCTIVE MAINTENANCE (TPM)

INTRODUCTION

The manufacturing of the twenty first century is characterized by high degree of automation and minimum or no human interference. A complete automated process with robots has become a reality. Now, in this completely automated environment, the quality depends upon the equipment as production equipments have become unmanageably sophisticated.

Though the operations/processes have been automated, maintenance still to a greater degree depends on human input. Automated and technologically advanced equipments demand the higher skills and competencies of maintenance worker or supervisor. The effectiveness of these equipments requires an appropriate maintenance organisation. TPM, which organizes all employees from top management to production line workers, is a company wide equipment maintenance system that can support highly sophisticated and technologically advanced equipments and production facilities.

The TPM has got twin objectives of zero breakdowns and zero defects (ZD). When breakdowns and defects are eliminated the equipment utilization will improve, costs are reduced and inventory is at its lowest level.

TPM is carried out on equipments on a company wide basis through small group activities.

Definition of TPM

TPM is defined as "Productive maintenance involving total participation".

The complete definition of TPM includes five elements :

- TPM aims to maximise equipment effectiveness.
- TPM establishes a thorough system of productive maintenance for entire life span of the equipment.
- TPM is to be implemented company wide (Engineering, Operations, Maintenance).
- TPM involves each and every employee.
- It is based on promotion of PM through motivation management.

The word "Total" in total productive maintenance refers to :

1. Total effectiveness indicates the pursuit of economic efficiency or profitability.
2. Total maintenance system includes maintenance prevention (MP) and corrective maintenance CM) and also the preventive maintenance.
3. Total participation of all employees.

To achieve overall equipment effectiveness, TPM works to eliminate the "six big losses" that are the obstacles to equipment effectiveness. The six big losses are represented in the table (S4.1).

Table S4.1: Six Big Losses

• **Down Time**
1. Equipment failure from breakdowns.
2. Set up and adjustments.
• **Speed Losses**
3. Idling and minor stoppages (abnormal operation of sensors, blockages etc.)
4. Reduced speed.
• **Defect**
5. Process defects (due to scrap and defects)
6. Reduced yields.

Stages of TPM

The period prior to 1950 can be referred to as "Breakdown maintenance period and in 1950 preventive maintenance was introduced and productive maintenance during 1960's". The development of TPM began from 1970 onwards.

The development stages of TPM in Japan can be identified as :

Stage I : Breakdown maintenance.
Stage II : Preventive maintenance.
Stage III : Productive maintenance.
Stage IV : TPM.

Equipment Effectiveness

Overall equipment effectiveness = Availability × Performance efficiency × Rate of quality products

$$\text{Availability} = \frac{\text{Operation time}}{\text{Loading time}}$$

$$= \frac{\text{Loading time} - \text{Down time}}{\text{Loading time}}$$

- Loading time or the net available time per day
  = Net available time per day – planned down time per day

(Planned down time refers to the amount of down time officially scheduled in production plan which includes down time for scheduled maintenance and management activities).

Operation time refers to the time during which the equipment actually operates.

Operation time = Loading time – Equipment down time.

- Equipment down time (non operation time) involves equipment stoppage losses, resulting from failure, and set up/adjustments, exchange of dies.

Availability is also called "Rate of operation time".

For example, assume a total loading time of 8 hours per day (480 min).

If the total down time per day = 80 minutes/day.

(Inclusive of setups/adjustment, breakdown etc.)

The operation time/day = 480 – 80 = 400 minutes/day.

$$\therefore \quad \text{Availability} = \frac{400 \text{ min}}{480 \text{ min}} \times 100 = 83.33\%$$

- **Performance Efficiency**

Performance efficiency is the product of the operating speed rate and net operating rate.

Operating speed rate of equipment refers to the difference between the ideal speed (based on equipments design capacity) and its actual operating speed.

Thus, $$\text{Operating speed} = \frac{\text{Theoretical cycle time}}{\text{Actual cycle time}}$$

Net operating rate measures the maintenance of a given speed over a given time.

It measures whether an operation remains stable despite of periods during which the equipment is operated at lower speed.

$$\text{Net operating rate} = \frac{\text{Actual processing time}}{\text{Operation time}}$$

Performance efficiency = Net Operating rate × Operating speed rate

$$= \frac{\text{Amount processed} \times \text{Actual cycle time}}{\text{Operation time}} \times \frac{\text{Ideal cycle time}}{\text{Actual cycle time}}$$

$$= \frac{\text{Processed amount} \times \text{Ideal cycle time}}{\text{Operation time}}$$

For example, if the rate of quality product is 98%, performance efficiency is 50%, and availability is

Then, Overall equipment effectiveness

= Availability × Performance efficiency × Rate of quality products

= 0.83 × 0.5 × 0.98 × 100 = 40.67%.

Though availability is 83.33 %, the overall equipment effectiveness when actually calculated is not even 50% i.e. the equipment is used at only half its effectiveness. The calculations of overall equipment effectiveness is represented in the table (S4.2).

1. Running time per day = 8 hrs × 60 minutes = 480 min.
2. Down time per day
3. Loading time per day (1 – 2)
4. Stoppage losses per day (Breakdown, setup etc.)
5. Operating time per day (3 – 4)
6. Output per day
7. Rate of quality products
8. Ideal cycle time
9. Actual cycle time
10. Actual processing time (6 × 9)
11. Availability (5/3)
12. Operating speed rate (8/9)
13. Performance efficiency (10/5)
14. Performance efficiency (12 × 13)
15. Overall equipment effectiveness (11 × 14 × 7)

Fig. S4.2. Overall Equipment Effectiveness Computation.

The ideal conditions are

Availability > 90 %

Performance efficiency > 95 %

Rate of quality products > 99 %

∴ Ideal overall equipment effectiveness should be

$$0.9 \times 0.95 \times 0.99 = 85 + \%$$

This is not a mere unachievable goal. The TPM prize-winning companies have equipment effectiveness greater than 85%. Thus, a TPM can be used to raise the overall equipment effectiveness upto 85%.

An increase in overall equipment effectiveness produces an increase in productivity.

Stages of TPM Development

Basically there are three stages and twelve steps of TPM development program

First Stage: Preparation Stage

In this stage, a suitable environment is created by establishing a plan for the introduction of TPM programme. This stage is analogous to the product design stage.

Second Stage: Preliminary Implementation Stage

This is analogous to the production stage of the product and this is also referred to as a "Stabilization Stage". Here the goals are to be set and also a specific time frame is mapped in order to boost the morale of the employees.

Third Stage: Implementation Stage

This is the stage of complete stabilization of TPM programme and a company must measure the actual results achieved against TPM targets.

These three stages can be implemented through twelve steps, which are represented in the fig (S4.1). If these twelve steps are implemented with right spirits, the company's overall effectiveness of equipments and hence the productivity and profits will be increased.

Details of Activities

1. **During Preparation Stage**
   - Organizing TPM lectures in a company and publishing articles on TPM in company's newsletters.
   - Educating the employees about the TPM. Seminars/conference as per the levels and presentation of success stories.
   - Formation of special committees at each level to promote TPM. Establish central head quarters and assign staff.
   - Existing situation analysis and goal setting.
   - Prediction of the likely results.
   - Preparation of detailed implementation plan for the preparation stage.

2. **Preliminary Implementation Stage**
   - Bring awareness among clients, affiliated and subcontracting companies.
   - Start battle against the six big equipment losses.
   - Create an atmosphere that increases workers morale and dedication.
   - Status reporting in meeting attended by clients, subcontractors highlighting the plans developed and works accomplished during the preparation phase.

3. **Implementation Stage**
   - Select the model equipments.
   - Form project teams.
   - Build diagnostic skills.
   - Establish worker certification procedure.
   - Include periodic and predictive maintenance, spare parts management, tools, blue prints and schedules.
   - Train the leaders who in turn train the group members.
   - Maintenance prevention (MP) design.
   - Life cycle cost analysis.
   - Evaluate the results and set higher goals.

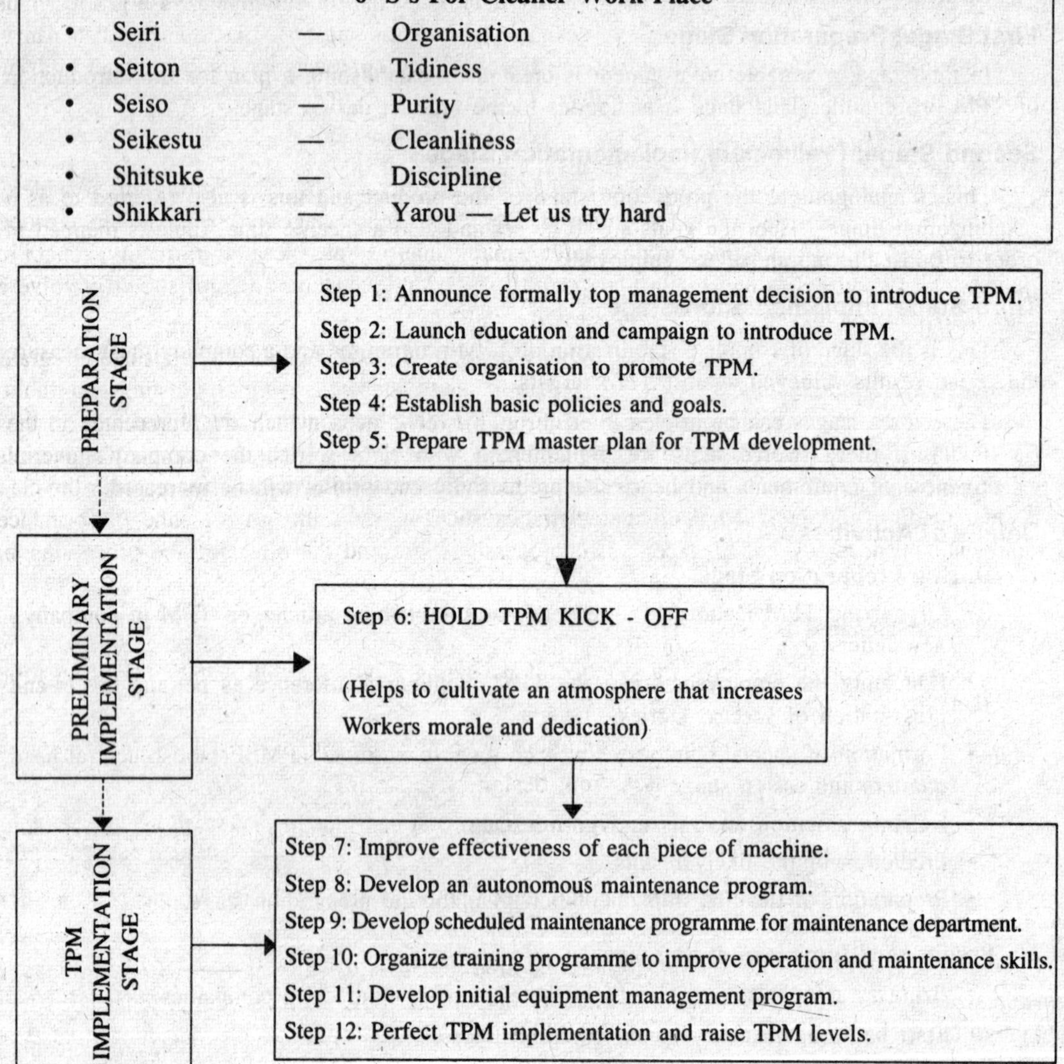

Fig. S4.1. Stages & Steps in TPM.

Supplement – Five

DESIGN FOR MANUFACTURING AND ASSEMBLY

The advancement in the design and production technology calls for interactive production theories like Design for manufacturing and Assembly (DFMA) or generally called "Design for Production". This concept reveals the fact that design department should work in co-ordination with production department resulting in economic production processes with the saving in time, labour and enhancement of quality of product.

Thus, design and manufacturing must be interrelated. Design and manufacturing should never be viewed as separate disciplines or activities. Each part or component of a product must be designed, so that it not only meets design requirements and specifications but also can be manufactured economically and with relative ease. This approach improves productivity and allows manufacturer to remain competitive.

Design for Manufacturing

In the design for manufacturing, all aspects of product design and planning are considered as totality. The issues that are to be addressed early in the design stage (or process) include proper material and process selection, product safety, maintainability produceability, quality, reliability and usability, aesthetics and human factors. All the people concerned (team) should involve at the start of planning stage of product design.

Design for manufacturing is a comprehensive approach to production of goods and integrates design with all down stream processes of manufacturing methods, assembly, testing and quality assurance etc. Effective implementation of design for manufacturing requires that designers have a fundamental understanding of the characteristics, capabilities and limitations of materials, manufacturing process and related operations, machinery and equipment. This includes the clear understanding and knowledge of characteristics such as variability in machine performance, dimensional accuracy and surface finish, processing time and the effect of the processing on quality.

Definition of DFM

Design for manufacturing is a practice of designing a product with manufacturing in mind so that they can be

- Designed in least time with minimum development cost.
- Makes the smooth transition from design phase to manufacturing.
- Can be assembled and tested with minimum cost and time have desired levels of quality and reliability.
- Helps to meet customer needs and hence helps to achieve better competitive position in the market.

Designers and product engineers must be able to assess the impact of design modification on manufacturing process selection, assembly inspection, tools and dies and product cost. It is essential to establish quantitative relationships in order to optimise the design for ease of manufacturing and assembly at minimum cost. Computer aided design; manufacturing and process planning have become indispensable tools for such analysis.

Guidelines for DFM

DFM guidelines are the statements of good design practices that have been developed through experience. These guidelines aid the designer to give design or product a manufacturing approach within the capability of the designer.

The various guidelines are :

1. **Minimise Total Numbers of Parts**

A part that is being eliminated results in great savings. It costs nothing to make, inspect and test. A part is a good candidate for elimination if there is no need for relative motion, no need for subsequent adjustment between parts. Best way to eliminate part is to make minimum part count a functional requirement of design at conceptual stage.

2. **Develop a Modular Design**

A module is self-contained component with a standard interface with other components in the system. Modular design is resistant to obsolescence as can be used elsewhere. It contributes to the simplification of final assembly.

3. **Minimise Part Variations**

The risk of quality problems will be reduced when part variations are kept at minimum. Minimizing part variation also minimizes information content required by the product. The use of standard parts is recommended to reduce parts.

4. **Design Parts to be Multifunctional**

A better way to reduce part count is to design a part that can carry out more than one function. *e.g.,* a part may be designed to provide a guiding, aligning or self-fixing feature in assembly.

5. **Design Parts for Multiple Use**

It is recommended to use the same part in more than one product.

6. **Design parts for ease of fabrication.**

Design Teams

It is challenge for the design engineer to design the product that could be manufactured without any difficulty in production department and there should be minimum changes in design once it is taken to manufacture. To achieve this, a design engineer instead of designing the product in isolation should adopt a collaborate or team approach. Team consisting of design, manufacturing engineer should be formed. They should meet at regular intervals and at critical stages of design to discus various issues regarding the manufacturing aspects of products.

The working of a design team is shown in fig (S5.1).

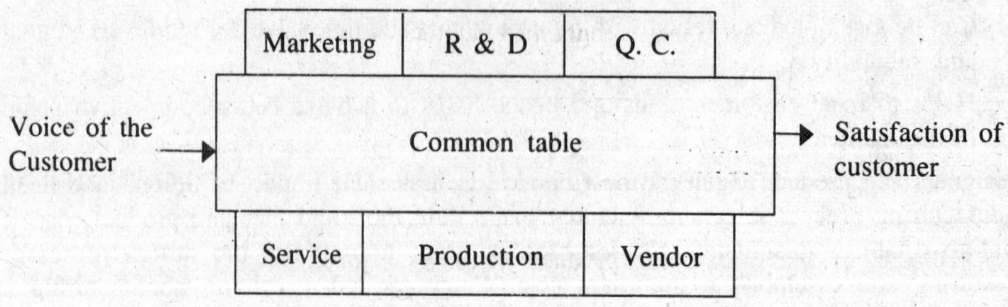

Fig. S5.1. Design Team.

Implementation of the DFM

For making DFM work, designers must know how to actually design products that are manufacturable. To implement the DFM, company should develop a programme. The DFM programme will accomplish a perfect documented details on manufacturability, a uniform practice/policy through out the organisation.

The various stages involved in DFM are represented in the table (S5.1).

For DFM to work, designers must know how to actually design products that are manufacturable. DFM is going to offer trouble free manufacturing and also results in following advantages.

1. Reduction in time to market.
2. Enhanced quality of products with less rejections and rework.
3. Can be assembled and tested with minimum cost and time.
4. Allows for quick and smooth transition in to production.
5. Satisfy customer's needs and compete well in the market.

Table S5.1: Stages in DFM

1. Obtain Management Support The support is assured if the motivation to DFM program comes from the top management.
2. Choose Participants for DPM Programme Participations should be such that they are aware of the role of manufacturing and design in their project.
3. Collect the Sources of DFM Information DFM information should be collected from all departments, *i.e.,* manufacturing, quality, purchasing, vendors and design.
4. Take feedback from production and quality.
5. Take additional information from machinery manufacturer and vendors.
6. Structure the information.
7. Define the level of importance and specify them as higher level or lower level importance.
8. Formulate rules and guidance.
9. Prepare checklists and public standard manuals checklists will be the written documents, which will become the part of formal procedure.
10. Training Training should be imparted to concerned persons to create familiarization amongst people about DFM.

Design for Assembly

The last step in manufacturing process is to assemble individual components into a final product minimizing the cost of assembly is clearly a design function.

The assembly of components into products can be categorized as

1. Manual assembly
2. Mechanically aided manual assembly (use of part feeders)
3. Automatic assembly
4. Automatic assembly with robots

To achieve higher productivity levels, design for ease of assembly must be given the highest priority. When automated assembly is considered one must determine whether the design of the product lends itself to such automation. This is particularly important in high technology areas where batch production is used, *e.g.*, computer hardware.

A study conducted recently revealed that reductions of 20 to 40% in manufacturing costs and increase of 100 to 200% in assembly productivity are readily obtainable through proper consideration of assembly at the design stage. This can be realized with simple techniques for analyzing even rough design and predicting assembly costs. The first step in these techniques is to identify the assembly process that is most likely to be economic for a particular product. Then, the product itself can be designed for that particular process.

Guidelines for Successful Assembly

1. Make available the parts having proper geometry and consistently uniform quality, make sure this at the design stage.
2. In case of automatic assembly, small parts are fed and oriented by vibrating belts, rotary disks, reciprocating arms etc. The parts should have sufficient rigidity to withstand feeding forces.
3. Minimise number of finished surfaces that must be protected from scratching or damage during assembly.
4. Avoid separate fasteners — the use of screws in assembly is expensive snap fits should be used whenever possible. Where ever necessary reduce the quality risk by minimizing number, size and variations of fasteners and by using standard fasteners.
5. Minimize assembly direction — All parts should be designed to enable the assembly from only one direction. The need to rotate consumes extra time and hence should be avoided.
6. Maximise compliance in assembly

   At design stage itself compliance features such as generous tapers, chamfers and radii are to be provided.
7. Minimise handling in assembly

   Parts should be designed to make the required position easy to achieve since the number of positions required in assembly enhances the risk of defects. Orientation can be assisted by design features, which help to guide and locate parts in proper position.

Design for Disassembly

This is an emerging concept and in this, the designer is confronted with the matter of creating products that are easy to take apart (disassemble) so that the product components can be conveniently recycled. This feature helps in ease of maintenance.

The features include :

1. Components can be easily separated, handled and cleaned to permit economic recycling.
2. Provide parts with two way snap fits or have break points on snap fits, which simplifies both assembly and disassembly.
3. Parts or components should be provided with clear identification for separation points, which also eases disassembly.
4. Components should have high tolerance design principles so that parts interlock there by substantially reducing the need for the fasteners.
5. Less number of parts which saves energy, time and money.

Computer Based DFM

To make DFM widely acceptable and adaptable in concurrent engineering requires the development of effective computer based design tools.

Many software tools are available for the design engineer. *e.g.*, a software package called "VSAS" (Variation Simulation Analysis Software) allows the designer to predict the assembly tolerance and manufacturing variation before the prototype is built. The model includes size and variation of each component and accuracy of each assembly component.

A DFM software developed by Boothroyd Dewhorst Inc. meets the needs of design engineer. This was originally based on analysis for assembly and it has been extended to the analysis of injection molded plastic parts. PCB assembly and also includes machining forging and sheet forming.

Some of the features of this software are :

- Evaluation of part complexity
- Calculation of part area and volume
- Estimation of mould cost based on part size, complexity, tolerance and appearance factors.
- Capabilities for instant cost recalculation when changes are made.

Supplement – Six

PROCESS DESIGN AND JOB DESIGN

At the strategic level, the major decisions concerning the production and operations are those concerned with design of physical processes for producing goods and services. These decisions encompass the selection of a process, choice of technology, process flow analysis and layout of the facilities. Once these decisions have been made, the process type, degree of automation, physical layout and job design have been largely determined. Process apart from technical aspects, also involves social, economic and environmental choices. Thus process design is a macroscopic decision making of an overall process route for converting the raw material in to finished product.

Functions of Process Design

The important decisions in process design are :

- To analyse the work flow for converting raw material in to finished product (flow analysis)
- To select the workstation for each included in the workflow.

These two aspects of process design are interrelated and thus needs the simultaneous decisions regarding the two.

Choice of Technology

Technology has become a dominant factor in business and industry. The technological advancement has greater impact on the industry as well as the society. The rate of obsolescence of technology is on the rise and the life of the technology is reducing at a faster pace. The new technologies that are coming will make the old ones less effective and redundant in terms of efficiency and effectiveness.

Technology is referred to as "The application of knowledge to solve human problems, technology in the narrower sense is defined as set of procedures, tools, methods, procedures and equipment used to produce goods/services". Managers should be aware of the knowledge of the performance characteristics of technologies that they manage. These performance characteristics include possible effects of technology on inputs; outputs process flow and costs, which can be evaluated by the managers.

Process Selection

Process selection is a strategic level decision of selecting what kind of production processes to have in the plant. Many of the times process selection is viewed as a layout problem of relatively lower level decision. But on the contrary, process selection is a higher level (strategic) decision as it affects the costs, quality and flexibility of operations. Process once selected, will bind the organisation with equipment, facilities and particular type of labour force, which limits the future strategic actions.

Types of Processes

At the basic level, the various types of processes are categorized as follows.

- Conversion Processes — conversion of raw materials in to the finished products, *i.e.*, conversion of iron ore into steel products.

- Fabrication Processes — Changing raw materials into some specific form, *i.e.*, making of a sheet metal into a storage box.
- Assembly Processes — mainly join parts already processed in to subassemblies or final assemblies.
- Analytic, Synthetic and Modifying Processes

An analytic process breaks down the raw materials into its constituent elements, *e.g.*, refining of crude oil into gasoline, diesel and other products.

A synthetic process combines parts into larger products. *e.g.*, Automobile, refrigerator.

Modifying processes are used in metal working industries. *e.g.*, a casting or forging is machined to get precise dimensions and shapes.

Process Flow Characteristics

A process flow structure refers to how a factory organizes material flow using one of the process types listed.

Basically, there are three types of flows :

- Line flow (Continuous and flow production)
- Intermittent flow (Batch production)
- Project (or job type) flow.

Another critical dimension affecting the choice of a process is whether the product is made to order or made to stock.

The types of manufacturing systems and their characteristics are explained in chapter 8 pp 112-113. Process characteristics are represented in table (S6.1).

Table S6.1: Process Characteristics

Characteristics	*Line flow*	*Intermittent flow*	*Project or job flow*
Product	Large batch or continuous	Batch type	Single or very few
Variety	Low	Medium	Very large
Market	Mass	Customized	Unique
Volume	High	Medium	Single or very low
Labour skill	Low	High	Very high
Job nature	Repetitive	May repeat after some time	Non repetitive
Investment	High	Medium	Low
Plant & machinery	SPM	General purpose	General purpose
Flexibility	Low	Medium	High
Cost of product	Low	Medium	High
Type of layout	Product	Process/Group layout	Functional
Planning & Control	Easy	Difficult	Difficult

Process Selection Decisions

Processes are categorized and selected based upon two characteristics or dimensions — product flow and the type of the customer order. The customer order is generally of two types make to stock and make to order.

Make to stock aims to produce products in advance of market requirement and helps to have a ready stock when demanded. Make to order aims to manufacture products only when the customer demands the products.

The process characteristics matrix is represented as shown in fig (S6.2).

Table S6.2: Process Characteristics Matrix

Flow type ↓	Make to Stock	Make to Order
Line Flow	I • Cement and fertilizer • Tooth paste and soaps	II • Automobile assembly line • Railway coaches
Intermittent	III • Pharmaceutical products • Fastners • Furniture	IV • Machine shop • Hospital
Project	V • Real estate homes • Commercial paintings	VI • Buildings, Dams, Bridges

The following factors will be considered while selecting the process

- Market conditions
- Investment requirement
- Labour
- Management skills
- Raw material
- Technology

A good selection decision requires the detailed analysis of each of the above factors. A market research study should be conducted to assess the potential demand and the other market conditions. The economic analysis of the each alternatives should be considered for decision. The cost structures such as fixed costs, revenues, present worth of cash flows, return on investment, breakdown point etc should be analyzed for alternatives processes to take decision regarding the process selection.

Product Process Strategy

Process selection is not a static decision but actually it is dynamic in nature. The relationship between process structure and product volume are represented on a product-process matrix represented in the fig (6.1).

The product life cycle of a firm is represented on one side of the matrix and the process side of the matrix represents the type of process. Process also follows a process life cycle like products. Firms are positioned along the diagonal of the matrix. The product-process matrix helps to describe the relationship between process and product strategy.

The upper left hand corner represents the job shop environment with general-purpose equipment and jobs made to order. Down the diagonal products are manufactured in batches and higher volume and product line has been standardized still down the line is the line flow process with few product lines, specialized equipments and highly structured a continuous flow type of operation with high volume and a no flexibility situation. Thus, the product-process strategy depends upon the firm's competitive position and the corporate strategy. All the members of the business do not necessary move down the diagonal together. Some manufactures might choose to be low volume intermittent process, which provides greater flexibility and customized products. Other companies might move down the diagonal and stress on standard products and low cost. A major issue in today's manufacturing strategy is to seek benefits of flexibility of job shop structure along with the cost advantages that go with assembly line or continuous flow structures. Thus, the new flexible manufacturing systems (FMS) make possible for a firm to operate over a wider range of product choices for the same process. This computerised technology allows the business to produce standardized products of large volumes and partially customized lower volume products from the same process. Thus, a firm can operate on a horizontal line, which cuts partly across the product-process matrix. But this is at an extra cost and heavy investment.

Process Flow Design

Process flow design focuses on the specific processes that raw materials, parts and sub-assemblies follow as they move through the plant. Process flow design and facility design are the micro level process design decisions. These micro level decisions affect the decisions on the other part of operations including scheduling decisions, inventory levels, the type of jobs designed and the method for quality control used.

The most common production management tools used in planning the process flow are assembly drawings, assembly charts, and route sheets and flow process charts. Each of these charts is a useful diagnostic tool and can be used to improve the operations during the steady state of production system.

The process flow can be viewed as a series of flows connecting inputs to outputs. The system approach helps in a better analysis of process flows, which includes in the system. Customer, outputs, inputs, supplier, boundaries and transformation.

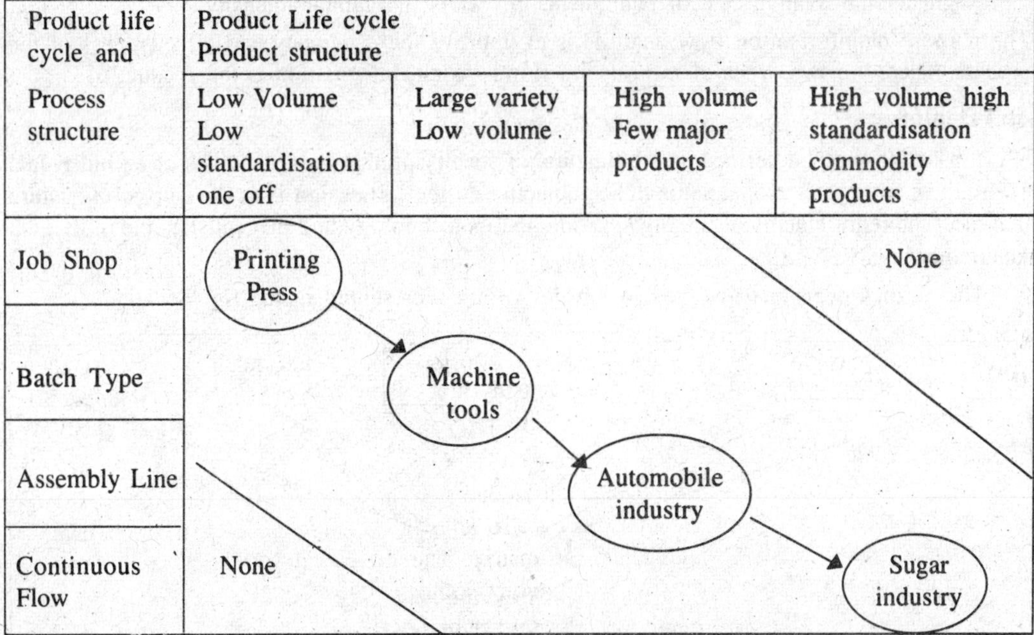

Fig. S6.1. Product-Process Matrix.

• **Material Flow Analysis**

Material flow analysis emphasis on reducing manufacturing throughput time (cycle time), time to order, manufacture and distribute a product from start to finish. This is achieved by reducing the waste. The material flow is described in greater detail for analysis using four types of principal types of documents - assembly drawing, assembly charts, route sheets and flow process chart.

• **Information Flow Analysis**

The purpose of information flow analysis is to improve the efficiency and effectiveness of the process. There are two types of information flows - First, information is the product of operations and second the information flow is used for management or control purposes. Process flow analysis should give due considerations to human elements not only in designing the new system but also in gaining acceptance for the new changes. The best way to accomplish this is to involve persons affected in every stage of diagnosis and design. This tends to increase the individual ownership of new systems and reduce the fear associated with change. Thus, a sociotechnical approach is needed to consider physical flow design simultaneously with the design of jobs. This approach should result in processes, which are both economically, and humanly rewarding.

• **Major Process Decisions**

Basically there are five common process decisions every managers should consider.

- Process choice determines whether resources are organised around products or processes in order to implement the flow strategy. The process choice decision depends on volumes and degree of customisation to be provided.
- Vertical Integration is the degree to which of firm's own production system or service facility handles the entire supply chain.
- Resource flexibility is the ease with which employees and equipment can handle wide variety of products, output levels, duties and functions.
- Customer involvement reflects the ways in which customers become part of the production process and the extent of their participation.
- Capital intensity is the mix of equipment and human skills in the production process; the greater the relative cost of equipment, greater is the capital intensity.

The purpose of information flow analysis is to improve the efficiency and effectiveness of the process. There are two types of information flows - First, information is the product of

Job Design

Job design may be defined as the function of specifying the work activities of an individual or group in an organizational setting. The objective of the job design is to develop job structure that meet the requirements of the organisation and its technology and that satisfies the individual requirements.

The various decisions involved in job design are represented in the fig (S6.2).

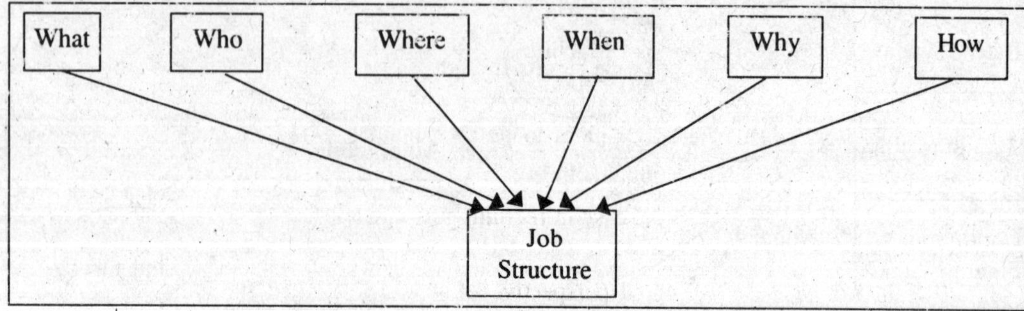

Fig. S6.2. Various Decisions Involved in Job Design.

The following trends are going to influence the job design.

- Workers concern and responsibility for quality — concept of quality at source or self-certification concept.
- Multiskilling of workers.
- Workers involvement in designing and organizing work.
- Extensive use of temporary or contract labours.
- Education of work force and ability to take challenges.
- Automation of heavy and hazardous work.

Various Aspects of Job Design

A. Socio-Technical Approach

This approach attempts to develop jobs that adjust the needs of the production process technology to the needs of the worker and the workgroup. It emphasis on both technical and social variables in relation to job design. This approach helps in designing the jobs that take into account the possible costs of turnover, absenteeism and boredom in relation to technology.

The individual or work group requires a logically integrated pattern of work activities that incorporates the following job design principles.

1. Task Variety

   Try to provide an optimal variety of tasks within each job. Too much variety is frustrating and too little leads to monotony and boredom. Optimal level is one, which allows the employee to take a rest from high level of attention or effort, while working on another jobs.

2. Skill Variety

   Employees derive satisfaction from applying number of skill levels.

3. Feed Back

   A means for informing employees regarding their achievement. Fast feedback aids in learning process.

4. Task Identity

   A particular set of tasks should be separated from other tasks by some clear boundary. A group or individual should have responsibility for some set of tasks that can be clearly defined, visible and meaningful.

5. Task Anatomy

   Employees should be able to exercise some control over their work i.e. they should be involved in decision-making.

B. Degree of Labour Specialization

Specialization of labour on one side results in high speed, low cost production and on other side it has some serious adverse effects on workers as well as on production system. The problem is to determine how much specialization is to be accounted for ?

The advantages of specialization include :

Rapid training of the workforce, ease in recruiting new work force, lower wages due to ease of substitutability and higher outputs.

The main limitations are limited perspective of workers, limited flexibility in workforce, repetitive nature of jobs and boredom.

C. Job Enrichment Approach

Job enrichment generally entails adjusting a specialized job to make it more interesting to the job holder. A job is said to be enlarged horizontally if the worker performs a greater number of variety of tasks and it is said to be enlarged vertically if the worker involves in planning, organizing and inspecting his own work.

Vertical enlargement referred to as job enrichment attempts to broaden workers influence in the transformation process by giving them certain managerial powers over their own activities. Thus, job enrichment is an approach to job design, which stresses the motivating potential in the work itself.

The organizational benefits of job enrichment occur both in quality and productivity. Quality in particular improves drastically because when individuals are responsible for their work output, they take ownership of it and simply do a better job. Productivity improvement also occurs from job enrichment.

D. Physical Consideration in Job Design

- The Work Physiology — incorporates the cost of moderate to heavy work in job design. Work physiology sets work rest cycles according to energy expended in various parts of the job.
- Ergonomics — is the term used to describe the study of physical arrangement of the workspace together with the tools used to perform a task. In ergonomics, attempt is made to fit the work to the body rather than forcing the body to conform to the work.
- Work Method — the principal approach to the study of work methods is the construction of charts such as operation process chart, flow process chart, Man machine charts and multiple activity chart with time study or standard time date. The choice of charting method depends on tasks activity level, *i.e.*, whether focus is on production process, worker at fixed place a worker interacting with equipment and worker interacting with other workers.

Supplement – Seven

LEARNING (OR EXPERIENCE) CURVES

Learning (or experience) curves has wider applications in the business world. Learning curves are integral part in planning corporate strategy such as decisions concerning pricing, capital investments and operating costs based upon the experience.

When a task or job is done more number of times, the time required to produce the job may decrease markedly. This reduction is sizable enough that it should be taken into account in pricing decisions, delivery-scheduling rates and in planning capacity utilization. The reasons for decrease in time (or labour hours) with increase in more number of units are :

1. Skill enhancement of workers for a particular set of tasks
2. Improvement in production methods and tooling
3. Improvement in layout and streamlining of flows
4. Economics of scale, in terms of time are also achieved.

The learning curves can be applied to individuals as well as to organisations. Individual learning refers to the improvement that results when people repeat a process (again and again) and gain skill or efficiency from their own experience. In organizational learning, learning results from practice but also comes from changes in administration, equipment, and product design. Both kinds of learning are expected in the organizational settings. Normally, one expects that the tasks that is being undertaken second time requires less time than the first, because considerable study analysis and thought may be involved in doing the task for the first time. This phenomenon is called a learning curve or some times an improvement curve, a progress curve etc.

Learning curve theory is based on three assumptions :

1. The time required to complete a given task or unit of a product will be less each time the task is repeated.
2. The unit time will decrease at a decreasing rate.
3. The reduction in time will follow a predictable pattern.

A learning curve is a graph or equation that expresses rate of improvement in productivity as more units are produced. The term learning suggests that the reduction in production time occurs because of improved dexterity of workers over time. Also the employees suggestions for improved work methods, designs for new tooling to assist in performing the work, redesign of product to make it easier or other innovative work methods or technological improvements may account for improvement in work. Unit production times may be improved in a few large steps or through number of small steps. Therefore, the progress or learning effect may be somewhat erratic rather than a smooth progression.

For example, the direct labour hours per unit may be reduced by a fairly consistent percentage each time the cumulative number of units are produced, *i.e.*, as output is doubled; there is a 20% reduction in direct labour hours per unit. For instance, with an 80% learning curve, the second unit requires 80% of the direct labour hours required to produce first unit. The fourth unit requires 80% of the time required to produce the second and the hundredth unit requires 80% of the direct labour hours required by 50th unit and so on.

This is represented with an example as follows :

Suppose the time required for one unit output of a machine tool is 1,00,000 hours for first machine, it would take 80,000 hours for machine 2,64,000 hours for machine 4 and so on.

Unit No.	*Unit direct labours hrs*	*Cum direct labour hrs*	*Cum Avg. Direct labour hrs*
1	1,00,000	1,00,000	1,00,000
2	80,000	1,80,000	90,000
4	64,000	3,14,210	78,553
8	51,200	5,34,591	66,824
16	40,960	8,92,014	55,751
⋮	⋮	⋮	⋮

The line connecting the co-ordinates of output and time was referred to as an 80 percent learning curve.

A mathematical expression for learning curve may be given as :

$$Y_n = Kx^n \quad \text{...(1)}$$

where Y_n = Number of direct labour hours required to produce nth unit.

K = Number of direct labor hours required to produce first unit.

x = Number of the unit for which the time is estimated.

n = Logarithm of the ratio of production time for a doubled quantity unit to production time for the base unit divided by 2.

$n = \dfrac{\log b}{\log 2}$ where 'b' is learning percentage.

For example, to find the labor hour requirement for eighth unit, and if K = 1,00,000 hours. Then, substituting the values in the above formula,

$$Y_8 = 1{,}00{,}000\,(8)^{\frac{\log 0.8}{\log 2}}$$

$$= 1{,}00{,}000\,(8)^{-0.322} = \frac{100000}{8^{0.322}}$$

$$= 51{,}192 \text{ hours.}$$

Therefore, it will take 51,192 hours to make the 8th unit.

Equation (1) is an exponential equation. When this type of curve is plotted on arithmetic scale co-ordinates, the rate of reduction in direct labour hours declines as more units are produced. The table (S7.1) gives the hours required for doubled quantities units for 90 percent and an 80 percent learning curve. The fig (S7.1) shows the learning curves plotted on arithmetic scales. A learning curve expressed in equation (1) is straight line when it is plotted on logarithmic scaled paper.

Table S7.1
Direct labour hours required to produce double quantity unit for 90 and 80 percentage learning curve

Cumulative Units	*Direct labour hours*	
	80%	*90%*
1	100.00	100.00
2	80.00	90.00
4	64.00	81.00
8	51.20	72.90
16	40.96	65.61
32	32.77	59.05
64	26.21	53.14
128	20.97	47.83
256	16.78	43.05

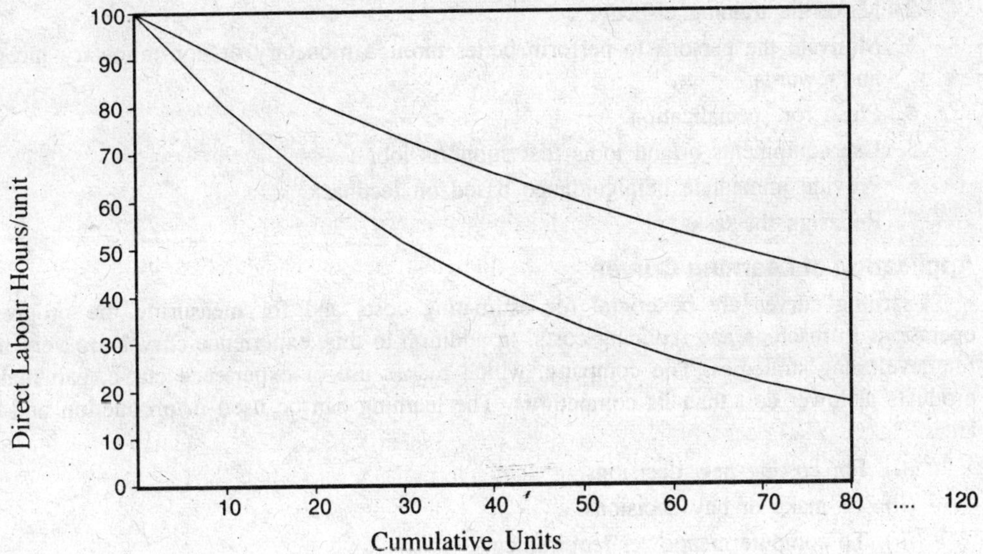

Fig. S7.1. Learning Curve.

Learning Curve Tables

When the learning percentage is known, the learning curves can be easily used to calculate estimated labour hours for a specific unit or for cumulative groups of units. We need only to multiply the initial labour unit value by the appropriate value in the table. The tables are given in appendix.

Estimation of Learning Percentage

The learning percentage can be easily obtained from the production records. Generally, longer the production history, more accurate the estimate. Usually, because of the initial problems during the start of the production, many companies do not begin collecting data for learning curve analysis until the production is stream lined and some units are produced. Extensive use of statistical analysis should be made. An exponential learning curve can be fitted to find out how well the curve fits past data.

If the production is yet to be started (no past information), then there are three options :

1. Assume that the learning percentage, which will be same as it, has been for previous applications within the same industry.
2. Assume that it is same as it for other similar products.
3. Analyze the similarities and differences between the proposed and the present start up and develop a revised percentage to fit the situation.

The basic question is how long does learning continue ? Whether the output stabilizes or there is a continuous improvement ? There are some areas, which show the trends to improvement on continuous basis. On the other hand, highly automated systems may have a near zero learning curve because after installation they quickly reach constant volume.

Factors Influencing Learning

Number of factors influence an individual's performance and rate of learning. Both the initial starting level and rate of learning are important.

The general guidelines to improve individual performance through learning curves are

1. Proper selection and orientation of the workers.
2. Make the training effective.
3. Motivate the persons to perform better through monetary or non-monetary incentives and rewards.
4. Plead for specialization.
5. Use equipments or/and tools that supports job.
6. Provide immediate help/guidance based on feedback.
7. Redesign the tasks.

Application of Learning Curves

Learning curves are beneficial for estimating costs and for measuring the progress of operations in machine and reducing costs. In addition to this, experience curves are very useful for developing strategies. The company, which makes use of experience curve, can make its products at lower cost than its competitors. The learning can be used in production and other areas.

(*i*) For costing new decisions
(*ii*) To make or buy decisions
(*iii*) To compute manpower requirements
(*iv*) To determine delivery schedules
(*v*) For purchase negotiations.

Supplement – Eight

SYSTEMATIC LAYOUT PLANNING (SLP) AND SYSTEMATIC HANDLING ANALYSIS (SHA)

SYSTEMATIC LAYOUT PLANNING (SLP)

Systematic Layout Planning (SLP) is an organized approach developed by Richard Muther (1973) which has got maximum practical applications in determining the best layout plan. SLP Consists of the following four phases

Phase I : Plant location — determining the plant site to be laid out.

Phase II : General overall Layout — establishing the general arrangement of the area to be laid out from the basic flow patterns.

Phase III : Detailed Layout Plans — establishing the detailed actual placement of each specific machine or equipment.

Phase IV : Installation — executing the layout plan.

Among the four phases, general overall layout and detailed layout planning which will be made in sequence in SLP, both according to the SLP procedure. The SLP procedure is represented in fig (S8.1).

Following five key factors are considered in this layout design process.

P — Product [What is to be produced]

Q — Quantity [How much of each item will be produced]

R — Routing [How will each item be produced]

S — Services or supporting activities [What will be support services for production]

T — Time [When will each item be produced].

The SLP procedure is based directly on three fundamentals (heart of any layout planning).

1. Relationships — the relative degree of closeness desired or required among things.
2. Space — the amount, kind and shape or configuration of the things being laid out.
3. Adjustments — the arrangement of things in a realistic best fit.

Steps Involved in SLP Procedure

The steps involved in SLP procedure are described as follows :

1. **Definition of objectives and scope of layout.**
2. **Input Data**

The preliminary planning step of SLP is the five key input factors (P, Q, R, S and T). An analysis of P and Q individually and in their relationship is specially important.

3. **Flow of Materials**

This determines the most effective process or routing by selecting the operations and sequences that will optimally produce P and Q in the specified time period (T) which is [Process planning]. For flow analysis, techniques of process charting for single product and for multi product output are useful. Flow of material analysis also includes the frequency or magnitude of material movement. Flow process charts and travel charts (To-From charts) are employed for flow analysis.

4. **Activity Relationships**

The flow of materials is a common basis for layout arrangements. But, in practical situations, all the support services must be integrated with the layout to ensure productive and effective activities. A systematic way of relating main activities is essential. The activity relationships diagram (REL) is often used for this purpose and indicates the relative importance and closeness between the two activities. REL chart is represented in fig (S8.2).

5. **Flow or Activity Relationship Diagram**

Once the analysis of flow is carried out and activity relationships are established through charting, both are now combined together and diagrammed us flow and/or activity relationship diagram. This diagram relates the various activities or departments,geographically with their relative closeness and intensity to each other without any regard to actual spaces required.

Here, the use of operation symbols for activities, identification numbers or letters for activities and number of lines is made to express the relationship and frequency or closeness value between the two activities.

6. **Space Determination**

The space or area of each piece of machinery, equipment and service facilities required for producing the products is determined. This is processed by first determining the standard area needed for installing each facility, then multiplying this by number of facilities required and finally, adding an extra space which may be required. The total sum of space requirements must be balanced against space available.

7. **Space Relationship Diagram**

The area to be allowed for each activity is represented in the flow/or activity relationship diagram, resulting in the space relationship diagram. This is a rough layout plan, which is generated from REL diagram.

8. **Adjusting the Diagram**

The space relationship diagram represents a theoretically ideal arrangement and as it is cannot be used. It is adjusted and manipulated to include modifying considerations such as, material handling methods, storage facilities, site conditions and surroundings personnel requirements, utilities, auxiliaries, detailed activity layouts etc. and practical limitations like cost, safety and employee preference.

9. **Analysis for Optimization**

When a certain evaluation criterion is decided for layout planning, optimization analysis is made such that the relative location areas for activities involved is optimized or satisfied, *e.g.*, by repeatedly replacing the relative allocation of two or three departments until no further reduction in transportation costs are generated.

10. **Evaluating and Determining the Best Layout**

After analysis, only few layout proposal plans will remain. Each has its own particular advantages and disadvantages. Now, it is required to select the best plan balancing advantages against disadvantages by using factor analysis or cost comparison method. Factor-analysis method is most effective general method of evaluating the alternative layouts. Finally, the procedure is set out for selecting the best alternative, and a scale model or virtual reality model of the selected layout may be used to aid visualization of the selected layout.

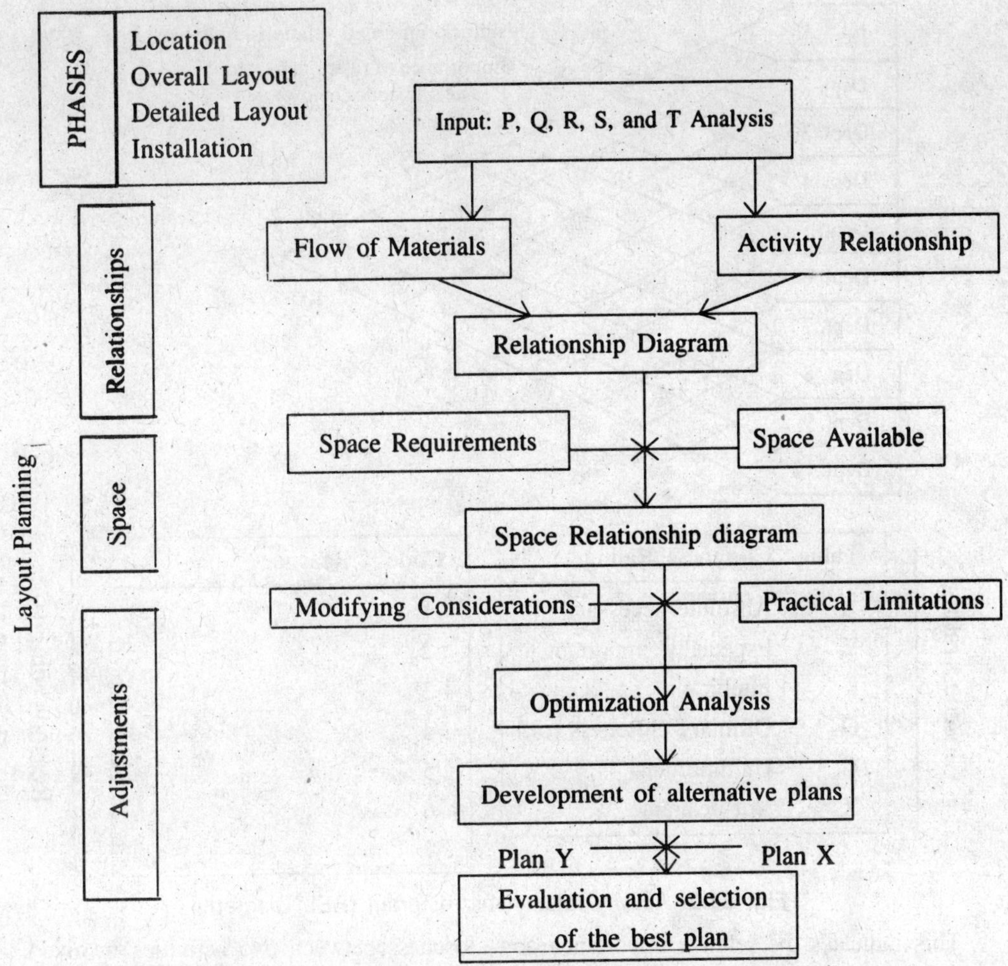

Fig. S8.1. Systematic Layout Planning (SLP) Procedure.

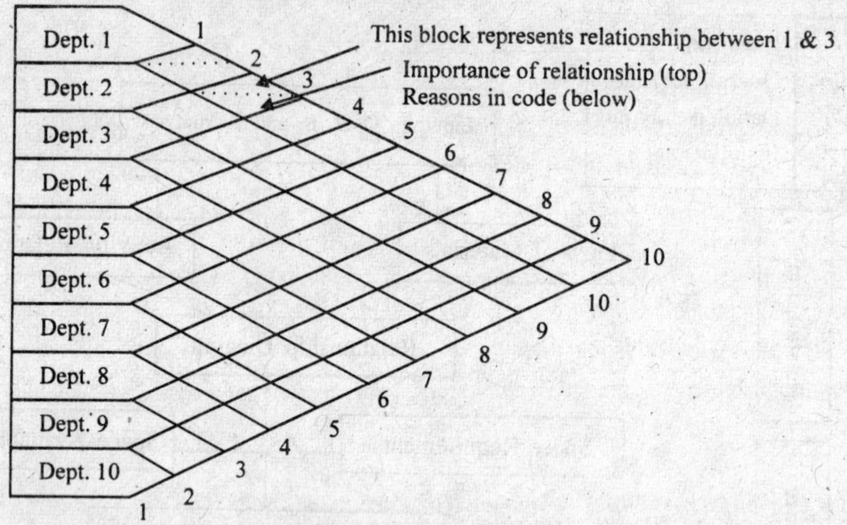

Dept. 1
Dept. 2
Dept. 3
Dept. 4
Dept. 5
Dept. 6
Dept. 7
Dept. 8
Dept. 9
Dept. 10

This block represents relationship between 1 & 3
Importance of relationship (top)
Reasons in code (below)

Closeness rating

Value	Closeness Raling
A	Absolute necessary
E	Especially important
I	Important
O	Ordinary closeness (ok)
U	Unimportant
X	Not desirable

Reasons behind closeness rating

Code	Reason
1	
2	
3	
4	
5	
6	
7	

Fig. S8.2. Activity relationship diagram (REL Diagram)

This indicates the relative importance and closeness between two activities involved.

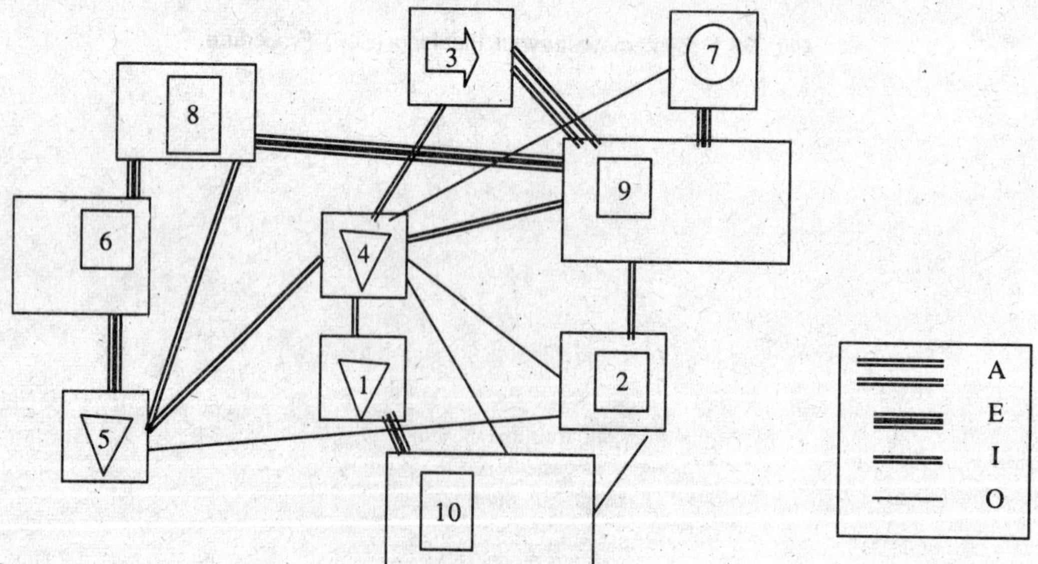

Fig. S8.3. Space relationship Diagram.

It is rough layout plan generated from REL Diagram.

SYSTEMATIC HANDLING ANALYSIS (SHA)

Phases of systematic Handling Analysis :

Phase I : **External Integration**

Examine all movements to and from the area in question preparatory to integration with the overall handling plans.

Phase II : **Overall Handling Plan**

Plan the methods of moving materials between areas make major decisions about system, type of equipment and container.

Phase III : **Detailed Handling Plan**

Plan the methods of moving materials within each area between work places and/or storage points.

Phase IV : **Installation**

Prepare drawings and specifications, obtain and install equipment. Train workers and supervisors.

SHA Procedure

The SHA procedure shown in fig (S8.5) and is flow chart fig (S8.4) is briefly described as follows.

1. **Classification of Materials**

Materials classification has the greatest impact on material handling. The basic classification may be one product (one material) situation or many product (or many material) situations. In many material situations, materials are grouped into material classes, and the similar class materials will be handled in similar way. This simplifies the analysis.

- Basic Classification of Materials

  Materials are classified as solids, liquids and gases. They may also be classified as individual piece, contained item or bulk.
- Materials are also classified on the basis of transportability physical and others. Physical characteristics of items that affect transportability are size, weight, shape risk of damage and condition.
- Quantity — Payload, Fast/slow movers batch.
- Timings — regular/intermittent, seasonal or no rush.

Materials classification can be done using the following procedure

- List all the items and item groups.
- Record all the physical characteristics of the item.
- Analyse the characteristics and identify the dominant characteristics.
- Based upon the similarity of characteristics, combine the material classes.

2. **Layout**

Layout and material handling are inseparable. Layout establishes the distances between the origin and destination. Generally, layout is considered as a constraint. The types of layouts are fixed position, patterns are straight, L-shape, U shape etc.

The information from the layout is :

- Origin — destination.
- The route and known method of material handling on this route.

- Kind of space — Floor load, ceiling height, column space etc.
- Detailed layout of each area.

3. **Analysis of Moves**

In this step, analysis of the moves is carried out based on physical and other characteristics of the materials, the route the distance of move (straight, rectilinear, horizontal and vertical). Also the physical situation of the route like :

- Directness and straightness
- Congestion — obstructions
- Climate and surroundings.

The analysis of flow refers to the following aspects :

- Intensity of flow
- Condition of flow
- Quality condition — batch, continuous/intermittent.
- Service.
- Timings.

The moves can be analyzed using process charts and route chart. The movement summary regarding all routes, all material classes is presented and also for each move the intensity, transportation condition of move and importance of move is to be presented.

5. **Flow Visualization**

Flow diagram and distance intensity charts are used for flow visualization. The flow diagram shows the movement of material over various routes, material class and intensity, the distance and directions of moves.

6. **Understanding MH Method**

The various sources should be referred/consulted in order to gain insight in to the advances in MH methods. The details like material handling equipment - Classification, Cost Movement System, the MH cost system (fixed and variable), unit load and palletizations are analyzed thoroughly and understand.

7. **Preliminary Handling Plans**

Here more than one handling plans are prepared.

8. **Modifications and Limitations**

The account is taken concerning peoples problems, procedural problems. The modifications are carried out based on :

- Production programme — long-range plan, expansion.
- Integration with external handling.
- Space limitation.
- Safety considerations.
- Quality and availability of existing equipments.
- Capital available for investment.

9. **Evaluation of Alternatives**

- Based upon the cost considerations (Fixed costs and operating costs).
- Intangible factors — flexibility, ease of future expansion, space utilization, working conditions etc.

10. **Selection of the Handling Plan**

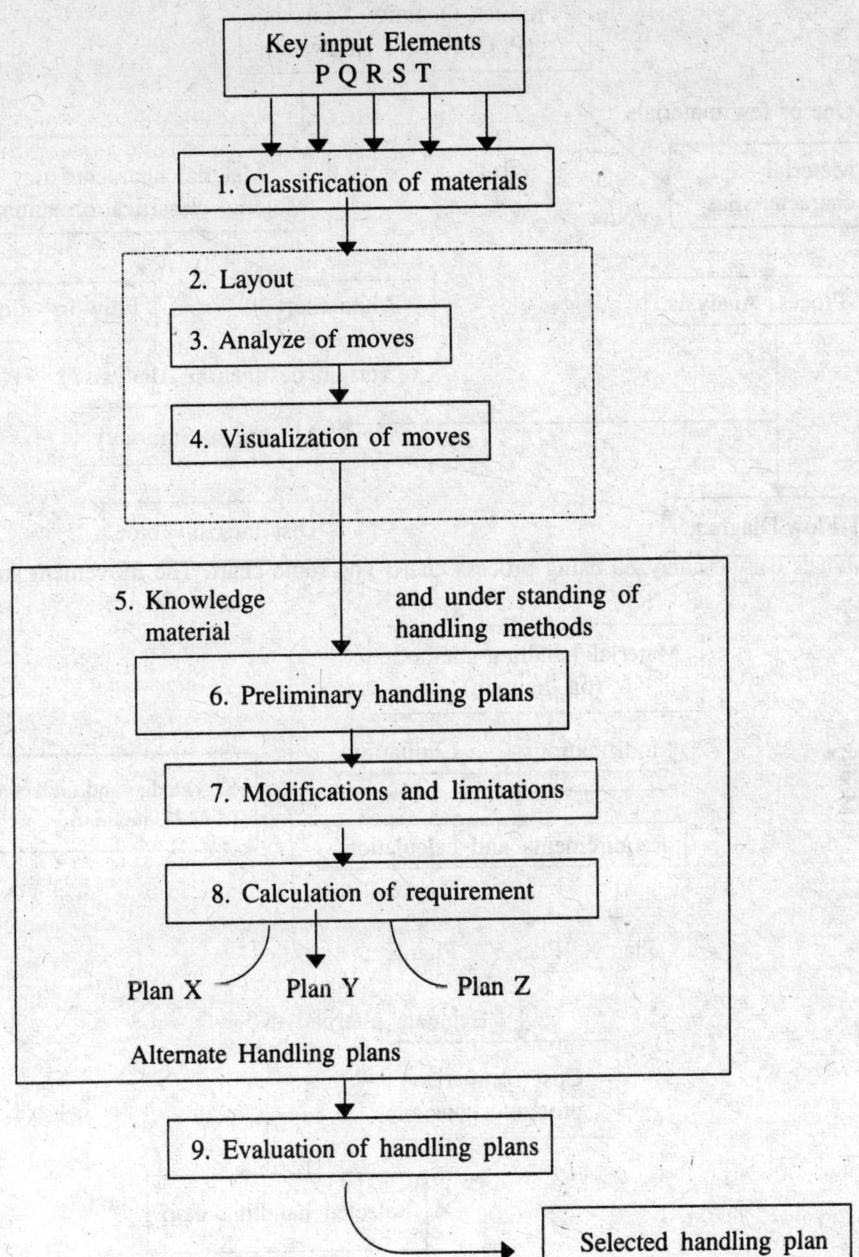

Fig. S8.4. Flow Chart for Systematic Handling Analysis (SHA).

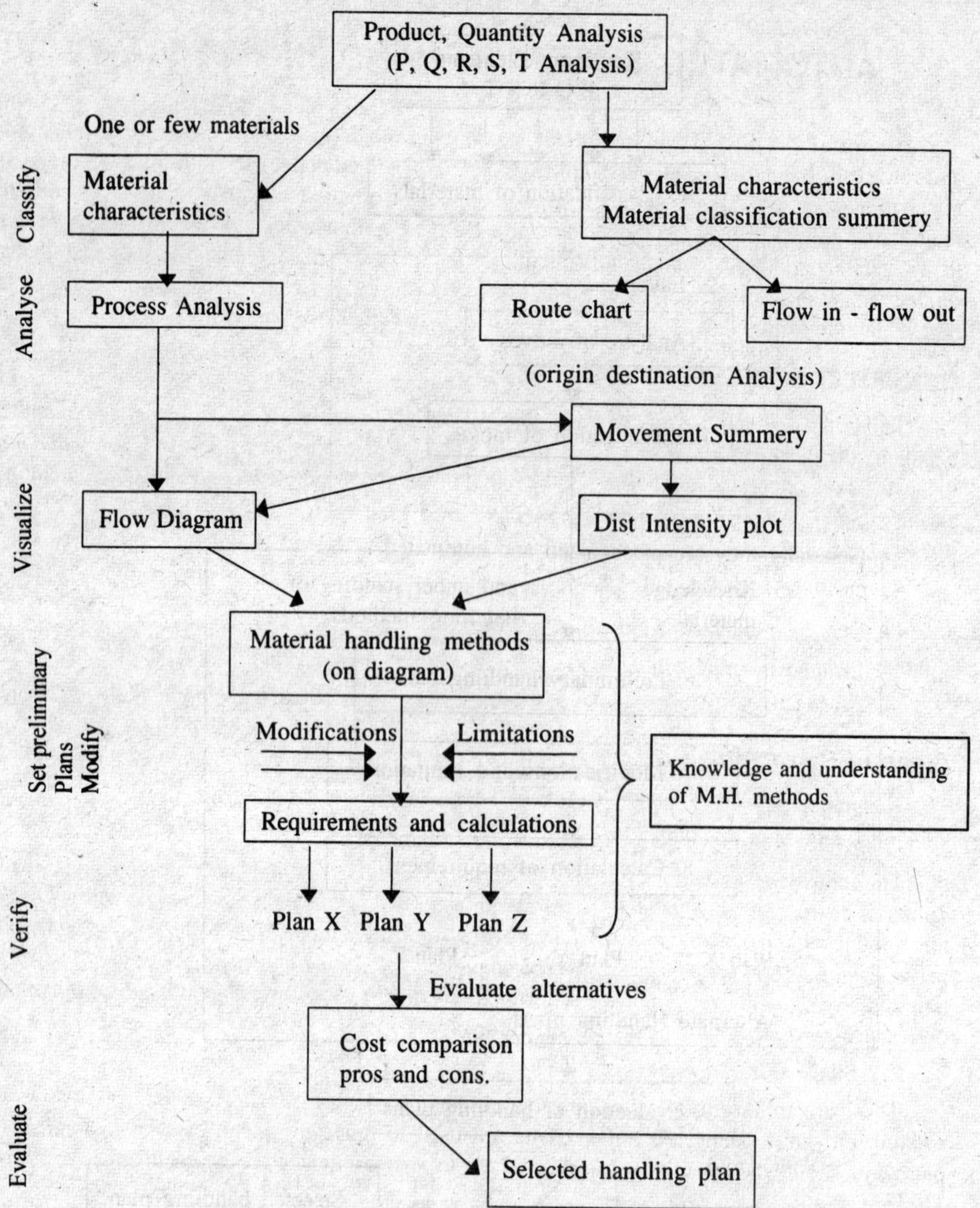

Fig. S8.5. SHA Pattern of Material Handling.

Supplement – Nine

AUTOMATED MATERIAL HANDLING SYSTEMS

Automated Materials Handling (AMH) Systems improve efficiency of transportation, storage and retrieval of materials. The examples include computerised conveyors, Automated storage and retrieval systems (AS/RS) in which computers direct automatic loaders to pick and place items. Automated guided vehicle (AGV) systems use embedded floor wires to direct driverless vehicles to various locations in the plant. Benefits of automated material handling systems include quicker material movement. Lower inventories and storage space, reduced product damage and higher labour productivity.

CONVEYOR SYSTEMS

Conveyor system is used when materials must be moved in relatively large quantities between specific locations over a fixed path. Usually they are powered to move the loads along the path ways and also some times gravity may cause the load to travel from one elevation to another. The characteristics of conveyors are

- Generally they are mechanized and automated.
- They are fixed in position to establish paths.
- May be mounted on the floor or overhead.
- Limited to unidirectional flow of materials.
- Generally use discrete loads but certain types can be used to move bulk loads only.

Types of Conveyors

The major types of conveyors are :

1. **Roller Conveyors**

This conveyor system consists of a series of rollers that are fixed perpendicular to the direction of travel. The rollers are contained in a fixed frame, which elevates the pathway above the floor level. The loads are moved forward as the roller rotate. The roller conveyors may be powered or may use gravity. The powered conveyors are driven by the different mechanisms like belt and chains, gears etc. They are used for delivering loads between manufacturing operations, delivery to and from storage and distribution applications.

2. **Skate Wheel Conveyors**

These are similar in operation to roller conveyors. However, instead of rollers, skate wheels rotating on shafts connected to the frame are used to move the pallet or containers along the pathway. The loads that can be transported are usually be lighter compared to roller conveyors.

3. **Belt Conveyors**

The materials are placed on the belt surface and travel along the moving oath. The belt is supported by a frame that has rollers spaced every few distances (feet). At each end of the conveyor there are driver rolls that power the belt.

4. **Chain Conveyors**

These are made of loops of endless chain in an over and under configuration around powered sprockets at the ends of the pathway. There may be one or more chains operating in parallel to from the conveyor.

5. **Slat Conveyors**

It uses an individual plat form called slats, that are connected to a continuously moving chain. Although its drive mechanism is the powered chain, it operates much like a belt conveyor. Loads are placed on the flat surface of the slates and are transported along with them.

6. **Overhead Trolley Conveyors**

A trolley is a wheeled carriage running on an overhead rail from which loads can be suspended. A trolley conveyor consists of multiple trolleys, usually equally spaced along the rail system by means of an endless chain or cable. The chain or cable is attached to a drive wheel that supplies power to the system. The path is determined by the configuration of the rail system. It has turns and changes in elevation to form an endless loop. Usually the hooks or baskets are suspended from the trolley to carry loads.

AUTOMATED GUIDED VEHICLE (AGV) SYSTEMS

An automated guided vehicle is a robot type vehicle that is used to carry objects from one place to another and can be programmed to travel in predetermined path. Automated guided vehicle systems (AGVS) is a material handling system that uses independently operated, self-propelled vehicles that are guided along defined pathways in the floor. They are usually powered by a battery and the path ways are normally defined by the wires embedded in the floor or reflective paint on the floor surface.

Components of AGV

The essential components of AGV's are :

1. Mechanical structure
2. Actuators for driving and steering mechanism
3. Servo controllers
4. Servo amplifiers
5. The computing facility and power system (on board)
6. Feed back components

Functions of AGV

- **Guidance**

  Allows the vehicle to follow a predetermined route, which is optimized for the material flow pattern of a given application.

- **Routing**

  Ability to make decisions along the guidance path in order to select optimum routes to specific applications.

- **Traffic Handling**

  It is the ability of the vehicle to avoid collisions with other vehicles at the same time maximising vehicle flow and therefore the load management throughout the system.

- **Load Handling and Transfer**

  Pickup and delivery method for AGVs, which may be simple and integrated with other subsystems.

- **System Management**

  Should be able to control the system for efficient system operation.

Types of AGV

Various types of Automated Guided Vehicles are :

(*i*) **Wire Guided Vehicles**

These types use a network of burried cables, which are arranged, in the form of complex closed loops. These are widely used for industrial applications.

(*ii*) **Painted Line Guided Vehicles**

These AGV's follow line on the floors, which have been painted using a fluorescent dye. Used widely in light engineering industries and office automation.

(*iii*) **Free Ranging AGV's**

The path of these vehicles is software programmable and can be altered easily. They provide more flexibility than guided vehicles and task of rescheduling is easy.

The AGV's can be categorized as :

- **Driverless Trains**

  This system consists of the towering vehicle that pulls one or more trailers to from a train. This is most popular type of AGV and is used where heavy payloads must be moved over large distances in factory or warehouses with intermediate pick up and drop off points along the route.

- **AGV Pallet Trucks**

  AGV's are used to move palletized loads along predetermined paths.

- **AGV Unit Load Carriers**

  These are used to move unit loads from one station to another station. They are often equipped for automatic loading and unloading by means of a powered rollers, moving belts, mechanized left platforms etc.

Application of AGV

1. Driverless Train Operation.

   Movement of large quantities of materials over relatively large distances.

2. Distribution Systems

   Unit load carriers and unit trucks are used for storage and distribution systems. Here the working of AGV's interface with other systems like automated storage and retrieval systems.

3. Assembly Line Operations.
4. Application in flexible manufacturing systems.
5. Other general application in office, hospitals etc.

AUTOMATED STORAGE AND RETRIEVAL SYSTEMS (AS/RS)

Automated Storage/Retrieval Systems (AS/RD) is defined by materials handling institute as "A combination of equipments and controls, which handles, stores and retrieves materials with precision, accuracy and speed under a defined degree of automation." The AS/R systems can be customised to meet specific applications and the AS/R systems will range from simple mechanized systems that are manually controlled to very large computerised systems, which are totally integrated with factory and warehouse applications.

Basic Components of an AS/RS

The components of AS/RS are :

1. Storage Structure.
2. Storage/Retrieval (S/R) Machine.
3. Storage Modules.
4. Pick up and Deposit Stations.

The storage structure supports the loads in the AS/RS and it is a strong and rigid fabricated structure where industrial compartments must be designed to contain the stored materials. The storage structure may also be used to support the roof and siding of the building in which AS/RS is installed. It also a supports the aisle hardware required to align the S/R machines with respect to individual storage compartments of AS/RS. This hardware includes the guide rails at the top and bottom of the structure as well as the end stops and also helps for safe operation of the S/R machines.

The S/R machine is used to carry out storage function, delivering loads from the input station in to storage or retrieving loads from storage and delivering them to output station. To perform these functions, the S/R machine should be capable of travelling in vertical and horizontal directions to align the carriage with the storage compartment in the storage structure. The carriage consists of shuttle mechanism to place and pickup loads from storage compartments. To get the desired motions of the S/R machine, the horizontal, vertical and shuttle drive systems are required.

The storage modules are the containers of the stored material. The storage modules may be pallets, steel wire baskets and container storage bins and special drawers. The sizes are standardized so that it permits the storage compartments of AS/RS.

The pickup and deposit stations are used to transfer loads to and from the AS/RS. They are located at the end of aisles for access by the S/R machines and the external handling system that brings loads to AS/RS and takes loads away. Pickup stations and deposit stations may be located at opposite ends of the storage aisle.

TYPES OF AS/RS

The important types of AS/RS are :

1. **Unit Load AS/RS**

   These are designed to handle large unit loads stored on pallets or other standard containers.

2. **Mini Load AS/RS**

   This is used to handle small loads that are contained on bins or drawers within the systems.

3. **Man-on Board AS/RS**

   This system permits individual items to be picked directly at their storage locations.

4. **Automated Item Retrieval System**

   These are designed for retrieval of individual items or small unit loads in a distribution warehouse. The items are stored in single file lanes rather than in bins or drawers.

5. **Deep Lane AS/RS**

   It is a high-density unit load storage system suitable when large quantities are to be stored. It stores number of loads in a single rack, one load behind the next. Each rack is designed for "Flow through" with input on one side and output on the other side.

Applications of AS/RS

Most applications of AS/RS systems are used in ware housing and distribution operations. The application areas include :

(*i*) Unit load storage and handling

(*ii*) Work in process storage system

(*iii*) Order picking

CAROUSEL STORAGE SYSTEMS

Carousel storage is a mechanized system where the load/unload station is manned by a human worker who activates the powered carousel to deliver a desired bin to the station. The carousel storage system consists of series of bins or baskets fixed to carries that are connected together and revolve around a long, oval track system. The track system is similar to a trolley conveyor system.

The controls used for modern carousel systems range from manual controls to the computer control. The manual controls include :

- Foot pedal control
- Hand control
- Key board control

Computer controls are implemented using various computer configurations from microprocessor based controllers for individual's carousels to centralized dedicated computers that control multiple carousels.

The carousel storage applications include — storage and retrieval operations, transport and accumulated and some special applications such as storing of items during electrical testing of components.

Supplement – Ten

FLEXIBLE MANUFACTURING SYSTEMS (FMS)

Flexibility has become a key consideration in the design of the manufacturing systems. Flexibility refers to those properties of manufacturing system that support changes in production capabilities or activities. Flexibility is referred to as the ability of a system to cope with changing circumstances. The flexibility is not a single homogeneous property but it consists of various different components such as machine flexibility, process flexibility, routing flexibility, volume flexibility etc.

A Flexible Manufacturing System (FMS) consists of a group of processing stations (usually CNC machines), interconnected by means of an automated handling and storage system and controlled by an integrated computer system. As this is capable of processing a variety of different types of jobs/parts simultaneously under NC program control at the various work-stations, it is referred to as FMS.

Components of FMS

The components of FMS are :

1. **Processing Stations**

Typically the processing stations are computer numerical control (CNC) machine tools that perform machining operations on families of parts. They are designed with other types of equipments including inspection stations, assembly work heads, and sheet metal processing. Most of the machines in FMS utilize randomly selectable heads to perform functions such as drilling, tapping etc.

2. **Material Handling Devices**

Various types of material handling equipment are used to transport the work parts and subassemblies between the processing stations. For work transportation individual conveys are used. The load and unload stations are used for loading and unloading of raw materials and finished parts.

3. **Auxiliary Equipments**

Apart from the machine tools, FMS can also include cleaning, automated measurement and inspection equipment.

4. **Computer Control System**

Computer control is used to co-ordinate the activities of the processing stations and material handling system. It oversees the operation of an entire FMS.

Some of the main functions of the controller are :

- Machine control (CNC, DNC)
- Production control
- Tools control
- Traffic and shuttle control
- Work handling system control
- Monitoring, controlling and reporting system performance

The structure of FMS is shown in fig (S10.1).

Types of FMS

Flexible manufacturing systems can be broadly grouped in to flexible transfer lines, flexible manufacturing cells and stand alone NC machines. All FMS consist of similar components. The number and types of machine tools may differ. The level of flexibility is the strategic decision in the development and implementation of FMS. Generally, the term cell can be used to refer to a machine grouping that consists of either manually operated or automated machines or combinations of the two. The cell may or may not include automated material handling system and it may or may not be computer controlled. The term flexibility manufacturing system generally means a fully automated system consisting of automated workstations, automated material handling and computer control. Manufacturing cell is used largely in connection with group technology but both the cells and FMS rely on GT approach in their design.

FMS can be described as either dedicated FMS or a random order FMS. A dedicated FMS is used to a much more limited variety of parts configurations hence their is a certain amount of process specialization and the machines can be designed for the specific processes required to make the limited part family random order FMS is appropriate if the part family is large and there are substantial variations in part configurations. It is more flexible than FMS. It is equipped with general-purpose machines to deal with the variations in product and is capable of processing parts in various sequences.

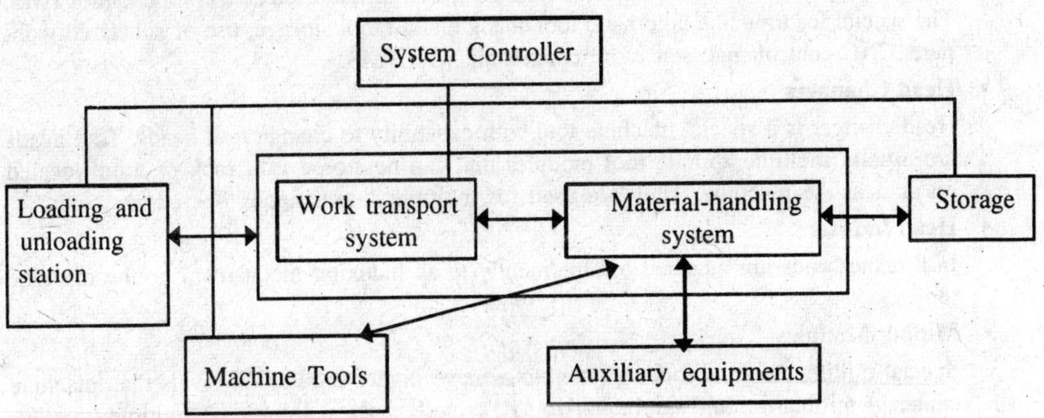

Fig. (S10.1). Structure of FMS.

Suitability of FMS

Flexible manufacturing systems normally fill the gap between high production transfer lines and low production NC machines. The position of FMS is shown in fig (S10.2).

Transfer lines are recommended for high volume output rates but the limitation being the minimum variety or configuration. The stand-alone NC or CNC machines can accommodate changes in part configurations but production rates are substantially lower and parts are usually made in batches.

The flexible manufacturing systems are used for the mid volume range between high volume transfer lines and the low volume stand alone NC-CNC machines. The production rates can be achieved. Various products can be simultaneously processed on the system. The set up time for changeover is minimised and economic batch size almost reduces to one and at the same time the average production rate increases. The advantage of FMS over transfer line is flexibility and FMS can be used for variety of part configurations. Where as transfer lines produce only one or limited variety parts.

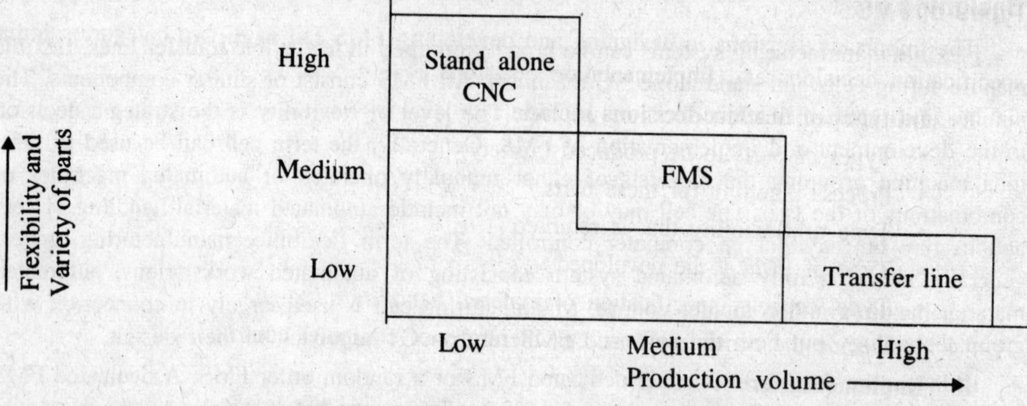

Fig. (S10.2). Applications of FMS Characteristics.

FMS Work Stations

The various types of machines used in FMS are :

- **Machining Centers**

  It is a highly automated stand-alone machine, which can be used as a component of FMS. The special features like automatic tool changing and tool storage, use of palletized work parts CNC control makes it more compatible with FMS.

- **Head Changers**

  Head changer is a special machine tool with capability to change tool heads. Tool heads are usually multiple spindle tool modules that can be stored in a rack or drum located on or near the machine. They are used for multiple simultaneous operations.

- **Head Indexers**

  In this the heads are attached semipermently to an indexing mechanism on the machine tool.

- **Milling Modules**

  Special milling machine modules help to achieve higher production levels than machine center. A milling module can be horizontal, spindle, vertical spindle or multiple spindles.

- **Turning Modules**

  Many of the parts made on FMS are held in a pallet fixture throughout processing on the FMS, thus the turning module must be designed to rotate the single point tool around the work.

- **Assembly Workstations**

  Industrial robots are usually considered to be the most appropriate for automated assembly workstations in FMS. They can be programmed to perform tasks with variation in sequence and motion patterns to accommodate the different products made on the system.

- **Inspection Stations**

  Inspection can be included in FMS either at the workstations or by designing special inspection stations. Co-ordinate measuring machines, special inspection probes that can be used in machine tool spindle and machine vision.

- **Sheet metal Processing Machines**

  Includes processing machines for operations such as punching, shearing, bending, forming etc.

- **Forging Stations**

  FMS are developed to automate the forging.

Designing FMS

The important decisions in designing and developing FMS fall in to two categories Initial specification decisions and Implementation decisions.

I. **Initial specification decisions include :**

- Types of parts to be produced (variety)
- Process planning for these parts
- Type of flexibility that is required
- Type of FMS to be developed
- Type, capacity specification of material handling systems
- Decide the control system, *i.e.*, hierarchy of computer control.

II. **Implementation decisions :**

- Determination of layout of FMS
- Determine the number of pallets required
- Determine the type and number of each type of fixture and design of these fixtures.
- Specification of strategies for FMS.

These decisions are taken after a thorough analysis of the above aspects to achieve maximum flexibility and efficiency.

Advantages and Limitations of FMS

Advantages

1. **Greater Flexibility**

   It provides manufacturing facilities which are flexible for family of work pieces i.e. there will be greater potential to make changes in terms of products, technology etc. as needed.

2. **Reduction in direct labour**

   Reduced (manual) material handling, and automation control of machines make it possible to operate an FMS with less direct labour in many instances.

3. **Reduced Capital Investment**

   Machine utilization with FMS can be about three times as high as with conventional machining. So, fever machines are required. Since there are fewer machines, less toolings are required to equip them. So, less inventory and lower space and hence reduced capital investment.

4. **Shorter Response Time**

   Setup time is relatively low with FMS as majority of the work is done automatically. The lead-time of production is hence very low and the response time will be shorter.

5. **Consistent Quality**

   Human error is minimised, as there is maximum automation. In the absence of human interference, the quality is consistent.

6. Better control of work.

Limitations

(*i*) Some sort of standardization of part design is required to reduce the tooling.

(*ii*) FMS requires long planning and development cycle to ensure the success of the system.

Supplement – Eleven

ROBUST DESIGN

Product design is not only a function of using latest technology or ensuring co-operation across cross-functional lines. Products and processes can be designed which are less affected by environmental variability. Tauguchi has given an improved definition of quality, which is stated as "a product has the ideal quality when it delivers on target performance each time its user uses the product, under all intended operating conditions and throughout its intended life." Tauguchi suggested that a product imparts a loss to society when its performance is not in target. This loss includes any inconvienience and monetary or other loss the customer incurs when using the product. Product quality is determined by the economic losses imposed upon the society from the time a product is released for shipment. For example, a loss caused by functional variation - the deviation of one of the principal functional characteristics of the product from the specified nominal (target) value of the product design specification. Product design may contribute a great deal of loss by including tolerances that may lead to assembly misfits. Similar deviations are also possible in other phases of product life cycle such as market development, maintenance of product, packaging etc.

Two types of undesirable and uncontrollable factors can cause deviations from the desired performance. These are the external and internal noise factors. Operating environmental variables such as temperature, humidity are examples of external noise factors. The problems caused by deterioration such as wearing of parts, manufacturing imperfections like machine setting etc. are internal factors.

Tauguchi suggested a robust design method — which is a systematic method for identifying process setting regions or parameters that are most sensitive to inherent process variation and minimise the effect of causes of variation. The primary goal of robust design is to evaluate these losses and effects and determine — the process conditions that would assure the product manufactured is initially on target and the characteristics of a product, which would make performance insensitive (Robust) to environmental and other factors. The basic concern of quality engineering is to assess and control variations and their effects. Factors causing variability in product or service are usually referred to as noise factors or errors. There are three categories of noise factors :

1. External Noise Factors — includes variables in the environment or conditions of use that disturb the functions of the product such as temperature, dirt, dust, humidity etc.
2. Internal Noise or deterioration noise, which includes the changes that have been taking place during its storage or wear out during use and thus, it, fails to deliver the target function.
3. Unit to unit noise or variational noise indicates the difference between individual products, which are produced to same specifications.

DESIGNING PERFORMANCE IN TO PRODUCT

Tauguchi has recommended a three-stage process for building performance and quality in to the product. The three stages are :

1. **System (Primary or Functional) Design**

This is the first step in design and it makes use of technical knowledge to reach the initial design of the product that delivers the basic, desired functional performance. This includes the design of the system, sub system and finally at the elemental, level design. This is basically a technology issue. This requires research in to concepts, technologies and specialized fields. Many innovations occur at this stage to result in concept design.

2. **Parameter Design**

This step aims at finding the optimum setting of the design parameters. At this stage, to obtain the optimum parameters, a physical or mathematical prototype is built for the product based on the functional design and this prototype is subjected to the efficient and well-designed statistical experiments. This gives the parameter values such that at which performance is optional. Two types of experiments are conducted. The first experimental aims at identifying process parameter values or setting such that the product manufactured by the process performs on target. The second aims at the type of experiments determining. The effect of the uncontrolled, environmental and other design parameters to find out the design parameters such that performance suffers minimum deviation from target. Parameter design identifies the optimum nominal values of the design parameter. If the product is so designed or the parameters are so chosen that its output characteristics are resistant to both external and internal noise, then it will function satisfactorily in spite of variability in its components and its cost will below.

3. **Tolerance Design**

In this step, the tolerances on the product design parameters are determined considering the loss that would be caused to society when the performance deviates from the target. Once the system has been designed along with the values for parameters, the designer has to set the tolerance of the parameters. Here, the environmental factors along with system parameter must be considered.

In tolerance design, the manufacturing tolerances that minimise the effect of noise factors and manufacturing costs are determined. The Tauguchi loss function is used here. The objective in tolerance design is to achieve a judicious trade off between quality loss attributable to performance variation and any increase in manufacturing cost.

TAUGUCHI LOSS FUNCTION

The loss function recognize the customer's desire to have products that are more consistent, part to part, and producer's desire to make a low cost product. Loss to society consist of

(*i*) Costs incurred in the production process.

(*ii*) Costs encountered during use by customer such as repair, lost business.

Dr. Tauguchi defined quality in terms of total loss imparted to the society from the time the product is shipped. Any user of product is likely to suffer at any arbitrarily time during the life span of the product due to performance variation. Other losses may also occur to the society as a result of harmful effects to society like pollution, presence of hydrocarbons etc.

Tauguchi defined the loss function as a quantity proportional to the deviation from the target value of quality characteristics. The loss function is used to evaluate the effect of quality improvement. The relationship between quality loss and performance deviation from target is represented as shown in fig (S11.1).

If y is the performance characteristic measured on a continuous scale when the ideal or target performance level is τ, then according to Tauguchi the loss caused $L(y)$ can be effectively modelled by a quadratic function.

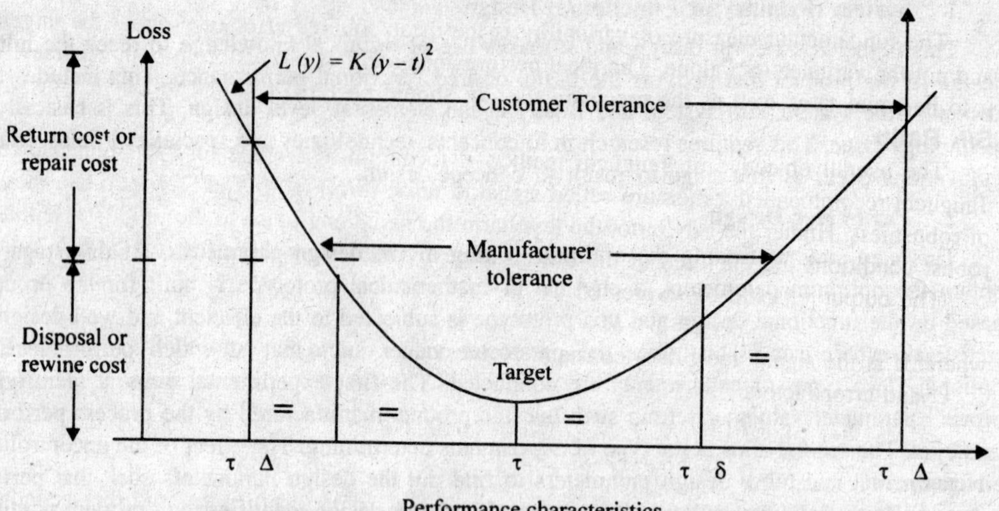

Fig. S11.1. Relationship between quality loss and performance deviation from target.

$$L(y) = K(y - \tau)^2$$

and K here is the proportionality constant and can be determined by

$$K = \frac{\text{Cost of a defective product}}{(\text{Tolerance})^2}$$

The loss function relates quality to a monetary loss. The quadratic loss function provides the necessary information (through signal to noise ratio) to achieve effective quality improvement. Loss function also indicates why it is just sufficient enough for products to be within the specification limits. Parts and components that must fit together to function are best made at their normal (mid point specification) dimensions than merely within their respective tolerance limits.

When the performance varies, the average loss can be determined by statistically averaging the quadratic loss. The average loss is proportional to the mean square of error of y about its target value τ.

Suppose the 'n' units are produced giving the performances $y_1, y_2, ..., y_n$ respectively, then the average loss caused by these units because of their not being on target τ is

$$\frac{1}{n}\left[L(y_1) + L(y_2) + ... + L(y_n)\right]$$

$$= \frac{K}{n}\left[(y_1 - \tau)^2 + (y_2 - \tau)^2 + ... + (y_n - \tau)^2\right]$$

$$= K\left[(\mu - \tau)^2 + \frac{n-1}{n}\sigma^2\right]$$

where $\mu = \dfrac{\Sigma y_i}{n}$ and $\sigma^2 = \dfrac{(y_i - \mu)^2}{(n-1)}$

Thus, the average loss caused by variability has two components

(*i*) The average performance (μ) being different from the target τ, contributes to the loss $K(\mu - \tau)^2$.

(*ii*) Loss $K\sigma^2$ results from performance (y_i) of the individual items being different from their own average μ.

The fundamental measure of variability is the mean squared error of y about the target (τ) and not the variance $(\sigma)^2$ alone. The ideal performance requires perfection in both accuracy (*i.e.*, μ to be equal τ) as well as the precision (*i.e.*, σ^2 should be zero).

S/N Ratio

The overall objective of Tauguchi method is to minimise average loss function. However, Tauguchi recommended a measure called signal to noise (S/N) ratio. The S/N ratio is a measure of robustness. Higher the S/N ratio, the less harm the variations cause to the product. Hence, the robust conditions are explored at the design stage by increasing the effect of the S/N ratio.

The output y is ideally expressed as

$$y = \beta \,.\, M$$

where M is the signal and b is a constant. Noise occurs due to the controllable variables x_1, x_2, ..., x_m and error factors z_1, z_2, ..., z_m

$$y = f\left(M_1, x_1, x_2, \ldots, x_m, z_1, z_2, \ldots, z_m\right)$$

Hence, $$y = \beta \cdot M + \left[f\left(M, x_1, x_2, \ldots, x_m, z_1, z_2, \ldots, z_m\right) - \beta M\right]$$

The first term of the RHS (βM) represents the ideal state (useful part) and the second part, deviation from ideal state (harmful part), meaning the robustness of y. An improved state is specified by increasing the former and decreasing the latter. A measure, which represents this improved state, is the S/N ratio the ratio between the useful part and the harmful part expressed as

$$\eta = \frac{\left(\dfrac{\partial f}{\partial M}\right)^2}{\displaystyle\sum_{i=1}^{n}\left(\dfrac{\partial f}{\partial x_i}\right)^2 \sigma x_i^2}$$

where σx_i is the observed standard deviation for x_i.

If the harmful part is expressed by a single error and its standard deviation is σ,

$$\eta = \frac{\beta^2}{\sigma^2}$$

Thus, the objective of robust design is to seek optimum settings of desired performance to achieve a particular target performance value under the noise conditions. Suppose that in a set of statistical experiments one finds the average quality characteristic to be μ, and the standard deviation caused by noise factors is σ. As delivering on target performance is the goal, one must make an adjustment in the design to get performance on target and also the loss after one has adjusted the process is due to the variability remaining. The loss after adjustment $\dfrac{\text{constant}}{\left(\mu^2 - \sigma^2\right)}$.

The factor (μ^2/σ^2) reflects the ratio of average performance μ^2 (signal) and σ^2 variance in performance (noise). Maximizing μ^2/σ^2 (S/N ratio) therefore becomes equivalent to minimising loss after adjustment. Thus, S/N ratio is a predictor of quality loss that isolates the sensitivity of the products function to noise factors. In robust design, one minimises this sensitivity to noise be seeking combinations of the DP setting that maximise the S/N ratio.

References For Further Reading

1. **Ross P.J. (1996),** "Tauguchi Techniques for Quality Engineering", (2nd Edn.), McGraw Hill, New York.
2. **Tapan Bagchi, (1993),** "Tauguchi Methods Explained Practical Steps to Robust Design", Prentice Hall of India, New Deihi.
3. **Tauguchi G., (1986),** "Introduction to Quality Engineering", Asian Productivity Organisation, Tokyo.

Supplement – Twelve

ENTERPRISE RESOURCE PLANNING (ERP)

INTRODUCTION

Information technology is revolutionizing the way the business is being done. For any organization to succeed, all business units or departments should work in harmony towards a common goal. Enterprise Resource Planning (ERP) is an extremely powerful tool, which provides seamless information system to support the various functional business modules of an enterprise. Most organizations are turning to available ERP package for solution to their information management problems. ERP package if chosen correctly, implemented judiciously and used efficiently will raise the productivity and profit of companies dramatically.

To reap the full benefits of technology such as information technology (IT) one has to devise a system with a holistic view of the enterprise which has to work around core activities of an organization and should facilitate uninterrupted flow of information across departmental barriers such systems can optimally plan and manage all the resources of the organization and hence called as ERP systems.

ERP cover the techniques and concepts employed for the integrated management of business as a whole, from the point of view of the effective use of resources to improve the efficiency of an enterprise. ERP system includes a commercial software package that provides for the seamless integration of all the information flowing through a company- financial, accounting, human resources, supply chain and customer information. The successful deployment of ERP results in an enterprise that has streamlined the data flow between different parts of the business. The main reason for popularity of ERP is the efficiency that an ERP system forces the organization to implement, the analysis and the reporting that can be used for long term planning and the efficient use of applications and system resources.

ERP system employs client / server technology, which means that a client (user) system runs an application (finance, human resources, production etc.) that accesses information from a common ERP database management system which is on the server. ERP operates through a common database at the core of the system and the database interacts with all of the various applications in the system

Need of ERP

ERP covers the techniques and concepts employed for the integrated management of business as a whole with an objective of effective use of management resources to improve the efficiency of the organization. ERP packages are integrated software packages that support the ERP concepts. ERP software is designed to model and automate many of the basic processes of the company from finance to the shop floor with a goal of integrating information across the company and eliminating complex expensive links between computer systems. ERP systems produce the dramatic improvements when used to connect parts of an organization and integrate its various processes. The ultimate benefit goes to the customer who will get better products and better services at affordable prices.

Benefits of ERP

The ERP system extends both direct and indirect benefits to the organization. The direct advantages include improved efficiency, information integration for better decision-making.

Faster response time to customer queries etc. The direct benefits include better corporate image, improved customer good will, customer satisfaction etc.

The benefits of ERP include,

1. Reduction of procurement lead time
2. On time delivery of goods and services
3. Reduction in cycle time from receipt of order to delivery of goods
4. Improved resource utilization
5. Better customer satisfaction
6. Increased flexibility
7. Reduces quality costs
8. Improvement and accuracy of decision-making capability.

Evolution of ERP

1. **Manufacturing — Pre 1960**

   Manufacturing system was based on traditional production planning and control system. The problems, which were faced, include,

   - Working of different functions in isolation
   - Reactive to the situation
   - Long manufacturing lead time
   - Inefficient inventory management
   - Chasing of work order and duplication of efforts.

2. **Manufacturing during 1960's**

   The evolution of CNC machines and computer integrated systems in manufacturing industries in USA as Materials Requirement Planning (MRP). MRP involves determining when to order raw materials and components for assembled products. It can also be used to reschedule orders in response to changing production priority and condition.

   This system is widely used in describing computer-based systems for time phased planning of raw materials, work-in-process and finished goods. It is based on :

   - Master production schedule (MPS)
   - Bill of materials (BOM)
   - Inventory status file

   MRP system computes the order in which they are assembled in to final products.

3. **Manufacturing during 1970s and 1980s**

   During this period, the techniques that help to plan capacity requirement were used along with MRP. Tools were developed to support the planning of aggregate production levels and development of anticipated production schedules. The closed loop MRP to provide the ability to translate the operating plan expressed in manufacturing terms such as units and Kgs into financial terms — rupees. This system is often referred to as "Manufacturing Resource Planning" (MRP-II). It is an expansion of closed loop MRP for managing entire manufacturing. MRP-II system provides information that is useful to all the functional areas. MRP-II supports sales and marketing by providing an order promising capability. Order promising is a method of tying customer's orders to finished goods in the MPS. MRP-II systems increase a company's efficiency by providing a central source of management information.

4. **Manufacturing during 1990s**

   The limitations associated with MRP systems led to the development of a new concept. Enterprise Resource Planning (ERP) by considering MPS as a key to integrate customers and suppliers. The integration of enterprise resource with suppliers and customers as a part of overall supply chain approach to planning and control systems is referred to as ERP.

5. **Manufacturing beyond 2000**

   In the future, manufacturing companies must have glass walls, if they are to survive. They must be able to see in to their supplier's factories, to view the status of their orders on the suppliers plant, equally your customers must be able to see in to your business to understand the status of their orders. It will be a world of total visibility where inefficiency will be visible, unacceptable and will be driven from the system.

   E-manufacturing is a new approach to manufacture and distribution. E-manufacturing is essentially a shrink-wrap manufacturing system infrastructure built on IT standards.

ERP Implementation Strategy

ERP project implementation has to go through different phases. Some of the phases can be done parallel.

The various phases of ERP implementation are :

1. **Pre-evaluation screening**

   The search for the perfect package starts, once the company has to decided to go for ERP. Amongst the large number of vendors and packages, initially it is better to limit the number of packages to less than five by carrying out pre-evaluation screening.

2. **Package evaluation**

   This is one of the important phases of ERP implementation, as the selection of the package will decide the success or failure of the project. None of the available packages are perfect in all respects. The objective here is to select the package that meets requirement of the company.

3. **Project planning phase**

   This phase deals with the design of the implementation process. This step deals With developing the plan and identifying roles and responsibilities to be assigned. The implementation team is selected and task is allocated. This phase will decide when to start the project, how to go about it and when it is going to be completed. Monitoring the project and exercising control at various control points are decided.

4. **Gap analysis**

   This is the critical step, which leads to the success of the project. This is a process through which the companies create a model of where they are now and in which direction it wants to move in the future.

5. **Reengineering**

   This phase is concerned with human resources. Implementation of ERP leads to reduction in number of manpower (downsizing efforts) and the implementation also requires some changes in job content and responsibilities. The BPR approach to an ERP implementation implies that there are really two separate but closely linked implementation involved in an ERP, a technical implementation and a business process implementation. The BPR approach emphasizes the human element of necessary change within organizations.

6. **Configuration**

   This is the main functional area of ERP implementation. During the process mapping companies cannot shut down their operations and hence the simulation of the actual business processes of the company will be used. When ERP consultants configure and test the prototype, they are able to solve any logistical problems before the actual go live configuration.

7. **Training of Implementation team**

   Concurrently along with the configuration, the implementation team is trained as to how to implement it. This is the phase where the company trains its employees to implement and latter run the system. The company to be self sufficient in running the ERP system, the company has to select those employee who have the right attitude and good functional knowledge.

8. **Testing**

   This phase consists of testing the system under extreme scenario's and conditions such as multiple user logging in at the same time, putting the same query by the multiple users, entering the invalid data, hackers trying to access restricted areas and so on.

   The tests are specifically designed to find weak links in these systems and are to be corrected before going live.

9. **Actual operation (going live)**

   All the technical work is almost completed, database is ready and on the functional side the prototype is fully configured and tested and ready to go operational. The system officially becomes ready and the old system is removed.

10. **Training of end users**

    In this phase, the actual users of the system will be imparted on how to use this system. This phase may start much before the system goes live. The employees who are going to use the new system are identified, their current skills are noted and based on the current skill level, and they are divided in to groups. Each group is given the training on the new system. The training is very important as the success of ERP depends on its end users.

11. **Post-implementation**

    This is the maintenance stage, which is very critical. To get the full benefits of ERP system, it is essential that the system should get the company-wide acceptance. The post ERP implementation will need a different set of skills than those whose less integrated system. The training will be an ongoing process, as new people will enter the organization.

ERP modules

All ERP packages contain many modules. The number and features of the modules vary with the ERP package.

Common modules available in almost all packages are :

- Finance
- Data base
- Sales and distribution
- Plant maintenance
- Inventory management
- Quality management

- Materials requirement planning
- Scheduling
- Shop floor control
- Capacity planning

Some packages will have additional modules and features. The product literature will give the detailed information regarding the same.

Integration of ERP modules

To survive and grow in a present competitive global market scenario, it is necessary for the industry to continue delivering consistently high quality products and services to its customers. A flexible cost efficient and responsive environment is therefore necessary to ensure that those challenges are met successfully and the best way to do this is to integrate the entire enterprise through an information backbone, *i.e.*, the ERP system. The flow chart for integrating the modules is shown in the fig. (S12.1)

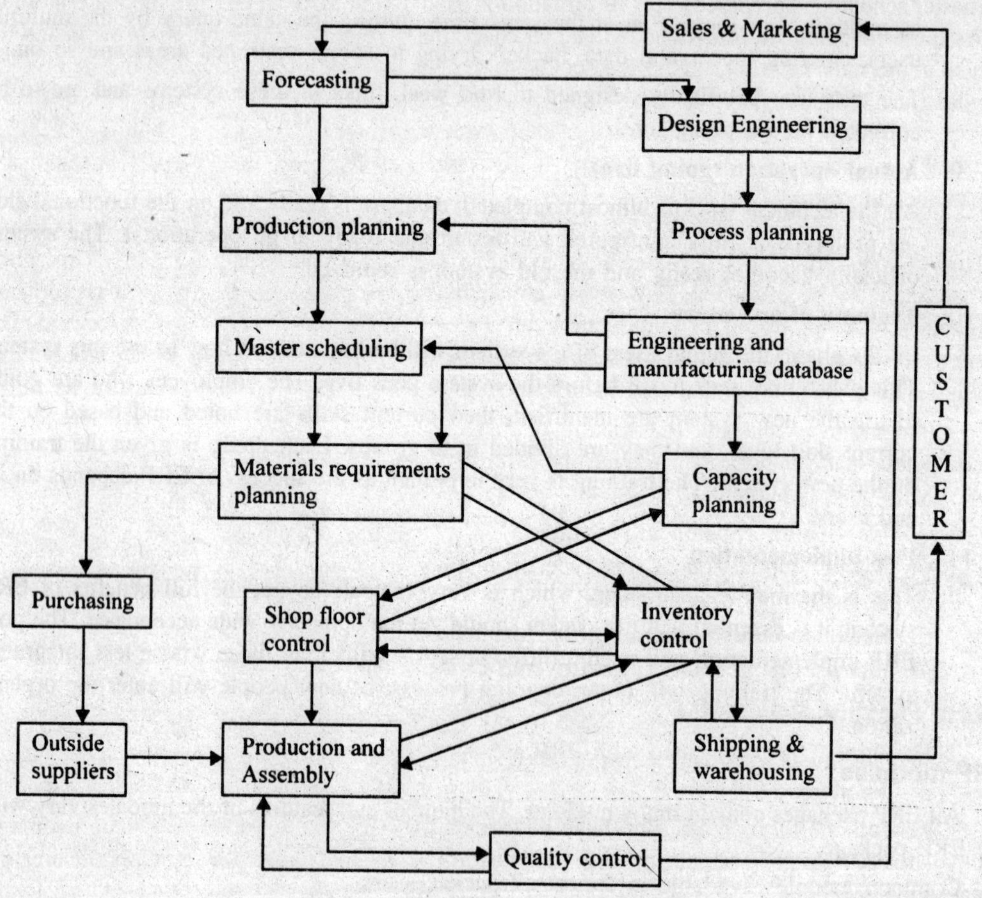

Fig. S12.1. Integration of ERP Modulus.

The starting point of information is customer. The flow of information starts and ends with customer. The product, which is manufactured by the company as per the requirement of the customer and supplied to him. Sales department works to increase the interest of the customer in the product and marketing identifies the new requirements of the customers. Based on the information from marketing, the design department prepares the specifications and drawing of the product and its components. As all the modules are integrated, the same product/component drawing is made available at all the departments and hence the duplication is avoided. The drawing is then analyzed and different operations are to be carried out are planned and are given to manufacturing and engineering database. Here the data available and data supplied are compared and the component is processed. This is integrated with production planning and materials requirement planning. Depending on this data, master schedule is prepared. Raw material required is processed at the right time and if necessary purchased from outside suppliers so that there should not be a shortage of materials.

Of all the modules MRP, shop floor control (SFC), capacity planning and inventory control are more important from manufacturing point of view. Due to integration of these modules the required information is available at right time and at right place. This can be illustrated as follows :

Once the design department decides the product features and components, this is supplied to the engineering and manufacturing database. Taking in to account capacity of the organization master schedule is prepared. This information is used by the MRP to procure the material in time. As these modules are integrated, MRP can check the inventory for the product/raw material and also at the same time can obtain the data from the manufacturing data base and decide how much quantity of material is to be purchased so that the required material can be made available at required time. This is then integrated to shop floor control and SFC is related with order release, order scheduling and order progress. It also takes information from inventory control and provides the information regarding materials required for the production. This order released is then supplied to production and assembly departments, which actually carries out the production. Due to integration, this department is available with information that how much quantity of material is to be produced, how many are to be for this production and what should be the quantity to be produced per production run. Production is governed by quality control, which takes the data from design engineering to update the information, and it always looks for the good quality of product that can be produced so that customer's requirements can be fulfilled.

Integration is thus the interlinking of these departments by computer networking. Use of computers made it possible to make the information flow in real time.

Advantages of integration are :

- Flow of information in real time
- Avoids duplication of any type of data
- Reduction in documentation
- Better co-ordination among the functional areas
- Increase in productivity and overall efficiency of the organization.

ERP Package Selection

The three terms connected with ERP are — vendors, consultants, and users.

1. **Vendors**

Vendors are the peoples who have developed the ERP packages. They are the people who have invested huge amounts of time and efforts in research and development to create the packaged solutions ?

The role of the vendor :

- Supply of product and its documentation
- Training and testing environment for implementation team
- Initial training for the company key users
- Project support function and quality control function of the product supplied
- Customizing the product to suit the actual process requirements of customer.

2. **Consultants**

Consultants are professionals who specialize in developing techniques and methodologies for dealing with implementation and solving various problems that eropup during implementation. They are the project experts. The role of the consultant is very valuable. They required to bring the know how about the packages and about the implementation which is not included in the documentation. The consultant should understand the total context and scope of the envisioned work and to know when to alert the company management about actions and decisions that must be undertaken so that the job will not be compromised and implementation should be appro-priate.

3. **End users**

These are the people who will be using ERP system once it is installed and are doing functions that are being automated or computered by ERP systems. With the implementation of the ERP system, the nature of the jobs will change drastically and the people will have to adapt to new challenging and more responsibilities. The success of ERP lies in how best the company can succeed in making employees accept this fact and assist in making the transformation.

Selection of packages

The evaluation/selection process is one of the most important phases of the ERP implementation, because the package that is selected will decide the success or failure of the project. Select right package first time is the befitting proposition as it is not easy to switch from one package to another easily.

The criteria like key benefits, the process and process steps are used in the pre-evaluation of the packages.

Some of the important points that are kept in mind in evaluating ERP software are:

1. Functional fit with the company's business processes
2. Degree of integration between various components of ERP systems
3. Flexibility and scalability
4. User friendliness
5. Ability to support multi-site planning and control
6. Technology — client capabilities, data base independence and security
7. Amount of customization required
8. Availability of regular up grades
9. Total costs including cost of license, training, implementation, maintenance, customization and hard ware requirements.

Normally, it is recommended to form an evaluation committee that will carry out evaluation process. This committee should comprise of people from various functional areas, the top management and the consultants.

ERP vendors

ERP market is very competitive and fast growing market. It is estimated that the growth potential in enterprise applications and enabling technologies is quite optimistic. Vendors in the ERP market are segmenting in to two tiers and focusing on expanded product functionality, new target markets and higher penetration rates. The top tier consists of five vendors — SAP – AG, Baan, People soft, Oracle applications and J.D. Adwards.

The Indian scenario, according to Data Quest Survey April 1999. SAP is the market leader with 20% market share. According to the survey, ERP does not appear to be new in the Indian market. While SAP R/3 and QAD's MFG/PRO continue to dominate the Indian market scene, and there is also a process of lesser known breeds like J. D.Edwards, and SSAS BPCS. Other familiar stalwarts like Oracle Financial, Ramco Marshal and Baan also dominate the second and third rungs of the domestic ERP market.

ADDITIONAL SOLVED PROBLEMS

PRODUCTIVITY AND WORKSTUDY

Problem 1

The following information is available for a factory :

Daily working hours	8
No. of working days in a week	6
No. of operators	20
Std. Hours per unit of production	4

During a particular week

Number of units produced	48
Absentee man days	40
Idle time due to load shedding	30 mandays

Find : (*i*) *Absenteeism percentage*

(*ii*) *Labour utilisation percentage*

(*iii*) *Productive efficiency of labour*

(*iv*) *Overall productivity of labour in terms of units produced/week/employee.*

Solution.

(*i*) Absenteeism percentage

$$= \frac{\text{Absentee mandays / week}}{\text{Total operator mandays / weeks}}$$

$$= \frac{40}{20 \times 6} \times 100 = 33.33\%$$

(*ii*) Labour utilisation percentage

$$= \frac{\text{Working mandays / week}}{\text{Attended mandays / weeks}} \times 100$$

$$= \frac{\text{Total operator mandays / week} - \text{Absenteeism} - \text{Idle time}}{\text{Attended mandays}} \times 100$$

$$= \frac{(20 \times 6) - 40 - 30}{(20 \times 6) - 40} \times 100$$

$$= 62.50\%$$

(*iii*) Productive efficiency of labour

$$= \frac{\text{Std. man hours produced / week}}{\text{Actual man hours worked / week}} \times 100$$

$$= \frac{4 \times 48}{[(20 \times 6) - 40 - 30] \times 8} \times 100$$

$$= 48\%$$

(*iv*) Overall productivity of labour

$$= \frac{\text{Total production / week}}{\text{Total operators}}$$

$$= \frac{48}{20} = 2.4 \text{ units / week / operator}$$

Problem 2

A factory can manufacture two products A and B by using either of two materials P or Q. A is expected to sell at Rs. 70 per unit and product B at Rs. 30 per unit. The operating data are as follows :

		Material P	*Material Q*
Output	*A*	*200 units*	*400 units*
	B	*300 units*	*200 units*
Quantity of raw material usage		*1,000 kg*	*1,000 Kg*
Labour usage		*300 man hrs.*	*250 man hrs.*
Electric energy consumption		*1000 KWhr*	*1500 KWhr*
Cost of raw material/kg		*Rs. 20*	*Rs. 30*
Labour per manhour		*Rs. 5*	*Rs. 5*
Electrical energy/KWhr		*Rs. 1.5*	*Rs. 1.5*

Compare the productivity of material, labour and electrical energy in using materials P and Q. Comment on the relative advantage of using either of the materials.

Solution.

$$\text{Productivity} = \frac{\text{Value of output}}{\text{Value of input}}$$

Sales value of output with material P

$= \text{Output of product } A \text{ in units} \times \text{Rate/unit of } A + \text{output of product } B \times \text{Rate/unit of } B$

$= 200 \times 70 + 300 \times 30 = \text{Rs. } 23{,}000$

Sales value of output with material Q

$= 400 \times 70 + 200 \times 30 = \text{Rs. } 34{,}000$

The partial productivity of different factors of production are computed as follows :

	Material P	*Material Q*
1. Productivity of raw materials $= \dfrac{\text{sales value of output}}{\text{value of raw material used}}$	$\dfrac{23000}{1000 \times 20} = 1.150$	$\dfrac{34000}{30000} = 1.133$
2. Labour productivity, *i.e.*, $= \dfrac{\text{sales value of output}}{\text{value of labour}}$	$\dfrac{23000}{300 \times 5} = 15.333$	$\dfrac{34000}{250 \times 5} = 27.200$
3. Electrical energy productivity, *i.e.*, $= \dfrac{\text{sales value of output}}{\text{value of electrical energy}}$	$\dfrac{23000}{1000 \times 1.5} = 15.333$	$\dfrac{34000}{1500 \times 1.5} = 15.111$

Comments

The productivities of (1) and (3) is nearly same by using either material P or Q. If labour is the key factor it is better to use material Q as labour productivity, for Q is higher, *i.e.*, 27.200 > 15.333.

Problem 3

The following data is available for a machine in a manufacturing unit.

Number of hours worked per day	8
Working days per month	25
Number of operators	1

Standard time per unit of production,

Machine time	22 *min*
Operator time	08 *min*
Total time/unit	30 *min*

(*i*) *If plant is operated at 75% efficiency, and the operator is working at 100% efficiency, what is the output per month ?*

(*ii*) *If the machine productivity is increased by 10% over the existing level, what will be the output per month.*

(*iii*) *If the operator efficiency is reduced by 20% over the existing level, what will be the output per month ?*

Solution.

(*i*) Plant is operated at 75% efficiency

$$\therefore \quad \text{Machine time} = 22 \times \frac{100}{75} = 22 \times \frac{4}{3} \text{ min}$$

$$\text{Operator time} = 8 \text{ min}$$

$$\text{Total time/unit} = \frac{22 \times 4}{3} + 8 = \frac{112}{3} \text{ min}$$

Total time available = 8 × 60 × 25 min

$$\therefore \quad \text{Number of units produced per month} = \frac{8 \times 60 \times 25}{(112/3)} = 321.43 \approx 322 \text{ units.}$$

(*ii*) If the machine productivity is increased by 10%, *i.e.*, plant efficiency = 75 + 10 = 85%.

$$\therefore \quad \text{Machine time} = \frac{22}{0.85} = 25.88 \text{ min}$$

Operator time = 8 min (assuming 100% efficiency)

Total time/unit = 33.88 min.

$$\therefore \quad \text{Number of units produced/month} = \frac{8 \times 60 \times 25}{33.88} = 354.19 \approx 355 \text{ units.}$$

(*iii*) If the operator efficiency is reduced by 20%, plant efficiency is 75%.

$$\therefore \quad \text{Machine time} = \frac{22}{0.75} = 29.33 \text{ min}$$

$$\text{Operator time} = \frac{8}{0.8} = 10 \text{ min/unit} \quad \text{(operator efficiency 80\%)}$$

Total time/unit = 29.33 + 10 = 39.33 min

$$\therefore \quad \text{Number of units produced/month} = \frac{8 \times 60 \times 25}{39.33} = 305.38 \approx 306 \text{ units.}$$

Problem 4

Continuous stop watch study observations for a job are given. Compute the standard time for the job, if the total allowances are 15%.

Ele. No.	Description	Cycle time (min)										P.R.
		1	2	3	4	5	6	7	8	9	10	
A	Loosen vice	0.09	0.49	0.89	1.31	1.70	2.09	2.50	2.88	3.29	3.71	90
B	Set bar length	0.16	0.56	1.38	1.38	1.76	2.16	2.57	2.95	3.36	3.78	110
C	Switch m/c	0.28	0.67	1.49	1.49	1.88	2.28	2.68	3.07	3.40	3.90	120
D	Unlock arm & set saw	0.41	0.80	1.61	1.61	2.00	2.41	2.80	3.20	3.62	4.03	100

Solution.

The individual elemental cycle timing are computed from the cumulative cycle times as shown in table below :

Ele. No.	Cycle time (min)										Avg. time	Normal time
	1	2	3	4	5	6	7	8	9	10		
A	0.09	0.08	0.09	0.10	0.09	0.09	0.09	0.08	0.09	0.09	0.089	0.080
B	0.07	0.07	0.06	0.07	0.06	0.07	0.07	0.07	0.07	0.07	0.068	0.075
C	0.12	0.11	0.12	0.11	0.12	0.12	0.11	0.12	0.13	0.12	0.118	0.142
D	0.13	0.13	0.14	0.12	0.12	0.13	0.12	0.13	0.13	0.13	0.128	0.128
											Total	0.425

$$\text{Standard time} = \frac{0.425}{(1-0.15)} = 0.500 \text{ minutes.}$$

Problem 5

The work study engineer carries out the work sampling study. The following observations were made for a machine shop.

The duration of study	*120 hrs*
Total number of observations	*7000*
No. Working activities	*1200*
Ratio between manual to machine elements	*2 : 1*
Average rating factor	*120 %*
Total number of jobs produced during study	*800 units*
Rest and personal allowances	*17 %*

Compute the standard time for the job.

Solution.

(*i*) Overall time per unit (To) $= \dfrac{\text{Duration of study}}{\text{Number of jobs produced during study}}$

$$= \frac{120 \times 60}{800}$$

$= 9$ min.

(*ii*) Effective time per piece (Te) $= \text{To} \times \dfrac{\text{Productive observations}}{\text{Total observations}}$

$$= 9 \times \frac{5800}{7000} = 7.46 \text{ min}$$

The effective time is to be segregated into manual time and machine element time.

Machine controlled time per piece (Tm) $= 7.46 \times \dfrac{1}{3} = 2.49$ min

Hand controlled time per piece (Th) $= 7.46 \times \dfrac{2}{3} = 4.97$ min

$\therefore$ Normal time per piece = Tm + Th × performance rating

= 2.49 + 4.97 × 1.2

= 4.86 min.

Standard time per piece = 8.46 (1 + 0.17) = 9.9 minutes.

PLANT LOCATION AND LAYOUT

Problem 6

An automobile air conditioner manufacturer currently manufactures its RB-300 line at three different places (locations), Plant A, Plant B and Plant C respectively. Recently, management has decided to build all compressors, a major component in a separate dedicated facility Plant D. Using center of gravity method and the information given determine the best location for plant D.

Assume linear relationship between volumes shipped and shipping costs. Quantity of compressors required by each plant.

Plant	*Compressors required/year*
A	*6,000*
B	*8,200*
C	*7,000*

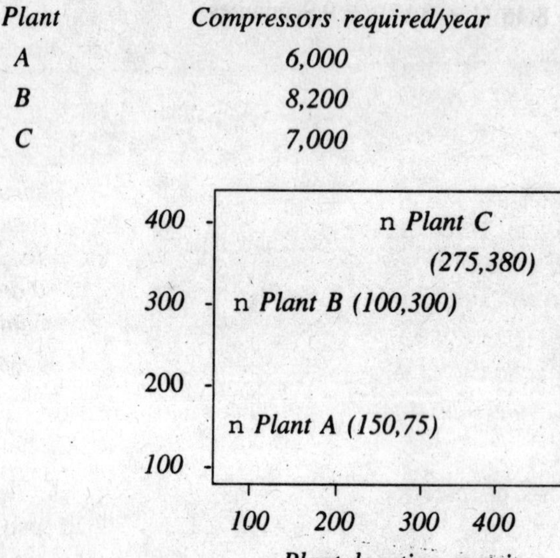

Plant location matrix

Solution.

The center of gravity method is a technique for locating facilities that considers the existing facilities, the distances between them and the volumes of goods to be shipped. This method assumes the inbound and outbound transportation. This method begins by placing the existing locations on a co-ordinate grid system. The choice of co-ordinate systems in entirely arbitrary. The purpose is to establish relative distances between locations, using longitude and latitude of a grid layout.

The center of gravity is found by calculating the x and y co-ordinates that result in the minimal transportation cost. The formula to calculate x and y co-ordinate.

$$C_x = \frac{\Sigma d_{ix} V_i}{\Sigma V_i}$$

$$C_y = \frac{\Sigma d_{iy} V_i}{\Sigma V_i}$$

where C_x = x co-ordinate of the center of gravity

C_y = y co-ordinate of the center of gravity

d_{ix} = x co-ordinate of the ith location

d_{iy} = y co-ordinate of the ith location

V_i = Volume of goods moved to or from the ith location.

$$d_{1x} = 150, \quad d_{1y} = 75, \quad V_1 = 6{,}000$$

$$d_{2x} = 100, \quad d_{2y} = 300, \quad V_2 = 8{,}200$$

$$d_{3x} = 275, \quad d_{3y} = 380, \quad V_3 = 7{,}000$$

$$C_x = \frac{\Sigma d_{ix} V_i}{\Sigma V_i} = \frac{(150 \times 6000) + (100 \times 8200) + (275 \times 7000)}{6000 + 8200 + 7000} = 172$$

$$C_y = \frac{\Sigma d_{iy} V_i}{\Sigma V_i} = \frac{(75 \times 6000) + (300 \times 8200) + (380 \times 7000)}{6000 + 8200 + 7000} = 262.7$$

The location for plant D is

$$D\,(C_x, C_y) = D\,[172, \; 263].$$

Problem 7

A company is planning to undertake the production of medical testing equipments has to decide on the location of the plant. Three locations are being considered, namely, Pune, Ahmednagar and Miraj. The fixed costs of three locations are estimated to be Rs. 300 Lakhs, 500 Lakhs and 250 Lakhs respectively. The variable costs are Rs. 3000, Rs. 2000 and Rs. 3500 per unit respectively. The average sales price of the equipment is Rs. 7000 per unit. Find

(*i*) *The range of annual production/sales volume for which each location is most suitable.*

(*ii*) *Select the best location, if the sales volume is of 18,000 units.*

Solution.

(*i*) Determination of total costs of three locations.

Total cost = Fixed cost + [volume or quantity produced] × [variable cost]

$= F + x.v$ where 'x' is the quantity to be produced.

a. Total cost at Pune = 300,00,000 + 3000 x ...(1)

b. Total cost at Ahmednagr = 500,00,000 + 2000 x ...(2)

c. Total cost at Miraj = 200,00,000 + 3500 x ...(3)

For the various volumes of production, *i.e.*, 5,000, 10,000, 15,000, 20,000 and 25,000 units, the total costs are computed at the three locations and they are plotted as shown in figure.

Table : Total costs at different volumes for three locations

(Rs. in Lakhs)

Volume (Nos)	5,000	10,000	15,000	20,000	25,000
Pune	450	600	750	900	1,050
Ahmednagar	600	700	800	900	1,000
Miraj	425	600	775	950	1,125

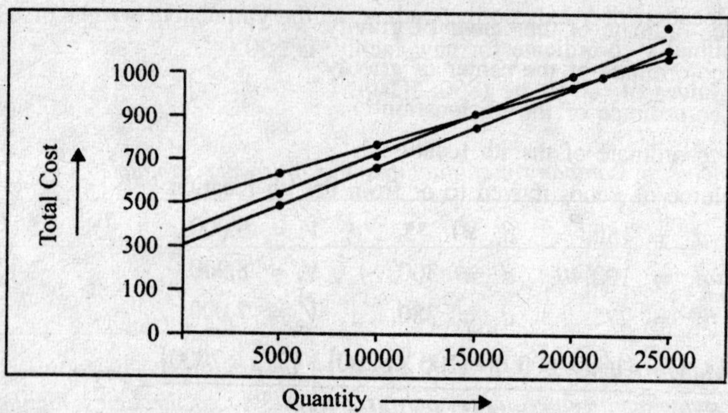

Decision Rules

For quantities upto 20,000 units, Miraj is the most economical location. For quantities above 22,000, Pune is the preferred location.

Problem 8

A new plant is to be located which will supply raw materials to a set of existing plants in a group of companies. There are five existing plants which require material movement with new plant. The locations of the existing plants are (400,200), (800,500), (1100,800), (200,900) and (1300,300). The material transported (in tons) per year from the new plant to various existing plants are 450, 1200, 300, 800 and 1500 respectively. Determine the optimum location for the new plant such that the distance moved is minimum.

Solution. Let (x, y) be the co-ordinates of the new plant.

(*i*) Determination of optional x co-ordinate for the new plant. The data of existing plants arranged in the increasing order of their x co-ordinates and their weights are represented in the table below.

Existing plant	*x co-ordinate*	*weight*	*cumulative weight*
4	200	800	800
1	400	450	1250
2	800	1200	2450
3	1100	300	2750
5	1300	1500	4250
Total			4250 tones

Thus, the median location corresponds to a cumulative weight of 4250/2 = 2125. From the table, the corresponding x co-ordinate value of 800. Since the cumulative weight first exceeds 2125 at $x = 800$.

(*ii*) Determination of y co-ordinate

Existing plant	*x co-ordinate*	*weight*	*cumulative weight*
1	200	450	450
5	300	1500	1950
2	500	1200	3150
3	800	300	2450
4	900	800	4250
Total			4250 tones

The median location of y axis corresponding, to the cumulative weight of 4250/2 = 2125 is 500. Hence optimal y co-ordinate for new facility is 500.

The optimal values of (x, y) are (800, 500).

Problem 9

Excel manufacturing company manufactures the following product.

Products	*P*	*Q*	*R*	*S*	*T*	*U*	*V*
Monthly volume of production (units)	*140*	*50*	*175*	*115*	*75*	*100*	*90*

There are six departments 1, 2, 3, 4, 5 and 6 in the plant, each performing a single type of operation. The sequence of operations each product has to go through a number of different departments and area requirements of each product is given.

Product	*Sequence*
P	*1 - 3 - 4 - 5 - 6*
Q	*2 - 4 - 5*
R	*1 - 4 - 6 - 3*
S	*6 - 2 - 4 - 5*
T	*4 - 1 - 2 - 5*
U	*1 - 4 - 5 - 6*
V	*1 - 3 - 4 - 5 - 6*

Area requirements of various departments are :

Departments	*1*	*2*	*3*	*4*	*5*	*6*
Area required (m^2)	*1000*	*2500*	*1500*	*2000*	*2000*	*1000*

The total available space is 100 m × 100 m. suggest suitable arrangement of the department.

Solution. Load summary is the basic input to the process layout. Load summary is computed combining the given product sequence and the production volume. For example, the frequency of transport between departments 1 and 3 are found as follows.

The movement 1 → 3 occurs for products P and V. Product P is produced in the volume of 140 + 90 = 230 units per month. The load summary matrix is computed as follows.

From the figure it is noticed that there is only one flow between the non adjacent department (3 and 6) and it is of significant magnitude of 175 units. Now, the layout is to be improved by reducing this flow.

(*i*) Exchange the places of the departments 2 and 6 in a grid. The resultant layout is shown below in fig. (P_1).

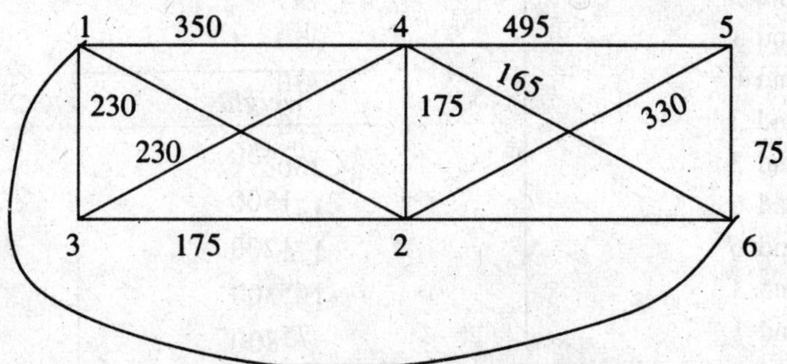

Fig. (P_1) Improved Layout.

This layout is an improvement over the earlier layout because of the non-adjacent flows have been reduced from 175 to 75. No further improvement is possible. The layout on the grid is transformed into actual areas by putting department squares on the grid and are arranged in the available space of 100 × 100 m².

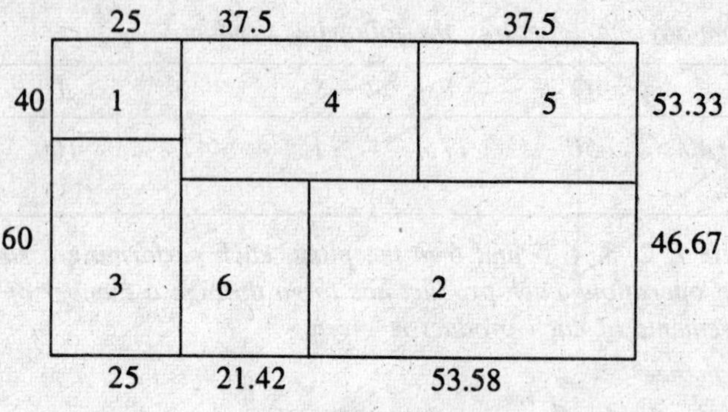

Final Layout

Load Summary

Dept.	1	2	3	4	5	6
1	–	(75)	140 + 90 (230)	175 + 75 + 100 (350)	(0)	(0)
2		–	(0)	50 + 115 (165)	(75)	(115)
3			–	140 + 90 (230)	(0)	(175)
4				–	140 + 50 + 115 + 100 + 90 (495)	(175)
5					–	140 + 100 + 90 (330)
6						–

Ranking the closeness requirements of pair of departments as per the load summary.

Pairs of Department	*Load Summary*	*Rank*
4 and 5	495	I
4 and 1	350	II
5 and 6	330	III
4 and 3	230	IV
1 and 3	230	IV
3 and 6	175	VI
4 and 6	175	VI
2 and 4	165	VIII
2 and 1	75	IX
2 and 5	75	IX

The rank I indicates that it is highly preferable to keep departments (D and E) close together. The increasing rank indicates the less and less preference. Based on the closeness rankings the departments are arranged on a grid along with their flows.

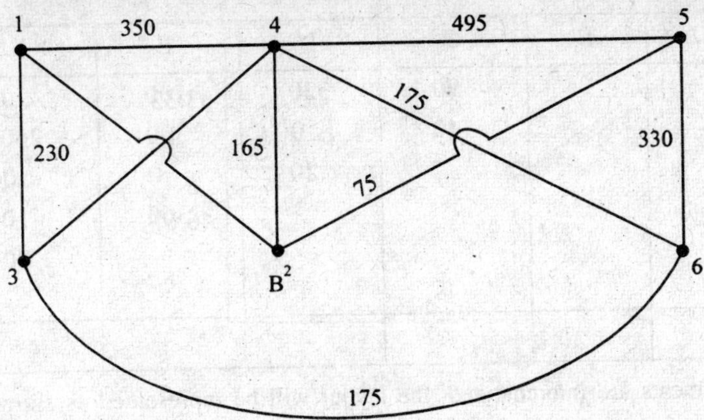

Problem 10

The present layout is shown in the figure. The manager of the department is intending to interchange the departments C and F in the present layout. The handling frequencies between the departments is given. All the departments are of the same size and configuration. The material handling cost per unit length travel between departments is same. What will be the effect of interchange of departments C and F in the layout ?

A	C	E
B	D	F

From / To	A	B	C	D	E	F
A	–	0	90	160	50	0
B	–	–	70	0	100	130
C	–	–	–	20	0	0
D	–	–	–	–	180	10
E	–	–	–	–	–	40
F	–	–	–	–	–	–

Solution. (*i*) The distance matrix for the existing layout considering only the departments that share boundaries with adjacent department.

Distance matrix for the existing layout.

From/To	A	B	C	D	E	F
A	–	1	1	2	2	3
B			2	1	3	2
C				1	1	2
D					2	1
E						1
F						–

(*ii*) Computation of total cost matrix (combining the inter departmental material handling frequencies and distance matrix.

Total cost matrix for initial layout.

From/To	A	B	C	D	E	F	Total
A		0	90	320	100	0	510
B			140	0	300	260	700
C				20	0	0	20
D					360	10	370
E						40	40
F							–
Total							1640

If the departments are interchanged, the layout will be represented as shown below.

A	F	E
B	D	C

The distance matrix and the cost matrix is represented as shown.

From/To	A	B	C	D	E	F
A	–	1	3	2	2	1
B			2	1	3	2
C				1	1	2
D					2	1
E						1
F						–

Total cost matrix for the modified layout.

From/To	A	B	C	D	E	F	Total
A		0	270	320	100	0	690
B			140	0	300	260	700
C				20	0	0	20
D					360	10	370
E						40	40
F							–
Total							1820

The interchange of departments C and F increases the total material handling cost. Thus, it is not a desirable modification.

Problem 11

A defence contractor is evaluating its machine shops current process layout. The figure below shows the current layout and the table shows the trip matrix for the facility. Health and safety regulations require departments E and F to remain at their current positions.

E	B	F
A	C	D

Current Layout

Dept.	A	B	C	D	E	F
A	–	8	3		9	5
B		–		3		
C			–		8	9
D				–		3
E					–	3
F						–

Trips between departments

Improve the layout based on trail and error method and also evaluate using load distance (ld) score ?

Solution.

Keep the departments E and F at the their current locations. Because C must be as close as possible to both E and F, put C between them. Place A directly south of E, and B next to A. All of the heavy traffic concerns have been accommodated. Department D is located in the remaining place. The proposed layout is shown in figure below. The load distance (ld) scores for the existing and proposed layout are shown below. As ld score for proposed layout is less, the proposed layout indicate improvement over existing.

E	C	F
A	B	D

Proposed Layout

Dept. Pair	*No. of Trips* (1)	*Existing plan*		*Proposed plan*	
		Distance (2)	*Load × Distance* (1 × 2)	*Distance* (3)	*Load × Distance* (1 × 3)
A – B	8	2	16	1	8
A – C	3	1	3	2	6
A – E	9	1	9	1	9
A – F	5	3	15	3	15
B – D	3	2	6	1	3
C – E	8	2	16	1	8
C – F	9	2	18	1	9
D – F	3	1	3	1	3
E – F	3	2	6	2	6
Total			92		67

DEMAND FORECASTING

Problem 12

There exists a relationship between expenditure on research and its annual profit. The details of the expenditure for the last six years is given below. Estimate the profit when the expenditure is 6 units.

Year	*Expenditure for research* (x)	*Annual Profit* (y)
1994	2	20
1995	3	25
1996	5	34
1997	4	30
1998	11	40
1999	5	31
2000	6	?

(*One unit corresponds to 1 Crore Rs*)

Solution.

Year	*Expenditure for research* (x)	*Annual Profit*	xy	x^2
1994	2	20	40	4
1995	3	25	75	9
1996	5	34	170	25
1997	4	30	120	16
1998	11	40	440	121
1999	5	31	155	25
Total	30	180	1000	200

$$\bar{x} = \frac{30}{6} = 5$$

$$\bar{y} = \frac{180}{6} = 30$$

The values a and b are computed as follows : for a linear regression equation

$$y = a + bx$$

$$b = \frac{\Sigma xy - n.\bar{x}.\bar{y}}{\Sigma x^2 - n\bar{x}^2}$$

$$b = \frac{1000 - 6 \times 5 \times 30}{200 - 6 \times 5 \times 5} = \frac{1000 - 900}{200 - 150} = 2.$$

$$a = \bar{y} - b\bar{x} = 30 - 2 \times 5 = 20$$

Thus, the model is $y = 20 + 2x$.

The profit when the expenditure is 6 units is

$$y = 20 + 2 \times 6 = 32 \text{ units of Rs}$$

Problem 13

Starwars Co. Ltd., uses simple exponential smoothing with smoothing constant α = 0.2 to forecast the demand. The forecast for the first week of March was 400 units and the actual demand turns out to be 450 units.

(i) *Estimate the demand for the second week of March.*

(ii) *If the actual demand for the second week of March is 460 units, forecast the demand upto April second week. Assume that the demand for subsequent weeks are 465, 434, 420, 498, 462 and 470 units.*

Solution.

(i) The forecast for the period (F_t) is given by

$$F_t = F_{t-1} + \alpha\left(D_{t-1} - F_{t-1}\right)$$

The forecast demand for March second week is,

$$F_t = 400 + 0.2\ (450 - 400)$$
$$= 410 \text{ units.}$$

(ii) The forecasted demand for the third week of March to April second week is worked out as shown in table below.

Week		*Demand* D_{t-1}	*Old forecast* F_{t-1}	*Forecast error* $D_{t-1} - F_{t-1}$	*Correction* $\alpha\left(D_{t-1} - F_{t-1}\right)$	*New forecast* (*Ft*) $F_{t-1} + \alpha\left(D_{t-1} - F_{t-1}\right)$
March	Week 2	460	410	50	10	420
	Week 3	465	420	45	9	429
	Week 3	434	429	5	1	430
April	Week 1	420	430	– 10	-2	428
	Week 2	498	428	70	14	442
	Week 3	462	442	20	4	446
	Week 4	470	446	24	4.8	500.8

Problem 14

The demand for different weeks of November and December month of Blue Herald Company is given below. Assume the initial forecast, i.e., for first week of November as 600 and corresponding initial trend is zero. Assume α = 0.1 and β = 0.2.

	November				*December*			
Week	*1*	*2*	*3*	*4*	*1*	*2*	*3*	*4*
Demand	*650*	*600*	*550*	*650*	*650*	*625*	*700*	*710*

Solution.

In this problem trend adjusted smoothed forecast is used to forecast the demand. Actually this model projects the next period forecast by adding a trend component to current period smoothed forecast, *i.e.*,

$$F_{t+1} = F_t + T_t$$

$$F_t = \alpha\, D_{t-1} + (1-\alpha)\left(F_{t-1} + T_{t-1}\right)$$

$$\therefore \quad T_t = \beta\left(F_t - F_{t-1}\right) + (1-\beta)\, T_{t-1}$$

where α and β are smoothing constant. The trend adjustment T_t uses a second smoothing coefficient β.

For First week of November,

$$F_t = \alpha\, D_{t-1} + (1-\alpha)\left(F_{t-1} + T_{t-1}\right)$$

The initial forecast for November first week is given as 600 units and initial trend (T_0) is zero. Now, substituting the given values in Ft equation, for first week of November.

$F_t = 0.1 \times 650 + 0.9\ (600 + 0) = 605$ units

$T_t = \beta\ (F_t - F_{t-1}) + (1 - \beta)\ T_{t-1}$

$= 0.2\ (605 - 600) + (1 - 0.2) \times 0$ (As initial trend T_{t-1} is assumed zero)

$F_{t+1} = F_t + F_t$

$= 605 + 1 = 606.$

The adjusted smoothed forecast for second week of November is 606 units.

For second week of November.

$F_t = 0.1 \times 600 + 0.9\ (605 + 1) =$ 605.4 Units.

$T_t = 0.2\ (605.4 - 605\ ) + 0.8 \times 1.0 = 0.88.$

$F_{t+1} = 605.4 + 0.88 = 606.28.$

The trend adjusted exponential smoothing forecaster is represented in the table.

Table : Trend adjusted exponential smoothing forecaster to November and December

Week	*Last period forecast* F_{t+1}	*Actual demand* D_{t-1}	*Smoothed Average* F_t	*Smoothed trend* T_t	*Next period forecast* F_{t+1}
Nov. Week 1	600.00	650	605.00	1.000	606.00
Week 2	605.00	600	605.40	0.880	606.38
Week 3	605.40	550	600.65	-0.246	600.40
Week 4	600.65	650	605.36	0.742	606.10
Dec. Week 1	605.36	625	607.99	1.120	609.11
Week 2	607.99	675	615.70	2.440	618.14
Week 3	615.70	700	626.33	4.080	630.41
Week 4	626.33	710	738.37	5.670	644.04

Problem 15

The Forecast and the actual demand for the 6 months is given below. Compute the various Forecast errors and tracking signal and comment on the forecast accuracy.

Month	*Actual Demand* (D_t)	*Forecasted Demand* (F_t)	*Deviation* $D_t - F_t$	*Cum deviation* $\Sigma (D_t - F_t)$
1	71	78	– 7	– 7
2	80	75	5	– 2
3	101	83	18	16
4	84	84	0	16
5	60	88	– 28	– 12
6	73	85	– 12	– 24

Solution.

The deviations and cumulative deviations are shown in table above.

(*i*) Mean Absolute Deviation (MAD)

MAD is the mean of absolute deviation of forecast demands from actual demand values.

$$\text{MAD} = \sum_{t=1}^{n} | D_t - F_t |$$

where D_t = actual demand for period '*t*'

F_t = forecast for period '*t*'

n = number of time periods used.

$$\therefore \qquad \text{MAD} = \frac{7+5+18+0+28+12}{6} = 11.7$$

(*ii*) Cumulative deviations or Running Sum of Forecast Error (RSFE)

This is the algebraic sum of the forecasting error i.e. both positive and negative signs are considered.

$$\text{RSFE} = -7 + 5 + 18 + 0 - 28 - 12 = -24$$

$$(iii) \text{ Tracking signal } = \frac{\text{RSFE}}{\text{MAD}} = \frac{-24}{11.7} = -2.05 = |2.05|$$

The tracking signal tells how well the forecast is predicting actual values, for it yields a measurement of the consistent difference between actual and forecasted values by expressing the cumulative deviation in terms of number of average deviations.

(*iv*) Mean Square Error (MSE)

It is the mean of the squares of the deviations of forecast demands from the actual demand values.

$$\text{MSE} = \sum_{t=1}^{n} \frac{(D_t - F_t)^2}{n}$$

$$= \frac{(-7)^2 + (5)^2 + (18)^2 + (-28)^2 + (-12)^2}{6}$$

$$= \frac{1326}{6} = 221.$$

(*v*) Mean Forecast Error (MFE)

Mean forecast errors is the mean of the deviations of the forecast demands from the actual demand.

$$\text{MSE} = \sum_{t=1}^{n} \frac{(D_t - F_t)}{n}$$

$$= \frac{-7+5+18+0-28-12}{6} = -4$$

(*vi*) Mean Absolute Percentage Error (MAPE)

Mean absolute percentage error is the mean of the percentage deviation of the forecast demands from the actual demands.

$$\text{MAPE} = \frac{1}{n} \sum_{t=1}^{n} \frac{|D_t - F_t|}{D_t} \times 100 = 0.294\%$$

Problem 16

The number of people visiting an arts exhibition appears to depend upon (a) the facilities available in the exhibition, (b) the distance of the location from the residential area (c) the entrance fee charged, Since these factors are difficult to measure directly, they are measured by means of a score for factors on 0-10 scale. The organisers believe that there is a direct and linear relationship between the number of people and these factors. The following data is collected on various exhibitions from cities. Estimate the relationship between the variables ?

Exhibition No.	*Facilities score*	*Distance score*	*Entrance fee score*	*Avg. of people/ week (000)*
1	8	4	8	5
2	10	5	2	4
3	2	6	4	1.5
4	4	6	6	2.5
5	6	8	9	7
6	3	10	7	5
7	4	5	4	3
8	10	7	9	9
9	1	5	5	1
10	1	7	4	2

Solution. Multiple Regression Model

The merit of the multiple regression model is that, it considers the effect of number of causative factors explicitly. Therefore by manupulating one of the independent variables or due to change in one of the value of the independent variable, the value by which the dependent variable will change can be computed and thus forecasted.

The regression equation

$$X_1 = b_{1,234}\, N + b_{12,34}\, X_2 + b_{13,24}\, X_3 + b_{14,23}\, X_4$$

The normal equations are

$$\Sigma X_1 = b_{1,234}\, N + b_{12,34}\, \Sigma X_2 + b_{13,24}\, \Sigma X_3 + b_{14,23}\, \Sigma X_4$$

$$\Sigma X_1 X_2 = b_{1,234}\, X_2 + b_{12,34}\, \Sigma X_2^2 + b_{13,24}\, \Sigma X_2 X_3 + b_{14,23}\, \Sigma X_2 X_4$$

$$\Sigma X_1 X_3 = b_{1,234}\ X_3 + b_{12,34}\ \Sigma X_2 X_3 + b_{13,24}\ \Sigma X_3^2 + b_{14,23}\ \Sigma X_3 X_1$$

$$\Sigma X_1 X_4 = b_{1,234}\ X_4 + b_{12,34}\ \Sigma X_2 X_4 + b_{13,24}\ \Sigma X_3 X_4 + b_{14,23}\ \Sigma X_4^2$$

Note that the partial regression co-efficient represents the slope of the linear relationship between variables X_1 and X_2 keeping the variable X_3 and X_4 constant and the constant that pertains to the relationship between the dependent variable X_1 and independent variables X_2, X_3 and X_4. N is the number of data points.

Thus, the given problem can be modelled as

$$X_1 = b_{1,234}\ N + b_{12,34}\ X_2 + b_{13,24}\ X_3 + b_{14,23}\ X_4$$

where X_1 = Average number of people per week

X_2 = Facilities index

X_3 = Distance index

X_4 = Entrance fee index

and $b_{1,234}$, $b_{12,34}$, $b_{13,24}$, and $b_{14,23}$ are constants. The normal equations are

$$\Sigma X_1 = b_{1,234}\ N + b_{12,34}\ \Sigma X_2 + b_{13,24}\ \Sigma X_3 + b_{14,23}\ \Sigma X_4$$

$$\Sigma X_1 X_2 = b_{1,234}\ X_2 + b_{12,34}\ \Sigma X_2^2 + b_{13,24}\ \Sigma X_2 X_3 + b_{14,23}\ \Sigma X_2 X_4$$

$$\Sigma X_1 X_3 = b_{1,234}\ X_3 + b_{12,34}\ \Sigma X_2 X_3 + b_{13,24}\ \Sigma X_3^2 + b_{14,23}\ \Sigma X_3 X_1$$

$$\Sigma X_1 X_4 = b_{1,234}\ X_4 + b_{12,34}\ \Sigma X_2 X_4 + b_{13,24}\ \Sigma X_3 X_4 + b_{14,23}\ \Sigma X_4^2$$

Compute the various ΣX_2, ΣX_2^2, $\Sigma X_2 X_3$ etc. from the given data points. These computations are represented in the table.

X_1	X_2	X_3	X_4	X_2^2	X_2X_3	X_2X_4	X_3^2	X_3X_4	X_4^2	X_1X_2	X_1X_3	X_1X_4
5	8	4	8	64	32	64	16	32	64	40	20	40
4	10	5	2	100	50	20	25	10	4	40	20	8
1.5	2	6	4	4	12	6	36	24	16	3	9	6
2.5	4	6	6	16	24	24	36	36	36	10	15	15
7	6	8	9	36	48	54	64	72	81	42	56	63
5	3	10	7	9	30	21	100	70	49	15	56	35
3	4	5	4	16	20	16	25	20	16	12	15	12
9	10	7	9	100	70	90	49	63	81	90	63	81
1	1	5	5	1	5	5	25	25	25	1	5	5
2	1	7	4	1	7	4	49	28	16	2	14	8
40	49	63	58	347	298	304	425	380	388	255	267	273

Therefore, the normal equations are

$$40 = 10\ b_{1,234} + 49\ b_{12,34} + 63\ b_{13,24} + 58\ b_{14,23}$$

$$255 = 49\ b_{1,234} + 347\ b_{12,34} + 298\ b_{13,24} + 304\ b_{14,23}$$

$$267 = 63\ b_{1,234} + 298\ b_{12,34} + 425\ b_{13,24} + 380\ b_{14,23}$$

$$273 = 58\ b_{1,234} + 304\ b_{12,34} + 380\ b_{13,24} + 388\ b_{14,23}$$

Solving the above equations, we get

$b_{1\ 234} = 0.522$, $b_{13,24} = 0.493$

$b_{14,23} = 0.443$, $b_{12,34} = -\ 4.222$.

Substituting these values, the regression equation is,

$$\boxed{X_1 = 0.522\ X_2 + 0.493\ X_3 + 0.443\ X_4 - 0.4222}$$

CAPACITY PLANNING AND PROCESS PLANNING

Problem 17

A manufacturing company has a product line consisting of five work stations in series. The individual workstation capacities are given. The actual output of the line is 500 units per shift.

Calculate (i) System capacity

(ii) Efficiency of the production line

Workstation. No.	*A*	*B*	*C*	*D*	*E*
Capacity/shift	*600*	*650*	*650*	*550*	*600*

Solution.

(*i*) The capacity of the system is decided by the workstation with minimum capacity/shift, *i.e.*, the bottleneck. In the given example, the work station 'D' is having a capacity of 550 units/shift which is a minimum.

Therefore, the system capacity = 550 units/shift.

(*ii*) The actual output of the line = 500 units/shift.

$$\text{Therefore, the system efficiency } = \frac{\text{Actual output}}{\text{System capacity}} \times 100$$

$$= \left(\frac{500}{550}\right) \times 100 = 90.91\%$$

Problem 18

A company intends to buy a machine having a capacity to produce 1,70,000 good parts per annum. The machine constitutes a part of the total product line. The system efficiency of the product line is 85%.

(*i*) *Find the system capacity.*

(*ii*) *If the time required to produce each part is 100 seconds and the machine works for 2000 hours per year. If the utilisation of the machine is 60% and the efficiency of the machine is 90%, compute the output of the machine.*

(*iii*) *Calculate the number of machines required ?*

Solution.

$$(i) \text{ System capacity } = \frac{\text{Actual output / annum}}{\text{System efficiency}}$$

$$= \frac{1,70,000}{0.85} = 2,00,000 \text{ units / annum}$$

$$= \frac{2,00,000}{2,000} = 100 \text{ units / hours}$$

(*ii*) Output per annum = Unit capacity × % utilisation × efficiency

$$\text{Unit capacity} = \frac{60 \times 60 \text{ sec}}{100 \text{ sec per unit}} = 36 \text{ units}$$

$$\text{Output per hour} = 36 \times 0.6 \times 0.9$$

$$= 19.44 \text{ units} \approx 20 \text{ units.}$$

(*iii*) $$\text{Number of machines required} = \frac{\text{System capacity}}{\text{Output per hour}}$$

$$= \frac{100}{20} = 5 \text{ machines}$$

Problem 19

The following activities constitute a work cycle.

(*i*) *Find the total time, theoretical output obtained from the machine.*

(*ii*) *Calculate the number of machines required to produce the three components from the information given below.*

Sr. No.	*Activity*	*Time (min)*
1.	*Unloading*	*0.25*
2.	*Inspection*	*0.35*
3.	*Loading job on machine table*	*0.40*
4.	*Machine operation time*	*0.90*

Components	*A*	*B*	*C*
1. Setup time per batch	25 min	55 min	45 min
2. Operation time (min/piece)	1.75	3.0	2.1
3. Batch size	350	550	575
4. Production per month	2450	4400	2875

Solution.

(*i*) Total cycle time (T) = 0.25 + 0.35 + 0.40 + 0.90 = 1.90 min.

(*ii*) Output of the machine

$$\text{Output} = \frac{60}{1.9} = 31.5 \approx 31 \text{ units.}$$

(*iii*) Number of machines required

Assume that the plant works on the single shift basis per day of 8 hours each.

The total time required for the processing the components is given by

Total time required = Setup time + operation time.

For component A,

Total time required = Setup time + operation time

$$= \left(\frac{\text{Production quantity}}{\text{Batch size}} \times \frac{\text{Setup time}}{\text{Batch}}\right) + \text{Operation time}$$

$$= \left[\frac{2450}{350} \times \frac{25}{60}\right] + \left[2450 \times \frac{1.75}{60}\right] = 74.374 \text{ hrs.}$$

For component *B*,

$$\text{Total time required} = \left[\frac{4400}{550} \times \frac{55}{60}\right] + \left[4400 \times \frac{3}{60}\right] = 227.33 \text{ hrs.}$$

For component *C*,

$$\text{Total time required} = \left[\frac{2875}{575} \times \frac{45}{60}\right] + \left[2875 \times \frac{2.1}{60}\right] = 104.375 \text{ hrs.}$$

Total time (hrs) required to process all the three components

$$= 74.374 + 227.33 + 104.375 = 406.079 \text{ hrs.}$$

Total number of hours available (assuming 25 working days) per month

$$= 8 \times 25 = 200 \text{ hrs.}$$

$$\therefore \text{ Number of machines required} = \frac{\text{Total number of machine hours required}}{\text{Total number of hours available}}$$

$$= \frac{406.079}{200} = 2.030 \approx 2 \text{ machines}$$

Assuming a machine efficiency of 85% and operator efficiency of 75%, the number of machines required are :

$$\text{Total hours required per month} = \frac{406.079}{0.85 \times 0.75} = 636.98 \text{ hrs.}$$

$$\text{Number of machines required} = \frac{636.98}{200} = 3.18 \approx 4 \text{ machines}$$

Problem 20

Three components are to be manufactured on three machines i.e. Center lathe, Milling machine and Cylindrical grinding machine.

(*i*) *Calculate the number of machines required of each kind to produce the components if the plant works for 48 hours per week.*

(*ii*) *Calculate the number of machines required assuming the machine efficiency of 75%.*

(*iii*) *How do you reduce the number of machines.*

The following information is given :

Machines	*Component A setup*	*Component A operation*	*Component B setup*	*Component B operation*	*Component C setup*	*Component C operation*
1. Center lathe	*30 min*	*2 min*	*55 min*	*2.5 min*	*40 min*	*1.5 min*
2. Milling machine	*45 min*	*8 min*	*30 min*	*4 min*	–	–
3. Cylindrical grinding	*50 min*	*10 min*	*60 min*	*8 min*	*60 min*	*10 min*
Other details						
Lot size	*350*		*400*		*600*	
Quantity demanded/month	*1750*		*4000*		*3000*	

Solution.

The total time required to process the required components on the machines.

***(i)* Center Lathe**

(*a*) Total time required for Component $A = \left[\frac{1750}{350} \times \frac{30}{60}\right] + \left[1750 + \frac{2}{60}\right] = 62.83$ hours

(*b*) Total time required for Component $B = \left[\frac{4000}{400} \times \frac{55}{60}\right] + \left[4000 + \frac{2.5}{60}\right] = 175.832$ hours

(*c*) Total time required for Component $C = \left[\frac{3000}{600} \times \frac{40}{60}\right] + \left[3000 + \frac{1.5}{60}\right] = 78.33$ hours

Total time required to process the components on center lathe

$= a + b + c = 62.83 + 175.83 + 78.333$

$= 314.995$ hrs/month.

Available time per machine per month $= 48 \times 4 = 192$ hours.

$\therefore$ No. of Lathe machines required $= \frac{\text{Total hours required / month}}{\text{Total hours available / month}}$

$= \frac{314.995}{192} = 1.64 \approx 2$ numbers

If the machine efficiency is considered as 85%, then

No. of lathes required $= \frac{\text{Total No. of hours required / month}}{\text{Total No. of hours available / month} \times \text{machine efficiency}}$

$= \frac{314.995}{192 \times 0.75} = 2.18 \approx 3$ machines

***(ii)* Milling Machine**

Total time required to process all the components per month

$= \left[\frac{1750}{350} \times \frac{45}{60} + 1750 \times \frac{8}{60}\right] + \left[\frac{4000}{400} \times \frac{30}{60} + 4000 \times \frac{4}{60}\right]$

$= 508.749$ hours.

Number of milling machines required

$= \frac{\text{Total hours required / month}}{\text{Total hours available / month}}$

$= \frac{508.749}{192} = 2.64 \approx 3$ numbers

If the machine efficiency is considered as 75%,

Number of milling machine required $= \frac{508.749}{192 \times 0.75} = 3.53 \approx 4$ numbers

Cylindrical Grinding Machines

Total time required to process all the components per month

$= \left[\frac{1750}{350} \times \frac{50}{60} + 1750 \times \frac{10}{60}\right] + \left[\frac{4000}{400} \times \frac{60}{60} + 4000 \times \frac{8}{60}\right] + \left[\frac{3000}{600} \times \frac{60}{60} + 3000 \times \frac{10}{60}\right]$

$= [4.166 + 291.666] + [10 + 533.333] + [5 + 500]$

$= 1344.165$ hours.

$$\text{Number of milling machines required} = \frac{\text{Total hours required / month}}{\text{Total hours available / month}}$$

$$= \frac{1344.165}{192} = 7 \text{ machines}$$

If the machine efficiency is considered as 75%,

$$\text{Number of milling machine required} = \frac{1344.165}{192 \times 0.75} = 9.33 \approx 10 \text{ machines}$$

(*iii*) Reduction in number of machines

(*a*) By introducing the second and third shift, the number of hours available will be increased and hence the number of machines required will be reduced.

(*b*) By increasing the utilisation of the machine. The availability of the machine will be increased by proper maintenance which reduces the break down and hence the down time. The production time will be increased and hence the plant utilisation.

Problem 21

The component can be processed on either of the two machines — Turrette lathe and Center lathe. The time and cost details are given below. Computer the breakeven quantity and state the decision rules.

Particulars	*Turrette-Lathe*	*Center Lathe*
1. Setup time (hrs)	*4.5*	*0.2*
2. Operation time/piece (min)	*4.0*	*35*
3. Setup cost/hour (Rs.)	*350*	*15*
4. Machining cost/hour (Rs.)	*45*	*25*

Solution.

Breakeven quantity refers to the quantity at which both the alternatives are equally feasible and the total cost of both alternatives are equal

Let 'Q' be the breakeven quantity

Total cost of production on Turrette lathe = Total cost of production on center lathe

$$= [\text{Setup cost + operation cost}]_{\text{Turret}} = [\text{Setup cost + operation cost}]_{\text{Center}}$$

$$= [4.5 \times 350] + \left[\frac{4}{60} \times Q \times 45\right] = [0.2 \times 15] + \left[\frac{35}{60} \times Q \times 25\right]$$

$$= 1575 + 3\,Q = 3 + 14.58\,Q$$

$$\therefore \quad 1572 = 11.58Q$$

$$Q = 136 \text{ units.}$$

If the quantity is ≤ 136, Center lathe is preferred.

If the quantity is ≥ 136, Turrette Lathe is preferred.

Problem 22

A customer is processed through each of the three operations A, B and C, in sequence. The process is designed to handle 100 customers a day. The average rate at which each operation can process customers is shown in the figure below :

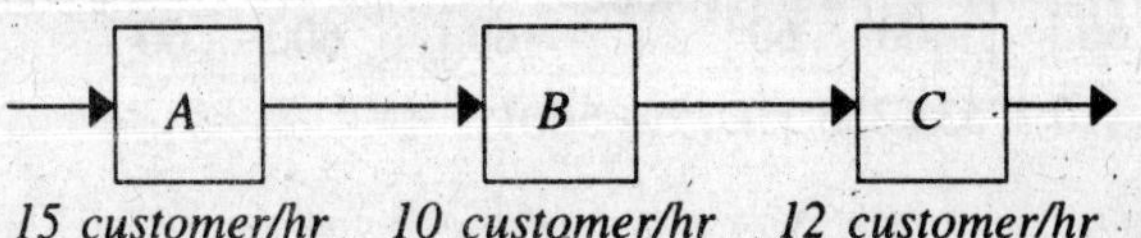

(a) For an 8 hour day identify any bottlenecks in the process.

(b) What effect will bottlenecks have on overall output and other operations.

(c) If all the processes are operated for 10 hours per day, is there still a bottleneck ?

(d) How will random arrivals affect the output and processing rate of this process ?

Solution.

(*a*) Identification of bottleneck for an 8 hour a day operation.

Operation A, Output for 8 hrs/day = 15 × 8 = 120 customers

Operation B, Output for 8 hrs/day = 10 × 8 = 80 customers

Operation C, Output for 8 hrs/day = 12 × 8 = 96 customers

Looking into the above calculations, the operations B which process 10 customer/hr and 80 customer a day becomes the bottleneck.

(*b*) Effect of the bottleneck on the overall output

The bottleneck operation decides the capacity and output. Here the capacity of the unit is 80 customer/day (due to bottleneck B) eventhough the process A and C have the output of 120 and 96 respectively.

(*i*) Effect of the bottleneck on other operation

Bottleneck operation B, makes the waiting time higher for C. Because of difference in out (C > B), facility has to wait fill the customer comes from B.

(*ii*) In case A, the operation will be completed and because of difference in processing rates, operation A will be idle for 25% time unless it is used for other purpose and the full capacity of A if used creates an inventory between A and B.

(*c*) If all the processes are operated for 10 hrs, then the designed output of 100 customers/hr are achieved still the process line is unbalanced as the processing capacities are unequal.

(d) Random arrivals affect the output and processing of customers.

The random arrivals affect the capacity utilisation also both capacity idle time and shortage.

Problem 23

An estimated demand per annum for a chemical is given below.

Demand in Kgs ('000)	*100*	*110*	*120*	*130*	*140*
Probability	*0.10*	*0.20*	*0.30*	*0.30*	*0.10*

(*i*) *If the capacity is set at 1,30,000 Kgs/annum, how much capacity cushion is existing?*

(*ii*) *What is the probability of idle capacity ?*

(*iii*) *What is the capacity utilisation of the plant at, 1,30,000 Kgs/annum capacity ?*

(*iv*) *If it is estimated that the lost business costs Rs. 1,00,000 per thousand Kgs of capacity, how much capacity should be built to minimise the total costs? The cost to build 1,000 Kgs of capacity is Rs. 50,000.*

Solution.

(*i*) The Capacity Cushion Available = Capacity – Average annual demand

Average annual demand $= \Sigma$ expected demand × probability

$= (100 \times 0.1 + 110 \times 0.2 + 120 \times 0.3 + 130 \times 0.3 + 140 \times 0.1)$

$= 121.$

∴ Capacity Cushion Available = 130 – 121

= 9,000 kgs.

(*ii*) Probability of idle capacity indicates the probability of demand less than 1,30,000 kgs.

∴ Probability of idle capacity = Σ probability of demand < 1,30,000 Kgs.

= 0.10 + 0.20 + 0.30 = 0.60

(*iii*) Average capacity utilisation

$$= \left(0.1 \times \frac{100}{130}\right) + \left(0.2 \times \frac{110}{130}\right) + \left(0.3 \times \frac{120}{130}\right) + \left(0.4 \times \frac{130}{130}\right)$$

$$= 92.31\%$$

(*iv*) Amount of capacity to minimise total costs

(*a*) Total costs to build 1,00,000 Kgs of capacity

= capacity cost + penalty cost

= (100 × 5000) + (1,00,000 × (0 × 0.1)+ (10 × 0.2) + (20 × 0.3) + (30 × 0.3) + (40 × 0.1))

= Rs. 2,60,000

(*b*) Total costs to build 1,10,000 Kgs of capacity

= capacity cost + penalty cost

= (110 × 5000) + (1,00,000 × (10 × 0.1) + (0 × 0.2) + (10 × 0.3) + (20 × 0.3) + (30 × 0.1))

= Rs. 1,75,000

(*c*) Total costs to build 1,20,000 Kgs of capacity

= (120 × 5000) + (1,00,000 × (0 × 0.1) + (0 × 0.2) + (0 × 0.3) + (10 × 0.3) + (20 × 0.1))

= Rs. 1,10,000

(*d*) Total costs to build 1,30,000 Kgs of capacity

= (130 × 5000) + (1,00,000 × (10 × 0.1)

= Rs. 7,50,000

(*e*) Total costs to build 1,40,000 kgs of capacity

= (140 × 5000) + (1,00,000 × 0)

= Rs. 7,00,000

It is therefore economical to build a capacity of 1,40,000 kgs per annum which results in minimum cost.

Problem 24

Machines A and B are both capable of processing the product. The following informations is given

Particulars	*Machine A*	*Machine B*
Investment	*Rs. 75,000*	*Rs. 80,000*
Interest on capital invested	*10 %*	*15 %*
Hourly charge (wages+power)	*Rs. 10*	*Rs. 8*
Pieces produced per hour	*5*	*8*
Annual operating hours	*2000*	*2000*

Which machine will give the lower cost per unit of production, if run for the whole year? If only 4000 pieces are to be produced in a year, which machine would give the lower cost per piece.

Solution.

Computation of cost per unit of production of machines

Particulars	*Machine A*	*Machine B*
Interest (fixed cost)	Rs 7,500	Rs 12,000
Variable cost (hourly charge × annual operating hours)	Rs 20,000	Rs 16,000
Total cost	Rs 27,500	Rs 28,000
Total output	5 × 2000 = 10,000	8 × 2000 = 16,000
Unit cost	Rs 2.75	Rs 1.75.

If the output is 4000 units per annum

Particulars	*Machine A*	*Machine B*
Interest (fixed cost)	Rs 7,500	Rs 7,500
Variable cost	Rs 8,000	Rs 4,000
(10 × 800)		(8 × 500)
Total cost	Rs 15,500	Rs 16,000
Unit cost	Rs 3.89	Rs 4.00.

Note : 800 hours will be required to produce the commodity with machine A and 500 hrs on machine B.

Problem 25

The Gamma corporation has developed a forecast for the group of items that has the following demand pattern.

Quarter	*Demand*	*Cumulative demand*
1	*•270*	*270*
2	*220*	*490*
3	*470*	*960*
4	*670*	*1630*
5	*450*	*2080*
6	*270*	*2350*
7	*200*	*2550*
8	*370*	*3920*

(*a*) *The firm estimates that it costs Rs. 150 per unit to increase production rate Rs. 200 per unit to decrease the production rate, Rs. 50 per unit per quarter to carry the items in inventory and Rs. 100 per unit if subcontracted. Compare the costs of the pure strategies.*

(*b*) *Given the cost figures in (a), evaluate the following mixed strategies company maintains a constant production rate of 250 units per quarter and permits 20% overtime when demand exceeds the production rate. The incremental cost of overtime is Rs. 25 per hour. The company plans to meet the excess demand by hiring and firing of workers.*

Solution.

(*a*) Different pure strategies are

Vary the work force size
Changing inventory levels
Sub contracting

Plan I — Varying the Size of Work Force

In this pure strategy, the actual demand is met by varying the work force size. This means that during the period of low demand, the company must fire the workers and during the period of high demand the company must hire workers. These two steps involve associated costs. In this strategy, the production units will be equal to the demand and values in each period. The cost of the plan is computed in the table below.

Quarter	*Demand*	*Cost of increasing production level (Rs)*	*Cost of decreasing production level (Rs)*	*Total cost of plan (Rs)*
1	270	–	–	–
2	220	–	50 × 200 = 10,000	10,000
3	470	250 × 150 = 37,500	–	37,500
4	670	200 × 150 = 30,000	–	30,000
5	450	–	220 × 200 = 44,000	44,000
6	270	–	180 × 200 = 36,000	36,000
7	200	–	70 × 200 = 14,000	14,000
8	370	170 × 150 = 25,500	–	25,500
	Total			1,97,000

Plan II — Changing Inventory Levels

In this plan, the company computes the average demand and sets its production capacity to this average demand. This results in excess of units in some periods and also shortage of units during some other periods. The excess units will be carried as inventory for future use and shortage of units can be fulfilled using future inventory. The cost of the plan II is computer in the table. The plan incurs a maximum shortage of 255 units during 5 periods. The firm might decide to carry 255 units from the beginning of period 1 to avoid shortage. The total cost of the plan is Rs. 96,000.

Quarter	*Demand forecast*	*Cumulative demand*	*Production level*	*Cumu. prod. level*	*Inventory*	*Adjusted inventory with 255 at beginning of period 1*	*Cost of holding inventory Rs*
1	270	270	365	365	95	350	17,500
2	220	490	365	730	240	495	24,750
3	470	960	365	1095	135	390	19,500
4	670	1630	365	1460	– 170	85	4,250
5	450	2080	365	1825	– 255	0	0
6	270	2350	365	2190	– 160	95	4,750
7	200	2550	365	2555	5	260	13,000
8	370	3920	365	2920	0	255	12,750
						Total	96,500

Plan III — Subcontracting

The additional demand other than the normal capacity is met by subcontracting. The cost of the plan III amounts to Rs. 1,32,000 as shown in table below.

Quarter	*Demand forecast*	*Production units*	*Subcontract units*	*Incremental cost @ Rs.100/unit*
1	270	200	70	70 × 100 = 7,000
2	220	200	20	20 × 100 = 2,000
3	470	200	270	270 × 100 = 27,000
4	670	200	470	470 × 100 = 47,000
5	450	200	250	250 × 100 = 25,000
6	270	200	70	70 × 100 = 7,000
7	200	200	0	0
8	370	200	170	170 × 100 = 17,000
			Total	1,32,000

The total cost of pure strategies is given below. On observation Plan II (Changing inventory levels) has the least cost.

Plan	*Total cost (Rs)*
Plan I	1,97,000
Plan II	96,500
Plan III	1,32,000

Table : Computation of the Cost Mixed Strategy

Quarter	*Demand forecast*	*Regular time prod. units*	*Additional units needed after R.T.*	*Over time production*	*Additional units needed after RT + OT*	*Cost of inventory*	*Cost of O.T.*	*Cost of changing work force*	*Total cost (Rs)*
1	270	250	20	50	– 30 (– 30)	1500	1250	0	2750
2	220	250	– 30	0	– 30 (– 60)	3000	0	0	3000
3	470	250	220	50	170 (110)	0	1250	16,500	17750
4	670	250	420	50	370 (370)	0	1250	39,000	40250
5	450	250	200	50	150 (150)	0	1250	44,000	45250
6	270	250	20	50	– 30 (– 30)	1500	1250	30,000	32750
7	200	250	– 50	0	– 50 (– 80)	4000	0	0	4000
8	370	250	120	50	70 (– 10)	5000	1250	0	1750
								Total	1,47,500

Note :

1. Negative quantities in bracket indicate inventory and positive quantities denote quantity to be produced by changing capacity.
2. Inventory at the end of period 2 is 60 units.
3. During period 3 increase in only 110 units are required after utilising 60 units of beginning inventory.

(*b*) Mixed Strategy

The components of mixed strategy are :

- Maintain a constant production rate of 250 units per quarter.
- Permit 20% overtime when the demand exceeds the production rate. The incremental cost of overtime is Rs. 25 per hour.
- To meet any further demand, it chooses to hire and fire workers.

Computation of the cost of strategy is shown in the table — with reference to table.

When there is a change of work force, the quantity of increase or decrease in the work force is to be computed and the same is to be multiplied with the cost of hiring and firing respectively.

During 3rd quarter, the additional number of units needed after utilising regular time capacity and overtime capacity is 110. Therefore, the cost of change in workforce (cost of hiring) in this period is (110 × 50 =) Rs. 16,500.

During 4th quarter, the additional number of units needed after utilising regular time capacity and overtime capacity is 370 units. This quantity is more than the corresponding quantity of the previous period by 260 units.

∴ Cost of hiring in this period is (260 × 150 =) Rs. 39,000.

During 5th quarter, the additional number of units needed is 150. This quantity is less than the corresponding quantity of the previous period by 220 units. Therefore, the cost of change in work force (firing) in this period is (220 × 200 =) Rs. 44,000.

During 6th quarter, there is an excess of 30 units, so the size of work force is brought to average regular time production quantity of 250 units in this period so, the change of work force costs (cost of firing) (150 × 200 =) Rs. 30,000.

Thus, the cost of mixed strategies = Rs. 1,47,000.

Problem 26

A company manufactures the consumer durable products and the company intends to develop an aggregate plan for six months starting from January through June. The following information is available.

Demand and working days.

Month	*Jan*	*Feb*	*Mar*	*Apr*	*May*	*June*
Demand	*500*	*600*	*650*	*800*	*900*	*800*
Working days	*22*	*19*	*21*	*21*	*22*	*20*

Cost details

Materials	*Rs. 100/unit*
Inventory carrying cost	*Rs. 10/unit/month*
Cost of stockout	*Rs. 20/unit/month*
Cost of subcontracting	*Rs. 200/unit*
Hiring and training cost	*Rs. 50/ worker*
Lay off cost	*Rs. 100/ worker*
Labour hours required	*Rs. 4/unit*
Regular time cost (for first 8 hrs)	*Rs. 12.50/hour*
Overtime cost	*Rs. 18.75/hr*
Beginning inventory	*200 units*
Safety stock required	*Nil*

Work out the cost of the following strategies

1. *Produce exactly to meet demand — vary the work force.*
2. *Constant work force — vary inventory and allow shortages*
3. *Constant work force and use subcontracting.*

Solution.

Strategy I : Produce exactly to meet demand by varying work force.

Assumption : Opening workforce equals the first month's requirements.

Table : Aggregate production planning requirements

	Jan	*Feb*	*Mar*	*Apr*	*May*	*June*	*Total*
Beginning inventory	200	0	0	0	0	0	
Forecasted demand	500	600	650	800	900	800	
Production requirement (demand + safety stock – beginning inventory	300	600	650	800	900	800	
Ending inventory (Beginning inventory + production requirement – Demand forecast)	0	0	0	0	0	0	

Plan I — Exact Production, vary Work Force

	Jan	*Feb*	*Mar*	*Apr*	*May*	*June*	*Total*
Production requirement	300	460	650	800	900	800	
Production hours required (production requirement × 4 hr/unit)	1200	2400	2600	3200	3600	3200	
Working days per month	22	19	21	21	22	20	
Hours per month per worker (working days × 8 hrs/day)	176	152	168	168	176	160	
No. of workers required (production hrs required ÷ hrs per month per worker	7	16	15	19	20	20	
New worker hired (assuming opening work force equal to first months requirement of 7 workers)	0	9	0	4	1	0	
Hiring cost (workers hired × Rs. 50)	0	450	0	200	50	0	700
Workers laid off	0	0	1	0	0	0	
Lay off cost (workers laid off × 100)	0	0	100	0	0	0	100
Regular production cost (production hrs required × 12.50 Rs./hrs)	15000	30000	32500	40000	45000	40000	202500
Total							203300

Plan II — Constant Work Force, vary Inventory and Stockout

* Assume a constant work force of 10.

	Jan	*Feb*	*Mar*	*Apr*	*May*	*June*	*Total*
Beginning inventory	200	140	– 80	– 310	– 690	– 1150	
Working days per month	22	19	21	21	22	20	
Production hrs available (working days/month × 8 hrs/day x 10 workers)	1760	1520	1680	1680	1760	1600	
Actual production (production hrs available ÷ 4 hours/unit)	440	380	420	420	440	400	
Forecasted demand	500	600	650	800	900	800	
Ending inventory (beginning inventory + actual)	140	– 80	– 310	– 690	– 1150	– 1550	
Shortage cost (unit short × Rs.20/unit)	0	1600	6200	13800	23000	31000	75600
Units excess (ending inventory – safety stock)	140	0	0	0	0	0	
Inventory cost (unit excess × 10)	1400	0	0	0	0	0	0
Regular production cost (production hrs required × 12.50 Rs./hrs)	22000	19000	21000	21000	22000	20000	125000
Total							202000

Plan III — Constant Work Force Subcontract

	Jan	*Feb*	*Mar*	*Apr*	*May*	*June*	*Total*
Production requirement	300	460	650	800	900	800	
Working days per month	22	19	21	21	22	20	
Production hrs available (working days × 8 hrs/day × 10 workers)	1760	1520	680	1680	1760	1600	
Actual production (productn. hrs available ÷ 4 hours per unit)	440	380	420	420	440	400	
Unit subcontracted (production requirements – actual production)	0	220	230	380	460	400	
Subcontracting cost (units subcontracted × Rs. 100)	0	8000	23000	38000	46000	4000	155000
Regular production cost (production hrs required × 12.50 Rs./hrs)	22000	19000	21000	21000	22000	20000	125000
Total							280000

Note : Assume a constant work force of 10.

600 – 140 = 460 units of beginning inventory in February.

Summary of the Plans

Plan	*Hiring*	*Lay off*	*Subcon-tract*	*RT prod.*	*Short-age*	*excess inven.*	*Total cost*
Plan I – Exact production vary work force	700	100	–	202500	–	–	203300
Plan II – Constant work force vary inventory and shortages	–	–	–	125000	75600	1400	202000
Plan III – Constant work force Subcontract	–	–	155000	125000	–	–	280000

Problem 27

A company manufactures seasonal products. The information regarding the seasonal demand pattern, available production capacities during regular time, overtime and other details are as follows :

Forecasted demand

Period	*1*	*2*	*3*	*4*
Demand (units)	*700*	*1000*	*2000*	*1200*

Available production capacity

Period	*Regular time*	*Over time*	*Subcontracting*
	(RT)	*(OT)*	*(SC)*
1	*900*	*350*	*600*
2	*1000*	*350*	*600*
3	*1100*	*350*	*600*
4	*700*	*350*	*600*

Initial inventory = *200 units*

Desired final inventory = *150 units*

Regular time production cost/unit = *Rs 125*

Overtime production cost/unit = *Rs 150*

Subcontracting cost/unit = *Rs 175*

Inventory cost/unit/period = *Rs 25*

Formulate this problem as a transportation model to determine the optimum production levels and means of production for the next four quarters.

Solution.

The problem is represented as a transportation model as shown in figure.

1. Initial inventory – Initial inventory of 200 units is available at period 1 at no additional cost. Carrying cost is Rs. 25 per unit per period, if it is carried for period 2, and period 3 and so on.

2. R.T. Production – Cost per unit is Rs. 125. If the units are produced and used in the same period. A carrying cost of Rs. 25/unit per period is added on for each period for the units retained.

3. Overtime Production (OT) – Cost per unit is Rs. 150 units if the units are used in the same month of their production otherwise a carrying cost of Rs. 25 per unit per period is applicable.

4. Subcontracting (SC) cost per unit is Rs. 175 per unit.

The initial allocations are made so as to use regular time production as fully as possible. Overtime and subcontracting can be allocated on a minimum cost basis.

- Cost of unused capacity during regular time is Rs. 50 /unit
- Final inventory – Final inventory requirement is 150 units at the end of period 4 and has been added to demand of period 4 (i.e. 1200 units) to obtain a total of 1350 units.

Table : Transportation Model

Period	*Supply of units from*	*Period in which products are demanded*				*Unused capacity*	*Total available capacity*
		1	2	3	4		
	Initial inventory	0	25 (200)	50	75	0	200
1	RT	125 (700)	150 (200)	175	200	50	900
	OT	150	175	200	225	0 (350)	350
	SC	175	175	175	175	0 (600)	600
2	RT	–	125 (600)	150 (400)	175	50	1000
	OT	–	150	175	200	0 (350)	350
	SC	–	175	175 (150)	175 (300)	0 (150)	600
3	RT	–	–	125 (1100)	150	.50	1100
	OT	–	–	150 (350)	175	0	350
	SC	–	–	175	175	0 (600)	600
4	RT	–	–	–	125 (700)	50	700
	OT	–	–	–	150 (350)	0	350
	SC	–	–	–	175	0 (600)	600
Demand		700	1000	2000	1350	2650	7700

The optimum solution to the problem is represented as below :

Period	*Means of production*	*Quantity*
0	Beginning inventory	200
1	Regular time	900
2	Regular time	1000
	Subcontracting	450
3	Regular time	1100
	Over time	350
4	Regular time	700
	Over time	350

LINE BALANCING

Problem 28

A company is setting an assembly line to produce 192 units per eight hour shift. The information regarding work elements in terms of times and immediate predecessors are given

Work element	*Time (Sec)*	*Immediate predecessors*
A	40	None
B	80	A
C	30	D, E, F
D	25	B
E	20	B
F	15	B
G	120	A
H	145	G
I	130	H
J	115	C, I
Total	720	

1. *What is the desired cycle time ?*
2. *What is the theoretical number of stations ?*
3. *Use largest work element time rule to workout a solution on a precedence diagram.*
4. *What are the efficiency and balance delay of the solution obtained ?*

Solution.

The precedence diagram is represented as shown below :

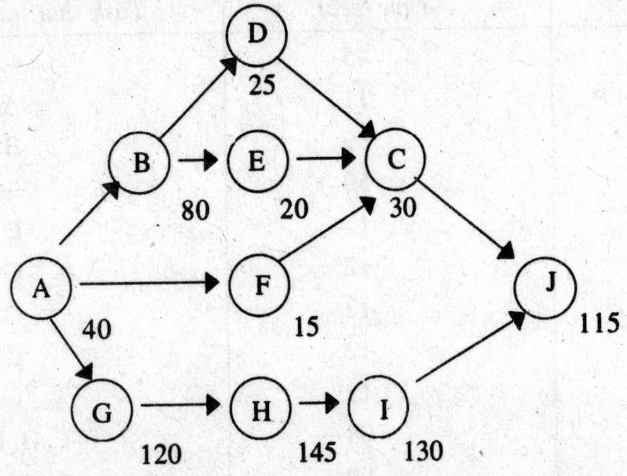

(*a*) Cycle time $= \frac{1}{r} = \frac{8 \text{ hours}}{192 \text{ units}} = 150 \text{ sec / unit}$

where r = output.

(*b*) Sum of the work elements is 720 seconds, so minimum number of work stations

$$= \frac{\Sigma t}{\text{cycle time}} = \frac{720 \text{ seconds / unit}}{150 \text{ sec / unit station}} = 4.8 \approx 5 \text{ stations}$$

Assignment of work elements to work stations.

Station	*Elements*	*work element time in sec*	*Cumulative time (Sec)*	*Idle time for station*
S_1	A B D	40 80 25	40 120 145	05
S_2	G E	120 20	120 140	10
S_3	H	145	145	05
S_4	I F	130 15	130 145	05
S_5	C J	30 115	30 145	05

(*d*) Efficiency $= \frac{\Sigma t}{n\,Ct} \times 100 = \frac{720}{5 \times 150} \times 100 = 96\%$

Thus, the balance delay is [100 – 96] 4 percent only.

Problem 29

The company is engaged in the assembly of a wagon on a conveyor. 500 wagons are required per day. Production time available per day is 420 minutes. The other information is given below regarding assembly steps and precedence relationships. Find the minimum number of work stations, balance delay and line efficiency.

Solution. The element times and precedence relationships

Task	*Time (sec)*	*Task that must precede*
A	45	–
B	11	A
C	09	B
D	50	–
E	15	D
F	12	C
G	12	C
H	12	E
I	12	E
J	08	F, G, H, I
K	09	–
Total	195	–

(*i*) Precedence diagram is constructed as per the given details.

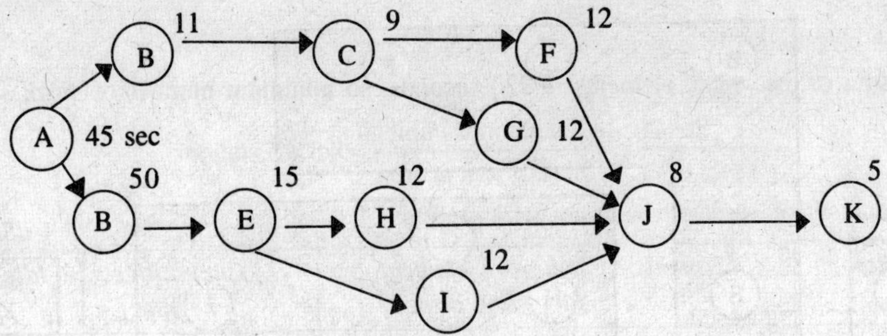

(*ii*) Determination of cycle time

$$C_t = \frac{\text{Production time / day}}{\text{Output / day}} = \frac{420 \times 60}{500} = \frac{25{,}200}{500} = 50.4 \text{ sec.}$$

(*iii*) Theoretical number of work stations required

$$N = \frac{\text{Total time}}{\text{Cycle time}} = \frac{195}{50.4} = 3.87 \approx 4 \text{ workstations}$$

(*iv*) Assign the elements to work stations based on the largest element time and the work stations to be required are 5.

Balance made according to largest number of followers task rule.

Station	*Task*	*Task time (sec)*	*Idle time (sec)*
S_1	A	45	5.4
S_2	D	50	0.4
S_3	B E C F	11 15 09 12	3.4
S_4	G H I J	12 12 12 8	6.4
S_5	K	9	41.4

(*iv*) Efficiency $= \dfrac{T}{\text{No. of stations} \times C} = \dfrac{195}{5 \times 50.4} \times 100 = 77.77\%$

(*v*) Balance delay = 100 – 77.77 = 22.23%

Precedence Graph

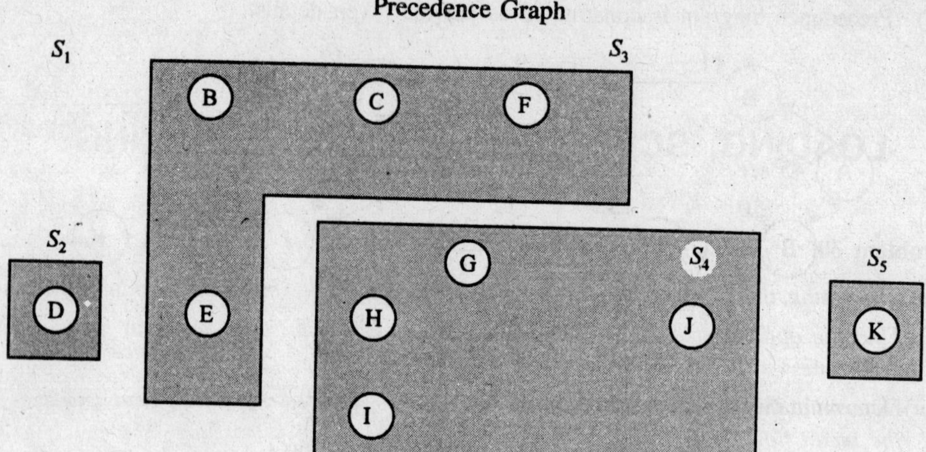

LOADING, SCHEDULING AND SEQUENCING

Problem 30

The processing times (t_j) in hrs for the five jobs of a single machine scheduling is given.

(*a*) *Find the optimal sequence which will minimise the mean flow time and find the mean flow time.*

(*b*) *Determine the sequence which will minimise the weighted mean flow time and also find the mean flow time.*

Job (j)	*1*	*2*	*3*	*4*	*5*
Processing time (t_j) *hrs*	*30*	*8*	*10*	*28*	*16*

Solution.

(*a*) First arrange the jobs as per the shortest processing time (SPT) sequence

Job (j)	2	3	5	4	1
Processing time (t_j) hrs	8	10	16	28	30

Therefore, the job sequence that minimises the mean flow time is 2-3-5-4-1.

Computation of minimum flow time (Fmin)

The flow time is the amount of time the job '*j*' spends in the system. It is a measure which indicates the waiting of jobs in the system. It is the difference between the completion time (C_j) and ready time (R_j) for job *j*.

$$F_j = C_j - R_j$$

Job (j)	2	3	5	4	1
Processing time (t_j) hrs	8	10	16	28	30
Completion time (C_j)	8	18	34	62	92

Since the ready time (R_j) = 0 for all *j*, the mean flow time $(\overline{Fj})$ is equal to C_i for all *j*.

Mean flow time $(\overline{F}) = \frac{1}{n}\sum_{j=1}^{n} F_j$

$$= \frac{1}{5}[8 + 18 + 34 + 62 + 92]$$

$$= \frac{1}{5}[214] = 42.8 \text{ hours.}$$

(*b*) The weights are given as follows :

Job (j)	1	2	3	4	5
Processing time (t_j) hrs	30	8	10	28	16
Weight (W_j)	1	2	1	2	3

The weighted processing time $= \dfrac{\text{Processing time } (t_j)}{\text{Weight } (W_j)}$

The weighted processing time is represented as

Job (j)	1	2	3	4	5
Processing time (t_j) hrs	30	8	10	28	16
Weight (W_j)	1	2	1	2	3
t_j / W_j	30	4	10	14	5.31

Thus, arranging the jobs in the increasing order of t_j / W_j (weighted shortest processing time WSPT).

We have the optimal sequence that minimises the weighted mean flow time is 2-5-3-4-1.

The weighted mean flow time $(\overline{F}w)$

$$\overline{F}w = \frac{\sum_{j=1}^{n} W_j \cdot F_j}{\sum_{j=1}^{n} W_j}$$

The weighted mean flow time is computed as follows for optimal sequence.

Job (j)	2	5	3	4	1
Processing time (t_j) hrs	8	16	10	28	30
$F_j = (C_j - R_j)$	8	24	34	62	92
W_j	2	3	1	2	1
$F_j \times W_j$	16	72	34	124	92

Weighted mean flow time $(\overline{F}w)$ is computed as

$$\overline{F}w = \frac{(16+72+34+124+92)}{(2+3+1+2+1)} = 37.55 \text{ hrs.}$$

Problem 31

The processing times and due dates of jobs for a single machine scheduling is given.

Job (j)	*1*	*2*	*3*	*4*	*5*	*6*	*7*
Processing time (t_j)	*10*	*8*	*8*	*7*	*12*	*15*	*18*
Due date (d_j)	*15*	*10*	*12*	*11*	*18*	*25*	*30*

Determine the sequence which will minimise the maximum lateness and also determine the maximum lateness with respect to the optimal sequence.

Solution.

Due date (d_j) is the time at which job 'j' is to be completed and Lateness (L_j) is the amount of time by which the completion time of the job j differs from the due date ($L_j = C_j - d_j$). Lateness is a measure which gives set of due dates of times. Lateness can be positive or negative. Positive lateness indicates the completion of job after its due date. It is therefore often desirable to optimize positive lateness.

Arranging the jobs as per Earliest Due Date (EDD), *i.e.*, increasing order of their due dates. The sequence is 2-4-3-1-5-6-7. This sequence gives the minimum value of maximum lateness (L_{max}).

Computation of L_{max}

Job (j) as per EDD sequence	2	4	3	1	5	6	7
Processing time (t_j)	8	7	8	10	12	15	18
Completion time (C_j)	8	15	23	33	45	60	78
Due date (d_j)	10	11	12	15	18	25	30
Lateness (L_j)	(– 2)	4	11	18	27	35	48

From the above table, the maximum L_j is 48. This is the (optimised) value of L_{max}.

Problem 32

The processing times for five jobs and their due dates are given for a single machine scheduling below.

Job (j)	*1*	*2*	*3*	*4*	*5*
Processing time (t_j) hrs	*9*	*7*	*5*	*11*	*6*
Due date (in days) (d_j)	*16*	*20*	*25*	*15*	*40*

(a) Determine the sequence (b) Total completion time (c) Average completion time

(d) Average number of jobs in the system and average job lateness using the following priority sequencing rules

(i) Shortest Processing Time (SPT)

(ii) Earliest Due Date (EDD)

(iii) Longest Processing Time (LPT)

(e) Compare the above characteristics for the three sequencing rules.

Solution.

(*i*) Shortest Processing Time (SPT) sequence.

As per this rule, the job with the shortest processing time is scheduled first and immediately followed by next lowest processing time and so on.

Job sequence	*Processing time* (t_j) *days*	*Flow time* (F_i) *days*	*Due date* *(dj) days*	*Job lateness* *(days)*
3	5	5	25	0
5	6	11	40	0
2	7	18	20	0
1	9	27	16	11
4	11	38	15	23
Total	38	99		34

The various characteristics are :

(*i*) Total completion time (flow time) = 38 days

(*ii*) Average completion time $= \dfrac{\text{Total flow time}}{\text{No. of jobs}} = \dfrac{99}{5} = 19.8$ days

(*iii*) Average number of jobs in the system

$$= \frac{\text{Total flow time}}{\text{Total process time (completion)}} = \frac{99}{38} = 2.61 \text{ jobs}$$

(*iv*) Average job lateness $= \dfrac{\text{Total job lateness}}{\text{No. of jobs}} = \dfrac{34}{5} = 6.8$ days

(*ii*) Earliest Due Date (EDD) rule

As per this rule priority is given to the job with earliest due date. Arranging the jobs as per EDD sequence gives the sequence as 4-1-2-3-5.

Job sequence *(j)*	*Processing time* (C_j)	*Flow time* (F_i)	*Due Date* (D_j)	*Job lateness* *(Days)*
4	11	11	15	0
1	9	20	16	4
2	7	27	20	7
3	5	32	25	7
5	6	38	40	0
Total	38	128		18

Characteristics :

(*a*) Total completion time = 38 days

(*b*) Average completion time $= \dfrac{\text{Flow time}}{\text{No. of jobs}} = \dfrac{128}{5} = 25.6$ days

(*c*) Average number of jobs in the system $= \dfrac{\text{Flow time}}{\text{Completion time}} = \dfrac{128}{38} = 3.37$ jobs

(*d*) Average job lateness $= \dfrac{18}{5} = 3.6$ days

(*iii*) Longest Processing Time (LPT) Rule

The job sequence and computations as per this priority sequencing rule is given as follows :

Job sequence (j)	*Processing time* (C_j)	*Flow time* (F_i)	*Due Date* (D_j)	*Job lateness* (*Days*)
4	11	11	15	0
1	9	20	16	4
2	7	27	20	7
5	6	33	40	0
3	5	38	25	13
Total	38	128		24

Characteristics :

1. Total completion time = 38 days
2. Average completion time $= \dfrac{\text{Flow time}}{\text{No. of jobs}} = \dfrac{129}{5} = 25.8$ days
3. Average number of jobs in the system $= \dfrac{\text{Flow time}}{\text{Completion time}} = \dfrac{129}{38} = 3.39$ jobs
4. Average job lateness $= \dfrac{24}{5} = 4.8$ days

(*b*) Comparison of Priority rules

Priority rule	*Total completion time (days)*	*Avg. Completion time (days)*	*Avg. No.of jobs in the system*	*Avg. Job lateness*
SPT	38	19.8	2.61	6.8
EDD	38	25.6	3.37	3.6
LPT	38	25.8	3.39	4.8

Problem 33

The following jobs are waiting to be processed in a turning shop today (July, 23). The estimates of the time needed to complete the jobs are as follows :

Jobs (j)	*Due date*	*Processing time* (t_j) *in days*
1.	*July, 31*	*9*
2.	*August, 2*	*6*
3.	*August, 16*	*24*
4.	*July, 29*	*5*
5.	*August, 30*	*30*

Sequence the jobs based on the minimum critical ratio.

Solution. The critical ratio is computed as

$$\text{Critical Ratio (CR)} = \frac{\text{Time remaining for due date of the job}}{\text{Time needed to complete the job}}$$

$$= \frac{\text{Time remaining}}{\text{Work remaining}}$$

As per the critical ratio rule, a job with the minimum critical ratio is given the first preference, *i.e.*, the lower is the critical ratio, higher is its priority.

The denominator of CR, *i.e.*, the time needed to complete the job includes the processing time remaining plus the transfer times plus the estimated waiting times remaining for the job to go through before it is completely processed.

The table below gives the calculations of time remaining and time needed and the critical ratio.

Jobs (j)	*Due date*	*Processing time* (t_j) *in days*	*Time needed to complete the job in days*	*Critical ratio* CR = *Tr / Tn*
1.	July, 31	8	9	8/9 = 0.89
2.	August, 2	10	6	10/6 = 1.167
3.	August, 16	24	24	24/24 = 1.00
4.	July, 29	6	5	6/5 = 1.20
5.	August, 30	38	30	38/30 = 1.27

Note :

Critical Ratio (CR) of less than one means that the job is already late.

The CR value of one indicates that the job is on schedule and greater than one indicates that the job has some slack available to it.

From the table, job 1 has the lowest critical ratio and has to be processed first and job 2 has the highest critical ratio and it is scheduled last.

The sequence is

1	3	4	5	2

Problem 34

A company has 8 large machines which receive preventive maintenance. The maintenance team is divided into 2 crews A and B. Crew A takes the machine power and replaces parts as per given maintenance schedule. The second crew resets the machine and puts back into operation. At all times the no passing rule is considered to be in effect. The servicing time for each machine is given below.

Machine	*a*	*b*	*c*	*d*	*e*	*f*	*g*	*h*
Crew A	5	4	22	16	15	11	9	4
Crew B	6	10	12	8	20	7	2	21

Determine the optimal sequence of scheduling the factory maintenance crews to minimise their idle times and represent it on the Gantt chart.

Solution.

Step I

The minimum processing time on crew A is kept first in sequence while machine with processing times minimum on crew B is kept last in sequence.

Optimal sequence :

b	h	a	e	c	d	f	g

Step II : Calculation of total elapsed time

Optimal sequence	*Crew A*		*Crew B*		*Idle time on Crew B*
	S	*F*	*S*	*F*	
b	0	4	4	4	4
h	4	8	14	35	0
a	8	13	35	41	0
e	13	28	41	61	0
c	28	50	61	73	0
d	50	66	73	81	0
f	66	77	81	88	0
g	77	86	88	90	0

Total elapsed time is 90 hrs.

Gantt Chart :

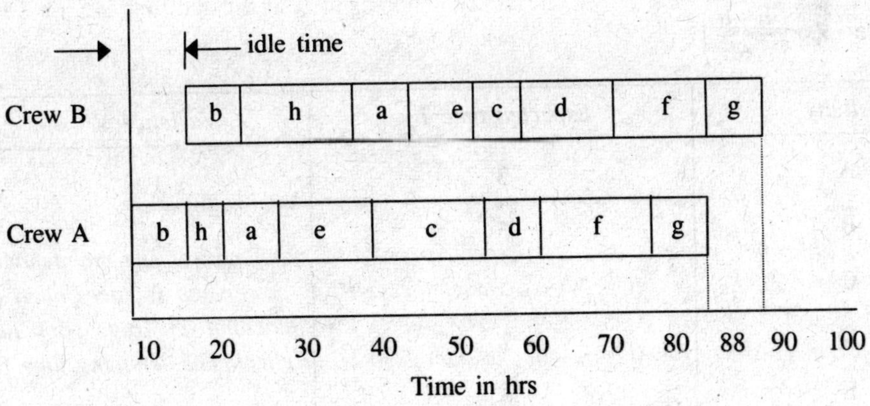

CPM AND PERT

Problem 35

A small project is composed of time activities whose time estimates are given below :

Activity	*A*	*B*	*C*	*D*	*E*	*F*	*G*	*H*	*I*
Optimistic time	*2*	*2*	*4*	*2*	*2*	*3*	*2*	*5*	*3*
Most likely time	*2*	*5*	*4*	*2*	*5*	*6*	*5*	*8*	*6*
Pessimistic time	*8*	*8*	*10*	*2*	*14*	*15*	*8*	*11*	*15*

Activities A, B and C can start simultaneously. Activity D follows activity A while E follows B. Activity D and E are followed by activity G while F is dependent on C, H depends on D and E, while I depends on F and G.

(i) Construct the network.

(ii) Find the expected duration and variance of each activity.

(iii) Calculate the slack for each event.

(iv) What is the critical path and expected project duration of the project.

(v) If the project due date is 28 days, what is the probability of not meeting the due date.

(vi) What should be the project duration for the probability of completion of 95%.

Solution.

Expected time $T_e = \dfrac{a+b+4m}{6}$

Variance $V = \left(\dfrac{b-a}{6}\right)^2$

Activity	*Expected time* T_e	*Variance V*
A	3	1
B	5	1
C	5	1
D	2	0
E	6	4
F	7	4
G	5	1
H.	8	1
I	7	2

Construction of Network

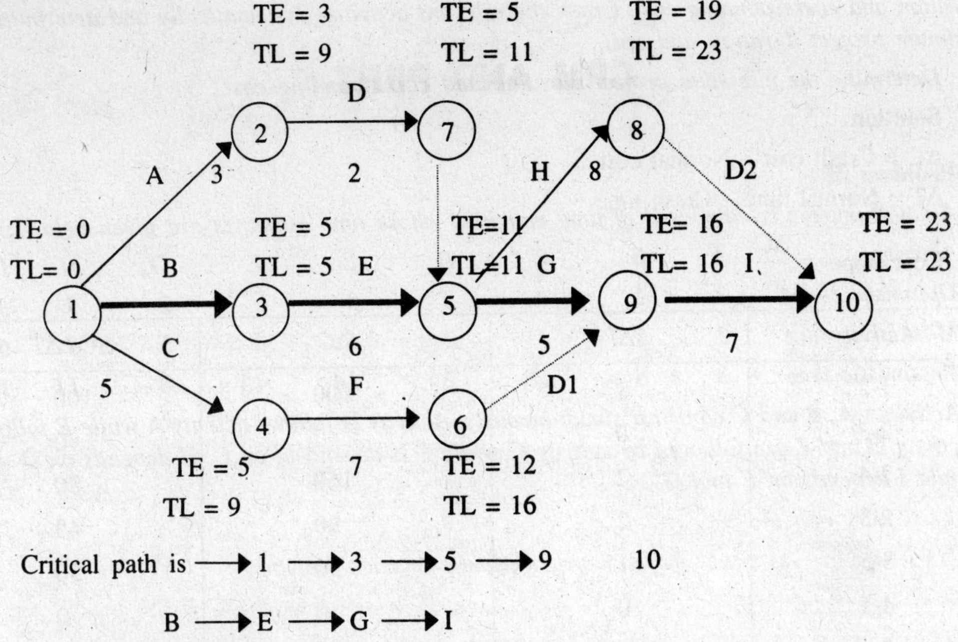

Critical path is ⟶ 1 ⟶ 3 ⟶ 5 ⟶ 9 10

B ⟶ E ⟶ G ⟶ I

Project duration 23 days.

Standard deviation $\sigma = \sqrt{V_{ij}} = 2.83$

(*i*) $Z = \dfrac{T_s - T_e}{6} = \dfrac{28 - 23}{6} = 0.833$

Probability of completion 78.8%.

(*ii*) With 95% probability

$$Z = 1.65 = \frac{T_s - 23}{6}$$

∴ Schedule time = 33 days

Problem 36

The following table gives data on normal cost and time, crash cost and time for a project. The indirect cost is Rs. 50/week.

Activity	*Time (week)*	*Normal cost (Rs)*	*Crash Time*	*Cost (Rs)*
1-2	3	300	2	400
2-3	3	30	3	30
2-4	7	420	5	580
2-5	9	720	7	810
3-5	5	250	4	300
4-5	0	0	0	0
5-6	6	320	4	410
6-7	4	400	3	470
6-8	13	780	10	900
7-8	10	1000	9	1200

Draw the network diagram and label it. Identify critical path and find out normal project duration and corresponding cost. Crash the relevant activities systematically and determine the optimum project duration and cost.

Determine the minimum project duration and corresponding cost.

Solution.

ΔC = Crash cost – Normal cost

ΔT = Normal time – Crash time

$$\text{Cost slope} = \frac{\Delta C}{\Delta T}$$

Activity	ΔT	ΔC	$\Delta C / \Delta T$
1-2	1	100	100
2-3	0	0	0
2-4	2	160	80
2-5	2	90	45
3-5	1	50	50
4-5	0	0	0
5-6	2	90	45
6-7	1	70	70
6-8	3	120	40
7-8	1	200	200

The network has been drawn for normal conditions and the times shown along the arrows are normal duration times.

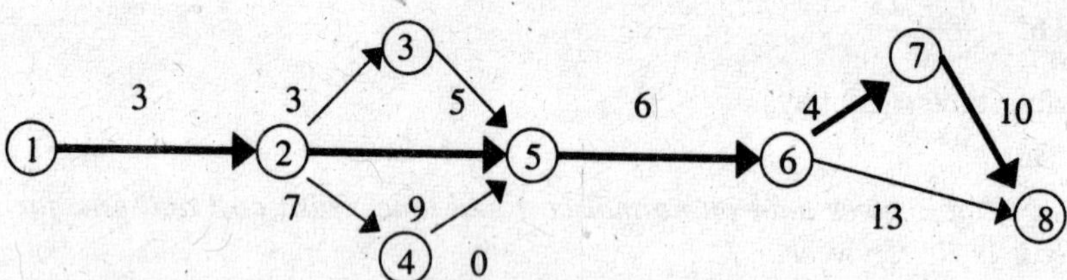

The critical path is $1 \longrightarrow 2 \longrightarrow 5 \longrightarrow 6 \longrightarrow 7 \longrightarrow 8$

The project duration is 32 weeks. Total cost of the normal project is the sum of the direct and indirect cost.

$\therefore$ Total project cost = Direct cost of all activities + Indirect cost of 32 weeks

$= 4220 + 32 \times 50$

$= 5820$ Rs.

Next step is to identify those activities along the critical path which can be crashed. Crash the least expensive activity, *i.e.*, the activity with the least cost slope.

Crashing activity 5-6 by 2 days with cost slope 45.

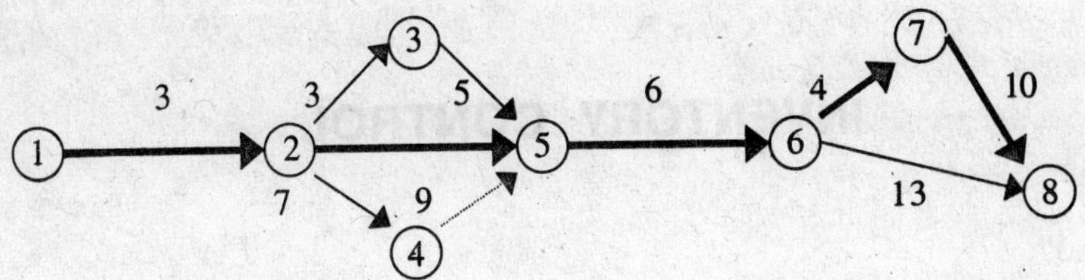

Critical path will remain as it is costs.

Project direct cost = 4220 + 2 × 45 = 4310.

Indirect cost = 30 × 50 = 1500.

Total project cost = 5810.

Step II : Crashing the activity 2-5 by 2 days and cost slope is 45.

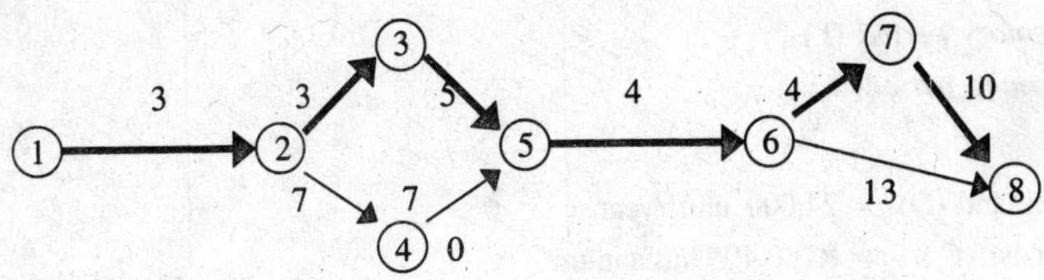

Now (1-2-3-5-6-7-8) paths are critical. Hence project duration is 28 days.

∴ Direct cost = 4310 + 2 × 45 = 4400

Indirect cost = 29 × 50 = 1450

Total project cost = 4400 + 1450 = 5850.

Step III : Crash the activity 3-5 by 1 day with cost slope 50.

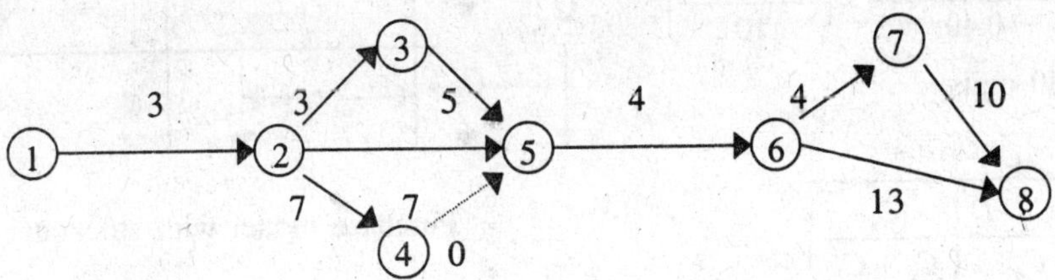

All the paths are critical.

∴ Direct cost = 4400 + 1 × 50 = 4450

Indirect cost = 28 × 50 = 1400

Total project cost = 4450 + 1400 = 5850.

Project duration is 28 days.

INVENTORY CONTROL

Problem 37

The annual demand for a machine component is 24,000 units. The carrying cost is Rs. 0.40 unit/year, the ordering cost is Rs. 20.00 per order and the shortage cost is Rs. 1,000 /unit/year. Find the values of the following :

(*i*) *Economic order quantity*

(*ii*) *Maximum inventory*

(*iii*) *Maximum shortage quantity*

(*iv*) *Cycle time*

(*v*) *Inventory period* (t_1)

(*vi*) *Shortage period* (t_2).

Solution.

Annual demand (D) = 24,000 units/year

Carrying cost (C_h) = Rs. 0.40/Unit/annum

Ordering cost (C_o) = Rs. 100/Unit/year

Shortage cost (C_s) = Rs. 10.00/Unit/year.

Economic order quantity (EOQ)

$$Q^* = \sqrt{\frac{2\,C_o \cdot D}{C_h}\left(\frac{C_s + C_h}{C_s}\right)}$$

$$Q^* = \sqrt{\frac{2 \times 20 \times 24000}{0.40} \times \left(\frac{10 + 0.40}{10}\right)}$$

$= 1580$ units.

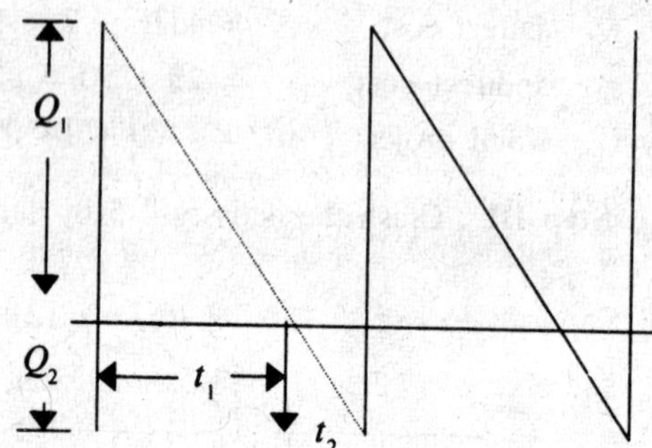

Purchase model with stockout

Q – Economic order quantity

Q_1 – Maximum inventory

Q_2 – Maximum stockout

$$Q_1^* = \sqrt{\frac{2\,C_o \cdot D}{C_h}\left(\frac{C_s}{C_s + C_h}\right)}$$

$$= \sqrt{\frac{2 \times 20 \times 24000}{0.40} \times \left(\frac{10}{10 + 0.40}\right)}$$

$= 1520$ units

Q_2 = Maximum stockout

$= Q^* - Q_1^*$

$= 1580 - 1520 = 60$ units

$$t = \frac{Q^*}{D} = \frac{1580}{24000} \times 365 = 24 \text{ days (cycle time)}$$

$$t_1 = \frac{Q_1^*}{D} = \frac{1520}{24000} \times 365 = 23 \text{ days (inventory period)}$$

$t_2 = t - t_1 = 24 - 23 = 1$ day (shortage period)

Problem 38

The demand for an item is 18,000 per year. Production rate is 3000 units/month. The carrying cost is Rs. 0.15/unit/month and the setup cost is Rs. 500 per setup. The shortage cost is Rs. 20.00 per unit per year. Find the following parameters.

1. *Economic Batch Quantity*
2. *Maximum Inventory*
3. *Maximum Stockout*
4. *Cycle time*

Solution.

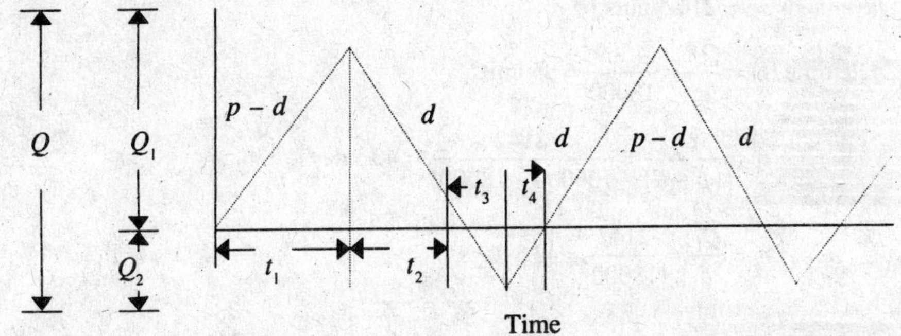

Production rate (p) = 3000 × 12 = 36,000 units/annum.

Demand (d) = 18,000 units/annum

Ordering cost (C_o) = Rs. 500 /setup

Carrying cost (C_h) = Rs (0.15 × 12) = Rs 1.8/unit/year.

Shortage cost (C_s) = Rs.20/unit/year.

Economic Order Quantity (Q^*)

$$Q^* = \sqrt{\frac{2\,C_o \cdot D_o}{C_h\,(1 - d/p)}\left(\frac{C_h + C_s}{C_s}\right)}$$

$$= \sqrt{\frac{2 \times 500 \times 18000}{1.8\,(1 - 18000/36000)} \times \left(\frac{1.80 + 20}{20}\right)}$$

= 4669 units

Q_1^* = Maximum inventory

$$= \sqrt{\frac{2\,C_o \cdot C_h}{C_s\,(C_s + C_h)}\left(\frac{p - d}{p}\right)\left(\frac{C_s}{C_s + C_h}\right)}$$

$$= \sqrt{\frac{2 \times 500 \times 18000\,(36000 - 18000)}{1.8 \times 36000} \times \frac{20}{20 + 1.8}}$$

= 2142 units

Q_2^* = Maximum Shortage

$$= \sqrt{\frac{2\,C_o \cdot C_h}{C_s\,(C_s + C_h)} \frac{d\,(p-d)}{p}}$$

$$= \sqrt{\frac{2 \times 500 \times 1.8}{20\,(20+1.8)} \times \frac{18000\,(36000-18000)}{36000}}$$

= 193 units

$$\left[\text{Check } Q_1^* = \left(\frac{p-d}{p} Q^*\right) - Q_2^*\right]$$

$$Q_1^* = \left[\frac{36000-18000}{36000} \times 4669\right] - 193$$

= 2142 units.

$$\text{Cycle time } (t) = \frac{Q^*}{d} = \frac{4669}{18000} = 95 \text{ units}$$

$$t_1 = \frac{Q_1^*}{(p-d)} = \frac{2142}{(36000-18000)} = 43.5 \text{ days}$$

$$t_2 = \frac{Q_1^*}{d} = \frac{2142}{18000} = 43.5 \text{ days}$$

$$t_3 = \frac{Q_2^*}{d} = \frac{193}{18000} = 4 \text{ days}$$

$$t_4 = \frac{Q_2^*}{(p-d)} = \frac{193}{(36000-18000)} = 4 \text{ days.}$$

Problem 39

A manufacturing company requires special gears at the rate of 300 numbers per year. Each gear costs Rs. 36. The procurement cost and inventory carrying costs are estimated at Rs. 30 and 20% respectively. If the supplier offers a discount of Rs. 2 per gear on an order of 200 or above, will it be advisable to avail the discount? What should be the order quantity ?

Solution.

Annual demand (D) = 300 units

Ordering cost/order (C_o) = Rs. 30

Basic price/unit (C_1) = Rs. 36

Discounted price/unit (C_2) = Rs. 34

Inventory carrying cost (i) = 0.20

The above prices are applicable to following quantities

Price	*Range of quantity*
Rs. 36	$0 < Q < 200$
Rs. 34	$200 \le Q$

(*i*) Compute EOQ (Q_2*) at discounted price C_2.

$$Q_2^* = \sqrt{\frac{2\,C_o \cdot D}{C_2\,i}}$$

$$= \sqrt{\frac{2 \times 30 \times 300}{34 \times 0.2}}$$

$$= 200 \text{ units.}$$

As the EOQ at discounted price C_2 is ≥ the price break quantity, it is economical to purchase the quantity (QB) equal to 200 units and avail the discount.

Problem 40

Annual demand for an item is 4800 units. Ordering cost is Rs. 500/order. Inventory carrying cost is 24% of the purchase price per unit, per year. The price breaks are given as follows.

Quantity	*Price (Rs) / unit*
$0 \le Q_1 \le 1200$	10
$1200 \le Q_2 \le 2000$	9
$2000 \le Q_3$	8

Find :

(a) Optimal Order Quantity

(b) If the ordering cost is changed to Rs.300 per order, find the optimal order quantity.

Solution.

Annual demand (D) = 4800 units. Let C_1 = 10

Ordering Cost (C_o) = 500 C_2 = 9

Inventory carrying cost (i) = 0.24 C_3 = 8

1. Find the economic order quantity (Q_3*) at the Price C_3

   EOQ price C_3 (Q_3*)

$$Q_3^* = \sqrt{\frac{2\,C_o \cdot D}{C_3\,i}}$$

$$= \sqrt{\frac{2 \times 500 \times 4800}{8 \times 0.24}}$$

$$= 1581 \text{ units.}$$

Since, Q_3* is less than the price break quantity (QB_2) (2000), go to the next step (2).

2. Find the economic order quantity (Q_2*) at price C_2.

   EOQ (Q_2*) at price C_2 is given by

$$Q_2^* = \sqrt{\frac{2\,C_o \cdot D}{C_2\,i}}$$

$$= \sqrt{\frac{2 \times 500 \times 4800}{9 \times 0.24}}$$

$$= 1491 \text{ units.}$$

Since Q_2* > price break quantity (QB_1), go to the next step.

3. Find out the total cost at Q_2^*

$$TC\left(Q_2^*\right) = 9 \times 4800 + \frac{500 \times 4800}{1491} + \frac{0.24 \times 9 \times 1491}{2}$$
$$= \text{Rs. } 46{,}420.$$

Find the annual total cost at QB_2 (2000)

$$TC\left(QB_2\right) = 8 \times 4800 + \frac{500 \times 4800}{2000} + \frac{0.24 \times 8 \times 2000}{2}$$
$$= \text{Rs. } 46{,}496/\text{annum}.$$

Comparing $TC\ (Q_2^*)$ and $TC\ (QB_2)$, the least cost is Rs. 46,420. Hence, the optimal order size is Q_2^* (*i.e.*, economic order quantity at C_2) which is equal to 1491.

(*B*) If the ordering cost (C_o) is reduced to Rs. 300/order.

(*i*) Calculate EOQ at the least price (C_3).

$$Q_3^* = \sqrt{\frac{2\, C_o \cdot D}{C_3\, i}} = \sqrt{\frac{2 \times 300 \times 4800}{8 \times 0.24}} = 1125 \text{ units.}$$

Since Q_3^* > the price break quantity (QB_2), proceed to step (*ii*).

(*ii*) Calculate economic order quantity at price C_2 (Q_2^*)

$$Q_2^* = \sqrt{\frac{2\, C_o \cdot D}{C_2\, i}} = \sqrt{\frac{2 \times 300 \times 4800}{9 \times 0.24}} = 1155 \text{ units.}$$

Since $Q_2^* < QB_1$ [price break quantity (1200)], proceed to step (*iii*).

(*iii*) Calculate EOQ (Q_1^*) at price C_1

$$Q_1^* = \sqrt{\frac{2\, C_o \cdot D}{C_1\, i}} = \sqrt{\frac{2 \times 300 \times 4800}{10 \times 0.24}} = 1096 \text{ units.}$$

As $Q_2^* < QB_1$, compute the total annual costs at Q_1^* and QB_1, QB_2

$$TC\left(Q_1^*\right) = 10 \times 4800 + \frac{300 \times 4800}{1096} + \frac{0.24 \times 10 \times 1096}{2} = \text{Rs } 50{,}629.$$

$$TC\left(QB_1\right) = 9 \times 4800 + \frac{300 \times 4800}{1200} + \frac{0.24 \times 9 \times 1200}{2} = \text{Rs } 45{,}696.$$

$$TC\left(QB_3\right) = 8 \times 4800 + \frac{300 \times 4800}{2000} + \frac{0.24 \times 8 \times 2000}{2} = \text{Rs } 41{,}040.$$

Since $TC\ (QB_2)$ is the minimum cost, the optimal order size is equal to 2000 units.

Problem 41

Daily usage of drug follows a normal distribution with a mean of 500 gm and a standard deviation of 50 gm. If the lead time for the procurement is 7 days and the drug store wants a risk of only 2% determine.

(*a*) *Re-order point*

(*b*) *Safety stock.*

The drug store practices 'Q' system of inventory.

Solution. Mean demand during lead time (DL) = 500 × 7 = 3500 gms per day. The weekly demand is also normally distributed and the variation of weekly demand equals the sum of daily variances.

Variance during lead time $= 7 \times (50)^2$

Standard deviation during lead time $= \sqrt{7 \times (50)^2} = 50\sqrt{7}$

The risk level being 2%, the service level is (100 – 2) = 98%.

Referring to normal distribution table, the service level of 0.98, corresponds to $Z = 2.05$.

and $$Z = \frac{x - \text{Mean}(\mu)}{\text{Std. deviation}} = \frac{x - 3500}{50\sqrt{7}} = 2.05$$

$\therefore$ $x = 3771$ gms.

This is the re-order point.

Safety Stock $(S_s) = x - \mu = 3771 - 3500 = 271$ gms.

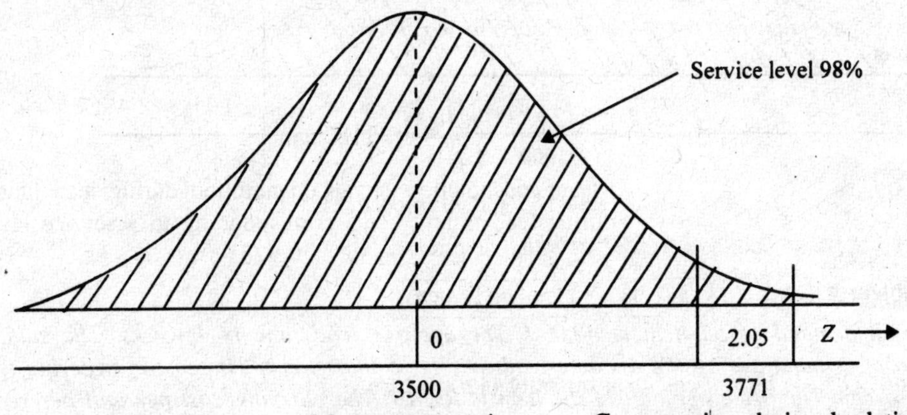

Problem 42

A company buys an item for its assembly. The usage pattern of this bought out items follows a normal distribution with a mean of 1000 items per week and a standard deviation of 200. The buying process takes one week. The inventory holding cost is Rs. 5 per unit per year and the cost of ordering is Rs. 200 per order. The company allows for only 2 stockout situations in a year. Compute the safety stock required.

Solution.

Service level is expressed here in terms of 2 stock-outs per year.

$\therefore$ It is required to find out the number of orders per annum.

$$\text{No. of ordering cycles per year} = \frac{\text{Annual demand }(D)}{\text{EOQ }(Q^*)}$$

$$\text{Economic Order Quantity }(Q^*) = \sqrt{\frac{2\,C_o\,D}{C_h}} = \sqrt{\frac{2 \times (52 \times 1000) \times 200}{5}}$$

$= 2039$ units.

$$\text{Number of Ordering cycles/year} = \frac{D}{D^*} = \frac{52000}{2039} = 26.$$

$$\therefore \text{ Service level is } \frac{26 - 2}{26} = 0.9231$$

The service level of 0.9231 corresponds to Z value of 1.425. (from normal distribution table).

Since the standard deviation of usage during a week (also lead time) is 200 units, the safety stock required is

Safety stock = $Z \cdot \sigma = 1.425 \times 200 = 285$ units.

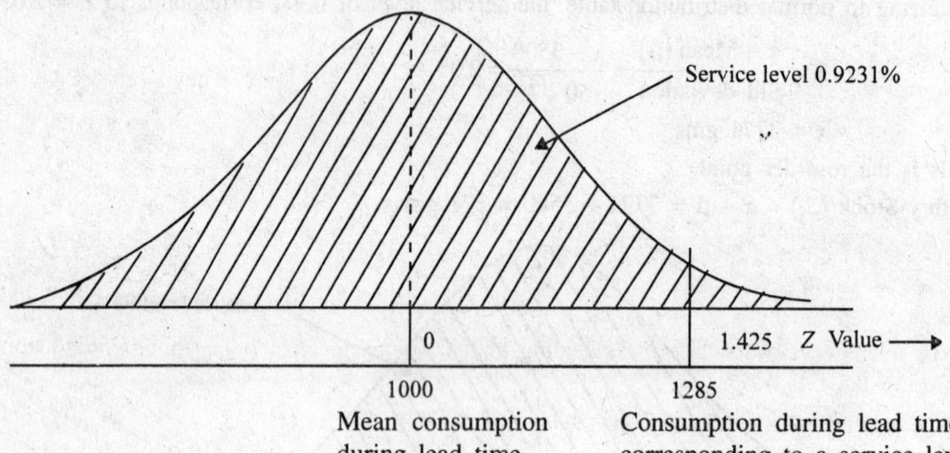

Problem 43

Annual demand of an item is 48,000. The average lead time is 4 weeks. The standard deviation of demand during the average lead time is 75 units/week. The cost of ordering is Rs. 400 per order. The price per unit of the item is Rs. 10. The carrying cost per unit per year is 15% of the purchase price. The maximum delay in lead time is 2 weeks and the probability of this delay is 0.25. Consider the service level of 0.95.

Find

(*i*) *Re-order level, if 'Q' system is followed.*

(*ii*) *The maximum inventory level, if 'P' system is followed.*

Solution.

Annual demand (D)	= 48,000 units
Ordering Cost per order (C_o)	= Rs. 400
Purchase Price per unit (C_1)	= Rs. 10
Carrying Cost per unit per annum (C_h)	= 0.15×10 = Rs. 1.50
Average lead time (L)	= 4 weeks
Standard deviation of demand during average lead time per week	= 75 units
Maximum delay	= 2 weeks
Probability of having maximum delay	= 0.25
Service level	= 0.95

Order Quantity in case of 'P' system equals EOQ

$$\text{Order Quantity } (Q^*) = \sqrt{\frac{2\,C_o\,D}{C_h}} = \sqrt{\frac{2 \times 400 \times 48000}{1.5}}$$

$$= 5060 \text{ units}$$

(*i*) *Reorder level for 'Q' system*

Demand during lead time (D_L) = Demand/week × Lead time (L)

$$= \frac{48000}{52} \times 4$$

= 3692 units

Standard deviation in demand during lead time $= \left(\sqrt{L}\right) \times$ standard deviation per week

$$= \sqrt{4} \times 75$$

= 150 units

Safety Stock during lead time (S_S) = $Z . \sigma$

= 1.64 × 150

= 246 units.

The average value of 'Z' corresponding to the service level of 0.95 from standard normal distribution table is 1.64.

Average demand during delivery delays (Reserve stock)

$$= \frac{\text{Demand }(D) \times \text{maximum delay}}{\text{No. of weeks per annum}} \times \text{probability of maximum delay}$$

$$= \frac{48{,}000}{52} \times 2 \times 0.25 = 462 \text{ units}$$

Re-order level (ROL) = Demand during lead time (D_L) + Safety stock (S_S)

Average demand during delivery delays (Reserve stock)

= 3692 + 246 + 462 = 4400 units.

(*ii*) Maximum inventory level for 'P' system

$$\text{Review period} = \frac{\text{EOQ }(Q^*)}{\text{Annual demand }(D)} = \frac{5060}{48{,}000} = 0.105 \text{ year} = 5.46 \text{ weeks.}$$

The review period is either 5 weeks or 6 weeks. The selection of review period is based on the total cost. The weeks for which the total cost is minimum is selected as review period.

(*a*) When review period is 5 weeks;

Total cost = Ordering cost + Carrying cost

$$= \left(\frac{52}{5}\right) \times 400 + \left(\frac{48000}{52} \times 5\right)\frac{1}{2} \times 1.50$$

= 4160 + 3462 = Rs. 7622.

(*b*) When review period is 6 weeks

Total cost = Ordering cost + Carrying cost

$$= \left(\frac{52}{6}\right) \times 400 + \left(\frac{48000}{52} \times 6\right)\frac{1}{2} \times 1.50$$

= 3467 + 4154 = Rs. 7621.

The total cost is minimum when the review period is 6 weeks. Hence, the review period of 6 weeks is selected.

Demand during lead time and review period $= \dfrac{48000}{52} \times (4 \times 6)$

$= 9231$ units

Safety stock during lead time and review period $= \sqrt{(4+6)} \times \sigma \times k$

$= \sqrt{10} \times 75 \times 1.64$

$= 389$ units

Average demand during delivery delays (Reserve stock)

$$= \frac{\text{Annual demand } (D) \times \text{maximum delay}}{\text{No. of weeks per annum}} \times \text{Probability of maximum delay}$$

$$= \frac{48000}{52} \times 2 \times 0.25 = 462 \text{ units}$$

Maximum Inventory level

= Demand during lead time and review period + safety stock during lead time and review period + Average demand during delivery delays

= 9231 + 389 + 462

= 10,182 units.

MAINTENANCE MANAGEMENT

Problem 44

Indian Electronics, manufactures TV sets and carries out the picture tube testing for 2000 hours. A sample of 100 tubes were put through this quality test during which two tubes failed. If the average usage of TV by the customer is 4 hours/day and if 10,000 TV sets were sold, then in one year how many tubes were expected to fail and what is the mean time between failures for these tubes ?

Solution.

The total test time = (100 tubes) × 2000 hours

= 200,000 tube-hours.

There are two tubes which have failed and hence the total time is to be adjusted for the number of hours lost due to the failures during the testing.

The lost hours are computed as = $2 \times \frac{2000}{2} = 2000$ hours.

The assumption is made here is that each of the failed tubes have lasted an average of half of the test period.

Therefore, the test shows that there are two failures during (2,00,000 – 2000) = 1,98,000 tube hours of testing.

During 365 days a year (four hours a day) for 10,000 tubes the number of expected failures are $\frac{2}{1,98,000} \times 10,000 \times 365 \times 4$

$= 147.47 \approx 148$ tubes approximately.

$$\text{Mean time between failures} = \frac{\text{1,98,000 tubes hrs. of testing}}{\text{2 failures}}$$

= 99,000 tubes hours per failure

$$= \frac{99,000}{4 \times 365}$$

= 67.68 tubes year per failure

Problem 45.

A company has 40 dumpers for extraction work which can be serviced on a preventive maintenance schedule at 5,000 Rs. Each. If the dumpers fail, it costs 12,500 to get them back in to service. Records show that probability of failures after maintenance are shown below :

Months after maintenance	*Probability of breakdown*
1	2
2	1
3	3
4	4

should the preventive maintenance policy be followed ? If so, how often should the vehicles be serviced ?

Solution.

Cost calculation of preventive and breakdown maintenance.

(*i*) Preventive Maintenance

Preventive maintenance cost = Servicing cost + Breakdown cost

= [No.of vehicles serviced] × Service cost/vehicle +

[Expected no.of breakdowns between service] [Breakdown cost/vehicle]

The cumulative expected number of breakdowns *B* in *M* months is given by,

$$B_n = N \sum_{1}^{n} p_n + B_{n-1}\, p_1 + B_{n-2}\, p_2 + \ldots + B_1\, p_{n-1}$$

where N = Number of vehicles

p = Probability of breakdown during a given month after maintenance

n = Maintenance period

Thus, $B_1 = N\,(p_1) = 40 \times 0.2 = 8.0$

$B_2 = N\,(p_1 + p_2) + B_1 p_1 = 40 \times (0.2 + 0.1) + 8\,(0.2) = 13.6$

$B_3 = N\,(p_1 + p_2 + p_3) + B_2 p_1 + B_1 p_2$

$= 40\,(0.2 + 0.1 + 0.3) + 13.6\,(0.2) + 8\,(0.1) = 27.52$

$B_4 = N\,(p_1 + p_2 + p_3 + p_4) + B_3 p_1 + B_2 p_2 + B_1 p_3$

$= 40\,(0.2 + 0.1 + 0.3 + 0.4) + 27.52\,(0.2) + 13.6\,(0.1) + 8\,(0.3) = 49.26$

The difference between the monthly cumulative totals, then represent an individual period breakdown, *i.e.*, the expected number of breakdowns during period 2 is 13.6 – 8 = 5.6.

The preventive maintenance cost analysis is shown in the table below.

	1 month	2 month	3 month	4 month
Cumulative break downs during PM period	8.0	13.60	27.52	49.26
Cost at 12,500 each	1,00,000	1,70,000	3,44,000	6,15,750
Preventive maintenance cost at Rs. 5,000 each	40,000	68,000	1,37,600	2,46,300
Total cost of maintenance	1,40,000	2,38,000	4,81,600	8,62,050

(*b*) Breakdown Maintenance Cost

Let C_p be the expected cost of breakdown

$$C_p = \frac{N\,C_r}{\Sigma\, T_n\,(p_n)}$$

where C_r = Cost of repairing the cells

T_n = Number of time periods after repair

p_n = Probability of breakdown during the time period '*n*'

Thus, $\Sigma T_n (p_n) = 1\ (0.2) + 2\ (0.1) + 3\ (0.3) + 4\ (0.4)$

$= 2.9$ months between breakdown

$$C_p = \frac{N\,C_r}{\Sigma T_n (p_n)} = \frac{40 \text{ vehicles} \times 12{,}500 / \text{breakdown}}{2.9 \text{ months} / \text{breakdown}}$$

$= 1{,}72{,}413.5$ Rs./month

The preventive maintenance policy of both 2nd month and 3rd month are preferred to the breakdown maintenance as the breakdown maintenance cost is higher than preventive maintenance cost for these two months.

Problem 46

The manager of a tool crib has to decide how many attenders he should have to operate the tool crib. The demand for service come at a mean rate of 24 per hour. An attender can service 10 requests. The demands show a negative exponential inter arrival time distribution and the service rates show a poison distribution. If the tool crib attenders are paid Rs. 8 per hour and the mechanics who come with requests are paid Rs. 15 per hour. Find the optimum number of attendents.

Solution.

Arrival rate $(\lambda) = 24$

Service rate $(\mu) = 10$

$$\therefore \quad \frac{\lambda}{\mu} = 2.40$$

Compare the average number of mechanics waiting for different number of attenders. Convert this into the cost of waiting of the mechanics (Idle/waiting time). The total cost of any alternative would be the cost of mechanics idle time and the cost of having the number of attendents. The least total cost alternative is chosen.

In computing the average number of mechanics waiting, make the use of appendix. Which gives the values of N_w (average number waiting) for various values of S (number of attendants).

The starting value of the S has to be greater than $\frac{\lambda}{\mu}$ (which is 2.4). So, the starting values of S is to be 3.

Computation for optional number of attendants

No. of attendants (*S*)	*Avg. no. of waiting Mechanics* (N_w)	*Avg. no. in system* $N_s = N_w + \frac{\lambda}{\mu}$	*Cost of Mechanics per hour* ($N_s \times 15$) *Rs*	*Cost of attendants* ($S \times 8$) *Rs*	*Total cost/hr Rs*
3	2.1261	4.521	67.89	24.00	91.89
4	0.4305	2.8305	42.86	32.00	**74.46**
5	0.1047	2.5047	57.57	40.00	77.57
6	0.0266	2.4266	36.40	48.00	84.40
7	0.0065	2.4065	36.10	56.00	92.10

Therefore, optimal number of tool crib attendants is 4.

Problem 47

The following table gives the operation cost, maintenance cost of a machine and the salvage value at the end of every year whose purchase price Rs. 2000.

(i) Find the economic life (best age of replacement) assuming that the money value will not change.

(ii) Find the economic life of the machine assuming interest rate of 15%.

End of your (n)	*Operation cost at the end of year*	*Maintenance cost at the end of year*	*Salvage value at the end of year*
1	*2000*	*200*	*10000*
2	*3000*	*300*	*9000*
3	*4000*	*400*	*8000*
4	*5000*	*500*	*7000*
5	*6000*	*600*	*6000*
6	*7000*	*700*	*5000*
7	*8000*	*800*	*4000*
8	*9000*	*900*	*3000*
9	*10000*	*1000*	*2000*
10	*11000*	*1100*	*1000*

Solution.

(*a*) **Without Considering Time Value of Money**

Since, the operation and maintenance cost occur at the end of the year they are added together to get the total at the end of each year.

$$\text{The average annual cost} = \frac{\text{Cumulative operation + Purchase price} \\ - \text{Salvage value and maintenance cost}}{\text{End of the year}}$$

The values of average annual cost are computed as shown in table. The average annual cost is minimum for $n = 4$. Hence the economic life of the machine is 4 years.

(*b*) **Here the Change in Value is Considered $i = 15\%$**

The steps to get the economic life are as follows :

(*i*) Discount the running and maintenance costs to the beginning of year 1.

(*ii*) Find the annual equivalent running and maintenance cost through year given.

(*iii*) Find the annual equivalent first cost through given years.

(*iv*) Find the total annual equivalent cost through n years.

Identify the end of the year for which the total annual equivalent cost is minimum.

Total annual equivalent cost =

$$\left[\begin{array}{l}\text{Cumulative sum of present} \\ \text{worth as of beginning of} \\ \text{year 1 of operation and} \\ \text{maintenance cost}\end{array} + \begin{array}{c}\text{First} \\ \text{cost}\end{array} - \begin{array}{l}\text{Present worth as} \\ \text{of beginning of} \\ \text{year 1 of salvage} \\ \text{value}\end{array}\right] \times (A/P,\ 15\%,\ n)$$

Table : Computation of Annual Equivalent Cost (i = 15%)

End of year (n)	Operation cost at end of year	Maintenance cost at end of year	Sum of operation and maintenance cost	P/F 15%, n	Present worth at beginning off year 1 of sum of maintenance and running costs	Cum sum of column F through years	Salvage value at the end of year	Present worth of salvage values	Total present work	A/P 15%, n	Total annum equivalent
A	B	C	$D = B + C$	E	$F = D \times E$	$G = \sum F$	H	$I = H \times E$	$J = G + 2000 - I$	K	$L = J \times K$
1	2,000	200	2,200	0.8696	1913.12	1913.12	10000	8696.00	13217.12	1.1500	15200
2	3,000	300	3,300	0.7562	2495.46	4408.58	9000	6805.80	17602.78	0.6151	10828
3	4,000	400	4,400	0.6575	2893.00	7301.58	8000	5260.00	22041.58	0.4380	9654
4	5,000	500	5,500	0.5718	3144.90	10446.48	7000	4002.60	26443.88	0.3503	9263
5	6,000	600	6,600	0.4972	3281.52	13728.00	6000	2983.20	30744.80	0.2983	9171
6	7,000	700	7,700	0.4323	3328.71	17056.71	5000	2161.50	34895.21	0.2642	9219
7	8,000	800	8,800	0.3759	3307.92	230364.63	4000	1503.60	38861.03	0.2404	9342
8	9,000	900	9,900	0.3269	3236.31	23600.94	3000	9807.70	42620.24	0.2229	9500
9	10,000	1,000	11,000	0.2843	3127.30	26728.24	2000	568.60	46159.64	0.2096	9675
10	11,000	1,100	12,100	0.2472	2991.12	29719.36	1000	247.20	49472.16	0.1993	9860

Economic life 5 years.

Total annual equivalent (Column L) = [Column G + 20,000 – Column I] × Column K

= Column J × Column K

Table showing average annual cost computation (i = 0%)

First cost = Rs. 20,000

End of year (n)	*Operation cost at end of year*	*Maintenance cost at end of year*	*Summation operation & maintenance cost at end of year*	*Cumulative sum of operation & maintenance at end of year*	*Salvage value at end of year*	*Total cost at the beginning of year 1*	*Average annual cost through year*
A	B	C	$D = B + C$	$E = \sum D$	F	$G = E + 2000 - F$	G/A
1	2000	200	2200	2200	10000	12200	12200
2	3000	300	3300	5500	9000	16500	8250
3	4000	400	4400	9900	8000	21900	7300
4	5000	500	5500	15400	7000	28400	7100
5	6000	600	6600	22000	6000	36000	7200
6	7000	700	7700	29700	5000	44700	7450
7	8000	800	8800	38500	4000	54500	7786
8	9000	900	9900	48400	3000	65400	8175
9	10000	1000	11000	59400	2000	77400	8600
10	11000	1100	12000	71500	1000	90500	9050

Problem 48

Following failure rates have been observed for a particular type of transistors in a computer.

End of week	*1*	*2*	*3*	*4*	*5*	*6*	*7*
Probability of failure	*0.07*	*0.18*	*0.30*	*0.48*	*0.69*	*0.89*	*1.00*

The cost of replacing a failed transistor individually is Rs. 10. If all the transistors are replaced simultaneously it would cost Rs. 2.5 per transistor. Any one of the following two options can be followed to replace the transistors.

(a) Replace the transistors individually when they fail.

(b) Replace all the transistors simultaneously at fixed intervals and replace the individual transistors as they fail in service during the fixed internal (group replacement policy)

Find out the replacement policy, i.e., individual or group replacement policy. If group replacement is optimal, then what equal intervals should all the transistors be placed ?

Solution.

(*a*) To start with, make an assumption that there are 100 transistors in use. Let p_i be the probability that a transistor which was new when placed in position for use, fails during ith week of its life.

Hence, $p_1 = 0.7$, $p_2 = 0.11$, $p_3 = 0.12$, $p_4 = 0.18$, $p_5 = 0.21$, $p_6 = 0.2$, $p_7 = 0.11$. Since, the sum of p_i is equal to 1 at the end of 7th week, the transistors are sure to fail during the 7th week. Assume that (*a*) transistors that fail during a week are replaced just before the end of the week end.

(*b*) The actual percentage of failures during a week for a sub group of transistors with the same as the expected percentage of failures during the week for that sub group of transistors.

Let, N_i be the number of transistors replaced at the end of ith week.

N_0 = Number of transistors replaced at the end of the week 'O' (*i.e.*, at the beginning of the 7th week) = 100.

N_1 = Number of transistor replaced at the end of the 1st week
$= N_0 \times p_1 = 100 \times 0.07 = 7.$

N_2 = Number of transistor replaced at the end of the 2nd week
$= N_0 \times p_2 + N_1 \times p_1 = 100 \times 0.11 + 7 \times 0.07 = 12.$

N_3 = Number of transistor replaced at the end of the 3rd week
$= N_0 \times p_3 + N_1 \times p_2 + N_2 \times p_1$
$= 100 \times 0.12 + 7 \times 0.11 + 12 \times 0.07$
$= 14$

N_4 = Number of transistor replaced at the end of the 4th week
$= N_0 \times p_4 + N_1 \times p_3 + N_2 \times p_2 + N_3 \times p_1$
$= 100 \times 0.18 + 7 \times 0.12 + 12 \times 0.11 + 14 \times 0.07$
$= 21$

N_5 = Number of transistor replaced at the end of the 5th week
$= N_0 \times p_5 + N_1 \times p_4 + N_2 \times p_3 + N_3 \times p_2 + N_4 \times p_1$
$= 100 \times 0.21 + 7 \times 0.18 + 12 \times 0.12 + 14 \times 0.11 + 21 \times 0.07$
$= 27$

N_6 = Number of transistor replaced at the end of the 6th week
$= N_0 \times p_6 + N_1 \times p_5 + N_2 \times p_4 + N_3 \times p_3 + N_4 \times p_2 + N_5 \times p_1$
$= 100 \times 0.2 + 7 \times 0.21 + 12 \times 0.18 + 14 \times 0.12 + 21 \times 0.11 + 27 \times 0.07$
$= 30$

N_7 = Number of transistor replaced at the end of the 7th week
$= N_0 \times p_7 + N_1 \times p_6 + N_2 \times p_5 + N_3 \times p_4 + N_4 \times p_3 + N_5 \times p_2 + N_6 \times p_1$
$= 100 \times 0.11 + 7 \times 0.2 + 12 \times 0.21 + 14 \times 0.18 + 21 \times 0.12 + 27 \times 0.11 + 30 \times 0.07$
$= 25.$

Cost of individual replacement

$$\text{Expected life of each transistor} = \sum_{i=1}^{7} i \times p_i$$

$= 1 \times 0.07 + 2 \times 0.11 + 3 \times 0.12 + 4 \times 0.18 + 5 \times 0.21 + 6 \times 0.2 + 7 \times 0.11$

$= 4.39$ weeks.

$$\text{No. of failures/week} = \frac{100}{4.39} = 23.$$

∴ Cost of individual replacement
= (No.of failures/week x individual replacement cost/transistor)
= 23 × 10 = Rs. 230.

Determination of group replacement cost

Cost per transistor when replaced simultaneously = Rs. 2.5.
Cost per transistor when replaced individually = Rs. 10.

The cost of transistor when replacement for different weeks is represented in the table below.

Table : Group Replacement Cost

End of week A	*Cost of replacing 100 transistors at a time (Rs)* B	*Cost of replacing transistors during given replacement period* C	*Total cost* D = B+C	*Average cost/ week* E = D/A
1	250	7 × 10 = 70	320	320
2	250	(7+12) × 10 = 190	440	220
3	250	(7+12+14) × 10 = 330	580	193
4	250	(7+12+14+21) × 10 = 540	790	198
5	250	(7+12+14+21+27) × 10 = 810	1060	212
6	250	(7+12+14+21+27+30) × 10 = 1110	1360	227
7	250	(7+12+14+21+27+30+25) × 10 = 1360	1610	230

From the above table, it is clear that the average cost per week is minimum for the third week.

Hence, group replacement cost is 3 weeks.

Individual replacement Cost/week = Rs. 230.

Minimum group replacement cost/week = Rs. 193.

Since the minimum group replacement cost per week as lesser than the individual replacement cost/week. Group replacement policy is best and hence all transistors are to be replaced once in three week and the transistors which fail during this three weeks period to be replaced individually.

Problem 49

The failure rate of machines follow a poisson distribution with a mean failure rate of 16 machines/day. The maintenance time of breakdown follows a negative exponential distribution with a mean service time of 20 minutes by a single mechanics. The cost of down time of machine is Rs. 3000 per hour. Maintenance mechanic is paid Rs. 100/Day. Find the optimal number of mechanics to be downtime and cost of mechanics is minimised. Assume 8 hrs. Shift per day.

Solution.

Machine breakdown rate = 16 per day.

Case 1. The number of mechanics is assumed as one.

Service rate of machines $(\mu) = \frac{1}{20} \times 60 \times 8 = 24$ per day

Utilisation on factor $(P) = \frac{16}{24} = 0.67$

Average waiting time of breakdown machines/day

$$W_1 = \frac{1}{\mu(1-P)} = \frac{1}{24(1-0.67)} = 0.126 \text{ hrs.}$$

Cost of downtime of machines = 0.126 × 3000 = Rs. 378.00

Cost of One mechanics per day = 1 × 100 = Rs. 100.00

Total cost = Rs. 478.00

Case 2. The number of mechanics is assumed as two.

Service rate of machines $(\mu) = \dfrac{1}{20/2} \times 60 \times 8 = 48$ per day

Utilisation on factor $(P) = \dfrac{16}{48} = 0.333$

Average waiting time of breakdown machines/day

$$W_1 = \frac{1}{\mu(1-P)} = \frac{1}{48(1-0.333)} = 0.031 \text{ hrs.}$$

Cost of downtime of machines = 0.031 × 3000 = Rs. 93.00

Cost of One mechanics per day = 2 × 100 = Rs. 200.00

Total cost = Rs. 293.00

Case 3. The number of mechanics is assumed as three.

Service rate of machines $(\mu) = \dfrac{1}{20/3} \times 60 \times 8 = 72$ per day

Utilisation on factor $(P) = \dfrac{16}{72} = 0.222$

$$W_1 = \frac{1}{\mu(1-P)} = \frac{1}{72(1-0.222)} = 0.01785 \text{ hrs.}$$

Cost of downtime of machines = 0.01785 × 3000 = Rs. 53.55

Cost of One mechanics per day = 3 × 100 = Rs. 300.00

Total cost = Rs. 353.55

The total cost of all these three cases is represented as follows.

No.of mechanics	*Total cost/day Rs*
1	478.00
2	293.00
3	353.55

From the above table, it is clear that optimal size of maintenance personnel is 2.

LINEAR PROGRAMMING PROBLEM

Problem 50

The annual hand made furniture show and sales occurs next month and the school of vocational studies is playing to make furniture for the sale. There are three wood working classes - I year, II year, III year at the school and they have decided to make three styles of chairs A, B and C. Each chair must receive work in each class and the time in hours for each chair in each class is given.

Chair	*I year*	*II year*	*III year*
A	*2*	*4*	*3*
B	*3*	*3*	*2*
C	*2*	*1*	*4*

In the next month there will be 120 available in I year class, 160 in the second year class and 100 hours in third year class to produce chairs. The teacher of the wood working class feels that a maximum of 40 chairs can be sold at the show. The teacher has determined that the profit from each type of chair will be A – Rs. 40, B – Rs. 35 and C – Rs. 30.

Formulate a linear programming model to determine how many chairs should be produced to maximise profit.

Solution.

Let x_1 are the chairs produced of A type

x_2 are the chairs produced of B type

x_3 are the chairs produced of C type

Formulation

$$2x_1 + 3x_2 + 2x_3 \le 120$$

$$4x_1 + 3x_2 + x_3 \le 160$$

$$3x_1 + 2x_2 + 4x_3 \le 100$$

$$x_1, x_2, x_3 \ge 0$$

Objective function

Maximise $Z = 40x_1 + 35x_2 + 30x_3$

The problem is solved by simple method, convert inequalities into equalities by adding slack variables.

S.T.

$$2x_1 + 3x_2 + 2x_3 + x_a = 120$$

$$4x_1 + 3x_2 + x_3 + x_b = 160$$

$$3x_1 + 2x_2 + 4x_3 + x_c = 100$$

Rewritting objective function as

$$Z - 40x_1 - 35x_2 - 30x_3 = 0$$

Selection of non basic variables

$n - m = 6 - 3 = 3.$

where n = No. of variables

m = No. of constraint equation.

where x_a, x_b and x_c are slack variables.

Non basic variables $x_1 = x_2 = x_3 = 0$ gives.

$\therefore \quad x_a = 120$

$\therefore \quad x_b = 160$

$\therefore \quad x_c = 100$

The initial solution is represented in starting table.

Starting table

Basics	Z	x_1	x_2	x_3	x_a	x_b	x_c	RHS	Ratio
Z	1	– 40	– 35	– 30	0	0	0	0	–
x_a	0	2	3	2	1	0	0	120	60
x_b	0	4	3	1	0	1	0	160	40
x_c	0	3	2	4	0	0	1	100	33.33

Entering variable x_1 leaving variable = x_c Iteration No. 1

Basics	Z	x_1	x_2	x_3	x_a	x_b	x_c	RHS	Ratio
Z	1	0	– 25/3	70/3	0	0	40/3	4000/3	–
x_a	0	0	5/3	– 2/3	1	0	– 2/3	160/3	32
x_b	0	0	1/3	– 13/3	0	1	– 4/3	80/3	80
x_c	0	1	2/3	4/3	0	0	1/3	100/3	50

Entering variable x_2 leaving variable = x_a Iteration No. 2

Basics	Z	x_1	x_2	x_3	x_a	x_b	x_c	RHS	Ratio
Z	1	0	0	20	5	0	10	1600	–
x_a	0	0	1	– 2/5	3/5	0	– 2/5	32	
x_b	0	0	0	42/10	– 3/25	1	– 94/75	20	
x_c	0	1	0	19/9	– 2/5	0	17/15	12	

As Z row has all positive values hence optimal solution has reached

$\therefore \quad Z = 1600$

$x_1 = 12$ units

$x_2 = 32$ units

$x_3 = 0$

Problem 51

Maximise $Z = 12x_1 + 20x_2 + 18x_3 + 40x_4$

Subjected to

$$4x_1 + 9x_2 + 7x_3 + 10x_4 \leq 6000$$
$$x_1 + x_2 + x_3 + x_4 \leq 4000$$
$$x_1, x_2, x_3, x_4 \geq 0$$

Solution.

Step I. Convert inequalities into equalities.

Maximise $Z - 12x_1 - 20x_2 - 18x_3 - 40x_4 = 0$

such that

$$4\,x_1 + 9x_2 + 7x_3 + 10x_4 + x_5 = 6000$$
$$x_1 + x_2 + x_3 + x_4 + x_6 = 4000$$

Step II. Initial solution

$$n - m = 6 - 2 = 4.$$

Putting $x_1, x_2, x_3, x_4 = 0$ gives

$\therefore \quad x_5 = 6000$

$\therefore \quad x_6 = 4000$

$\therefore \quad Z = 0$

The results of initial solution are represented in starting table.

Starting table

Basics	Z	x_1	x_2	x_3	x_4	x_5	x_6	RHS
Z	1	– 12	– 20	– 18	– 40	0	0	0
x_5	0	4	9	7	10	1	0	6000
x_6	0	1	1	3	40	0	1	4000

Iteration No. 1

Leaving variable x_6, Entering variable x_4

Basics	Z	x_1	x_2	x_3	x_4	x_5	x_6	RHS
Z	1	– 11	– 19	– 15	0	0	1	4000
x_5	0	15/4	35/4	25/4	0	1	– 1/4	5000
x_4	0	1/40	1/40	3/40	1	0	1/40	100

Iteration No. 2

Leaving variable x_2, Entering variable x_5

Basics	Z	x_1	x_2	x_3	x_4	x_5	x_6	RHS
Z	1	– 20/7	0	– 10/7	0	76/35	16/35	52000/35
x_2	0	3/7	1	5/7	0	4/35	– 1/35	20000/35
x_4	0	1/70	0	2/35	1	– 1/350	1/1400	3000/35

Iteration No. 3

Leaving variable x_1. Entering variable x_2

Basics	Z	x_1	x_2	x_3	x_4	x_5	x_6	RHS
Z	1	0	20/3	10/3	0	10780/3675	3220/8575	56000/3
x_1	0	1	7/3	5/3	0	4/15	– 1/15	20000/15
x_4	0	0	– 1/30	0.046	1	-6.66×10^{-3}	-1.66×10^{-3}	200/3

As in iteration no. 3. All the values in Z row are positive.

Hence, optimal solutions has reached hence solution is

Maximum Value is 56000/3.

$$x_1 = 4000/3$$

$$x_4 = 200/3$$

$$x_2 = x_3 = 0.$$

Substituting the values in Z equation.

$$Z = \frac{12 \times 4000}{3} + 20 \times 10 + 7 \times 0 + 40 \times \frac{200}{3}$$

$$= \frac{48000}{3} + \frac{8000}{3} = \frac{56000}{3}$$

APPENDICES

APPENDIX A1 : Normal Distribution

	.00	.01	.02	.03	.04	.05	.06	.07	.08	.09
.0	.5000	.5040	.5080	.5120	.5160	.5199	.5239	.5279	.5319	.5359
.1	.5398	.5438	.5478	.5517	.5557	.5596	.5636	.5675	.5714	.5753
.2	.5793	.5832	.5871	.5910	.5948	.5987	.6026	.6064	.6103	.6141
.3	.6179	.6217	.6255	.6293	.6331	.6368	.6406	.6443	.6480	.6517
4.	.6554	.6591	.6628	.6664	.6700	.6737	.6772	.6808	.6844	.6879
.5	.6915	.6950	.6985	.7019	.7054	.7088	.7123	.7157	.7090	.7224
.6	7257	.7291	.7324	.7357	.7389	.7422	.7454	.7486	.7517	.7549
.7	.7580	.7611	.7642	.7673	.7704	.7734	.7764	.7794	.7823	.7852
.8	.7881	.7910	.7939	.7967	.7995	.8023	.8015	.8078	.8106	.8133
.9	.8159	.8186	.8212	.8238	.8264	.8289	.8315	.8340	.8365	.8389
1.0	.8413	.8438	.8461	.8485	.8508	.8531	.8554	.8577	.8599	.8621
1.1	.8643	.8665	.8686	.8708	.8729	.8749	.8770	.8790	.8810	.8830
1.2	.8849	.8869	.8888	.8907	.8925	.8944	.8962	.8980	.8997	.9015
1.3	.9032	.9049	.9066	.9082	.9099	.9115	.9131	.9147	.9162	.9177
1.4	.9192	.9207	.9222	.9236	.9215	.9265	.9279	.9292	.9306	.9319
1.5	.9332	.9345	.9357	.9370	.9382	.9394	.9406	.9418	.9429	.9441
1.6	.9452	.9463	.9474	.9484	.9495	.9505	.9515	.9525	.9535	.9545
1.7	.9554	.9564	.9573	.9582	.9591	.9599	.9608	.9616	.9625	.9633
1.8	.9641	.9649	.9656	.9664	.9671	.9678	.9686	.9693	.9699	.9706
1.9	.9713	.9719	.9726	.9732	.9738	.9744	.9750	.9756	.9761	.9767
2.0	.9772	.9778	.9783	.9788	.9793	.9798	.9803	.9808	.9812	.9817
2.1	.9821	.9826	.9830	.9834	.9834	.9838	.9842	.9846	.9850	.9857
2.2	.9861	.9864	.9868	.9871	.9875	.9878	.9881	.9884	.9887	.9890
2.3	.9893	.9896	.9898	.9901	.9904	.9906	.9909	.9911	.9913	.9916
2.4	.9918	.9920	.9922	.9925	.9927	.9929	.9931	.9932	.9934	.9936
2.5	.9938	.9940	.9941	.9943	.9945	.9946	.9948	.9949	.9951	.9952
2.6	.9953	.9955	.9956	.9957	.9959	.9960	.9961	.9962	.9963	.9964
2.7	.9965	.9966	.9967	.9968	.9969	.9970	.9971	.9972	.9973	.9974
2.8	.9974	.9975	.9976	.9977	.9977	.9978	.9979	.9979	.9980	.9981
2.9	.9981	.9982	.9982	.9983	.9984	.9984	.9985	.9985	.9986	.9986
3.0	.9987	.9987	.9987	.9988	.9988	.9989	.9989	.9989	.9990	.9990
3.1	.9990	.9991	.9991	.9991	.9992	.9992	.9992	.9992	.9993	.9993
3.2	.9993	.9993	.9994	.9994	.9994	.9994	.9994	.9995	.9995	.9995
3.3	.9995	.9995	.9995	.9996	.9996	.9996	.9996	.9996	.9996	.9997
3.4	.9997	.9997	.9997	.9997	.9997	.9997	.9997	.9997	.9997	.9998

APPENDIX A2 : Tables of Random Numbers

71509	68310	48213	99928	64650	13229	36921	58732	13459	93487
21949	30920	23287	89514	58502	46185	00368	82613	02668	37444
50639	54968	11409	36148	82090	87298	41396	71111	00076	60029
47837	76716	09653	54466	87987	82362	17933	52793	17641	19502
31735	36901	92293	19293	57582	86043	69502	12601	00535	82697
04174	32342	66532	07875	54445	08795	63563	42295	74646	73120
96980	68728	21154	56181	71843	66134	52396	89723	96435	17871
21823	04027	76402	04655	87276	32593	17097	06913	05136	05115
25922	07122	31485	52166	07645	85122	20945	06369	70254	22806
32530	98882	19105	01769	20276	59401	60426	03316	41438	22012
00159	08461	51810	14650	45119	97920	08063	70819	01832	53295
66574	21384	75357	55888	83429	96916	73977	87883	13249	28870
00995	28829	15048	49573	65277	61493	44031	88719	73057	66010
55114	79226	27929	23392	06432	50200	39054	15528	53483	33972
10614	25190	52647	62580	51183	31338	60008	66595	64357	14985
31359	74469	58126	59192	23371	25190	37841	44386	92420	42965
09736	51873	94595	61367	82091	63835	86858	10677	58209	59820
24709	23224	45788	21426	63353	29874	51058	29958	61220	61199
79957	67598	74102	49824	39305	15069	56327	26905	34453	53964
66616	22137	72805	64420	58711	68435	60301	28620	91919	96080
01413	27281	19397	36231	05010	42003	99865	20924	76151	54089
88238	80731	20777	45725	41480	48277	45704	96457	13918	52375
57457	87883	64273	26236	61095	01309	48632	00431	63730	18917
21614	06412	71007	20255	39890	75336	89451	88091	61011	38072
26466	03735	39891	26361	86816	48193	33492	70484	77322	01016
97314	03944	04509	46143	88908	55261	73433	62538	63187	57352
91207	33555	75942	41668	64605	38741	86189	38197	99112	59649
46791	78974	01999	78891	16177	95746	78076	75001	51309	18791
02376	40372	45077	73705	56076	01853	83512	81567	55951	27156
33994	56809	58377	45976	01581	78389	18268	90057	93382	28494
92588	92024	15048	87841	38008	80689	73098	39001	10907	88092
73767	61534	66197	47147	22994	38197	60844	86926	27595	29907
51517	39870	94094	77092	94595	37904	27553	02229	44993	10468
33910	05156	60844	89012	21154	68937	96477	05867	95809	72827
09444	93069	61764	99301	55826	78849	26131	28201	91417	98172
96896	43760	72890	78682	78243	24061	55449	53587	77574	51580
97523	54633	99656	08503	52563	12099	52479	72374	79581	57143
42568	30794	32613	21802	73809	60237	70087	36650	54487	43718
45453	33136	90246	61953	17724	42421	17611	95369	94208	95369
52814	26445	73516	24897	90622	35018	70087	60112	09025	05324
87318	33345	14546	15445	81588	74561	12246	47858	08983	18205
08063	83575	25294	93027	09988	04487	88364	31078	22200	91019
53400	62078	52103	25650	75315	18916	06809	88217	12245	33053
90789	60614	20862	34475	11744	24437	55198	55219	74730	59820
73684	25859	86858	48946	30914	79017	53776	72537	83638	44680
82007	12183	89326	53713	77782	50368	01748	39033	47042	65758
80208	30920	97774	41417	79038	60531	32990	57770	53441	58732
62434	96122	63019	58439	89702	38657	60049	88761	22785	66093
40718	83199	65863	58857	49886	70257	27511	99426	53985	84077

APPENDIX A3 : Cumulative Poisson Probabilities

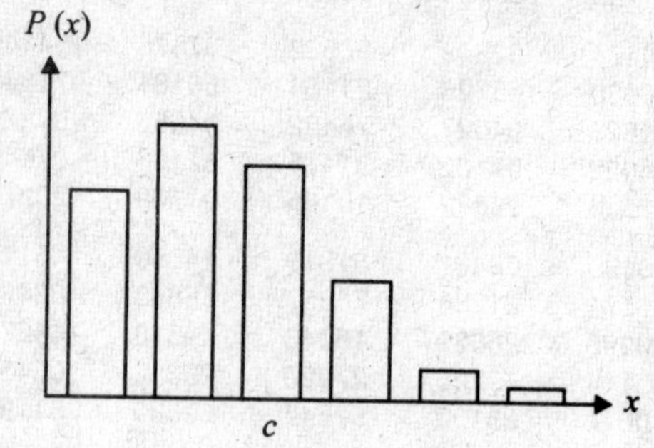

	x												
np	**0**	**1**	**2**	**3**	**4**	**5**	**6**	**7**	**8**	**9**	**10**	**11**	**12**
.05	.951	.999	.1000										
.10	.905	.995	1.000										
.15	.861	.990	.999	.1000									
.20	.819	.982	.999	.1000									
.25	.779	.974	.998	.1000									
.30	.741	.963	.996	.1000									
.35	.705	.951	.994	.1000									
.40	.670	.938	.992	.999	.1000								
.45	.638	.925	.989	.999	.1000								
.50	.607	.910	.986	.998	.1000								
.55	.577	.894	.982	.998	.1000								
.60	.549	.878	.977	.997	.1000								
.65	.522	.861	.972	.996	.999	.1000							
.70	.497	.844	.966	.994	.999	.1000							
.75	.472	.827	.959	.993	.999	.1000							
.80	.449	.809	.953	.991	.999	.1000							
.85	.427	.791	.945	.989	.998	.1000							
.90	.407	.772	.937	.987	.998	.1000							
.95	387	.754	.929	.984	.997	.1000							
1.0	.368	.736	.920	.981	.996	.999	.1000						
1.1	.333	.699	.900	.974	.995	.999	.1000						
1.2	.301	.663	.879	.966	.992	.998	.1000						
1.3	.273	.627	.857	.957	.989	.998	.1000						
1.4	.247	.592	.833	.946	.986	.997	.999	.1000					
1.5	.223	.558	.809	.934	.981	.996	.999	.1000					
1.6	.202	.525	.783	.921	.976	.994	.999	.1000					
1.7	.183	.493	.757	.907	.970	.992	.998	.1000					
1.8	.165	.463	.731	.891	.964	.990	.997	.999	.1000				
1.9	.150	.434	.704	.875	.956	.986	.997	.999	.1000				
2.0	.135	.406	.677	.857	.947	.983	.995	.999	.1000				
2.2	.111	.355	.623	.819	.928	.975	.993	.998	.1000				
2.4	.091	.308	.570	.779	.904	.964	.988	.997	.999	.1000			
2.6	.074	.267	.518	.736	.877	.951	.983	.995	.999	.1000			
2.8	.061	.231	.469	.692	.848	.935	.986	.992	.998	.999	.1000		
3.0	.050	.199	.423	.647	.815	.916	.966	.988	.996	.999	.1000		
3.2	.041	.171	.380	.603	.781	.875	.955	.983	.994	.998	.1000		
3.4	.033	.147	.340	.558	.744	.871	.972	.977	.992	.997	999	.1000	
3.6	.027	.126	.303	.515	.706	.844	.927	.969	.988	.996	999	.1000	
3.8	.022	.107	.269	.473	.668	.816	.909	.960	.984	.994	.998	.999	.1000
4.0	.018	.092	.238	.433	.629	.785	.889	.949	.979	.992	.997	.999	.1000

P (x)

c

x

$$P\ x \le c = \sum_{x=0}^{x=c} \frac{\lambda^{x} \cdot e^{-\lambda}}{x!}$$

	x													
np	**0**	**1**	**2**	**3**	**4**	**5**	**6**	**7**	**8**	**9**	**10**	**11**	**12**	**13**
4.2	.015	.078	.210	.395	.590	.753	.867	.936	.972	.989	.996	.999	1.000	
4.4	.012	.066	.185	.359	.551	.720	.844	.921	.964	.985	.994	.998	.999	1.000
4.6	.010	.056	.163	.326	.513	.683	.818	.905	.995	.980	.992	.997	.999	1.000
4.8	.008	.048	.143	.294	.476	.651	.791	.887	.994	.975	.990	.996	.999	1.000
5.0	.007	.040	.125	.265	.440	.616	.762	.867	.932	.968	.986	.995	.998	.999
5.2	.006	.034	.109	.238	.406	.581	.732	.845	.918	.960	.982	.993	.997	.999
5.4	.005	.029	.095	.213	.373	.546	.702	.822	.903	.951	.977	.990	.996	.999
5.6	.004	.024	.082	.191	.342	.512	.670	.797	.886	.941	.972	.988	.995	.998
5.8	.003	.021	.072	.170	.313	.478	.638	.771	.867	.929	.965	.984	.993	.997
6.0	.002	.017	.062	.151	.285	.446	.606	.744	.847	.916	.957	.980	.991	.996
6.2	.002	.015	.054	.134	.259	.414	.574	.716	.826	.902	.949	.975	.989	.995
6.4	.002	.012	.046	.119	.235	.384	.542	.687	.803	.886	.939	.969	.986	.994
6.6	.001	.010	.040	.105	.213	.355	.511	.658	.780	.869	.927	.963	.982	992
6.8	.001	.009	.034	.093	.192	.327	.480	.628	.755	.850	.915	.955	.978	.990
7.0	.001	.007	.030	.082	.173	.301	.450	.599	.729	.830	.901	.947	.973	.987
7.2	.001	.006	.025	.072	.156	.276	.420	.569	.703	.810	.887	.937	.967	.984
7.4	.001	.005	.022	.063	.140	.253	.392	.539	.676	.788	.871	.926	.961	.980
7.6	.001	.004	.019	.055	.125	.231	.365	.510	.648	.765	.854	.915	.954	.976
7.8	.000	.004	.016	.048	.112	.210	.338	.481	.620	.741	.835	.902	.945	.971
8.0	.000	.003	.014	.042	.100	.191	.313	.453	.593	.717	.816	.888	.936	.966
8.2	.000	.003	0.12	.037	.089	.174	.290	.425	.565	.692	.796	.873	.926	.960
8.4	.000	.002	.010	.032	.079	.157	.267	.399	.537	.666	.774	.857	.915	.952
8.6	.000	.002	.009	.028	.070	.142	.246	.373	.509	.640	.752	.840	.903	.945
8.8	.000	.001	.007	.024	.062	.128	.226	.348	.482	.614	.729	.822	.890	.936
9.0	.000	.001	.006	.021	.055	.116	.207	.324	.456	.587	.706	.803	.876	.926
9.2	.000	.001	.005	.018	.049	.104	.189	.301	.430	.561	.682	.783	.861	.916
9.4	.000	.001	.005	.016	.043	.093	.173	.279	.404	.535	.658	.763	.845	.904
9.6	.000	.001	.004	.014	.038	.084	.157	.258	.380	.509	.633	.741	828	.892
9.8	.000	.001	.003	.012	.033	.075	.143	.239	.356	.483	.608	.719	.810	.879
10.0	0	.000	.003	.010	.029	.067	.130	.220	.333	.458	.583	.697	.792	.864
10.2	0	.000	.002	.009	.026	.060	.118	.203	.311	.433	.558	.674	.772	.849
10.4	0	.000	.002	.008	.023	.053	.107	.186	.290	.409	.533	.650	.752	.834
10.6	0	.000	.002	.007	.020	.048	.097	.171	.269	.385	.508	.627	.732	.817
10.8	0	.000	.001	.006	.017	.042	.087	.157	.250	.363	.484	.603	.710	.799
11.0	0	.000	.001	.005	.015	.038	.079	.143	.232	.341	.460	.579	.689	.781
11.2	0	.000	.001	.004	.013	.033	.071	.131	.215	.319	.436	.555	.667	.762
11.4	0	.000	.001	.004	.012	.029	.064	.119	.198	.299	.413	.532	.644	.743
11.6	0	.000	.001	.003	.010	.026	.057	.108	.183	.279	.391	.508	.622	.723
11.8	0	.000	.001	.003	.009	.023	.051	.099	.169	.260	.369	.485	.599	.702
12.0	0	.000	.001	.002	.088	.020	.046	.090	.155	.242	.347	.462	.576	.682

	x													
np	**0**	**1**	**2**	**3**	**4**	**5**	**6**	**7**	**8**	**9**	**10**	**11**	**12**	**13**
12.2	0	0	0.000	0.002	0.007	0.018	0.041	0.081	0.142	0.225	0.327	0.439	0.553	0.660
12.4	0	0	0.000	0.002	0.006	0.016	0.037	0.073	0.131	0.209	0.307	0.417	0.530	0.639
12.6	0	0	0.000	0.001	0.005	0.014	0.033	0.066	0.120	0.194	0.288	0.395	0.508	0.617
12.8	0	0	0.000	0.001	0.004	0.012	0.029	0.060	0.109	0.179	0.269	0.374	0.485	0.595
13.0	0	0	0.000	0.001	0.004	0.011	0.026	0.054	0.100	0.166	0.252	0.353	0.463	0.573
13.2	0	0	.000	.001	.003	.009	.023	.049	.091	.153	.235	.333	.441	.551
13.4	0	0	.000	.001	.003	.008	.020	.044	.083	.141	.219	.314	.420	.529
13.6	0	0	.000	.001	.002	.007	.018	.039	.075	.130	.204	.295	.399	.507
13.8	0	0	.000	.001	.002	.006	.016	.035	068	.119	.189	.277	.378	.486
14.0	0	0	0	.000	.002	.006	.014	.032	.062	.109	.176	.260	.358	.464
14.2	0	0	0	.000	.002	.005	.013	.028	.056	.100	.163	.244	.339	.443
14.4	0	0	0	.000	.001	.004	.011	.025	.051	.092	.151	.228	.320	.423
14.6	0	0	0	.000	.001	.004	.014	.032	.062	.084	.139	.213	.302	.402
14.8	0	0	0	.000	.001	.003	.009	.020	.042	.077	.129	.198	.285	.383
15.0	0	0	0	.000	.001	.003	.008	.018	.037	.070	.118	.185	.268	.363
15.2	0	0	0	.000	.001	.002	.007	.016	.034	.064	.109	.172	.251	.344
15.4	0	0	0	.000	.001	.002	.006	.014	.030	.058	.100	.160	.236	.326
15.6	0	0	0	.000	.001	.002	.005	.013	.027	.053	.092	.148	.221	.308
15.8	0	0	0	0	.000	.002	.005	.011	.025	.048	.084	.137	.207	.291
16.0	0	0	0	0	.000	.001	.004	.010	.022	.043	.077	.127	.193	.275
16.2	0	0	0	0	.000	.001	.004	.009	.020	.039	.071	.117	.180	.259
16.4	0	0	0	0	.000	.001	.003	.008	.018	.035	.065	.108	.168	.243
16.6	0	0	0	0	.000	.001	.003	.007	.016	.032	.059	.100	.156	.228
16.8	0	0	0	0	.000	.001	.002	.006	.014	.029	.054	.092	.145	.214
17.0	0	0	0	0	.000	.001	.002	.005	.013	.026	.049	.085	.135	.201
17.2	0	0	0	0	.000	.001	.002	.005	.011	.024	.045	.078	.125	.188
17.4	0	0	0	0	.000	.001	.002	.004	.010	.021	.041	.071	.116	.176
17.6	0	0	0	0	0	.000	.001	.004	.009	.019	.037	.065	.107	.164
17.8	0	0	0	0	0	.000	.001	.003	.008	.017	.033	.060	.099	.153
18.0	0	0	0	0	0	.000	.001	.003	.007	.015	.030	.055	.092	.143
18.2	0	0	0	0	0	.000	.001	.003	.006	.014	.027	.050	.085	.133
18.4	0	0	0	0	0	.000	.001	.002	.006	.012	.025	.046	.078	.123
18.6	0	0	0	0	0	.000	.001	.002	.005	.011	.022	.042	.082	.115
18.8	0	0	0	0	0	.000	.001	.002	.004	.010	.020	.038	.066	.106
19.0	0	0	0	0	0	.000	.001	.002	.004	.009	.018	.035	.061	.098
19.2	0	0	0	0	0	0	.000	.001	.003	.008	.017	.032	.056	.091
19.4	0	0	0	0	0	0	.000	.001	.003	.007	.015	.029	.051	.084
19.6	0	0	0	0	0	0	.000	.001	.003	.006	.013	.026	.047	.078
19.8	0	0	0	0	0	0	.000	.001	.002	.006	.012	.024	.043	.072
20.0	0	0	0	0	0	0	.000	.001	.002	.005	.011	.021	.039	.066

APPENDIX A4
Learning Curve : Table of Unit Values

	Learning Rate							
Unit	60%	65%	70%	75%	80%	85%	90%	95%
1	1.0000	1.0000	1.0000	1.0000	1.0000	1.0000	1.0000	1.0000
2	0.6000	0.6500	0.7000	0.7500	0.8000	0.8500	0.9000	0.9500
3	0.4450	0.5052	0.5682	0.6338	0.7021	0.7729	0.8462	0.9219
4	0.3600	0.4225	0.4900	.0.5625	0.6400	0.7225	0.8100	0.9025
5	0.3054	0.3678	0.4368	0.5127	0.5956	0.6857	0.7830	0.8877
6	0.2670	0.3284	0.3977	0.4754	0.5617	0.6570	0.7616	0.8758
7	0.2383	0.2984	0.3674	0.4459	0.5345	0.6337	0.7439	0.8659
8	0.2160	0.2746	0.3430	0.4219	0.5120	0.6141	0.7290	0.8574
9	0.1980	0.2552	0.3228	0.4017	0.4930	0.5974	0.7161	0.8499
10	0.1832	0.2391	0.3058	0.3846	0.4765	0.5828	0.7047	0.8433
12	0.1602	0.2135	0.2784	0.3565	0.4493	0.5584	0.6854	0.8320
14	0.1430	0.1940	0.2572	0.3344	0.4276	0.5386	0.6696	0.8226
16	0.1296	0.1785	0.2401	0.3164	0.4096	0.5220	0.6561	0.8145
18	0.1188	0.1659	0.2260	0.3013	0.3944	0.5078	0.6445	0.8074
20	0.1099	0.1554	0.2141	0.2884	0.3812	0.4954	0.6342	0.8012
22	0.1025	0.1465	0.2038	0.2772	0.3697	0.4844	0.6251	0.7955
24	0.0961	0.1387	0.1949	0.2674	0.3595	0.4747	0.6169	0.7904
25	0.0933	0.1353	0.1908	0.2629	0.3548	0.4701	0.6131	0.7880
30	0.0815	0.1208	0.1737	0.2437	0.3346	0.4505	0.5963	0.7775
35	0.0728	0.1097	0.1605	0.2286	0.3184	0.4345	0.5825	0.7687
40	0.0660	0.1010	0.1498	0.2163	0.3050	0.4211	0.5708	0.7611
45	0.0605	0.0939	0.1410	0.2060	0.2936	0.4096	0.5607	0.7545
50	0.0560	0.0879	0.1336	0.1972	0.2838	0.3996	0.5518	0.7486
60	0.0489	0.0785	0.1216	0.1828	0.2676	0.3829	0.5367	0.7386
70	0.0437	0.0713	0.1123	0.1715	0.2547	0.3693	0.5243	0.7302
80	0.0396	0.0657	0.1049	0.1622	0.2440	0.3579	0.5137	0.7231
90	0.0363	0.0610	0.0987	0.1545	0.2349	0.3482	0.5046	0.7168
100	0.0336	0.0572	0.0935	0.1479	0.2271	0.3397	0.4966	0.7112
120	0.0294	0.0510	0.0851	0.1371	0.2131	0.3255	0.4830	0.7017
140	0.0262	0.0464	0.0786	0.1287	0.2038	0.3139	0.4718	0.6937
160	0.0237	0.0427	0.0734	0.1217	0.1952	0.3042	0.4623	0.6869
180	0.0218	0.0397	0.0691	0.1159	0.1879	0.2959	0.4541	0.6809
200	0.0201	0.0371	0.0655	0.1109	0.1816	0.2887	0.4469	0.6757
250	0.0171	0.0323	0.0584	0.1011	0.1691	0.2740	0.4320	0.6646
300	0.0149	0.0289	0.0531	0.0937	0.1594	0.2625	0.4202	0.6557
350	0.0133	0.0262	0.0491	0.0879	0.1594	0.2532	0.4105	0.6482
400	0.0121	0.0241	0.0458	0.0832	0.1453	0.2454	0.4022	0.6419
450	0.0111	0.0224	0.0431	0.0792	0.1399	0.2387	0.3951	0.6363
500	0.0103	0.0210	0.0408	0.0758	0.1352	0.2329	0.3888	0.6314
600	0.0090	0.0188	0.0372	0.0703	0.1275	0.2232	0.3782	0.6229
700	0.0080	0.0171	0.0344	0.0659	0.1214	0.2152	0.3694	0.6158
800	0.0073	0.0157	0.0321	0.0624	0.1163	0.2086	0.3620	0.6098
900	0.0067	0.0146	0.0302	0.0594	0.1119	0.2029	0.3556	0.6045
1,000	0.0062	0.0137	0.0286	0.0569	0.1082	0.1980	0.3499	0.5998
1,200	0.0054	0.0122	0.0260	0.0527	0.1020	0.1897	0.3404	0.5918
1,400	0.0048	0.0111	0.0240	0.0495	0.0971	0.1830	0.3325	0.5850
1,600	0.0044	0.0102	0.0225	0.0468	0.0930	0.1773	0.3258	0.5793
1,800	0.0040	0.0095	0.0211	0.0446	0.0895	0.1725	0.3200	0.5743
2,000	0.0037	0.0089	0.0200	0.0427	0.0866	0.1683	0.3149	0.5698
2,500	0.0031	0.0077	0.0178	0.0389	0.0806	0.1597	0.3044	0.5605
3,000	0.0027	0.0069	0.0360	0.0360	0.0760	0.1530	0.2961	0.5530

Source : Albert N. Schrieber, Richard A. Johnson, Robert C. Meier, William T. Newell, Henry C. Fischer, *Cases in Manufacturing Management*, New Delhi, McGraw-Hill Book Company, 1965, p. 464.

APPENDIX A5
Present Value Factors for a Single Payment

Number of Periods (n)	Interest Rate (r) 0.01	0.02	0.03	0.04	0.05	0.06	0.08	0.10	0.12	0.14	0.16	0.18	0.20	0.22	0.24	0.26	0.28	0.30
1	.9901	.9804	.9709	.9615	.9524	.9434	.9259	.9091	.8929	.8772	.8621	.8475	.8333	.8197	.8065	.7937	.7812	.7692
2	.9803	.9612	.9426	.9246	.9070	.8900	.8573	.8264	.7972	.7695	.7432	.7182	.6944	.6719	.6504	.6299	.6104	.5917
3	.9706	.9151	.9151	.8890	.8638	.8396	.7983	.7513	.7118	.6750	.6407	.6086	.5787	.5507	.5245	.4999	.4768	.4552
4	.9610	.8885	.8885	.8548	.8227	.7921	.7350	.6830	.6355	.5921	.5523	.5158	.4823	.4514	.4230	.3968	.3725	.3501
5	.9515	.8626	.8626	.8219	.7835	.7473	.6806	.6209	.5674	.5194	.4761	.4371	.4019	.3700	.3411	.3149	.2910	.2693
6	.9420	.8880	.8375	.7903	.7462	.7050	.6302	.5645	.5066	.4556	.4104	.3704	.3349	.3033	.2751	.2499	.2274	.2072
7	.9327	.8706	.8131	.7599	.7107	.6651	.5835	.5132	.4523	.3996	.3538	.3139	.2791	.2486	.2218	.1983	.1776	.1594
8	.9235	.8535	.7894	.7307	.6768	.6274	.5403	.4665	.4039	.3506	.3050	.2660	.2326	.2038	.1789	.1574	.1388	.1226
9	.9143	.8368	.7664	.7026	.6446	.5919	.5002	.4241	.3606	.3075	.2630	.2255	.1938	.1670	.1443	.1249	.1084	.0943
10	.9053	.8203	.7441	.6756	.6139	.5584	.4632	.3855	.3220	.2697	.2267	.1911	.1615	.1369	.1164	.0922	.0847	.0725
11	.8963	.8043	.7224	.6496	.5847	.5268	.4289	.3505	.2875	.2366	.1954	.1619	.1346	.1122	.0938	.0787	.0662	.0558
12	.8874	.7885	.7014	.6246	.5568	.4970	.3971	.3186	.2567	.2076	.1685	.1372	.1122	.0920	.0757	.0625	.0517	.0429
13	.8787	.7730	.6810	.6006	.5303	.4688	.3677	.2897	.2292	.1821	.1452	.1163	.0935	.0754	.0610	.0496	.0404	.0330
14	.8700	.7579	.6611	.5775	.5051	.4423	.3405	.2633	.2046	.1597	.1252	.0985	.0779	.0618	.0492	.0393	.0316	.0254
15	.8613	.7430	.6419	.5553	.4810	.4173	.3152	.2394	.1827	.1401	.1079	.0835	.0649	.0507	.0397	.0312	.0247	.0195
16	.8528	.7284	.6232	.5339	.4581	.3936	.2919	.2176	.1631	.1229	.0930	.0708	.0541	.0415	.0320	.0248	.0193	.0150
17	.8874	.7142	.6050	.5134	.4363	.3714	.2703	.1978	.1456	.1078	.0802	.0600	.0451	.0340	.0258	.0197	.0150	.0116
18	.8787	.7002	.5874	.4936	.4155	.3503	.2502	.1799	.1300	.0946	.0691	.0508	.0376	.0279	.0208	.0156	.0118	.0089
19	.8700	.6864	.5703	.4746	.3957	.3305	.2317	.1635	.1161	.0829	.0596	.0431	.0313	.0229	.0168	.0124	.0092	.0068
20	.8613	.6730	.5537	.4564	.3769	.3118	.2145	.1486	.1486	.0728	.0514	.0365	.0261	.0187	.0135	.0098	.0072	.0053
21	.8114	.6598	.5375	.4388	.3589	.2942	.1987	.1351	.0926	.0638	.0443	.0309	.0217	0.154	.0109	.0078	.0056	.0040
22	.8034	.6468	.5219	.4220	.3418	.2775	.1839	.1228	.0826	.0560	.0382	.0262	.0181	.0126	.0088	.0062	.0044	.0031
23	.7954	.6342	.5067	.4057	.3256	.2618	.1703	.1117	.0738	.0491	.0329	.0222	.0151	.0103	.0071	.0049	.0034	.0024
24	.7876	.6217	.4919	.3901	.3101	.2470	.1577	.1015	.0659	.0431	.0284	.0188	.0126	.0085	.0057	.0039	.0027	.0018
25	.7798	.6095	.4776	.3751	.2953	.2330	.1460	.0923	.0588	.0378	.0245	.0105	.0105	.0069	.0046	.0031	.0021	.0014

Number of Periods (n)	Interest Rate (r)																	
	0.01	0.02	0.03	0.04	0.05	0.06	0.08	0.10	0.12	0.14	0.16	0.18	0.20	0.22	0.24	0.26	0.28	0.30
26	.7720	.5976	.4637	.3607	.2812	.2198	.1352	.0839	.0525	.0331	.0211	.0135	.0087	.0057	.0037	.0025	.0016	.0011
27	.7644	.5859	.4502	.3468	.2678	.2074	.1252	.0763	.0469	.0291	.0182	.0115	.0073	.0047	.0030	.0019	.0013	.0008
28	.7568	.5744	.4371	.3335	.2551	.1956	.1159	.0693	.0419	.0255	.0157	.0097	.0061	.0038	.0024	.0015	.0010	.0006
29	.7493	.5631	.4243	.3207	.2429	.1846	.1073	.0630	.0374	.0224	.0135	.0082	.0051	.0031	.0020	.0012	.0008	.0005
30	.7419	.5521	.4120	.3083	.2314	.1741	.0994	.0573	.0334	.0196	.0116	.0070	.0042	.0026	.0016	.0010	.0006	.0004
35	.7059	.5000	.3554	.2534	.1813	.1301	.0676	.0356	.0189	.0102	.0055	.0030	.0017	.0009	.0005	.0003	.0002	.0001
40	.6717	.4529	.3066	.2083	.1420	.0972	.0972	.0221	.0107	.0053	.0026	.0013	.0007	.0004	.0004	.0001	.0001	.0000

$$P = \frac{F}{(1+r)^n} = F(pf)$$

where P = present value of a single investment

F = future value of a single payment

n = number of periods for which P is to be invested

r = periodic interest rate

pf = present value factor for \$ 1 = $1/(1 - t)^n$